# 16th International Conference on Positron Annihilation 2012

# (ICPA-16)



Journal of Physics: Conference Series Volume 443

## Bristol, United Kingdom
## 19-24 August 2012

### Editors:

Ashraf Alam
Stephen Dugdale
Paul Coleman
Mina Roussenova

ISBN: 978-1-62748-675-0
ISSN: 1742-6588

**Printed from e-media with permission by:**

Curran Associates, Inc.
57 Morehouse Lane
Red Hook, NY 12571

**Some format issues inherent in the e-media version may also appear in this print version.**



Printed by Curran Associates, Inc. (2013)

For permission requests, please contact the Institute of Physics
at the address below.

Institute of Physics
Dirac House, Temple Back
Bristol BS1 6BE UK

Phone: 44 1 17 929 7481
Fax: 44 1 17 920 0979

techtracking@iop.org

**Additional copies of this publication are available from:**

Curran Associates, Inc.
57 Morehouse Lane
Red Hook, NY 12571 USA
Phone: 845-758-0400
Fax: 845-758-2634
Email: curran@proceedings.com
Web: www.proceedings.com

## Papers

### Fundamentals



### $e^+$/Ps interactions with matter



### Anti-hydrogen physics



### Metals, oxides and theoretical

**Semi-conducting solids**



**Polymers and soft matter**

**Liquids**



**Porous and layered structures**

**$e^+$ beam developments**

**Advances in techniques**

## PREFACE

The $16^{th}$ International Conference on Positron Annihilation (ICPA-16) was held at the University of Bristol, United Kingdom during 19-24 August, 2012. This triennial conference is the foremost gathering of the Positron Annihilation Physics community and it was hosted in the UK for the first time since the series of meetings first started back in 1965. The University of Bristol, the *Alma Mater* of Paul Dirac, is situated at the heart of the city, and it has established a worldwide reputation in research and teaching. Many of the topics which were discussed during ICPA-16 form an integral part of the research themes in the schools of Physics, Chemistry and Engineering of this University.

ICPA-16 attracted a diverse audience, both from academic and industrial institutions, with over 200 participants from 29 countries. It continued the long held tradition of showcasing novel research in the field of positron annihilation and a total of 170 papers were presented as talks and posters. The papers reported studies of metallic and semi-conducting solids, polymers and soft matter, porous materials, surfaces and interfaces, as well as advances in experimental, analytical and biomedical applications. The high quality of the presented work, coupled with the enthusiastic exchange of ideas, provided an invaluable forum, especially for younger researchers and postgraduate students. The excellence of student presentations was acknowledged by the award of prizes for the best student posters, which were received by David Billington (University of Bristol, UK), Moussa Sidibe (CEMHTI, France) and Hongxia Xu (Tohoku University, Japan).

All papers published in the Conference Proceedings were reviewed by ICPA-16 participants. We are indebted to all reviewers who contributed their time and intellectual resources, allowing the refereeing and editing process to move smoothly toward the compilation of the Proceedings. Our sincere thanks and gratitude go to everyone who contributed to the success of the conference. We are grateful to all participants for their informative talks, poster presentations and fruitful discussions; the session chairs for keeping to the tight time schedule and for making sure the oral presentation sessions ran smoothly; Maria Dugdale for her time and effort in organising the social programme for the accompanying persons; the student volunteers from the Bristol Positron Group for all their help and time before, during and after the conference; the Bath positron group for helping with the organisation of the excursion and last, but not least, the University of Bristol Conference Office staff for their help with the organisation of the conference. We are also very grateful for the financial and logistical help from the University of Bristol and financial support from our sponsors and exhibitors, Ortec and Canberra.

We conclude by wishing the Organising Committee of ICPA-17 all the best for a successful conference. We look forward to seeing everyone in China in 2015.

*Ashraf Alam, Paul Coleman, Stephen Dugdale and Mina Roussenova*

Guest Editors
Bristol, April 2013

**Local organising committee**

M A Alam, S Dugdale and M Roussenova
University of Bristol, UK
P Coleman and S Townrow
University of Bath, UK
M Charlton
University of Swansea, UK
G Laricchia
University College London, UK
D Keeble
University of Dundee, UK

**Steering committee**

M A Alam (UK, Secretary)
P Coleman (UK)
B Ganguly (India)
Y Kobayashi (Japan)
P Mascher (Canada)
H Schut (Netherlands)
R Krause-Rehberg (Germany)

**International Advisory committee**

P Coleman (UK, Chairman)
R Krause-Rehberg(Germany, Vice-chairman)
M A Alam (UK)
G Laricchia (UK)
R Brusa (Italy)
M Doyama (Japan)
B Ganguly (India)
C Hugenschmidt (Germany)
Zs Kajcsos (Hungary, deceased)
Y Kobayashi (Japan)
J Kuriplach (Czech Republic)
P Mascher (Canada)
A Mills (USA)
Y Nagashima (Japan)
M Puska (Finland)
H Schut (Netherlands)
A Seeger (Germany)
Y Shirai (Japan)
A Somoza (Argentina)
A Stewart (Canada)
Z Tang (China)
A Weiss (USA)

**Conference Sponsors and Exhibitors**

ORTEC® CANBERRA

University of BRISTOL

Journal of Physics: Conference Series **443** (2013) 012001 doi:10.1088/1742-6596/443/1/012001

# Positronium formation in molecular oxygen

**J. Archer, S.M. Trilov and P.G. Coleman**

Department of Physics, University of Bath, Bath BA2 7AY, UK

E-mail: p.g.coleman@bath.ac.uk

**Abstract**. New measurements of the positronium (Ps) formation cross section in molecular oxygen have been made using the 'positron loss' technique used in earlier measurements in the noble gases – the reduction in beam intensity being attributed to Ps formation. The results are generally in agreement, in terms of dependence on incident positron energy, with earlier measurements, but are lower in magnitude than those deduced from total ionization measurements below the first threshold for direct ionization, which employed ion detection. $Q_{Ps}$ for molecular oxygen exhibits an interesting peak-like dependence on incident positron energy just above threshold, and possible reasons for this structure are discussed. These include the possibility of a dissociative attachment process; a unique peak at similar projectile energies has long been observed in electron-oxygen scattering. The most likely explanation, however, lies in coupling between Ps formation and excitation to the continuum, the latter having previously been shown also to have a peaked energy dependence above threshold.

## 1. Introduction

Total cross sections for positron scattering by molecular oxygen, $Q_{tot}$, have been measured by a number of laboratories since 1975 [eg 1-3]. $Q_{tot}$ has an essentially smooth dependence on incident positron energy, and there is reasonable agreement between the different measurements.

In 1980 the positronium (Ps) formation cross section for $O_2$ was reported by Charlton et al [4]. The measurements were made by detecting the three annihilation gamma photons resulting from ortho-Ps decay in coincidence as the positrons passed through a gas cell. The resulting cross section, $Q_{Ps}$, exhibited a peak-like structure in the few eV above threshold, in common with most other gases studied. However, this structure was later found to be a result of premature pick-off annihilation of the o-Ps on the gas cell walls.

Interestingly, however, measurement of the total ionisation cross section for positrons on $O_2$ (ie, Ps formation plus direct ionization), involving ion detection, by the same group in 1993 showed that in $O_2$, unlike other gases, the near-threshold peak was real [5]. The authors linked this feature to coupling of the Ps formation channel with that for excitation to the continuum, to which we shall return.

In 2005 Marler and Surko [6] measured $Q_{Ps}$ (not $Q_{Ps} + Q_{ion}$) using a trap-based system, by a method based on positron loss from the beam (as used by Fornari et al [7]). They confirmed the presence of a peak in $Q_{Ps}$ above threshold and again invoking coupling with excitation to the continuum.

The present work was initially performed to attempt to settle the apparent differences in absolute magnitudes of the $Q_{Ps}$ values reported in refs. [5] and [6], particularly at the near-threshold peak. This original aim was augmented by the need to try to understand more fully the reasons for the existence of the structure in the energy dependence of $Q_{Ps}$.

## 2. Experimental technique

The methodology used to measure $Q_{Ps}$ is essentially that described by Thornton et al [8], and thus will only be briefly summarised here. Positrons from a ~20MBq $^{22}Na$ source are moderated by two annealed 50% transmission tungsten meshes and their energy distribution narrowed to ~ 800meV by a 92% tungsten cut-off mesh mounted immediately above the moderator. This latter mesh also improves the reflection of positrons scattered back towards the source. The moderated positrons, in a beam of 4mm diameter, are accelerated to the desired mean energy (in the range 2-40eV) by a potential applied to the moderator and are guided by a 7mT axial magnetic field through a 70mm-long gas cell to the 10mm-diameter cone of a channel electron multiplier (CEM) detector 360mm away. The front end of the CEM is held at -2kV, with a fine mesh cover to prevent the 'sucking out' of electrons from its cone and to reduce the background due to fast secondary electrons from the moderator meshes. Between the gas cell and detector is mounted a cylindrical retarding field analyser (RFA) to which a potential can be applied in order to discriminate against scattered positrons.

The positron signal count rate $I_0$ was determined by measuring the difference between the total CEM count rate and that recorded when the beam was *just* stopped by raising the potential on the cut-off mesh in front of the moderator. With gas continuously bled into the cell – at a density for which the maximum total attenuation was no more than 15% - the count rate is $I$. When a potential is applied to the RFA which only allows unscattered positrons to pass through (and thus effectively stops those positrons which have been scattered through elastic or inelastic channels), $(I_0-I)/I_0$ is the 'total' attenuation of the beam, $A_{tot}$. When no potential is applied to the RFA all positrons which survive the scattering process can pass through, and the only loss is via neutralisation by Ps formation (assuming that direct annihilation is negligible); $(I_0-I)/I_0$ is then designated $A_{Ps}$. $Q_{Ps}$ is then deduced from $(A_{Ps}/A_{tot})Q_{tot}$, where $Q_{tot}$ is the total scattering cross section measured previously [1-3].

## 3. Results

The total scattering cross section $Q_{tot}$ is given by the Beer-Lambert expression $(-1/nL)\ln(1-A_{tot})$, where $nL$ is the atomic number density-path length product for the gas cell. As $nL$ was held constant for all positron energies, the measured $A_{tot}$ values could be used to obtain the dependence on positron energy $E$ of $Q_{tot}$ for positron-oxygen scattering, which by normalisation could then be compared with earlier measurements – with which reasonable agreement was obtained, as shown in Fig.1.

Fig. 2 shows the results for $Q_{Ps}$ from the current measurements and those of refs. [5] and [6]. The former used the thin-target approximation $Q_{Ps} = (1/nL)A_{Ps}$. If we use this expression then agreement between our results and those of ref. [6] is excellent for energies $E$ up to 20eV. However, the disagreement in the magnitude of $Q_{Ps}$ below the threshold for direct ionisation between the current results and those of ref. [5] is unexplained.

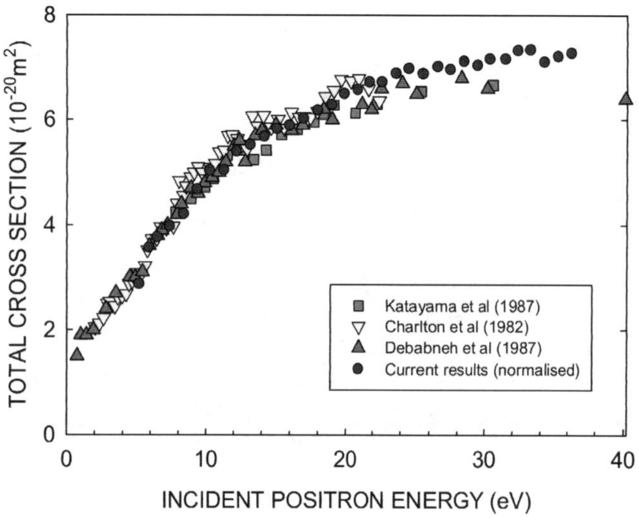


**Figure 1.** Total scattering cross sections for positron scattering by molecular oxygen.

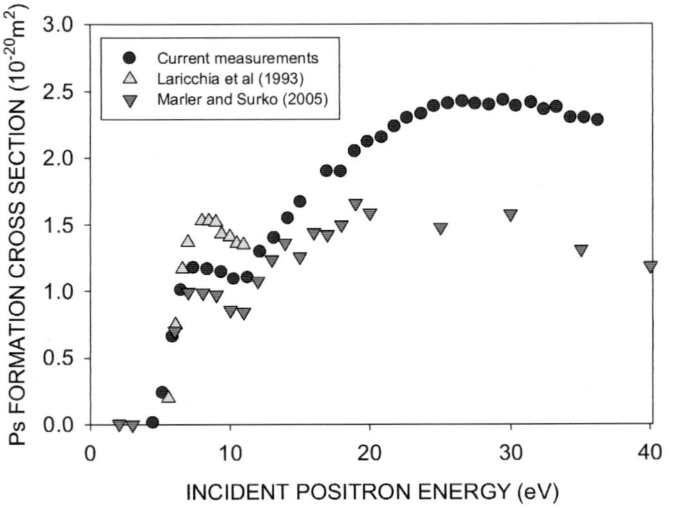


**Figure 2.** Cross sections for Ps formation in $O_2$. The results of Laricchia et al. are those for total positron-$O_2$ ionisation below the threshold for direct ionisation.

The divergence of the current results and those of ref. [6] at higher energies could be explained by an increasing underestimation of $A_{tot}$ as $E$ increases; but this implies a lack of discrimination against small-forward-angle elastic scattering, whereas inelastic processes (against which it is easy to discriminate) become increasingly important at higher $E$. Rather than discuss further this discrepancy in the magnitude of $Q_{Ps}$ at higher energies between the current results and those of ref. [6], we shall turn instead to the mechanism for the peak in $Q_{Ps}$ in the few eV above threshold.

## 4. Discussion

Coupling between scattering processes near the threshold for Ps formation has been seen in the noble gases (here between $Q_{Ps}$ and the elastic cross section $Q_{el}$) [9,10]. It is thus appropriate to consider whether the peak in $Q_{Ps}$ in $O_2$ above threshold is also evidence of channel coupling.

It is interesting to note that in electron-$O_2$ scattering there is a pronounced peak in the cross section for the formation of $O^-$ ions (Fig. 3) which occurs at exactly the same energy as the peak in $Q_{Ps}$ [11]. It is tempting to attempt to correlate this unusual peak with that in $Q_{Ps}$. One possible scenario would involve the dissociative attachment of a positron ($e^+ + O_2 \rightarrow Oe^+ + O$) or Ps ($e^+ + O_2 \rightarrow PsO + O^+$), whereby the positron would be lost directly (via annihilation) or after reorganisation into Ps + $O^+$ + O. Either route would leave an $O^+$ ion, which would be essential in the measurements of Laricchia et al [5]. However, Cheng et al [13] and others have shown that positrons do not bind to O atoms and, although PsO is possible there is an energy deficit against its formation of ~11.2eV in $e^+$-$O_2$ scattering.

Finally, note that the cross section at the peak in Fig. 3 is only about 1% of $Q_{Ps}$ at the peak; we consequently should rule out any dissociative attachment mechanism to explain the structure in $Q_{Ps}$.

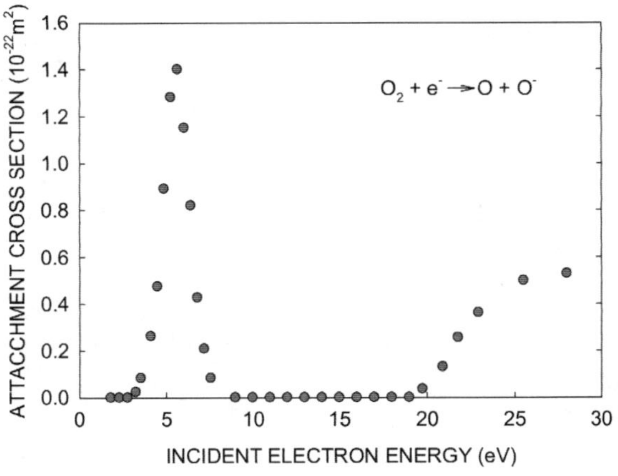


**Figure 3.** Cross section for the formation of $O^-$ ions via electron-$O_2$ impact. Data taken from ref.[11] normalised to absolute values using ref [12].

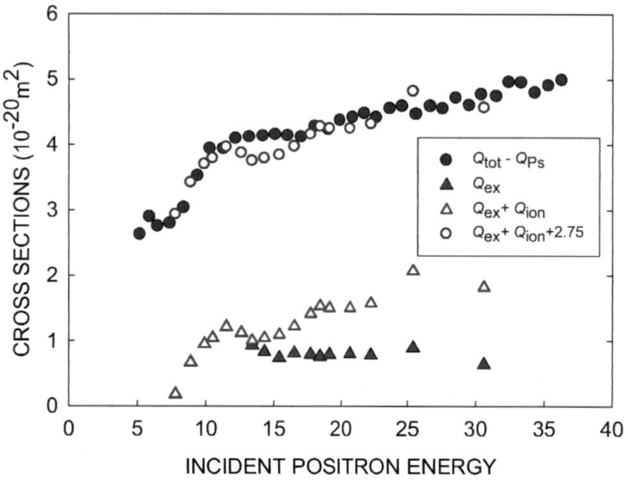


**Figure 4.** Solid circles: current (normalised) total minus Ps formation cross sections. Open circles: addition of the excitation and ionisation cross sections of Katayama et al [1] to a constant $2.75x10^{-20}m^2$ to represent elastic scattering.

Laricchia et al and Marler and Surko [5,6] both linked the effect to coupling with excitation to the (Schumann-Runge) continuum – ie dissociative excitation. Katayama et al [1] measured the excitation cross section $Q_{ex}$ by a positron time-of-flight method – ie the scattered positrons survive the interaction. They saw a significant peak in $Q_{ex}$ above the Schumann-Runge threshold at 7.1eV, implying a broad resonance-like behaviour for positron (but not electron) excitation (see Fig. 4). However, they invoked a coupling between Ps formation and excitation which possibly involves the formation of a $PsO_2^+$ complex, which in the light of recent calculations does not now seem likely [14].

The question remains as to why $Q_{tot}$ appears to vary smoothly with $E$, whereas the partial cross sections $Q_{Ps}$ and $Q_{ex}$ have structure. In Fig. 4 the current ($Q_{tot}$ – $Q_{Ps}$) is compared with ($Q_{ex}$+$Q_{ion}$) of Katayama et al [1] added to a constant 2.75 $x10^{-20}m^2$ to represent $Q_{el}$ in this energy range. Apart from the small discrepancy in the 10-15eV region, the two data sets are remarkably similar, suggesting that the structure in $Q_{Ps}$ should not be regarded as a peak centred at ~ 8eV, but a dip at ~ 11eV, where there is strong coupling with dissociative excitation.

Thanks are due to David Schrader and Gleb Gribakin for stimulating discussions on positron and Ps binding with atoms and molecules.

## References


[1] Katayama Y, Sueoka O and Mori S 1986 *J. Phys.* B*: At. Mol. Phys.* **20** 1654

[2] Charlton M, Griffith TC, Heyland GR and Wright GL 1982 *J. Phys.* B*: At. Mol. Phys.* **16** 333

[3] Debabneh MS, Hsieh YF, Kauppila WE, Kwan CK, Smith SJ, Stein TS and Uddin MN 1987 *Phys. Rev.* A **38** 1210

[4] Charlton M, Griffith TC, Heyland GR, Lines KS and Wright GL 1980 *J. Phys.* B*: At. Mol. Phys.* **13** L757

[5] Laricchia G, Moxom J and Charlton M 1993 *Phys. Rev. Lett.* **70** 3229

[6] Marler J P and Surko C M 2005 *Phys. Rev.* A **72** 3

[7] Fornari LS, Diana LM and Coleman PG 1983 *Phys. Rev. Lett.* **51** 2276-9

[8] Thornton MJ and Coleman PG 2011 *J. Phys. B: At. Mol. Opt. Phys.* **44** 145201

[9] Jay PM and Coleman PG 2010 *Phys. Rev.* A **82** 012701

[10] Jones ACL, Caradonna P, Makochekanwa C, Slaughter DS, McEachran RP, Machacek JR, Sullivan JP and Buckman SJ 2010 *Phys. Rev. Lett.* **105** 073201

[11] Hagstrum HD and Tate JT 1941 *Phys. Rev.***59** 354

[12] Lindsay BG and Mangan MA 2003 *Landolt-Bornstein: Photon and Electron Interactions with Atoms, Molecules and Ions* vol 1/17C ed Y Itakawa (New York: Springer) pp 1-77

[13] Cheng X, Babikov D and Schrader DM 2011 *Phys. Rev. A* **83** 032504

[14] Schrader D M (private communication) and Cheng X, Babikov D and Schrader D M 2012 *Phys. Rev.* A **85** 021503

# The sub-THz direct spectroscopy of positronium hyperfine splitting

**A Miyazaki[1], T Yamazaki[1], T Suehara[1], T Namba[1], S Asai[1], T Kobayashi[1], H Saito[2], Y Tatematsu[3], I Ogawa[3] and T Idehara[3]**

[1]Department of Physics, Graduate School of Science, and International Center for Elementary Particle Physics (ICEPP), the University of Tokyo, 7-3-1, Hongo, Bunkyo-ku, Tokyo, Japan
[2]Institute of Physics, Graduate School of Arts and Sciences, the University of Tokyo, 3-8-1, Komaba, Meguro-ku, Tokyo, Japan
[3]Research Center for Development of Far-Infrared Region, University of Fukui (FIR-FU), 3-9-1, Bunkyo, Fukui-shi, Fukui, Japan

E-mail: `miyazaki@icepp.s.u-tokyo.ac.jp`

**Abstract.** The positronium hyperfine splitting is a good target to study Quantum Electrodynamics in the bound state. There is a discrepancy between precision measurements and a theoretical calculation. We are planning to directly measure the positroinum hyperfine structure for the first time. A gyrotron oscillator is used as a novel radiation source in terahertz region. A Fabry-Pérot resonator is also developed to increase photon density. We have already observed the direct transition at 202.9 GHz. The direct measurement of the order of 100 ppm will be performed within about a year.

## 1. Introduction

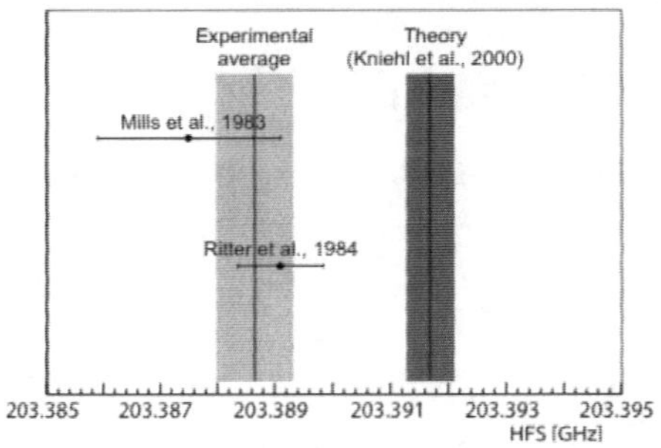


**Figure 1.** Discrepancy of Ps-HFS.

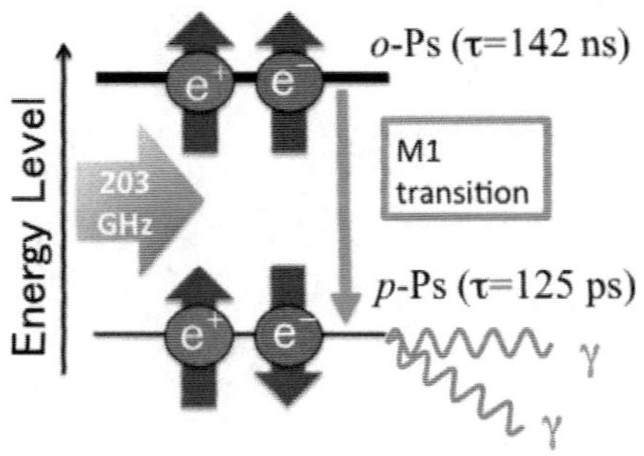


**Figure 2.** Schematic of the direct transition between Ps-HFS levels.

The hyperfine splitting of ground-state positronium (Ps-HFS) is a notable physics quantity to study Quantum Electrodynamics (QED) in the bound state. Some measurements of Ps-HFS were performed in 1980s. Precise calculations have been developed recently. The average of previous two experiments is 203.388 65(67) GHz (3.3 ppm) [1][2], whereas the theoretical value of $\mathcal{O}\left(\alpha^3 \ln \alpha^{-1}\right)$ is 203.391 69(41) GHz (2.0 ppm) [3][4][5]. Figure 1 shows this discrepancy of $3.9\sigma$ (15 ppm) between measurements and the QED calculation. In order to validate the QED theory, common systematic uncertainties in all previous experiments have to be checked.

All the foregoing studies measured the Zeeman interval under a static magnetic field of about 1 T. The positronium hyperfine splitting is indirectly calculated by the Zeeman splitting of about 3 GHz. Therefore, we point out inhomogeneity of the static magnetic field might be one of the underestimated systematic errors. A direct measurement of Ps-HFS without any magnetic fields can be free from this suspicion.

Figure 2 schematically shows how to measure Ps-HFS directly. An electromagnetic wave of about 203 GHz stimulates the transition from *o*-Ps to *p*-Ps. Since *p*-Ps promptly (125 ps) decays into two $\gamma$ rays, this transition can be probed by an increase of two $\gamma$-ray annihilation ratio which draws a Lorentzian curve with changing frequency. The natural rate (the spontaneous emission rate) of this direct transition is $A^{theo} = \frac{g'^2 \alpha \hbar^2 \omega^3}{3 m_e^2 c^4} = 3.37 \times 10^{-8}\,\mathrm{s}^{-1}$, where $g'$ is the g-factor of the bound-state electron, $\omega$ is $2\pi \times 203.4\,\mathrm{GHz}$. The rate is very small because this is a magnetic dipole transition (M1 transition). High power terahertz radiation of over 10 kW is essential for this experiment.

## 2. Terahertz Optics

High power 203 GHz radiation is achieved by developing two novel terahertz devices: *gyrotron oscillator* and *Fabry-Pérot resonator.*

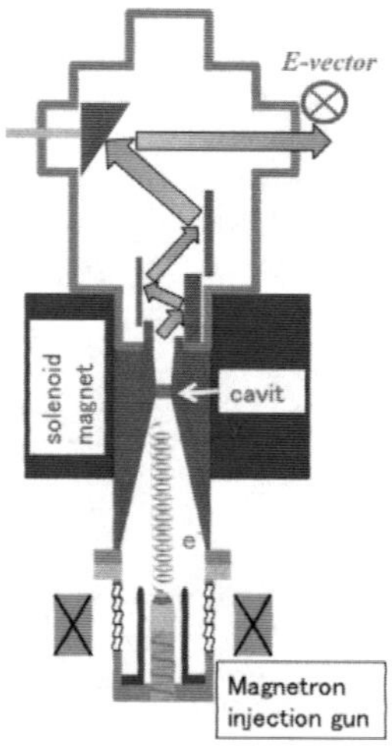


**Figure 3.** Schematic view of a gyrotron oscillator.

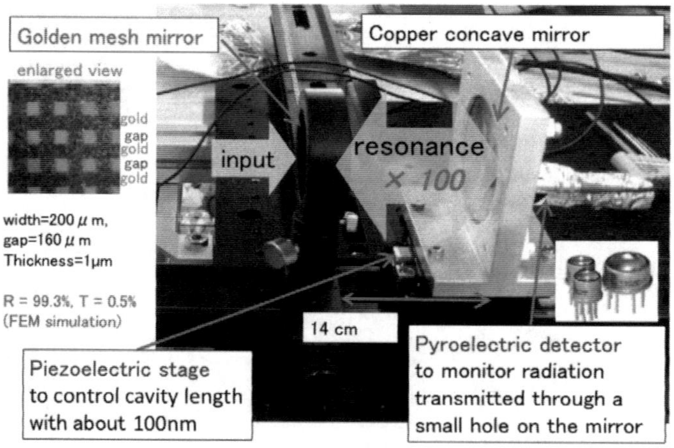


**Figure 4.** Photograph of the Fabry-Pérot cavity.

### 2.1. *Gyrotron oscillator*

A gyrotron oscillator developed in the field of nuclear fusion is the most powerful radiation source in terahertz region. It consists of a magnetron injection gun (MIG), a cavity in a superconducting solenoid and an internal mode-converter as shown in Fig. 3. Electrons emitted from MIG move in a cyclotron motion under a strong magnetic field. The interaction between electrons and the cavity self-consistently bunches the cyclotron motion of many electrons to extract a coherent electromagnetic wave upward from the cavity. The internal mode-converter shapes $TE_{52}$ into a linearly polarized gaussian beam [6]. Output power of our gyrotron is about 300 W stably when acceleration voltage and current are 18 kV and 500 mA, respectively. Oscillation frequency is more or less monochromatic ($\Delta f$ =1 MHz).

*2.2. Fabry-Pérot resonator*

Gyrotron output of 300 W is insufficient to drive the Ps-HFS transition. A Fabry-Pérot resonator is used to increase photon density. Figure 4 shows a photograph of the Fabry-Pérot resonator. A golden mesh mirror is developed to efficiently introduce a gaussian beam into the resonator. An opposite side is a concave Copper mirror on which a pyroelectric detector monitors transmitted radiation through a very small hole. The cavity length is controlled by a piezoelectric stage with accuracy of about 100 nm. The equivalent power in this resonator is measured about 10 kW, which can sufficiently stimulate the direct transition of positronium.

## 3. Positronium assembly and $\gamma$-ray detectors

The Fabry-Pérot resonator is in a chamber filled with nitrogen gas. We produce positronium there with a positron source $^{22}$Na as shown in Fig. 5. A thin plastic scintillator exists next to the source so as to tag time of the positron emission. Four $LaBr_3$ (Ce) crystal scintillators surrounds the resonator. Transiting *p*-Ps decays into two back-to-back $\gamma$ rays of 511 keV, while *o*-Ps decays into three $\gamma$ rays. Since these crystal scintillators have high energy resolution (FWHM 4% @511 keV), the signal of the direct transition can be efficiently separated from the *o*-Ps background.

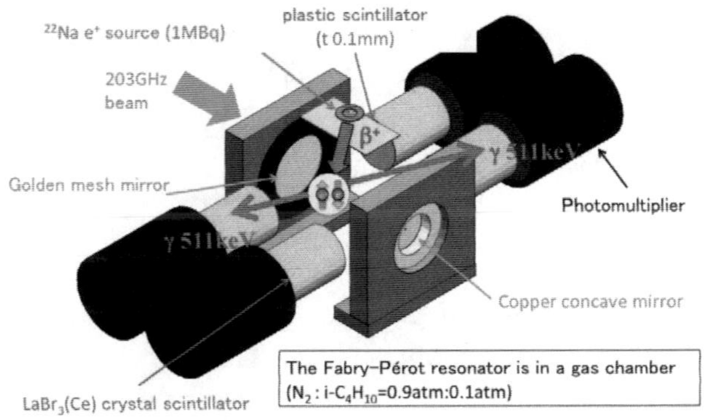


**Figure 5.** Arrangement of the positronium assembly and $\gamma$-ray detectors. Isobutane is mixed so as to reduce slow positrons.

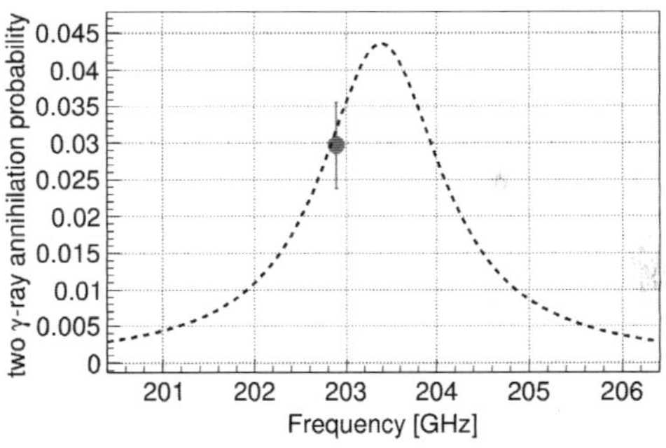


**Figure 6.** The direct transition at 202.9 GHz. The solid point shows the measured transition; the dotted line does a theoretical prediction.

## 4. Current Result and Improvements

The experiment of the direct transition at 202.9 GHz was performed [7]. Figure 6 shows the observed transition compared with a theoretical calculation. The direct transition is obtained with significance of 5.4$\sigma$. The measured natural transition rate $A^{exp} = 3.1^{+1.6}_{-1.2} \times 10^{-8}\,s^{-1}$ is consistent with the theoretical value. The error of this experimental rate is mainly due to the accuracy of the absolute power calibration of terahertz radiation.

In order to measure the Ps-HFS directly, the frequency must be swept around 203.4 GHz. Unfortunately, frequency of the gyrotron oscillator is fixed. A gyrotron cavity is a waveguide-like cavity (open cavity) in which electrons from the MIG move in a cyclotron motion. The oscillation frequency can be changed within the quality factor of the cavity, while it is too narrow (100 MHz) to draw the whole resonance curve of the Ps-HFS (FWHM 1.3 GHz; correspoing to *p*-Ps lifetime). We decided to create some gyrotron cavities with different diameters to change oscillation frequency. We will replace the gyrotron cavity when we measure different frequency points. All the necessary cavities have been already prepared.

The Fabry-Pérot cavity has a problem. The golden mesh mirror is melted away under accumulated power of over 10 kW. Since it was evaporated on a quartz substrate whose thermal conductivity is very small (5 $WK^{-1}m^{-1}$), the mesh pattern cannot be cooled by water. High resistance silicon (150 $WK^{-1}m^{-1}$) was selected as a new substrate. Figure 7 shows a new mesh mirror with a water cooling system. Figure 8 shows an obtained resonance of the new Fabry-Pérot resonator. The resonance occurs when cavity length matches half-integer multiple of wavelength (1.47 mm). This figure shows one of these resonances. Equivalent power in the resonator becomes over 20 kW. Water cooling works very well.

**Figure 7.** Photograph of the silicon mesh with water cooling.

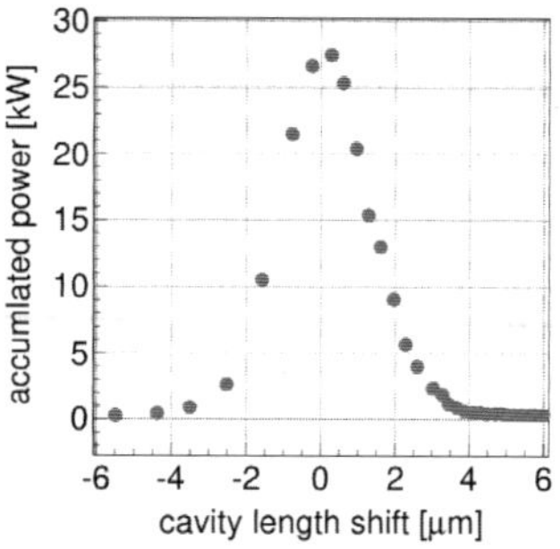


**Figure 8.** Measured resonance of the new Fabry-Pérot resonator.

## 5. Conclusion

The measured value of the Ps-HFS significantly differs from a theoretical calculation. The direct measurement of the Ps-HFS has a potential to study this problem in the future. We firstly observed the direct transition from *o*-Ps to *p*-Ps using a gyrotron oscillator and a Fabry-Pérot resonator. The positronium hyperfine splitting will be measured with accuracy of the order of 100 ppm in about one year for the first time. In order to improve the accuracy to ppm level, we need following four improvements. Firstly, an accurate power detector (relative accuracy of 0.1% level) of a terahertz wave should be developed. Secondly, the linewidth of the gyrotron oscillation should be improved from 1 MHz to 100 kHz. Thirdly, a positron beam should be used to increase statistic. Finally, the experiment should be performed in vacuum, so that the thermalization effect of positronium would be eliminated.

## References

[1] A P Mills 1983 *Phys. Rev. A* **27** 262
[2] M W Ritter, P O Egan, V W Hughes, and K A Woodle 1984 *Phys. Rev. A* **30** 1331
[3] B A Kniehl and A A Penin 2000 *Phys. Rev. Lett.* **85** 5094
[4] K. Melnikov and A. Yelkhovsky 2001 *Phys. Rev. Lett.* **86** 1498.
[5] R. J. Hill 2001 *Phys. Rev. Lett.* **86** 3280.
[6] Y Tatematsu, Y Yamaguchi, T Idehara, T Ozeki, R Ikeda, T Kanemaki, I Ogawa and T Saito 2012 *J. Infrared Milli. Terahertz Waves* **33** 292
[7] T Yamazaki, A Miyazaki, T Suehara,T Namba, S Asai, T Kobayashi, H Saito, I Ogawa, T Idehara, ans S Sabchevski 2012 *Phys. Rev. Lett.* **108** 253401

Journal of Physics: Conference Series **443** (2013) 012003 doi:10.1088/1742-6596/443/1/012003

# A simple shape-free model for pore-size estimation with positron annihilation lifetime spectroscopy

**Ken Wada and Toshio Hyodo**

Institute of Materials Structure Science, High Energy Accelerator Research Organization (KEK), 1-1 Oho, Tsukuba, Ibaraki 305-0801, Japan

E-mail: ken.wada@kek.jp

**Abstract.** Positron annihilation lifetime spectroscopy is one of the methods for estimating pore size in insulating materials. We present a shape-free model to be used conveniently for such analysis. A basic model in classical picture is modified by introducing a parameter corresponding to an effective size of the positronium (Ps). This parameter is adjusted so that its Ps-lifetime to pore-size relation merges smoothly with that of the well-established Tao-Eldrup model (with modification involving the intrinsic Ps annihilation rate) applicable to very small pores. The combined model, *i.e.*, modified Tao-Eldrup model for smaller pores and the modified classical model for larger pores, agrees surprisingly well with the quantum-mechanics based extended Tao-Eldrup model, which deals with Ps trapped in and thermally equilibrium with a rectangular pore.

## 1. Introduction

The positron annihilation lifetime spectroscopy (PALS) is used for estimating size of pores in insulating materials. The lifetime of the spin-triplet Ps, *o*-Ps, trapped in pores (defects, voids or free spaces) is shorter than the value in vacuum owing to its quenching through the annihilation with an electron on the pore surface (pick-off annihilation). Thus on assumption that the pickoff annihilation rate is practically independent of the material comprising the pore wall, the size of the pore can be deduced from the measured lifetime. The above assumption of the insensitivity of the pick-off annihilation rate to the material species is supported by a recent observation that the probabilities of the pick-off quenching of the thermalized *o*-Ps per collision with all the atoms and molecules in gas phase observed so far are almost constant [1].

## 2. Models for estimation of pore sizes by PALS analysis

### *2.1. Models based on quantum mechanics*

The original model for pore-size estimation with PALS analysis, the Tao-Eldrup (TE) model [2–4], is applicable to size estimation of very small pores ($<\sim 1\,\mathrm{nm}$). The TE model regards the *o*-Ps in the pore as a single particle with a mass twice as large as the electron mass in the ground state in a spherical potential well with radius $R$ (pore radius). To simplify the picture of the penetration of the Ps wave function into the wall, the pore is modeled by an infinitely deep well of radius $R + \Delta R$ with the position of the abrupt potential step recessed inside the wall. It is assumed that the fraction of the Ps wave function within the thickness $\Delta R$ contributes to the pick-off annihilation with the rate of the spin-averaged Ps annihilation in vacuum, $\lambda_{\rm A} = (\lambda_{\rm S} + 3\lambda_{\rm T})/4 = 2\,{\rm ns}^{-1}$; here $\lambda_{\rm S}$ and $\lambda_{\rm T}$ are the intrinsic annihilation rates in vacuum for spin-singlet Ps (*p*-Ps) and *o*-Ps, respectively. A straightforward calculation of the overlap gives the annihilation rate of the trapped *o*-Ps as

$$\lambda_{\rm TE}(R) = \lambda_{\rm A}\left[1 - \frac{R}{R+\Delta R} + \frac{1}{2\pi}\sin\left(\frac{2\pi R}{R+\Delta R}\right)\right] \quad . \tag{1}$$

The parameter $\Delta R$ has been empirically determined to be 0.165 nm by fitting to data acquired in variety of well-characterized small-pore materials.

In figure 1, the relation between *o*-Ps lifetime to the mean free length of the pore is plotted (relation between the mean free length and the pore radius is described in the next section).

The TE-model is insufficient, however, for pores larger than $\sim 1\,{\rm nm}$. Several attempts to modify this model for larger pores have been reported [5–10]. It is trivial to add the vacuum annahilation rate of *o*-Ps [6], as

$$\lambda_{\rm MTE}(R) = \lambda_{\rm TE}(R) + \lambda_{\rm T} \tag{2}$$

which we call modified Tao-Eldrup (MTE) model in the present paper. In addition, inclusion of the (translational) excited states in the pores is considered [5]; the separations between energy levels become narrower as the pore size increases, resulting in larger fraction of the Ps atoms annihilated from the excited states. Gidley *et al.* employed a model of rectangular pore geometry [7,8] to avoid calculation difficulty for a spherical pore [5]. This extended Tao-Eldrup (ETE) model gives the annihilation rate of *o*-Ps trapped in and thermally equilibrium with a rectangular well with sides $a$, $b$, $c$ at temperature $T$ as

$$\lambda_{\rm RTE}(a,b,c) = \lambda_{\rm A} - \frac{\lambda_{\rm S} - \lambda_{\rm T}}{4} F(a,\delta,T)F(b,\delta,T)F(c,\delta,T)\,, \tag{3}$$

where

$$F(x,\delta,T) = 1 - \frac{2\delta}{x+2\delta} + \frac{\sum_{i=1}^{\infty}\frac{1}{i\pi}\sin\left(\frac{2i\pi\delta}{x+2\delta}\right)\exp\left\{-\beta i^2/(x+2\delta)^2 kT\right\}}{\sum_{i=1}^{\infty}\exp\left\{-\beta i^2/(x+2\delta)^2 kT\right\}}\,. \tag{4}$$

Here, $\beta = h^2/16m = 0.188\,{\rm eV\,nm}^2$ and $m$ is the electron mass (the Ps mass is $2m$). The value of the adjustable parameter $\delta$ has been determined by fitting to experimental data as 0.18 nm. Note that the pore dimension in the original RTE model was $(a-2\delta)(b-2\delta)(c-2\delta)$ [7,8], but we have redefined it as $abc$ and the potential wall dimension as $(a+2\delta)(b+2\delta)(c+2\delta)$ to be consistent with the picture of the traditional TE model.

### *2.2. Models based on classical mechanics*

Classical models are applicable where the Ps mean free path between collisions, $\overline{L}_{\rm Ps}$, is much larger than the thermal Ps de Broglie wavelength $\lambda_{\rm dB}$ $(= h/\sqrt{4\pi mkT} = 3.05\,{\rm nm}$ at room temperature). The pick-off quenching rate of *o*-Ps is given by the product of pick-off annihilation probability per collision with the cavity wall, $P_{\rm A}$, and the collision frequency, $v_{\rm th}/\overline{L}_{\rm Ps}$; the total annihilation rate of *o*-Ps is described as [7]

$$\lambda_{\rm CM} = \frac{v_{\rm th}P_{\rm A}}{\overline{L}_{\rm Ps}} + \lambda_{\rm T}\,, \tag{5}$$

where $v_{\rm th}$ is the thermal velocity of the Ps. Gidley et. al deduced the product of $v_{\rm th}P_{\rm A}$ at room temperature to be $0.021 \pm 0.002\,{\rm nm/ns}$ for silica powders of two different grain sizes for which $\overline{L}_{\rm Ps}$ was systematically varied by controlling the packing from 35 nm to 600 nm [11]. This model

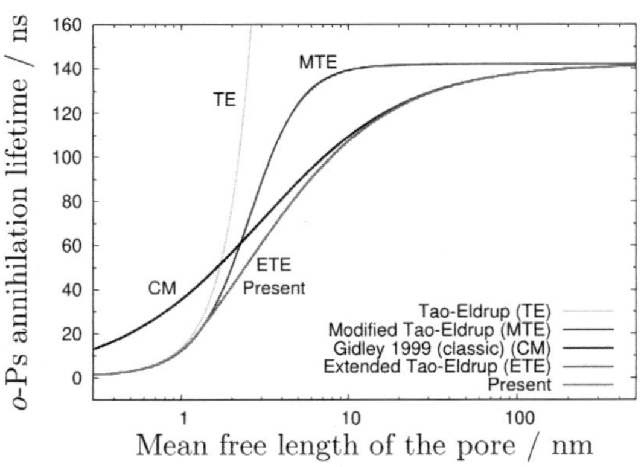


**Figure 1.** Models for pore-size estimation with PALS analysis. The curves of the present model and ETE model are indistinguishable.

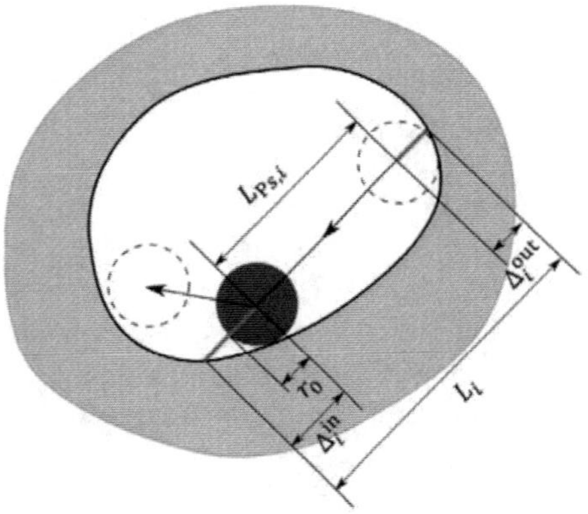

**Figure 2.** The difference between flight length of the Ps and that of an infinitely small particle

successfully relates the $o$-Ps annihilation rate and the mean free length for free spaces larger than several tens nm but fails for smaller; at $\sim 1\,\mathrm{nm}$, this model predicts annihilation rate about an order of magnitude lower than (lifetime longer than) the MTE model (figure 1).

Another classical-mechanics based model, specific for PALS measurement in air, has also been reported [10].

## 3. A new model based on a classical picture

The failure of the classical model in the smaller mean free length region may be removed by taking into account the effect of the size of Ps atom.

A free flight length of the Ps before the $i$-th collision with wall surface, $L_{\mathrm{Ps},i}$, is schematically shown in figure 2, or $L_{\mathrm{Ps},i} = L_i - \left(\Delta_i^{\mathrm{out}} + \Delta_i^{\mathrm{in}}\right)$, where $L_i$ is the free flight length of the infinitely small particle in the same trajectory, and $\Delta_i^{\mathrm{out}} + \Delta_i^{\mathrm{in}}$ represents the effect of the finite size. The average length $\overline{L}_{\mathrm{Ps},i}$ is written as

$$\overline{L}_{\mathrm{Ps},i} = \sum_{i=1}^{n} \frac{L_i - \left(\Delta_i^{\mathrm{out}} + \Delta_i^{\mathrm{in}}\right)}{n} = \overline{L} - 2\overline{\Delta}\,. \tag{6}$$

Here, $\overline{L}$, represents the mean free length of the pore. Substituting $\overline{L}_{\mathrm{Ps}}$ in (5) with the right-hand side of eq. (6), the $o$-Ps annihilation rate in this modified classical model (MCM), $\lambda_{\mathrm{MCM}}$, is given as

$$\lambda_{\mathrm{MCM}}(\overline{L}) = \frac{v_{\mathrm{th}} P_{\mathrm{A}}}{\overline{L} - 2\overline{\Delta}} + \lambda_{\mathrm{T}}\,. \tag{7}$$

The adjustable parameter $\overline{\Delta}$ is set so that the $o$-Ps-lifetime to cavity-size relation merges smoothly with that of the MTE model. Since the MTE model is described as a function of radius $R$ of the spherical pore, we transformed $R$ into $\overline{L}$ with a procedure using the general relation, $\overline{L} = 4V/A$ [12] for a cavity of volume $V$ with surface area $A$, giving $\overline{L} = 4R/3$. Two functions touch at $\overline{L} = 1.28\,\mathrm{nm}$ when $2\overline{\Delta} = 0.76\,\mathrm{nm}$. A combined model is practical: MTE model for pores less than $\overline{L} = 1.28\,\mathrm{nm}$ or measured $o$-Ps lifetime less than 21.1ns, and MCM for pores greater than $\overline{L} = 1.28\,\mathrm{nm}$ or lifetime longer than 21.1 ns, i.e.,

$$\lambda_{o-\mathrm{Ps}}^{\overline{L}} = \begin{cases} \lambda_{\mathrm{MTE}}(R = 3\overline{L}/4) & (\overline{L} < 1.28\,\mathrm{nm} \text{ or } 1/\lambda_{o\text{-Ps}} < 21.1\,\mathrm{ns}) \\ \lambda_{\mathrm{MCM}}(\overline{L}) & (\overline{L} \geq 1.28\,\mathrm{nm} \text{ or } 1/\lambda_{o\text{-Ps}} \geq 21.1\,\mathrm{ns}) \end{cases}, \tag{8}$$

where $v_{\mathrm{th}}P = 0.021 \pm 0.002\,\mathrm{nm/ns}$ and $2\overline{\Delta} = 0.76\,\mathrm{nm}$. When pores are assumed to be spherical with radius $R$, the following expression where $\overline{L}$ is transformed into $R$ is used:

$$\lambda^{R}_{o\text{-Ps}} = \begin{cases} \lambda_{\mathrm{MTE}}(R) & (R < 0.96\,\mathrm{nm} \ \text{or} \ 1/\lambda_{o\text{-Ps}} < 21.1\,\mathrm{ns}) \\ \dfrac{3}{4}\dfrac{v_{\mathrm{th}}P}{R - r_0} + \lambda_{\mathrm{T}} & (R \geq 0.96\,\mathrm{nm} \ \text{or} \ 1/\lambda_{o\text{-Ps}} \geq 21.1\,\mathrm{ns}) \end{cases}, \quad (9)$$

where $2r_0 = 3\overline{\Delta} = 1.14\,\mathrm{nm}$, which corresponds to an effective size of the Ps. This value is one order of magnitude larger than the diameter of the Ps (quantum mechanically expected value for the electron-positron distance in Ps), 1.058 nm, and about one thirds of the thermal de Broglie wave length of the Ps at room temperature, $\lambda_{\mathrm{dB}} = 3.05\,\mathrm{nm}$.

When applied to cubic pores, the present model (MTE model + MCM) agrees surprisingly well with the RTE model throughout entire region. These are indistinguishable in figure 1.

Figure 3 shows the present model applied to spherical pores (9) and experimental data cited in ref. [6, 8] and those reported in references [10, 13]. The function (9) in the region $R < 0.96\,\mathrm{nm}$ and $R \geq 0.96\,\mathrm{nm}$ is plotted with blue and red lines, respectively. The sets of data in refs. [10] and [14] which was obtained in the normal atmosphere have been corrected for the spin-conversion quenching of $o$-Ps with a rate of $0.0048 \pm 0.004\,\mathrm{ns}^{-1}$ [15, 16] by $O_2$ in the air and shown with uncertainties. The data from ref. [11] is converted from $\overline{L}$ into $R$. Other data reported without measurement conditions specified are plotted at the values of "$R$" stated in the references.

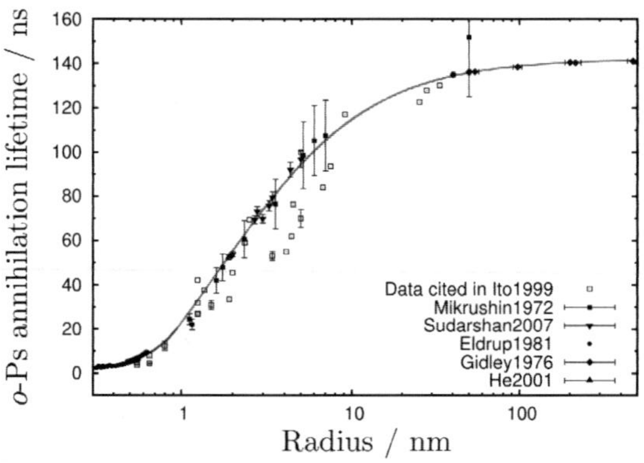


**Figure 3.** The present model applied to spherical pores, (9), and experimental data

As an application of the present shape-free model, we have measured the $o$-Ps annihilation rates in several silica aerogels having very complex open-pore structures. The mean free lengths obtained are consistent with other physical dimensions [17].

## References

[1] Wada K, Saito F, Shinohara N and Hyodo T 2012 *Eur. Phys. J.* D **66** 108.
[2] Tao S J 1972 *J. Chem. Phys.* **56** 5499.
[3] Eldrup M, Lightbody D and Sherwood J N 1981 *Chem. Phys.* **63** 51.
[4] Wang Y Y, Nakanishi Y, Jean Y C *et al.* 1990 *J. Polym. Sci., Part B: Polym. Phys.* **28** 1431.
[5] Goworek T, Ciesielski K, Jasinska B *et al.* 1998 *Chem. Phys.* **230** 305.
[6] Ito K, Nakanishi H and Ujihira Y 1999 *J. Phys. Chem.* B **103** 4555.
[7] Gidley D W, Frieze W E, Dull T L *et al.* 1999 *Phys. Rev.* B **60** 8. The pore dimension was taken to be the position of the infinite potential step in this paper. It has been corrected on the web page of the lab (url: http://positrons.physics.lsa.umich.edu/current/nanopos/conversion/index.html) and the discrepancy in the physical picture with the other works does not exist anymore.
[8] Dull T L, Frieze W E, Gidley D W *et al.* 2001 *J. Phys. Chem.* B **105** 4657.
[9] Dutta D, Ganguly B N, Gangopedhyay D *et al.* 2004 *J. Phys. Chem.* B **108** 8947.
[10] Sudarshan K, Dutta D, Sharma S K *et al.* 2007 *J. Phys.: Condens. Matter* **19** 386204.
[11] Gidley D W, Marko K A and Rich A 1976 *Phys. Rev. Lett.* **36** 395.
[12] Kosten C W, 1960 *Acustica* **10** 245.
[13] He Y, Zhang H, Chen Y *et al.* 2001 *J. Phys.: Condens. Matter* **13** 2467.
[14] Mikrushin A D, Levin B M, Goldanskii V I *et al.* 1972 *Russ. J. Phys. Chem.* **46** 368.
[15] Goldanskii V, Mokrushin A, Tatur A *et al.* 1975 *Appl. Phys* **5** 379.
[16] Shinohara N, Suzuki N, Chang T *et al.* 2001 *Phys. Rev.* A **64** 042702.
[17] Paper in preparation

16th International Conference on Positron Annihilation (ICPA-16) IOP Publishing
Journal of Physics: Conference Series **443** (2013) 012004 doi:10.1088/1742-6596/443/1/012004

# Exchange and correlation effects in the strongly interacting He-Ps system

**A. Zubiaga, F. Tuomisto and M. J. Puska**

E-mail: asier.zubiaga@aalto.fi
Department of Applied Physics, Aalto University, P.O. Box 11100, FIN-00076 Aalto Espoo, Finland

**Abstract.** The density functional theory has been used to calculate the total energy of the system comprising a He and a positronium atom. The single particle orbitals have been used to calculate the non-interacting kinetic energy, while the electron and positron densities were used for the Hartree energy, the electron exchange-correlation energy and the external potential energies. The electron exchange-correlation energy has been calculated within the Local Density Approximation and the General Gradient Approximation (PBE). For the electron-positron correlation energy the formula by Boronski and Nieminen has been used. The results have been compared to many-body wavefunction calculations employing the exact diagonalization of an explicitly correlated gaussians basis. While the many-body value result predicts that HePs is not bound, the density functional approach predicts a bound state with mean nucleus-positron distance of ~9 a.u.. To discuss the origin of this discrepancy the exchange-correlation energy has been deduced from the many-body result and it has been compared to the density functional scheme.

## 1. Introduction

The ortho-positronium (o-Ps) atom is the bound state of an electron and a positron with the total spin $S = 1$. It is formed in molecular matter and in some insulators, such as $SiO_2$, where the electron density is low in the interstitial region [1]. Its annihilation with an electron of the matter occurs typically through the two-gamma channel. The resulting pick-off annihilation rate depends on the overlap of the o-Ps with the electron density of the matter and it can reduce the positron lifetime remarkably [2]. The pick-off annihilation lifetime spectroscopy of o-Ps has a rather unique role as a method capable to study open volumes in polymers [3] and biostructures [4; 5].

The understanding of the experimental results would benefit from calculations of o-Ps states in molecular soft-matter. However both the electron and the positron in the Ps atom are light quantum-mechanical particles and the non-adiabatic correlation effects have to be properly taken into account; something that has inhibited progress in this field. For delocalized positrons in metals and semiconductors, where Ps is not formed and the density of positrons can be assumed to be low, the quantum-mechanical state of a single positron [6] can be represented within the two component density functional theory (DFT). Accurate lifetime values have been obtained, and even the lifetimes of positrons trapped at vacancies can be calculated with good accuracy [7]. Using a DFT approach for the simulation of the positronium atom would also be desirable but the description of the electron-positron bound state is challenging.

Journal of Physics: Conference Series **443** (2013) 012004 doi:10.1088/1742-6596/443/1/012004

In this work, the many-body wavefunction and the interaction energy of the unbound HePs system has been calculated by using explicitly correlated gaussian and the stochastical variational method (ECG-SVM) [8; 9] for a set of configurations with the nucleus-positron mean distance ranging between 2.0 a.u. and 40 a.u.. The DFT exchange-correlation energy has been deduced using the electron and positron densities of the many-body calculation. The interaction and the exchange-correlation energies have been calculated using the Local Density Approximation (LDA) and the General Gradient Approximation implementation by Perdew et al. (PBE) for the electron-electron exchange correlation.

## 2. Computational methods

The many-body wavefunction is written as a linear combination of properly antisymmetrized ECG functions. The 5-particle system comprising the heavy nucleus (treated as a single particle), the electrons and the positron has been described by the non-relativistic Hamiltonian $\widehat{H} = \sum_i \frac{\vec{p}_i^2}{2m_i} - T_{cm} + \sum_{i<j} \frac{q_i q_j}{4\pi\epsilon_0 r_{ij}}$, where $\vec{p_i}$, $m_i$, and $q_i$, are the momenta, masses, and charges of the particles, respectively, $r_{ij}$ is the distance between the $i^{\text{th}}$ and $j^{\text{th}}$ particles, and $T_{cm}$ is the center-of-mass (CM) kinetic energy. More details about the calculations can be found in reference [9].

The energy of the HePs system under the two component DFT with electron density $n_-$ and positron density $n_+$ is [10]:

$$E \;=\; F[n_-] + F[n_+] + Z\int d\mathbf{r}\frac{n_+(\mathbf{r}) - \mathbf{n}_-(\mathbf{r})}{r} - \int d\mathbf{r}d\mathbf{r}'\frac{n_-(\mathbf{r})n_+(\mathbf{r}')}{|\mathbf{r}-\mathbf{r}'|} + E_c^{e-p}[n_+, n_-] \quad (1)$$

where $E_c^{e-p}$ is the electron-positron correlation energy and $F[n]$ the hamiltonian for one component DFT ($F[n] = T[n] + \frac{1}{2}\int d\mathbf{r}\int d\mathbf{r}'\frac{n(\mathbf{r})n(\mathbf{r}')}{|\mathbf{r}-\mathbf{r}'|} + E_{xc}[n]$). The LDA [11] and PBE [12] functionals have been used for the exchange-correlation energy ($E_{xc}$) while for $E_c^{e-p}$ the parametrization by Boronski and Nieminen (BN) [10] has been used. The non-interacting kinetic energies have been calculated from the single particle orbitals. The positron single particle orbital is the square root of the positron density. For the electrons, the orbitals are also approximated as the square root of the components of the electron density forming a singlet and a triplet with the positron, $\rho_i^e =< \Psi|\hat{P}_i^{e-p}\hat{\rho}^e|\Psi >$. $\hat{P}_i^{e-p}$ projects the singlet or triplet component of the electron density operator and $\Psi$ is the ECG-SVM many-body wavefunction.

## 3. Results & Discussion

The values of the interaction energy $E_I$=$E_{HePs}$-$E_{He}$-$E_{Ps}$ of HePs are shown in table 1 ordered according to the increasing nucleus-positron mean distance $\langle r_p \rangle$. The ECG-SVM result, $E_I^{ED}$, is always positive and approaches zero for large $\langle r_p \rangle$ values. The energy values decrease from 0.4054 Ha in the strongest constrained system to 0.0017 Ha in the less constrained system and it further decreases going towards zero at larger separations [9]. The positron non-interacting kinetic energy $E_{kin}^p$ agrees with the zero point energy of positronium $E_{conf}^{sq}$ in an infinite square potential with R=2$\langle r_p \rangle$. $E_{kin}^p$ is ~37% of the total interaction energy for the most confined system but for the system with the largest $\langle r_p \rangle$ it is ~83%. The DFT interaction energy has been calculated using the density obtained from the ECG-SVM wavefunction and the LDA ($E_I^{LDA}$) and PBE ($E_I^{PBE}$) functionals for the indistinguishable particle exchange-correlation energy. The interaction energies are similar in LDA and PBE. At $\langle r_p \rangle$ larger than 2 a.u., the interaction energy is negative and it shows a minimum at $\langle r_p \rangle$~9 a.u. (~ -0.031 Ha). At very long separations $E_I^{LDA}$ and$E_I^{PBE}$ increases towards zero. In the most confined system, the interaction energy obtained from the density is also lower than the ECG-SVM value.

In order to gain a deeper insight into the origin of the discrepancy between the ECG-SVM and the DFT results, the exchange-correlation energy has been calculated by subtracting from

**Table 1.** Values of the interaction energy from the ECG-SVM calculation ($E_I^{ED}$), DFT ($E_I^{LDA}$ and $E_I^{PBE}$) and non-interacting kinetic energy of the positron in Hartrees. The zero-point energy inside an infinite square potential of $R=2\langle r_p\rangle$ is also shown. The values are ordered according to $\langle r_p\rangle$ (in a.u.).

| $\langle r_p\rangle$ | 2.07 | 5.00 | 10.11 | 14.55 | 19.41 |
|---|---|---|---|---|---|
| $E_I^{ED}$ | 0.4054 | 0.0492 | 0.0079 | 0.0033 | 0.0017 |
| $E_I^{LDA}$ | 0.2422 | -0.0087 | -0.0307 | -0.0231 | -0.0129 |
| $E_I^{PBE}$ | 0.2354 | -0.0110 | -0.0313 | -0.0232 | -0.0129 |
| $E_{kin}^{p}$ | 0.1504 | 0.0267 | 0.0057 | 0.0026 | 0.0014 |
| $E_{conf}^{Sq}$ | 0.1435 | 0.0246 | 0.0060 | 0.0029 | 0.0016 |

the many-body result the DFT single particle kinetic and potential (External and Hartree terms) energies. The values shown in figure 1 are compared to the corresponding DFT values. The ECG-SVM exchange-correlation energy has an almost constant value of -1.31 Ha for systems with $\langle r_p\rangle$ larger than 10 a.u. and it is lower ( $E_{xc}^{ED}$ ~-1.45 Ha is lower, in the more confined systems. While the total PBE+BN exchange-correlation energies are lower than ECG-SVM values for all $\langle r_p\rangle$, the total LDA+BN exchange-correlation energies are larger than those from ECG-SVM when $\langle r_p\rangle$ is larger than 15 a.u.. It keeps on growing till $\langle r_p\rangle\sim$35 a.u. and the increase is markedly stronger, -0.3 Ha, than in the ECG-SVM result, -0.14 Ha. The LDA results are ~0.06 Ha larger than the PBE values in all the calculated systems and the interaction energy is similar under both approximations. The PBE exchange-correlation energies are lower than the ECG-SVM values in all the systems but the LDA value are larger for $\langle r_p\rangle\geq$15 a.u..

The difference between the ECG-SVM and the DFT total interaction energies come from the low values of the DFT exchange-correlation energy in the more confined systems, in which the electron-electron exchange-correlation energy is the dominating contribution. The exchange-correlation functionals are known to have severe difficulties in describing the energetics of closed-shell molecules and atoms with HOMO-LUMO gaps. DFT is also known to be unable to describe the attractive dispersion energy [13] so that the predicted interaction energies are typically too large. In addition, in the HePs system the attractive dispersion interaction is very weak, because the polarizability of He is low. The BN parametrization of the electron-positron correlation energy, obtained from many-body results for a single positron in an infinite homogeneous electron gas, is also a source of inaccuracy. It should be noted though that the electron-positron correlation energy for a finite positron density would be smaller and it cannot account for the observed difference [14].

## 4. Conclusions

The He-Ps interaction energy obtained from many-body ECG-SVM results and that arising from a DFT scheme have been compared. The interaction energies are similar in magnitude but show important differences in their behaviors as the confinement of the system is varied. While the ECG-SVM total energy decreases monotonically as the mean positron increases, the DFT results show a minimum at ~9 a.u. and it incorrectly predicts a bound state for the HePs system. In the most confined systems the ECG-SVM energies are more repulsive than the DFT energies. The origin of the weakened repulsive in DFT has been linked to the electron-electron exchange term which fails to account for the strong repulsion when the system is confined. The electron-positron correlation energy can also be a source of inaccuracy but it cannot explain the low values of the interaction energy in the DFT result.

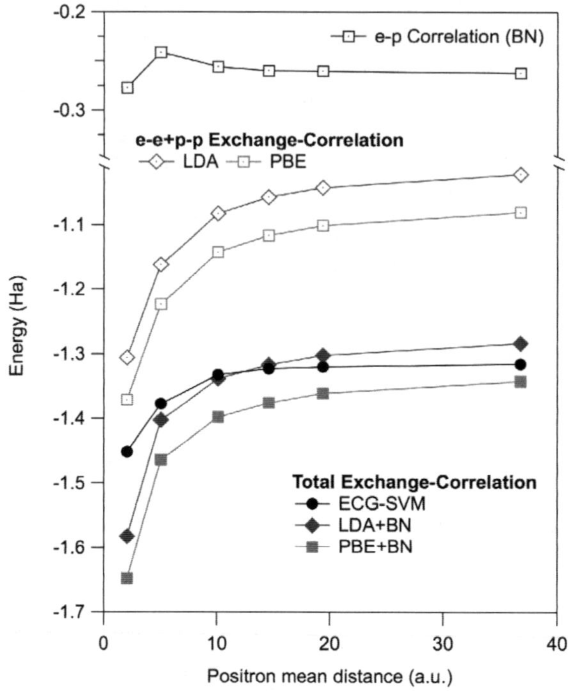


**Figure 1.** (Color online) The total exchange-correlation energy terms for the ECG-SVM (black full circles) and the DFT LDA (purple full diamonds) and PBE (red full squares) energies versus $\langle r_p \rangle$. See the text for the definition of the ECG-SVM exchange-correlation energy. Also the DFT electron-electron exchange correlation energies (purple dotted diamonds for LDA and red dotted square for PBE) and the BN electron-positron correlation energy (black dotted squares) are shown.

## 5. Acknowledgments
This work was supported by the Academy of Finland through individual fellowships and the centre of excellence program.

## References

[1] Stepanov S V and Byakov V M 2003 *Principles and application of positron & positronium chemistry* 5 (World Scientific)

[2] Mogensen O E 1995 Springer series in chemical physics *Positron Annihilation in Chemistry* vol 58 ed Lotsch H (Springer-Verlag) chap Principles of Positron Annihilation in Molecular Solids, p 193

[3] Uedono A, Suzuki R, Ohdaira T, Uozumi T, Ban M, Kyoto M, Tanigawa S and Mikado T 1998 *J. Polym. Sci.* Part B **36** 2597

[4] Sane P, Salonen E, Falck E, Repakova J, Tuomisto F, Holopainen J and Vattulainen I 2009 *J. Phys. Chem.* B *Letters* **113** 1810

[5] Dong A W, Pascual-Izarra C, Pas S J, Hill A J, Boyd B J and Drummond C J 2009 *J. Phys. Chem.* B **113** 84

[6] Puska M J and Nieminen R M 1994 *Rev. Mod. Phys.* **66** 841

[7] Makkonen I and Puska M J 2007 *Phys. Rev.* B **76** 054119

[8] Varga K and Suzuki Y 1995 *Phys. Rev.* C **52** 2885
[9] Zubiaga A, Tuomisto F and Puska M J 2012 *Phys. Rev.* A **85** 052707
[10] Boroński E and Nieminen R M 1986 *Phys. Rev.* B **34** 3820
[11] Perdew J P and Wang Y 1992 *Phys. Rev.* B **45** 13244
[12] Perdew J P, Burke K and Ernzerhof M 1996 *Phys. Rev. Lett.* **77** 3865
[13] Johnson E R, Mackie I D and DiLabio G A 2009 *J. Phys. Org. Chem.* **22** 1127
[14] Lantto L J 1987 *Phys. Rev.* B **36** 5160

Journal of Physics: Conference Series **443** (2013) 012005 doi:10.1088/1742-6596/443/1/012005

# Trapped antihydrogen: A new frontier in fundamental physics.

**N. Madsen**

Department of Physics, College of Science, Swansea University, Swansea SA2 8PP, UK

**The ALPHA Collaboration**[1]

**Abstract.** Antihydrogen, the bound state of an antiproton and a positron, has been produced at low energies at CERN since 2002. Antihydrogen is of interest for use in precision tests of nature's fundamental symmetries. The charge conjugation/parity/time reversal (CPT) theorem, a crucial part of the foundation of the standard model of elementary particles and interactions, demands that hydrogen and antihydrogen have the same spectrum. Given the current experimental precision of measurements on the hydrogen atom, subjecting antihydrogen to rigorous spectroscopic examination would constitute a compelling, model-independent test of CPT. Antihydrogen could also be used to study the gravitational behaviour of antimatter.

However, until recently, experiments have produced antihydrogen that was not confined, precluding detailed study of its structure. Experimenters working to trap antihydrogen have faced the challenge of trapping and cooling relativistic antiprotons and using them to make antihydrogen cold enough to be trapped in a magnetic minimum trap with a depth of only 50 $\mu$eV. In November 2010 the ALPHA collaboration demonstrated the first trapping of antihydrogen, thus opening the door to precision measurements on anti-atoms which can soon be subjected to many of the same techniques as developed for atoms. The prospect for such measurements improved further with ALPHA's demonstration of 1000 s confinement of the anti-atoms in the summer of 2011 and the recent first detection of resonant quantum interactions in antihydrogen.

## 1. Introduction

Antihydrogen ($\bar{\mathrm{H}}$), the bound state of an antiproton ($\bar{\mathrm{p}}$) and a positron ($e^+$) is the simplest anti-atom and also the simplest neutral antimatter system. As such it holds the promise of precise comparisons between matter and antimatter using spectroscopic techniques [1]. Depending on the particular aspect of the (anti)atom investigated, one can study various aspects of the fundamental CPT-theorem which states that all physical laws must be unchanged under the simultaneous application of Charge conjugation, Parity inversion and Time reversal. Antihydrogen, due to its neutrality, furthermore facilitates directly investigating the gravitational interaction of antimatter.

[1] C. Amole, M. D. Ashkezari, M. Baquero-Ruiz, W. Bertsche, P. D. Bowe, E. Butler, A. Capra, C. L. Cesar, M. Charlton, A. Deller, S. Eriksson, J. Fajans, T. Friesen, M. C. Fujiwara, D. R. Gill, A. Gutierrez, J. S. Hangst, W. N. Hardy, M. E. Hayden, A. J. Humphries, C. A. Isaac, S. Jonsell, L. Kurchaninov, J. T. K. McKenna, S. Menary, P. Nolan, K. Olchanski, A. Olin, P. Pusa, F. Robicheaux, E. Sarid, D. M. Silveira, S. Stracka, C. So, R. I. Thompson, D. P. van der Werf & J. S. Wurtele

The first low energy $\bar{\mathrm{H}}$ suitable for scrutiny was made by the ATHENA collaboration working at the CERN Antiproton Decelerator (AD) [2] in 2002 [3]. Following this success, the ALPHA collaboration was started in 2005 with the intermediate goal of trapping $\bar{\mathrm{H}}$ and ultimate goal of performing precision comparisons of H and $\bar{\mathrm{H}}$. ALPHA succeeded in trapping $\bar{\mathrm{H}}$ in 2010 [4] and further in holding on to it for 1000 s in 2011 [5]. These results led to the first detection of a resonant quantum transition in an anti-atom in 2012 [6], thus kicking off the field of $\bar{\mathrm{H}}$ spectroscopy. This paper discusses the challenges that were faced and met to achieve all of these results, and gives a brief glimpse of the future of $\bar{\mathrm{H}}$ physics, in particular in ALPHA.

## 2. Making cold antihydrogen

Antihydrogen can be formed by merging plasmas of $e^+$ and $\bar{p}$. To date, this is the most efficient method available, and in a typical experiment in ALPHA the conversion efficiency of $\bar{p}$, the rarest of the two species, is about 30% [4]. The basic scheme is illustrated in Figure 1. All the charged particles used ($e^+$, electrons ($e^-$)and $\bar{p}$) are held in a Penning-Malmberg trap where an axial magnetic field ensures transverse confinement and axial electric field ensures axial confinement. A key difficulty, which was identified already in the ATHENA experiment [7], is that one must ensure that the $\bar{p}$ that form $\bar{\mathrm{H}}$ are cold as they possess the bulk of the kinetic energy of the synthesised $\bar{\mathrm{H}}$. This may seem obvious, but in initial experiments the focus was on ensuring that the $e^+$ were cold, and it was assumed that the $\bar{p}$ would then come to equilibrium with them before forming $\bar{\mathrm{H}}$. However, the key parameter for recombination is the relative velocity, not the temperature, and as the $\bar{p}$ are much more massive than the $e^+$, they may form $\bar{\mathrm{H}}$ with cold $e^+$ even with an order of magnitude larger kinetic energy [7].

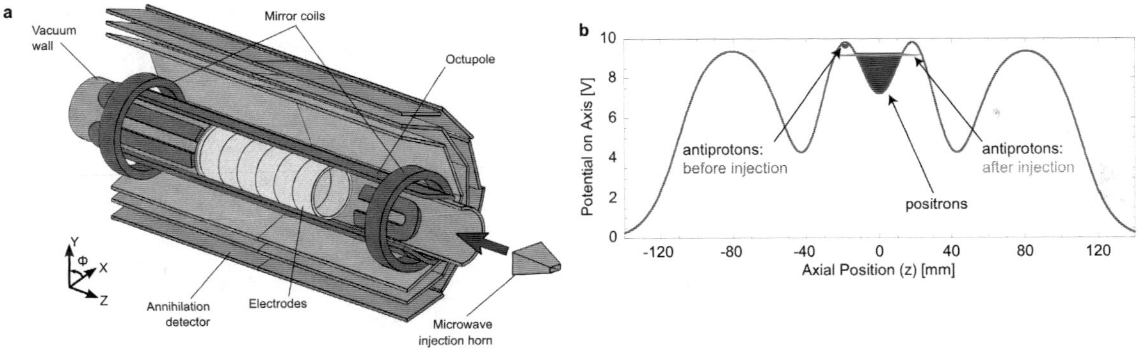


**Figure 1.** (a) Cut-away, drawing of the $\bar{\mathrm{H}}$ synthesis and trapping region. The super conducting atom-trap magnets, the annihilation detector, and some of the Penning-Malmberg trap electrodes are shown. An external solenoid (not shown) provides a 1-T magnetic field for the Penning-Malmberg trap. (b) The electrical potentials used for merging cold $\bar{p}$ and positrons to make cold $\bar{\mathrm{H}}$. See text for details.

The route taken by ALPHA to improve upon this situation was to (a) work towards making all participating species colder and (b) develop methods to merge the two species without jeopardising the achieved low temperatures.

### *2.1. Ensuring cold particles for antihydrogen formation*

The $\bar{p}$ are delivered by the CERN AD at 5.3 MeV [2] and directed through a thin aluminium foil ($\sim$0.2 mm), which ensures that a maximum of particles can be captured dynamically using electric fields switched at $\sim$5 kV [8]. The $\bar{p}$ are then cooled through collisions with an $e^-$ plasma which has been pre-loaded. The $e^-$ plasma cools through emission of cyclotron radiation in the strong axial magnetic field of the Penning-Malmberg trap, cooling towards the temperature of the surroundings which are held at 7 K. In ALPHA, the $e^-$ typically achieve final temperatures of 100-200 K. The neutral trap magnetic fields, to be discussed later, introduce a strong azimuthal asymmetry which is observed to both induce dynamic aperturing and lead to heating of the plasmas [9, 10, 11, 12]. This means that to keep the plasmas cold we must keep them radially small and axially short. The $\bar{p}$ plasma is therefore compressed through the rotating wall (RW) technique [13] which in ALPHA in set to act on the combined $e^-$ $\bar{p}$ plasma, in order to ensure $e^-$ cooling also after compression [14]. The $e^-$ are then released from the mixture by carefully, but quickly pulsing one of the confining electrodes. Using this methodology ALPHA routinely prepares $\bar{p}$ clouds of 40000 particles with a radius of 0.25 mm and a temperature of $\sim$400 K, the $e^-$ ejection process causing some heating. The $\bar{p}$ are then further cooled using evaporative cooling [15] and a typical final $\bar{p}$ cloud consists of 15000 particles with a radius of 0.4 mm and a temperature of 100 K.

The $e^+$ are accumulated and transferred using well established techniques [16, 17]. These techniques also include the RW compression of the $e^+$. Using these techniques as well as evaporative cooling of the $e^+$ ALPHA typically obtains $e^+$ plasmas with a radius of 0.8 mm containing $1\times10^6$ $e^+$ at a density of $5\times10^7$ cm$^{-3}$ and a temperature of $\sim$40 K.

### *2.2. Merging antiprotons and positrons*

As mentioned above, the traditional technique of launching eV $\bar{p}$ into a cold $e^+$ plasma generated $\bar{H}$ that is too energetic to be trapped [7]. ALPHA first attempted to solve this issue by merging the plasmas by slowly moving the potentials holding the $\bar{p}$ and $e^+$ in neighbouring wells [18]. This led to some new loss mechanisms in the neutral trap magnetic fields [19]. However, the technique did not result in trapped $\bar{H}$, so a novel technique using autoresonance [20] to excite the $\bar{p}$ collectively in their well until they enter the $e^+$ plasma was introduced [21]. This technique makes use of the fact that a anharmonic oscillator will lock to an external drive under certain conditions and by carefully chirping the drive frequency energy can be transferred to the oscillator in a deterministic fashion. To trap $\bar{H}$ ALPHA prepares the $\bar{p}$ and $e^+$ in neighbouring wells, energises the magnetic trap, evaporatively cools the $e^+$ and finally merges the two by autoresonantly exciting the $\bar{p}$ into contact with the $e^+$ (Figure 1b).

## 3. Trapped antihydrogen and its detection

As there are no readily available means for rapidly cooling the $\bar{H}$ they must be trapped directly after formation. Trapping can be done using the forces exerted by a magnetic field on the magnetic dipole moment of the (anti)atom that arises mainly due to the $e^+$ spin. The potential energy $U$ of a dipole in a magnetic field $\mathbf{B}$ is given by $U = -\boldsymbol{\mu}\cdot\mathbf{B}$, where $\boldsymbol{\mu}$ is the magnetic dipole moment.

In order to reduce the perturbative influence of the transverse multipole magnetic fields necessary to create the three dimensional minimum of magnetic field strength that forms the atom trap, ALPHA chose to use a transverse octupole rather than a quadrupole configuration common in the atom-trapping community [22].

To trap $\bar{H}$ it is synthesised with the neutral trap energised following the recipe given above. After autoresonant excitation the merged species are held for one second before all charged particles are ejected from the trap. Following this the super conducting neutral trap is de-energised rapidly with an e-folding time of 9 ms by dissipating its stored energy in external resistors. The trap depth is reduced to less than 1% of the full value in about 30 ms [22]. We look for signs of trapped $\bar{H}$ in this time window.

When searching for trapped $\bar{H}$ two potential false signals contribute. The first is from the cosmic rays passing through the apparatus. To deal with this background ALPHA uses a unique annihilation detector constructed with three layers of double-sided silicon strip detectors in a barrel geometry around the core of the apparatus [23]. The detector allows reconstruction of the tracks of the charged pions that appear from $\bar{p}$ annihilations, and by combining these tracks an annihilation vertex can be found, localising the exact position in space where the $\bar{p}$ annihilated. Cosmic rays can be distinguished by their very different tracking geometry. Using these techniques we obtain a detection efficiency of $\bar{p}$ annihilations of 57±6% with a background rate of 47±2 mHz. The effective background for detection of trapped $\bar{H}$ is further reduced by only searching for annihilation signals in the 30 ms window of de-energising the neutral trapping fields. The background fake rate is thus $1.4\pm0.1 \times 10^{-3}$ counts per experiment or one event in 700 experiments, thus effectively eliminating this issue [24, 5].

However the annihilation detector detects only $\bar{p}$ annihilations. In the ATHENA experiment, a similar detector was used, but with the additional capacity of detecting $e^+$ annihilations [3]. With the added material for the neutral trap magnets $e^+$ detection was not an option for ALPHA, and instead an extra layer of silicon was added to allow the track curvature to be measured [23]. The second type of false positives is thus bare trapped $\bar{p}$.

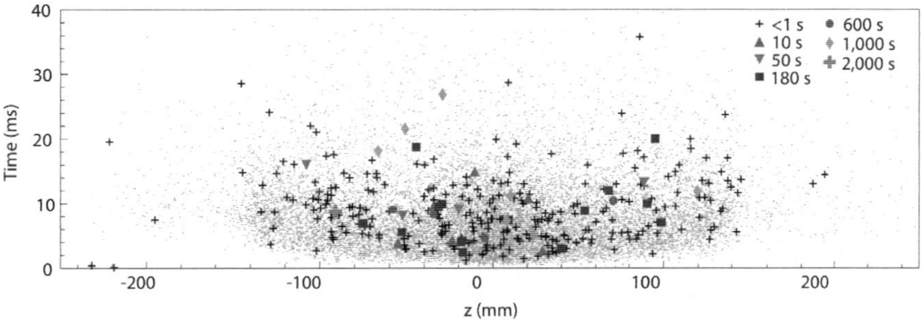


**Figure 2.** Time and axial $z$ distribution of annihilations on release of $\bar{H}$ from the neutral trap for different confinement times (see legend) and comparison with simulation (grey dots).

Before de-energising the neutral trap, charged particles are cleared from the trap by a series of electric pulses. Charged particles, in particular $\bar{p}$, may be trapped on their motional magnetic moment, so-called mirror-trapping. Trapping depends on the relative difference between their transverse (to the magnetic field) and longitudinal velocities. The clearing electric fields applied allow us to eject $\bar{p}$ with transverse energy less than 20 eV. As most of the manipulating potentials leading up to the $\bar{H}$ formation are lower than this it is unlikely that any $\bar{p}$ remain after clearing [25]. However, experimental errors and unforeseen circumstances could lead to a population of energetic $\bar{p}$ in the trap whose signal could be confused for trapped $\bar{H}$. To avoid these issues ALPHA applied three controls for fake trapped $\bar{H}$ signals. The first was to carry out the full $\bar{H}$ trapping process with all fields and manipulations but while heating the $e^+$ plasma to 1000 K and thus suppressing the $\bar{H}$ formation, and therefore any true $\bar{H}$ trapping signal. The second consisted of applying axial bias fields across the atom trap volume in order to cause any mirror-trapped $\bar{p}$ still present to be preferentially located either on the left or the right side of the trapping volume. The third and final control was to simulate the release of both $\bar{p}$ and $\bar{H}$ atoms from the trapping volume. Due to their different dynamics and energies their behaviour during de-energisation can easily be distinguished with the ALPHA annihilation detector [25]. Figure 2 shows 309 annihilation events recorded in 2010, with most of them taken with a wait time after $\bar{H}$ synthesis of less than one second. For most of the data presented here, a static electric bias field of 500 V $m^{-1}$ was applied during the confinement and shutdown stages to deflect bare $\bar{p}$ that may have been mirror-trapped. The bias field ensured that annihilation events could only be produced by neutral $\bar{H}$. We observed significant signal at holding times of up to 1000 s, sufficient to envisage the first measurements on ground state $\bar{H}$ [5].

## 4. Resonant quantum transitions in antihydrogen

Figure 3 shows the Breit-Rabi diagram for the ground state of (anti)hydrogen. For the ground state of hydrogen in a magnetic field, magnetic dipole transitions exist between the low-field seeking (trapped) states with the $e^+$ spin antiparallel to the local magnetic field ($|c>$ and $|d>$) and the high-field seeking (untrapped) states with the spin parallel. By inducing a spin-flip transition from low to high fields seeking state one may thus eject trapped $\bar{H}$ from the trap.

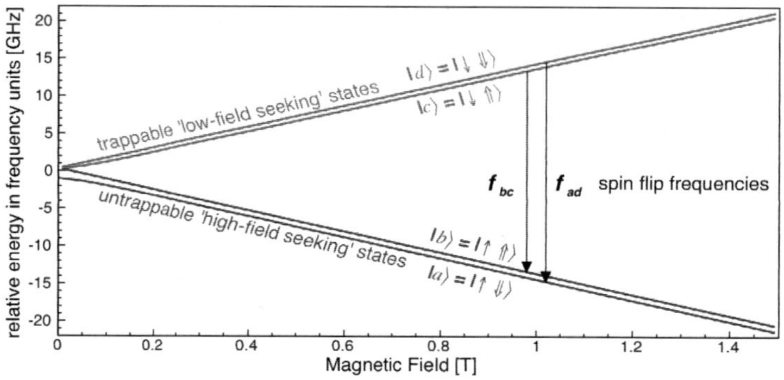


**Figure 3.** The Breit-Rabi diagram, showing the relative hyperfine energy levels of the ground state of the hydrogen (and $\bar{H}$, assuming CPT invariance) atom in a magnetic field. In the state vectors shown (for the high-field limit), the single arrow refers to the $e^+$ spin and the double arrow refers to the $\bar{p}$ spin.

In a series of experiments we illuminated the trapped $\bar{H}$ with microwave radiation in order to induce such transitions. As the $\bar{p}$ spin state is unknown we illuminated both the (bc) and the (ad) transition frequencies to ensure maximum effect [6]. To ensure frequency overlap with the transitions we swept the transition frequency +10/-5 MHz around the expected frequency of ~28 GHz. In order to further ensure interaction, we illuminated the two transitions for 15 s each six times repeated, for a total of 180 s of illumination. We made a series of measurements with the microwaves on and off resonance. To ensure against effects of standing waves in the apparatus, we shifted the microwaves on and off resonance by either changing the magnetic field during illumination by 3.5 mT or shift the microwaves by +100 MHz [6].

We searched for an effect of the microwaves in two distinct ways: (a) In so-called disappearance mode, we looked for a suppression of the trapping rate. We observed a clear effect with an decrease in trapping rate of a factor 10.5±2.0 from off to on resonance. (b) In appearance mode, we looked for signal during the microwave illumination. As the standard annihilation discrimination algorithms would result in too high a background from cosmic radiation for this, we refined the analysis to increase the signal to background. We reduced the background by an order of magnitude and the sensitivity by 25%. Figure 4 shows the results, clearly showing a peak in each of the two first 15 s windows where the two transitions are consecutively illuminated. Little to no signal is present when the microwaves are off resonance, with the possible exception of a small signal in the second 15 s period, which would be consistent with resonance with the tail of the lower transition. We thus observed resonant quantum transitions in trapped $\bar{H}$.

## 5. Outlook

With trapped $\bar{H}$, held for up to 1000 s, and the first observation of resonant quantum transitions, the field of $\bar{H}$ spectroscopy is taking its first steps. As the successful spin-flip measurement demonstrated, long confinement times and sensitive detection makes it feasible to make measurements with only a single anti-atom trapped on average. In the future we envisage

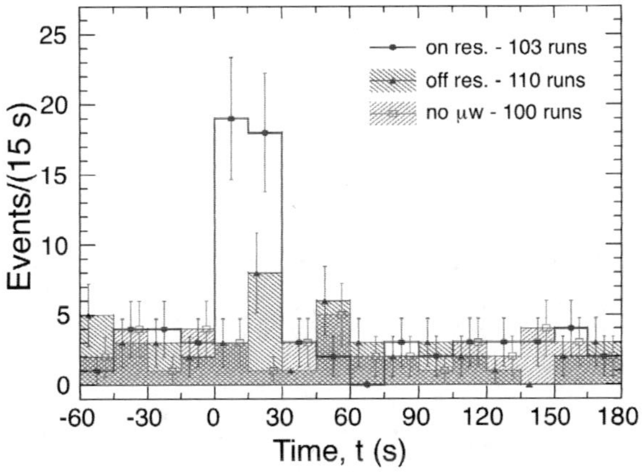


**Figure 4.** The number of appearance mode annihilation events as a function of time between the end of $\bar{\text{H}}$ production and the trap shutdown. Microwave power is first applied at time t = 0 s. The expected cosmic background per bin per run is 0.026 ± 0.005 events. The error bars are due to counting statistics.

experiments on the $\bar{\text{p}}$ spin-flip transition $|c>$ to $|d>$ which could give new bounds on the $\bar{\text{p}}$ magnetic moment. Furthermore, the ALPHA apparatus is currently being upgraded to allow laser-access and the first laser-spectroscopy of $\bar{\text{H}}$. Finally, trapped $\bar{\text{H}}$ may also serve as a good starting point for measurements of the gravitational influence on $\bar{\text{H}}$, in particular in combination with our unique annihilation position resolution.

## References

[1] Madsen N 2010 *Phil. Trans. Roy. Soc. A* **368** 3671
[2] Maury S 1997 *Hyp. Int.* **109** 43
[3] Amoretti M, *et al.*, (ATHENA Collaboration), 2002 *Nature* **419** 456
[4] Andresen G B, *et al.*, (ALPHA Collaboration) 2010 *Nature* **468** 673
[5] Andresen G B, *et al.*, (ALPHA Collaboration) 2011 *Nature Physics* **7** 558
[6] Amole C, *et al.*, (ALPHA Collaboration) 2012 *Nature* **483** 439
[7] Madsen N, *et al.*, (ATHENA Collaboration), 2005 *Phys. Rev. Lett* **94** 033403
[8] Gabrielse G, *et al.*, (TRAP Collaboration) 1989 *Phys. Rev. Lett.* **63** 1360
[9] Fajans J, *et al.* 2005 *Phys. Rev. lett.* **95** 155001
[10] Andresen G B, *et al.*, (ALPHA Collaboration), 2007 *Phys. Rev. Lett.* **98** 23402
[11] Andresen G B, *et al.*, (ALPHA Collaboration), 2008 *Phys. Plas.* **15**, 032107
[12] Fajans J, Madsen N and Robicheaux F, 2008 *Phys. Plas.* **15**, 032108
[13] Huang X-P *et al.* 1997 *Phys. Rev. Lett.* **78** 875
[14] Andresen G B, *et al.*, (ALPHA Collaboration) 2008 *Phys. Rev. Lett.* **100** 203401
[15] Andresen G B, *et al.*, (ALPHA Collaboration) 2010 *Phys. Rev. Lett.* **105** 013003
[16] Surko C M & Greaves R G 2004 *Phys. Plasmas* **11** 2333
[17] Jørgensen L V, *et al.*, (ATHENA Collaboration), 2006 *Phys. Rev. Lett.* **95** 025002
[18] Andresen G B, *et al.*, (ALPHA Collaboration) 2010 *Phys. Lett. B* **685** 141
[19] Andresen G B, *et al.*, (ALPHA Collaboration) 2009 *Phys. Plas.* **16** 100702
[20] Fajans J and Friedland L 2001 *Am. Jour. Phys.* **69** 1096
[21] Andresen G B, *et al.*, (ALPHA Collaboration) 2011 *Phys. Rev. Lett.* **106** 025002
[22] Bertsche W, *et al.*, (ALPHA Collaboration), 2006 *Nucl. Inst. & Meth. A* **566** 746
[23] Andresen G B, *et al.*, (ALPHA Collaboration) 2012 *Nucl. Inst. & Meth. A* **684** 73
[24] Andresen G B, *et al.*, (ALPHA Collaboration) 2011 *Phys. Lett. B* **695** 95
[25] Andresen G B, *et al.*, (ALPHA Collaboration) 2011 *New J. Phys.* **14** 015010

Journal of Physics: Conference Series **443** (2013) 012006 doi:10.1088/1742-6596/443/1/012006

# Linac-based positron source and generation of a high density positronium cloud for the GBAR experiment

**L. Liszkay[1], P. Comini[1], C. Corbel[2], P. Debu[1], P. Dupré[3], P. Grandemange[3], P. Pérez[1], J-M. Rey[1], N. Ruiz[1], Y. Sacquin[1]**

[1] CEA, Irfu, Centre de Saclay, F-91191 Gif-sur-Yvette, France

[2] Laboratoire des Solides Irradiés, Ecole Polytechnique, 91128 Palaiseau, France

[3] CSNSM-IN2P3/CNRS, Université de Paris Sud, 91406 Orsay, France

E-mail: laszlo.liszkay@cea.fr

**Abstract**. The aim of the recently approved GBAR (Gravitational Behaviour of Antihydrogen at Rest) experiment is to measure the acceleration of neutral antihydrogen atoms in the gravitational field of the Earth. The experimental scheme requires a high density positronium cloud as a target for antiprotons, provided by the Antiproton Decelerator (AD) - Extra Low Energy Antiproton Ring (ELENA) facility at CERN. We introduce briefly the experimental scheme and present the ongoing efforts at IRFU CEA Saclay to develop the positron source and the positron-positronium converter, which are key parts of the experiment. We have constructed a slow positron source in Saclay, based on a low energy (4.3 MeV) linear electron accelerator (linac). By using an electron target made of tungsten and a stack of thin W meshes as positron moderator, we reached a slow positron intensity that is comparable with that of $^{22}$Na-based sources using a solid neon moderator. The source feeds positrons into a high field (5 T) Penning-Malmberg trap. Intense positron pulses from the trap will be converted to slow ortho-positronium (o-Ps) by a converter structure. Mesoporous silica films appear to date to be the best candidates as converter material. We discuss our studies to find the optimal pore configuration for the positron-positronium converter.

The aim of the GBAR experiment [1] is the direct measurement of the gravitational acceleration of the simplest antiatom, antihydrogen, in the gravitational field of the Earth. Free fall of extremely cold $\overline{\mathrm{H}}$ atoms will be observed and $\overline{\mathrm{g}}$ will be determined with better than 1 % precision. The weak equivalence principle for antimatter has never been tested directly. Precision measurement of the gravitational acceleration on charged antiparticles (e.g. positron, antiproton) is rendered impossible by the fact that the residual electromagnetic forces are far stronger than the gravitational one in any conceivable measuring apparatus. On the other hand, deceleration and cooling of neutral particles to the energy range where the measurement of $\overline{\mathrm{g}}$ can be performed in the laboratory length scale is very difficult. GBAR is based on an experimental scheme [2] originally suggested by Walz and Hänsch [3]. Positively charged antihydrogen ions ($\overline{\mathrm{H}}^+$, which consist of an antiproton and two positrons) are cooled to nanoelectronvolt energy (10 μK temperature range). The cold ions are then neutralized by photodetachment and the time between the photodetachment and the annihilation of the antiatoms at the bottom (or top) of the experimental chamber is used to determine $\overline{\mathrm{g}}$. The experiment poses a number of challenges: creation of the $\overline{\mathrm{H}}^+$ ion, its capture and cooling to extreme low energy are steps that have never been achieved before. The present paper, after a brief presentation of the GBAR experimental scheme, gives a short summary of the efforts made at CEA/IRFU to develop the equipment needed to generate the dense positronium cloud, which is a prerequisite for the creation of the $\overline{\mathrm{H}}^+$ ions.

## 1. The GBAR experimental scheme

A simplified scheme of the GBAR experiment is shown in Figure 1. The antiprotons are provided by the Antiproton Decelerator (AD) facility at CERN. The particles will be further decelerated from 5.3 MeV to 100 keV kinetic energy in the ELENA antiproton storage ring. Finally, the $\bar{\mathrm{p}}$ pulses reach a few keV kinetic energy in the switched electrostatic decelerator of the experiment. The antiprotons hit a target consisting of a dense positronium (Ps) cloud. Its density should be sufficiently high so that both charge exchange reactions

$$\bar{\mathrm{p}} + Ps \rightarrow \overline{\mathrm{H}} + e^- \qquad (1)$$
$$\overline{\mathrm{H}} + Ps \rightarrow \overline{\mathrm{H}}^+ + e^- \qquad (2)$$

that are necessary for the production of the positive ion, take place with sufficient efficiency. The positronium states for the highest possible formation efficiency of $\overline{\mathrm{H}}^+$ in the target cloud will be set by laser excitation. The cross sections of the two reactions are being studied by numerical simulations [4].

The created $\overline{\mathrm{H}}^+$ ions are captured and decelerated to a kinetic energy in the electronvolt range. Cooling to extreme low energy will be achieved in two steps. In the first one the ions will reach a temperature in the mK range by sympathetic cooling with a laser Doppler cooled $^9Be^+$ cloud. In the final cooling step a $\overline{\mathrm{H}}^+$-$^9Be^+$ atom pair will be cooled to the 10 μK temperature range by Raman sideband cooling.

For the photodetachment of the $\overline{\mathrm{H}}^+$ ions a well tuned, horizontally polarized laser beam will be used, in order to minimize the initial velocity of the antihydrogen atoms. The time and place of the annihilation of the antihydrogen atom will be detected by position sensitive detectors, surrounding the experimental chamber. The chamber will provide typically 30 cm distance for the free fall.

## 2. The positron source

Production of the positronium cloud at every antiproton pulse (typically every 110 s) requires a primary slow positron ($e^+$) source with more than $10^8$ $e^+$/s intensity. This intensity is beyond the capability of the currently available isotope-based, solid neon moderated slow positron sources. The planned positron source of the GBAR experiment is based on a linear electron accelerator (linac) with low (up to 20 MeV) beam energy. Positrons will be created by pair production in a cooled tungsten target. In the baseline design, a stack of thin tungsten meshes is used as positron moderator. The moderated positrons will be transported in a weak (typically 8 mT) magnetic field to the positron accumulator, a multi-electrode Penning-Malmberg trap with 5 T magnetic field. At the low transport energy used, most positrons could not pass the magnetic mirror at the entrance of the trap. A special injection system, which forms short (typically 100 ns) bunches from the approximately 3 μs long primary slow positron pulses and accelerates $e^+$ to kiloelectronvolt kinetic energy before injection into the trap, is being developed to overcome the magnetic mirror effect. The cooling scheme [5,6] uses a cold electron plasma, which is trapped in a separate potential well in the trap. Positrons will be cooled by the trapped electron plasma, and collected in a second potential well. Eventually, approximately $2 \times 10^{10}$ positrons will be accumulated in the trap. The particles will be ejected in a short (<100 ns) pulse and will hit a positron-positronium converter target.

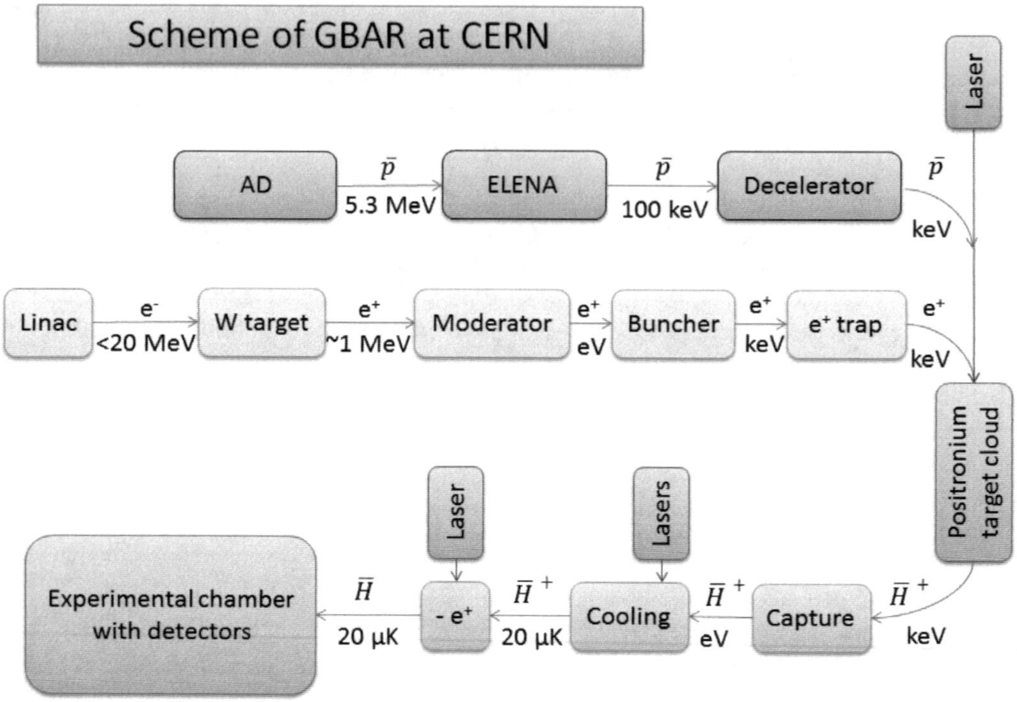


**Figure 1** The scheme of the GBAR experiment (AD=Antiproton Decelerator, ELENA= (Extra Low Energy Antiproton Ring).

## 3. The positronium target cloud

A positronium target cloud with up to $10^{12}$ Ps/cm$^3$ density is needed for the efficient formation of $\overline{H}^+$. Positronium will be formed from the intense positron pulse in a positron/positronium converter structure. The converter should re-emit positronium upon irradiation with $e^+$. Only ortho-positronium can be used, as the 125 ps lifetime of the singlet para-positronium state is too short for the reaction with the antiproton pulse. In order to increase the effective density of the positronium cloud, the converter has to emit Ps at low energy. Furthermore, it has to withstand the irradiation with intense $e^+$ pulses without loss of efficiency. On the basis of the literature and our previous studies [7], the most promising candidate for a $e^+$/Ps converter is thin (typically 300-1000 nm) mesoporous silica film, grown on silicon. Ps is created in the film and escapes from the pore system with high (>40 %) efficiency and relatively low energy (<50 meV).

The positronium cloud will be created in a small chamber (Figure 2), typically 1x1x10 mm size. The $e^+$/Ps converter forms one side of the chamber. The other three sides are made of a material that reflects Ps with high efficiency. The slow positron pulse will enter the cavity on an open side. It will be then deviated by an electrostatic or magnetic field and the positrons will be implanted into the converter film at approximately 3 keV energy. Ortho-positronium, created in the film and escaping into the cavity, will be confined in the cavity by the reflecting walls. The chamber should give access to the antiproton beam and the laser beam used to excite Ps, and leave place for the exit of the created $\overline{H}^+$ (and also the much more numerous neutral antihydrogen atoms).

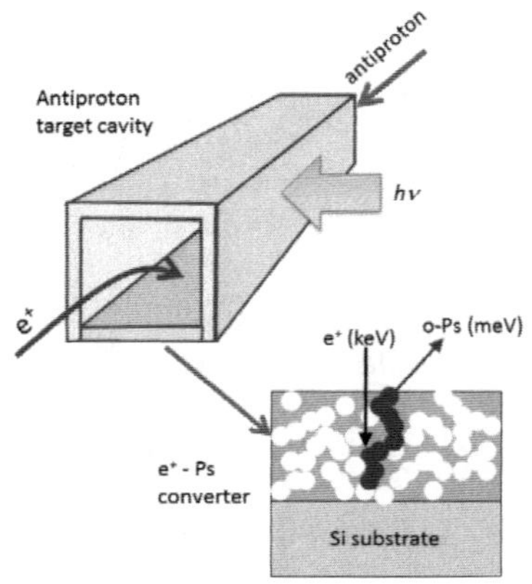


**Figure 2** Scheme of the positronium cloud production chamber and the positron-positronium converter film.

## 4. Development of the $e^+$/Ps converter

As the positronium cloud is a key ingredient of the GBAR setup, positronium creation, escape probability and energy should be studied in detail to ensure that a Ps cloud is formed with the density needed for the ion production. Various positron annihilation spectroscopic methods can be used to study positronium creation and escape in thin porous silica films. The most frequently available slow positron beam based detection method uses the measurement of the energy distribution of the annihilation gamma photons. The fraction of positrons which annihilate with three gamma photons can be deduced from these measurements. As most ortho-positronium in vacuum annihilate with three gamma photons, this quantity gives information on the ortho-positronium escape fraction. However, this detection method is not able to separate o-Ps annihilation in the film (i.e., in pores) from the annihilation events in vacuum. Only positron lifetime spectroscopy can provide full information on the o-Ps escape process. In the design of the lifetime spectrometer and in the analysis of the spectra, care should be taken that the different detection efficiency of different annihilation sites (annihilation in vacuum and annihilation in the film) and the difference between the detection efficiency of 2 gamma and 3 gamma annihilation events do not distort the results. Measurements on mesoporous silica [7,8] have shown that a combination of lifetime and 3 gamma fraction method is an efficient and reliable way to study o-Ps escape in these systems.

The minimum energy of the re-emitted o-Ps is determined by the quantum confinement in the pores [9,10,11]. One way to decrease the energy is by increasing the pore size of the silica, as the zero-point energy decreases with increasing pore size. Measurements on macroporous silica films have shown that films with as large as 32-77 nm pore size exhibit significant (up to 25 %) o-Ps escape fraction [12]. Further studies are underway to find the pore size and porous fraction that gives the optimal o-Ps escape fraction/o-Ps energy combination for the use of the o-Ps converter.

## 5. The slow positron test source at CEA/IRFU

We developed a slow positron source [13] as a test system for the essential parts of the GBAR setup and to provide slow positrons for positron spectroscopy. It is based on a compact, low energy (4.3 MeV) linac with 200 Hz repetition frequency, 4 μs pulse length and 100 μA mean current. The slow positron intensity is approximately $3 \times 10^6$ $e^+$/s. The slow positron transfer line can supply $e^+$ for the 5 T Penning-Malmberg trap or to a beam line where slow positron spectroscopy will be performed. A magnetic beam switch is used to move the beam between the two beam lines. To provide a quasi-continuous positron intensity for spectroscopy use, a pulse stretcher has been installed and tested.

## 6. Summary and outlook

Extrapolation of the measured slow positron yield for higher electron energies shows that the desired positron intensity can be reached at 18-20 MeV linac energy. Further improvement of the positron moderator structure is possible by improving the target-moderator geometry. Attempts will be made to develop a solid neon moderator, which would presumably increase the positron yield significantly. The main elements of the positron cloud production target have been developed; the target cavity has to be tested first with a low intensity positron beam, then with intensities near the final one. A major challenge is to accumulate $2x10^{10}$ positrons in the high field trap and eject it into the target cavity within 100 ns. The positron source, together with the antihydrogen ion cooling system and the experimental chamber, has to be installed at the experimental site at CERN by 2016, the expected date of the commissioning of the ELENA ring.

## 7. Acknowledgement

This work is supported in part by Conseil Général de l'Essonne under ASTRE 2006 funding and by Research National Agency under contract number ANR-05-BLAN-0380-01.

## References

[1] Pérez P and the GBAR collaboration 2011 Proposal to measure the Gravitational Behaviour of Antihydrogen at Rest GBAR, CERN-SPSC-2011-029, Geneva, Switzerland
[2] Pérez P and Rosowsky A 2004 *Nucl. Inst. Meth. A* **532** 523
[3] Walz J and Hänsch T 2004 *General Relativity and Gravitation* **36** 561
[4] Comini P and Hervieux P-A, in this volume
[5] Oshima N, Kojima T M, Niigaki M, Mohri A, Komaki K, Iwai Y and Yamazaki Y 2003 *Nucl. Inst. Meth. B* **205** 178
[6] Dupré P 2010 Ph. D. thesis, Université de Pierre et Marie Curie Paris VI, Orsay, France
[7] Liszkay L, Corbel C, Pérez P, Desgardin P, Barthe M-F, Ohdaira T, Suzuki R, Crivelli P, Gendotti U, Rubbia A, Etienne M and Walcarius A 2008 *Appl. Phys. Lett.* **92** 063114
[8] Liszkay L, Corbel C, Raboin L, Boilot J-P, Pérez P, Brunet-Bruneau A, Crivelli P, Gendotti U, Rubbia A, Ohdaira T, and Suzuki R 2009 *Appl. Phys. Lett.* **95** 124103
[9] Mariazzi S, Salemi A and Brusa R S 2008 *Phys. Rev. B* **78** 085428
[10] Crivelli P, Gendotti U, Rubbia A, Liszkay L, Pérez P, Corbel C 2010 *Phys. Rev. A* **81** 052703
[11] Cassidy D B, Crivelli P, Hisakado T H, Liszkay L, Meligne V E, Pérez P, Tom H W K and Mills A P Jr. 2010 *Phys. Rev. A* **81** 012715
[12] Liszkay L, Guillemot F, Corbel C, Boilot J-P, Gacoin T, Barthel E, Pérez P, Barthe M-F, Desgardin P, Crivelli P, Gendotti U and Rubbia A 2012 *New Journal of Physics* **14** 065009
[13] Rey J-M, Coulloux G, Debu P, Dzitko H, Hardy P, Liszkay L, Lotrus P, Muranaka T, Noel C, Pérez P, Pierret O, Ruiz N and Sacquin Y, in this volume.

# $\bar{\mathrm{H}}$ and $\bar{\mathrm{H}}^+$ production cross sections for the GBAR experiment

**P Comini**[1] **and P-A Hervieux**[2]

[1]CEA Saclay, DSM/Irfu/SPP, 91191 Gif sur Yvette Cedex, France
[2]IPCMS/DON, 23 rue du Loess, BP 43, 67034 Strasbourg Cedex, France

E-mail: pauline.comini@cea.fr

**Abstract**. The production and cooling of the $\bar{\mathrm{H}}^+$ ion is the key point of the GBAR experiment (Gravitational Behaviour of Antihydrogen at Rest), which aims at performing the free fall of antihydrogen atoms to measure $\bar{g}$, the acceleration of antimatter on Earth. $\bar{\mathrm{H}}^+$ ions will be obtained from collisions between a positronium cloud and antiprotons delivered by the AD/ELENA facility at CERN, with intermediate formation of antihydrogen atoms. In order to optimise the experimental production of $\bar{\mathrm{H}}^+$ ions, we computed the total cross sections of the two corresponding reactions, within the same theoretical framework of the Continuum Distorted Wave – Final State (CDW-FS) model. The different contributions of the $\bar{\mathrm{H}}$ excited states have been systematically investigated for different states of Ps. The results exhibit an increase of the $\bar{\mathrm{H}}$ production toward low kinetic energies, in agreement with experimental data and previous calculations, whereas the largest $\bar{\mathrm{H}}^+$ production is obtained with low energy ground-state antihydrogen atoms. These theoretical predictions suggest that the overall production of $\bar{\mathrm{H}}^+$could be optimal for 2 keV antiproton impact energy, using positronium atoms prepared in the 2p state.

## 1. Introduction

The GBAR experiment (Gravitational Behaviour of Antihydrogen at Rest) has been recently approved by the CERN Research Board [1]. It aims at performing the free fall of antihydrogen atoms in order to measure $\bar{g}$, the gravitational acceleration of antimatter on Earth, and thus test the Weak Equivalence Principle for antimatter. The main source of uncertainty in this experiment is the initial antihydrogen ($\bar{\mathrm{H}}$) velocity. This is why ultra-cold antihydrogen atoms of a few neV are required. The key idea of the GBAR experiment is to produce $\bar{\mathrm{H}}^+$ ions that can be cooled using techniques developed in cold-atom physics [2]. They will be obtained from a two-step process: (i) creation of antihydrogen from collisions between keV antiprotons ($\bar{\mathrm{p}}$) and a gas of positronium atoms (Ps), (ii) production of $\bar{\mathrm{H}}^+$ from further collisions between these antihydrogen atoms and the positronium cloud. The two reactions are:

$$\bar{p} + Ps(n_{Ps}, l_{Ps}) \longrightarrow \bar{H}(n_H, l_H) + e^- \quad (1)$$

$$\bar{H}(n_H, l_H) + Ps(n_{Ps}, l_{Ps}) \longrightarrow \bar{H}^+ + e^- \quad (2)$$

where the positronium atoms can be optically prepared in any state ($n_{Ps}$, $l_{Ps}$), and $\bar{\mathrm{H}}$ can be produced from reaction (1) in any ($n_H$, $l_H$) state. Experimentally, the antiproton kinetic energy (delivered by the AD/ELENA at CERN), the positronium excited state and its fraction with respect to the ground state

Journal of Physics: Conference Series **443** (2013) 012007 doi:10.1088/1742-6596/443/1/012007

can be chosen to increase the number of $\overline{\mathrm{H}}^+$ produced per $\bar{\mathrm{p}}$ pulse. The aim of the present theoretical study is to optimise these parameters by computing the total cross sections of reactions (1) and (2).

Although the three-body reaction (1) has been already widely theoretically studied [3,4] this is not the case for the four-body reaction (2) for which only a few theoretical results can be found, often limited to the simplest cases (H(1s) + Ps($n_{Ps}$=1,2)) [5]. In the present study, the cross sections of the two reactions have been computed at the same level of approximation within the framework of the Continuum Distorted Wave – Final State model (CDW-FS).

In the following, we give a short description of the CDW-FS model and present our theoretical predictions for both reactions. Atomic units are used unless otherwise specified.

## 2. Theoretical model

Reactions (1) and (2) are equivalent, according to the principle of microreversibility and using charge conjugation, to the two following reactions:

$$e^+ + H(n_H, l_H) \longrightarrow p + Ps(n_{Ps}, l_{Ps}) \tag{3}$$

$$e^+ + H^- \longrightarrow H(n_H, l_H) + Ps(n_{Ps}, l_{Ps}) \tag{4}$$

Cross sections for reactions (1) and (3) (respectively (2) and (4)) are related by a simple kinetic factor [6].

The common theoretical framework chosen here to describe reactions (3) and (4) is called Continuum Distorted Wave – Final State (CDW-FS). It is indeed closely related to CDW methods since special attention is given to the treatment of the boundary conditions, hence the use of Coulomb wave functions to describe exactly the asymptotic states. The presence of a charged target (respectively charged target fragment) in the initial (resp. final) channel requires a particular treatment. Originally introduced to study fast positron/hydrogen-like collisions [7], the CDW-FS model includes distortions in the final channel (*prior* version) related to the Coulomb continuum states of both the positron and the electron in the field of the residual target. The 3-body formulation of the model, well-suited for reaction (3), has been later adapted to 4-body reactions [8], and applied in particular to the case of positronium formation from positron collisions with metastable helium [9], which is very similar to reaction (4). We closely followed the formalism developed in references [7] and [9], where all the details can be found.

In the interest of the GBAR experiment, we want to investigate the low energy regime (down to 1 eV in the centre of mass). We are well-aware that our collisional model is normally only valid for intermediate and high impact energies and that the *post/prior* discrepancy inherent to the CDW approximation increases toward low energies [10]. If the values of the cross sections are certainly not reliable close to the energy thresholds, we can however extract meaningful and valuable information from their relative behaviours since the two reactions are investigated using the same model.

The treatment of the $H^-$ ion can be performed at different levels of approximation. $H^-$ is a weakly bound system (ionisation potential: 0.76 eV) which only exists in its ground state and exhibits strong electronic correlation effects. A large variety of wave functions is available in the literature. At present, the most accurate is proposed by Le Sech [11]. For the sake of computational simplicity, we used the uncorrelated Chandrasekhar wave function [12], which includes radial correlations (but not angular).

## 3. Results

We systematically computed the cross sections of reactions (3) and (4) for Ps(1s) to Ps(3d) and H(1s) to H(4f). The contribution of the higher hydrogen excited states ($n_H > 4$) can be estimated using the well-known $n^{-3}$ rule. We have checked that this correction is usually small. All cross sections are given as a function of the initial antiproton kinetic energy in the laboratory frame, where the positronium atom is supposed to be at rest.

Journal of Physics: Conference Series **443** (2013) 012007 doi:10.1088/1742-6596/443/1/012007

### 3.1. Reaction (1)

In Figure 1, we have chosen to present in detail our theoretical predictions for the case of Ps(1s). For the sake of clarity, for a given quantum number $n_H$, the cross sections have been summed over the $l_H$ states. Due to the energy threshold values for $n_H \geq 2$ and to the very low cross sections for $n_H = 1$, almost no $\overline{H}$ are expected to be produced below 5 keV. Above thresholds, the production of higher excited states of $\overline{H}$ is favored (the same behaviour is found for the other states of Ps). This agrees with other calculations [3]. We also compared our cross sections summed over the $\overline{H}$ states with the experimental results (inclusive measurements) of Merrison *et al.* [13], where hydrogen atoms were formed by impact of protons on a positronium target. Although our theoretical predictions slightly overestimates the $\overline{H}$ production, the agreement with the experimental data is rather good, thus validating the use of the CDW-FS model for evaluating the cross sections of the 3-body reaction (1).

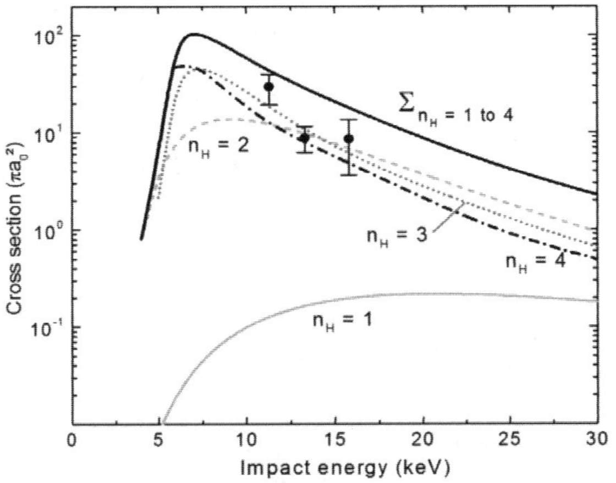


**Figure 1.** Cross sections for reaction (1) with Ps(1s) and $\overline{H}$ labelled by $n_H$ (see text) as a function of the antiproton impact energy. Available experimental data are represented by black dots with error bars [13].

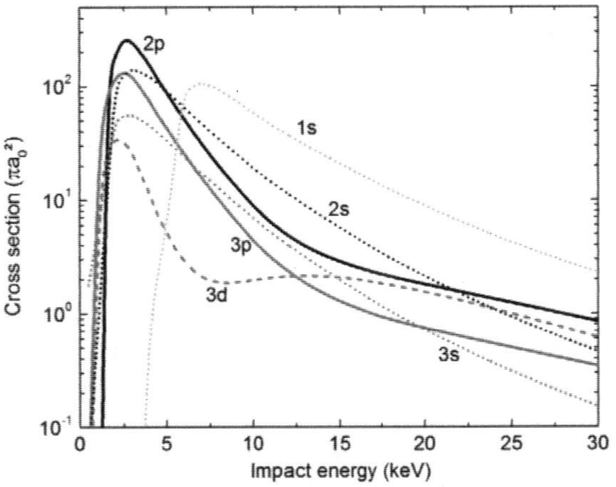


**Figure 2.** $\overline{H}$ production cross sections (summed over $n_H$ from 1 to 4) as a function of the antiproton impact energy, for Ps(1s) to Ps(3d).

In Figure 2, we compare the total cross sections summed over all the computed $\overline{H}$ states, for Ps(1s) to Ps(3d). Except for Ps(1s), the production of $\overline{H}$ is maximal in the range 2-4 keV, Ps(2s), Ps(2p) and Ps(3p) exhibiting the highest cross sections.

### 3.2. Reaction (2)

The most accurate value of the $H^-$ ground-state energy is -0.5277 a.u. [14]. This happens to be almost equal to the sum of the binding energies of hydrogen in its ground state (-0.5 a.u.) and of the positronium atom in a state $n_{Ps} = 3$ (-0.02778 a.u.). Therefore, in that peculiar case, we can expect the $\overline{H}^+$ formation to be highly efficient (nearly resonant) almost at zero impact kinetic energy. Indeed, as can be seen on Figure 3, a sharp increase of the cross section at zero impact energy is observed for Ps(3d) (also for Ps(3p), not shown on the figure). However, it is worth noticing that reaction (2) with Ps(2p) in the entrance channel is once again dominant. This might be attributed to the high polarisability of the 2p orbital. Another very important observation is that the contribution of the excited states of $\overline{H}$ to the production of $\overline{H}^+$ is always negligible, except in the case of Ps(1s). For ground state $\overline{H}$, Ps(2s) and Ps(3s), not shown on Figure 3, are just below Ps(1s).

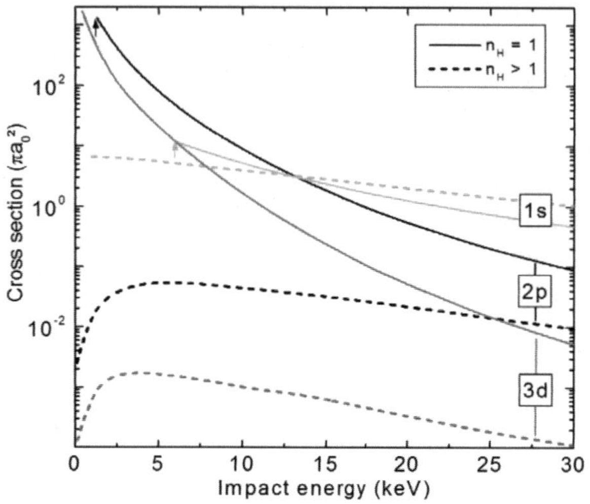


**Figure 3.** Cross sections for reaction (2) as a function of the antiproton impact energy. Only results for Ps(1s), Ps(2p) and Ps(3d) are presented. Solid curves represent antihydrogen ground-state in the entrance channel, while the dashed curves correspond to the contribution of all other $\bar{H}$ excited states. The arrows indicate the threshold energies.

## 4. Conclusion

Excitation of positronium atoms can indeed lead to a higher rate of $\bar{H}^+$ production for GBAR. Despite the dramatic increase of the cross section close to the threshold of reaction (2) with Ps(3d), this might not be the best option since below 1 keV antiproton impact energy, very few antihydrogen atoms will be produced through reaction (1). Our calculations rather suggest that Ps(2p) would be the most attractive solution and the optimal antiproton impact energy around 2 keV. However, Ps(2p) (and also Ps(3p)) is a short-lived state requiring to set-up a sophisticated laser system: these experimental constraints may lead to favour Ps(3d) over the p-states. Nonetheless, one should not forget that in reality, most of the positronium atoms will remain in their ground state (as they are produced).

The present study also highlights the need to de-excite the $\bar{H}$ produced through the first reaction. It leads also to important experimental constraints. Therefore, in order to accurately handle the evolution of the different Ps and $\bar{H}$ populations over time, a simulation of the interaction region is deeply needed. Moreover, work on the influence of the angular correlations in $H^-$ is also currently underway.

Once again, it must be emphasised that our collisional model (CDW-FS) might not yield quite reliable results at very low impact energy (near threshold) where more sophisticated calculations (e.g. coupled channel models) are needed. Nevertheless it can provide valuable information on the relative behaviour of the cross sections relevant for the design of the GBAR experiment.

## References

[1] The GBAR Collaboration 2011 CERN-SPSC-2011-029, SPSC-342
Debu P 2012 *Hyperfine Interact.* **212** 51
[2] Walz J and Hänsch T W 2004 *Gen. Relativ. Gravit.* **36** 561
[3] Mitroy J 1995 *Phys. Rev.* A **52** 2859
[4] Ghanbari-Adivi E 2010 *J. Phys. B: At. Mol. Opt. Phys.* **43** 135202 and references therein
[5] Roy S and Sinha C 2008 *Eur. Phys. J.* D **47** 327 and references therein
[6] Humberston J W, Charlton M, Jacobsen F M and Deutch B I 1987 *J. Phys.* B **20** L25
[7] Fojón O A, Rivarola R D, Gayet R, Hanssen J and Hervieux P-A 1996 *Phys. Rev.* A **54** 4923
[8] Fojón O A, Hanssen J, Hervieux P-A and Rivarola R D 2000 *J. Phys. B: At. Mol. Opt. Phys.* **33** 3093
[9] Hanssen J, Hervieux P-A, Fojón O A and Rivarola R D 2000 *Phys. Rev.* A **63** 012705
[10] Brandsen B H, Joachain C J and McCann J F 1992 *J. Phys. B: At. Mol. Opt. Phys.* **25** 4965
[11] Le Sech C 1997 *J. Phys. B: At. Mol. Opt. Phys.* **30** L47
[12] Chandrasekhar S 1944 *Astrophys. J.* **100** 176
[13] Merrison J P, Bluhme H, Chevallier J, *et al.* 1997 *Phys. Rev. Lett.* **78** 2728
[14] Pekeris C L 1958 *Phys. Rev.* **112** 1649

Journal of Physics: Conference Series **443** (2013) 012008 doi:10.1088/1742-6596/443/1/012008

# Vacancy clusters in ultra fine grained metals prepared by severe plastic deformation

**J. Čížek[1], O. Melikhova[1], Z. Barnovská[1], I. Procházka[1], R. K. Islamgaliev[2]**
[1]Charles University in Prague, Faculty of Mathematics and Physics,
V Holešovičkách 2, 180 00 Prague 8, Czech Republic
[2]Institute of Physics of Advanced Materials, Ufa State Aviation Technical University,
12 K. Marx Street, 450000 Ufa, Russian Federation

E-mail: jakub.cizek@mff.cuni.cz

**Abstract.** Severe plastic deformation of metals introduces not only dislocations, but also a high concentration of vacancies. In the present work we employed positron lifetime spectroscopy for investigation of deformation-induced vacancies in ultra fine grained metals prepared by high pressure torsion. It was found that in all metals studied deformation-induced vacancies agglomerate into small vacancy clusters. Experimental positron lifetime results were combined with *ab-initio* theoretical calculations of positron parameters for vacancy clusters of various sizes. This new approach described in this paper enables to determine the size distribution of vacancy clusters.

## 1. Introduction

Severe plastic deformation (SPD) enables to achieve an extreme grain refinement down to nanoscopic dimensions leading to formation of ultra fine grained (UFG) structure. SPD introduces a high density of dislocations into the strained material. Moreover, non-conservative movement of dislocations leads to formation of vacancies. An indirect evidence for deformation-induced vacancies in UFG metals comes from comparison of differential scanning calorimetry (DSC) and electrical resistivity with X-ray diffraction (XRD) line profile analysis [1,2]. Dislocation densities determined by DSC from the total stored energy and by electrical resistivity are significantly higher than those obtained by XRD [2]. This discrepancy can be explained by excess vacancies, which also contribute to the total stored energy and electrical resistivity, but do not cause broadening of XRD profiles. Direct evidence for deformation-induced vacancies in UFG Cu and Ni was obtained by positron annihilation spectroscopy [3,4]. Since in majority of metals vacancies are mobile at room temperature, some fraction of deformation-induced vacancies disappear by diffusion to sinks at grain boundaries, while other vacancies agglomerate and form small vacancy clusters which are more stable than single vacancies. Driving energy for this process is reduction of the surface energy of vacancies.

Among the variety of SPD-based techniques, the high pressure torsion (HPT) [5] leads to the strongest grain refinement. For this reason HPT was chosen for preparation of UFG samples studied in the present work. The deformation-induced vacancy clusters in various UFG metals with fcc, bcc and hcp structure were investigated by positron lifetime (LT) spectroscopy combined with *ab-initio* theoretical calculations. This approach enables to determine the size distribution of deformation-induced small vacancy clusters.

## 2. Experimental

A digital spectrometer [6] with excellent time resolution of 145 ps (FWHM $^{22}Na$) was employed for LT investigations. At least $10^7$ annihilation events were accumulated in each LT spectrum. A $^{22}Na_2CO_3$ positron source with activity of ~1 MBq deposited on a 2 μm thick mylar foil was always sealed between two identical pieces of sample. Diameter of the source spot was ~ 3 mm. The source contribution which comes from positrons annihilated in the source spot and the covering mylar foil consisted of two weak components with lifetimes of 0.368 and 1.5 ns.

Specimens of well annealed Cu (99.95%), Al (99.9999%), Fe (99.99%), Nb (99.9%), W(99.95%), and Ti (99.7%) were subjected to quasi-constrained HPT straining under pressure of 6 GPa. The samples were strained by 5 HPT revolutions performed at 450°C (W), 150°C (Nb) and at room temperature (Cu, Al, Fe, Ti). The grain size determined by transmission electron microscopy falls in the range 100-250 nm for all samples studied except of Al which exhibits the mean grain size of ~ 0.5 μm. HPT-deformed specimens were disk shaped, with a diameter of ~ 10 mm and a thickness of ~ 0.3 mm. LT investigations were always performed in the centre of the sample disk except of HPT-deformed Cu where LT characterizations were done not only in the centre but also at the periphery with positron source positioned at the radial distance $r$ = 3 mm from the centre.

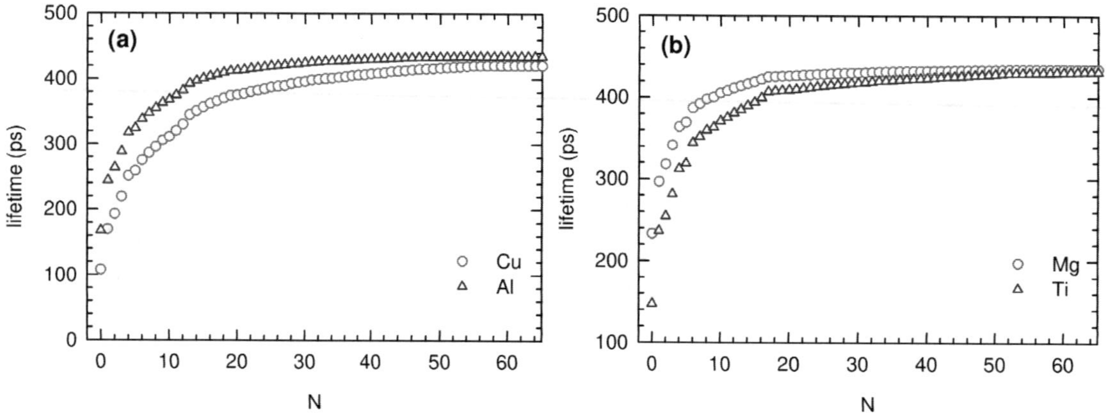


**Figure 1.** Calculated dependence of the positron lifetime on the size of vacancy clusters consisting of $N$ vacancies: (a) fcc metals Cu, Al; (b) hcp metals Mg, Ti.

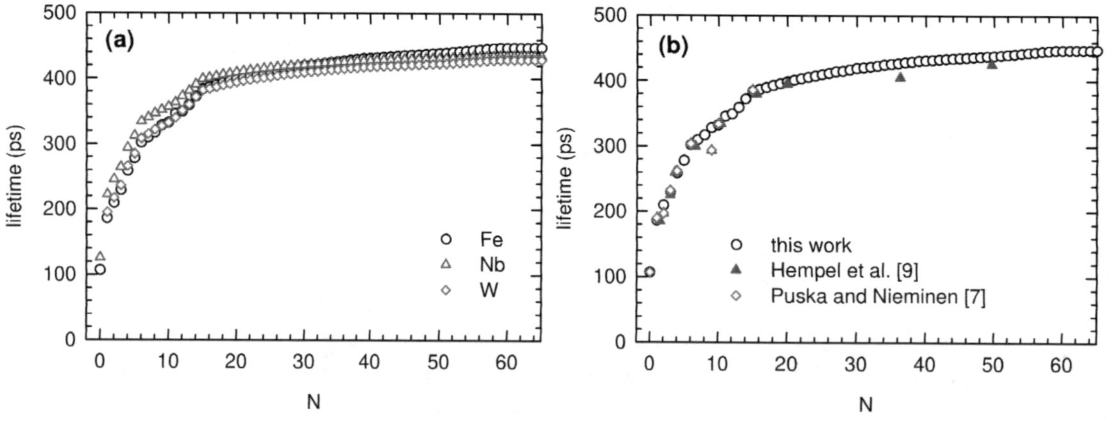


**Figure 2.** (a) Calculated dependence of the positron lifetime on the size of vacancy clusters consisting of $N$ vacancies for bcc metals Fe, Nb, W; (b) Comparison of the lifetimes calculated in this work for Fe with data existing in literature.

Journal of Physics: Conference Series **443** (2013) 012008 doi:10.1088/1742-6596/443/1/012008

## 3. Theoretical calculations

Theoretical calculations of positron lifetimes were performed within the so-called standard scheme employing the atomic superimposition (ATSUP) method for construction of electron density [7]. The electron–positron correlations were treated according to Boroński and Nieminen [8]. The calculations have been done in 512 atom-based supercells. Vacancies were introduced simply by removing corresponding atoms from the supercell. Equilibrium geometry of vacancy clusters was obtained by minimizing their surface. No atomic relaxations were considered in the present stage of calculations.

Calculated lifetimes of positrons trapped at vacancy clusters consisting of $N$ vacancies are plotted in Fig. 1(a) and 1(b) for fcc metals (Al, Cu) and hcp metals (Mg, Ti), respectively. Calculated positron lifetimes for metals with bcc structure (Fe, Nb, W) are plotted in Fig. 2(a). Positron lifetime firstly strongly increases for very small clusters, while for larger clusters it gradually saturates. Theoretical calculations of positron lifetimes for selected vacancy clusters in Fe were performed in Refs. [7,9]. These results which are plotted in Fig. 2(b) are in a satisfactorily agreement with our calculations.

**Table 1.** Results of LT measurements of HPT deformed metals.

*Fit using discreet components only*

| Sample | $\tau_1$ (ps) | $I_1$ (%) | $\tau_2$ (ps) | $I_2$ (%) | $\tau_3$ (ps) | $I_3$ (%) | $\chi^2/n$ |
|---|---|---|---|---|---|---|---|
| Cu | - | - | 164(1) | 76.0(6) | 256(2) | 24.0(5) | 1.06 |
| Al | 149(8) | 71(5) | 252(9) | 26(1) | 450(50) | 3(1) | 1.03 |
| Fe | - | - | 150.8(6) | 91.1(4) | 360(9) | 8.9(4) | 1.07 |
| Nb | - | - | 173.9(8) | 94(1) | 300(10) | 6(1) | 1.10 |
| W | - | - | 160.8(6) | 90.8(4) | 367(7) | 9.2(7) | 1.04 |
| Ti | - | - | 185.2(5) | 98.4(7) | 430(70) | 1.6(7) | 1.05 |

*Fit using size distribution of vacancy clusters, Eq. (1)*

| Sample | $\tau_1$ (ps) | $I_1$ (%) | $\tau_2$ (ps) | $I_2$ (%) | $v_d$ | $I_d$ (%) | $\chi^2/n$ |
|---|---|---|---|---|---|---|---|
| Cu | - | - | 164(1) | 73.3(8) | 3.7(2) | 26.7(8) | 1.03 |
| Al | 158(1) | 86.5(4) | 240(3) | 10(1) | 40(3) | 3.5(4) | 1.02 |
| Fe | - | - | 150.8(8) | 89.8(2) | 11.3(5) | 10.2(3) | 1.03 |
| Nb | - | - | 174(1) | 91.2(3) | 3.0(2) | 8.8(5) | 1.04 |
| W | - | - | 161.6(6) | 90.5(5) | 13(1) | 9.5(7) | 1.02 |
| Ti | - | - | 182(1) | 96.3(5) | 39(2) | 3.7(7) | 1.03 |

## 4. Results and discussion

The LT spectra of HPT-deformed metals with exception of Al consist of two components, see Table 1. The lifetime $\tau_2$ of the shorter component is slightly shorter compared to that for a monovacancy [10] which is typical for dislocations. It is well known that dislocation line itself is only a shallow positron trap [11]. Positron pre-trapped at a dislocation diffuses quickly along the dislocation line until it becomes finally confined at a deep trap provided by a vacancy anchored in the compressive elastic field of dislocation [11]. Hence, positron is finally annihilated in a vacancy squeezed by the elastic field of dislocation resulting in lifetime slightly shorter than that for a bare monovacancy. In case of HPT-deformed Al a short lived free-positron component with lifetime $\tau_1$ was observed in addition to the dislocation component. This indicates a dynamic recovery of UFG structure during HPT processing, which was observed also in HPT-deformed Mg [12]. In other metals studied the concentration of defects is high enough to cause saturated positron trapping.

All HPT-deformed metals studied exhibit a long-lived component with lifetime $\tau_3$ (see Table 1) which comes from positrons trapped at vacancy clusters. The mean size of vacancy clusters can be estimated by comparison of the lifetime $\tau_3$ with the theoretical calculations in Figs. 1, 2. However, HPT-deformed metals contain obviously certain size distribution $P(N)$ of deformation-induced vacancy clusters. Hence LT spectrum of a HPT-deformed metal can be assumed in a general form

$$S(t)=\left\{\sum_i \frac{I_i}{\tau_i} e^{-\frac{t}{\tau_i}} + I_d \left(\sum_N \nu_N\right)^{-1} \sum_N \frac{\nu_N}{\tau_N} P(N)\, e^{-\frac{t}{\tau_N}}\right\} \otimes R(t) + B, \quad (1)$$

where $\tau_i$, $I_i$ denotes the lifetimes and the relative intensities of discreet exponential components, here the free positron component ($i$ = 1), if present, and the dislocation component ($i$ = 2). The symbol $I_d$ denotes the relative intensity of the cluster contribution given by a superposition of exponential components with lifetimes $\tau_N$ of positrons trapped at vacancy clusters consisting of $N$ vacancies. These lifetimes were obtained from theoretical calculations in Figs. 1, 2. Each component is weighted by a factor $\nu_N P(N)$, where $\nu_N$ is the specific positron trapping rate accounting for varying cross-section for positron trapping at vacancy clusters of various size. For small vacancy clusters ($N \leq 10$) $\nu_N$ is proportional to the cluster size, while for larger clusters the slope of this dependence decreases. Nieminen and Laakkonen [13] calculated $\nu_N$ for vacancy clusters of various sizes in Al. From interpolation of these results the dependence of $\nu_N$ on the cluster size can be approximated as $\nu_N / \nu_1 \approx 9.4\,(1 - e^{-0.13\,N})$. The ideal spectrum in brackets in Eq. (1) is convoluted with the resolution function $R(t)$ of the spectrometer and a constant background $B$ from random coincidences is added. The relative intensities and size distribution of vacancy clusters always satisfy the normalization conditions

$$\sum_i I_i + I_d = 1 \quad \text{and} \quad \sum_N P(N) = 1. \quad (2)$$

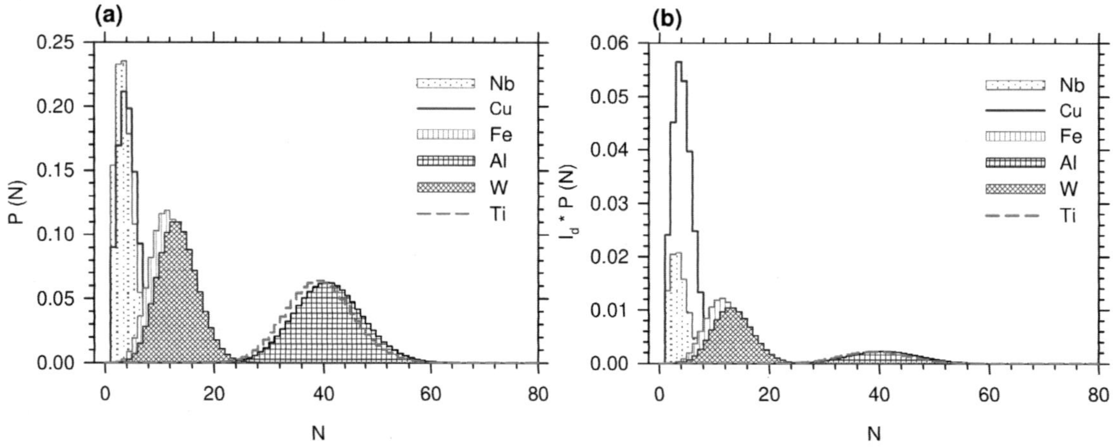


**Figure 3.** (a) The size distribution *P(N)* of vacancy clusters in various HPT-deformed metals; (b) the size distribution *P(N)* of vacancy clusters multiplied by the relative intensity $I_d$ of the cluster contribution. All results were measured at the centre of the sample disks.

Vacancy clusters are created by agglomeration of deformation-induced vacancies. Movement of a screw dislocation with a jog results in creation of a row of vacancies or interstitial atoms. The creation of vacancies dominates due to their lower formation energy. Hence, vacancies are introduced in rows created by moving dislocation. Since vacancy clusters are formed by agglomeration of vacancies in these rows, the size of vacancy clusters depends predominantly on the number of vacancies in these rows. Drag of a dislocation jog is result of many attempts to move the jog under applied stress. Hence the number of vacancies in the row created by a jog dragged by moving dislocation can be described by the Poisson distribution

$$P(N) = \frac{\nu^N e^{-\nu_d}}{N!}, \quad (3)$$

where $\nu_d$ is the mean size vacancy clusters. The second part of Table 1 shows result of fitting of LT spectra by Eq. (1) with *P(N)* given by Eq. (3). It was found that such model function describes the experimental data accurately in all metals studied. Since we use the Poisson distribution *P(N)* with the only fitting parameter $\nu_d$ (the variance of the distribution equals to the mean) the model function given by Eq. (1) does not introduce any new fitting parameter compared to the conventional approach with discreet components. Last column in Table 1 shows the $\chi^2$ coefficient per the number of degrees-of-freedom *n*. Obviously fitting of LT spectra by Eq. (1) improved $\chi^2$ coefficient with respect to the conventional model with discreet components in case of all samples except of Al where it remains the same. Moreover, in Al the fitting by Eq. (1) resulted in lifetime $\tau_2$ which agrees better with lifetime of positrons trapped at dislocations in Al [14]. Fig. 3(a) shows the distributions *P(N)* (normalized to the unit area) obtained from fitting of LT spectra. The distributions multiplied by the relative intensity $I_d$ of the cluster contribution are plotted in Fig. 3(b). The latter distributions give information not only about the size distribution of vacancy clusters, but also about their concentration. Cu and Nb exhibit very narrow size distribution of vacancy clusters with the mean size of ~ 4 vacancies, Fe and W contain larger vacancy clusters with the mean size of ~ 10 vacancies, Al and Ti contain the biggest vacancy clusters with the mean size of ~ 40 vacancies having a wide size distribution. The size distribution of vacancy clusters in HPT-deformed metals is not directly related to their crystalline structure, e.g. fcc Cu exhibits very small clusters while fcc Al contains big clusters. Systematic investigations on a broader set of UFG metals are required to identify the factors influencing the size distribution of deformation-induced clusters. However, it seems to be clear that the majority of vacancy clusters is not formed directly by deformation, but rather by aggregation of deformation-induced vacancies. Since the accumulation rate is influenced by the mobility of vacancies the big clusters in Al and Ti are likely formed due to low activation energy $E_m$ for migration of vacancies in these two metals. Theoretical calculations revealed $E_m$ = 0.43 eV [15] and 0.42 eV [16] for Ti and Al, respectively, which is less than half of the vacancy migration energy in Cu, $E_m$ = 1.00 eV [17].

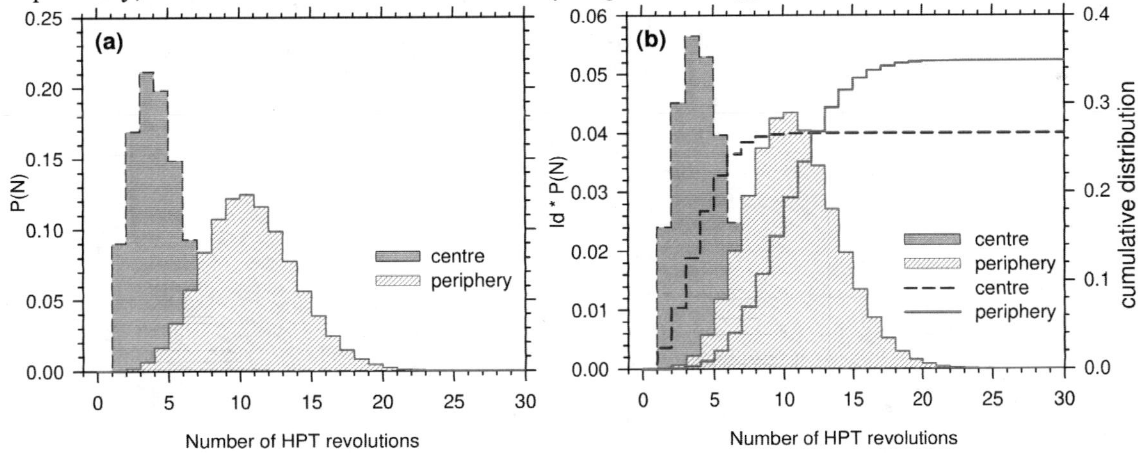


**Figure 4.** Comparison of the size distribution of vacancy clusters in UFG Cu in the centre of the sample ($r$ = 0) and at the periphery ($r$ = 3 mm): (a) comparison of the normalized size distribution *P(N)*; (b) the size distribution *P(N)* multiplied by the relative intensity $I_d$ of the cluster component. Thick lines show cumulative distribution $I_d \, \Sigma \, P(N)$.

The vacancy production rate at low temperature plastic deformation depends on the strain rate $\dot{\varepsilon}$ [18]

$$\Pi = \alpha \frac{\sigma \Omega_0}{E_f} \dot{\varepsilon} , \tag{4}$$

where the coefficient $\alpha \approx 0.1$ [19], $\sigma$ is the applied stress, $\Omega_0$ the atomic volume and $E_f$ the vacancy formation energy. In torsion deformation the strain rate increases with the radial distance $r$ from the centre of the sample (corresponding to the rotation axis). The results given in Table 1 and Fig. 3 were obtained with the positron source positioned at the centre of the sample. Since all samples studied were deformed in the same way, the strain rate was identical for all samples. The diameter of the central area of the sample probed by positrons is approximately 3 mm due to finite size of the source spot. To examine the influence of strain rate on the size distribution of vacancy clusters UFG Cu sample was measured not only at the centre but also at the "periphery" with the positron source located at the distance $r = 3$ mm from the centre. A comparison of the size distribution of vacancy clusters measured at the centre and at the periphery is given in Fig. 4(a). Obviously the vacancy clusters at the periphery are larger compared to those in the centre due to higher production rate of vacancies. The size distribution $P(N)$ multiplied by the relative intensity $I_d$ of the cluster distribution is plotted in Fig. 4(b) together with the cumulative distribution $I_d\, \Sigma_N\, P(N)$. The total concentration of deformation-induced vacancies is clearly higher at the periphery because of higher strain rate.

## 5. Conclusions

SPD introduces a lot of vacancies which subsequently agglomerate into clusters. Size distribution of these deformation-induced clusters in various UFG metals prepared by HPT was determined by LT spectroscopy combined with *ab-initio* theoretical calculations. It was found that size distribution of vacancy clusters is considerably different in various UFG metals. A high mobility of vacancies in Al and Ti leads to formation of rather big clusters with mean size of ~ 40 vacancies and wide size distribution, Fe and W exhibit smaller vacancy clusters with size of ~ 10 vacancies, very small clusters with size of ~ 4 vacancies and narrow size distribution were found in Cu and Nb. Vacancy clusters at the periphery of the sample disk are larger compared to those in the centre due to higher strain rate.

## 6. References

[1] Schafler E, Steiner G, Korznikova E, Kerber M and Zehetbauer M 2005 *Mater. Sci .Eng. A* **410-411** 169
[2] Ungar T, Schafler E, Hanák P, Bernstorff S and Zehetbauer M 2007 *Mater. Sci. Eng. A* **462** 398
[3] Čížek J, Janeček M, Srba O, Kužel R, Barnovská Z, Procházka I and Dobatkin S 2011 *Acta Mater.* **59** 2322
[4] Oberdorfer B, Steyskal E-M, Sprengel W, Puff W, Pikart P, Hugenschmidt Ch, Zehetbauer M, Pippan R and Würschum R 2010 *Phys. Rev. Lett.* **105** 146101
[5] Zhilyaev A P and Langdon T G 2008 *Prog. Mater. Sci.* **53** 893
[6] Bečvář F, Čížek J, Procházka I and Janotová J 2005 *Nucl. Instrum. Methods A* **539** 372
[7] Puska M J and Nieminen R M 1983 *J Phys F* **13** 333
[8] Boroński E and Nieminen R M 1983 *Phys Rev B* **34** 3820
[9] Hempel A, Hasegawa M, Brauer G, Plazaola F, Saneyasu M and Tang Z 1999 *Proc. of the 9th Int. Conf. on "Enviromental Degradation of Materials in Nuclear Systems-Water Reactors"*, (Aug. 1-5, 1999, Newport Beach, CA, USA)
[10] Campillo Robles J M, Ogando E and Plazaola F 2007 *J. Phys.: Condens. Matter* **19** 176222
[11] Smedskjaer L, Manninen M and Fluss M 1980 *J. Phys. F: Met. Phys.* **10** 2237
[12] Čížek J, Procházka I, Smola B, Stulíková I, Kužel R, Matěj Z, Cherkaska V, Islamgaliev R K and Kulyasova O 2005 *Mater. Sci. Forum* **482** 183
[13] Nieminen R M and Laakkonen J 1979 *Appl. Phys.* **20** 181
[14] Petersen K, Repin I A and Trumpy G 1996 *J. Phys.: Condens. Matter* **8** 2815
[15] Verite G, Willaime F and Fu Ch-Ch 2007 *Solid State Phenomena* **129** 75
[16] Da Fano A and Jacucci G 1977 *Phys. Rev. Lett.* **39** 950
[17] Plishkin Yu M and Podchinenov I E 2006 *Phys. stat. sol. (a)* **38** 51
[18] Militzer M, Sun W P and Jonas J J 1993 *Acta Mater.* **42** 133
[19] Gottstein G, Bewerunge J, Mecking H and Wollenberger H 1975 *Acta Met.* **23** 641

**Acknowledgments**

This work was supported by the Czech Science agency (project P108/12/G043).

Journal of Physics: Conference Series **443** (2013) 012009 doi:10.1088/1742-6596/443/1/012009

# High-temperature cuprate superconductors studied by x-ray Compton scattering and positron annihilation spectroscopies

**Bernardo Barbiellini**

Department of Physics, Northeastern University, Boston MA 02115, USA

E-mail: bba@neu.edu

**Abstract.** The bulk Fermi surface in an overdoped ($x = 0.3$) single crystal of $La_{2-x}Sr_xCuO_4$ has been observed by using x-ray Compton scattering. This momentum density technique also provides a powerful tool for directly seeing what the dopant Sr atoms are doing to the electronic structure of $La_2CuO_4$. Because of wave function effects, positron annihilation spectroscopy does not yield a strong signature of the Fermi surface in extended momentum space, but it can be used to explore the role of oxygen defects in the reservoir layers for promoting high temperature superconductivity.

## 1. Introduction

Discovered by Bednorz and Müller in 1986 [1], high temperature superconductivity in cuprates has mobilized scientists around the world, both to understand its origin and to find new compounds with higher superconducting transition temperature $T_c$.

The lanthanum compounds $La_{2-x}M_xCuO_4$ have variable impurity content $x$ and M = Ca, Ba, Sr. Their crystal structure, shown in Fig. 1, belongs to the $K_2NiF_4$ family. It is built up from Cu-O superconducting layers alternating with rock salt type layers, which provide the charge reservoirs. When $x = 0$, these materials have antiferromagnetic insulating phases. It is only with increasing $x$ that the metallic state and superconductivity set in. A naive chemical argument finds that, when $x = 0$, copper has a valency +2 and thus contains one hole (missing electron). The increase of $x$ creates excess of holes that causes the oxides to conduct.

However, the physics and chemistry of doping high-temperature cuprate superconductors still remain something of a mystery. For instance, standard theories cannot readily explain the sudden destruction of superconductivity at high doping concentrations. For this reason, we have undertaken [2] Density Functional Theory (DFT) calculations for nano-sized supercells of $La_2CuO_4$ with dopants. These calculations show that weak ferromagnetism appears around clusters of high Ba concentration and suggest that ferromagnetism and superconductivity compete in overdoped samples. Our studies on doping have also shown how by scattering high energy x-rays from single crystals of $La_{2-x}Sr_xCuO_4$, one can directly image the character of holes doped into this material [3]. At low doping the hole O $2p$ character is much stronger than the Cu $3d$ character. However, at high doping the Cu $3d$ character becomes dominant. These observations are highly significant because via x-ray scattering we have new tools for directly seeing what the dopant atoms are doing in these materials [4]. The same advanced x-ray characterization could be used to study cathode materials for lithium batteries [5], where the octahedron formed by a transition metal atom and six oxygen atoms plays a key role like the $CuO_6$ building block in the $La_2CuO_4$ compound (see Fig. 1).

Since high $T_c$ superconductivity appears to be near a metal-insulator transition there is a particular significance attached to the existence of the Fermi Surface (FS) in the normal state. Currently, the only direct probe of FS is angle-resolved photoemission (ARPES) [6]. While ARPES has many strengths, this has created the problem that most of our understanding of cuprates is based on measurements from a single, surface sensitive technique that has been applied only to a limited number of materials that cleave. Thus, the risk is that experimental artifacts can be interpreted as fundamental physics. Therefore, it is important to develop new approaches to measuring both electron momentum density and FS applicable to a much wider class of materials.

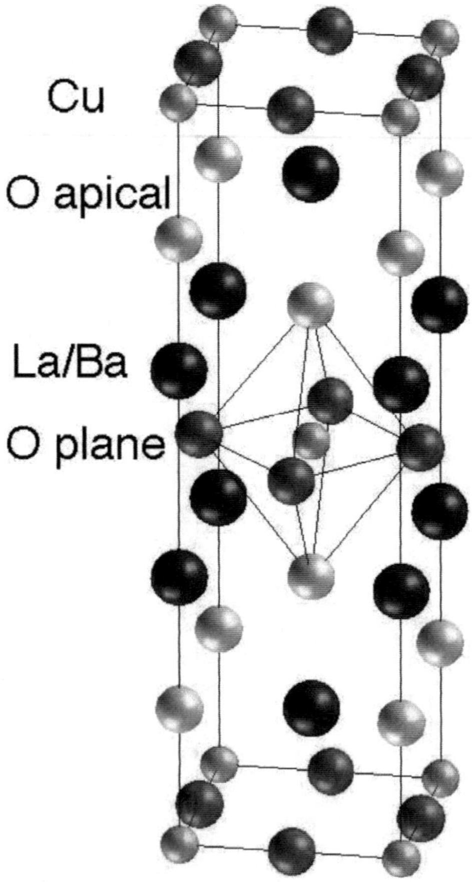


**Figure 1.** The structure of $La_{2-x}Ba_xCuO_4$. The La and Ba atoms (blue spheres) are in the charge reservoir layers, which contains all the dopant atoms. The so-called apical O atoms (yellow spheres) are located in the same layers while the Cu atoms (green spheres) and the planar O atoms (red spheres) belong to the superconducting layers. The $CuO_6$ octahedron (shown in the middle) is formed by a $CuO_4$ plaquette and two apical O atoms and determines the important $x^2 - y^2$ and $z^2$ features of the electronic structures.

## 2. 2D ACAR spectroscopy

The two dimensional angular correlation of positron annihilation radiation (2D ACAR) [7, 8, 9] depends, by momentum and energy conservation, on the two photon momentum density $R(p)$ of an annihilating electron-positron pair [10, 11]. Given a typical electronic momentum $|p| \sim 10^{-2}mc$, the angle between the two photons deviates only by a few milliradians from

Journal of Physics: Conference Series **443** (2013) 012009 doi:10.1088/1742-6596/443/1/012009

anticollinearity, since each photon has momentum $\sim mc$ (energy $\sim mc^2$). Thus angles $\theta_x$ and $\theta_y$ are directly related to the $p_x = mc\theta_x$ and $p_y = mc\theta_y$ components of $p$. The two dimensional (2D) ACAR spectrum is measured by $\gamma$ detectors at a large distance from the sample (about 10 meters). Then one obtains the projection of $R(\mathbf{p})$

$$N(p_x, p_y) = \int dp_z \, R(\mathbf{p}). \tag{1}$$

Actually, the distribution $N(p_x, p_y)$ is convoluted with the resolution function of the experimental setup and the typical experimental resolution is of the order 0.03 a.u. of momentum. The 2D ACAR has been successful for a determination of the FS in many metallic systems [11], but similar studies of the high-$T_c$ oxides have met difficulties [4, 9]. Although the electron-positron momentum density measured in a positron-annihilation experiment contains FS signatures, the amplitude of such signatures is controlled by the extent to which the positron wave function overlaps with the states at the Fermi energy. Calculations of the positron density distribution indicates that the positron does not probe well the FS contribution of the Cu-O planes in $La_{2-x}Sr_xCuO_4$ [12]. Indeed, we see little evidence of FS signatures in either the computed or the measured 2D ACAR distributions.

## 3. X-ray Compton scattering spectroscopy

Compton scattering refers to inelastic x-ray scattering [13, 14, 15] in the deeply inelastic limit in which the non-resonant scattering cross-section is well-known to become proportional to the ground state momentum density [16]. Compton scattering thus provides a uniquely powerful probe of the many body ground state wave function. The technique is also well suited for investigating disordered alloys because, being a ground state probe, it does not require long electron mean free paths. By contrast, transport type experiments such as the De Haas-van Alphen effect are limited to low defect and impurity concentrations. Like other light scattering techniques, Compton scattering is genuinely a bulk probe, which is not complicated by the presence of surface effects involved in ARPES. Historically, the capabilities of the Compton technique have been slow to realize in practice because only a limited momentum resolution ($\sim 0.4$ a.u.) was possible with gamma ray sources. However, with the availability of second and third generation synchrotron light sources and improvements in detector technology, it has become possible now to achieve momentum resolution of about 0.1 a.u. in wide classes of materials, and of about 0.01 a.u. in low-Z systems [13, 17]. These advances have sparked a renewed interest in Compton scattering as a tool for investigating Fermi liquid and bonding related issues and also as a unique window on hitherto inaccessible correlation effects in the electron gas. We have developed extensive methodology for obtaining momentum density and highly accurate Compton profiles (CPs) within the Fermi liquid framework [11, 18] and generalized it to the correlated Resonating valence bond (RVB) ground state [19, 20]. These studies have shown, for example, the surprising degree to which Li deviates from the Fermi liquid paradigm [21]. These results have also prompted us to look for ways of going beyond the standard DFT framework by using antisymmetric geminal products (AGP) of pair wavefunctions to construct the many-body state in a solid. The AGP is an important building block of the RVB. This scheme can be applied to highly correlated electron systems such as high temperature superconductors. In the ground state of these materials, multi-configuration effects lead to non-integer occupancies of one-electron states. The AGP and RVB calculations have shown that the deviation from integer occupancy becomes stronger as correlations grow. However, the important finding of our recent work in an overdoped ($x = 0.3$) single crystal of $La_{2-x}Sr_xCuO_4$ [12] is that strong correlation effects that ordinarily reduce the Compton anisotropy amplitudes are no longer effective at this doping. The quantitative agreement between DFT calculations and momentum density experiments (2D ACAR and 2D Compton scattering reconstruction) shown

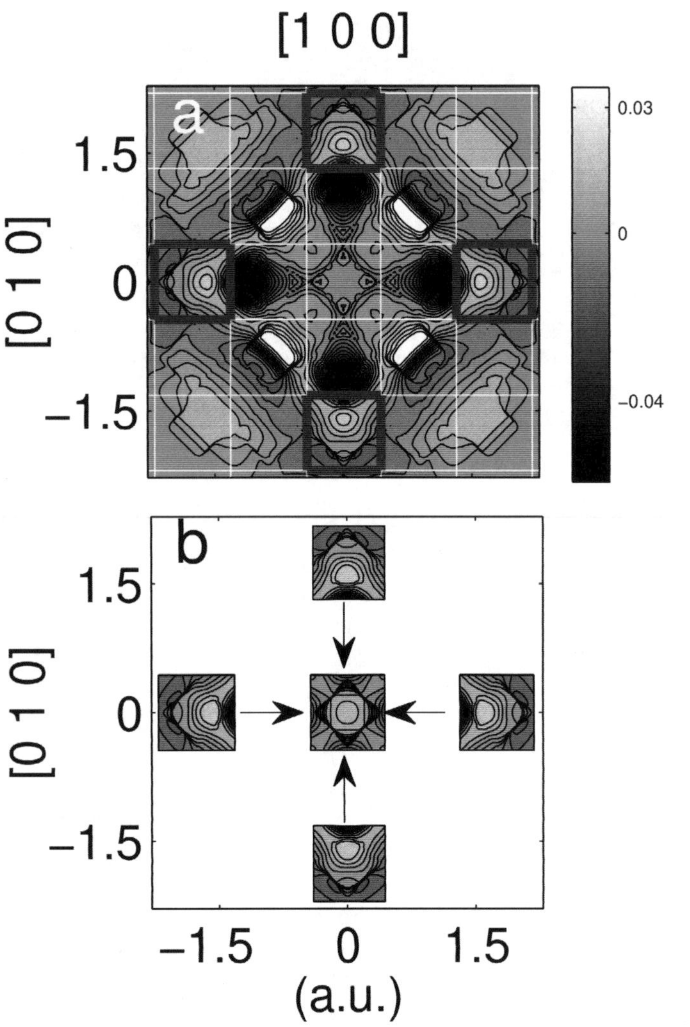


**Figure 2.** FS analysis from the 2D electron momentum density distribution: a) Cylindrical anisotropy for theoretical electron momentum density. The white lines define the Brillouin zones, while blue squares indicate a family of zones where the FS features are particularly strong. (b) The blue squared regions are isolated from the rest of the spectra and folded back to the central region. The color scale is in units of the 2D momentum density at the origin. The total amplitude of the 2D momentum density anisotropy is of the order of 8% in excellent agreement with the experiment.

in Ref. [12] suggests that Fermi liquid physics is restored in the overdoped regime. A FS signal was also clearly observed by Compton scattering in the third Brillouin zone along [100] and the calculated FS topology was found to be in good accord with the corresponding experimental data. In particular, Fig 2 (a) shows the cylindrical anisotropy of the theoretical 2D electron moment density. Because the FS is periodic, a complete FS must exist in each Brillouin zone, but with its intensity modulated by electronic wave functions effects. For a predominantly $d$ character at the FS, these wave functions will strongly suppress spectral weight near the origin, so the FS breaks are most clearly seen in higher Brillouin zones. These FS breaks appear superimposed on the momentum density in the form of discontinuities which can occur in any Brillouin zone. In Fig.2 (a), the Fermi surface breaks are the regions where the contours run closely together. This means that the electron momentum density varies rapidly at these locations. Fig.2 (a) shows the calculated Fermi breaks in several Brillouin zones. In particular, in the third zones framed with blue squares, the arc-like features are theoretically predicted FSs associated with

Cu-O planes. Due to the tetragonal symmetry, a rotation of the spectrum by $\pi/2$ will generate symmetry-related regions (blue squares) with equivalent strong FS features. These regions are isolated in Fig.2 (b). By folding back only these regions as shown by the arrows of Fig.2 (b), we produce a full FS, where strong wave functions effects are substantially circumvented.

## 4. Oxygen defects in high temperature superconductors

The positron lifetime is an ideal tool to detect low concentrations of defects in materials [10]. It is determined by the positron-electron contact density, which is remarkably higher than the unperturbed electron density [9]. The ratio of these two densities is called the *enhancement factor* and it is used in ACAR calculations as well. The agreement between the positron annihilation theory and the experiment is usually excellent within the Generalized Gradient Approximation (GGA) [22, 23]. The GGA can also be safely applied for the calculation of positron annihilation rates at defects in cuprates [24].

Curiously, it has been difficult to obtain cuprates single crystals of a sufficient size and perfection. In fact, some of these substances have atomic defects such as oxygen vacancies and interstitials, others contain charge density waves and some others present themselves with a number of different phases. Are all these only accidental complications of minor importance to high temperature superconductivity? Or, are they on the contrary connected to essential properties of high temperature superconductivity? An empirical law [25] (only valid for high temperature superconductors) suggests that

$$k_B T_c \tau \sim \hbar, \tag{2}$$

where $k_B$ is the Boltzmann's constant, $\hbar$ is the Planck constant, $T_c$ is the superconducting critical temperature and $\tau$ is the electron relaxation time in the normal state. Therefore, in order to increase $T_c$, one can increase the electron scattering (i.e. lower $\tau$) by introducing certain impurities or defects. The intense level of scattering given by Eq. 2 corresponds to the so-called *Planckian dissipation limit*, beyond which Bloch-wave propagation becomes inhibited, i.e. the quasi-particle states themselves become incoherent. The fact that the relaxation rate $1/\tau$ scales with $T_c$ implies that the interaction causing this anomalous scattering is also associated with the superconducting pairing mechanism [26].

Interestingly, a different route to hole doping is to introduce excess of oxygen in the $La_2CuO_4$ compound, which is possible because there are extra sites in the reservoir layers for the oxygen atoms to occupy. $La_2CuO_{4+\delta}$ when doped in the range of $0.1 \leq \delta \leq 0.12$ achieves the best $T_c$, however the form of the oxygen interstitials distribution [27] can change both $T_c$ and $\tau$ dramatically. So far as the technique of positron spectroscopy is concerned, it has become clear that the propensity to oxygen defects [24] has an essential influence on positron annihilation measurements. For example, we have shown that positrons are trapped at oxygen vacancies in the superconducting compound $YBa_2Cu_3O_{7-\delta}$ while this trapping becomes negligible in the non-superconducting sample where Y has been replaced by Pr [28]. Our goal is therefore to show via positron annihilation spectroscopy that oxygen defects (both interstitials and vacancies) are connected to essential properties of high temperature superconductivity [29].

## 5. Conclusion

The quantitative agreement between the calculations and the experiment for ACAR as well as momentum density anisotropies suggests that $x = 0.3$ overdoped $La_{2-x}Sr_xCuO_4$ can be explained within the conventional Fermi-liquid theory. The FS analysis from Compton scattering confirms previous ARPES FS measurements showing an electron-like FS in the overdoped regime. This validation is important since we provide via deep inelastic x-ray scattering experiments a truly bulk-sensitive image of momentum density maps of electrons near the

Fermi level. In general, this momentum density information is difficult to extract from ARPES experiments due to difficulties associated with matrix element effects and the well-known surface sensitivity of ARPES. However, improvements in the momentum resolution are still needed to bring Compton scattering into the fold of mainstream probes for the cuprates FS. Higher momentum resolution can also enable the study of the FS smearing due to the superconducting energy gap opening [9]. Positron annihilation spectroscopy cannot probe well the FS of the Cu-O layers, but it can be extremely useful to explain why oxygen defects at reservoir layers play an important role in high temperature superconductivity [29].
*Acknowledgments:* This work is supported by the USDOE, Grants No. DE- FG02-07ER46352 and No. DE-FG02-08ER46540 (CMSN) and benefited from the allocation of supercomputer time at NERSC and Northeastern University Advanced Scientific Computation Center (ASCC).

## References

[1] J.C. Bednorz and K.A. Müller, Z. Phys. B **64** (1986) 180
[2] B. Barbiellini and T. Jarlborg, Phys. Rev. Lett. **101** (2008) 157002
[3] Y. Sakurai, M. Itou, B. Barbiellini, P. E. Mijnarends, R. S. Markiewicz, S. Kaprzyk, J.-M. Gillet, S. Wakimoto, M. Fujita, S. Basak, Yung Jui Wang, W. Al-Sawai, H. Lin, A. Bansil, and K. Yamada, Science **332** (2011) 698
[4] B. Barbiellini and P.M. Platzman, Mat. Sci. Forum **255-257** (1997) 189
[5] Xiaosong Liu, Jun Liu, Ruimin Qiao, Yan Yu, Hong Li, Liumin Suo, Yong-sheng Hu, Yi-De Chuang, Guojiun Shu, Fangcheng Chou, Tsu-Chien Weng, Dennis Nordlund, Dimosthenis Sokaras, Yung Jui Wang, Hsin Lin, Bernardo Barbiellini, Arun Bansil, Xiangyun Song, Zhi Liu, Shishen Yan, Gao Liu, Shan Qiao, Thomas J. Richardson, David Prendergast, Zahid Hussain, Frank M. F. de Groot and Wanli Yang, J. Am. Chem. Soc. **134** (2012) 13708
[6] A. Damascelli, Z. Hussain, Z.-X. Shen, Rev. Mod. Phys. **75** (2003) 473
[7] S. Berko, M. Hagooie and J.J. Mader, Phys. Lett. A **63** (1977) 335
[8] P.E. Bisson, P. Descouts, A. Dupanloup, A.A. Manuel, E. Perreard, M. Peter and R. Sachot, Helv. Phys. Acta **55** (1982) 110
[9] M. Peter, T. Jarlborg, A.A. Manuel, B. Barbiellini, S.E. Barnes, Z. Naturforsc. **48a** (1993) 390
[10] B. Barbiellini, in *New Directions in Antimatter Chemistry and Physics*, edited by C. M. Surko and F. A. Gianturco (Kluwer Academic Publishers, The Netherlands, 2001)
[11] B. Barbiellini, S. B. Dugdale and T. Jarlborg, Computational Materials Science **28** (2003) 287
[12] W. Al-Sawai, B. Barbiellini, Y. Sakurai, M. Itou, P. E. Mijnarends, R. S. Markiewicz, S. Kaprzyk, S. Wakimoto, M. Fujita, S. Basak, H. Lin, Yung Jui Wang, S. W. H. Eijt, H. Schut, K. Yamada, and A. Bansil, Phys. Rev. B **85** (2012) 115109
[13] E.D. Isaacs, P.M. Platzman, Physics Today (February 1996) 40
[14] Yung Jui Wang, B. Barbiellini, Hsin Lin, Tanmoy Das, Susmita Basak, P. E. Mijnarends, S. Kaprzyk, R. S. Markiewicz, and A. Bansil, Phys. Rev. B **85** (2012) 224529
[15] B. Barbiellini, P. Nicolini, Phys. Rev. A **84** (2011) 022509
[16] I. G. Kaplan, B. Barbiellini, A. Bansil, Phys. Rev. B **68** (2003) 235104
[17] M. J. Cooper, P. E. Mijnarends, N. Shiotani, N. Sakai, and A. Bansil, *X-Ray Compton Scattering* (Oxford University Press, Oxford, 2004)
[18] A. Bansil, B. Barbiellini, S. Kaprzyk and P.E. Mijnarends, J. Phys. Chem. Solids **62** (2001) 2191
[19] B. Barbiellini, J. Phys. Chem. Solids **61** (2000) 341
[20] B. Barbiellini, A. Bansil, J. Phys. Chem. Solids **62** (2001) 2181
[21] D. Nissenbaum, L. Spanu, C. Attaccalite, B. Barbiellini and A. Bansil, Phys. Rev. B **79** (2009) 03541
[22] B. Barbiellini, M. J. Puska, T. Torsti and R. M. Nieminen, Phys. Rev. B **51** (1995) 7341
[23] B. Barbiellini, M. J. Puska, T. Korhonen, A. Harju, T. Torsti and R. M. Nieminen, Phys. Rev. B **53** (1996) 16201
[24] B. Barbiellini, M. J. Puska, A. Harju and R. M. Nieminen, J. Phys. and Chem. of Solids **56** (1995) 1693
[25] J. Zaanen, Nature **430** (2004) 512
[26] N. E. Hussey, R. A. Cooper, Xiaofeng Xu, Y. Wang, I. Mouzopoulou, B. Vignolle and C. Proust, Phil. Trans. R. Soc. A **369** (2011) 369
[27] M. Fratini, N. Poccia, A. Ricci, G. Campi, M. Burghammer, G. Aeppli and A. Bianconi, Nature **466** (2010) 841
[28] A. Shukla, L. Hoffmann, A.A. Manuel, E. Walker, B. Barbiellini and M. Peter, Phys. Rev. B **51** (1995) 6028
[29] T. Jarlborg, B. Barbiellini, R.S. Markiewicz and A. Bansil, Phys. Rev. B **86** (2012) 235111

Journal of Physics: Conference Series **443** (2013) 012010 doi:10.1088/1742-6596/443/1/012010

# Microstructural Characterization of Ferritic/Martensitic Steels by Positron Annihilation Spectroscopy

**S. Hari Babu, G. Amarendra[1], R. Rajaraman and C.S. Sundar**

Materials Science Group, Indira Gandhi Centre for Atomic Research, Kalpakkam-603102, TN, India

[1]E-mail: amar@igcar.gov.in

**Abstract.** Positron annihilation (PA) studies are carried on two ferritc/martensitic steels (modified 9Cr-1Mo and EUROFER97) and Fe-9Cr binary alloy. Normalized modified 9Cr-1Mo steel is subjected to isochronal heat treatments between 300 K – 1273 K with PA studies at different intervals. Due to changes in the concentration of positron trapping open volume defects associated, different stages of the microstructural changes are identified. These results are corroborated with scanning electron microscopy studies. In EUROFER97 steel, the relative increase in precipitate number density due to the additional cold-work (after normalization) is brought out. In binary Fe-9Cr alloy, which is the model alloy of 9Cr ferritic/martensitic steels, the effect of dislocations on Cr segregation is brought out by studying alloys with two different initial conditions of varying dislocation densities.

## 1. Introduction

Ferritic/martensitic steels are considered as important candidate structural materials in fast nuclear reactors pertaining to good void swelling resistance. Void swelling in ferritic steels is around 1% while for the austenitic stainless steels it is 10-28 % for the same fluence of neutron irradiation (18 x $10^{22}$ n/cm$^2$, E > 0.1 MeV) at 693 K [1]. Ferritc steels (in particular, 9Cr-1Mo steels) have strength higher than 304 austenitic stainless steel up to 973 K. However, the conventional ferritic steels have poorer high temperature creep strength as compared to austenitic steels [2]. The strength and long term performance of ferritic steels critically depend on the initial stable microstructure for the given operating conditions. The microstructure consists of lath structure with a variety of precipitates dispersed [3]. The Fe-9Cr binary alloy is a model system for 9Cr ferritic steels, which could shed light on the nature of Cr segregation, known as α' phase, in Fe-Cr based steels. From a fundamental point of view, the Fe-Cr phase diagram continues to evoke a lot of interest and getting revised with more precise experimental results and ab-initio calculations, as evident from a recent critical review [4]. Earlier positron annihilation spectroscopy (PAS) was extensively employed for characterization of steels [5, 6] and model alloys [5, 7, 8]. In the present work, PAS is employed on modified 9Cr-1Mo steel, EUROFER97 steel and Fe-9Cr binary alloy to find out the different microstructural changes occurring as a function of heat treatment. Electron microscopy studies are used for obtaining corroborative information. The microstructural changes as seen by PAS are compared with previous studies.

## 2. Experimental Details

The chemical composition of the steels and the binary alloy used in the present work is given in Table-1. Mod. 9Cr-1Mo steel and EUROFER97 steel samples were normalized at 1313 K for 1h followed by heat treatment between 300 – 1027 K with a step size of 50 K and a holding time of 1 h. PAS measurements were carried at selected heat treatment steps. Fe-9Cr model alloys were considered in two different initial conditions viz. (a) heat treatment at 1423 K for 1 h followed by air-quenching (AQ) (b) heat treatment at 1073 K for 2 h (HT). Further details regarding thermomechancal treatments can be found in [8].

**Table 1.** The chemical composition of materials used in the present work.

| Element | Composition (wt %) | | |
|---|---|---|---|
| | **Mod. 9Cr-1Mo** | **EUROFER97** | **Fe-9Cr** |
| C | 0.10 | 0.11 | < 0.0001 |
| Cr | 8.50 | 8.96 | 9.50 |
| Mn | 0.40 | 0.49 | - |
| Si | 0.25 | 0.04 | - |
| Mo | 0.95 | - | - |
| W | - | 1.08 | - |
| V | 0.22 | 0.20 | - |
| Ni | 0.20 | 0.02 | - |
| Nb | 0.08 | - | - |
| Ta | - | 0.14 | - |
| N | 0.01 | 0.021 | - |
| P | <0.006 | <0.005 | - |
| S | <0.006 | 0.004 | - |
| Fe | Bal. | Bal. | Bal. |

Positron lifetime measurements were carried out at room temperature with a fast-fast lifetime spectrometer having a time resolution of 260 ps (FWHM), using a $^{22}$Na source. Source lifetime and intensity were deduced using annealed Fe as a reference sample. For each measurement, more than $10^6$ total counts were accumulated and the measured lifetime spectra were analyzed using the LT program [9]. In the present experiments, only a single mean lifetime is obtained for all the measurements over the entire annealing temperature range. Doppler broadening annihilation line shape measurements were carried out at room temperature using a HPGe detector having energy resolution of 1.29 keV at 662 keV. The defect sensitive S-parameter is defined as the ratio between the area under annihilation peak energy region of 511 ± 1 keV and the total area under the region of 511 ± 10 keV. Scanning electron microscopy (SEM) was carried out using a Carl Zeiss FE SEM with an INCA EDS analyser (M/s. Oxford) in secondary emission mode. Samples subjected to the same heat treatment procedures as described earlier were used. These samples for electron microscopy were prepared by standard metallographic methods of grinding and polishing.

## 3. Results and Discussion

The normalized ferritic/martensitic steel consists of martensitic lath structure with high concentration of dislocations. Martensitic steel in this state also contains vacancies and vacancy-carbon complexes as explained previously in reference [10]. Figure-1 shows the Doppler broadening S-parameter obtained for normalized sample as a function of heat treatment. It also presents positron lifetime values for corresponding heat treatments from previous work [10]. The results can be understood as three stages of microstructural changes i.e., (i) the S-parameter and lifetime show decrease, as compared to values in a normalized sample, upto 673 K (ii) an increase and decrease during heat treatment between 773 - 1073 K (iii) increase beyond 1073 K.

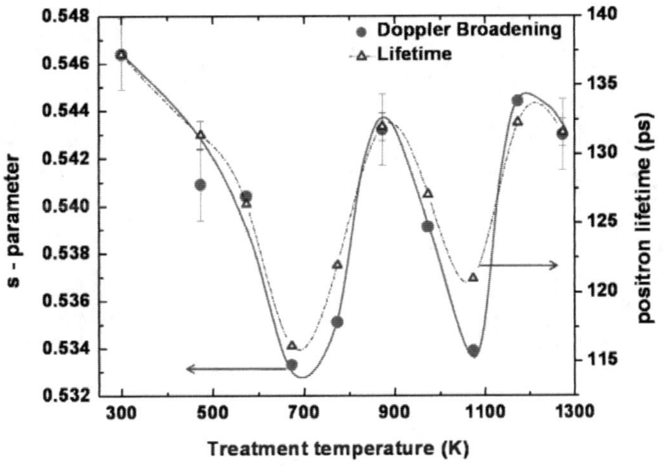


**Figure 1.** Doppler S-parameter and lifetime as a function of temperature in modified 9Cr-1Mo steel. Lifetime results are from reference [10]. The changes seen between 773 K – 1073 K correspond to precipitation.

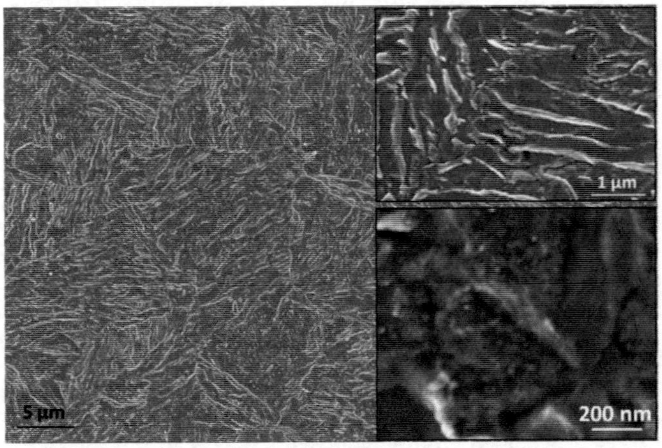


**Figure 2.** SEM micrograph of as-normalized modified 9Cr-1Mo steel. A Similar microstructure is observed in 673 K annealed sample.

In the first stage, as a function of heat treatment the S-parameter decreases till 673 K indicating defect (dislocations and/or vacancy complexes) annealing. While the PAS is sensitive to these changes, other techniques such as hardness and ultrasonic velocity did not show sensitivity [10]. A SEM micrograph of the normalized sample is shown in Figure-2. It is found to be similar up to 673 K [10]. Increase in positron lifetime above 773 K corresponds to the nucleation and growth of the metal carbide/nitrides precipitates with majority of the precipitates being $M_{23}C_6$ [3, 11]. With the increase in the number density and the size of precipitates, the associated precipitate-matrix interface area increases and hence, the interfacial open volume defects increases. It is possible that the increase in lifetime could also be due to vacancy defects inside the precipitates. It is known from earlier calculations that positron lifetime in metal vacancy is higher than the corresponding perfect metal carbide (or nitride) lattice [12]. However, the present experiments are inadequate to identify the exact defect site. Nevertheless, these results confirm that the sites are associated with precipitates.

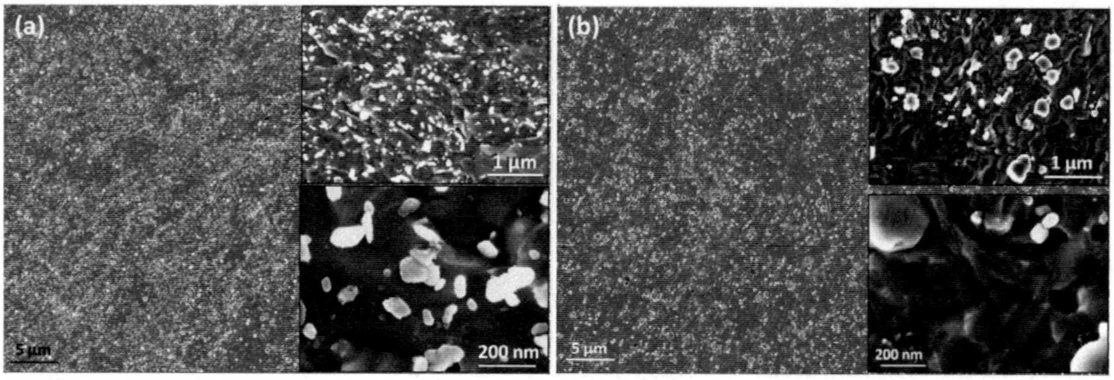


**Figure 3.** SEM micrographs of modified 9Cr-1Mo steel samples treated at (a) 873 K and (b) 1073 K. Magnified images are shown on the right side. In the 873 K treated sample, large concentration of precipitates with mean separation of ~150 nm are seen. In the 1073 K sample, the precipitate concentration decreases while the mean separation increases to > 0.5 µm.

Beyond 873 K, the lifetime decreases indicating decrease in positron trapping defects. Above 873 K, the precipitates coarsen at the expense of small precipitates resulting in less concentration of precipitates having bigger size. After heat treatment at 1073 K, the mean distance between precipitates increases to ~1 µm. The positron diffusion length being about 100 nm in metals and alloys, the fraction of positrons trapped at precipitates decreases with the increase in the mean separation length of the precipitates. Figure-3 shows the precipitate microstructure of 873 K and 1073 K heat treated samples at different magnifications. In the 873 K heat treated sample the precipitates are found to be high in concentration and small in size, while the sample heat treated at 1073 K shows lower concentration of precipitates of larger size. Hence, the microstructures at 873 K and 1073 K indicate growth and coarsening stages respectively. The third stage beyond 1073 K (Figure-1) is a mere reproduction of the microstructure similar to the initial condition of the normalized sample. This is because, heat treatment above 1073 K and subsequent quenching results in martensitic structure. However, there could be differences in the size of prior austenitic grain boundaries that would not be able to influence positron annihilation parameters significantly.

In summary, the microstructural changes in terms of quenched-in defects annealing below 673 K, the growth and the coarsening of precipitates between 773 K – 1073 K and the formation of martensitic phase around 1173 K are distinguished using positron annihilation. Our previous studies using ultrasonic velocity and hardness techniques [10] were found to be insensitive to the defect annealing stages below 673 K. While these techniques show sensitivities for the changes in lath structure and the associated precipitation stages, they show monotonic changes for both, growth and coarsening stages, indicating they cannot distinguish the precipitation stages. However, the martensitic phase formation is duly detected by ultrasonic velocity and hardness.

Lifetime studies were also carried out on EUROFER97 samples and they exhibit similar stages of defect annealing, precipitation growth and coarsening. Figure-4 shows the mean positron lifetime during the growth and the coarsening stages, with one of the samples further cold worked to 10 %, upon normalization, to increase the nucleation sites in the form of dislocations. As can be seen, the positron lifetime in the cold-worked sample is longer as compared to only normalized sample, throughout the range, indicating more positron traps in the cold worked sample. Since the positron lifetimes are almost the same before precipitation in both samples, the increase in trapping is expected to be associated with the variation in precipitate associated defects. Hence, the increase in positron lifetime is due to an increase in the number density of precipitates. As a summary, this study brought out the variation in precipitate concentrations using positron lifetime technique.

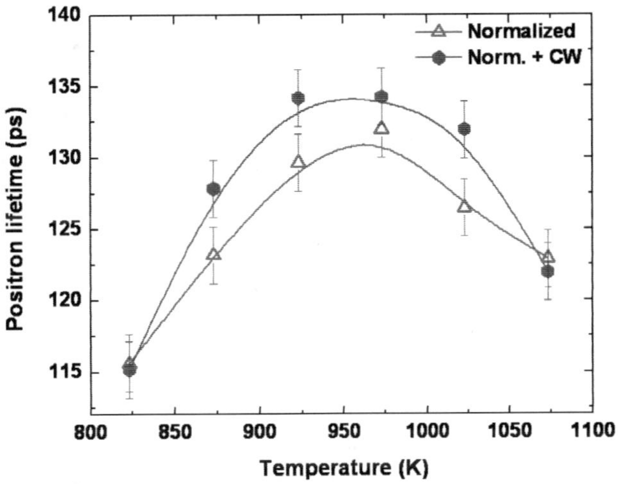


**Figure 4.** Selected temperature range in EUROFER97 steel showing the precipitation stage, under two different initial conditions. Due to cold-work the precipitate number density increases, resulting in an increased lifetime.

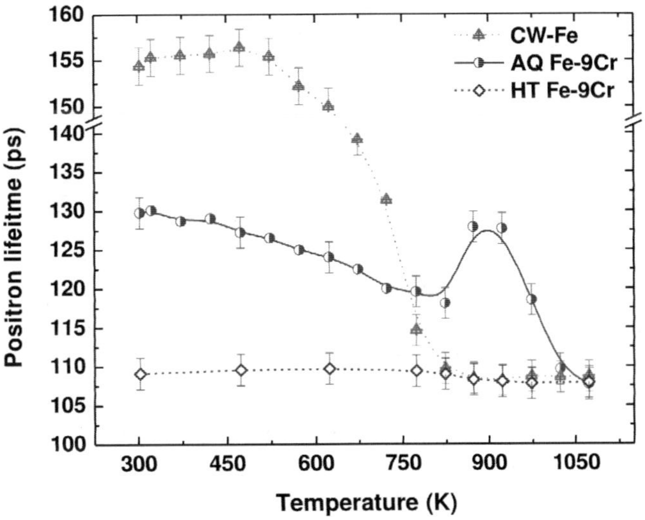


**Figure 5.** Positron lifetime in Fe-9Cr alloys with two different initial conditions (AQ & HT), showing the effect of dislocations on precipitation between 823 K – 973 K [8]. The onset of precipitation is correlated with dislocation annealing in the cold-worked (CW) pure Fe sample.

Figure-5 shows results of Fe-9Cr samples in comparison to cold-worked 99.99% pure Fe samples. AQ samples quenched from elevated temperature (FCC phase) contains random dislocations and they are found to be annealed out by 773 K [8]. Upon further heat treatment, they showed a longer lifetime between 823 K – 973 K while, the dislocation free HT sample showed no change in lifetime in the corresponding temperature range. Annealing of the majority of dislocations by 773 K in pure Fe sample clearly indicates that the Cr segregation in Fe-Cr sample is induced by dislocation annealing, indicating that the Cr segregation is driven by dislocations. Transmission electron microscopy confirmed the existence of precipitates at 900 K and their dissolution by 1073 K [8]. A similar kind of Cr segregation upon helium implantation was reported previously [13]. Further, the nature of precipitates is found, indicatively, as Fe-Cr intermetallic.

## 4. Summary

PAS studies were carried out on two different ferritic steels and a Fe-9Cr binary alloy to investigate the microstructural changes as a function of heat treatment. The effect of dislocations on Cr precipitation in Fe-9Cr alloys is brought out. The effect of cold work on the precipitation in EUROFER97 steel is presented. The growth and coarsening stages in Modified 9Cr-1Mo steels are distinguished using PAS and the results are corroborated with SEM studies.

## 5. Acknowledgements

One of the authors (SHB) acknowledges the Department of Atomic Energy (DAE), India for awarding research fellowship.

## References

[1] Klueh R L, Nelson A T 2007 *J. Nucl. Mater.* **371** 37
[2] Booker M K, Sikka V K, Booker B L P 1981 *Proc. of ASM International conference on ferritic steels for High Temperature Applications* Warrendale. PA. (Ashok K. Khare Eds.. ASM, Metals Park, OH) 257
[3] Fujio Abe 2008 *Sci. Technol. Adv. Mater.* **9** 013002
[4] Wei Xiong, Malin Selleby, Qing Chen, Joakim Odqvist and Yong Du 2010 *Crit. Rev. Solid. State* **35** 125
[5] Lambrecht M and Malerba L 2011 *Acta Mater.* **59** 6547
[6] Rajaraman R, Padma Gopalan, Viswanthan B, Venkadesan S 1994 *J. Nucl. Mater.* **217** 325
[7] Banhart J, Lay M D H, Chang C S T, Hill A J 2011 *Phys. Rev. B* **83** 014101
[8] Hari Babu S, Rajaraman R, Amarendra G, Govindaraj R, Lalla N P, Arup Dasgupta, Gopal Bhalerao and Sundar C S 2012 *Philos. Mag.* **92** 2848
[9] Kansy J 1996 *Nucl. Instrum. Meth A* **374** 235
[10] Hari Babu S, Rajkumar K V, Hussain S, Amarendra G, Sundar C S and Jayakumar T 2013 *J. Nucl. Mater.* **432** 266
[11] Wendell B. Jones, Hills C R and Polonis D H 1991 *Metall. Trans. A* **22A** 1049
[12] Puska M J, Sob M, Brauer G, Korhonen T 1994 *Phys. Rev. B* **49** 00947
[13] Krsjak V, Slugen V, Miklos M, Petriska M and Ballo P 2009 *Appl. Surf. Sci.* **255** 153

Journal of Physics: Conference Series **443** (2013) 012011 doi:10.1088/1742-6596/443/1/012011

# What we can learn from magnetic Compton scattering: application to the determination of spin polarization

**J.A. Duffy**

Department of Physics, University of Warwick, Coventry, CV4 7AL, UK

E-mail: j.a.duffy@warwick.ac.uk

**Abstract.** Studies of spin-resolved electron momentum densities involve the measurement of the so-called magnetic Compton profile. This is a one-dimensional projection of the electron momentum distribution of only those electrons that contribute to the spin moment of a sample. The technique is applicable to ferri- and ferromagnetic materials. The profile is obtained via the inelastic 'Compton' scattering of high energy X-rays. Since electrons originating from different atomic orbitals have specific momentum densities, it is often possible to determine the origin of the magnetism present. Typically, interpretation requires the use of electronic structure calculations using molecular orbital and band structure approaches. Here, we highlight the application of the technique to the determination of the Fermi level spin polarization, the knowledge of which is important to the development of novel spintronic materials.

## 1. Introduction to magnetic Compton scattering

Compton scattering studies involve the non-resonant inelastic scattering of high energy photons. Two specific applications of the technique are the measurements of Fermi surfaces and spin resolved electron momentum densities. Incident photon energies of 90-200keV are normally used, and, for magnetic measurements, a degree of circular polarization is required. In order to meet these requirements, the experiments are usually performed at the ESRF and SPring-8 synchrotron X-ray sources.

When the X-rays are inelastically scattered by the electrons in the sample, the scattered photon energy distribution is Doppler broadened because of the electrons' momentum distribution. If the scattering event is described within the impulse approximation [1, 2], the measured spectrum represents a 1-dimensional projection of the electron momentum distribution, $n(\mathbf{p})$, resolved along the $z$ direction, which is parallel to the scattering vector. This is the Compton profile:

$$J(p_z) = \iint n(\mathbf{p})\, dp_x dp_y, \quad (1)$$

where the integral of $J(p_z)$ is taken over occupied electron states.

Fermi surface studies typically involve an incident monochromatic photon energy of around 90keV. The high X-ray energies used in the experiments mean that the bulk electronic structure is measured. The method requires a high-resolution energy dispersive measurement in order to obtain the momentum resolution necessary to resolve Fermi surface features.

Examples of such studies include the superconductors $La_{2-x}Sr_xCuO_4$ [3], $Na_xCoO_2.yH_2O$ [4] and $Ba(Fe_{1-x}Co_x)_2As_2$ [5]. Whilst several methods are widely used for Fermi surface measurements (including the closely-related positron technique [6]), there are a number of circumstances where Compton scattering is the technique of choice. Positron annihilation, for example, requires a very low defect concentration, and can suffer from positron-electron correlation effects [7]. Quantum oscillatory methods require long electronic mean-free-paths, and photoemission is inherently surface sensitive.

If the incident photon beam has a component of circular polarization, the scattering cross-section contains a term which is spin dependent [8]. This term may be isolated from the charge scattering by either flipping the direction of the sample magnetization or the photon helicity parallel and antiparallel with respect to the scattering vector, resulting in a *magnetic* Compton profile (MCP), $J_{mag}(p_z)$. This MCP is defined as the 1D projection of the spin-polarized electron momentum density:

$$J_{mag}(p_z) = \iint [\, n^{\uparrow}(\mathbf{p}) - n^{\downarrow}(\mathbf{p}) \,] \, dp_x dp_y. \tag{2}$$

Here $n^{\uparrow}(\mathbf{p})$ and $n^{\downarrow}(\mathbf{p})$ are the momentum densities of the majority and minority spin bands. Since the MCP is the difference between two measured Compton profiles, components arising from spin-paired electrons cancel, as do most sources of systematic error. Because only those electrons that contribute to the spin moment of the sample contribute to the integral of this MCP, it is then possible to determine the spin magnetic moment, usually by comparison with a measurement, under the same experimental conditions, of a sample with known spin moment. Such straightforward spin magnetometry has, for example, permitted the clarification of the temperature dependence of the spin and orbital moments in magnetite [9].

Crucially, from more detailed analysis, it is normally possible to determine the origin of the spin moment via comparison with calculated profiles. This is because electrons in different orbitals have different momentum density distributions: more localised electrons (for example the 4f contribution to the magnetism in Gd) have a correspondingly broad contribution whereas more itinerant electrons have a narrower momentum density distribution. This has been illustrated clearly in materials such as Gd [10], GdY [11], the magnetocaloric material $Gd_3Pd_7$ [12] and the ferromagnetic superconductor $RuSr_2GdCu_2O_8$ [13], where the itinerant 5d contribution is clearly observable in addition to the localised 4f moment. Such an approach can also be applied to 4f-3d systems again because the 3d electrons have a narrower momentum density distribution, as demonstrated in $SmMn_2Ge_2$, where the 4f and 3d spin moments were isolated [14]. Here, the total spin moment approaches zero, but the measurement revealed this to arise from large antiparallel Sm 4f and Mn 3d contributions. Here, the ability to differentiate between the 4f and 3d contributions, together with the sensitivity to only the spin moment, also revealed a subtle field dependence of the individual moments even when the bulk magnetisation was constant [15].

Electronic structure calculations are particularly useful for the interpretation of magnetic Compton profiles. Direct predictions of the MCPs can be made by calculating the resultant spin-dependent momentum distributions. For example, in Ni, as has been described by Dixon *et al.* [16], the experimental results measured along four crystallographic directions provided a stringent test of three different calculations. In such metals, features associated with the Fermi surface arising from bands crossing the Fermi level may be clearly visible. In this case, the theory generally reproduced the experimental results well, with good agreement with the experimental spin moment, and consistency with the Fermi surface features that were observed. However, the results demonstrated a characteristic problem with the *ab initio* calculations: there is a distinct discrepancy at low momentum, where the theoretical results typically overestimate the spin density. This problem is related to the precise location of the spin-polarized bands relative to the Fermi level. Because in the band structure there is a marked energy dependence of the density of states (DOS) at the Fermi level, very small changes in the calculation have a significant effect on the resultant spin density. It has been demonstrated previously [17, 18], that the agreement with experiment may be optimised by rigidly shifting the bands contributing to the spin moment, as will be demonstrated in this paper.

More recent research has required the use of further sophisticated approaches, relying on either molecular orbital methods (GAMESS) or a variety of electronic structure calculations. In $NbFe_2$, comparison with electronic structure calculations for different possible ground states revealed the magnetic order to be ferrimagnetic [19]. Such work shows how electronic structure calculations can be used as the basis for fitting to the data to reveal the pertinent features of the underlying electronic structure, again demonstrating the benefit of not simply relying on purely *ab initio* calculations.

In this paper, we report on recent work that demonstrates the possibility of determining the Fermi level spin polarization in highly spin-polarized materials. We apply the method to Ni, which provides an excellent test of the approach, demonstrating comparison with previous experimental measurements and theory. Our work shows that an accurate determination of the Fermi level spin polarization is indeed possible, and that the band shifting and fitting described above is essential to the technique's success.

## 2. Experimental details and considerations

Magnetic Compton profiles are normally measured on the beam line BL08W at the SPring-8 synchrotron or on ID15 at the ESRF. The SPring-8 facility has a sample environment consisting of a superconducting solenoid to provide the applied magnetic field of up to 2.5T and temperatures down to approximately 6K. An incident X-ray energy of 175keV is normally used. The energy spectrum of the scattered flux is measured using a ten-element Ge detector at a mean scattering angle of 173°, giving a momentum resolution of the magnetic Compton spectrometer, taken as the full-width at half-maximum (FWHM) of the instrument response function, of 0.50 a.u. of momentum (where 1 a.u.$=1.99 \times 10^{-24}\,\mathrm{kg\,m\,s^{-1}}$). At the ESRF, an Oxford Instruments Spectromag cryomagnet is currently used to obtain fields up to 9 tesla. However, the experimental results presented in this paper were collected with a rotating 1 tesla permenent magnet at room temperature. At the ESRF a thirteen-element Ge detector is available. The incident energy of 220keV and scattering angle of 172° result in a 0.44 a.u. resolution. In both experiments, the data need to be corrected for energy-dependent detector efficiency, sample absorption, and the relativistic scattering cross-section. The profiles are then corrected for multiple scattering using the technique described by Felsteiner [20].

## 3. Determination of spin polarization at the Fermi level

The degree of spin polarization at the Fermi level is a measure of the majority and minority density of states (DOS) at this energy. Hence, it contains information concerning the relative contribution to conduction provided by the two spin channels. The degree of spin polarization, $P$, can be defined simply as follows:

$$P = \frac{N_\uparrow - N_\downarrow}{N_\uparrow + N_\downarrow}, \tag{3}$$

where $N_\uparrow$ and $N_\downarrow$ are the spin up and spin down Fermi level densities of states respectively.

Ideally, materials for applications would be fully spin-polarized 'half-metals' (HMs), $P = 1$, where the density of states at the Fermi level is finite for one spin, but zero for the other, such that carriers of only one spin exist at the Fermi level. Such spin-polarized transport is central to the concept of 'spintronics'. In these materials and devices, the goal is to exploit the electron spin for applications such as data storage and read-heads. For example, magnetic random access memory devices (MRAM) are based on the concept of spin-dependent tunnelling magnetoresistance [21]. Ferromagnets typically have a degree of Fermi level spin-polarization, but in general this is only partial, and not high enough for applications. Only a few promising candidates, such as $CrO_2$ [22, 23], optimally doped $La_{0.7}Sr_{0.3}MnO_3$ [24], $Co_2MnAl_{(1-x)}Sn_x$ [25, 26] and possibly $Co_{(1-x)}Fe_xS_2$ [27, 28, 29, 30] have been found to exhibit high polarizations.

The direct and unambiguous experimental measurement of $P$ is difficult. Techniques such as saturation magnetization or transport measurements can only give an indication of half-metallicity, and techniques which give access to the magnitude of $P$ directly such as spin-polarized photoemission or point-contact Andreev reflection (PCAR) are heavily dependent on the surface quality of the sample. Moreover, the quantity that is actually measured by transport-based probes is weighted by the Fermi velocity. As pointed out by Mazin [31], when comparing such data with theoretical results, the precise definition of $P$ is problematic, resulting in several definitions [31, 32]. In most studies, the polarization is defined in terms of the spin-dependent density of states, $N_{\uparrow/\downarrow}$ at the Fermi level ($E_F$), and the spin-dependent Fermi velocity, $v_{F,\uparrow/\downarrow}$. Following Mazin [31], $P_n$ is defined such that,

$$P_n = \frac{N_\uparrow v^n_{F,\uparrow} - N_\downarrow v^n_{F,\downarrow}}{N_\uparrow v^n_{F,\uparrow} + N_\downarrow v^n_{F,\downarrow}}, \tag{4}$$

where $v_{F,\uparrow/\downarrow}$ is raised to the power of $n$. For $n = 0$, as quoted by most theoretical studies, $P_0$ is solely defined by the DOS. Taking $n = 1$ or $n = 2$, the polarization is weighted with $v_F$ in the ballistic or $v_F^2$ in the diffusive regime. For HMs, $P_n$ is unambiguously 100% while in the case of non-HMs, the values of $P_n$ will generally differ according to the definition adopted. However, the exact weighting of $P_n$ is difficult to extract experimentally and it is thus not clear which regime is appropriate for particular classes of material [33]. Consequently the comparison with other techniques and theory is notoriously difficult [31, 33].

$P_n$ can be calculated directly from band structure calculations, but, as is clear from magnetic Compton scattering measurements such as those on Ni, *ab initio* calculations can have a characteristic difficulty in correctly describing the correct Fermi energy electronic structure. In this paper, we demonstrate a solution to this problem. By performing a rigid-band fit of the calculated MCPs to experiment, we can make an accurate determination of the Fermi level DOS, and it becomes possible to extract $P_n$ for $n = 0$ as well as the weighted values for $n = 1, 2$.

In order to demonstrate this method, the spin polarization of Ni has been calculated and compared with previous experimental [34] and theoretical results [31]. The experimental data gave a spin polarization of $P_2 = 23 \pm 3\%$, using spin-polarized electron tunnelling. Whilst this technique gives excellent results, it relies on high-quality films grown on a superconducting substrate, which is not possible for many materials. For our magnetic Compton scattering approach, experimental data for four crystallographic directions from Dixon *et al.* [16] were used for the fitting procedure. The results of this fitting are shown in Fig. 1. For more details of the method of producing these fitted data, and intracacies involved, see references [17, 18]. It is important to note that several directional MCPs, measured on a single crystal, should be fitted simultaneously in order to have confidence in the optimised DOS and hence $P$ values.

The self-consistent results for the spin polarization $P_n$ are shown in Fig. 2a. Our data agree well with those of Mazin [31]. It can be seen that the calculated value of $P_2$ is 0%, whereas the experimentally determined value from the spin-polarized tunnelling measurements is 23±3% [34]. Tuning our calculations by fitting the resultant magnetic Compton profiles to the experimental data as shown in Fig. 1, produces the polarization plots shown in Fig. 2b. Here, the fitting had only a small effect on $P_0$ and $P_1$. On the other hand, the fitting increased the value of $P_2$ from 0% to $20 \pm 2\%$. This fitted value compares very well with the previous experimental result. This result demonstrates how the method provides a useful measure of the spin polarization values. Further evidence of the methods efficacy can be seen in the study of the highly spin-polarized

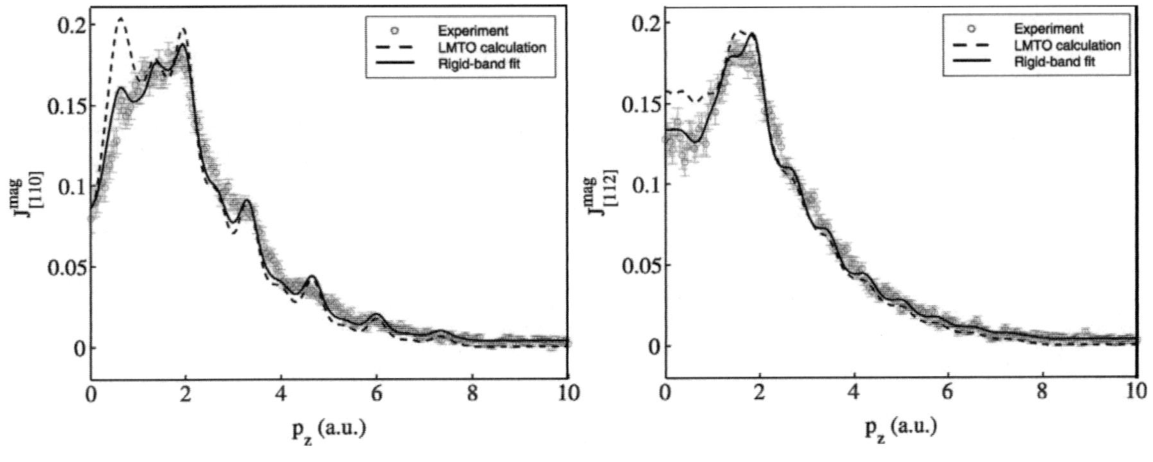


**Figure 1.** The magnetic Compton profiles for Ni, measured along the [100] and [110] directions. Also shown are the results of the raw and tuned calculations. Note that the agreement at low momentum is much better for the fitted band structure.

$Co_{(1-x)}Fe_xS_2$ system [35].

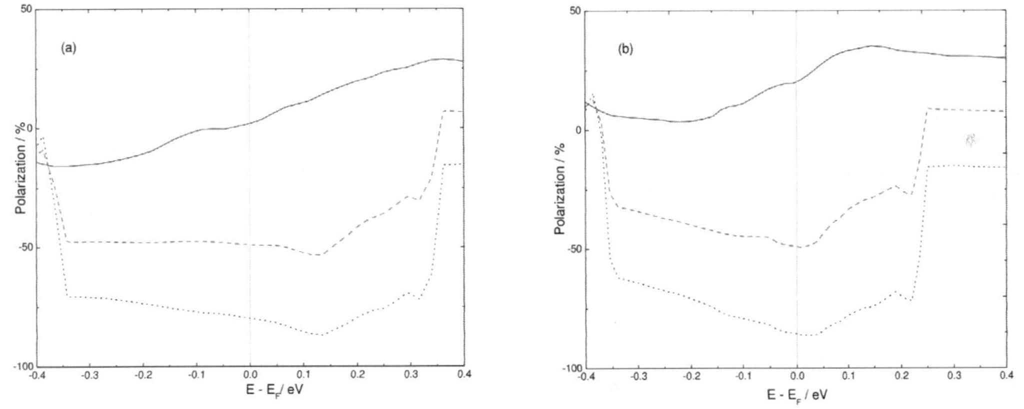


**Figure 2.** The value of the spin polarization across the Fermi level for (a) the raw and (b) the fitted calculations. Results are shown for $P_0$ (blue dots), $P_1$ (red dashes) and $P_2$ (green line). It can be seen that the fitting changes the Fermi level value of $P_2$ from 0% to $20 \pm 2\%$

.

## 4. Outlook

Magnetic Compton scattering has a wide range of applications. Current developments include the improvement of the sample environment available. Our new 'Spectromag' cryomagnet can provide magnetic fields up to 9 tesla, and temperatures down to 1.5K. A newly developed microfurnace insert enables temperatures up to approximately 800K to be obtained. This facility is extending the range of materials that can be studied experimentally.

## Acknowledgments

I would like to thank many people for their contributions to this research, including C. Utfeld, S.R. Giblin, J. W. Taylor, C. Shenton-Taylor, J. Laverock, S.B. Dugdale, J.E. McCarthy, M.A.G. Dixon and M.J. Cooper. I am grateful for the support of T. Buslaps, V. Honkimaki at the ESRF, and acknowledge the ESRF for providing the beamtime used for the nickel measurements.

## References

[1] Platzman P M and Tzoar N 1970 *Phys. Rev.* **2** 3556
[2] Holm P 1988 *Phys. Rev. A.* **37** 3706
[3] Sakurai Y *et al.* 2011 *Science* **332** 698
[4] Laverock J *et al.* 2007 *Phys. Rev. B* **76** 052509
[5] Utfeld C *et al.* 2010 *Phys. Rev. B* **81** 064509
[6] Crowe S J *et al.* 2004 *Europhysics Letters* **65** 235
[7] Laverock J *et al.* 2010 *Phys. Rev. B* **82** 125127
[8] Bell F, Felsteiner J and Pitaevskii L P 1996 *Phys. Rev. A.* **53** R1213
[9] Duffy J A 2010 *et al.*, *Phys. Rev. B* **81** 134424
[10] Duffy J A *et al.* 1998 *J. Phys.: Condens. Matter* **10** 10391
[11] Duffy J A *et al.* 2000 *Phys. Rev. B* **61** 14331
[12] Shenton-Taylor C *et al.* 2007 *J. Phys.: Condens. Matter* **19** 186208
[13] Banfield Z F *et al.* 2005 *J. Phys.: Condens. Matter* **17** 5533
[14] McCarthy J E *et al.* 2000 *Phys. Rev. B* **62** R6073
[15] Bebb A M *et al.* 2005 *Phys. Rev. B* **71** 024407
[16] Dixon M A G *et al.* 1998 *J. Phys.: Condens. Matter* **10** 2759
[17] Major Zs *et al.* 2006 *J. Phys. Chem. Sol.* **65**, 2011
[18] Major Zs *et al.* 2004 *Phys. Rev. Lett.* **92**, 107003
[19] Haynes T D *et al.* 2012 *Phys. Rev. B* **85**, 115137
[20] Felsteiner J 1997 private communication
[21] Wolf S A *et al.* 2001 *Science* **294** 1488
[22] Ji Y *et al.* 2001 *Phys. Rev. Lett.* **86** 5585
[23] Parker J A *et al.* 2002 *Phys. Rev. Lett.* **88** 196601
[24] Soulen R J *et al.* 1998 *Science* **282** 85
[25] Fecher G H *et al.* 2005 *J. Phys.: Condens. Matter* **17** 7237
[26] Nakatani T M *et al.* 2007 *J. Appl. Phys.* **102** 033916
[27] Zhao G L, Callaway J and Hayashibara M 1993 *Phys. Rev. B* **48** 15781
[28] Wang L *et al.* 2005 *Phys. Rev. Lett.* **94** 056602
[29] Wang L *et al.* 2006 *Phys. Rev. B* **73** 144402
[30] Leighton C *et al.* 2007 *J. Phys.: Condens. Matter* **19** 315219
[31] Mazin I I 1999 *Phys. Rev. Lett.* **83** 1427
[32] Müller G M *et al.* 2009 *Nat. Mater.* **8** 56
[33] Woods G T *et al.* 2004 *Phys. Rev. B* **70** 054416
[34] Meservey R and Tedrow P M 1994 *Phys. Rep.* **238** 173
[35] Utfeld C *et al.* 2009 *Phys. Rev. Lett.* **103** 226403

Journal of Physics: Conference Series **443** (2013) 012012 doi:10.1088/1742-6596/443/1/012012

# PALS combined with Charpy-V tests at WWER reactor pressure vessel steels

V Slugeň[1], A Kryukov[2], M Petriska[1], J Veterníková[1], S Sojak[1], V Sabelová[1], R Hinca[1]
[1]Institute for Nuclear and Physical Engineering, Slovak University of Technology, Ilkovičova 3, 81219 Bratislava, Slovakia
[2]European Commission, Joint Research Centre - Institute for Energy, PO Box 2, 1755 ZG Petten, The Netherlands

**Abstract.** This paper presents results from our long-term studies of irradiated, commercially used WWER reactor pressure vessel steels. Results from Charpy-V tests and positron annihilation spectroscopy techniques are compared and discussed in details, having in mind actual state of art and other microstructural studies in this area. The optimal region for annealing of irradiation induced defects was analyzed. It was shown that WWER steel with low impurity contents has good radiation stability and operation these reactor pressure vessels could be extended beyond a design lifetime.

## 1. Introduction

The basic reason of RPV mechanical properties degradation is the neutron irradiation, resulting in hardening and embrittlement of the steel which the RPV is made of. The prediction of radiation embrittlement is performed usually in accordance with relevant codes and standards that are based on large amount of information from surveillance and test irradiation programmes. Considerable data exists regarding the effect of neutron irradiation on pressure vessel steels; from both mechanical properties and microstructure features, [1].

In the last years, several reliable and predictive modelling approaches have been developed (thanks also to EC-framework projects PERFECT, REWE and PERFORM60) [2], according to which the assumed RPV-steel damage is generated by three major contributions and their synergisms:

i) copper rich radiation induced nano-precipitates,
ii) phosphorus segregation at different internal surfaces and
iii) the basic damage of the material matrix.

The neutron embrittlement of RPV steel was a serious problem in first generation of WWER reactors: WWER-440/230. The major part of the WWER-440/230 pressure vessels were made of steel with high phosphorous and copper content in the welds. The radiation embrittlement caused by the influence of these impurities shorted the safe operational lifetime below the design value.

Undoubtedly, changes in the structure could be basically registered using several methods. Material properties of the RPV steels and influence of thermal and neutron treatment on these properties are routinely investigated by macroscopic methods such as Charpy V-notch and tensile tests. A number of semi-empirical laws, based on macroscopic data, have been established, but, unfortunately, these laws are not completely consistent with all measured data and do not provide the desired accuracy. Therefore, many additional test methods, excellent summarised in [3] have been developed to unravel the complex microscopic mechanisms responsible for RPV steel embrittlement. The possible contribution of non-destructive techniques as Mössbauer spectroscopy (MS), positron annihilation spectroscopy (PAS) and

Journal of Physics: Conference Series **443** (2013) 012012 doi:10.1088/1742-6596/443/1/012012

transmission electron microscopy (TEM) was analysed in [4-6]. In this paper we focused mostly on comparison of Charpy-V results and PAS studies performed on irradiated WWER RPV steels.

## 2. Expermental

In the framework of the "Extended Surveillance Specimen Program", started in 1995 at the nuclear power plant (NPP) Bohunice (Slovakia), several specimens, which were prepared originally for Mössbauer spectroscopy measurements, but because of the proper size (10x10x0,05 mm) and the polished surface also suitable for positron annihilation studies using the conventional lifetime set-up as well as the pulsed low energy positron system (PLEPS) measurement, were selected and measured before their placement into the special irradiation chambers, near the core of the operated nuclear reactor, and after 1, 2 and 3 years resistance there (neutron fluence in the range from 7,8 $10^{23}$ $m^{-2}$ up to 2,5 $10^{24}$ $m^{-2}$). The chemical composition and the irradiation conditions of the studied RPV-steel specimens are shown in Table 1 and the irradiation conditions in Table 2.

*Table 1 – The chemical composition of the studied RPV-steel specimens.*

| Code | Type of steel | Contents of alloying elements in RPV specimens (wt.%) | | | | | | | | | | | |
|---|---|---|---|---|---|---|---|---|---|---|---|---|---|
| | | C | Si | Mn | Mo | Ni | Cr | Cu | P | S | V | Co | Total |
| ZM - Base metal WWER-440 | 15Kh2MFA | 0.14 | 0.31 | 0.37 | 0.58 | 0.20 | 2.64 | 0.08 | 0.011 | 0.017 | 0.27 | 0.019 | 4.651 |
| ZK - Weld metal WWER-440 | Sv10KhMFT | 0.048 | 0.37 | 1.11 | 0.39 | 0.12 | 1.00 | 0.104 | 0.01 | 0.013 | 0.13 | 0.020 | 3.347 |

*Table 2 – Irradiation conditions of specimens at the 3rd unit of nuclear power plant Bohunice (Slovakia).*

| | Material | Code of sample | Time of irrad. [eff. days] | Neutron Fluency [$m^{-2}$] | Damage [dpa] | Thickness of sample [μm] |
|---|---|---|---|---|---|---|
| 1 | Base material – non-irradiated | ZM | 0 | 0 | 0 | 60 |
| 2 | Base material – 1 year irradiated | ZM1Y | 280,8 | 7,81E23 | 0,08 | 50 |
| 3 | Base material – 2 year irradiated | ZM2Y | 578,5 | 1,64E24 | 0,17 | 40 |
| 4 | Base material – 3 year irradiated | ZM3Y | 894,3 | 2,54E24 | 0,26 | 30 |
| | | | | | | |
| 5 | Weld – non-irradiated | ZK | 0 | 0 | 0 | 55 |
| 6 | Weld – 1 year irradiated | ZK1Y | 280,8 | 7,81E23 | 0,08 | 45 |
| 7 | Weld – 2 years irradiated | ZK2Y | 578,5 | 1,64E24 | 0,17 | 25 |

| | | | | | | |
|---|---|---|---|---|---|---|
| 8 | Weld – 3 years irradiated | ZK3Y | 894,3 | 2,54E24 | 0,26 | 47 |

PLEPS technique at University of Bundeswehr (Neubiberg, Germany) [7] was used in 2000-2004 for the investigation of neutron-irradiated RPV-steels. This system enables the study of the micro structural changes in the region from 20 to 550 nm (depth profiling) with small and very thin (<50 μm) specimens, therefore reducing the disturbing $^{60}$Co radiation contribution to the lifetime spectra to a minimum. Such a disturbance is the limiting factor for the investigation of highly-irradiated RPV specimens with conventional positron lifetime systems. In comparison to a triple coincidence setup of positron-lifetime spectroscopy, used in our laboratory in Bratislava, PLEPS reduces the time for the measurements by about a factor 500, resulting in qualitatively comparable spectra and enables in addition the estimation of the defect concentrations. Results were reported in [8].

Actually, we would like to compare this output to Charpy V tests performed on the same type of steels. Our current results show that the Charpy shift increases with the level of irradiation dose (Fig.1). Also in this case values for welds are slightly lower than at base metal. It is necessary to note that the irradiation damage over 1 dpa is behind expected damage during expected reactor operating lifetime.

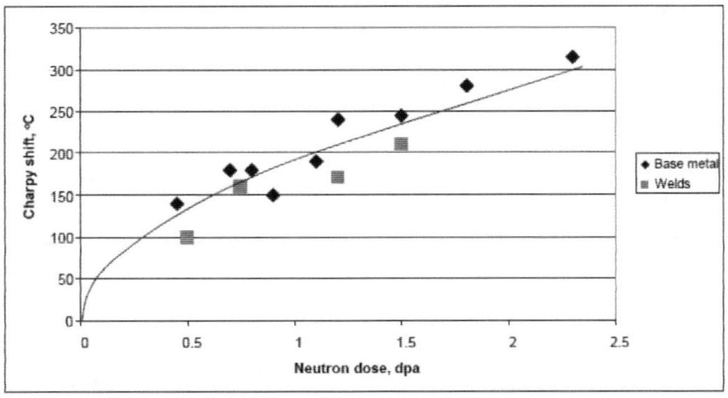


*Fig.1 – Charpy shift versus neutron dose for WWER-440 steels.*

An excellent correlation between Charpy test and positron annihilation results can be observed in case of post-irradiation annealing experiments, comparing Fig. 2 to Fig.3.

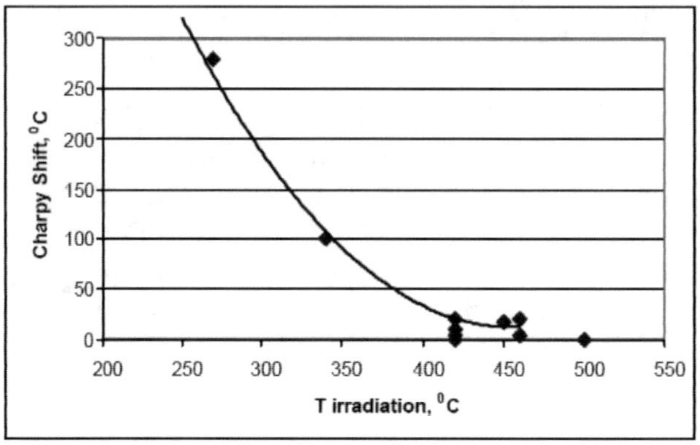


*Fig.2 – Charpy shift of irradiated specimens in dependence to irradiation temperature.*

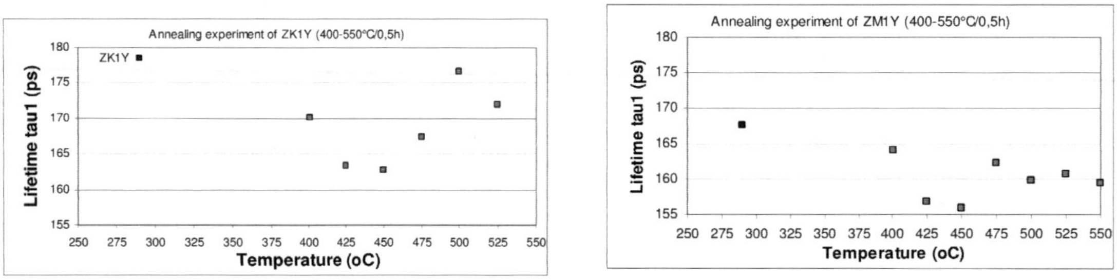


*Fig. 3 - Positron lifetime in defects (Tau1) of irradiated and annealed WWER-440 base (ZM1Y) and weld metals (ZK1Y) to level of neutron fluence 1.25x10$^{24}$ m$^{-2}$ (~0,13 dpa). The error bars were less than 3ps.*

Positron annihilation lifetime results are presented in Fig.3. Intensities of this "defects component" assigned as tau1 changed during isochronal annealing minimal (± 2%). Increase of the positron lifetime values after temperature of about 475 °C could be caused by the carbides precipitation or recombination (mostly VC, $Cr_7C_3$, $Cr_{23}C_6$ are created after irradiation at about 270 °C) and this increase was observed at all types of WWER steels also in the past. This precipitation should not be significant in case of Western type of steel, where zero Vanadium and less chromium content is present (up to 0,7 wt.% in contrast to WWER base metal – up to 3 wt.%).

## 3. Conclusion

Positron Annihilation Spectroscopy is one of the non-destructive spectroscopic methods which can contribute to the complex evaluation of the RPV-steels microstructure and can in this way contribute to the nuclear safety of NPP. Actual EURATOM framework projects (LONGLIFE, NULIFE) are focused on studies towards ensuring of longer NPP operation including PAS analyses.

The second generation of WWER RPV steels of WWER-440 (V-213) is comparable with RPV-steels used in Western Europe and their quality enables prolongation of NPP operating lifetime over projected 40 years. The embrittlement of CrMoV steel is very low due to the low phosphorus and copper content.

Clear correlation between PAS results and Charpy shifts measurements performed on irradiated WWER commercially used RPV-steels was presented and discussed in details. Neutron irradiation at the reactor operating temperatures causes dominantly point defects, which are well detectable for positron annihilation lifetime techniques. The post-irradiationn annealing experiments shown that the optimal region for removing of irradiation caused defects is 425-475 °C. In contrast to RPV steels commercially used in Western Europe, where the positron lifetime permanently decrease up to 700 °C, different chemical composition (V and Cr content) is responsible for this effect. Based on PAS results, due to creation of additional defects in RPV steels over 475 °C, we recommend not overrunning this temperature by annealing.

In the future, PAS techniques can be applied effectively also for evaluation of microstructural changes caused by extreme external loads, simulating irradiation by proton implantation and for the evaluation of the effectiveness of post-irradiation thermal treatments. Therefore, we would like to use our results collected during last 20 years from measurement of different RPV-steels in "as received", irradiated and post-irradiation annealed and compare them to results where the real neutron irradiation will be replaced by proton implantation. This challenge has several open questions having in mind the fact that protons have positive charge and their behaviour in metals can be different. Nevertheless, advantages connected with much comfortable handling, transport and measurement with "not active" specimens seems to be very promising and valuable for those who would like to apply additional non-destructive techniques in this study.

## 4. References

[1] Davies L M 1999 *Int. J. Press. Vess, & Piping.* **76** 163
[2] Jumel S and Van Duysen J C 2007 *J. Nucl. Mat.* **366** 256
[3] Phythian W J and English C A 1993 *J. Nucl. Mater.* **205** 162
[4] Miller M K et al 2000 *J. Nucl. Mater.* **282** 83
[5] Brauer G et al 1991 *Nucl. Eng. & Desg.* **127** 47
[6] Slugen V and Magula V 1998 *Nucl. Eng. & Desg.* **186/3** 323
[7] Bauer-Kugelmann W et al 2001 *Mat. Sci. Forum* **363-365** 529
[8] Slugen et al 2002 *Appl. Surf. Sci.* **194** 150

Journal of Physics: Conference Series **443** (2013) 012013 doi:10.1088/1742-6596/443/1/012013

# Positron annihilation sites in nano lead sulfide powders

**A A Rempel[1,2], A A Valeeva[1,2], K Sato[3], N S Kozhevnikova[1]**
[1]Institute of Solid State Chemistry, the Ural Branch of the Russian Academy of Sciences, Pervomaiskaya 91, Ekaterinburg 620990, Russia
[2]Ural Federal University named after First President of Russia B.N. Yeltsin, Mira 19, Ekaterinburg, 620002, Russia
[3]Department of Environmental Sciences, Tokyo Gakugei University, Nukuikita 4-1-1, Koganei, Tokyo 184-8501, Japan

E-mail: rempel@ihim.uran.ru

**Abstract.** The coincident Doppler broadening of the positron-electron annihilation radiation has been applied for the identification of vacancies in nano lead sulfide (nano-PbS) powders. The results of the positron annihilation studies on nano-PbS with different particle size are presented. Nano-PbS specimens have been synthesized by the chemical bath deposition in aqueous medium. Independently of the size of nanoparticles (from 12 to 16 nm) the positron lifetime (PLT) for all specimens has been found to be the same and equal to 391±1 ps. Independently of the size of nanoparticles the chemical surrounding of positron annihilation sites identified by coincident Doppler broadening measurements was the same, namely lead (Pb).

## Introduction

Coincident measurements of the Doppler broadening of the positron-electron annihilation radiation is a powerful tool for the site detection of vacancies in complex substances. The coincident Doppler broadening of positron annihilation radiation (CDBPAR) gives access to the electron momentum distribution of the core electrons (CE) of the atoms surrounding the annihilation site. In case of positron trapping at vacancies in binary compounds this element-specific information allows for the identification of the sublattice on which the vacancies are located [1].

Lead sulfide (PbS) has been used as a functional material in infrared photon detectors [2]. Since decade after development of the synthetic routes for quantum dots this material has been applied for studies of quantum-confinement effects in nanoparticles [3]. Such studies might extend the detection application to near infrared wave range in near future. That is why a lot of recent scientific activities are devoted to nano-PbS synthesis, structure, and properties [4, 5, 6]. Valuable information on the positron lifetimes for a set of chalcogenide semiconductors is available in [7, 8, 9, 10].

In this paper we present our recent results on vacancies identification in nano-PbS powders. It should be noted that the method of synthesis of nano-PbS is of great importance, because it depends on the method what kind of defects the compound will contain, i.e. structural vacancies in metal or nonmetallic sublattices, agglomerates of vacancies, vacancies in interfaces etc.

---

[1] Corresponding author Andrey A Rempel

**Experimental**

Positron lifetime spectra with 1-2×$10^6$ counts were measured by means of a fast-slow $\gamma\gamma$ spectrometer with a time resolution of 205 ps. The Doppler broadening of the positron-electron annihilation radiation was measured by coincident detection of the two annihilation quanta for background suppression using two Ge detectors with high energy resolution (FWHM = 1.2 keV). Two-dimensional energy spectra ($N(E1,E2)$) of the pairs of annihilation photons were recorded with a total number of 3-5×$10^7$ counts within 3 days. One-dimensional CDBAR spectra with optimum statistics were obtained by cutting the two-dimensional energy spectra along diagonal $E1 + E2 = 1022$ keV with an energy width of 1 keV. Since the one-dimensional spectra were symmetric with respect to the 511 keV, the right- and left-hand sides of the spectra were superimposed. The one-dimensional CDBAR spectra with low background (< $10^{-5}$) extend up to an energy shift $|\Delta E|$ = 12.8 keV corresponding to an electron momentum of $50\times10^{-3}m_0c$ , where $m_0$ is the electron rest mass and $c$ is the velocity of light.

The chemical bath deposition (CBD) of nano-PbS was performed by the direct reaction between water-soluble lead acetate $Pb(OAc)_2$ and sodium sulfide $Na_2S$ in aqueous medium:

$$Pb(OAc)_2 + Na_2S = PbS\downarrow + 2\ NaOAc$$

Primary concentrations of the precursors for nano-PbS synthesis by chemical bath deposition and size of nanoparticles are listed in Table 1. By XRD, SEM and TEM it was determined that the average size of crystallites for different nano-PbS was in the range between 12 and 16 nm [11].

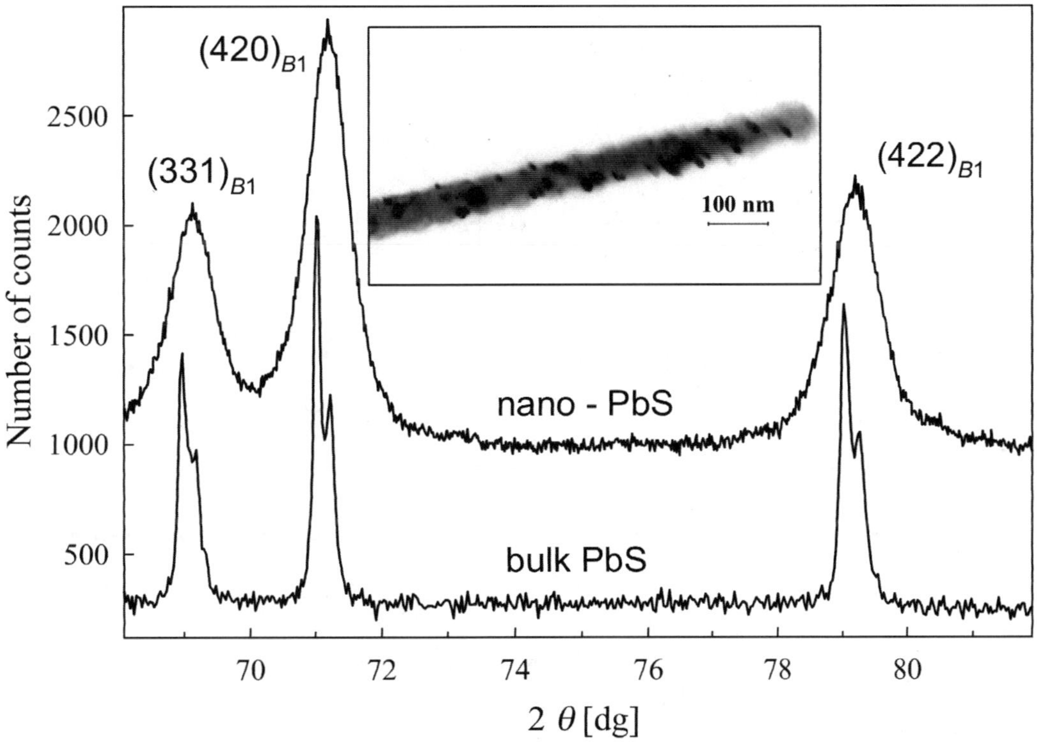


**Figure 1.** XRD spectra showing a broadening of lines due to small nanosized grains of nano-PbS as compared with coarse-grained bulk PbS. Insert shows TEM of nano-PbS powder [11]

According to XRD analysis the structure of synthesized samples of nano-PbS corresponds to $B1$ (rock salt) structure. Nano-PbS is homogeneous and contains only one phase. The XRD spectrum of nano-PbS (Figure 1) shows a huge broadening of diffraction doublet lines (331), (420), (422) as compared with the XRD spectrum of bulk PbS with narrow diffraction reflections split into Cu$K\alpha_{1,2}$ doublet. The average size of nanoparticles (Table 1) was calculated from the broadening of the lines by means of the Williamson-Hall method [12, 13]. For positron annihilation studies the nano-PbS powders were compacted by uniaxial pressing of $P = 20$ kgf/cm$^2$ into disk-shaped pellets.

**Table 1.** Lattice constant, average size of nanoparticles and primary concentrations of the precursors for nano-PbS synthesized by CBD

| Specimen | $a_{B1}$, pm | <$D$>, nm | Synthesis conditions | |
|---|---|---|---|---|
| | | | $[Pb(OAc)_2]$, mol/l | $[Na_2S]$, mol/l |
| PbS 1 | 591.8 | 12 | 0.05 | 0.05 |
| PbS 2 | 594.0 | 15 | 0.025 | 0.025 |
| PbS 3 | 593.5 | 16 | 0.25 | 0.25 |

**Results and discussion**
In present paper the positron annihilation studies (PAS) results on nano-PbS with different size of particles synthesized by CBD are presented.

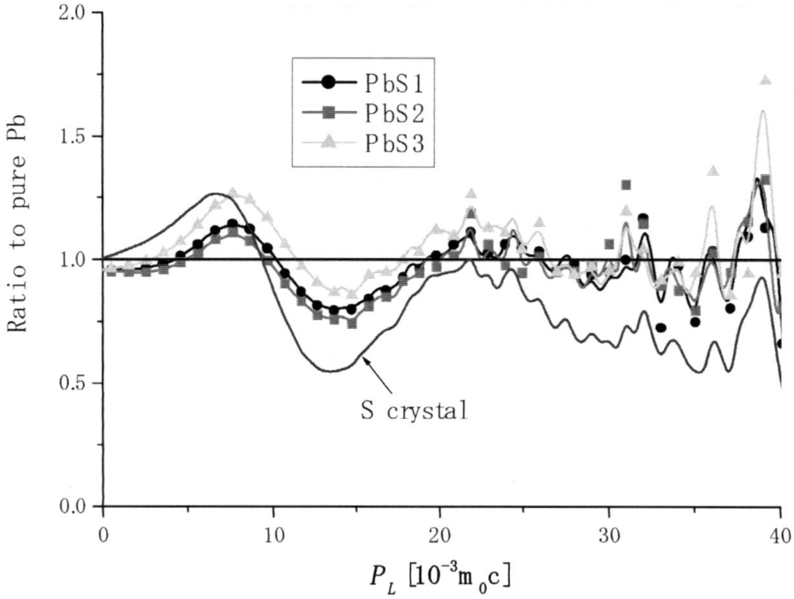


**Figure 2.** Ratios of the coincidence Doppler broadened spectra for the PbS specimens (PbS 1, PbS 2, PbS 3 and for sulfur (S crystal) to the Doppler broadened spectrum of defect-free pure lead (pure Pb)

According to the positron lifetime measurements in nano-PbS powders an exclusively single positron lifetime component was observed. Independently of the size of nanoparticles the positron lifetime for all specimens was the same and equal to 391±1 ps.

The attribution of the positron lifetimes to defects is made by the analysis of CDBAR spectra, analogous to that suggested earlier in [1, 14]. It could be seen from Figure 2 that the position and the slope of the all three curves of nano-PbS in the CE range (from $20\times10^{-3}m_0c$ to $40\times10^{-3}m_0c$ ) are very similar to each other. The curve for sulfur (S crystal) has very different position and slope in the CE range. The nearly zero slope of the curves ratio and ratio values close to 1 for the Doppler broadening spectra ratio to bulk lead (pure Pb) spectrum means that positron annihilation sites in nano-PbS are surrounded by, mostly, Pb atoms and to a small extent by S atoms. These results are compared with the results of our previous studies on nano-PbS synthesized using another methodology of CBD described in [15]. In our previous study we observed two components in PLT spectrum, which were ascribed to positron trapping in the lattice vacancies on lead sublattice (204 ps, 20 %) and in interfaces (340 ps, 80 %). The contribution by interfaces in case of nanopowders with smaller particle size (from 12 to 16 nm) was much larger (about one order of magnitude), than in case of nanopowders with larger particle size (180 nm [15]). Newly obtained and previous data will be consistent if positron trapping rates in lattice vacancies and interfaces do not change drastically upon particle size in nanoscale regime. In this case for nano-PbS specimens with smaller particles almost all the positrons are trapped at interfaces. A little bit larger (15 %) positron lifetime in interfaces (391 ps) with smaller particles is due to higher contribution of free volumes in triple junctions in such nanomaterial.

**Summary**

The CDBPAR techniques reported here are expected to be of interest for an identification of defects and specific investigation of defects and atomic processes in complex condensed matter in general, etc. It has been found that independently of the size of nanoparticles (from 12 to 16 nm) the positron lifetime (PLT) for all specimens of nano-PbS is the same 391±1 ps. The chemical surrounding of positron annihilation sites identified by coincident Doppler broadening measurements is mostly Pb and to small extent S.

**Acknowledgements**

The authors would like to thank S.I. Sadovnikov for his assistance. This work was partially financially supported by the Russian Foundation for Basic Research (RFBR No. 11-08-00314) and by the Ural Branch of the RAS (12-P-234-2003).

**References**

[1] Rempel A A, Sprengel W, Blaurock K, Reichle K J, Major J and Schaefer H-E 2002 *Phys. Rev. Letters* **89** 185501, Strite S and Morkoc H 1992 *J. Vac. Sci. Technol.* B **10** 1237
[2] Liu X and Zhang M 2000 *Int. J. Infrared Millimeter Waves* **21** 1697
[3] Mcdonald S A, Konstantatos G, Zhang S, Cyr P W, Klem E J D, Levina L and Sargent E H 2005 *Nat. Mater.* **4** 138
[4] Zagorac D, Doll K, Schön J C and Jansen M 2011 *Phys. Rev. B* **84** 045206
[5] Rempel A A , Kozhevnikova N S, Leenaers A J G and van den Berghe S 2005 *J. Cryst. Growth* **280** 300
[6] Sadovnikov S I and Rempel A A 2009 *Phys. Solid State* **51** 2375
[7] Polity A, Krause-Rehberg R, Zlomanov V, Stanov V, Chatchaturov A and Mäkinen S 1993 *J. Cryst. Growth* **131** 271
[8] Plazaola F, Seitsonen A P and Puska M J 1994 *J. Phys.: Condens. Matter* **6** 8809
[9] Eijt S W, Mijnarends P E, van Schaarenburg L C, Houtepen A J, Vanmaekelbergh D, Barbiellini B and Bansil A 2009 *Applied Physics Letters* **94** 091908
[10] Sharma S K, Sudarshan K, Maheshwari P, Dutta D, Pujari P K, Shah C P, Kumar M and Bajaj P 2011 *European Physical Journal B* **82** 335
[11] Sadovnikov S I, Kozhevnikova N S, Pushin V G and Rempel A A 2012 *Inorg. Mater.* **48** 21
[12] Hall W H 1949 *Proc.Phys. Soc., Sect. A* (London) **62** 741.
[13] Hall W H and Williamson G K 1951 *Proc. Phys. Soc. Sect. B.* (London) **64** 937
[14] Valeeva A A, Rempel A A, Sprengel W and Schaefer H-E 2003 *Phys. Chem. Chem. Phys.* **5** 2304
[15] Rempel A A, Valeeva A A and Kozhevnikova N S 2010 *JETP Letters* **92** 146

Journal of Physics: Conference Series **443** (2013) 012014 doi:10.1088/1742-6596/443/1/012014

# Dielectric properties and structural defects in $BaTi_{1-x}Sn_xO_3$ ceramics

**Zhang Lijuan**[1], **Wang Lihai**[2], **Liu Jiandang**[1], **Cheng Bin**[1], **Zhao Minglei**[2], **Ye Bangjiao**[1,*]

[1] Department of Modern Physics, University of Science and Technology of China, Hefei 230026, P. R. China

[2] School of Physics, Shandong University, Jinan 250100, P. R. China

[*] Email: bjye@ustc.edu.cn

**Abstract.** As a function of Sn doping concentration x, dielectric properties and X-ray diffraction measurements were carried out on $BaTi_{1-x}Sn_xO_3$ ($BTS_x$) ceramics fabricated by the solid-state reaction route. Positron annihilation lifetime spectroscopy and coincident Doppler-broadening spectroscopy were also measured for the evaluation of defects in the $BTS_x$ ceramics. Dielectric properties measurement reveals that the permittivity of $BTS_x$ ceramic gradually increases with increasing Sn dopant content for $x \leq 3\%$, and then decreases. This change of permittivity is found to agree well with the relative defect concentration estimated using two positron annihilation techniques. The S-W plot indicates that the defect species do not change with Sn doping. The variation correlations between defects and dielectric properties further proves that $BTS_x$ ceramics with the higher relative defect concentration present a lower permittivity.

## 1. Introduction

As one of the most important lead-free dielectric ceramics, barium titanate ($BaTiO_3$)-based ceramics have been attracting extensive interests for about 50 years, owing to their excellent dielectric properties and wide range of applications.[1-3] However, in order to meet the broad application in advanced electronics nowadays, the constant research of its structure defect and modification is in demand.[4] The $BaTiO_3$-based ceramics are of perovskite structures with an $ABO_3$ formula. Unique electric/dielectric properties of $BaTiO_3$-based materials can be carried out by varying the preparation conditions and by doping with various isovalent cations on both A (Ba) and B (Ti) sites.[5-6]

Recently, solid solution $BaTi_{1-x}Sn_xO_3$ ($BTS_x$) is attracting enormous attention due to the applications of capacitors, relaxors, sensor and the advantages of low internal stresses, low production costs and the possibility to design systems from lead-free materials.[7-8] Wei et al. reported the dielectric relaxation in paraelectric phase of $Ba(Ti,Sn)O_3$ ceramics.[9] Markovic et al reported $BTS_x$ materials have high values of dielectric permittivity, which depends on the Sn content.[6] It is shown that the partial replacement of titanium by tin improve the dielectric behavior.[4] Besides, partial isovalent substitution of $Ti^{4+}$ by $Sn^{4+}$ leads to a shifting of the paraelectric–ferroelectric phase transition (Curie point, Tc) to lower temperatures.[6, 10] These researches mainly aimed at the influences of doping and preparation conditions on the dielectric properties. It is known that the properties of $BaTiO_3$-based ceramics might strongly depend on defects that exist in large concentrations.[11] The substitution of $Ti^{4+}$ by $Sn^{4+}$ ions in the $BTS_x$ lattice might cause the lattice mismatch due to the different atomic radii. The research on the variation of structure defects and its influence on the dielectric properties induced by Sn dopants were necessary. Unfortunately, the microstructure defects doped with Sn content observation, to our knowledge, still required further investigation.

Positron annihilation spectroscopy (PAS), due to its sensitivity and selectivity to crystal defects, is well suited to the nondestructive study of structural defects in solid materials.[12-13] In recent years, a lot of significant results have been achieved by applying positron annihilation techniques into ceramics defects studies.[14-15] In this work, the structural defects and dielectric properties in the $BTS_x$ ceramics were studied. We tried to understand the relationship between the defect mechanisms and the dielectric properties with different Sn dopant content.

## 2. Experimental Procedures

All the $BTS_x$ ceramic samples in this literature were provided by the group of the Meilei Zhao in Shandong University. $BTS_x$ ceramics with x (Sn doping concentration) =0.00, 0.01, 0.03 and 0.05 were prepared by the conventional solid-state reaction route. The starting powder materials were $BaCO_3$ (99.9%), $TiO_2$ (99.9%) and $SnO_2$ (99.9%), which were of analytical grade and were weighed following the overall formula ($BaTi_{1-x}Sn_xO_3$) (x from 0 to 0.05). The starting powder materials were mixed and ball-milled with alcohol for 12 h, subsequently dried, sifted, and pressed into sheets and pre-sintered for 4 h in air at 1050 $^0$C. Then the sheets were again reground for 12 h, dried, sifted and pressed into sheets of φ13mm×2mm with the addition of a little organic adhesive; finally the sheets were sintered for 2 h in air at 1240 $^0$C and used for measurements.

The structure and composition of the samples were characterized by X-ray diffraction (XRD) technique using X' PERT PRO diffractometer. The dielectric constants at room temperature were determined at 1 kHz, 10 kHz and 1 MHz using an Agilent 4292A precision impedance analyzer with the environmental temperature was 300 K.

The positron annihilation spectroscopy (PAS) measurements were performed at room temperature. Two PAS techniques were carried out for the evaluation of defects in the $BTS_x$ ceramics. The positron lifetime spectra (PLS) were measured using a conventional ORTEC fast-slow coincidence system with a resolution of 185 ps. A 20 μCi $^{22}$Na positron source was sandwiched between two pieces of identical samples. Each spectrum contains total counts of $2\times10^6$. In order to obtain more information about defects, the coincident Doppler-broadening spectroscopy (CDBS) measurements were undertaken using two HPGe Detectors. Each spectrum was collected with a total count over $8\times10^6$.

## 3. Results and discussion

Figure 1 shows the X-ray diffraction patterns of the $BTS_x$ ceramic samples with x=0.00, 0.01, 0.03 and 0.05, respectively. The structural analysis mainly indicates that the samples consist of single perovskite phase without any other impurity phases. The diffraction patterns of $BTS_{0.00}$ (Pure $BaTiO_3$) ceramic sample shows that the peak nearby 2θ=45° splits into (002) and (200) reflections, however, a single (111) peak at 2θ=39°, which are characteristic to $BaTiO_3$ tetragonal phase. Compared with the $BTS_{0.00}$ ceramic sample, the diffraction patterns of $BTS_x$ ceramic samples with x=0.01, 0.03 and 0.05 shows that the (002)/(200) double peak of each sample still remains but is not obvious. This suggests that a low content of cubic phase is interfused accompanying the Sn doping. Besides, it can be observed from XRD patterns that increasing of Sn content causes slightly systematic shift of peaks towards lower 2θ angles. This shift is because that the substitution of $Ti^{4+}$ (0.068 nm) by $Sn^{4+}$ (0.071 nm) ions increases the unit cell volume.

A detailed study of the microstructural defects in the $BTS_x$ ceramics was performed by positron annihilation lifetime measurements. Each lifetime spectrum was fitted by LIFETIME9 program with three-component fitting procedure. The longest lifetime $\tau_3$ and its intensity $I_3$ were around 2.4 ns and below 1.2% respectively, without significant variation for all compositions. This component is attributed to the ortho-positronium (o-Ps) annihilation in the source or at the surfaces among the grains in the ceramics and is neglected in the following discussion. The shortest lifetime component $\tau_1$ is around 140-150 ps, while the second lifetime $\tau_2$ has a value between 206-248 ps. $\tau_1$ might be attributed to the average value of free positron lifetime and the lifetime in smaller cation vacancies like $V_O$. $\tau_2$ can be regarded as the weighted average lifetime of the positron in the mono-vacancies (like $V_{Ba}$, $V_{Ti}$)[16].

Average lifetime $\tau_m=(\tau_1 I_1+\tau_2 I_2)/(I_1+I_2)$ could synthetically reflect the material microstructure. It is because that the value of $\tau_m$ could present an overall reflection on the defect traps in crystal lattice of materials. Therefore, it could provide the details of the electronic structure and defect distribution in the materials. The variation of average positron lifetime $\tau_m$ as a function of the x for $BTS_x$ is presented in figure 2. It is seen that the average lifetime $\tau_m$ gradually decreases with the increase of x up to 3%. While the x exceeds 3%, the average lifetime $\tau_m$ begins to increase. This indicates that doping small amount of Sn content could reduce the defects of $BaTiO_3$. However, the structure defects in the $BTS_x$ increased when the Sn doping concentration exceeds 3%.

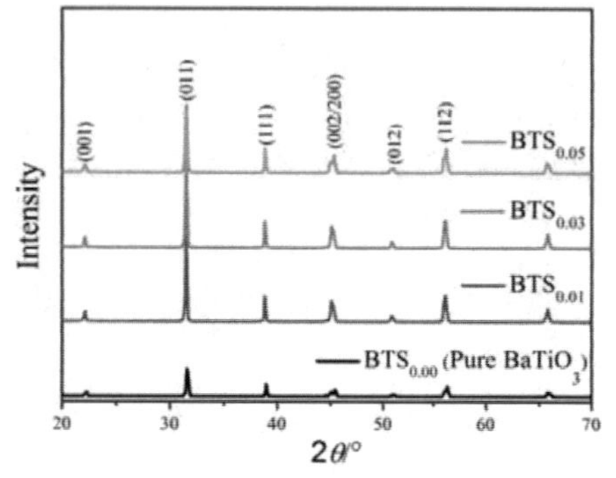


**Figure 1.** X-ray diffraction patterns of $BTS_x$ ceramics

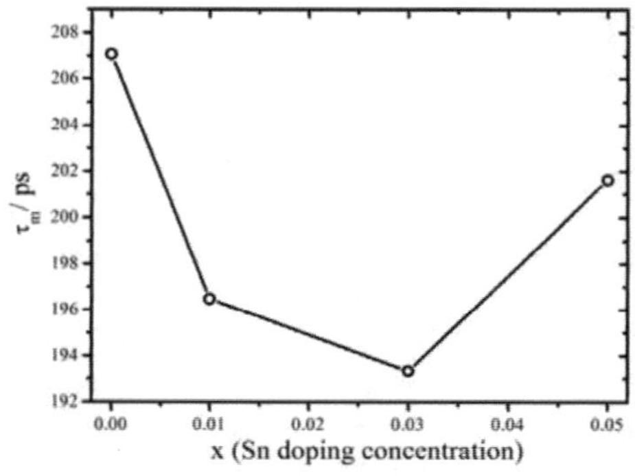


**Figure 2.** Average lifetime as a function of the x for $BTS_x$ ceramics

In order to obtain more information about structural defects, we carried out CDBS measurements in $BTS_x$ ceramics samples. The S-parameter as a function of the x is shown in figure 3. It is clear from figure 3 that the S-parameter drops with increasing Sn doping concentration when the x is below 3%. For samples with Sn doping concentration of exceeding 3%, the S-parameter increases slowly. The variation of S-parameter with x gives a preliminary idea of the changing defective state of the samples. The coincident Doppler-broadening is characterized by the S- and W-parameter. The S-parameter reflects the annihilation probability of positrons with the low-momentum valence electrons. Therefore, it suggests that the main origin of a decrease in S-parameter is due to a decrease in the relative concentrations of defects in the material. Therefore, the same conclusion about the defects evolution can be obtained in both PLS measurement and CDBS measurement.

Positron annihilation correlation plots of the S-parameter and its complementary W-parameter are found to be useful in distinguishing various defect species or the chemical surrounding of the defects. It can be seen in figure 4, the data points fall on a straight line, suggesting that the defect species do not change with Sn doping.

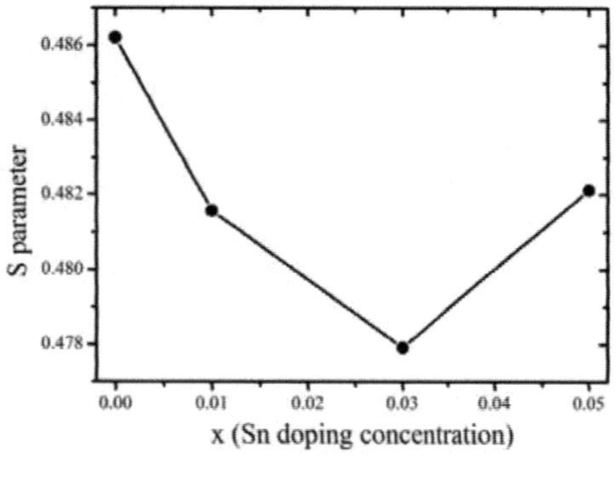


**Figure 3.** The S-parameter as a function of the x for $BTS_x$ ceramics

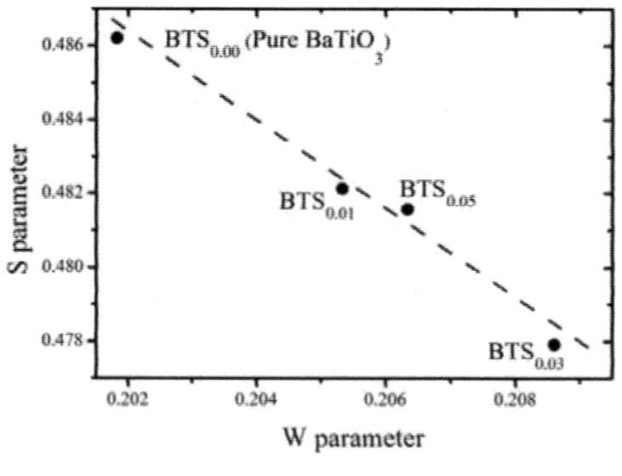


**Figure 4.** The variation of S- versus W-parameter for $BTS_x$ ceramics

The variations of relative permittivity of $BTS_x$ ceramic samples with Sn doping concentration x, at various frequencies (1 kHz, 10 kHz and 1 MHz.) are shown in figure 5. The permittivity of the sample decreases with the increasing frequency. The variation of the permittivity with the corresponding frequency can be qualitatively explained by the Debye's theory. At every frequency, the permittivity of $BTS_x$ ceramic samples gradually increases to about 4000 with Sn doping concentration (when Sn doping concentration is below 0.03). With more Sn doped ($x>0.03$), the permittivity of $BTS_x$ ceramics decreases. Compared figure 5 with figure 2 and figure 3, it can be discovered that $BTS_x$ ceramic samples with the higher relative defect concentration present a lower permittivity.

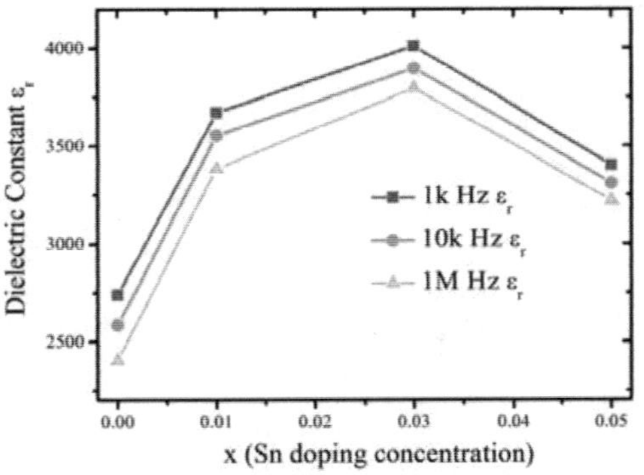


**Figure 5.** Permittivity for $BTS_x$ ceramic versus x at various frequencies

## 4. Conclusion

In summary, a detailed study of the variation of the X-ray diffraction, dielectric constant and positron annihilation spectra have been carried out on $BTS_x$ ceramics for x from 0.00 to 0.05 fabricated by the conventional solid-state reaction route. The following conclusions have been observed:

(1) A low content of cubic phase is interfused into $BaTiO_3$ tetragonal phase accompanying the Sn doping. Besides, increasing Sn content causes slightly systematic shift of peaks towards lower 2θ angles. This shift is because that the substitution of $Ti^{4+}$ (0.068 nm) by $Sn^{4+}$ (0.071 nm) ions increases the unit cell volume.

(2) The permittivity of $BTS_x$ ceramics in the same x decreases with increasing frequency. However, at each measuring frequency, the permittivity of $BTS_x$ ceramic gradually increases with Sn dopant content for $x \leq 3\%$ and decreases after it.

(3) The results of the positron annihilation lifetime spectroscopy and coincident Doppler-broadening spectroscopy agree well with the variation of the dielectric property. The S-W plot indicates that the defect species do not change with Sn doping. The variation correlations between defects and dielectric properties further proves that $BTS_x$ ceramic with the higher relative defect concentration present a lower permittivity.

**Acknowledgements** This work is supported by the State Key Program of National Science Foundation of China (Grant No. 10835006).

## References

[1] Macchi C, Somoza A, Dupasquier A *et al* 2001 *J. Phy.: Condens. Matter* **13** 5717
[2] Yongping Pu, Kai Chen, Haidong Wu 2011 *Journal of Alloys and Compounds* **509** 8561
[3] Miriam S C, Walter S, Alberto S 2007 *Journal of Physics and Chenistry of Solids* **68** 1315
[4] Mitic V V, Paunovic V, Purenovic J *et al* 2012 *Ceramics Internationa*l **38** 1295
[5] Natesan B, Hua C 2001 *Journal of materials in electronics* **12** 527
[6] Markovic S, Mitric M, Cvjeticanin N *et al* 2007 *J Eur Ceram Soc* **27** 505
[7] Alena M, Olga M, Gunnar S *et al* 2008 *J Electroceram* **20** 43
[8] Gheorghe A, Marin C, Paul G 2010 *J Mater. Sci.* **45** 2606
[9] Xiaoyong W, Xing W, Xi Y 2008 *J Electroceram* **21** 226
[10] Smilja M, Cedomir J, Ljiljana V *et al* 2010 *J Eur Ceram Soc* **30** 1427
[11] Tang C Q, Zhang D M, Xue Q Z 1995 *Cryst. Res. Technol.* **30** 837
[12] Krause R R, Leipner H S 1998 *Positron Annihilation in Semiconductors* (Springer Verlag, Berlin)
[13] Coleman P G 2000 *Positron Beams and Their Applications* (World Scientific, Singapore)
[14] Chen Z Y, Chen Z Q, Wang D D *et al* 2011 *Applied Surface Science* **258** 19
[15] Zhang L J, Wang T, Wang L H *et al* 2012 *Scripta Materialia* **67** 61
[16] V. J. Ghosh, B. Nielsen 2000 Phys. Rev. B **61**,270

Journal of Physics: Conference Series **443** (2013) 012015 doi:10.1088/1742-6596/443/1/012015

# Positron Annihilation Study of Cs-Deficient Pollucite

**P Guagliardo[1], E R Vance[2], K Sudarshan[1], J Davis[2], J F Williams[1], C Ranganathaiah[1] and S Samarin[1]**

[1]Centre for Antimatter-Matter Studies, School of Physics, The University of Western Australia, 35 Stirling Highway, Crawley, WA 6009, Australia
[2]Institute for Materials Engineering, Australian Nuclear Science and Technology Organisation, Menai, NSW 2234, Australia

E-mail: paul.guagliardo@uwa.edu.au

**Abstract**. Positron annihilation has been applied to study a series of non-stoichiometric pollucite samples with the composition $Cs_{(1-x)}Al_{(1-x)}Si_{(2+x)}O_6$ for x = 0.0-0.25. Lifetime results showed a monotonic increase in the o-Ps lifetime and intensity as the Cs concentration was reduced. This is consistent with literature suppositions that vacancies are created as the Cs content is reduced.

## 1. Introduction

Pollucite is a caesium aluminosilicate with the composition $CsAlSi_2O_6$. Its structure is a rigid three-dimensional framework consisting of 48 corner-sharing (Si, Al)$O_4$ tetrahedra and 16 $Cs^+$ ions occupying its large 12-coordinate cavities [1-3]. The Cs-deficient form of pollucite has found utility in applications requiring high thermal stability, such as in gas turbine components, as its thermal expansion behaviour can be tuned by the amount of $Cs^+$ in the unit cell [4]. In addition, pollucite has been considered as a possible material for the immobilization of $^{137}Cs$ radioactive waste, owing to the fact that its structure can incorporate up to 40 wt% of Cs, which creates a highly dense wasteform superior to glass-ceramic or zeolite alternatives [5, 6]. Kobayashi et al. [3] have shown by powder x-ray diffraction measurements that the lattice parameters of $Cs_{(1-x)}Al_{(1-x)}Si_{(2+x)}O_6$ showed a nearly linear decrease with a reduction in the amount of $Cs^+$ ions (increasing x), consistent with the replacement of $Cs^+$ ions with 12-fold vacancies in the structure. To investigate this we have used positron annihilation, x-ray diffraction and scanning electron microscopy to study a series of non-stoichiometric pollucite samples.

## 2. Experimental

X-ray diffraction (XRD) was performed with a Siemens D500 instrument, using Co Kα radiation, as well as a PANalytical diffractometer, using Cu Kα radiation. Scanning Electron Microscopy (SEM) was carried out on samples mounted in epoxy resin and polished to a 1-micron diamond finish. Approximately 50 Å of carbon was evaporated onto the surfaces under vacuum to prevent charging. The SEM was a Zeiss Ultra Plus instrument with an attached Oxford Instruments X-Max 80mm$^2$ SDD X-ray microanalysis system. The SEM was operated at an accelerating voltage of 15 kV.

The positron lifetime spectrometer was a fast-fast coincidence spectrometer, with detectors consisting of a BC418 plastic scintillator coupled to a Burle 8850 photomultiplier tube. A 30 μCi $^{22}$NaCl source was used for collecting spectra; the source was encapsulated in 8 μm Kapton foil.

Experiments were carried out in the usual sandwich geometry between two identical samples. The time resolution of the instrument was determined to be 220 ps from spectra of a high-purity annealed nickel sample. Each spectrum had 3-4x$10^6$ counts and was analysed using PALSfit Version 1.64 [7]. For all samples, a three component analysis gave satisfactory fits (a variance close to unity). Measurements were conducted at room temperature and in air. Doppler broadening measurements were carried out with a HPGe detector with an energy resolution of approximately 1.5 keV at 511 keV.

Two series of Cs-deficient pollucite samples were produced by slightly varying methods, giving rise to differing levels of phase purity. The first series of samples with the composition $Cs_{(1-x)}Al_{(1-x)}Si_{(2+x)}O_6$ for x = 0.0, 0.1, 0.2 and 0.25 were made from mixed Al/Cs nitrate solutions and Ludox, a 40 wt% colloidal silica solution. The chemicals were accurately weighted, stir-dried, calcined, ball-milled in non-polar cyclohexane, and then sintered for 5 hours at 1200°C. Samples of $CsAlSi_5O_{12}$ and $Cs_{1.05}Al_{1.05}Si_{1.95}O_6$ were also produced by the same method. For the $Cs_{(1-x)}Al_{(1-x)}Si_{(2+x)}O_6$ samples, XRD showed major pollucite phase in all samples; however these samples contained a substantial contribution from $CsAlSi_5O_{12}$ (over 10%), with the amount increasing with x. The second series of samples ($Cs_{(1-x)}Al_{(1-x)}Si_{(2+x)}O_6$ for x = 0, 0.1, 0.2) was produced in a similar manner; however sintering was performed at a higher temperature of 1450°C. This gave rise to a higher degree of pollucite phase, as determined by XRD analysis. The x = 0 sample contained only a small amount of $CsAlSiO_4$, with the strongest lines being only 2% of the intensity of the strongest pollucite line. The x = 0.1 sample exhibited only pollucite, insofar as any other X-ray lines were less than 1% of the intensity of the strongest pollucite line. The x = 0.2 sample did exhibit some $CsAlSi_5O_{12}$, but the strongest line was only 6% of the strongest pollucite line.

SEM on the pollucite phases in the Cs-deficient samples also indicated Cs deficiencies corresponding approximately to the nominal sample stoichiometries.

## 3. Results and Analysis

Lifetime spectra could be successfully analysed in terms of three discrete components. The shorter component $\tau_1$ is associated with p-Ps annihilation and positron annihilation in the bulk material; $\tau_2$ can be attributed to positron annihilation in defects, and the nanosecond component $\tau_3$ is indicative of o-Ps pick-off annihilation in structural open volume. Fitting results for these samples and the resulting pore sizes calculated from $\tau_3$ using the Tao-Eldrup (TE) model [8] are given in table 1. The $\tau_3$ lifetimes correspond to pores in the range of 0.33-0.38 nm; this is close to the ionic size of Cs (0.38 nm) [9].

**Table 1.** Fitting results and pore diameters calculated from the TE model for set 1 of the pollucite samples.

| Composition | $\tau_1$ (ps) | $\tau_2$ (ps) | $\tau_3$ (ps) | $I_1$ (%) | $I_2$ (%) | $I_3$ (%) | d (nm) |
|---|---|---|---|---|---|---|---|
| $CsAlSi_2O_6$ | 198 (10) | 435 (9) | 980 (20) | 24 (2) | 65 (1) | 11 (1) | 0.33 |
| $Cs_{0.9}Al_{0.9}Si_{2.1}O_6$ | 192 (10) | 447 (8) | 1060 (10) | 23 (2) | 62 (1) | 15.3 (0.7) | 0.35 |
| $Cs_{0.8}Al_{0.8}Si_{2.2}O_6$ | 168 (8) | 453 (6) | 1120 (9) | 19 (1) | 60 (1) | 20.8 (0.5) | 0.37 |
| $Cs_{0.75}Al_{0.75}Si_{2.25}O_6$ | 153 (10) | 440 (8) | 1120 (11) | 16 (1) | 61 (1) | 22.9 (0.6) | 0.37 |
| $CsAlSi_5O_{12}$ | 150 (10) | 457 (7) | 1170 (12) | 13.8 (0.9) | 63.4 (0.5) | 22.7 (0.6) | 0.38 |
| $Cs_{1.05}Al_{1.05}Si_{1.95}O_6$ | 188 (8) | 419 (6) | 1045 (13) | 24 (1) | 64 (1) | 12.0 (0.5) | 0.35 |

It can be seen that as the Cs content decreased, both the lifetime and intensity of the o-Ps component increase monotonically. This could suggest an increase in the structural open volume initiated by Cs vacancy creation. However, as noted above, the amount of $CsAlSi_5O_{12}$ increases with x. Lifetime results for $CsAlSi_5O_{12}$ are also shown in table 1; it can be seen that this sample has the longest lifetime with high intensity. This is consistent with reports that $CsAlSi_5O_{12}$ can exhibit a substantial Cs-deficiency [10]. Thus, the monotonic increase in $\tau_3$ and $I_3$ with x could simply be due to the increasing concentration of $CsAlSi_5O_{12}$ in the samples. A sample with excess Cs was also produced, which gave rise to results closest to that of stoichiometric pollucite.

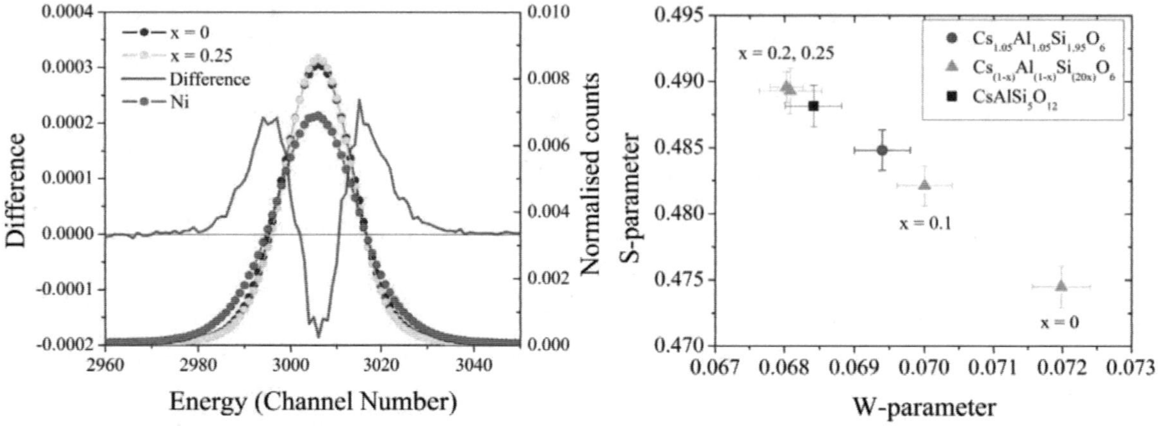


**Figure 1.** Left: 511 keV photo peak (taken with HPGe detector) for pollucites x = 0.0 and 0.25, and nickel; the difference between the pollucite spectra is also shown; right: S- vs. W-parameter plot.

In figure 1, Doppler broadening spectra are shown for the stoichiometric pollucite $CsAlSi_2O_6$ and for $Cs_{0.75}Al_{0.75}Si_{2.25}O_6$ (annealed Ni is also shown as a reference). For the Cs-deficient sample there is an increase in the peak intensity relative to the stoichiometric pollucite, which is apparent in the difference spectra (also shown in figure 1). An S vs. W plot is shown in figure 1 where the line shape parameters for all samples are plotted; it can be seen that there is an increase in the S-parameter with increasing x. Again, this could indicate an increase in structural open volume (i.e. increased annihilation with low momentum electrons), or an increase in the concentration of $CsAlSi_5O_{12}$, which displays similar line shape parameters to the x = 0.2 and 0.25 cases.

PALS fitting results for the samples with a higher degree of phase purity are given in table 2. For these samples there is a monotonic increase in $\tau_3$ and $I_3$ with decreasing Cs content. As the content of $CsAlSi_5O_{12}$ does not increase monotonically for these samples then this provides an indication that it is Cs vacancy creation, rather than the development of a second phase, that is giving rise to the observed increase. This is consistent with the results of Kobayashi et al. [3]. The o-Ps lifetimes and intensities for both sets of samples are compared in figure 2. The variation in $\tau_3$ is significantly reduced for sample set 2, which displayed higher phase purity; however, the variation in $I_3$ with doping is comparable for both sets of samples.

**Table 2.** PALS fitting results and pore diameters calculated from the TE model for the final set of pollucite samples.

| x | $\tau_1$ (ps) | $\tau_2$ (ps) | $\tau_3$ (ps) | $I_1$ (%) | $I_2$ (%) | $I_3$ (%) | d (nm) |
|---|---|---|---|---|---|---|---|
| 0 | 188 (8) | 465 (5) | 1081 (13) | 17 (1) | 69.7 (0.6) | 13.4 (0.5) | 0.358 |
| 0.1 | 182 (7) | 467 (5) | 1108 (12) | 18.3 (0.9) | 66.6 (0.6) | 15.2 (0.5) | 0.366 |
| 0.2 | 180 (8) | 464 (5) | 1108 (8) | 15.7 (0.9) | 63.0 (0.5) | 21.3 (0.5) | 0.366 |

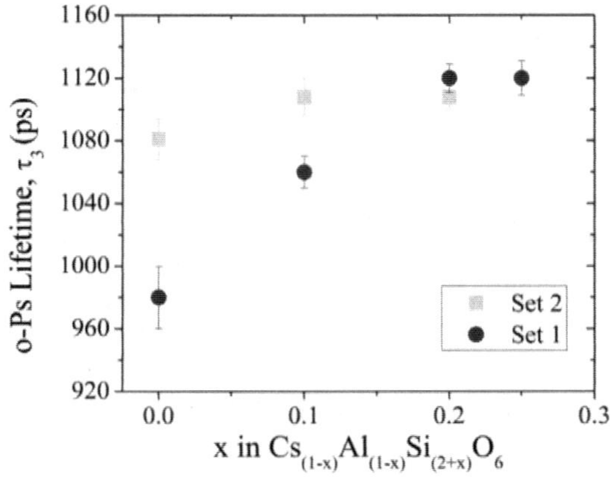


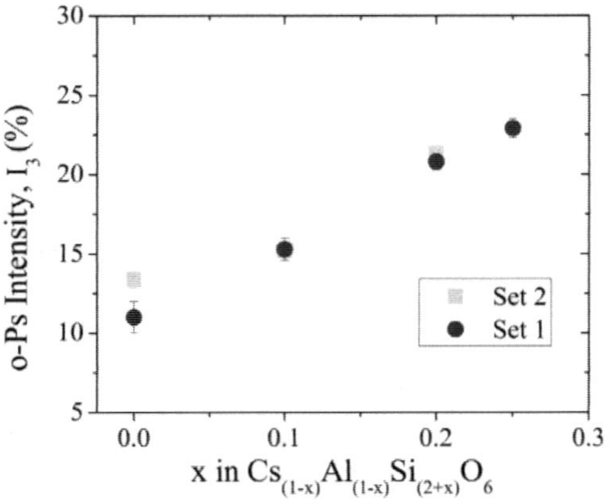


**Figure 2.** $\tau_3$ and $I_3$ vs. x for both sets of pollucite samples; for sample set 1 there is an in increase in the amount of $CsAlSi_5O_{12}$; set 2 is closer to single phase pollucite.

## 4. Conclusion

Two sets of Cs-deficient pollucite samples were studied with differing degrees of phase purity. Both sets of samples showed a monotonic increase in the o-Ps lifetime and intensity as the Cs concentration was reduced. This is consistent with literature suppositions that vacancies are created as the Cs content is reduced. However, in all samples there was a contribution from non-pollucite phases which could influence the positron annihilation parameters. In the future we will attempt to produce single phase pollucite so as to remove any ambiguity in the data.

## References

[1] Beger R M 1969 *Z. Kristallogr.* **129** 280
[2] Hirao K, Soga N and Kunugi M 1976 *J. Phys. Chem.* **80** 1612
[3] Kobayashi H, Sumino S, Tamai S and Yanase I 2006 *J. Am. Ceram. Soc.* **89** 3157
[4] MacLaren I, Cirre J and Ponton C B 1999 *J. Am. Ceram. Soc.* **82** 3242
[5] Komarneni S and Roy R 1983 *J. Am. Ceram. Soc.* **66** 471
[6] Hess N J, Espinosa F J, Conradson S D and Weber W J 2000 *J. Nucl. Mater.* **281** 22
[7] Olsen J V, Kirkegaard P, Pedersen N J and Eldrup M 2007 *Phys. Status Solidi (c)* **4** 4004
[8] Tao S J 1972 *J. Chem. Phys.* **56** 4754
[9] Shannon R 1976 *Acta Crystallogr. Sec. D* **32** 751
[10] Gatta G D, Rotiroti N, Fisch M, Kadiyski M and Armbruster T 2008 *Phys. Chem. Minerals* **35** 521

Journal of Physics: Conference Series **443** (2013) 012016 doi:10.1088/1742-6596/443/1/012016

# Positron study of alpha particles effect on oxide-dispersion-strengthened steels

**J Veternikova, V Slugen, S Stancek, J Degmova, V Sabelova, S Sojak, M Petriska, R Hinca**

Institute of Nuclear and Physical Engineering, Faculty of Electrical Engineering and Information Technology, Slovak University of Technology, Ilkovicova 3, 812 19 Bratislava, Slovakia

jana.veternikova@stuba.sk

**Abstract**. Microstructure of oxide-dispersion-strengthened steels with different chromium content – MA 956 (20% Cr), ODM 751 (16% Cr) and MA 957 (14% Cr) were studied by positron annihilation lifetime spectroscopy. Samples were measured before and after helium ion implantation ($He^{2+}$); therefore microstructure changes and radiation resistance to alpha particles of these steels were observed. Defect accumulation due to the radiation treatment was proven in all investigated materials. After ion implantation, mean lifetimes increased of about ~ 10 ps, which indicates significant change in microstructure. According to calculations of a defect volume from defect concentration and pre-dominant size of defects, ODM 751 is the most resistant steel in comparison to other investigated materials.

## 1. Introduction

Oxide-dispersion-strengthened alloys belong to perspective candidate materials for new reactors within the frame of the international program Generation IV. Therefore, it is necessary to ensure their integrity during their projected lifetimes. Construction steels applied for the advanced nuclear facilities could be exposed to high temperatures (~ 900°C) and high irradiation levels (~ 150 dpa) [1], which significantly deteriorate their mechanical properties (hardness, embrittlement and toughness) [2].

This paper deals with a study of microstructure of oxide-dispersion-strengthened steels with different chromium content – MA 956 (20% Cr), ODM 751 (16% Cr) and MA 957 (14% Cr). A radiation resistance to alpha irradiation, which transmits high energies in a short-range area, was investigated for samples of the studied steels via positron annihilation lifetime spectroscopy.

The effect of alpha particles was simulated by helium ion implantation with a kinetic energy up to 500 keV performed by using a linear accelerator at Slovak University of Technology. Radiation damage around 45 dpa was applied. Two states of the samples, before and after the implantation, were measured and results were compared. Defect accumulation due to the radiation treatment was proven in all investigated materials.

## 2. Experiment

Three following commercial high chromium ferritic ODS steels were measured: MA 956 and MA 957 (products of Incoloy) [3] as well as ODM 751 (product of Dour Metal) [4]. The chemical composition of investigated steels (See table 1) was observed by optical emission spectroscopy at Institute of Materials, Slovak University of Technology.

**Table 1.** Chemical composition of steels (in % wt.).

| Steels | C | Mn | Ni | Cr | Mo | Ti | Al | Cu | Si | Nb | N | $Y_2O_3$ |
|---|---|---|---|---|---|---|---|---|---|---|---|---|
| **MA 956** | 0.07 | 0.12 | 0.07 | 19.97 | 0.10 | 0.30 | 3.40 | 0.03 | 0.04 | 0.01 | 0.04 | 0.50 |
| **ODM 751** | 0.07 | 0.07 | 0.02 | 16.20 | 1.74 | 0.66 | 3.80 | 0.02 | 0.06 | 0.01 | 0.03 | 0.50 |
| **MA 957** | 0.03 | 0.09 | 0.13 | 14.40 | 0.03 | 1.00 | 0.03 | - | 0.04 | - | | 0.30 |

These investigated ODS alloys were produced by mechanical alloying, i.e. matrix materials were milled and mixed together with yttrium particles to form solid solutions with a uniform dispersion of oxide nano-particles, and the mixtures were then consolidated using Hot Isostatic Pressing (HIP) at 1150°C under a pressure of 103MPa. The recrystallization heat treatment resulted in a coarse columnar grain structure. Samples of investigated steels were prepared from as-received material by cutting the steel sheets into suitable pieces. After cutting, the sample surfaces were polished in order to remove surface impurities. Mechanical treatment of samples affects surface and subsurface layers, although samples are polished almost into a mirror level.

The samples were investigated by positron annihilation lifetime spectroscopy (PALS). The PALS [5] can determine concentration and size of vacancy-type defects in sample with very low concentration (from 0.1 to 500 ppm) [6]; therefore can describe area where microscopy techniques are not so sensitive. The measuring equipment [7] used in this work consists of two $BaF_2$ scintillation detectors connected in fast-fast mode. As positron source, $^{22}Na$ covered in kapton foil (POSK 22) was used. The resulting spectrum was evaluated by program LifeTime9 [8]. The value of FWHM parameter was close to 240 ps. Fit Variant (reduction of chi-square) achieved value in range (1; 1.1).

Samples were loaded by radiation damage performed at a linear accelerator belonging to Slovak University of Technology. Helium ions ($He^{2+}$) with a kinetic energy up to 500 keV were used. An implantation depth achieved approximately 1 μm according SRIM simulations [3]. The implantation level – surface charge was 0.16 C/cm$^2$ (~ 1x10$^{17}$ ions cm$^{-2}$) and the radiation damage was around 45 dpa. During the ion implantation, a defect accumulation occurred as result of atom knocking-out due to slowing down of ions. Therefore increase of defect concentration or defect size was assumed in the investigated samples [9].

## 3. Results and Discussion

PALS spectra were decomposed into three components according to the Standard trapping model [10]. The shortest lifetime (LT1) of MA 956 and ODM 751 achieved values up to 110ps, which describes positron annihilation in bulk, eventually values reduced due to presence of defects especially for MA 956. The LT1 of MA 957 was much higher (170 ps) which also indicates presence of dislocations or mono-vacancies.

The second positron lifetime (LT2) found within the range from 225 to 300 ps characterizes vacancy type defects and is dependent on the size of three dimensional vacancy clusters $V_n$ consisting n vacancies. This lifetime can be also affected by positron annihilation in $Y_2O_3$ nano-particles (theoretical value 240 ps [11]) or in similar precipitates or segregation.

Values of LT2 are presented and compared each other in figure 1. In this graph, the differences of microstructure are demonstrated via the presence of vacancy defects. Also evident increase of lifetimes due to implantation (See figure 1) was found for both groups of steels as was assumed. The lifetime defining the bulk (LT1) and also defects (LT2) grew due to the effect of ion-atom collisions and atom knocking-out.

The highest defect size was observed for steel ODM 751, which contains probably clusters with 5 vacancies ($V_5$) in pre-dominance. After implantation the defect size remained stable – unchanged. Five or four vacancy clusters ($V_4$ – $V_5$) were observed in the implanted sample of MA 957. This steel demonstrated the most increase of LT2 due to implantation, the non-implanted sample had only three-vacancy clusters ($V_3$). Lower defect growth, but also evident change of LT2, was recorded for MA 956. MA 956 contained three-vacancies ($V_3$). Due to radiation treatment, defects start to enlarge and

LT2 increased. The implanted MA 956 has probably three-vacancies with some amount of four-vacancies ($V_3$- $V_4$).

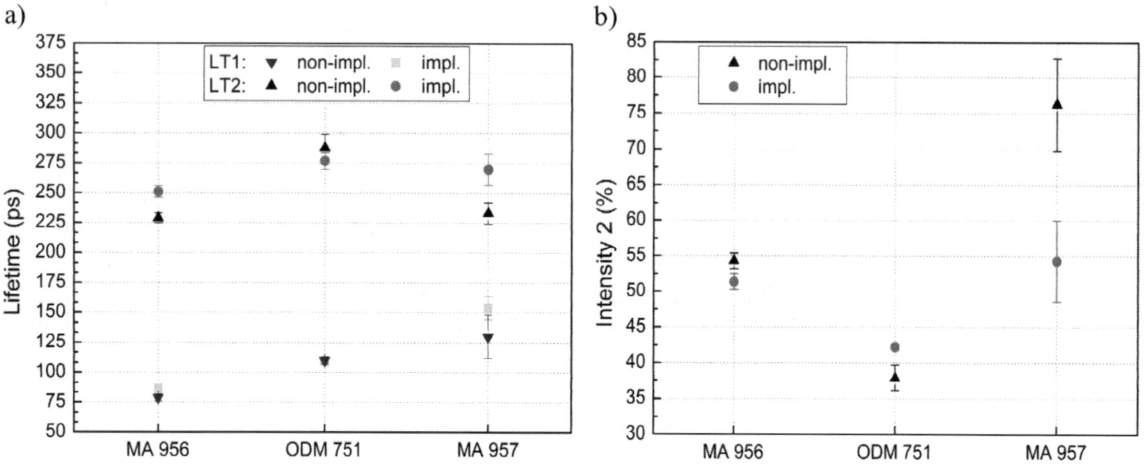


**Figure 1.** PALS results for non-implanted (non-impl.) and implanted (impl.) samples.

The intensities of positron annihilation in defects (I2) shown in figure 1b demonstrated a value decrease for all steels after implantation except of steel ODM 751. This indicated that during implantation the defect amount was lessened. Defects merged to the larger ones as LT2 values have already presented. Only steel ODM 751 has higher I2 value after implantation ($\Delta$I2= 4.3 %), which means that accumulation of new defects occurred along with the weak merging of defects in this material.

Results of positron study can be also treated into mean lifetime (MLT), which is not loaded by error formed during spectra decomposition into the lifetimes and intensities. The MLT has absolute deviation equal to 2 ps; therefore it introduces a quite accurate way for comparison of steels. The MLT of our investigated steels is presented in figure 2a, where also changes of MLT between samples before and after implantation are displayed. There, we can observe influence of implantation on defect accumulation.

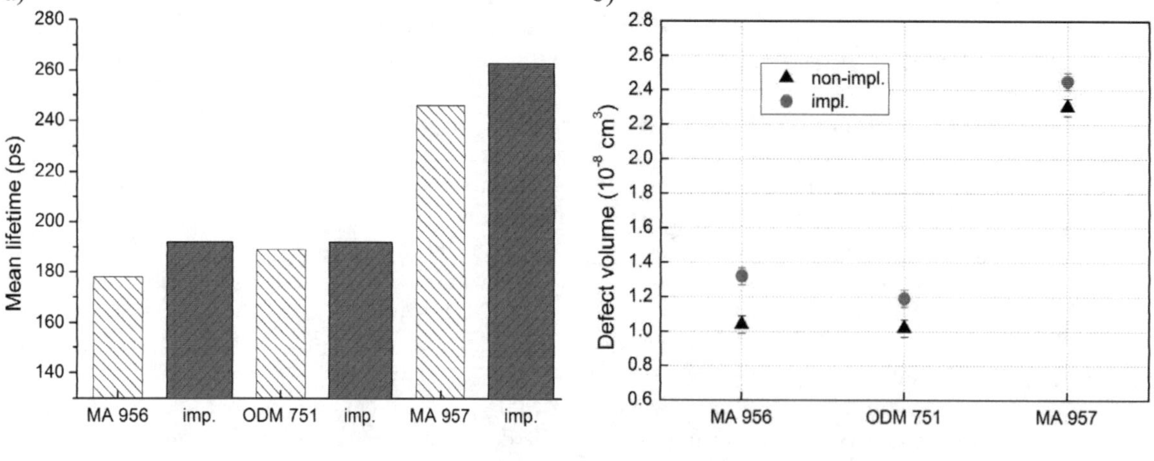


**Figure 2.** Positron mean lifetime values for samples before and after implantation (a).The change of defect volume after implantation (b).

The smallest MLT change was identified for the samples of ODM 751 (3 ± 2 ps). Enough high change due to implantation was proven for MA 956 (14 ± 2 ps) and MA 957 (17 ± 2 ps). ODM 751 seems to be the most radiation resistant material in compare to other investigated steels, although it contains the biggest defects.

With the use of LT2 and I2 values, defect concentration was calculated according to equations published in [12, 13]. The defect concentration also decreased after implantation due to defect merging and the growth of the defect size. However defect volume calculated from the defect concentration and the defect size of pre-dominance defects increased after implantation, which is evident proof of accumulated damage by alpha particles. The highest change of vacancy defect volume due to implantation was found for MA 956. MA 957 is typical by the highest volume of vacancy defects, but it is necessary to realize that dislocation or mono-vacancies found in this material were not included in the volume, which were calculated only from LT2 and I2 values.

## 4. Conclusion

Three different commercial oxide-dispersion strengthened steels were investigated by positron annihilation spectroscopy. Steels differ in chemical composition, mostly in chromium and aluminum content. They were formed by the same process of manufacture, but their sensitivity to defect accumulation during $He^{2+}$ ions (alpha particles) implantation is not similar. Material ODM 751 seems to be the most resistant, but it contains the biggest defects with the lowest concentration. Steel MA 956 indicated the highest radiation damage due to defect accumulation shown in the change of defect volume.

This study can supplement complex investigation of candidate materials for GEN IV reactors and can be helpful for final selection of the most appropriate construction material for application in environment with alpha radiation.

**Acknowledgement**
Financial contributions from the association EURATOM.CA and VEGA 1/0366/12 are acknowledged.

**References**

[1] Zinkle S J 2004 *Fusion Materials Science: Overview of Challenges and Recent Progress* 46th Annular Meeting Savannah USA.
[2] Rykalin N N, Bystrov L N, Ivanov L I 1973 *Radiation Effects* **17** 241.
[3] Gelles D S 1966 *J. Nuc. Mat.* **233-237** 293.
[4] Kazimierzak B, Prignon M, Lecomte-Mertens Ch, Coutsouradis D 1954 *Anti-Corrosion Methods and Materials* **37** 4.
[5] Ziegler J F, Ziegler M D, Biersack JP 2010 *Nucl. Instr. Meth. Phys. Res. B* **268** 1818.
[6] Slugen V 2006 *What kind of information we can obtain from Positron Annihilation Spectroscopy* Report of European Commission, Netherlands EUR 22468.
[7] Petriska M, Zeman A, Slugen V, Krsjak V, Sojak S 2009 *Phys. Stat. Solidi. C* **60** 2465.
[8] Kansy J 1996 *Nucl. Instr. Meth. Phys. Res. A* **374** 235.
[9] Eldrup M, Singh B N, Golubov S I 2010 *Final Report on Investigations of the Influence of Helium Concentration and Implantation rte on Cavity Nucleation and Growth during Neutron Irradiation of Fe and Eurofer 97* Report of National Laboratory for Sustainable Energy ISSN 0106-2840.
[10] Hautojarvi P, Corbel C 1995 *Positron spectroscopy of solid* IOS Press Amsterdam.
[11] Alinger M J, Glade S C, Wirth B D, Odette G R, Toyama T, Nagai Y, Hasegawa M 2009 *Mat. Sci. Eng. A* **518** 150.
[12] Krause-Rehberg R, Leipner S H 1998 *Positron annihilation in Semiconductors* Springer Berlin Germany ISBN 3-540-64371-0.
[13] Hautojarvi P, Pollanen L, Vehanen A, Yli-Kauppila J 1983 *J. Nucl. Matter* **114** 250.

Journal of Physics: Conference Series **443** (2013) 012017 doi:10.1088/1742-6596/443/1/012017

# Cu Precipitates in Fe Ion Irradiated Fe-Cu Alloys Studied Using Positron Techniques

**X.Z. Cao[1], P. Zhang[1], Q. Xu[2], K. Sato[2], H. Tsuchida[3], G.D. Cheng[1], H.B. Wu[1], X.P. Jiang[1], R.S. Yu[1], B.Y. Wang[1], L. Wei[1]**

[1] Key Laboratory of Nuclear Analysis Techniques, Institute of High Energy Physics, Chinese Academy of Sciences, Beijing 100049, China

[2] Research Reactor Institute, Kyoto University, Osaka 5900494, Japan

[3] Quantum Science and Engineering Center, Kyoto University, Uji 6110011, Japan

E-mail: caoxzh@ihep.ac.cn

**Abstract.** In this paper, we summarized our recent experimental results on Fe-Cu model alloys irradiated by Fe ion. Two kinds of Fe-Cu alloys with 0.3%Cu and 0.6%Cu were prepared and irradiated by 2.5 MeV Fe ion at 573 K. Irradiation dose is 0.1 dpa and 1.2 dpa for each type alloy respectively. Positron annihilation techniques of slow positron beam were used to investigate the irradiation induced defects. Results show that the S parameters are higher in the specimens with high irradiation dose, but the intensity of Cu peaks of CDB is lower. It indicates that the precipitation of Cu atoms formed easily as lower irradiation dose.

## 1. Introduction

The formation of Cu precipitates plays a key factor in the reactor pressure vessel (RPV) steels as irradiation induced by neutrons [1, 2]. Extensively studies focused to the formation and migration of the Cu precipitates at past decades, because the precipitation of Cu atoms can increase the hardness and decrease the ductility, especially it can leads to embrittlement in irradiation induced RPV steels [3-5]. As one of the menaces to the safety of nuclear power reactor operating, to understand the formation mechanism of irradiation induced defects and Cu precipitates in RPV steels is one of the important research subjects [6-9].

As the complicated irradiation effects in commercial RPV steels, such phenomena of radiation induced segregation and precipitation is important as the migration of Cu atoms excess point defect fluxes and super saturation during the displacement and energy deposition of implanted particles. It is well know that substantial recombination and defect clustering can be produced by neutron irradiation in metals as high energy recoils of fast neutrons within dense cascades [10, 11]. Recent work has suggested that remarkable precipitates formed during post-radiation procedure with temperature affection [3-5]. One assumed viewpoint is that embryos of Cu precipitates can also be nucleated and contribute to the hardening directly [11, 12]. Compared to the neutron irradiation effects in RPV steels, high-energy heavy particle can produce similar displacement damage within acceptable experiment research. Fe ion produced by accelerator was usually performed as the implanted particles, which can produce complex defects without impurity elements.

The mechanism of the formation of defects and nucleating and clustering of Cu atoms are still unclear under irradiation performed, and the evolution of defects with radiation dose and Cu contents still require further investigation, especially in the nucleation stage of the Cu precipitates [4]. But it is little know experimentally because the difficulty of observation method, such as transmission electron microscopy and atom probe tomography [1, 13, 14]. Positron annihilation techniques (PAT) have

[1] X.Z. Cao, E-mail: caoxzh@ihep.ac.cn

been used to investigate tiny Cu precipitates, defects and their complexes in Fe-Cu alloys [5], especially the positrons can be implanted to the interesting depth with appropriate energy using slow positron beam. Previously, we have published several papers related to Cu precipitates in Fe-Cu alloys. In the present study, PAT of slow positron beam were carried out to elucidate the effects of radiation dose and Cu concentration on defects and Cu precipitates formation in Fe ions irradiated Fe-Cu alloys.

## 2. Experimental

*2.1.* ***Materials*****:** Two series of Fe-Cu alloys with 0.3 and 0.6 mass% Cu were smelted from Fe (99.99% purity) and Cu (99.9% purity) in vacuum using a high-frequency induction furnace. Specimens, with a thickness of 0.2 mm and 8 mm square sheet, were irradiated with Fe ions after quenched in water from 1173 K. Fe ions with 2.5 MeV were implanted using a tandem Pelletron accelerator (model: 6SDH-2) of Quantum Science and Engineering Center, Kyoto University. Irradiation dose performed to 0.1 dpa and 1.2 dpa for each series at 573 K, respectively.

*2.2.* ***Positron annihilation techniques:*** PAT experiments were performed at slow positron beam facility of institute of high energy physics (IHEP). Slow positrons are generated by a 50 mCi (2005) $^{22}$Na radiation source. During the measurement of positron annihilation Doppler broadening spectroscopy (DBS), beam energy was changed from 0.18 keV to 25.0 keV. In this present study, two parameters were introduced to DBS to characterize the information of specimens, which are named S and W parameters respectively. S parameter was defined as the ratio of the counts in the centre area near the 511 keV (510.2-511.8 keV) of γ-rays to the total counts of spectrum (503.34-518.66 keV). And W parameter was defined as the ratio of the counts in the summed areas of 514.83–518.66 keV and 503.34–507.17 keV to the total counts of spectrum.

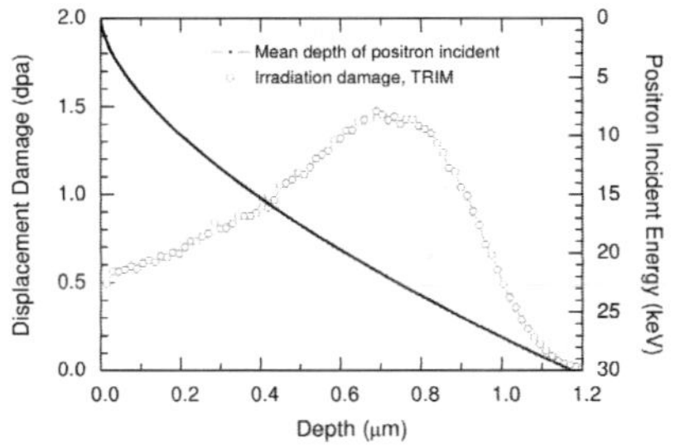


**Figure 1.** Depth distribution of damage calculated using the TRIM code for the Fe-Cu alloy irradiated with Fe ions to 1.2 x $10^{17}$ Fe/m$^2$.

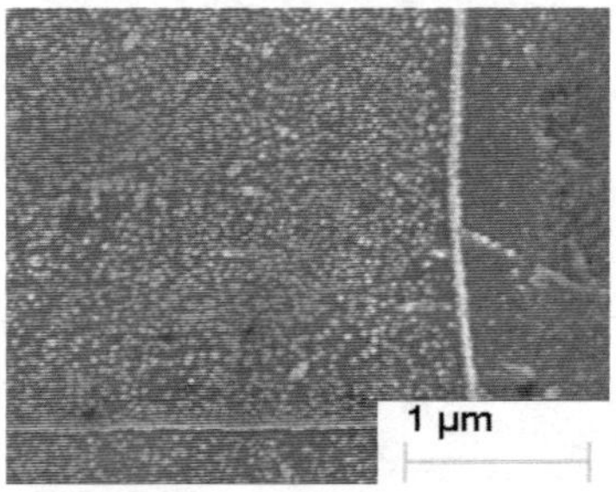


**Figure 2.** SEM image of Fe-0.3%Cu alloy with a irradiation dose as 0.1 dpa induced by 2.5 MeV Fe ions at 573 K.

Compared to the DBS, two HPGe detectors were used to detect the γ-ray pairs emitted by annihilation of positron and electron in coincident Doppler broadening (CDB) techniques. The information of elements around the positron trapping position can be described, because the momentum of atom-bound electron can be measured in CDB if it was annihilated [16, 17]. The damage peak of Fe ions in Fe-Cu alloys was about 0.7 μm from the irradiation surface, according to the simulation using the TRIM code. In order to investigate the precipitation of Cu atoms, CDB measurements with a mono-energetic positron beam were performed to the irradiation damage area in this present study. The implanted mean depth (nm) of positrons may be estimated by empirical equation as $R=40\times E^{1.6}/\rho$, using parameters of positron energy E with keV and materials density ρ with g/cm$^3$ [18-20]. The results of TRIM code simulation and the mean depth of positron incident are shown as Figure 1.

## 3. Results and discussion

*3.1.* ***SEM results:***The surface microstructure of Fe-Cu alloy was observed by SEM (Hitachi S-4700). Figure 2 showed the result of the Fe-0.3%Cu alloy with a radiation dose as 0.1 dpa induced by Fe ions at 573 K. It shows that some crystalline structure formed with many tiny clusters on the surface, especially aggregated at the grain boundary. The concentration of Cu distributed on these tiny clusters and the grain boundary is high than mean distribution, which confirmed by the energy dispersive spectrometer (EDS) of SEM. Some paper has also reported that Cu atoms migrated and aggregated at the grain boundaries with the diffusion of vacancies defects [21].

*3.2.* ***PAT results:*** Figure 3 showed the S parameters with positron energy changed from 0.18 keV to 20 keV. One un-irradiated Fe-Cu specimen was also measured as the reference specimens. According to the description in Figure 1, the positrons were implanted near 0.7 μm from surface and almost closing to the area of Fe ions deposition. Results show that the S parameters near to 20 keV are dramatically increased after irradiated, but there is no clearly difference affected by Cu content and irradiation dose. Because of closing to the Fe ions deposition area, the defects may be saturated to trapping positrons to annihilate. At the range of positron energy changed from 2 keV to 15 keV, the S parameters of specimens with 1.2 dpa is larger than those with 0.1 dpa. It indicated that more defects were produced at the area of ions track with high irradiation dose. The affection to S parameters contributed by Cu concentration can also be negligible. And the S parameters lower than 2 keV may be contributed by the surface defects, as shown in Figure 2.

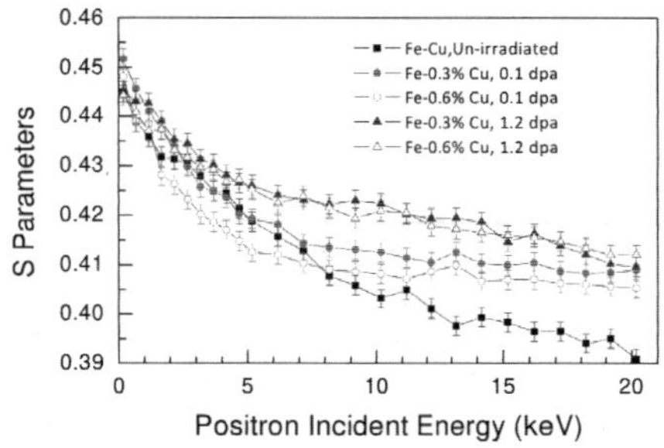


**Figure 3.** S parameters with positron energy for Fe-Cu alloys with different Cu content and irradiation dose.

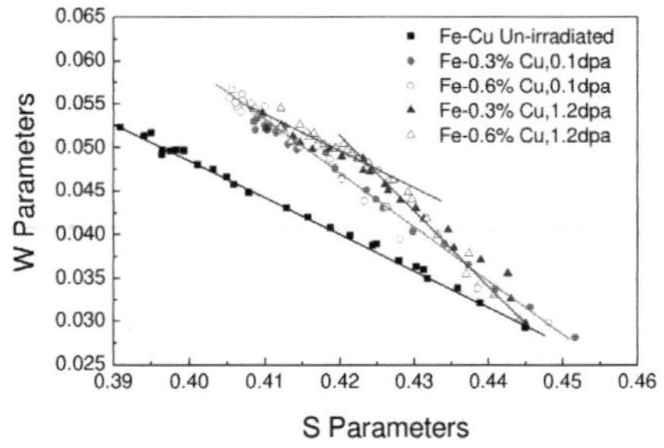


**Figure 4.** S-W curves for Fe-Cu alloys with Cu content and irradiation dose. The straight line was drawn only as guide eye.

As the W parameter was usually used to describe the information of core electrons annihilated with positrons, the slopes of S-W plots can present the annihilation mechanism of positrons after trapped [17]. The S-W plots of specimens are shown in Figure 4. It shows that the slopes changed clearly in irradiated alloys compare to un-irradiated one. And an inflection potion appeared clearly in the specimens with 1.2 dpa, which is corresponding to the 2~15 keV of positron energy range. It indicates that the mechanism of positron annihilation was changed in 1.2 dpa specimens. It may be contributed more by positrons annihilation trapping to defects. That can also be explained by the change of S parameters affected by irradiation dose, as shown in Figure 3.

According to the results calculated by TRIM, CDB experiments were performed as the positron energy with 13 keV, which related to the depth near 0.3 μm of Fe-Cu alloys from irradiation surface. The results with ratio to pure Fe are shown in Figure 5. It shows that a typical peak was observed near to 24 ($10^{-3}$ $m_0c$) in irradiated Fe-Cu alloys, same momentum with the peak of pure Cu. The difference at low momentum related to the formation of defects induced by irradiation [22], as shown in small graph. Comparing with the results of S parameters and S-W plots, more detailed information can be investigated using CDB experiments, especially related to high momentum as a range with 13~30 ($10^{-3}$ $m_0c$). In which, the intensity of Cu peaks are higher with 0.1 dpa dose than 1.2 dpa. It means that most of the positrons annihilated with electrons of Cu at 0.1 dpa. It can be attributed to that the Cu precipitates can formed easily within the small defects induced with 0.1 dpa, or that the increasing of counts related to free electrons affect to the decreasing of Cu peaks in 1.2 dpa specimens. The other reasonable explain is that many tiny Cu precipitates with vacancies type defects were produced in 0.1 dpa specimens, and positrons can be trapped easily by these vacancies type defects [15]. But in 1.2 dpa specimens, those tiny Cu precipitates and vacancies are formed to clusters and complex clusters, which lead to the increasing of S parameters and decreasing of Cu peaks [4, 23]. The change of Cu peaks with Cu content in 0.1 dpa specimens is few, but is increased with Cu content increasing in 1.2 dpa specimens. This can also attribute to the clustering of vacancies induced by high irradiation dose, and easier to form complex clusters in lower Cu content specimens [3, 24].

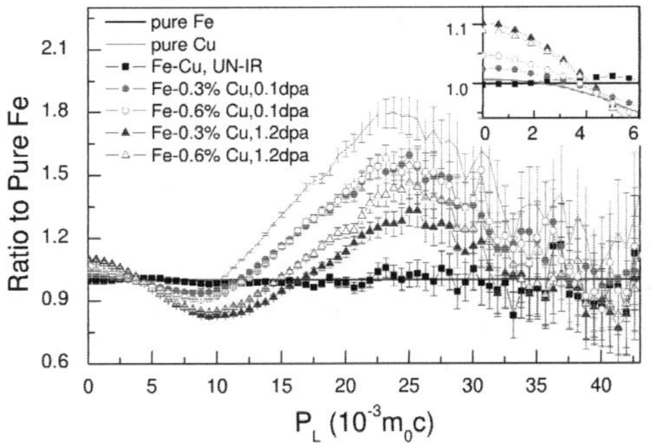


**Figure 5.** CDB results as the positron energy with 13 keV, including pure Cu, un-irradiated Fe-Cu alloy, Fe-0.3%Cu alloys with 0.1 dpa and 1.2 dpa irradiation dose, and Fe-0.6%Cu alloys with 0.1 dpa and 1.2 dpa irradiation dose, respectively. Small graph is the detailed information as the momentum lower than 6 ($10^{-3}$ $m_0c$).

## 4. Conclusion

The PAT were performed to study the formation of defects and precipitation of Cu atoms, especially the CDB based on the slow positron beam were used to investigate the Cu precipitates formed with complex defects induced by Fe ions irradiation. The defects formed in Fe-Cu alloys at the ions track area and deposition area, and more defects produced as the irradiation dose increased. Precipitation of Cu atoms is affected by the irradiation dose and the Cu content in Fe-Cu alloys. Further experiments will be performed to investigate the migration of defects and tiny Cu precipitates with post-annealed time and temperature.

**Acknowledgements** This work was partially supported by the NSFC (Grant Nos. 91026006).

## References

[1] Othen P J, Jenkins M L, Smith G D W, and Phythian W J 1991 *Philos. Mag. Lett.* **64** 383
[2] Phythian W J and English C A 1993 *J. Nucl. Mater.* **205** 162
[3] Nagai Y, Tang Z, Hassegara M, Kanai T and Saneyasu M 2000 *Phys. Rev. B* **63** 134110
[4] Xu Q, Yoshiie T and Sato K 2006 *Phys. Rev. B* **73** 134115
[5] Cao X Z, Xu Q, Sato K and Yoshiie T 2009 *Phys. Status Solidi C* **6** 2355
[6] Liu J Z, Walle A, Ghosh G F and Asta M 2005 *Phys. Rev. B* **72** 144109
[7] Golubov S I, Osesky Y N, Serra A and Barashev A V 1995 *J. Nucl. Mater.* **226** 252
[8] Ackland G J, Bacon D J, Calder A S and Harry T 1997 *Philos. Mag. A* **25** 713
[9] Lambrecht M and Almazouzi A 2009 *J. Nucl. Mater.* **385** 334
[10] Harry T and Bacon D J 2002 *Acta Mater* **50** 209
[11] Auger P, Pareige P, Akamatsu M and Blavette D 1994 *J. Nucl. Mater.* **225** 225
[12] Perlado J M, Victoria M, Caturla M J, Maria J, Gamez M L, Arevalo C, Martinez E, Mota F, Velarde M, Velarde G and Cepas P 2006 *J. Phys. IV France.* **133** 805
[13] Zhang C, Enomoto M, Yamashita T and Sano N 2004 *Metal .Mater. Trans, A* **35** 1263
[14] Pareige P, Radiguet B and Barbu A 2007 *J. Nucl. Mater.* **360** 104
[15] Onitsuka T, Takenaka M, Kuramoto E, Nagai Y and Hasegawa M 2002 *Phys. Rev. B* **65** 012204
[16] Wang B Y, Ma Y Y, Zhang Z, Yu R S and Wang P 2008 *Appl. Surf. Sci.* **255** 119
[17] Hao X P, Wang B Y and Yu R S 2008 *Chin. Phys. Lett.* **25** 1034
[18] Biersack J P and Haggmark L G 1980 *Nucl. Instrum. And Meth.***174** 257
[19] Vehanen A, Saarinen K Hautojarvi P and Huomo H 1987 *Phys. Rev.B* **35** 4606
[20] Asoka K P, Lynn K G and Welch D O 1994 *J. Appl. Phys.***76** 4935
[21] Odette G R 1983 *Scr. Metall.* **17** 1183
[22] He S M, van Dijk N H, Schut H and van der Zwaag S 2012 *Phys. Rev. B* **81** 094103
[23] Lambrecht M and Almazouzi A 2011 *J. Phys: Conference Series.* **265** 012009
[24] Nagai Y, Takadate K, Tang Z, Ohkubo H, Takizawa H and Hasegawa M 2011 *J. Phys: Conference Series.* **265** 012007

# Defect studies of ZnO films prepared by pulsed laser deposition on various substrates

**O. Melikhova[1], J. Čížek[1], I. Procházka[1], R. Kužel[1], M. Novotný[2], J. Bulíř[2], J. Lančok[2], W. Anwand[3], G. Brauer[3], J. Connolly[4], E. McCarthy[4], S. Krishnamurthy[4], J.-P. Mosnier[4]**
[1]Charles University in Prague, Faculty of Mathematics and Physics, V Holešovičkách 2, 180 00 Prague 8, Czech Republic
[2]Institute of Physics, Academy of Sciences of the Czech Republic, Na Slovance 2, 182 21 Prague, Czech Republic
[3]Institut für Strahlenphysik, Helmholtz-Zentrum Dresden-Rossendorf, Postfach 510119, D-01314, Dresden, Germany
[4]National Centre for Plasma Science and Technology, School of Physical Sciences, Glasnevin, Dublin 9, Ireland

E-mail: oksana.melikhova@mff.cuni.cz

**Abstract.** ZnO thin films deposited on various substrates were characterized by slow positron implantation spectroscopy (SPIS) combined with X-ray diffraction (XRD). All films studied exhibit wurtzite structure and crystallite size 20-100 nm. The mosaic spread of crystallites is relatively small for the films grown on single crystalline substrates while it is substantial for the film grown on amorphous substrate. SPIS investigations revealed that ZnO films deposited on single crystalline substrates exhibit significantly higher density of defects than the film deposited on amorphous substrate. This is most probably due to a higher density of misfit dislocations, which compensate for the lattice mismatch between the film and the substrate.

## 1. Introduction

Externally-pumped lasing observed in epitaxial ZnO films stimulated a great interest in the preparation of high-quality ZnO films [1]. Pulsed laser deposition (PLD) enables the production of high-quality ZnO films at lower temperatures than other methods, due to high energy of the ablated particles in the laser-produced plasma plume [2]. The quality of films deposited by PLD is influenced by laser fluence, substrate temperature, the atmosphere in the deposition chamber, and also by the type of substrate. In the present work, we aim to clarify the role played by the substrate. To this end we have carried out structural investigations of very thin ZnO films, deposited by PLD under identical conditions, on MgO (100), sapphire (0001) and fused silica (FS) substrates, respectively. We selected these substrates because of their different crystallographic structure; MgO crystallizes in a rocksalt structure in contrast to the hexagonal lattice of sapphire and the amorphous phase of FS.

## 2. Experimental

ZnO films were deposited by PLD using a frequency-quadrupled Nd:YAG laser providing 90 mJ of 266 nm laser light in 6 ns pulses at the repetition rate of 10 Hz, ablating an ultra-high purity ZnO ceramic target with a fluence on target of 2.8 $Jcm^{-2}$. Polished sapphire (0001), MgO (100) and FS substrates baked out in vacuum at 950°C were used for depositions. The ZnO films were grown in an oxygen atmosphere at a pressure of 10 Pa, with a substrate temperature of $T_s$ = 300°C as a result of 5,200 laser shots. A hydrothermally-grown bulk ZnO (0001) single crystal with an O-terminated surface supplied by MaTecK GmbH was used as a reference material in SPIS investigations.

XRD studies were performed with an X'Pert MRD diffractometer using Cu-$K_\alpha$ radiation. SPIS investigations were performed using a slow positron beam, SPONSOR [3], with the energy of incident positrons adjustable in the range from 0.03 to 36 keV. Doppler broadening (DB) of the annihilation photopeak was measured by a HPGe detector with an energy resolution of (1.09 ± 0.01) keV at 511 keV. Evaluation of DB was perfor using the line shape *S* parameter. The central energy region for the calculation of *S* was chosen as $|E - m_0c^2| < 0.93$ keV and all the *S* parameters presented in this paper are normalized to the bulk value $S_0$ = 0.5068(5) determined in the reference ZnO crystal.

## 3. Results and discussion

The XRD studies described in detail in Ref. [4] revealed that all films studied exhibit wurtzite ZnO structure with lattice parameters comparable with those given in literature and have crystallite size falling in the range 10-200 nm. The half-widths (FWHM) of the ZnO (0002) rocking curves of the films on MgO, sapphire and FS substrates were 1.2°, 1.3° and 10°, respectively. The ZnO films deposited on single-crystalline substrates (MgO, sapphire) exhibit local epitaxy, while the ZnO film deposited on FS exhibits (0001) fibre texture with a random lateral orientation of crystallites in the plane of substrate. Hence, the mosaic spread of planes is relatively small for the films grown on single crystalline substrates while it is substantial for the film grown on an amorphous FS substrate.

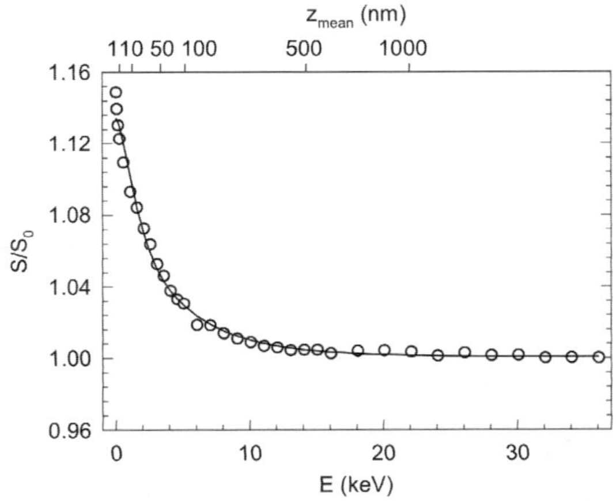


**Figure 1.** The dependence of the *S* parameter on the positron energy *E* for the reference ZnO (0001) single crystal. The mean positron penetration depth $z_{mean}$ is shown on the upper horizontal axis. The solid line shows the model curve calculated by VEPFIT.

SPIS measurements were performed on (i) a reference ZnO (0001) single-crystal in order to determine positron characteristics of bulk ZnO, (ii) bare substrates in order to characterize the defects in the substrates, and (iii) ZnO films deposited on the MgO, sapphire and FS substrate. The *S(E)* curve, showing the dependence of *S* on positron energy *E* for the reference ZnO single crystal, is plotted in Fig. 1. At the lowest energy, virtually all positrons annihilate on the ZnO crystal surface. With increasing energy, positrons penetrate progressively deeper into the ZnO crystal and the fraction of positrons diffusing back to the surface decreases, ultimately yielding the bulk values $S_0$ which correspond to all positrons annihilating in the ZnO bulk. The mean positron diffusion length $L_0$ = (47 ± 1) nm was obtained from the fitting of the *S(E)* curve by VEPFIT [5] assuming a single layer system. This value is in reasonable agreement with the positron diffusion length determined in a ZnO single crystal in Ref. [6], but is significantly shorter than the mean positron diffusion length of about 100-200 nm, typical for nearly defect-free semiconductors [7]. This can be explained by positron trapping in complexes consisting of a Zn-vacancy associated with a hydrogen atom ($V_{Zn}$+H) detected in the ZnO single crystal by positron lifetime spectroscopy [8].

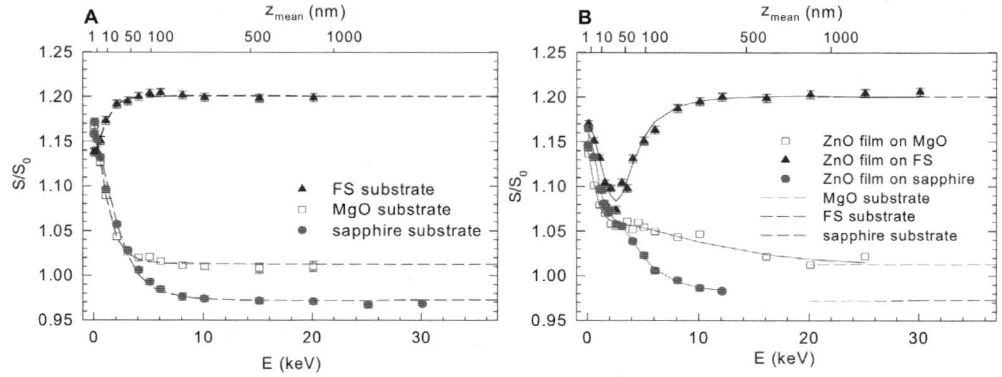


**Figure 2.** Dependence of the *S* parameter on positron energy for (A) bare substrates and (B) ZnO films. The mean positron penetration depth $z_{mean}$ is shown on the upper horizontal axis. Dashed and solid lines show model curves calculated by VEPFIT for bare substrates and ZnO films, respectively.

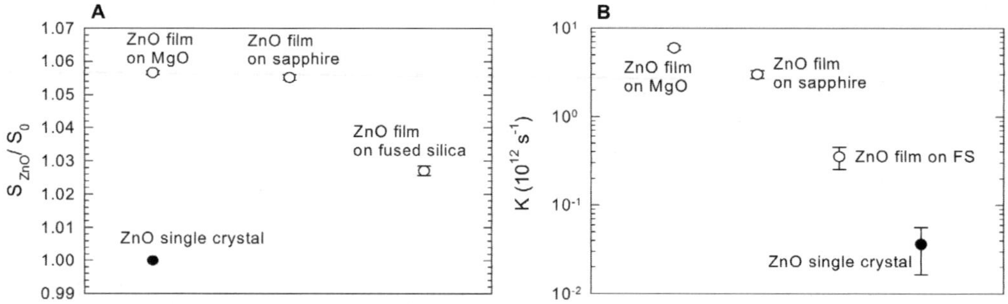


**Figure 3.** (A) The *S* parameter for ZnO layer obtained from fitting of *S(E)* curves for ZnO films in Fig. 2; (B) estimated positron trapping rate to defects in ZnO films and the reference crystal.

The *S(E)* curves for bare substrates are shown in Fig. 2(A). These curves were again fitted by VEPFIT (dashed lines) assuming a single layer structure for each substrate. The mean positron diffusion lengths for the substrates are: (16.9 ± 0.4) nm for MgO, (20 ± 1) nm for sapphire and (22 ± 1) nm for FS. Hence, all substrates exhibit positron diffusion length significantly shorter than expected for a perfect lattice [7]. This testifies that the substrates contain open-volume defects in the sub-surface layer, which are most probably dislocations introduced into the substrates by surface polishing.

The *S(E)* for various ZnO films are plotted in Fig. 2(B). At low energies positrons annihilate in the ZnO films, at higher energies $E > 4$ keV positrons are able to penetrate into the substrates, and *S* converges to the substrate value indicated by dashed line in Fig. 2(B). The model functions calculated by VEPFIT assuming a two-layered structure (ZnO film + substrate) are plotted in Fig. 2(B) by solid lines. The thicknesses of the ZnO layers obtained from fitting are (80 ± 2) nm, (45 ± 3) nm and (76 ± 1) nm for the films deposited on MgO, sapphire and FS substrates, respectively, and agree well with those obtained by XRD reflectivity [4]. The best-fit values for the *S* parameter of the ZnO layer are plotted in Fig. 3(A). Upon inspection of these results, it becomes clear that all ZnO films have a larger value of the *S* parameters than the reference ZnO single crystal. This shows that ZnO films contain concentrations of defects higher than that of the reference ZnO single crystal, independent of the substrate used in the deposition. The ZnO films likely contain point defects similar to the reference ZnO single crystal, i.e. $V_{Zn}$+H complexes. In addition, they will likely contain two other types of defects, namely: (i) misfit dislocations compensating the lattice mismatch between the film and the substrate, and (ii) open-volume defects at grain boundaries or interfaces between various crystallites.

The positron trapping rate $K$ in defects is proportional to the net concentration of defects in ZnO films. This quantity can be estimated from the expression [7] $K = 1/\tau_B\,[(L_{+,B} / L_+)^2-1]$, where $\tau_B = 154$ ps is the bulk positron lifetime [8], $L_{+,B}$ is the positron diffusion length in defect-free ZnO, and $L_+$ is the positron diffusion length measured in the ZnO film. Since defect-free ZnO crystals are not available $L_{+,B} \approx 150$ nm was estimated from expression $L_{+,B} = (D_+\ \tau_B)^{1/2}$ using the positron diffusion coefficient $D_+ \approx 1.6$ cm$^2$s$^{-1}$ estimated from the formula $D_+ = 1/3 <l> (<v^2>)^{1/2}$, where $<l> \approx 5$ nm is the mean free path between positron scattering events by longitudinal acoustic phonons determined for defect-free semiconductors GaAs and Ge in Ref. [9] and $<v^2>$ is the mean square thermal positron velocity [7]. From inspection of estimated trapping rates in Fig. 3(B), it becomes clear that all the ZnO films studied exhibit higher defect concentration than the reference ZnO single crystal. Moreover, the defect densities in the films deposited on the single crystalline substrates (MgO and sapphire) are higher than the concentration of defects in the film deposited on the FS substrate.

In the ZnO film deposited on the FS substrate, the mismatch between atomic positions in the substrate and in the film is, to some extent, compensated for by many differing orientations (tilting) of the ZnO crystallites. As a consequence, the density of misfit dislocations in ZnO crystallites is relatively low and positrons are trapped predominantly in open volume misfit defects at the interfaces between the crystallites. On the other hand, ZnO films deposited on MgO and sapphire single crystalline substrates exhibit local epitaxy. In these cases, the lattice mismatch between the film and the substrate accumulates mainly by misfit dislocations. Finally, it should be mentioned that the width of the rocking curve, which is often used for characterization of quality of epitaxial thin films, is a measure of the perfection of the structural relationship but is not directly correlated with the density of defects in the film. From comparison of XRD and SPIS results in this work, we can conclude that ZnO films deposited on single crystalline substrates (MgO and sapphire) exhibit narrow rocking curves, but concentration of defects in these films is substantially higher than in the film deposited on the amorphous FS, substrate which is characterized by significantly wider rocking curve.

## 4. Conclusions

Very thin ZnO films deposited by PLD on various substrates were characterized. All films exhibit higher concentrations of open volume defects than a bulk ZnO single crystal. The ZnO films deposited on the single crystalline substrates (MgO, sapphire) exhibit a higher concentration of defects than film deposited on an amorphous FS substrate. This is most probably due to a higher density of misfit dislocations, which compensate for the lattice mismatch between the film and the substrate.

## 5. References


[1] Bagnall B M, Chen Y F, Zhu Z, Yao T, Koyama S, Shen M Y and Goto T 1997 *Appl. Phys. Lett.* **70** 2230

[2] Hu W S, Liu Z G, Sun J, Zhu S N, Xu Q Q, Feng D and Ji Z M 1997 *J. Phys. Chem. Sol.* **58** 853

[3] Anwand W, Kissener H R and Brauer G 1995 *Acta Physica Polonica A* **88** 7

[4] Novotný M *et al.* 2012 *J. Phys. D*, 45 225101

[5] v Veen A, Schut H, Clement M, de Nijs J, Kruseman A and Ijpma M 1995 *Appl Surf Sci* **85** 216

[6] Koida T, Uedono A, Tsukazaki A, Sota T, Kawasaki M and Chichibu S F 2004 *phys. stat. sol. (a)* **201** 2841

[7] Krause-Rehberg R and Leipner H S 1999 Positron Annihilation in Semiconductors – Defect Studies, Springer, Berlin

[8] Brauer G *et al.* 2009 *Phys. Rev. B* **79** 115212

[9] Soininen E, Mäkinen J, Beyer D and Hautojärvi P 1992 *Phys. Rev. B* **46** 13104


**Acknowledgments**

This work was supported by the Czech Science agency (project P108/11/0958), the Czech Ministry of Education, Youths and Sports (project MEB101102) and DAAD (project 50755628).

Journal of Physics: Conference Series **443** (2013) 012019 doi:10.1088/1742-6596/443/1/012019

# Vacancy clustering in pure metals: some first principle calculations of positron lifetimes and momentum distributions

**C. R. Luna[1], C.Macchi[1§], A. Juan[2] and A. Somoza[3]**

[1]IFIMAT, UNICEN and CONICET, Pinto 399, Tandil, Argentina.
[2]IFISUR, Dpto. Física, UNS and CONICET, Av. Alem 1253, Bahía Blanca, Argentina
[3]IFIMAT, UNICEN and CICPBA, Pinto 399, Tandil, Argentina.

Email: cmacchi@exa.unicen.edu.ar

**Abstract**. First results of a systematic study on the vacancy clustering process in Al, Cu, Mg and Nb are presented. To this aim first principle calculation of positron lifetimes and positron-electron momentum distributions were performed. We test the reliability of the computational scheme used by comparing some of the calculated results with experimental ones.

## 1. Introduction

Vacancies have a predominant influence on the solid properties because of their important role in the atomic diffusion. When produced in supersaturation, for instance, during quenching, irradiation, or plastic deformation, besides monovacancies, vacancy clusters may be formed. These defects influence structural properties of the solids such as resistance to plastic deformation, solute diffusion, kinetics of thin films growth, etc.

It is well recognized that Positron Annihilation Spectroscopy (PAS) is a powerful tool to study vacancy-like defects. In fact, by measuring positron lifetimes and momentum distributions of the positron annihilation radiation it is possible to obtain information about the open volumes associated with defects and their chemical environment [1]. However, in some cases in which more than monovacancies or divacancies cannot be directly revealed by means of PAS it is necessary to perform *ab-initio* calculations that allow interpreting the experimental results.

In this work, first results of a systematic first principle calculation study on the vacancy clustering process in some pure metals are presented. Specifically, changes in the positron lifetimes and in the CDB positron-electron momentum distributions during vacancy clustering in pure Al, Cu, Mg and Nb were studied. These metals were selected since vacancy-like defects play an important role in: (i) Precipitation processes of Al-based age hardenable alloys; and (ii) Hydrogen sorption kinetic in Nb-doped nanostructured Mg films. These two topics are currently being studied in the Positron Group at IFIMAT (Argentina). In order to check the reliability of the results obtained, when possible some of the calculated results are compared with experimental ones.

## 2. Computational method

### 2.1. Positron calculations

A general introduction to *ab-initio* calculations of positron annihilation characteristics in solids can be found in [2]. In the present work, positron lifetimes and momentum distributions of the annihilating positron-electron pairs for positrons delocalized in the bulk or trapped in vacancy-like defects were calculated following the procedure developed by Puska and Makonnen based on the VASP code (see [3] and references therein). Positron *ab-initio* calculations have been performed within the frame of the density-functional theory (DFT) [4] which was implemented in the Vienna Ab-initio Simulation Package (VASP) code [5]. The projector augmented wave (PAW) pseudopotential was used to account the electron-ion core interaction, using the local density approximation (LDA) [6] for the exchange-correlation term. In the calculation, the Brillouin-zone of the different structures was sampled using a 6x6x6 Monkhorst-Pack ***k***-point mesh [7]. For the plane-wave basis set a cut-off of 650 eV was used. The self-consistent calculations were considered to converge when the difference in the total energy of the crystal between consecutive steps did not exceed $10^{-5}$ eV.

### 2.2. Solid modeling

The periodic structures of the studied elements were simulated using different supercell sizes depending on the element crystal structure. Information corresponding to the crystal structure and lattice parameters used for calculations was taken from experimental data reported in the literature [8]. The unit cell of the crystalline structure has been used as the supercell for bulk calculations. For the calculation in defected materials one or more atoms were removed. In such a case, the supercell should be large enough in order to avoid the defect interaction with its periodic image; therefore, the system describes an isolated vacancy quite well. However, in practice, the size of the supercell cannot be built arbitrarily large. In all cases, we checked whether such supercell size was sufficient to obtain positron characteristics independent of the supercell size. The maximum number of atoms per supercell used to reach convergence was: 256 atoms (64 unit cells) for the fcc crystal structure, 128 atoms (64 unit cells) for the bcc crystal structure and 240 atoms (60 orthorombic unit cells [9]) for the hcp crystal structure.

In these supercells, when possible, vacancy clusters were created maintaining a spherical or ellipsoidal geometry. Furthermore, for all the studied structures, the volume of the vacancy clusters was estimated. To this aim, an ellipsoid was fitted inside the open volume left by the vacancy cluster.

On the other hand, when generating vacancy clusters the possible relaxation of the crystal lattice was not considered, that is the ions neighboring vacancy agglomerates were not allowed to relax from their ideal position. A more precise calculation of the annihilation parameters could be obtained considering the relaxation of the cluster-neighboring ions and the forces exerted on these ions by the localized positron; however, the study of these effects goes beyond the aim of this work.

## 3. Experimental

Plates of well-annealed high-purity Al, Cu and Mg were used to characterize the bulk state of these elements. Defect signatures were obtained following the procedure described by Somoza *et al.* [10]. The positron lifetime spectrometer was a fast-fast timing coincidence system with a time resolution of 254 ps. A 20 μCi source of $^{22}$NaCl deposited on a thin Kapton foil was sandwiched between two identical specimens. After subtracting the source component, the spectra with about $10^6$ counts were satisfactorily analyzed using the POSITRONFIT program [11]. The CBD spectra were obtained using two HP-Ge detectors (90% efficiency) with an energy resolution of 1.2 keV at the 511 keV photopeak.

## 4. Result and Discussion

In figure 1, the evolution of the calculated positron lifetime ($\tau^{calc}$) as a function of the estimated vacancy cluster size for the four studied elements is presented. In this figure, the bottom abscissa indicates the estimated void volume while the top axis indicates the number of vacancies into the cluster. As can be seen in the four panels, all $\tau^{calc}$ evolutions show a similar behavior: an initial rapid increase of $\tau^{calc}$ for small volumes then reaching asymptotic values for large void volumes. Besides, it can be seen that in all cases, the transition between these two stages occurs for void volumes ~0.3nm$^3$. Depending on the crystal structure of the metal, this volume corresponds to different number of vacancies. The $\tau^{calc}$ values obtained for perfect lattices and for structures containing monovacancies are in very good agreement with those reported in the literature (see for example [8] and references therein). Moreover, in the case of Mg the evolution of $\tau^{calc}$ is in very good agreement with recently reported results obtained using the non-autoconsistent method [12]. In the same way, the observed increase of $\tau^{calc}$ with the growth of the vacancy cluster size for Al shows a good agreement with results reported in [2]. Although the values reported here are systematically lower than those calculated by Puska *et al.*, this difference can be assigned to the method of calculation used in these two works.

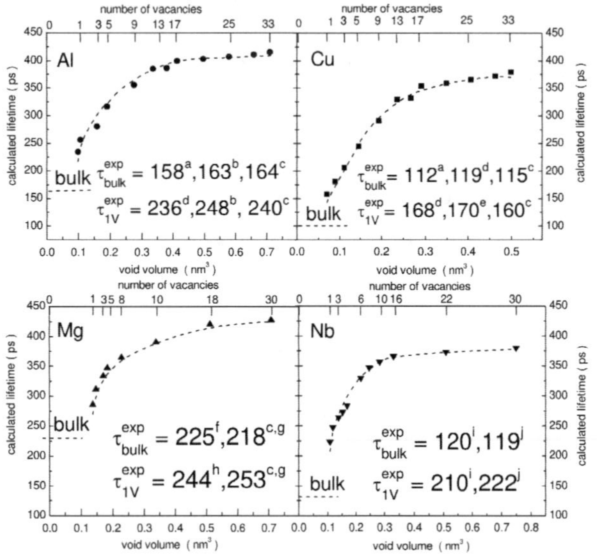


Figure 1: Evolution of the calculated positron lifetimes as a function of the estimated void volume (bottom) and the number of vacancies (top) for the four pure elements studied. Dotted lines are only eye guides.

[a][13] [b][14] [c]This work [d][15] [e][16] [f][17] [g][12] [h][18] [i][8] [j][19]

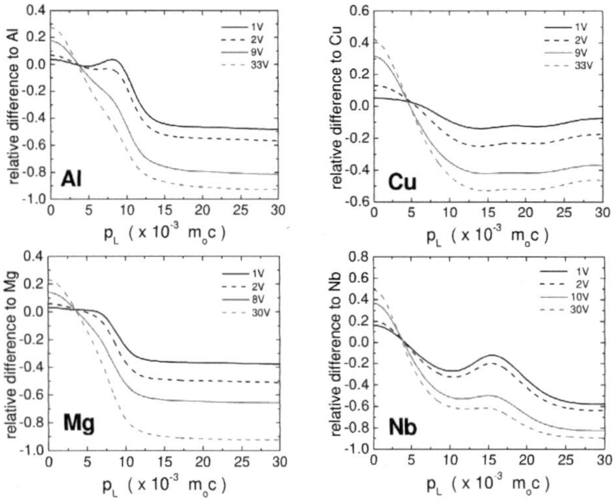


Figure 2: Evolution of the calculated momentum distributions of the annihilating electron-positron pairs in Al, Cu, Mg and Nb containing vacancy-clusters of different sizes. *To simulate the experimental CDB spectra, all the calculated curves were convoluted with a Gaussian function (FWHM = 1.2 keV; that is, 4.6x10$^{-3}$ $m_oc$).*

In figure 1, experimental lifetime values measured in pure well-annealed metals and in metals containing monovacancies are also given. As already known, the experimental values for the defect-free bulk and for pure metals containing monovacancies differ with the $\tau^{calc}$ in no more than 10%. However, in the case of Mg containing monovacancies this difference increases until about 20 %. The same behavior was reported by Folegati *et al.* [18].

In figure 2, the relative (with respect to perfect element) calculated momentum distributions of the annihilating electron-positron pairs in Al, Cu, Mg and Nb containing vacancy clusters of different sizes are shown. To check the goodness of the computational scheme used, when possible it is important to compare these distributions with CDB experimental curves. To this aim, we have compared calculated CDB distributions with the corresponding experimental ones measured in our laboratory. As an example, in figure 3, results of the calculated and experimental relative difference ratio curves (with respect to perfect Al) for perfect Cu and Mg are shown. On the other hand, in figure 4 experimental and calculated relative ratio curves for Cu and Mg containing monovacancies were plotted. From figures 3 and 4, it can be concluded that for Cu there is a very good agreement between the experimental and calculated CDB distributions. In the case of Mg this good agreement maintains until about $10x10^{-3}m_0c$, the calculated curves qualitatively show the same "structure" as than experimental ones but they are systematically above (or below) the experimental ratio curves. Work in progress will make it possible to go deeper into the analysis and discussion of the results briefly presented in this work.

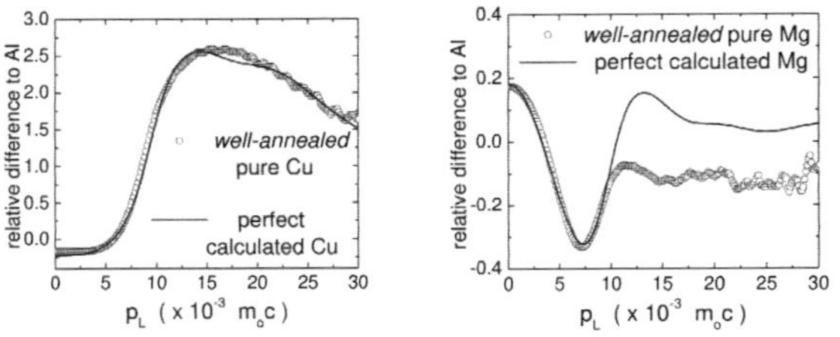


Figure 3: Calculated and experimental Cu/Al relative difference curve (left) and Mg/Al relative difference curve (right).

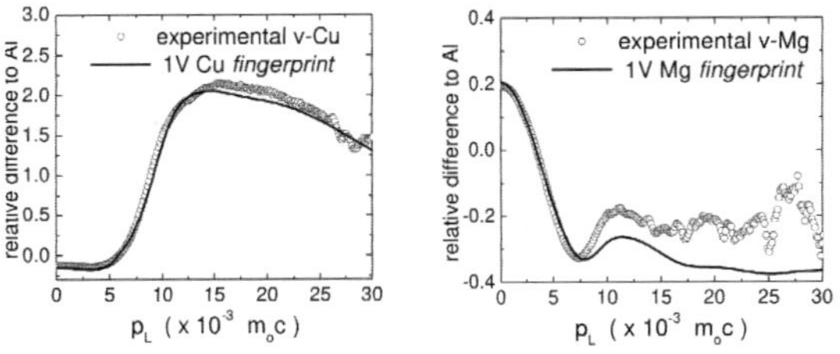


Figure 4: Calculated and experimental CDB curves for Cu monovacancy (left) and Mg monovacancy (right).

**References**

[1] Hautojärvi P and Corbel C 1995 *Positron Spectroscopy of Solids*, ed A Dupasquier and A P Mills Jr (Amsterdam: IOS Press) p 491
[2] Puska M J and Nieminen R M 1994 *Rev. Mod. Phys.* **66,** 841
[3] Makkonen I, Hakala M and Puska M J 2006 *Phys. Rev. B* **73**, 035103
[4] Kohn W and Sham L J 1965 *Phys. Rev.*, A1133
[5] Kresse G and Joubert D 1999 *Phys. Rev. B* **59**, 1758
[6] Blöchl P E 1994 *Phys. Rev. B* **50**, 17953
[7] Monkhorst H J and Pack J D 1976 *Phys. Rev. B* **13**, 5188
[8] Campillo J M, Ogando E and Plazaola F 2007 *J. Phys.: Condens. Matter* **19**, 176222
[9] Shin D and Wolverton C 2010 *Scripta Materialia*, 531
[10] Somoza A, Petkov M P, Lynn K G and Dupasquier A 2002 *Phys Rev B* **65**, 094107
[11] Kirkegaard P, Pedersen N J and Eldrup M, 1989 Roskilde, Denmark: PATFIT Program Risø-M-2740 RNL
[12] Checchetto R, Bazzanella N, Kale A, Miotello A, Mariazzi S, Brusa R S, Mengucci P, Macchi C, Somoza A, Egger W and Ravelli L 2011 Phys Rev B **84**,
[13] Staab T E M, Somieski B and Krause-Rehberg R 1996 *Nucl. Instrum. Methods Phys. Res. A* **381**, 141
[14] Schaefer H E, Gugelmeier R, Schmolz M and Seeger A 1986 *Proceedings of the Eighth International Conference on Vacancies and Interstitials in Metals and Alloys*, ed C Ambromeit and H Wollenberger (Aedermannsdorf: Trans Tech)
[15] Petters A K 1998, Diploma work, Martin-Luther Universität Halle-Wittenberg
[16] Staab T E M, Krause-Rehberg R and Kieback B 1999 *J. Mater. Sci.* **34**, 3833
[17] Hautojärvi P, Johansson J, Vehanen A, Yli-Kauppila J, Hillairet J and Tsanetakis P 1982 *Appl. Phys. A: Solids Surf.* **27**, 49
[18] Folegati P, Makkonen I, Ferragut R and Puska M J 2007 *Phys Rev B* **75** , 054201
[19] Čížek J, Procházka I, Daniš S, Melikhova O,Vlach M, Žaludová N, Brauer G, Anwand W, Mücklich A, Gemma R, Nikitin E, Kirchheim R and Pundt A 2007, *Phys Stat Sol (c)* **4**, 3485

Journal of Physics: Conference Series **443** (2013) 012020 doi:10.1088/1742-6596/443/1/012020

# Effects of oxide distributed in grain boundaries on microstructure stability of nanocrystalline metals

**Kai Zhou, Hui Li, JinBiao Pang, Zhu Wang** [1]

School of Physics, Wuhan University, Wuhan 430072, China

E-mail: wangz@whu.edu.cn

**Abstract.** Nanocrystalline copper and zinc prepared by high-pressure compaction method have been studied by positron lifetime spectroscopy associated with X-ray diffraction. For nanocrystalline Cu, mean grain sizes of the samples decrease after being annealed at 900 °C and increase during aging at 180 °C, revealing that the atoms exchange between the two regions. The positron lifetime results indicate that the vacancy clusters formed in the annealing process are unstable and decomposed at the aging time below 6 hours. In addition, the partially oxidized surfaces of the nanoparticles hinder the grain growth during the ageing at 180 °C, and the vacancy clusters inside the disorder regions which are related to $Cu_2O$ need longer aging time to decompose. In the case of nanocrystalline Zn, the open volume defect (not larger than divacancy) is dominant according to the high relative intensity for the short positron lifetime ($\tau_1$). The oxide (ZnO) inside the grain boundaries has been found having an effect to hinder the decrease of average positron lifetime ($\tau_{av}$) during the annealing, which probably indicates that the oxide stabilizes the microstructure of the grain boundaries. For both nanocrystalline copper and zinc, the oxides in grain boundaries enhance the thermal stability of the microstucture, in spite of their different crystal structures. This effect is very important for the nanocrystalline materials using as radiation resistant materials.

## 1. Introduction

Some computer simulations have been performed on nanocrystalline materials recently which reveal the self-healing ability of these materials for the radiation damage [1, 2]. These materials are highly promising for the use in the future nuclear reactors. When nanocrystalline materials are irradiated by particles, the grain boundaries play a key role which can serve as effective sinks for radiation-induced defects such as interstitials and vacancies. Moreover, these interstitials then are emitted to annihilation vacancies in the crystallites. Hence the stability of the microstructure is essential for the nanomaterials keeping the radiation resistance, especially at the elevated temperature.

Positron annihilaion technique can sensitively detect open volume defects in materials, and gives detailed information about the defects, such as, types and concentrations of defects, and the chemical identity of the elements surrounding the defects. Many positron annihilaion measurements have performed on nanocrystalline materials to investigate defects associated with the extraordinary structure of the grain boundaries in nanocrystalline materials [3-5]. However, interpretation of the positron lifetime results has to be different because of the obviously different structure of nanocrystalline materials prepared by different techniques.

[1] To whom any correspondence should be addressed.

In this work, positron lifetime spectroscopy associated with X-ray diffraction is used to study the microstructure of two typical nanocrystalline metals, Cu and Zn. The emphasis is to investigate the effects of oxide in grain boundaries on the microstructure thermal stability of nanocrystalline metals.

## 2. Experimental

Samples were prepared by high-pressure compaction method. Cu nanoparticles for samples A and B were produced by flow-levitation (FL) method [6] and chemical vapor deposition (CVD) [7], respectively. Zn nanoparticles for samples C and D were prepared by wire explosion method [8]. The mean grain size for all as-prepared samples is about 55 nm obtained by XRD. Sample B was compacted in air. Nanoparticles for sample D were oxidized in air at 100 °C for 100 minutes. These treatments were to increase the oxide in grain boundaries of samples B and D.

In order to reduce the gas impurities and the residual porosity, the samples A and B were annealed at 900 °C for 30 minutes under the argon atmosphere before ageing experiment. The ageing experiment was performed at 180℃ for aging time from 1 to 108 hours in a vacuum furnace. Isochronal annealing of sample C and D was performed in the Ar gas atmosphere from the temperature of 50 to 410 °C for 30 minutes at each temperature point. After each ageing or annealing, the samples were cooled slowly to room temperature.

Before positron lifetime and XRD measurements, the samples were polished with emery paper (800 mesh) to remove the possible surface oxide. X-ray diffraction measurements were performed on a BRUKER AXS D8 ADVANCE X-ray diffractometer using Cu Kα radiation. A digital positron lifetime spectrometer (for details see [9]) with a time resolution of about 240 ps was used to perform the lifetime measurements at room temperature. The positron source employed was $^{22}$NaCl encapsulated with Ti foil, which is sandwiched by two identical samples under investigation.

## 3. Results and discussion

### *3.1.* Nanocrystalline Cu

The mean grain size (obtained by XRD) versus ageing time is shown in figure 1. What is surprising is the decrease in the mean grain size of both samples after the annealing at 900 °C. The grain size obtained using XRD profile analysis is the ordered region size. After annealing the ordered regions decrease because the atoms in the edge of the ordered regions are activated thermally and move to the disordered regions. We also can see from figure 1 that the mean grain size of sample B increases slowly compared to that of sample A. Sample B was prepared in air, so the surface oxidation of nanoparticles could be expected [10]. This oxide ($Cu_2O$) (confirmed by XRD) distributed in the grain boundaries of the sample retard the increase of the mean grain size in the sample.

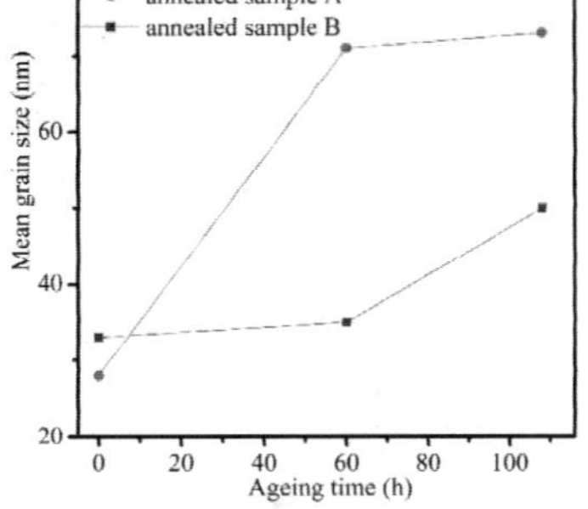


**Figure 1.** Mean grain sizes of samples A and B as a function of aging time at 180 °C. The ageing time '0' means as-annealed.

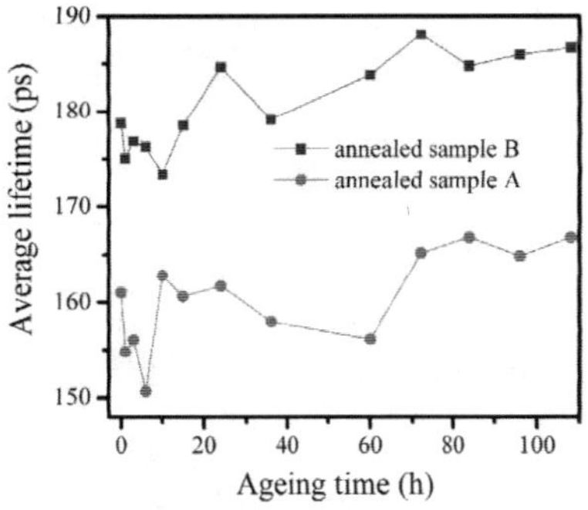


**Figure 2.** Average lifetimes for annealed samples A and B as a function of ageing time at 180 °C.

The spectra of nanocrystalline Cu samples were decomposed into three components $\tau_1$, $\tau_2$ and $\tau_3$, with relative intensities $I_1$, $I_2$, and $I_3$. The lifetimes $\tau_1$ and $\tau_2$ are roughly 130 ps and 310 ps, respectively. The lifetime $\tau_2$ is come from positron annihilation in vacancy defect containing about ten monovacancies[11]. The longest lifetime $\tau_3$ is between 1 and 2 ns. Detailed discussions on these components can be found in [10]. We here present the positron average lifetime $\tau_{av}$ ($\tau_{av}= \Sigma_i \tau_i I_i$) for a description of changes that occur during the ageing. From figure 2 we can see that the ageing behavior of average lifetime $\tau_{av}$ for annealed samples A and B is similar at the intervals of about 0-10 h and 70-108 h. In the former interval the average lifetimes decrease to their minimum values for samples A and B and in the latter interval increase slowly to their maximum values. Furthermore, the average lifetime for both samples may reach their saturation values when the ageing time is beyond 108 h, and then the microstructure of the samples would be in a stabilization state. The obvious difference in the ageing behavior of $\tau_{av}$ is at the mediate interval of the ageing time, where the average lifetime for sample A slightly reduces but for sample B increases in general. This difference may be caused by the partial surface oxidization of nanoparticles for sample B. The oxide impedes the decomposition of the vacancy clusters distributed in grain boundaries and causes the continuous increase in the size of the vacancy clusters. When the vacancy clusters become large enough, they become unstable and then might collapse. This deduction is supported by the changes of the lifetime $\tau_2$ (not shown here).

### *3.2.* Nanocrystalline Zn

For sample C, the lifetime components $\tau_1$ and $\tau_2$ with their relative intensities $I_1$ and $I_2$ as a function of annealing temperature are presented in figure 3. The longest lifetime component $\tau_3$ (about several nanoseconds) is not discussed because of their really small intensity (below 0.5%). The lifetime $\tau_2$ is attributed to positron annihilation at the large vacancy clusters containing dozens of monovacancies. Because the lifetimes of monovacancy and divacancy in bulk Zn are about 220 ps and 260 ps [12, 13] respectively, the lifetime $\tau_1$ for as-prepared sample C is an effective lifetime containing contributions from the positrons trapped both at defects containing one or two monovacancies and in disorder regions which will give a lifetime a little smaller than that of monovacancy [14].

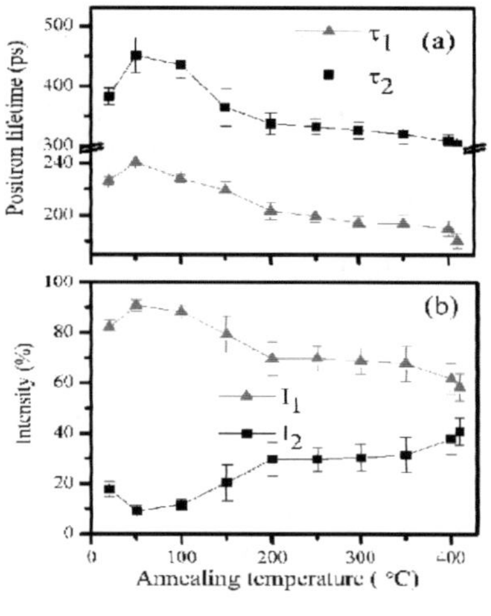


**Figure 3.** Positron lifetimes and their relative intensities for sample C with different annealing temperatures.

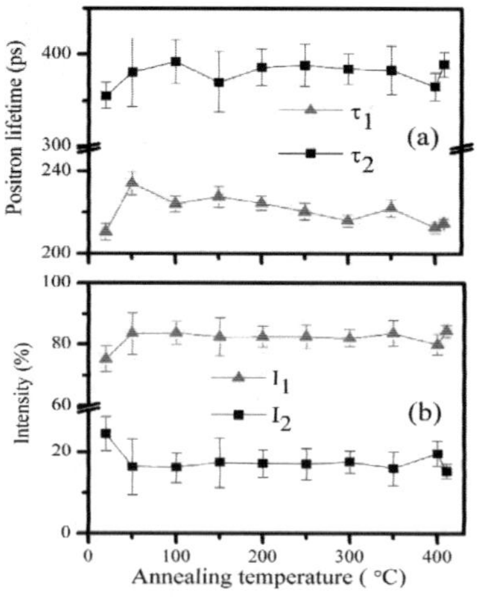


**Figure 4.** Positron lifetimes and their relative intensities for sample D with different annealing temperatures.

In order to clearly observe the influence of the oxide in grain boundaries on the thermal stability of the microstructure, sample D is annealed for comparison. The positron lifetime result for sample D is presented in figure 4. From this figure we can see that the lifetimes $\tau_1$ and $\tau_2$ both increase at the beginning of the annealing, which is due to the release of the stress induced in the compacting process.

After the temperature reaches 100 °C the lifetime $\tau_2$ is almost unchanged, indicating that the size of vacancy cluster changes very little. The lifetime $\tau_1$ declines very slowly above 50 °C. The recovery process of the open volume defects is suppressed. It is noteworthy that the relative intensities $I_1$ and $I_2$ are almost unchanged after the annealing temperature reaches 50 °C, which is very different to sample C. In the case of sample C, the intensity $I_2$ increases fast from 10% to 30% in the interval of 50-200 °C, and then changes slight. For sample D, above 50 °C the intensities $I_1$ and $I_2$ are almost unchanged and there is only a slow decrease of the lifetime $\tau_1$. Hence, we conclude that the oxide in grain boundaries have an active effect on the stability of the defect structure. If we consider the disordered regions in grain boundaries to be a kind of defect, the stability of the defect structure suggests the stability of the microstructure of grain boundaries. The typical positron diffusion length in crystal is about 100 nm. When grain size is larger than the diffusion length, positron may annihilate in the free state which will contribute a short lifetime component (about 160 ps) [15]. The lifetime $\tau_1$ does not below 220 ps during the whole annealing process probably indicating that the mean grain size of the sample is still below about 100 nm. This observation probably suggests that the oxide in grain boundaries suppress the grain growth.

## 4. Conclusions

Positron annihilation spectroscopy and XRD were used for investigating the influences of the oxide in grain boundaries on the microstructure thermal stability of nanocrystalline Cu and Zn prepared by compacting very small crystallites. For nanocrystalline Cu, the decrease of mean grain sizes after annealing at 900 °C and increase during ageing at 180 °C is observed by XRD. However, the increase of the grain size of the sample with oxide in grain boundaries is apparently slow. According to the positron lifetime results the vacancy clusters related to $Cu_2O$ need a longer aging time to decompose during the ageing process at 180 °C. For nanocrystalline Zn, the experiment results indicate that small open volume defects are dominant in the grain boundaries. The oxide in grain boundaries stabilizes the microstructure of the nanocrystalline Zn sample under the heat treatment. Although copper and zinc have different crystal structures, which are face-centred-cubic (fcc) and hexagonal-close-packed (hcp) respectively, oxides distributed in their grain boundaries have a similar effect to enhance thermal stability of the microstructure. However other properties may be degraded due to the oxides.

**Acknowledgements**
This work was financially supported by the Natural Science Foundation of China (Project No. 11275142).

## References

[1] Bai X -M, Voter A F, Hoagland R G, Nastasi M, Uberuaga B P 2010 *Science* **327** 1631.
[2] Ackland G 2010 *Science* **327** 1587.
[3] Jakub Čížek, Ivan Procházka, Miroslav Cieslar et al 2002 *Phys. Rev.* B **65** 094106
[4] Würschum R, Scheytt M, and Schaefer H -E 1987 *Phys. Stat. Sol.* A **102** 119
[5] Schaefer H -E, Würschum R, Birringer R, and Gleiter H 1988 *Phys. Rev.* B **38** 9545
[6] Li C-M, Lei H, Tang Y-J, Luo J-S, Liu W and Chen Z-M 2004 *Nanotechnology* **15** 1866
[7] Swihart M T 2003 *Curr. Opinion in Colloid and Interface Sci.* **8** 127
[8] Wang Q, Yang H B, Shi J L and Zou G T 2001 *Mater. Sci. Eng.* A **307** 190
[9] Li H, Shao Y D, Zhou K, Pang J B, Wang Z 2011 *Nucl. Instrum. Meth.* A **625** 29
[10] Zhou K, Li H, Pang J B, Wang Z 2011 *Physica B* **406** 760
[11] Ohkubo H, Tang Z, Nagai Y et al 2003Mater. Sci. Eng. A**350** 95
[12] Hidalgo C and Diego N de 1986 *Appl. Phys.* A **40** 25
[13] Campillo J M, Plazaola F and Diego N de 2000 *J. Phys.: Condens. Matter* **12** 9715
[14] Sui M L, Lu K, Deng W, Xiong L Y, Patu S and He Y Z 1991 *Phys. Rev.* B **44** (1991) 6466
[15] Zhou K, Li H, Pang J B, Wang Z 2012 *Physica B* **407** 1219

Journal of Physics: Conference Series **443** (2013) 012021 doi:10.1088/1742-6596/443/1/012021

# Vacancy Formation Enthalpy in Polycrystalline Depleted Uranium

**K R Lund[1†], K G Lynn[1], M H Weber[1] and M A Okuniewski[2]**
[1]Washington State University, Center for Materials Research, Pullman, WA, United States
[2]Idaho National Laboratory, Idaho Falls, ID, United States

E-Mail: kasey.lund@email.wsu.edu

**Abstract.** Positron Annihilation Spectroscopy was performed as a function of temperature and beam energy on polycrystalline depleted uranium (DU) foil. Samples were run with varying heat profiles all starting at room temperature. While collecting Doppler-Broadening data, the temperature of the sample was cycled several times. The first heat cycle shows an increasing S-parameter near temperatures of 400K to 500K much lower than the first phase transition of 941K indicating increasing vacancies possibly due to oxygen diffusion from the bulk to the surface. Vacancy formation enthalpies were calculated fitting a model to the data to be 1.6± 0.16 eV. Results are compared to previous work [3,4].

## 1. Introduction

Positron Annihilation Spectroscopy (PAS) can be used to determine vacancy formation enthalpies in metals [1]. Using Doppler-Broadening of the Annihilation Radiation (DBAR), one is able to observe changes in the line shape parameter, also known as the S-parameter, to calculate vacancy concentrations [2]. Uranium has three phases between temperatures of 300K to its melting point of 1407K [3-5]. α-U from ambient temperatures to 941K is orthorhombic, β-U from 941 to 1048K is a complex tetragonal, and γ-U from 1048K to its melting point is BCC.

Matter *et al* [3] and Kögel *et al* [4] used PAS to report phase transition formation enthalpies using bulk PAS measurements, however their results are conflicting with one another. Both Matter and Kögel annealed large polycrystalline uranium samples; 8mm diameter 15mm long cylinder for Reference 3, and 13mm diameter 6.5mm long cylinder for Reference 4 to remove stresses, strains, and impurities such as oxygen. It is possible that not all of the oxygen was removed from their samples; this could be the cause of the lower enthalpies that were reported..

Senanayake *et al* [6] used high-resolution X-ray photoelectron spectroscopy to study the formation of defects created by argon sputtering in $UO_2$. They report a decrease in the oxygen and uranium binding energy, along with an increase in the rate of oxygen diffusion from the bulk to the surface, at temperatures greater than 400K.

This experiment uses a variable energy PAS technique in an attempt to understand the significant discrepancy of the previously reported results for formation enthalpies from the α to the γ phase. The effects of oxygen migration from the bulk to the surface are found to be important to understand the difference in the reported results.

† To whom any correspondence should be addressed.

## 2. Procedure

2.1. *Doppler Broadening of the Annihilation Radiation.* For temperature dependent DBAR measurements, the positron energy was set to 66keV, corresponding to a mean positron implantation depth of ≈1.7μm into the bulk of the DU. This value was obtained using [1] <z>400E^1.6/ ρ , where ρ is the density in g/cm$^3$ and E is the energy of the positron beam, <z> is given in Å.

The formation enthalpy was calculated using the temperature dependant scans and the equations [2],

$$C_v = C_v^0 \exp\left(\frac{-H_v^f}{k_b T}\right) \tag{1}$$

and,

$$K = \mu C_v = \lambda_b \frac{S - S_b}{S_v - S} \tag{2}$$

Here $C_v^0$ is the pre-exponential factor containing the entropy of formation, K is the trapping rate, μ is the specific trapping coefficient, $C_v$ is the concentration of vacancy defects, $\lambda_b$ is the bulk annihilation rate, S is the line shape parameter (from here on referred to as the S-parameter), $S_b$ is the S-parameter for the bulk, $S_v$ is the S-parameter for vacancies, $H_v^f$ is the vacancy formation enthalpy, and $k_b$ is Boltzmann's constant. Combining equations 1 and 2, and solving for the S-parameter as a function of temperature we obtain,

$$S(T) = \frac{S_b(T) + S_v \xi \exp(\frac{-H_v^f}{k_b T})}{1 + \xi \exp(\frac{-H_v^f}{k_b T})}, \qquad S_b(T) = S_b(300K) + \Delta S(T - 300K) \tag{3}$$

where, $\xi = C_v^0 \mu \lambda_b^{-1}$. By measuring $S_v$ and $S_b$ from the data and assigning ξ and $H_v^f$ as fitting variables, Equation 3 was used to fit the temperature dependent data.

2.2. *Sample Preparation.* Depleted uranium foil 150μm thick was purchased from International Bio-Analytical Industries, with isotopic and elemental purities greater than 99.8%. A 4cm$^2$ rectangle was cut. It was mounted to a HeatWave Labs molybdenum body ultra high vacuum substrate-heating element, with tantalum wire to hold it in place. Tantalum and titanium foil was placed around the sample and used as heat shielding. K-type thermocouples were spot welded onto the sample and heater. Ta wire provided additional tension to securely hold them down.

The sample was placed in the positron beam line at Washington State University under a vacuum of $10^{-8}$ Torr, and analyzed with energy and temperature dependent Doppler-Broadening of the Annihilation Radiation (DBAR). The temperature of heating element was varied by using an AC HeatWave Labs control box (see Table 1). While the temperature was cycling, an HPGe detector with resolution of ≈1.34keV at 511keV collected the annihilation radiation spectrum. Total counts for each spectrum was greater than $2\times10^6$, and had total peak counts greater than $3\times10^5$. These spectra were

**Table 1.** Heating History

| Cycle # | Highest Temperature (K) | Measured Rate up (K/min) | Measured Rate Down (K/min) |
|---|---|---|---|
| 1 | 550 | 8.8 | -3.8 |
| 2 | 1000 | 1.5 | -23.2 |
| 3 | 1000 | 14.4 | -4.9 |
| 4 | 1150 | 2.7 | -33.2 |
| 5 | 1100 | 2.7 | -2.8 |
| 6 | 1150 | 2.7 | -12.3 |
| 7 | 1100 | 1.9 | -1.8 |
| 8 | 1150 | 3.2 | -2.7 |
| 9 | 1100 | 2.4 | -13.7 |
| 10 | 1270 | 2.7 | -8.0 |
| 11† | 1270 | 3.0 | -2.6 |

†Data reported in Figure 1 and 2.

compiled to show the temperature dependence of the S-parameter. The low momentum S-parameter limits were set to 510.3 and 511.7keV while the high momentum W-parameter limits were set symmetrically about the 511 peak to 513.4 to 514.4 and 506.0 to 508.6keV. These settings produce an average bulk S value (beam energy between 20 and 70keV) of 0.5009±0.0007 for Si(100).

## 3. Results and Discussion

Figure 1 shows the fit of Equation 3 over data of a sample after being annealed at 1200K. There is a linear dependence between the S-parameter and temperature from 300 to 800K because of thermal expansion or a change due to a positron-electron wave function overlap. The data from this region were fit linearly to find $S_b(T) = [(1.4\pm0.1)\times10^{-5}]T + (0.433\pm0.006)$. If we assume a linear dependence with temperature for $S_v$, we find $S_v(T) = [(1.4\pm0.5)\times10^{-5}]T + (0.441\pm0.006)$.

These relations were inserted into Equation 3 for $S_b$ and $S_v$ to find $H_v^f = 1.6\pm0.2$ eV and $\xi/10^{+7}=8.3 \pm17$. This $H_v^f$ is greater than previously reported γ-phase values that were obtained using other PAS techniques (see Table 2). If, however, $S_v$ was set to a constant, 0.459 or 0.464, the enthalpy ranged from 2.14 eV to 1.03 eV respectively.

**Table 2.** Formation Enthalpy Summary

| Source | $H_v^f$ (eV) | $\Delta H_v^f$ (eV) |
|---|---|---|
| This Work | 1.6 | 0.2 |
| Ref 3 | 1.2 | 0.3 |
| Ref 4[a] | 0.3 | --- |

[a] Error not reported

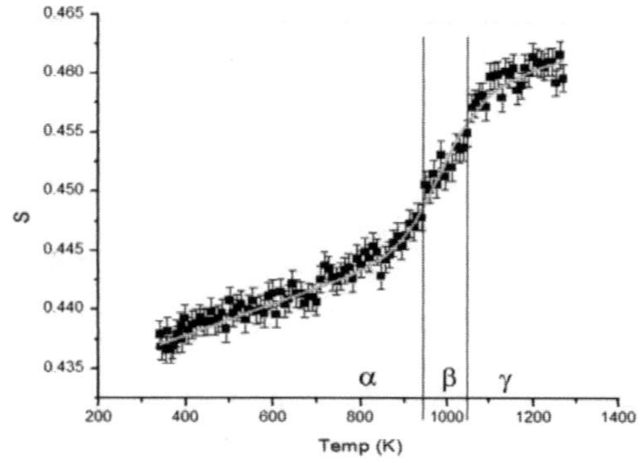


**Figure 1.** S vs. T data and fit to Equation 3. Vertical lines indicate the phase transition temperatures.

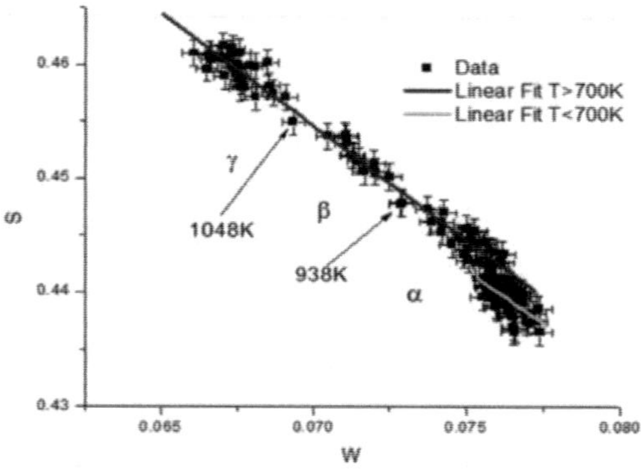


**Figure 2.** S vs. W plot corresponding to the data in Figure 1 showing a single defect type in the three phases.

Figure 2 shows the S vs. W plot for the same data in Figure 1. The lower temperatures are on the bottom right of the graph and the higher temperatures are displayed on the upper left corner. Two data points closest to the transition temperatures are marked with arrows and the corresponding phases are labelled. The data in Figure two fall on one of two lines, first for data ranging from 300 to 700K and then for data where T>700K. Because of the single slope from 700K and above, we see that there is only one defect type [7] as seen by these measurements spanning the three phase transitions.

It is important to note the differences between this work and Reference 3 and 4. First, this work used a variable energy beam of positrons where as the previous work placed the positron source in contact with the samples. Second, the depleted uranium sample used here is much smaller than those used in the earlier work. The last major difference between this work and that done before is that in the previous work, both publications report discontinuous jumps in the S data at or near the phase transition temperature. In this work those discontinuities are not observed. This is possibly due to the differences in sample preparation.

Figure 3 shows three heat cycles of a sample while the energy was set to 66 keV. The sample's surface was consisting of an oxide layer ($U_3O_8$ and $UO_2$)[8]. After heating past 800K, the samples turned from black to a metallic grey color. Heating the sample creates vacancies observed by an increase in the S-parameter until equilibrium is achieved. Because the S-parameter is related to the materials vacancy concentration [1] and the primary mechanism for oxygen diffusion at these temperatures is vacancy diffusion, the results in Figure 3 are interpreted as oxygen desorbing through the uranium and leaving vacancies in its stead. The increase in the S-parameter near 450K is in close agreement with the findings on oxygen out-diffusion reported by Senanayake [6]. These vacancies were noticeably reduced when heated past the first phase transition of 941K (see Figure 4).

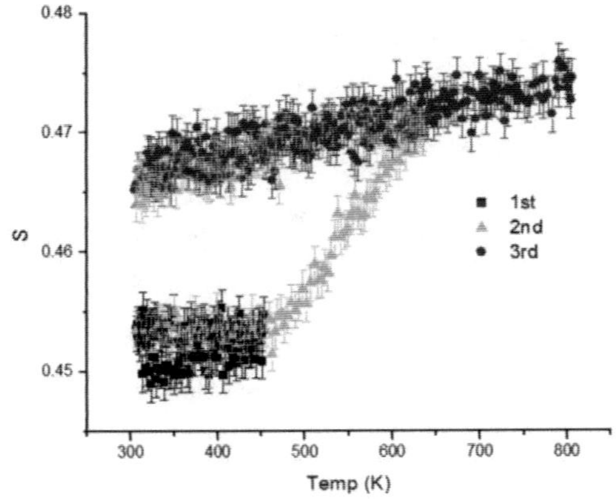


**Figure 3.** S vs. T (K) plot showing an as-received sample heated and cooled at 1 degree K per minute for three heat cycles.

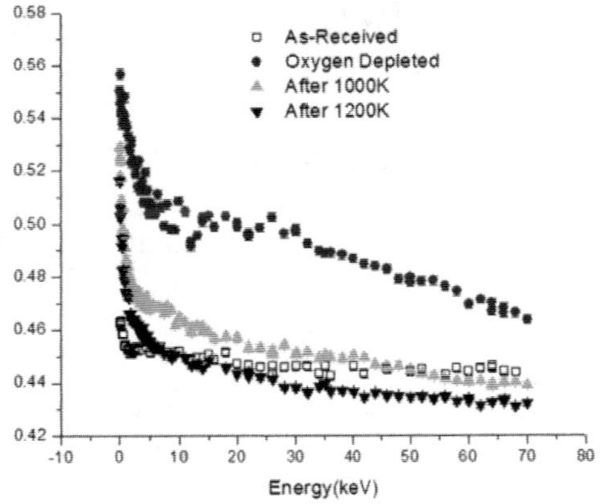


**Figure 4.** S vs. E plot showing the depth profile changes after different heat treatments.

## 4. Conclusion

PAS has been used to calculate a formation enthalpy of 1.6±0.16 eV for polycrystalline depleted uranium. Oxygen desorption from the bulk to the surface was observed near temperatures of 450K. After annealing, vacancies were removed and a single defect type was consistent with our data in all three phases of the uranium sample.

Future work could be used to investigate the effects oxygen has on the formation enthalpy. Different sample preparations, such as acid etching and annealing, will be compared to the as-received samples. A 2-detector coincidence measurement could be used to identify impurity-vacancy complexes. Investigation of uranium alloys and single crystals would be of considerable interest to analyze in the same fashion.

## Acknowledgments

We would like to acknowledge Idaho National Laboratory for funding under contract number 00014002, as well as Narendra Parmar and Ryan Stewart for their discussions and help in the lab.

## 5. References

[1] Schultz P J and Lynn K G 1988 *Rev. Modern Phys.* **60** No. 3
[2] Siegel R W 1980 *Ann. Rev Mater. Sci.* **10** 393-425
[3] Matter H, Winter J and Triftshäuser W 1980 *J. Nuclear Materials* **88** 273-278
[4] Kögel G, Sperr P, Triftshäuser W and Rothman S J 1985 *J. Nuclear Materials* **131** 148-157
[5] Touloukian T S 1975 *Thermophysical Properties of Matter 12* (New York: Plenum)
[6] Senanayake S D, Waterhouse G I N, Chan A S Y, Madey T E, Mullins D R and Idriss H 2007 *J. Phys. Chem. C* **111** 7963-7970
[7] Eichler S, Gebauer J, Börner F, Polity A, Krause-Rehberg R, Wendler E, Weber B, Wesch W and Börner H 1997 *Phys. Rev. B* **56** No. 3, 1393-1403
[8] Sullivan R 2011 *Chemical Forms of Uranium* (Chicago: Argon National Laboratory Report)

# Positron annihilation study of free volume in electron irradiated $Zr_{50}Cu_{40}Al_{10}$ bulk glassy alloy: Effects of thermal relaxation before irradiation

**N Onodera[1], A Ishii[1], A Iwase[1], Y Yokoyama[2], K Sato[3], Q Xu[3], T Yoshiie[3] and F Hori[1]**

[1] Department of Materials Science, Osaka Prefecture University, 1-1, Gakuen-cho, Naka-ku, Sakai, Osaka 599-8531, Japan

[2] Institute for Materials Research, Tohoku University, 2-1-1, Katahira, Aoba-ku, Sendai 980-8577, Japan

[3] Research Reactor Institute, Kyoto University, 2, Asashiro-Nishi, Kumatori-cho, Sennan-gun, Osaka 590-0494, Japan

E-mail: horif@mtr.osakafu-u.ac.jp

**Abstract**. Change in free volume reflects various properties such as hardness and ductility of glassy alloys. The electron- and ion- irradiations affect the free volume of ZrCuAl bulk glassy alloys that is associated with the mechanical properties. In this study, as-quenched $Zr_{50}Cu_{40}Al_{10}$ bulk glassy alloys and structural relaxed one by annealing for 5 hours at 673K below glass transition temperature ($T_g$) were irradiated by 8 MeV electrons with a maximum fluence of about $2\times10^{18}$ $e^-/cm^2$ at room temperature. The behaviour of free volume in these samples was investigated by positron annihilation techniques. X-ray diffraction measurements showed that no crystallization occurred after the irradiation. We observed the positron lifetime increased by the irradiation for both as-quenched and structural relaxed samples. The increases in positron lifetime at $2\times10^{18}$ $e^-/cm^2$ were 9 psec for as-quenched and 12 psec for relaxed sample, respectively. In addition, the increase in positron lifetime with irradiation fluence was clearly different for the two types of sample. These facts imply that the thermal relaxation before irradiation of $Zr_{50}Cu_{40}Al_{10}$ bulk glassy alloy influences the radiation effects, especially the free volume change.

## 1. Introduction

The Zr-based bulk glassy alloys are known to have a good glass forming ability, and attractive mechanical properties such as high strength and toughness [1]. These properties closely related to change in "free volume", which means frozen excess open volume containing in glassy structure [2]. The bulk glassy alloy has superior stabilized structure, while it is nonequilibrium state. On the other hand, various approaches have been studied to improve the poor ductility of these metallic glasses. One of the most popular is to anneal them. It is well known that annealing below glass transition temperature ($T_g$) results in "structural relaxation", with an increase in density, but without crystallization [2]. This structural relaxation has relation with the shrinkage of free volume [8]. According to recent studies, a mixture of local domains with liquid-like weakly bonding and regions of solid-like strongly bonding exist in glassy alloys [3,4]. These imply that structural relaxation affects some local structure change and thermal stabilities of the bulk glassy alloys. Other approaches to improve materials features we suggest are electron and ion irradiation. We have been studying the radiation effects of electron and ion irradiation on the free volume and mechanical properties of ZrCuAl bulk glassy alloy [5,6,7]. In general, electron irradiation introduces vacancies and interstitials by local atomic displacement in crystalline metals. However, the radiation damage by energetic particles irradiation for the "bulk" glassy alloys has not been clarified yet. In this study, therefore, in order to study the influence of structural relaxation on radiation effect, we performed electron irradiation to as-quenched and structural relaxed $Zr_{50}Cu_{40}Al_{10}$ bulk glassy alloys.

## 2. Experimental procedure

$Zr_{50}Cu_{40}Al_{10}$ bulk glassy alloy was fabricated by the tilt casting method in an arc furnace and cut into the size of ϕ 6-8 mm × 0.5 mm disc [9]. Some of these quenched samples were annealed for 5 hours at 673 K, which is below glass transition temperature $T_g$ [8]. All of these annealed and as-quenched samples were irradiated with 8 MeV electrons at room temperature by using an electron linear accelerator at the Research Reactor Institute, Kyoto University. The fluence of electron irradiation was from $2\times10^{17}$ to $2\times10^{18}$ $e^-/cm^2$. Positron annihilation lifetime and coincidence Doppler broadening (CDB) measurements were performed for these irradiated samples at room temperature. All the positron annihilation lifetime spectra were obtained by using a conventional fast-fast circuit with a time resolution of about 200 psec (FWHM) and analyzed by the POSITRONFIT program [10]. The positron annihilation spectra and CDB spectra consist of more than $10^6$ and $10^8$ counts, respectively. To confirm crystallinities of all samples due to irradiation, X-ray diffraction measurement was also performed using Rigaku Ultima IV.

## 3. Results and discussion

Figures 1 and 2 show X-ray diffraction spectra for as-quenched and structural relaxed $Zr_{50}Cu_{40}Al_{10}$ bulk glassy alloys before and after electron irradiation. From these figures it can be seen for the range of electron irradiation fluences studied no crystalline peak were observed. The amorphous state is retained for both as-quenched and structural relaxed samples by electron irradiation.

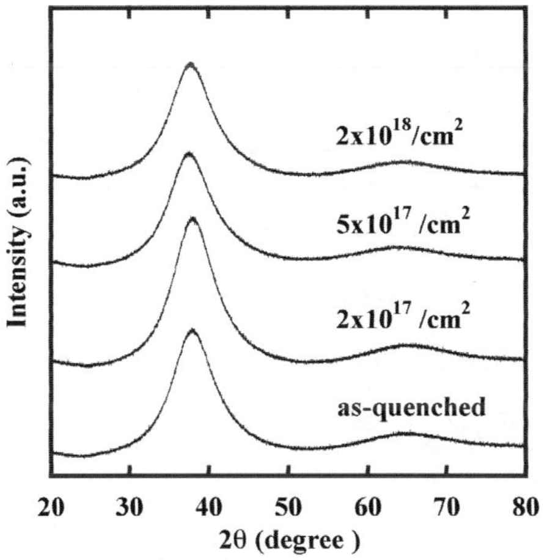


**Figure 1.** X-ray diffraction spectra for as-quenched $Zr_{50}Cu_{40}Al_{10}$ bulk glassy alloys after electron irradiation.

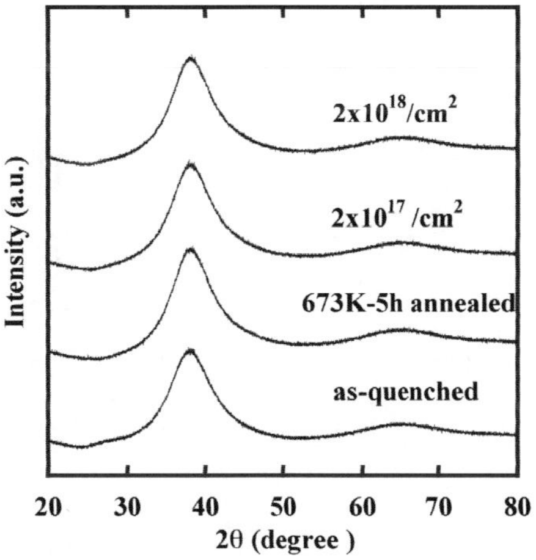


**Figure 2.** X-ray diffraction spectra for $Zr_{50}Cu_{40}Al_{10}$ bulk glassy alloys with annealing followed by electron irradiation.

The mean positron lifetime value τ of as-quenched and structural relaxed samples are 166 and 157 psec, respectively. The positron lifetime decreases 9 psec by the annealing. This decreasing of positron lifetime reveals shrinkage of free volume by structural relaxation [8]. Figures 3 and 4 represent the change in positron lifetime τ by electron irradiation as a function of irradiation fluence for as-quenched and structural relaxed $Zr_{50}Cu_{40}Al_{10}$ bulk glassy alloys. We fitted all the positron lifetime spectra with one, two, three components and the one component fits always had the lowest chi-squared. This indicates that the distribution of open volume size has single peak even though the samples are irradiated by electron. The positron lifetime increases with electron irradiation for both alloys. An increment of positron lifetime at maximum fluence of $2\times10^{18}$ $e^-/cm^2$ is about 10 psec for both cases. It can be estimated that electron irradiation introduces displacement defects for as-quenched and structural relaxed $Zr_{50}Cu_{40}Al_{10}$ bulk glassy alloys. However, the increasing trend of positron lifetime with increase of irradiation fluence is clearly different between as-quenched and structural relaxed samples. In the case of as-quenched sample, the positron lifetime increases gradually at about $5\times10^{17}$ $e^-/cm^2$ irradiation as shown in Figure 3. On the other hand, in the case of structural relaxed one, the positron lifetime increases at low fluence of electron irradiation and its value is almost constant. From these results, following explanations can be given. In the case of as-quenched $Zr_{50}Cu_{40}Al_{10}$ alloys, the increasing in the size of free volume at low irradiation fluence is suppressed by formation of chemical short-range ordering (CSRO) through irradiation displacement at mainly liquid-like region in glassy alloy [3]. Actually, the formation of CSRO due to structural relaxation has been reported [11]. Above $1\times10^{18}$ $e^-/cm^2$ irradiation, no more CSRO formation takes place and it becomes vacancy type displacement damage that is retained. On the other hand, in the case of structural relaxed $Zr_{50}Cu_{40}Al_{10}$ alloys, after the initial increase further increases in the radiation fluence do not change the positron lifetime. This may suggest that no structural relaxation results from the irradiation because the liquid-like regions no longer exists due to annealing before irradiation.

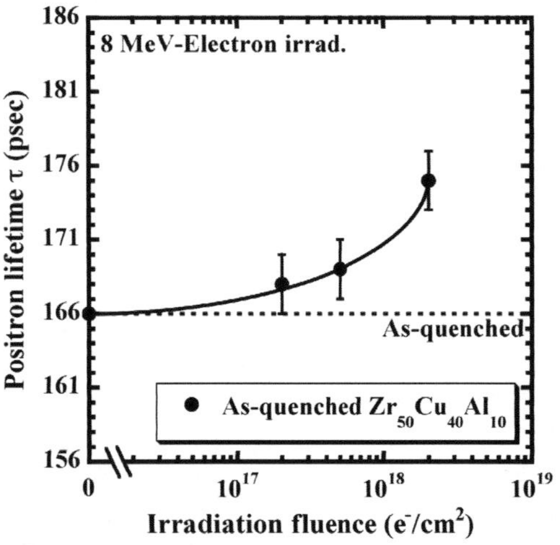


**Figure 3.** Change in positron lifetime τ by electron irradiation for as-quenched $Zr_{50}Cu_{40}Al_{10}$ as a function of irradiation fluence.

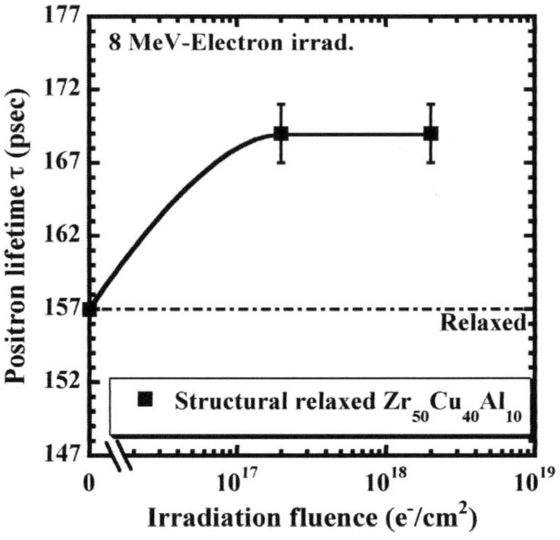


**Figure 4.** Change in positron lifetime τ by electron irradiation for structural relaxed $Zr_{50}Cu_{40}Al_{10}$ as a function of irradiation fluence.

Figures 5 and 6 show CDB profile before and after electron irradiation for as-quenched and structural relaxed $Zr_{50}Cu_{40}Al_{10}$ bulk glassy alloys in the form of ratio of the CDB intensity to that for Al metal. CDB profile for ZrCuAl bulk glassy alloy does not change by structural relaxation [8], but it clearly changes by crystallization with long range atomic reordering [12]. No significant change in the CDB ratio profile with electron irradiation was observed for either type of sample. This suggests that long-range atomic reordering around free volume did not occur.

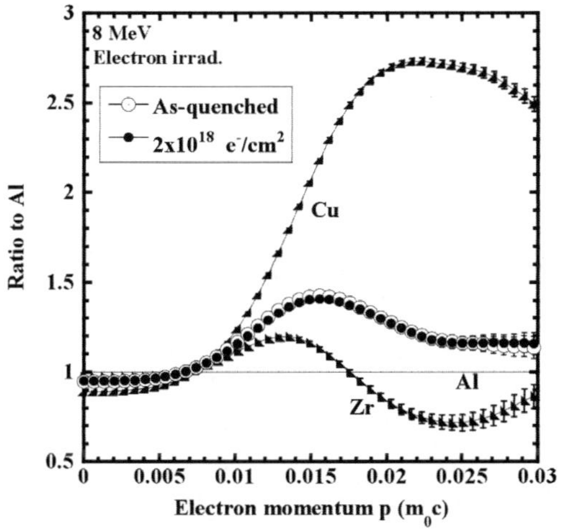


**Figure 5**. CDB ratio profile of as-quenched $Zr_{50}Cu_{40}Al_{10}$ with electron irradiation, expressed in the form of ratio to pure Al. CDB spectra of pure Zr and Cu are also plotted.

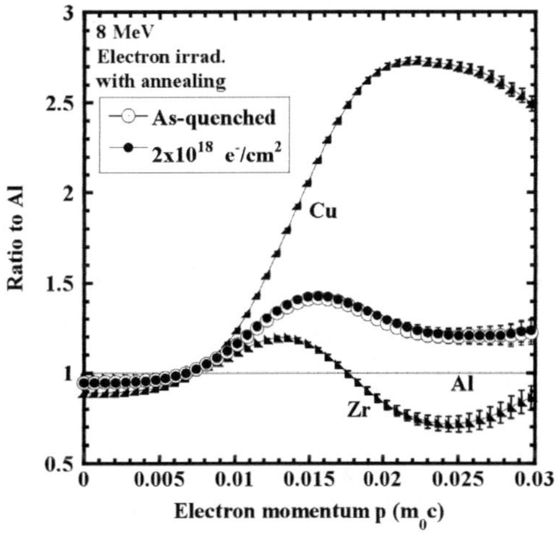


**Figure 6**. CDB ratio profile of structural relaxed $Zr_{50}Cu_{40}Al_{10}$ with electron irradiation, expressed in the form of ratio to pure Al. CDB spectra of pure Zr and Cu are also plotted.

## 4. Summary

The effect of structural relaxation on the radiation damage by electrons for $Zr_{50}Cu_{40}Al_{10}$ bulk glassy alloys was studied. We found that electron irradiation introduces the atomic displacements without crystallization. Moreover, we found that the structural relaxation by annealing before irradiation affects the radiation damage formation in this bulk glassy alloy by the electron irradiation especially at low fluence.

**Acknowledgment**

Authors would like to express cordial thanks to technical staffs for electron irradiation at Research Reactor Institute, Kyoto University. This work was also performed under the inter-university cooperative research program of the Advanced Research Center of Metallic Glasses, Institute for Materials Research, Tohoku University.

## References

[1] Inoue A 2000 *Acta. Mater.* **48** 279
[2] Yokoyama Y, Akeno Y, Yamasaki T, Liaw P K, Buchanan R A and Inoue A 2005 *Mater. Trans.* **46** 2755
[3] Ichitsubo T and Matsubara E 2009 *Mater. Sci. Eng.* A **521-522** 236
[4] Shimono M and Onodera H 2007 *Mater. Sci. Eng.* A **449-451** 717
[5] Hori F, Onodera N, Fukumoto Y, Iwase A, Kawasuso A, Yabuuchi A, Maekawa M and Yokoyama Y 2011 *J. Phys.: Conf. Ser.* **262** 012025
[6] Fukumoto Y, Ishii A, Iwase A, Yokoyama Y and Hori F 2010 *J. Phys.: Conf. Ser.* **225** 012010
[7] Onodera N, Ishii A, Fukumoto Y, Iwase A, Yokoyama Y and Hori F 2012 *Nucl. Inst. Meth.* B **282** 1
[8] Ishii A, Hori F, Iwase A, Fukumoto Y, Yokoyama Y and Konno T J 2008 *Mater. Trans.* **49** 1975
[9] Yokoyama Y, Fukaura K and Inoue A 2002 *Intermetallics* **10** 1113
[10] Kirkegaard P, Eldrup M, Mogensen O E and Pedersen N J 1981 *Comp. Phys. Commun.* **23** 307
[11] Komatsu T and Matusita K 1986 *J. Mater. Sci.* **21** 1693
[12] Hori F, Yano T, Yokoyama Y, Akeno Y and Konno T J 2007 *J. Alloys. Comp.* **434-435** 207

Journal of Physics: Conference Series **443** (2013) 012023 doi:10.1088/1742-6596/443/1/012023

# Cavity nucleation and growth during helium implantation and neutron irradiation of Fe and steel

**M. Eldrup and B.N. Singh**

Department of Energy Conversion and Storage, Technical University of Denmark, Risø Campus, DK-4000 Roskilde, Denmark

moel@dtu.dk

**Abstract**. The present work concerns investigations of damage accumulation during helium implantation of pure iron and the reduced activation ferritic-martensitic steel 'EUROFER 97' at 323K and 623K as well as during neutron irradiation with or without prior helium implantation. The defect microstructure, in particular the cavities, was characterized using Positron Annihilation Lifetime Spectroscopy (PALS) and Transmission Electron Microscopy (TEM). The PALS investigations reveal a clear difference between the He implantation effects in Fe and EUROFER 97 at both temperatures. For both materials the mean positron lifetime increases with He dose in the range 1 – 100 appm, although the increase is stronger for Fe than for EUROFER 97 and for both materials stronger for implantation at 323K than at 623K.

## 1. Introduction

Effects of neutron irradiation on physical and mechanical properties of reduced activation ferritic-martensitic (RAFM) steels are being extensively studied since they are considered to be candidate materials for fusion reactors [1, 2].

It has previously been demonstrated that in pure iron as well as in the RAFM steel called F82H [2] irradiation with fission neutrons causes void formation already at temperatures as low as 323 - 373K [3-5]. This is in agreement with general experimental observations that voids in bcc metals and alloys are formed at relatively low temperatures, i.e. close to the recovery stage III [6]. In a fusion reactor these materials are expected to be exposed to 14 MeV neutrons that introduces concurrent generation of helium atoms at a relatively high He-to-displacement ratio (~10-20 appm/dpa) and at damage rates of $\sim 10^{-6}$ dpa/s in the form of cascades and sub-cascades. This is a matter of serious concern from the point of view of the application of these alloys in structural components.

The main objective of the present work was to obtain experimental results on cavity evolution already during the He-implantation stage and then to follow the evolution of these cavities after post-implantation neutron irradiation to different displacement doses.

Details of the present experimental work can be found in two Risø Reports [7, 8]. Reference [8] also describes the modeling of vacancy cluster and helium-vacancy cluster evolution under a variety of different conditions.

## 2. Materials and Experimental Procedure

Specimens of pure (99.99 wt%) iron and the RAFM steel called EUROFER 97 [2] were investigated in the present work. Details of the sample preparation procedure can be found in [7-9].

Strip specimens of both pure iron and EUROFER 97 steel were implanted with helium at Forschungszentrum Jülich (Germany). A beam of 28 MeV α-particles, energy degraded by a rotating wheel equipped with Al foils, was used for implantation of the specimens to provide a homogeneous He concentration throughout a depth of about 100 μm (i.e. the specimen thickness) over an area of about 13 × 12 $mm^2$. The implantations were carried out at ~323K or at 623K to concentration levels of 1, 10 and 100 appm of helium at implantation rates in the range of $1.2\times10^{-3}$ to $1.2\times10^{-2}$ appm He/s [9]. The implantation gives rise to displacement damage doses of $1.5\times10^{-4}$, $1.5\times10^{-3}$ and $1.5\times10^{-2}$ dpa, respectively for the three concentrations of implanted helium, since on average, each implanted He atom gives rise to 154 atomic displacements [9]. (This is equivalent to ~6500 appm He/dpa or roughly 400 times higher than expected in a fusion reactor. Thus, the present implantation-irradiation conditions do not simulate those in a fusion reactor). Both un-implanted and He-implanted specimens were irradiated with fission neutrons in the BR-2 reactor at Mol (Belgium) at 323K or 623K with a displacement dose rate of ~$4.5\times10^{-8}$ dpa/s to doses in the range of 0.001 to 0.23 dpa.

Positron Annihilation Lifetime Spectroscopy (PALS) was used for measurements on a wide range of specimens and Transmission Electron Microscopy (TEM) to study fewer, selected specimens.

## 3. Results

### *3.1 Positron Annihilation Lifetime Spectroscopy*

A detailed analysis of the measured lifetime spectra was carried out: All spectra for implanted and/or irradiated specimens could be decomposed into two or three lifetime components using the PALSfit program [10]. In order to give a simple qualitative overview of the general trend of all the data obtained by PALS for iron and EUROFER 97 which have either been implanted with helium, neutron irradiated or neutron irradiated after helium implantation, Figure 1 shows mean positron lifetimes as functions of total displacement dose at both 323 K and 623 K.

*3.1.1 Iron.* For He implantation at both temperatures the mean lifetime increases with He concentration and hence displacement dose (Fig. 1a and b). Qualitatively, a mean-lifetime increase can be ascribed to an increase in density of defects and/or size of cavities. The difference in mean lifetime after implantation to the same He concentration at the two temperatures is mainly due to a much larger cavity density at 323 K than at 623 K. The figures also demonstrate that the influence of He implantation prior to neutron irradiation is quite different at the two temperatures. At 323 K the presence of He has only a moderate influence on the result of neutron irradiation, while at 623 K the presence of He has a clear effect on the cavity population after neutron irradiation.

*3.1.2 EUROFER 97.* Figures 1c and 1d show the mean positron lifetime as a function of total displacement dose for EUROFER 97 at 323 K and 623 K. The relatively small effect (compared to Fe) of neutron irradiation with or without prior He implantation is clearly displayed. As Fig. 1 shows, in the investigated He concentration range of 0 – 100 appm He, changes of the mean lifetime fall in the range up to 30 ps for EUROFER 97 at 623 K, while for Fe the range of mean lifetime variations is about 60 ps at the same temperature. At 323 K the changes of the mean lifetime are appreciably larger (about 95 ps and 135 ps, respectively).

The results of the detailed analyses of all measured lifetime spectra (lifetimes and intensities) showed an appreciable scatter for nominally identical specimens. This scatter makes it impossible to define any trend in these data that can be correlated with the He implantation rate. Also the mean lifetimes show a certain scatter. Thus, attempts to correlate mean lifetimes with implantation rates were also unsuccessful. On the other hand the mean lifetimes shown in Fig. 1 clearly correlate with total He concentration. In spite of the scatter of the data for the resolved lifetime components, it is possible to derive estimates of sizes and densities of the cavities in the irradiated and/or implanted samples (see details in [4, 7, 8]).

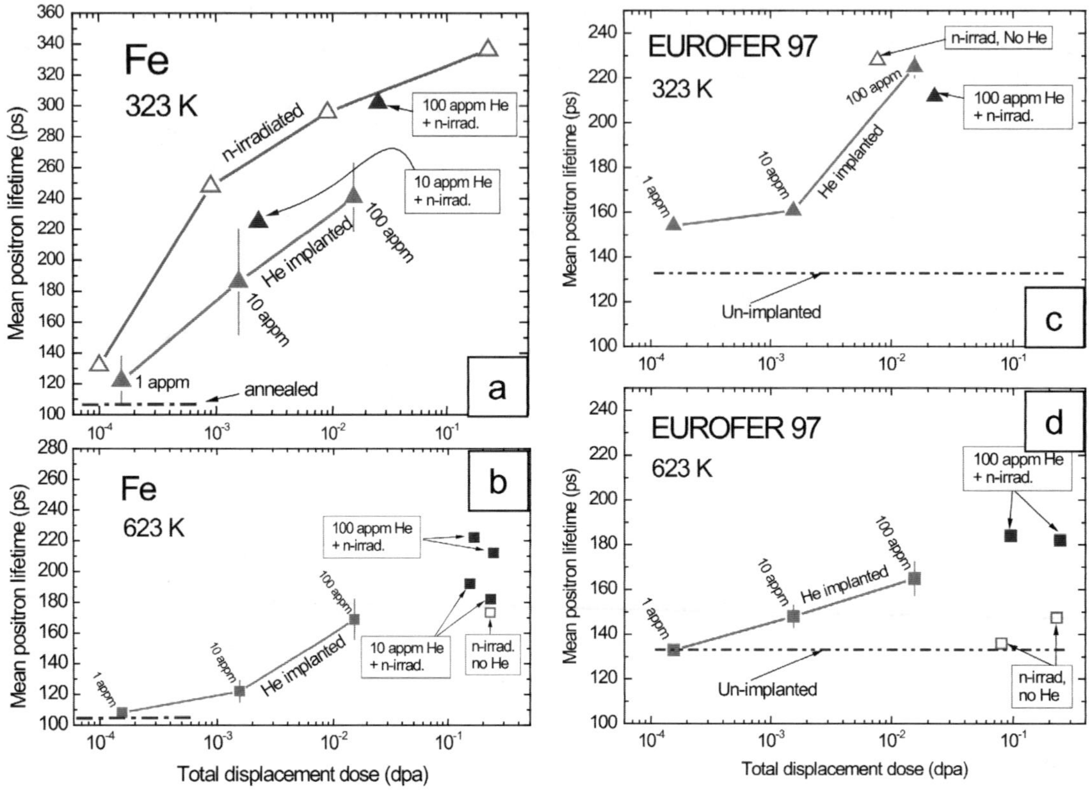


Figure 1. Mean positron lifetimes for iron (a, b) and EUROFER 97 (c, d) which have either been implanted with helium, neutron irradiated or neutron irradiated after helium implantation at 323K (upper frames: a and c) or at 623K (lower frames: b and d) [7, 8]. Implantation of 100 appm helium gives rise to a displacement dose of 0.015 dpa. The green open symbols are for neutron irradiation without helium implantation, while the blue filled symbols are results for neutron irradiation after implantation to a dose of 10 or 100 appm helium. An error bar associated with a red point indicates the scatter of results from several specimens implanted to the same nominal He dose.

*3.2 Transmission Electron Microscopy*

TEM could not resolve any cavities in Fe or EUROFER 97 after implantation of 100 appm He, neither at 323K nor at 623K. However, neutron irradiation of Fe at 623K to a dose level of 0.23 dpa leads to cavities (Fig. 2A). He implantation (100 appm) prior to neutron irradiation, results in a cavity density increase by a factor of ~7 (Fig. 2B). In EUROFER 97 a very inhomogeneous cavity distribution, formed at dislocations and interfaces, was observed after He implantation with subsequent neutron irradiation. In addition, a very low density of very large voids has been observed in Fe neutron irradiated at 323K (no He), even at a dose level of as low as 0.036 dpa.

## 4. Conclusion

The present experiments have shown some rather interesting effects of helium implantation and post-implantation neutron irradiation on cavity formation and growth in pure iron. The results presented in Fig. 1a suggest, for example, that at 323K He implantation prior to neutron irradiation does not make any substantial addition of cavities compared to neutron irradiation without prior He implantation. This implies that at this temperature the homogeneous nucleation of cavities is very efficient even in the absence of implanted helium.

At 623K (i.e. above the recovery stage V), on the other hand, the implanted helium does cause an increase in cavity nucleation and growth (Fig. 1b). This is because at this temperature, the implanted helium atoms help to stabilize the cavity nuclei against shrinkage by thermal evaporation of vacancies, which would take place in the absence of He. Consequently, both cavity nucleation and growth are enhanced.

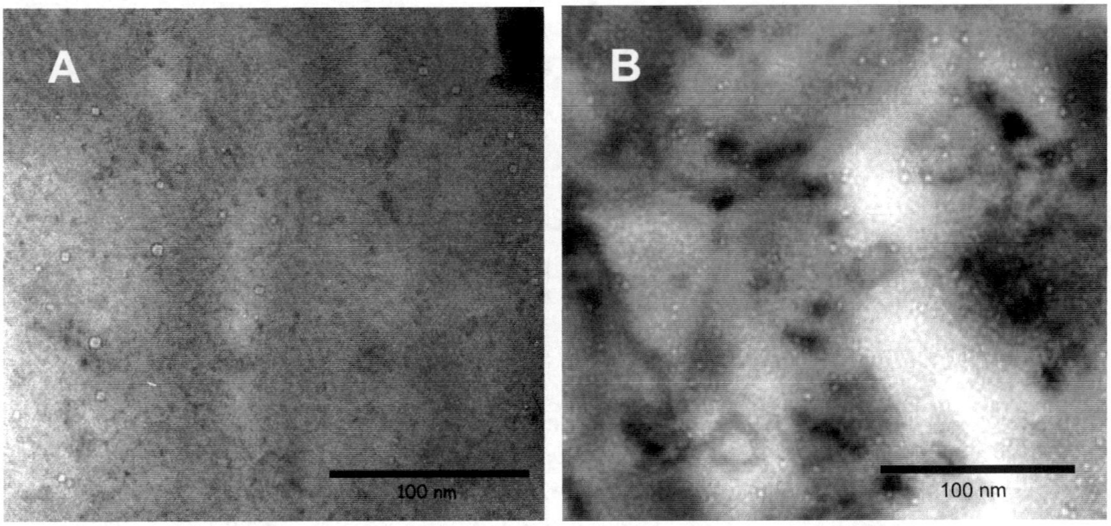


Figure 2. Transmission electron micrographs showing cavities in pure Fe after A) neutron irradiation to a dose of 0.23 dpa, giving an average void size of ~ 4.0 nm and density of $\sim 1.5 \times 10^{21}$ $m^{-3}$ and B) implantation of 100 appm He and subsequent neutron irradiation to a dose of 0.23 dpa, giving an average void size of ~ 3.5 nm and density of $\sim 1 \times 10^{22}$ $m^{-3}$. Irradiation and implantation temperatures were 623 K.

## 5. Acknowledgements
The work reported here was partly funded by the European Fusion Technology Programme. The authors wish to thank S.I. Golubov for discussions, P. Jung for providing numerous helium implanted specimens, P. Jacquet for organizing neutron irradiations in the BR-2 reactor at Mol, D. Edwards and E. Johnson for carrying out TEM examinations of some of the specimens and G. Christiansen, N.J. Pedersen and B. Olsen for technical assistance.

## References
[1] Klueh R L, Ehrlich K and Abe F 1992 *J. Nucl. Mater.* **191 – 194** 116.
[2] Baluc N, Gelles D S, Jitsukawa S, Kimura A, Klueh R L, Odette G R, van der Schaaf B and Yu Jinnan 2007 *J. Nucl. Mater.* **367-370** 33.
[3] Eldrup M and Singh B N 2000 *J. Nucl. Mater.* **276** 269.
[4] Eldrup M, Singh B N, Zinkle S J, Byun T S and Farrel K 2002 *J. Nucl. Mater.* **307 – 311** 912.
[5] Eldrup M and Singh B N 2003 *Risø Report No. Risø-R-1241(EN)* pp. 21. http://orbit.dtu.dk/fedora/objects/orbit:90110/datastreams/file_7712456/content
[6] Singh B N and Evans J H 1995 *J. Nucl. Mater.* **226** 29.
[7] Singh B N, Eldrup M, Golubov S I, Edwards D J and P. Jung 2005 *Risø Report No. Risø-R-1496(EN)* pp. 38. http://orbit.dtu.dk/fedora/objects/orbit:90096/datastreams/file_7710901/content
[8] Eldrup M, Singh B N and Golubov S I 2010 *Risø Report No. Risø-R-1619(EN)* pp. 64 http://orbit.dtu.dk/fedora/objects/orbit:82170/datastreams/file_4667852/content
[9] Jung P. 2004 *Helium implantation in pure iron and RAFM steels at <70 and 350℃* (Task:TW4-TTMS-001a, EFDA Technology Programme).
[10] Olsen J V, Kirkegaard P, Pedersen N J, Eldrup M 2007 *phys. stat. sol.* (c) **4** 4004 and www.PALSfit.dk

# Coincidence Doppler Broadening of Positron Annihilation Radiation in Fe

**E do Nascimento, V R Vanin, N L Maidana and O Helene**

Instituto de Física da Universidade de São Paulo, São Paulo, SP, Brasil

eduardon@if.usp.br

**Abstract**. We measured the Doppler broadening annihilation radiation spectrum in Fe, using $^{22}$NaCl as a positron source, and two Ge detectors in coincidence arrangement. The two-dimensional coincidence energy spectrum was fitted using a model function that included positron annihilation with the conduction band and *3d* electrons, *3s* and *3p* electrons, and in-flight positron annihilation. Detectors response functions included backscattering and a combination of Compton and pulse pileup, ballistic deficit and shaping effects. The core electrons annihilation intensity was measured as 16.4(3) %, with almost all the remainder assigned to the less bound electrons. The obtained results are in agreement with published theoretical values.

## 1. Introduction

Doppler broadening of electron-positron annihilation radiation is an important tool in the field of materials science and has been used since the 1970s in the investigation and characterization of point defects in materials [1,2]. This technique has been applied more recently in intermetallic alloys as prospective materials for new industrial applications [3]. In fact, vacancies and other lattice defects have a significant impact on mechanical as well as electrical properties of materials [3]. As a non-destructive method, upon penetration into the sample, positrons lose almost all of their kinetic energy, reach thermal equilibrium and diffuse into the material before annihilation. During the electron-positron annihilation process, the energy-momentum is conserved and, in most of the cases, two photons are emitted with energies given by $E_{1,2} = m_e c^2 - B_i/2 \pm p_z c/2$, where $m_e c^2$ is the electron rest mass energy, $B_i$ the electron binding energy, $c$ the speed of light, and $p_z$ the momentum component of the electron-positron pair in the direction of the emitted gamma-rays. The use of two Ge detectors allows the observation of both annihilation photons in the Doppler Broadening Spectroscopy [4], with great improvement in the combined energy resolution for the study of the annihilation peak and an important reduction in the observed background, when compared to measurements taken with one detector.

In this paper, we report an experiment of Coincidence Doppler Broadening Spectroscopy (CDBS) of positron annihilation in Fe. A model function was fitted to the obtained data in order to determine the distribution of electron momenta, in a procedure similar to that applied to CDBS for aluminum and silicon [5,6], using core electron binding energies taken from literature [7]. This procedure allows the experimental determination of the annihilation parameters and response function parameters along with their uncertainties.

## 2. Experimental Setup

A $3.7 \times 10^5$ Bq (10 μCi) $^{22}$NaCl carrier free Amersham positron source was deposited directly between two 2-mm-thick polycrystalline iron discs (99.99% pure) and positioned in the middle of two Ge detectors with end caps standing 10 cm apart, in coincidence arrangement. The detectors were calibrated in energy using the gamma-rays from an $^{192}$Ir source placed near the iron discs. A 2D-spectrum was taken for 76 h with about $1.8 \times 10^7$ counts in the annihilation photons coincidence peak. The resolutions of the detectors were 1.6 and 1.3 keV for 511 keV photons.

Figure 1 shows the obtained coincidence spectrum. When $E_1$ and $E_2$ represent the photons energies measured in the detectors, it is found that the coincidence peak is broader along the line $E_1+E_2 = 2m_e c^2$ due to the Doppler shift of the annihilation radiation. The ridges crossing below the annihilation peak are due to coincidences with annihilation photons or 1274 keV gamma-rays from $^{22}$Na decay that undergo Compton scattering inside the detector active volume. The spectrum also shows features that cannot be seen in this figure [8]: exponential tails due to ballistic deficit, pulse shaping problems, pileup, and events of backscattering of the annihilation photons. There is still a small fraction of positrons that annihilate before thermalizing, the so-called *in-flight positron annihilation*, and the corresponding events show a larger energy spread, of a few hundreds of keV. It is also seen in Figure 1 a continuum between the ridges, associated to events where both photons scatter in the detectors and escape the active volumes. There is more information in this spectrum, that cannot be retrieved by visual inspection as, for example, the various distributions related to annihilation with core electrons from different shells, which are displaced from the line $E_1+E_2 = 2m_e c^2$ by their binding energies $B_i$. Accidental coincidences were corrected using another matrix built with the time window displaced from the prompt peak.

## 3. Fitting – Function Model

A two dimensional function was fitted to the experimental histogram (Figure 1) in a 90 keV × 90 keV region around the two annihilation-photons coincidence peak. The main components of this model function are described below.

Positron annihilation with electrons of the conduction band was represented by:

$$f_c = \sum_{i=1}^{3} C_i (E_1 - E_2 - \alpha_i)(E_1 - E_2 + \alpha_i) \tag{1}$$

along the line $E_1+E_2 = 1022$ keV, where $\alpha_i$ are the cutoff parameters ($C_i = 0$ when $|E_1-E_2| > \alpha_i$). We fitted the empirical parameters $C_i$ and $\alpha_i$ to the experimental spectrum. The *3d* orbital was included in these terms because the detector resolutions did not allow to separate this orbital from the conduction band; hence, we have defined an effective band of less bound electrons, in a procedure similar to that adopted by Tang [8]. Eventually, this term will include also the contribution from the electrons within the defects.

Positron annihilation with core electrons was described by Gaussian functions:

$$f_{nl} = \frac{A_{nl} e^{-(E_1 - E_2)^2 / 2\sigma_{nl}^2}}{\sqrt{2\pi}\sigma_{nl}} \tag{2}$$

along the line $E_1+E_2+B_{nl} = 1022$ keV, where $B_{nl}$ is the *nl*-orbital binding energy [7]; here, *nl* stands for *3s* and *3p*. We fitted the empirical parameters $A_{nl}$ and $\sigma_{nl}$.

The in-flight positron annihilation was taken into account by a function that depends on the angle $\theta$ between the emitted annihilation photons,

$$f_f = A_f \; e^{-\lambda/d(\theta)} \tag{3}$$

where the constraint equation $d(\theta) = E_1^{-1} + E_2^{-1} = (1-\cos\theta)/m_e c^2$ comes from the conservation laws [6], and relation (3) extends over all $E_{1,2}$ consistent with this constraint. Here we fitted the empirical parameters $A_f$ and $\lambda$.

Finally, spectrum distortion by the ballistic deficit, pulse pileup, and incomplete charge collection were taken into account using internal ($E_{1,2} < m_e c^2$) and external ($E_{1,2} > m_e c^2$) exponential tails for each detector. The two internal and two external ridges were included in the model function to account for coincidence events where one of the annihilation photons is measured in the full energy peak and the other undergoes scattering in the detector before escaping the active volume. It was assumed that the continuum is proportional to the ridges and, thus, the only fitted parameter is the proportionality factor [5,6]. We fitted also the coincidence peak two-dimensional position, the inclination of the $E_1+E_2 = 2m_e c^2$ line with respect to the CDBS axes, and two parameters for the detector resolutions.

The functions $f_c$, $f_{nl}$, $f_f$ , the exponential tails, the ridges, and the background were added and convolved with the detector response functions given by two Gaussians. The fitted spectrum is shown in figure 2. The chi-squared value ($\chi^2$) was calculated using the expression:

$$\chi^2 = \sum_{i,j} \frac{(n_{ij} - F_{ij})^2}{F_{ij} + 2a_{ij}} \tag{4}$$

where $n_{ij}$ is the number of observed events in channel $(i,j)$ of the Doppler broadening real coincidence spectrum, $a_{ij}$ is the number of observed events in channel $(i,j)$ of the Doppler broadening accidental coincidence spectrum and $F_{ij}$ is the fitted function.

## 4. Results and Conclusion

With this simplified model, the obtained $\chi^2$ value was $3.98\times10^4$ for about $3.06\times10^4$ degrees of freedom. This chi-squared value corresponds to a low significance level, and shows that this simple model is unable to explain the phenomenon in its complexity.

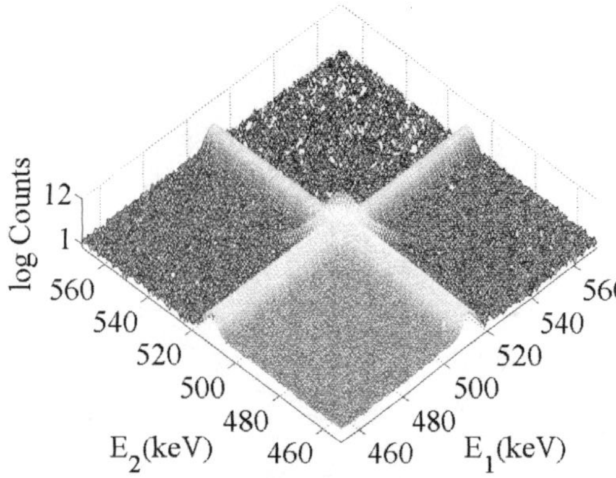


**Figure 1**. Experimental coincidence Doppler Broadening spectrum of the annihilation radiation in Fe.

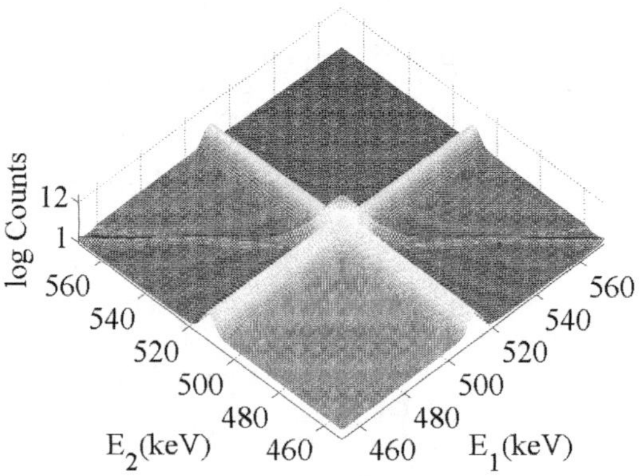


**Figure 2**. Fitted coincidence Doppler Broadening spectrum of the annihilation radiation in Fe.

Figure 3 shows the fitted and experimental Doppler broadening profile spectrum for Fe, where the *3p* plus *3s* electrons correspond to 16.4 (3) % of total annihilation events, with the remaining

83.6 (3) % assigned to conduction, *3d* electrons, and defects; these intensities can be compared to those calculated by Tang et al. [9], respectively ~ 15.6% and ~ 84.4%. In-flight positron annihilation amounts to a small fraction of percent, and the corresponding events are shown along with backscattering events in Fig. 2. The less bound electron band (conduction, *3d*, and defects) cutoff parameters obtained were 7, 12 and 17 × $10^{-3}$ $m_ec$. We call attention for the fact that our Fe sample was polycrystalline, therefore the measurement corresponds to an average of the orientations of the momentum distributions of electrons in the sample.

In conclusion, with the simplified model presented here – three parabolas and two Gaussian functions with a standard response function [5,6] – we were able to describe the main components of the electron-positron annihilation peak in Fe, as can be seen in Figure 3, and obtained annihilation intensities in agreement with theoretical results. Thus, it could be possible to identify the chemical element present in the annihilation site by CDBS experiment, with relatively few counts.

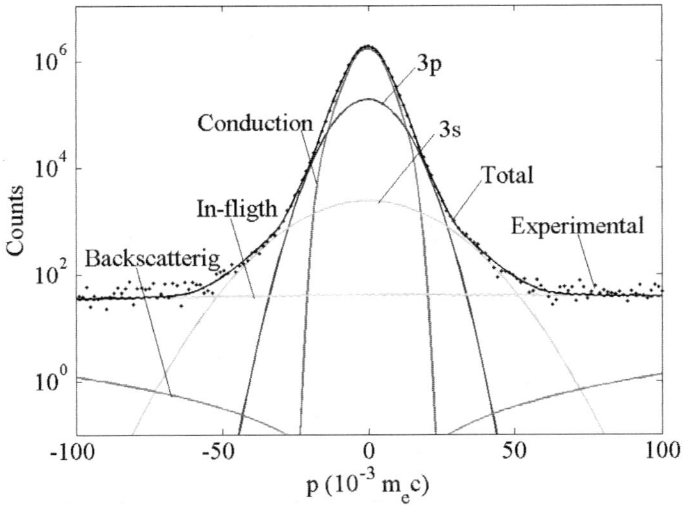


**Figure 3**. Fitted and experimental Doppler broadening profile spectrum for Fe, including the *3p*, *3s* and conduction electrons components of annihilation events. In-flight positron annihilation and backscattering events in the same region of interest are also shown.

**Acknowledgments**
We acknowledge the support of Conselho Nacional de Desenvolvimento Científico e Tecnológico – CNPq, Coordenação de Aperfeiçoamento de Pessoal de Nível Superior – CAPES and Fundação de Amparo à Pesquisa do Estado de São Paulo – FAPESP.

**References**
[1] Krause-Rehberg R and Leipner H 1999 *Positron Annihilation in Semiconductors* (Springer-Verlag, Berlin)
[2] Puska M J and Nieminen R M 1994 *Reviews of Modern Physics* **66** 841
[3] Melikhova O, Cizek J, Kuriplach J, Prochazka I, Cieslar M, Anwand W and Brauer G 2010 *Intermetallics* **18** 592
[4] Lynn K G, MacDonald J R, Boie R A, Feldman L C, Gabbe J D, Robbins M F, Bonderup E and Golovchenko J 1977 *Phys Rev Lett* **38** 241
[5] do Nascimento E, Helene O, Vanin V R and Takiya C 2005 *Nucl Instrum Meth A* **538** 723
[6] do Nascimento E, Helene O, Vanin V R, da Cruz M T F and Moralles M 2009 *Nucl Instrum Meth A* **609** 244
[7] Firestone R B and Shirley V S 1996 *Table of Isotopes,* eighth edition, vol II, John Willey & Son
[8] do Nascimento E, Fernández-Varea J M, Vanin V R, Helene O and Maidana N L 2011 *AIP Conf Proc* **1351** 216
[9] Tang Z, Hasegawa M, Nagai Y and Saito M 2002 *Phys Rev B* **65** 195108-1

Journal of Physics: Conference Series **443** (2013) 012025 doi:10.1088/1742-6596/443/1/012025

# Vacancy-induced hardening in Fe-Al alloys

**F Lukáč[1], J Čížek[1], I Procházka[1], Y Jirásková[2], D Janičkovič[3], W Anwand[4] and G Brauer[4]**

[1] Charles University in Prague, Faculty of Mathematics and Physics, V Holešovičkách 2, 18000 Praha 8, Czech Republic

[2] Institute of Physics of Materials, Academy of Science of the Czech Republic, Žižkova 513/22, Brno, 61662, Czech Republic

[3] Institute of Physics, Slovak Academy of Science, Dúbravská cesta 9, 84511 Bratislava, Slovak Republic

[4] Institut für Strahlenphysik, Helmholtz-Zentrum Dresden-Rossendorf, PO Box 510119, D-01314 Dresden, Germany

E-mail: lukac@mbox.troja.mff.cuni.cz

**Abstract.** Hardness of Fe-Al alloys shows a non-trivial dependence on chemical composition and thermal treatment of samples and cannot be fully explained by consideration of intermetallic phases formed according to the equilibrium phase diagram of Fe-Al system. Hardening in Fe-Al alloys caused by quenched-in non-equilibrium vacancies was studied in this work. Non-equilibrium vacancies introduced into the alloys by quenching from 1000 °C were detected by means of positron annihilation spectroscopy. The concentration of quenched-in vacancies was found to strongly increase with increasing Al content from $\approx 10^{-5}$ at.$^{-1}$ in the alloy with $c_{Al} = 18$ at. % up to $\approx 10^{-1}$ at.$^{-1}$ in the alloy with $c_{Al} = 45$ at. %. Comparison of the vacancy concentration and the Vickers microhardness revealed that hardening is proportional to square root of concentration of quenched-in vacancies.

## 1. Introduction

Iron aluminides are well-known for their excellent mechanical properties, high temperature corrosion resistance and their low density when comparing to industrial steels. The Fe-rich part of the equilibrium phase diagram of Fe-Al system [1] exhibits intermetallic phase regions of B2, $D0_3$ phases and disordered A2 phase region. The mechanical properties of Fe-Al alloys change with Al composition and also after various heat treatments. Hardening caused by quenched-in non-equilibrium vacancies was proposed by Chang et al. [2] stating that a vacancy-like point defect acts as an obstacle for dislocation movement.

Positron annihilation spectroscopy provides unique methods capable of determination of vacancy concentration, namely positron lifetime spectroscopy (LT), employing simple trapping model, and variable energy slow positron annihilation spectroscopy (VEPAS), using positron diffusion length obtained from backdiffusion of implanted positrons. Recently, it has been demonstrated that VEPAS enables precise determination of vacancy concentration even in samples containing very high density of defects, where LT is not applicable due to saturated positron trapping [3]. In this work we employed VEPAS for determination of the concentration of quenched-in vacancies in Fe-Al alloys with various composition and examined the effect of vacancies on hardness.

## 2. Experimental

Fe-Al alloys with Al concentration $c_{Al}$ falling in the range from 18 to 45 at.% were prepared by arc melting from high purity Fe (99.99%) and Al (99.99%) in Ti-gettered Ar atmosphere. Specimens were sealed in evacuated quartz ampoules and quench to the room temperature water after the one hour annealing at 1000 °C. The samples were investigated in the as-quenched state and after subsequent annealing at 520 °C for 1h.

A well annealed pure $\alpha$-Fe (99.99%) specimen was used as a reference sample. LT characterization of this sample resulted in a single component spectrum with lifetime of (107 $\pm$ 1) ps testifying that virtually all positrons annihilate in the free state and the sample can be considered as a defect-free material. Detailed description of LT setup is described elsewhere [3].

VEPAS measurements were performed using magnetically guided energy variable positron beam "SPONSOR" [4] with slow positrons of energy from 30 eV to 35 keV. Energy spectra of annihilation $\gamma$ rays were measured by HPGe detector having an efficiency of $\approx$ 30 % and an energy resolution of (1.06 $\pm$ 0.01) keV (FWHM at 511 keV). The Doppler broadening of annihilation profile was evaluated using the line-shape $S$-parameter. All $S$-parameters shown in this work were normalized to the bulk $S$-parameter $S_0 = 0.50784(5)$ measured in the well-annealed $\alpha$-Fe reference sample at the positron energy of 35 keV. The dependence of the $S$-parameter on the positron energy E (so-called $S(E)$ curve) for each alloy was analysed using the VEFPIT code [5].

The Vickers microhardness (HV) was measured by STRUERS Duramin-2 micro-tester. The load of 100 g was applied for 10 s.

## 3. Results and discussion

LT characterization of quenched alloys [3] revealed a strong contribution of positrons trapped at quenched-in vacancies. The concentration of vacancies increases with increasing Al content. In alloys with $c_{Al} > 25$ at.% the concentration of vacancies $c_V$ is so high that it exceeds the saturated trapping limit ($c_{V,max,LT} \approx 2 \times 10^{-4}$ at.$^{-1}$). Therefore, vacancy concentration in alloys with $c_{Al} > 25$ at.% cannot be determined from LT measurements. However, VEPAS backdiffusion measurements enable to determine the concentration of vacancies also in the alloys with $c_{Al} > 25$ at.%. The the dependence of the $S$-parameter on the positron energy measured in the quenched alloys and the alloys subsequently annealed at 520 °C are plotted in Fig. 1(a) and 1(b), respectively. Smooth lines in Fig. 1 represent model curves calculated by VEPFIT software [5] assuming two layered model consisting of (i) a thin oxide surface layer and (ii) a bulk Fe-Al layer. The thickness of the oxide layer obtained from fitting falls into the range of 5-40 nm. This surface layer exhibits high density of defects which is testified by a positron diffusion length of $\sim$ 6 nm.

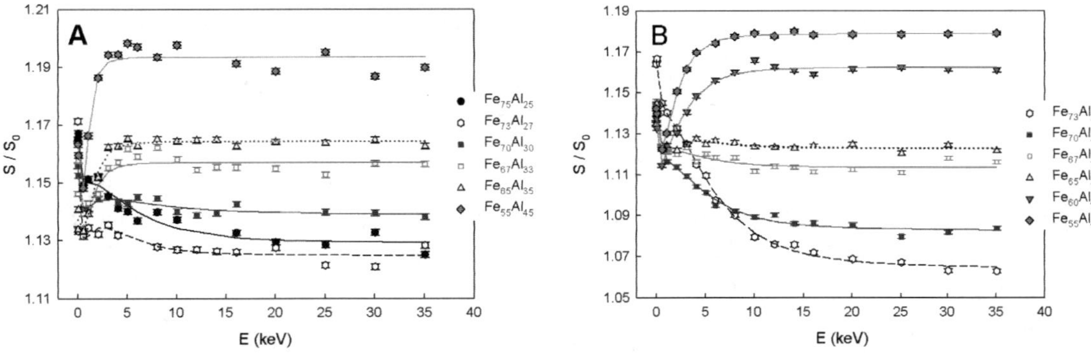


**Figure 1.** VEPAS results: $S(E)$ curves of various Fe-Al alloys in (a) the as-quenched state and (b) after annealing at 520 °C for 1 h.

The S-parameter and the positron diffusion length $L_+$ for Fe-Al bulk obtained from fitting of $S(E)$ curves are plotted in Fig. 2 for both quenched and annealed alloys. Assuming that Fe-Al alloys contain only a single type of positron traps, which was confirmed by LT spectroscopy [3], the vacancy concentration can be calculated from the formula [6]:

$$c_V = \frac{1}{\nu_V \tau_B}\left(\frac{L_{+,B}^2}{L_+^2} - 1\right), \tag{1}$$

where $\nu_V = (4 \times 10^{14})\ \mathrm{s}^{-1}$ is the specific positron trapping rate for Fe-vacancy [7], $\tau_B$ is the free positron lifetime in a defect-free alloys calculated theoretically in [3] and $L_{+,B}$ is the mean positron diffusion length in a defect-free alloy. This quantity can be calculated from equation

$$L_{+,B} = \sqrt{D_+ \tau_B}, \tag{2}$$

where $D_+$ is the room temperature positron diffusion coefficient estimated as a weighted average of positron diffusion coefficients for pure Fe and Al, i.e. $D_+ = (1 - c_{Al})D_{+,Fe} + c_{Al}D_{Al}$ . The positron diffusion coefficient for aluminium $D_{+,Al} = (1.7 \pm 0.2)\ \mathrm{cm^2s^{-1}}$ was determined in [8], for iron the value $D_{+,Fe} = (1.87 \pm 0.05)\ \mathrm{cm^2s^{-1}}$ was determined using Eq. (2) from measured positron diffusion length $L_{+,Fe} = (142 \pm 2)$ nm of well-annealed $\alpha$-Fe. Bulk positron diffusion lengths $L_{+,B}$ for Fe-Al alloys estimated by Eq. 2 fall into the interval of 141-145 nm.

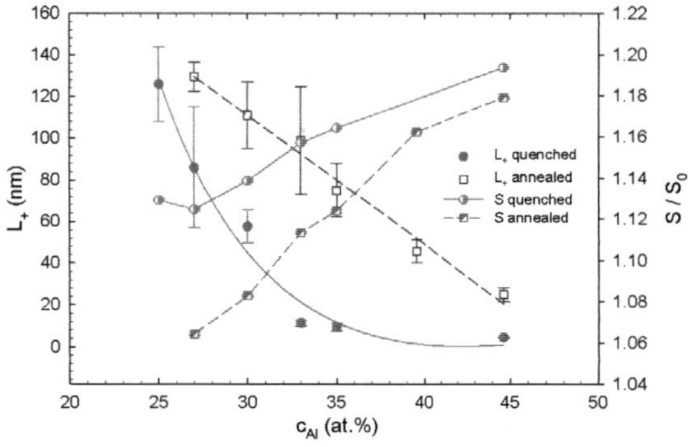


**Figure 2.** Positron parameters obtained from fitting of $S(E)$ curves for Fe-Al alloys: positron diffusion length $L_+$ and S-parameter.

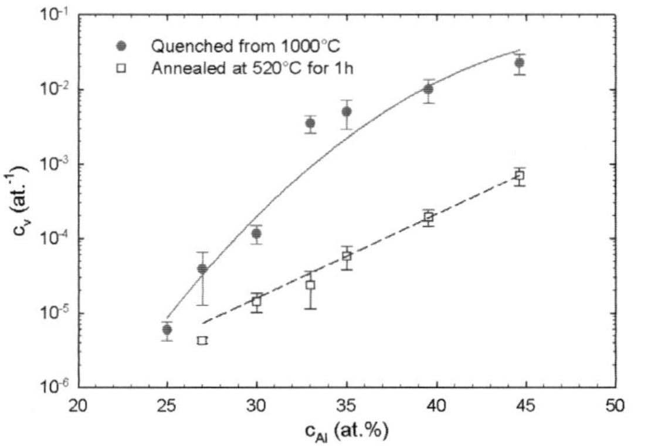


**Figure 3.** The vacancy concentration calculated from VEPAS results by Eq. 1.

The vacancy concentration calculated using Eq. 1 is plotted in Fig. 3 for the alloys quenched from 1000 °C and the alloys subsequently annealed at 520 °C. One can see in the figure that the concentration of quenched-in vacancies substantially increases with increasing Al content. The annealing at 520 °C leads to a significant drop of $c_V$ indicating annihilation of vacancies.

Fig. 4 shows HV of Fe-Al alloys in the as-quenched state and after subsequent annealing at 520 °C. In alloys with $c_{Al} > 27$ at.%, the annealing at 520 °C leads to a significant drop of HV due to annihilation of vacancies. On the other hand, in alloys with $c_{Al} < 27$ at.%, the concentration of vacancies is too small to cause a detectable increase of hardness. Hence, hardness of Fe-Al alloys is a superposition of two contributions $\mathrm{HV} = \mathrm{HV}_V(c_V) + \mathrm{HV}_c(c_{Al})$, where $\mathrm{HV}_V$ expresses hardening caused by vacancies while $\mathrm{HV}_c$ is the part of hardness influenced by composition. The former contribution is negligible in alloys with $c_{Al} < 27$ at.%, while in alloys with $c_{Al} > 27$ at.% both contributions are important. Since quenched-in vacancies are point obstacles pinning dislocations, $\mathrm{HV}_V = 6\gamma\mu\sqrt{c_V}$ [2], where $\mu \sim 100$ GPa is the shear modulus of FeAl [9] and $\gamma$ is a coefficient less than unity expressing the strength of the interaction of vacancies with a

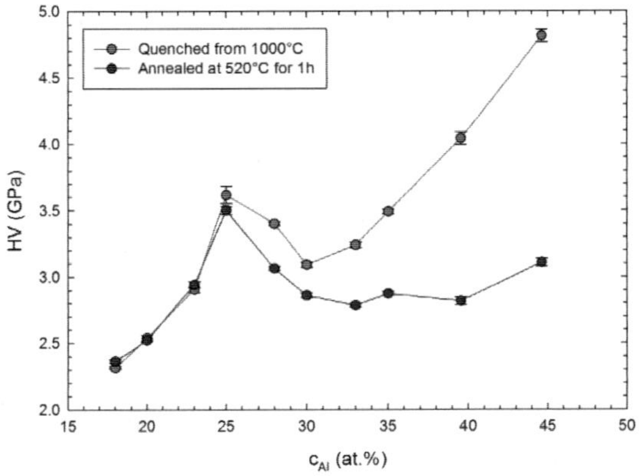


**Figure 4.** Microhardness of Fe-Al alloys with the Al content ranging from 18 to 45 at. %.

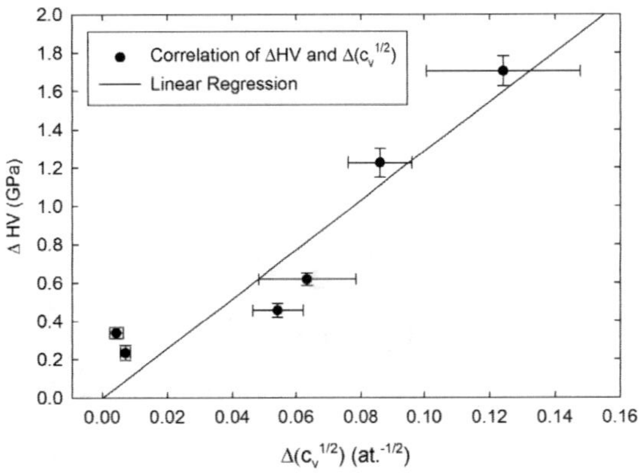


**Figure 5.** Difference of microhardness Δ HV between as-quenched and annealed alloy plotted as a function of the difference of $\sqrt{c_V}$.

dislocation. In order to eliminate the composition part $HV_c$ we calculate the difference of HV between the as-quenched and the annealed sample $\Delta$ HV $\sim \sqrt{c_{V,q}} - \sqrt{c_{V,a}}$, where $c_{V,q}$ and $c_{V,a}$ denote the concentration of vacancies in the as-quenched and the annealed state, respectively. Fig. 5 shows that $\Delta$ HV is indeed proportional to $\Delta\sqrt{c_V}$ taking into account experimental uncertainties. From the slope of this dependence we obtained $\gamma = 0.020 \pm 0.005$, which is in order of magnitude comparable with that obtained by Chang et al. [2] .

## 4. Conclusions

The influence of quenched-in vacancies on hardness of Fe-Al alloys was investigated in the present work. It was found that the concentration of quenched-in vacancies increases with increasing Al content. Subsequent annealing at 520 °C causes a remarkable decrease of vacancy concentration. In alloys with Al content higher than 27 at.% the concentration of vacancies becomes so high that causes a significant hardening. The hardness increment in these alloys was found to be proportional to the square root of the vacancy concentration.

## Acknowledgments

This work was supported by the Czech Science Foundation (Project No. P108/11/1350), Grant Agency of Charles University (Project No. 348111) and by the grant SVV-2012-265303, by the Ministry of Education, Youths and Sports of the Czech Republic (project MEB101102) and by the Deutscher Akademischer Austauschdienst DAAD (project 50755628).

## References

[1] Kubaschewski O 1982 *Iron-Binary Phase Diagrams* (Berlin: Springer-Verlag) p 5
[2] Chang Y A, Pike L M, Liu C T, Bilbrey A R and Stone D S 1993 *Intermetallics* **1** 107
[3] Čížek J, Lukáč F, Procházka I, Kužel R, Jirásková Y, Janičkovič D, Anwand W and Brauer G 2012 *Physica B: Condensed Matter* **407** 2659
[4] Anwand W, Kissener H R and Brauer G 1995 *Acta Phys. Polon. A* **88** 7
[5] Veen A, Schut H, Clement M, de Nijs J, Kruseman A and Ijpma M R 1995 *Applied Surface Science* **85** 216
[6] Veen A, Schut H, de Vries J, Hakvoort R A and Ijpma M R 1990 *AIP Conf. Proc.* **218** 83
[7] Schaefer H E, Damson B, Weller M, Arzt E and George E P 1997 *Physica status solidi (a)* **160** 531
[8] Soininen E, Huomo H, Huttunen P A, Mäkinen J, Vehanen A and Hautojärvi P 1990 *Phys. Rev. B* **41** 6227
[9] Wolfenden A and Marmouche M 1983 *J. Metals* **35** 90

Journal of Physics: Conference Series **443** (2013) 012026 doi:10.1088/1742-6596/443/1/012026

# Positron annihilation studies of zirconia doped with metal cations of different valence

**I Prochazka[1,4], J Cizek[1], O Melikhova[1], T E Konstantinova[2], I A Danilenko[2], I A Yashchishyn[2], W Anwand[3] and G Brauer[3]**

[1] Charles University in Prague, Faculty of Mathematics and Physics, Department of Low Temperature Physics, V Holesovickach 2, CZ 180 00 Praha 8, Czech Republic
[2] Donetsk Institute for Physics and Engineering named after O.O. Galkin, National Academy of Sciences of Ukraine, Luxemburg Street 72, 83114 Donetsk, Ukraine
[3] Institut für Strahlenphysik, Helmholtz-Zentrum Dresden-Rossendorf, POB 51 01 19, D-01314 Dresden, Germany
[4] Corresponding author. E-mail: ivan.prochazka@mff.cuni.cz

**Abstract**. New results obtained by applying positron annihilation spectroscopy to the investigation of zirconia-based nanomaterials doped with metal cations of different valence are reported. The slow-positron implantation spectroscopy combined with Doppler broadening measurements was employed to study the sintering of pressure-compacted nanopowders of tetragonal yttria-stabilised zirconia (*t*-YSZ) and *t*-YSZ with chromia additive. Positronium (Ps) formation in *t*-YSZ was proven by detecting 3γ-annihilations of ortho-Ps and was found to gradually decrease with increasing sintering temperature. A subsurface layer with enhanced 3γ-annihilations, compared to the deeper regions, could be identified. Addition of chromia was found to inhibit Ps formation. In addition, first results of positron lifetime measurements on nanopowders of zirconia phase-stabilised with MgO and $CeO_2$ are presented.

## 1. Introduction

Zirconium dioxide (zirconia) is the basic constituent of a variety of industrial materials suited, particularly, as ceramics for operation at high temperatures. Stabilisation of high-temperature zirconia phases is, however, needed in order to avoid material failure when passing phase transformation temperatures. This is achieved by adding a proper amount of a suitable phase-stabiliser agent. Trivalent yttrium oxide (yttria) is traditionally used as stabilizer of $ZrO_2$ lattice referred to as yttria-stabilised zirconia (YSZ). If the yttria addition amounts 2÷3 mol.% the lattice becomes tetragonal (*t*-YSZ). Current trends in prospecting zirconia-based materials with improved properties are focused on the following directions: (i) Nanopowders are utilised as starting substances for the preparation of materials by pressure-compaction and sintering. (ii) Unlike traditional, but relatively expensive yttria ($Y_2O_3$), other metal oxides are examined as phase-stabilisers, e.g. magnesia (MgO) or ceria ($CeO_2$). (iii) Another metal oxide admixture, like e.g. chromia ($Cr_2O_3$) [1], is introduced together with the phase stabiliser into the $ZrO_2$ lattice, giving rise to ternary oxide systems.

Over the past few years, we have employed positron annihilation spectroscopy in systematic investigations of YSZ nanomaterials [2]. The present contribution communicates our latest results obtained by means of slow-positron implantation spectroscopy (SPIS) on sintering of *t*-YSZ nanopowders as well as the first conventional positron lifetime measurements on zirconia nanopowders with various stabilising agents.

## 2. Experimental

### 2.1. Samples

Pure $ZrO_2$ (with monoclinic structure) and *t*-YSZ nanopowders (with 3 mol.% $Y_2O_3$), prepared using co-precipitation technique described in [3], were investigated. The nanopowders were compacted under a uniaxial pressure of 500 MPa into disks of 15 mm diameter and 5 mm thickness. Beside the binary *t*-YSZ, the ternary nanopowders (*t*-YSZC) containing 0.3 and 0.7 mol.% $Cr_2O_3$ were studied, too. The specimens underwent a sequence of 1 hour annealing at temperatures $T_S = 600$, 700, 800 and 900 °C in air followed by a slow cooling to room temperature. Magnesia-, ceria- and chromia-stabilised zirconia compacted nanopowders (MgSZ, CeSZ and CrSZ, respectively, with 10, 12 and 3 mol.% of stabiliser), were prepared by a similar co-precipitation procedure.

### 2.2. Measurements and data analysis

SPIS experiments were conducted using a monoenergetic positron beam [4] at positron energies $E_+$ covering a region from 27 eV to 35 keV. Annihilation of implanted positrons was observed by a single HPGe $\gamma$-spectrometer. The shape of the Doppler broadened (DB) $2\gamma$ annihilation peak was characterised with ordinary $S$ and $W$ parameters. In addition, the relative fractions, $F$, of implanted positrons annihilated as positronium (Ps) were deduced from counts in a valley left to the annihilation peak in a way closely similar to that of paper [5]. The measurement on *t*-YSZC at $E_+$=35 keV (the mean positron penetration depth $\bar{z} \approx 2$ µm) could serve as the reference case with no Ps formation [2].

Conventional positron lifetime measurements were carried out at ambient temperature in air using the apparatus and data processing methods described in [2] and references cited therein.

## 3. Results and discussion

### 3.1. Compacted nanopowders

In table 1, observed positron lifetimes and intensities ($\tau_i$ , $I_i$ , $\sum I_i$=100) for compacted nanopowders are presented. The mean crystalite sizes $d$ estimated from X-ray diffraction are shown in the table as well. Since crystallite size is substantially smaller than the mean positron diffusion length, positrons annihilate at GBs from trapped states in two kinds of defects: (i) vacancy-like misfit defects situated along grain boundaries (GBs), lifetime $\tau_1$, and (ii) larger defects at intersections of GBs (triple points) and comparable in size with few-vacancy clusters, lifetime $\tau_2$. The lifetimes $\tau_1$ and $\tau_2$ in MgSZ and CeSZ appeared to be considerably greater than in the other samples. This may consist in different influence of Mg and Ce stabiliser cations on electronic structure of the zirconia host lattice. Since in compacted nanopowders positrons are annihilated predominantly at GBs relatively strong variations of positron lifetimes $\tau_1$ and $\tau_2$ for various stabilisers indicate segregation of stabiliser cations at GBs.

**Table 1.** Crystalite size $d$ determined by X-ray diffraction and positron lifetime data on zirconia-based compacted nanopowders. The ortho-Positronium (oPs) lifetime and the total Ps intensity including para-Positronium contribution are shown. Amounts of additives are given in mol.%. In parentheses, standard deviations are indicated in units of the last significant digit.

| Sample | $d$ [nm] | $\tau_1$ [ps] | $I_1$ [%] | $\tau_2$ [ps] | $I_2$ [%] | $\tau_{oPs}$ [ns] | $I_{Ps}$ [%] |
|---|---|---|---|---|---|---|---|
| pure $ZrO_2$ (monoclinic) | 23 | 187(3) | 42(2) | 377(4) | 48(2) | 31(2) | 9.6(4) |
| *t*-YSZ ($ZrO_2$+3%$Y_2O_3$) | 18 | 174(3) | 27(2) | 373(3) | 63(2) | 30(2) | 9.7(4) |
| MgSZ ($ZrO_2$+10%MgO) | 11 | 261(3) | 52(2) | 441(4) | 41(2) | 18.1(2) | 6.0(5) |
| CeSZ ($ZrO_2$+12%$CeO_2$) | 9.1 | 254(4) | 22(1) | 408(1) | 76(1) | 19.7(4) | 2.0(5) |
| CrSZ ($ZrO_2$+3%$Cr_2O_3$) | | 178(3) | 16(1) | 374(3) | 84(1) | | |
| *t*-YSZ + 0.3%$Cr_2O_3$ | 18 | 191(3) | 32(1) | 381(2) | 69(1) | | |
| *t*-YSZ + 0.7%$Cr_2O_3$ | 15 | 191(3) | 26(1) | 390(2) | 74(1) | | |

In addition, Ps formation in pores of a few nanometer in size (nanovoids) was observed in all compacted nanopowders except of those containing chromia. Pores with radius of ≈ 1.5 nm were observed in pure $ZrO_2$ and binary *t*-YSZ compacted nanopowders, while MgSZ and CeSZ contain smaller nanovoids with radius of ≈ 0.9 nm. This correlates reasonably with smaller crystallite size in MgSZ and CeSZ compacted nanopowders and appears to be in accordance with expectations of an aggregate model [6] for the cavities between primary nanoparticles. Ps formation is suppressed below the detection limit as a result of Cr cations segregated along grain boundaries whenever a small amount of $Cr_2O_3$ is introduced into the $ZrO_2$ lattice [1,2].

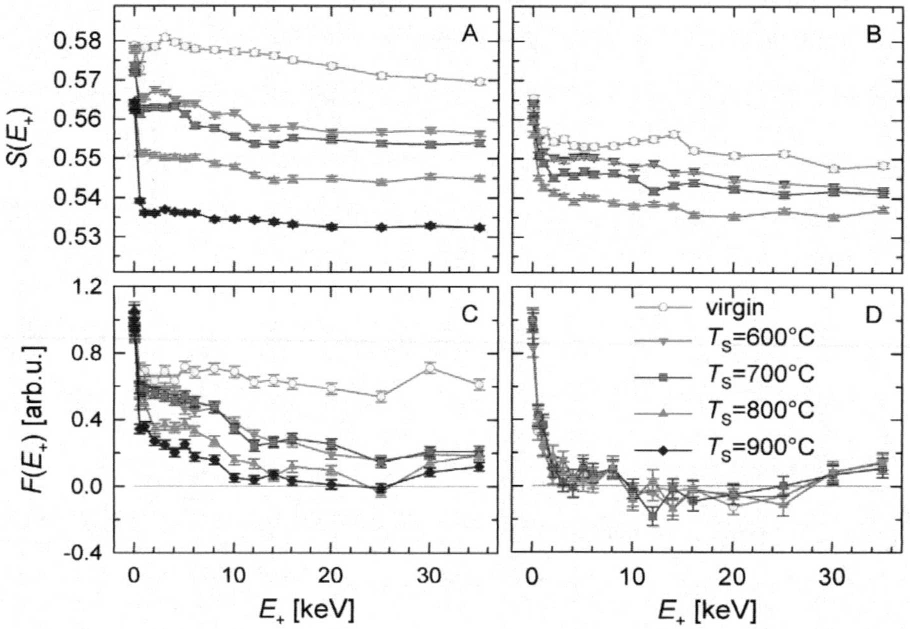


**Figure 1.** SPIS results (*S*- and *F*-parameters) for *t*-YSZ (left panels) and *t*-YSZC (right panels).

### *3.2. Sintering of t-YSZ and t-YSZC*

Binary *t*-YSZ, where Ps is formed, and ternary *t*-YSZC containing 0.7 mol.% $Cr_2O_3$, where Ps formaton is inhibited, were selected for investigation of the influence of sintering on defect structure. In figure 1, parameters *S* and *F*, respectively, measured in *t*-YSZ and *t*-YSZC nanomaterials are plotted versus the positron energy $E_+$. It can be immediately seen from the figure, that for energies $E_+ < 0.3$ keV ($\bar{z} < 2$ nm) a portion of positrons or Ps diffused back to the surface and annihilated there from a surface state. At positron energies $E_+ > 20$ keV ('bulk' region with $\bar{z} > 1$ μm), plateau-like dependences of both *S*- and *F*-values as functions of $E_+$ are observed. In the virgin *t*-YSZ nanopowder, Ps formation takes place in the whole positron energy range, i.e. at the surface and also in the bulk. This is reflected by *S* parameter in the bulk which is increased due to a sharp contribution of para-Positronium (cf. figures 1A,B) and *F* parameter in figure 1C which is substantially enhanced compared to ternary *t*-YSZC in figure 1D, where *F*-values consistent with zero were found above $E_+ \approx 0.5$ keV. An approximate absolute calibration of the *F*-parameter scale may be deduced from ortho-Ps lifetime measurements in *t*-YSZ bulk nanopowders [2] leading to the probability of ≈ 2 % for $3\gamma$ annihilation of Ps in bulk.

The plateau values of *S* exhibit a well-pronounced decrease with increasing sintering temperature both in the binary *t*-YSZ and the ternary *t*-YSZC, see figures 1A,B. This behaviour should be regarded as a consequence of the sintering-induced grain growth and disappearance of pores. The latter process seems to be dominating for the first annealing step ($T_S = 600$ °C) leading to a strong decrease of the *F* parameter in *t*-YSZ (see figure 1 C). In ternary *t*-YSZC, disappearance of pores takes place as well but since formation of Ps in this sample is inhibited it is not reflected by the *F* parameter. Above $T_S$ = 700 °C, the disappearance of open space defects at triple points dominates leading to a significant decrease of *S*-values, while changes of the *F* parameter are very small. The existence of an intermediate layer between the surface and the bulk region is, in addition, indicated by the data in figure 1. The layer is characterised by the *S*- and *F*- values which are remarkably lower than the respective surface parameters, but slightly enhanced compared to the bulk region. The upper boundary of the intermediate layer can be estimated to lie between ≈ 100 and 300 nm (respective positron energies of ≈ 6 to 10 keV). An origin of this layer may be tentatively assigned to sintering-induced moving of open-volume defects from the sample interior toward its surface.

## 4. Conclusions

Structural studies of zirconia-based nanomaterials stabilised by yttria, chromia, magnesia and ceria revealed three kinds of open volume defects: (i) vacancy-like defects along GBs, (ii) larger point defects with size comparable to a cluster of vacancies at intersections of GBs and (iii) pores between grains. Ps is formed in the pores in all samples except of those containing chromia, which inhibits Ps formation. Sintering of compacted nanopowders leads to disappearance of pores followed by grain growth.

## Acknowledgements

This work was financially supported by The Czech Science Foundation under project P108/11/1396, The Academy of Sciences of The Czech Republic (project KAN 300100801) and The Ministry of Education, Youth and Sports of The Czech Republic (project MEB 101102). Financial support obtained from The National Academy of Sciences of Ukraine (project 89/12-H) and German Academic Exchange Service (project 71 31 308 022) is acknowledged as well. One of authors (OM) would like to appreciate specific funding by Charles University in Prague within University Research Centre (programme UNCE “Physics of Condensed Matter and Functional Materials”).

## References

[1] Yashchishyn I A, Korduban A M, Trachevskiy V V, Konstantinova T E, Danilenko I A, Volkova G K and Nosolev I K 2010 *Functional Materials* **17** 306-10
[2] Prochazka I, Cizek J, Melikhova O, Kuriplach J, Anwand W, Brauer G, Konstantinova T E, Danilenko I A and Yashchishyn I A 2012 *Defects and Diffusion Forum* **331** 181-99
[3] Konstantinova T, Danilenko I, Glazunova V, Volkova G and Gorban O 2011 *J. Nanoparticle Research* **13** 4015-23
[4] Anwand W, Kissener H R and Brauer G 1995 *Acta Phys. Polonica* A **88** 7-11
[5] Wu Y C, Jiang J, Wang S J, Kallis A and Coleman P G 2011 *Phys. Rev.* B **84** 064123
[6] Ito K, Yagi Y, Hirano S, Miyayama M, Kudo T, Kishimoto A and Ujihira Y 1999 *J. Ceram. Soc. Japan* **107** 123-7

Journal of Physics: Conference Series **443** (2013) 012027 doi:10.1088/1742-6596/443/1/012027

# Effect of carbon on copper precipitation in deformed Fe-based alloys studied by positron annihilation spectroscopy

**S Zhang [1], H Schut [2], E Brück [1], S van der Zwaag [3] and N H van Dijk [1]**

[1] Fundamental Aspects of Materials and Energy, Faculty of Applied Sciences,

Delft University of Technology, Mekelweg 15, 2629JB Delft, The Netherlands

[2] Neutron and Positron Methods in Materials, Faculty of Applied Sciences,

Delft University of Technology, Mekelweg 15, 2629JB Delft, The Netherlands

[3] Novel Aerospace Materials group, Faculty of Aerospace Engineering,

Delft University of Technology, Kluyverweg 1, 2629 HS Delft, The Netherlands

Email: S.Zhang-1@tudelft.nl

**Abstract**. The role of deformation-induced defects and carbon addition on the copper precipitation during aging at 550 °C is investigated in high-purity Fe-Cu-B-N-C alloy samples by Coincidence Doppler Broadening. In samples with 0% and 8% cold pre-strain, the influence of tensile pre-deformation on the precipitation kinetics of copper is studied. The deformation-induced defects are found to enhance the Cu precipitation kinetics. A sharp reduction in open volume defects is accompanied with a strong increase of Cu signature during the initial stage of aging, implying that the open defects (mainly dislocations) act as nucleation sites for Cu precipitates. A comparison of the time evolution of *S-W* plots between Fe-Cu, Fe-Cu-B-N, and Fe-Cu-B-N-C alloys indicates that the addition of carbon does not alter the Cu precipitation mechanism but decelerates the kinetics.

## 1. Introduction

Recently, we studied the copper precipitation behavior in Fe-Cu and Fe-Cu-B-N alloys during aging by positron annihilation spectroscopy and neutron scattering to evaluate the potential contribution of Cu precipitation to self-healing of open volume defects created by prior plastic deformation [1,2]. Carbon, being the primary alloying element in steel, is widely used to raise the mechanical properties of the steel, such as tensile strength, hardness and resistance to wear and abrasion. For copper containing iron-based alloys, it is therefore of importance to clarify the influence of carbon on the Cu precipitation behavior in iron-based alloys. In earlier studies the carbon content was found to affect the activation energy of Cu diffusion in steel [3]. Positron annihilation has successfully been utilized to study the interaction of irradiation-induced vacancies and carbon in the Fe matrix [4,5]. Carbon was demonstrated to act as trapping site for vacancies and to form carbon-vacancy pairs, which favor the formation of precipitates. The effect of carbon on the deformation-induced Cu precipitation, however, has attracted limited attention. It is therefore desirable to investigate the influence of carbon on the copper precipitation behavior, which is assumed to lead to self-healing in Fe-Cu and Fe-Cu-B-N alloys under higher temperature loading conditions. In the present paper, we investigate the influence of carbon on the copper precipitation during thermal aging in deformed and un-deformed Fe-Cu-B-N-C alloys using positron annihilation spectroscopy.

## 2. Experimental

The composition of the Fe-Cu-B-N-C alloy produced by Goodfellow Ltd is in Table 1. The Ce was added to prevent S segregation at grain boundaries. Dog-bone shaped samples with a thickness of 0.5 mm were machined by spark erosion from the rolled sheet material. The samples were solution treated at 800 °C for 3 h in evacuated silica tubes filled with 200 mbar ultrahigh purity argon, and subsequently quenched into water at room temperature. Some samples were tensile pre-deformed at room temperature using a 2 kN microtensile tester (Deben). Vickers microhardness testing was carried out using a load of 500 g.

Table 1. The chemical composition of Fe-Cu-B-N-C alloy (in wt. %) with balance iron. The Ce concentration amounts to the nominal concentration.

| Cu | B | N | C | S | Ce |
|---|---|---|---|---|---|
| 1.00 | 0.050 | 0.019 | 0.087 | 0.002 | 0.015 |

The coincidence Doppler broadening (CDB) measurements were carried out by using a sandwich of two samples with a $^{22}$Na positron source in between. By measuring the Doppler shift in the energy of the two 511 keV annihilation $\gamma$-ray, one obtains the information on the momentum distribution of the electrons involved in the positron annihilation. In comparison with the conventional 1-detector measurement, the 2-detector technique results in a significant improvement in the peak-to-background ratio, with better energy resolution. The overall energy resolution was 1 keV at 511 keV, corresponding to a momentum resolution of $4\times10^{-3}$ $m_0c$ (full width at half maximum, FWHM).

## 3. Results and discussion

Figures 1 and 2 show the time evolution of the CDB ratio curves of the Fe-Cu-B-N-C alloy with 0% and 8% pre-strain, respectively. The momentum distribution is represented in relative terms as $(\rho - \rho_{Fe})/\rho_{Fe}$, where $\rho$ is the momentum spectrum for the positron annihilation in the samples and $\rho_{Fe}$ is the momentum spectrum measured on annealed (defect-free) pure iron. The CDB ratio curve for a pure bulk copper sample shows a broad peak at around $24\times10^{-3}$ $m_0c$, which is the characteristic feature for the 3*d* electrons of Cu. In the low momentum region ($< 5\times10^{-3}$ $m_0c$), the presence of open-volume defects as positron trapping sites is the dominant contribution to the variation in the momentum distribution [6].

As shown in figure 1, for the as-quenched Fe-Cu-B-N-C alloy, the Cu peak intensity continuously grows with aging time up to 1 h, followed by gradual decrease at longer aging times. Simultaneously, a continuous rise is observed in the low-momentum region, suggesting that the formation of new open-volume defects is associated with the formation of Cu precipitates. For the deformed Fe-Cu-B-N-C alloy (figure 2), a similar evolution of the CDB curves is observed. The broad peak in the low $p_L$ region before aging is attributed to the positron trapping at open-volume defects introduced by cold deformation (8% pre-strain). During the subsequent initial aging up to 1 h, the peak obviously decreases while the copper peak gradually increases, implying a simultaneous reduction of the open-volume defects with the formation of copper precipitates.

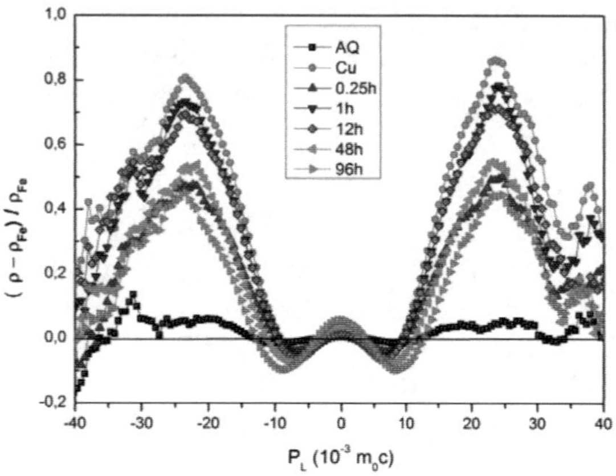


**Figure 1.** Coincidence Doppler Broadening relative to annealed iron $(\rho - \rho_{Fe})/\rho_{Fe}$, as a function of the electron momentum $p_L$ for the Fe-Cu-B-N-C alloy with 0% pre-strain.

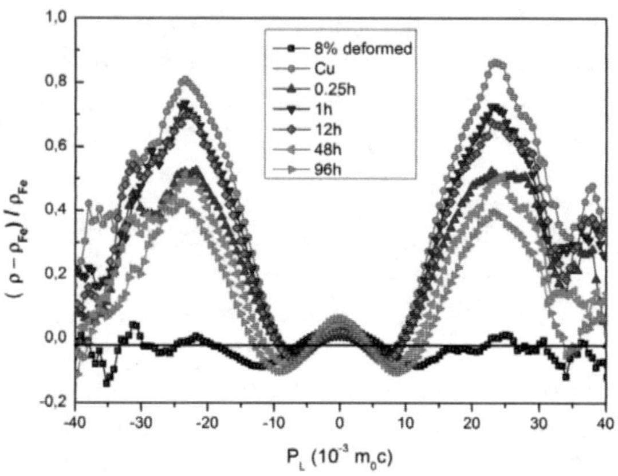


**Figure 2.** Coincidence Doppler Broadening relative to annealed iron $(\rho - \rho_{Fe})/\rho_{Fe}$, as a function of the electron momentum $p_L$ for the Fe-Cu-B-N-C alloy with 8% pre-strain.

Figure 3 shows the evolution of the *S*-*W* points for the as-quenched and the 8% deformed sample for ageing at 550 °C and times up to 96 h. The parameters *S* and *W* are defined as the ratio of low-momentum ($|p_L| < 3.1\times10^{-3}\ m_0c$) and high-momentum ($9.2\times10^{-3}\ m_0c < |p_L| < 24.3\times10^{-3}\ m_0c$) regions to the total region, respectively. All *S*-*W* points are normalized to those for annealed pure iron ($S_{Fe}$ = 0.461, $W_{Fe}$ = 0.120). The *S*-*W* points approach the value of the annealed pure Cu after an aging time of 2-4 h for the Fe-Cu-B-N-C alloy without deformation and after 1-2 h for the Fe-Cu-B-N-C alloy with 8% pre-strain. For longer aging times, a shift is observed towards a high *S* and a low *W* value, indicating that the interface between the copper precipitates and the matrix gradually loses coherency when the Cu precipitates grow during further aging [1,2]. For aging times from 4 to 96 h the evolution of the *S* and *W* parameters follows the same trajectory for the deformed and undeformed samples, where the undeformed sample lags behind with respect to the deformed sample. The faster evolution for the deformed sample indicates that the nucleation of the Cu precipitates is promoted by the open volume defects (mainly dislocations) introduced by the pre-strain. The subsequent coarsening shows a similar behavior for both alloys, while the overall precipitation kinetics is promoted by deformation.

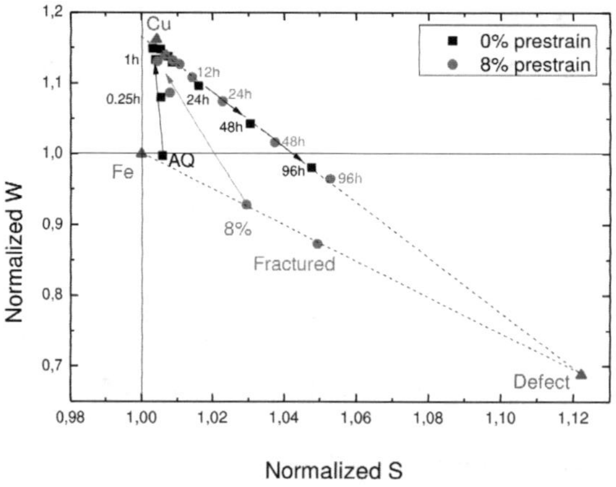


**Figure 3.** Time evolution of *S* and *W* parameters for the as-quenched Fe-Cu-B-N-C alloy with 0% and 8% pre-strain.

To clarify the role of carbon on the copper precipitation during aging, we compare the *S-W* plot of the deformed Fe-Cu-B-N-C alloy with that of the deformed Fe-Cu and the deformed Fe-Cu-B-N. It is interesting to note that the addition of carbon decelerates the Cu precipitation, which is opposite to the effect of boron and nitrogen. The *S-W* point of the deformed Fe-Cu-B-N-C alloy annealed for 1 h overlaps with that of the Fe-Cu alloy annealed for 1 h (figure 4), whereas the *S-W* point of the Fe-Cu-B-N alloy after 1 h annealing is quite close to that of the Fe-Cu-B-N-C alloy annealed for 12 h (figure 5). The shift of the *S-W* points for the Fe-Cu-B-N-C alloy is observed to be less than that of the Fe-Cu alloy and the Fe-Cu-B-N alloy for the aging time longer than 1 h. A possible explanation for the observation is that carbon reduces the solubility of B and N in the alloy via the precipitation of boron carbides. According to Kim and co-workers [7], in boron-containing low carbon steels, $Fe_{23}(B,C)_6$ particles containing N are found to precipitate at around 700 to 900 °C, which may reduce the levels of solute B and N in the matrix. Further study is required to clarify the structure of the carbides formed during the solution treatment and the following quench process.

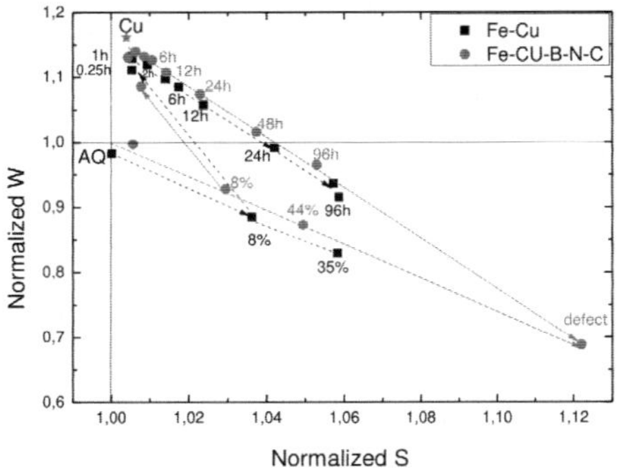


**Figure 4.** *S-W* plots of Fe-Cu and Fe-Cu-B-N-C alloys with 8% pre-strain during aging at 550 °C for 96 h.

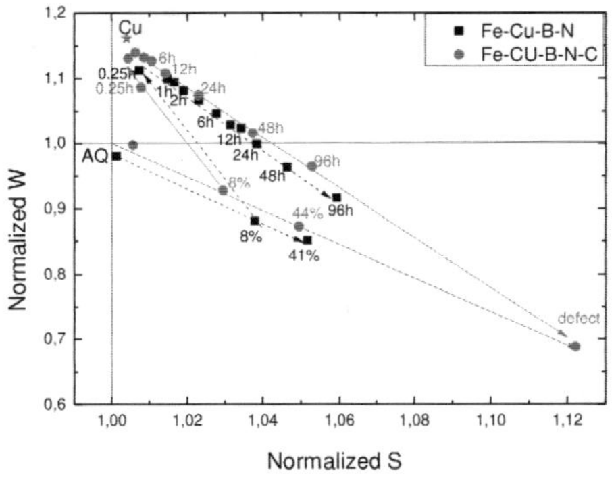


**Figure 5.** *S-W* plots of Fe-Cu-B-N and Fe-Cu-B-N-C alloys with 8% pre-strain during aging at 550 °C for 96 h.

## 3. Conclusions

Positron annihilation spectroscopy is a powerful technique to monitor the fine details of the Cu precipitation in a cold rolled Fe-Cu-B-N-C alloy during aging at 550 °C. During the initial aging stage, the deformed sample exhibits a sharp reduction in the defects accompanied with a strong copper signature, implying that the open defects act as nucleation sites for Cu precipitates and accelerate the precipitation. A comparison between the *S-W* plots for Fe-Cu, Fe-Cu-B-N, and Fe-Cu-B-N-C alloys indicates that the addition of carbon decelerates the kinetics of Cu precipitation but does not change the mechanism.

## References

[1] He S M, van Dijk N H, Schut H, Peekstok E R and van der Zwaag S 2010 *Phys. Rev.* B **81** 094103
[2] He S M, van Dijk N H, Paladugu M, Schut H, Kohlbrecher J, Tichelaar F D and van der Zwaag S 2010 *Phys. Rev.* B **82** 174111
[3] Rassoul E S 1997 *J. Mater. Sci.* **32** 6471
[4] Vehanen A, Hautojärvi P, Johansson J, Yli-Kauppila J and Moser P 1982 *Phys. Rev.* B **25** 762
[5] Hori F, Takenaka M, Kuramoto M and Aono Y 1993 *Scr. Metall. Mater.* **29** 243
[6] Nagai Y, Tang Z, Hassegawa M, Kanai T and Saneyasu M 2001 *Phys. Rev.* B **63** 134110
[7] Kim S K, Kim J S and Kim N J 2002 *Metall. Mater. Trans.* A **33** 701

Journal of Physics: Conference Series **443** (2013) 012028 doi:10.1088/1742-6596/443/1/012028

# Skyrmion-like Solitons and Topological Quasi-positronium in Graphite-potassium Intercalation Compounds

**M. Saito, T. Sasaki, and I. Kanazawa**

Department of Physics, Tokyo Gakugei University, Nukuikita-machi 4-1-1, koganei-shi, Tokyo, 184-8501, Japan

E-mail: f077121y@st.u-gakugei.ac.jp

**Abstract.** We have analyzed the narrow components of the angular correlation of positron annihilation radiation of $C_{24}K$, by using the topological charge density of the skyrmion-like soliton, and have estimated experimentally the configuration of the skyrmion-like soliton.

## 1. Introduction

Alkali-metal graphite intercalation compounds (AGICs) have been studied in understanding the electronic structures of these intercalations, which are nearly two-dimensional synthetic metals [1]. It is known that some of AGICs absorb a fairy amount of hydrogen in their intercalate layers and AGICs have catalytic activities for hydrogen layers of the compounds [2]. Hydrogens are physisorptively accommodated in the molecules in the interstices among the alkali-metal ions in intercalant layers of stage-2 compounds $C_{24}M$ (M=K, Rb, and Cs) at temperatures below 200 K [3]. Chemisorption of hydrogen takes place in both first and second stage compounds at higher temperature [4]. Enoki et al. [5,6] have studied the absorbing process of hydrogens in AGICs by ESR (Electron Spin Resonance). They have shown that in $C_8K$, $C_{24}K$, and $C_{24}Rb$, the hydride ions are stabilized after the dissociation of the hydrogen molecules into atoms and the subsequent charge transfer, and in $C_8Rb$ hydrogens are absorbed in the atomic form. In addition, hydrogen physisorption and chemisorption effects in AGICs have been studied by Doppler-broadening positron-annihilation radiation (DBPAR) [7-10]. The DBPAR spectral line-shape of $C_8K$ became sharp through hydrogen chemisorption at 300 K [7], while the spectral line-shape of $C_{24}Cs$ became broad through the physisorption of hydrogen molecules at 77 K [8,9]. The intensity of the narrow component of DBPAR spectrum of $C_8Rb$ was suppressed through hydrogen absorption at 300 K, while it decreased at first and then increased through the absorption at 373 K [10]. From the change in DBPAR spectral line-shape of $C_8Rb$, the hydrogen accommodated in $C_8Rb$ is considered to be atomic at 300 K and to be in the form of the hydride ion at 373 K. Cartier et al. [11-13] have reported very interesting results of the angular correlation of positron annihilation radiation (ACAR) for AGICs. That is, it is seen that a strong anisotropic and narrow component appears in the center of the ACAR of $C_8K$ and $C_{24}K$, after subtraction of a broad contribution which accounts for annihilations with the graphite $\pi$- and $\sigma$-electrons [14]. Features of electrons contributing to the narrow component in the ACAR might be in good agreement with the quasi-two dimensional band structure, which predicts an interlayer state with quasi-two dimensional free-electron character parallel to the carbon planes [15]. There exist some problems in the assumption that the narrow component of positron annihilation spectra corresponds to the interlayer state in AGICs. That is, the energy level of the interlayer state in the second-stage AGICs is above Fermi energy $E_F$. Furthermore, it looks like that the narrow component is not attributed to the simple positronium Ps in AGICs, because simple Ps cannot exist in the metallic state of AGICs with the screening effect. So far the origin of the narrow component in the positron annihilation spectra has not been confirmed. Recently the present authors [16] have analyzed the annihilation narrow components of the ACAR spectra [17] with the theoretical formula, which is extended from "topological quasi-positronium model" [18] and have estimated roughly a radius of the induced soliton-like density, by using a baby skyrmion model. In this study, we shall analyze the annihilation narrow components of the ACAR, by using the topological charge density of the skyrmion-like soliton, and have estimated experimentally the configuration of the skyrmion-like soliton.

## 2. Topological quasi-positroniums and the skyrmion-like soliton

The atomic density of potassium metal atoms intercalated between graphite layers in $C_{24}K$ is reduced to 2/3 of the density of the close-packed structure in $C_8K$. Posternak et al. [15] have shown that the existence of a new kind of electron state in graphite and in alkali graphite intercalation compounds is predicted with use of self-consistent local-density all-electron theory. These interlayer states, which exhibit quasi-two dimensional free-electron character parallel to the layers, form a band close to the Fermi energy and provide for alkali graphite intercalation compounds a new understanding of the origin of bands near the Fermi energy. Since the energy level of the interlayer state is close to the Fermi energy level, $SU(2)$ pseudospin structure can be introduced by assigning "up" and "down" pseudospins to the electron on the interlayer state level and the Fermi energy level, respectively. A skyrmion is a topological soliton spread over two energy levels. Now, we shall consider the quantum fluctuation in quasi $(2+1)$ electron state's system. A charged skyrmion-like soliton in quasi $(2+1)$ intercalate state in AGICs might be described as a topological soliton [18-22]. "Topological quasi-positronium" is regarded as the skyrmion-like soliton, which traps a positron, of the electron-density. The electron number associated with the skyrmion-like soliton can be derived from the induced current $J^{\mu}_{sky}$ on the $SU(2)$ pseudospin-like skyrmion configuration [20-22]. The electron number $N_{sky}$ is

$$N_{sky} = \int dxdy\, J^{\mu}_{sky} \, . \tag{1}$$

$$J^{\mu}_{sky}(r) = \frac{1}{\pi} \frac{4\xi^2}{[r^2 + 4\xi^2]^2} \, . \tag{2}$$

$$\text{where} \quad r^2 = x^2 + y^2. \tag{3}$$

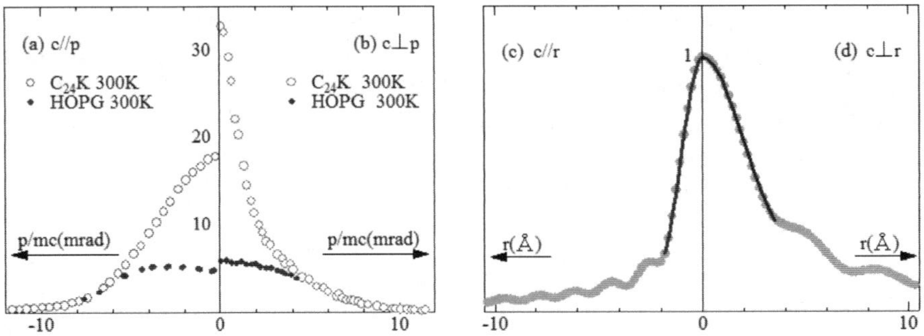


**Figure 1.** Figure 1 (a) and (b) show the annihilation spectrum for pair-momentum p parallel (perpendicular) to the crystallographic c-axis, p//c (p⊥c) in $C_{24}K$ [17].Figure 1 (c) and (d) show the autocorrelation function of the electron-wave function parallel to the c-axis and perpendicular to the c-axis , which is fourier-transformed from the narrow component for pair-momentum p parallel to the c-axis and perpendicular to the c-axis, respectively. The solid line in Figure 1 (d) shows the function $J^0_{sky}(r)$. The value of $\xi$ is estimated to be $\sim$ 2.1 Å. The solid line in Figure 1 (c) shows the Gaussian function. $\propto \exp(\frac{-r^2}{2\sigma^2})$. The value of $\sigma$ is estimated to be $\sim$ 1.1 Å.

In the case of the $SU(2)$ pseudospin-like skyrmion, the electron number $N_{sky}$ in eq(1) is estimated to be 1. The autocorrelation function, which is related to the electron distribution, parallel to the quasi-two dimensional electron layer in the interlayer states in $C_{24}K$ can represented by the $SU(2)$ pseudospin-like skyrmion configuration $J^0_{sky}(r)$ with $\xi \sim$ 2.1 Å.

## 3. Conclusion

We have analyzed the narrow components of the ACAR of $C_{24}K$ , by using the topological charge density of the skyrmion-like soliton. The topological charge density parallel to the graphite layer can be represented approximately by the $SU(2)$ pseudospin-like skyrmion configuration $J^0_{sky}(r)$ with $\xi$ $\sim$ 2.1 Å.

## References

[1] Dresselhaus M S, Dresselhaus G 1981 *Adv. Phys.* **30** 139
[2] Enoki T, Miyajima S, Sano M, Inokuchi H 1990 *J. Mater. Res.* **5** 435
[3] Watanabe K, Soma M, Onishi T, Tamaru K 1971 *Nature. Phys. Sci.* **233** 160
[4] Kondou T, Inokuchi M, Wakayama N 1965 *J. Chem. Phys.* **43** 3766
[5] Enoki T, Inokuchi H, Sano M 1982 *Chem. Phys. Lett.* **86** 285
[6] Enoki T, Sano M, Inokuchi H 1983 *J. Chem. Phys.* **78** 2017
[7] Murakami H, sano M, Kanazawa I, Enoki T, Kurihara T, Sakurai Y, Inokuchi H 1985 *J. Chem. Phys.* **82** 4728
[8] Kanazawa I, Sano M, Enoki T, Murakami H, Sakurai Y, Inokuchi H 1985 *Synth. Met.* **12** 225
[9] Sano M, Kanazawa I, Murakami H, Sakurai Y, Enoki T, Inokuchi H 1985 *Chem. Phys. Lett.* **122** 143
[10] Murakami H, Kanazawa I, Sano M, Enoki T, Inokuchi H 1988 *J. Phys. Chem. Solids* **49** 457
[11] Cartier E, Heinrich P, Pfluger P, Guntherodt H J 1981 *Phys. Rev. Lett.* **46** 272
[12] Cartier E, Heinrich P, Pfluger P, Guntherodt H J 1981 *Solid State Commun.* **38** 985
[13] Cartier E, Gunbler U M, Heinrich P, Geiser V, Pfluger P, Guntherodt H J 1983 *Synth. Met.* **8** 119
[14] Kanazawa I, Tanigawa S, Suzuki R, Sano M, Inokuchi H 1990 *Phys. Rev.* B **42** 11583
[15] Posternak M, Baldereschi A, Freeman A J, Wimmer E, Weinert M 1983 *Phys. Rev. Lett.* **50** 761
[16] Saito M, Yamada K, Kanazawa I 2013 *Mater. Sci. Forum* **733** 115
[17] Cartier E, Heinrich P, Geiser V, Guntherodt H J 1985 *Positron Annihilation* ed jain P C, Singra R M, Gopinathan K P (World Sci. Pub. Singapore) p 678
[18] Kanazawa I 1995 *Mater. Sci. Forum* **175-178** 695
[19] Polyakov A M 1997 *Nucl. Phys.* B **120** 429

[20] Wilczek F 1980 *Fractional Statistics and Anyon and Superconductor* (World. Sci. Pub. Singapore)
[21] Ezawa Z F 1997 *Phys. Rev.* B **55** 7771
[22] Manton N, Sutclitte P 2007 *Topological Solitons* (Cambridge Univ. Press. Cambridge)

Journal of Physics: Conference Series **443** (2013) 012029 doi:10.1088/1742-6596/443/1/012029

# Positron annihilation study of the Mg-Zn-Y alloys with long period stacking ordered (LPSO) structures

**H X Xu** [1,2,4], **K Inoue** [2], **Y Nagai** [2,5], **T Toyama** [2], **Y Matsukawa** [2], **A Kuramoto** [2], **D Egusa** [3], **E Abe** [3] **and B J Ye** [1,4]

[1] State Key Laboratory of Particle Detection and Electronics, University of Science and Technology of China, Hefei 230026, China
[2] The Oarai Center, Institute for Materials Research, Tohoku University, Oarai, Ibaraki 311-1313, Japan
[3] Department of Materials Engineering, University of Tokyo, Tokyo 113-8656, Japan
[4] Department of Modern Physics, University of Science and Technology of China, Hefei 230026, China

E-mail: nagai@imr.tohoku.ac.jp

**Abstract**. The Mg-Zn-Y alloys with long period stacking ordered (LPSO) structures have been studied by positron annihilation lifetime (PAL), coincidence Doppler broadening (CDB) and atom probe tomography (APT). The positron lifetime for all the Mg-Zn-Y alloys is in a range of 221~225 ps, very close to the positron lifetime for pure Mg bulk, 222 ps. Low temperature measurements of the positron lifetime also give no evidence for shallow positron trapping sites in the LPSO phases. The CDB shows that most of the positrons are annihilated with electrons of Mg. These results suggest that sub-nano scale open volumes, which were expected to exist in the Zn/Y enriched layers synchronized with stacking faults of the LPSO phases by the first principles calculations, are not present.

## 1. Introduction

Recently, long period stacking ordered (LPSO) structures formed in the Mg-Zn-Y alloys have attracted more attention, which are believed to play a critical role in improving the mechanical properties of the Mg-Zn-Y alloys [1-7]. The LPSO structures (four polytypes: 10H, 14H, 18R, 24R) are long period stacking variants of hcp-Mg by introducing the Zn/Y enriched layers synchronized with stacking faults periodically. The Zn/Y atoms are located on four close-packed layers, which have an ABCA stacking sequence and form a local face centered cubic (fcc) structure. Preceding first principles calculations [7] suggested that sub-nano open volumes with mono-vacancy size may be present in these Zn/Y enriched layers of the LPSO phases due to the large displacements of the constituent atoms. Thus it is necessary to confirm experimentally whether such open volumes exist or not in the LPSO phases, because these open volumes may affect the stability of the LPSO phases.

[5] To whom any correspondence should be addressed.

However, the local structure of the Zn/Y enriched layers of the LPSO phases is difficult to be observed even by using high resolution TEM and STEM.

Positron annihilation spectroscopy (PAS) is well known to detect open volumes in materials. Since the open volumes act as positron attractive centers with negative charge and low electron density, positrons can easily get trapped by the open volumes and show differences in annihilation characteristics, such as increased positron lifetime and annihilation ratio with electrons having low momentum. Positron annihilation lifetime (PAL) could provide us information on the open volumes. Coincidence Doppler broadening (CDB) could identify the chemical environment around the positron annihilation sites [8-11].

In order to confirm whether the open volumes as expected by the calculations exist or not in various types of the LPSO phases in the Mg-Zn-Y alloys, The PAL and CDB measurements were employed at a temperature range from 10 K to 300 K in the present study.

## 2. Experimental procedure

At present, it is very difficult to prepare the single crystal Mg-Zn-Y alloys, and the formation of the LPSO polytypes depends on the heat-treatments, thus the $Mg_{97}Zn_1Y_2$ and $Mg_{85}Zn_6Y_9$ alloys subjected to the present study are polycrystalline alloys going through various heat-treatments. Their master alloy ingots were prepared either by high-frequency induction melting or by conventional furnace melting. $Mg_{97}Zn_1Y_2$ ingots were solution-treated at 823 K for 8 h, followed by quenching into water and annealed at 473 K for 120 h, 573 K for 120 h and 673 K for 48 h, respectively, in order to obtain various types of the LPSO phases. The $Mg_{85}Zn_6Y_9$ ingot was extruded and annealed at 673 K for 72 h to obtain near single 18R-LPSO phase for giving a direct comparison. Details of the preparation procedure are described elsewhere [4,7]. The LPSO phases in each sample have already been confirmed by TEM [4] or STEM [7] and the results are listed in Table 1.

Atom probe tomography (APT), a powerful method for elemental mapping of materials in three dimensional real space with nearly atomic-scale resolution [12], was employed on the $Mg_{85}Zn_6Y_9$ in order to explore whether the chemical order of the 18R-LPSO phase is consistent with the structural model proposed by STEM [7].

The PAL measurements for all the specimens were performed by using a digital oscilloscope. The $^{22}$Na positron source was sandwiched between two pieces of identical pellets. The time resolution is about 190 ps in full width at half maximum (FWHM) and each spectrum contains more than $2\times10^6$ counts.

The CDB measurements for all the specimens were performed by using two HP-Ge detectors. The overall energy resolution is 1.0 keV in FWHM, which corresponds to a momentum resolution of $3.91\times10^{-3}$ $m_0$c, where $m_0$ is the electron or positron rest mass and c is the speed of light. The difference in energies of two γ-rays is $cp_L$, where $p_L$ is the longitudinal component of the positron-electron momentum along the direction of the γ–ray emission [10-11]. Each spectrum contains more than $3\times10^7$ accumulated events. The CDB ratio curves were obtained by normalizing the momentum distribution of each spectrum to pure Mg. The peak to background ratio is more than $10^5$, which enables us to observe positron annihilation with core electrons [8-11].

## 3. Results and discussion

### *3.1 Atom probe study of 18 R in the $Mg_{85}Zn_6Y_9$*

Atom map of Zn and Y in the 18R LPSO structure in the $Mg_{85}Zn_6Y_9$ was shown in Figure 1. The Zn/Y enriched layers of the 18R in the $Mg_{85}Zn_6Y_9$ alloy with an interval distance of ~1.6 nm are observed clearly, which is consistent with the 18R-LPSO structural model [6] proposed basing on STEM observation. The

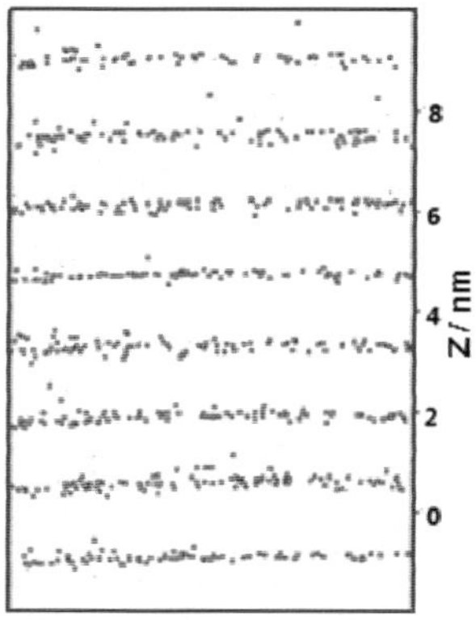


**Figure1.** Atom map of the Zn/Y enriched layers of the 18R in the $Mg_{85}Zn_6Y_9$. Pink and green dots represent Y and Zn atoms, respectively.

ratio of Y to Zn within four closed packed Zn/Y enriched layers was 1.33 from AP, in good agreement with that value for $Zn_6Y_8$ clusters 1.33 proposed by the first principles calculations [7].

*3.2 Positron annihilation lifetime of the Mg-Zn-Y alloys*

Table 1 shows the positron lifetime of the Mg-Zn-Y alloys measured at 300 K. After subtracting the contribution from source component ($\tau_s$=391 ps, $I_s$=14.8%) and background, all the positron lifetime spectra of the Mg-Zn-Y alloys comprised only a single lifetime component ($\tau_1$: 221~ 225 ps), very close to that for pure Mg bulk ($\tau_b$ ~ 222 ps). This result indicates positrons are not trapped by open volumes in the LPSO phases of the Mg-Zn-Y alloys.

**Table 1.** The positron lifetime of the Mg-Zn-Y alloys

| Sample | Mg | $Mg_{85}Zn_6Y_9$ extruded | $Mg_{97}Zn_1Y_2$ cast bulk | | | |
|---|---|---|---|---|---|---|
| Heat treatment | — | 673 K, 72 h | As quenched | 473 K, 120 h | 573 K, 120 h | 673 K, 48 h |
| Phase | α | 18R | α+18R+10H | — | α+18R | α+14H+18R |
| Positron Lifetime/ps | 221.5±0.2 | 222.8±0.2 | 221.0±0.3 | 222.4±0.2 | 225.5±0.2 | 222.4±0.3 |

If open volumes in the Zn/Y enriched layers of the LPSO phases are shallow trapping sites for positron, they could be detected by low temperature measurements, therefore the positron lifetime and CDB measurements were performed as a function of temperature (from 10 K to 300 K) for the $Mg_{85}Zn_6Y_9$ alloy. However, the positron lifetime of this alloy measured at 10 K, 100 K and 300 K was 224.2 ps, 228.0 and 222.8 ps, respectively. Nearly no shallow trapping effect is seen in this alloy, which suggests the LPSO phases in the Mg-Zn-Y alloy are free from the open volumes.

In addition to the PAL, the CDB measurements of the Mg-Zn-Y alloys were performed to study the chemical environment where the positrons are annihilated. Figure 2 (a) shows the CDB results of pure Y, pure Zn and the Mg-Zn-Y alloys performed at 300 K. Figure 2 (b) shows the CDB results of the $Mg_{85}Zn_6Y_9$ alloy performed at a temperature range from 10 K to 300 K.

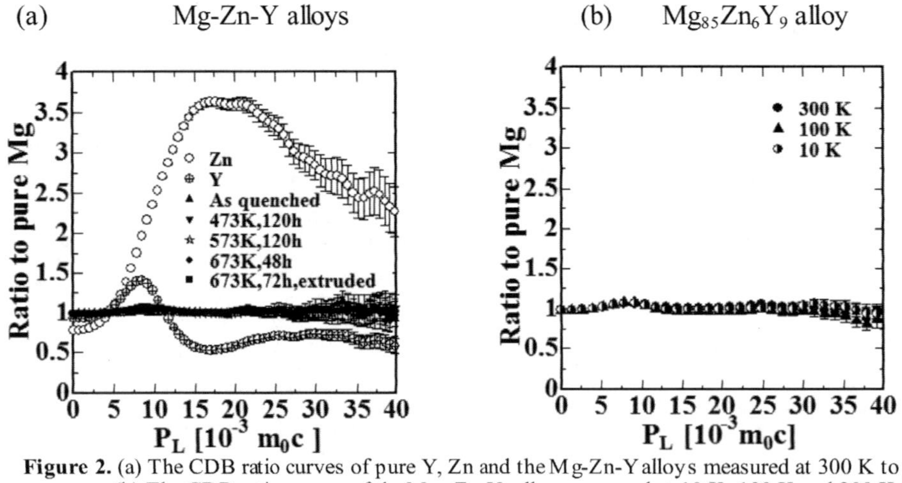


**Figure 2.** (a) The CDB ratio curves of pure Y, Zn and the Mg-Zn-Y alloys measured at 300 K to pure Mg, (b) The CDB ratio curves of the $Mg_{85}Zn_6Y_9$ alloy measured at 10 K, 100 K and 300 K to pure Mg.

As seen from Figure 2 (a), Zn showed a very high and wide annihilation peak located at 10~ 35×$10^{-3}$ $m_0$c, which was ascribed to the annihilation of positrons with 3d electrons of Zn. Y showed an annihilation peak and valley located at 10×$10^{-3}$ $m_0$c and 16×$10^{-3}$ $m_0$c, respectively. Although the Mg-Zn-Y alloys went through different heat-treatments and hot-extrusion process, their ratio curves obtained at 300 K exhibited common features: no difference either in shape or in height for each other over the whole momentum region, neither Zn nor Y annihilation peak/valley was found. From figure 2 (b), the ratio curves obtained at 10 K, 100 K and 300 K for the $Mg_{85}Zn_6Y_9$ alloy was quite similar to each other either. The CDB measurements show that most of positrons are annihilated with electrons of Mg and no other elements are detected by positrons in the Mg-Zn-Y alloys.

Positron annihilation results suggest that the LPSO phases are free from open volumes, in contrast to the prediction of the first principles calculations [7]. Very recently, another group [13] speculate that the clusters in the Mg-Zn-Y alloys become stable by introducing extra atoms to the sites where the open volumes were expected to form based on the first principles calculations, thus the open volumes should not exist by the occupation of extra atoms, consistent with the present positron annihilation result.

As very qualitative discussion, another possibility that the LPSO phases have open volumes but not detected by positrons is not completed excluded, because the positron affinity of Mg is ~1 eV lower than Zn and Y [14]. The positrons may preferably annihilate with electrons of Mg locating in intervals of the Zn/Y enriched layers, thus the open volumes within the Zn/Y enriched layers can not trap positrons. However, it is very difficult to estimate quantitatively, further study including the first principles calculations of the positron state in the LPSO phases is necessary to explain in the future.

## 4. Conclusions

The $Mg_{97}Zn_1Y_2$ cast bulk alloys and $Mg_{85}Zn_6Y_9$ extruded alloy with the LPSO phases were studied by positron annihilation spectroscopy. The Mg-Zn-Y alloys employed in this study went through different heat-treatments and hot-extrusion, and consequently various types of the LPSO phases were obtained. However, there are no differences between each Mg-Zn-Y alloy and pure Mg either in positron lifetime or in CDB curve, which suggest the LPSO phases in the Mg-Zn-Y alloys are free from open volumes.

**Acknowledgements**

This work is supported by Grant-in-Aid for Scientific Research on Innovative Areas (No. 23109002), by the China Scholarship Council (CSC) Foundation and National Natural Science of China (Grant No. 11175171 and 11105139).

**References**

[1] Kawamura Y, Hayashi K, Inoue A and Masumoto T 2001 *Mater. Trans.* **42** 1172
[2] Abe E, Kawamura Y, Hayashi K and Inoue A 2002 *Acta Mater.* **50** 3845
[3] Itoi T, Seimiya T, Kawamura Y and Hirohashi M 2004 *Scr. Mater.* **51** 107
[4] Ono A, Abe E, Itoi T, Hirohashi M, Yamasaki M and Kawamura Y 2008 *Mater. Trans.* **49** 990
[5] Zhu Y M, Morton A J and Nie J F 2010 *Acta Mater.* **58** 2936
[6] Abe E, Ono A, Itoi T, Yamasaki M and Kawamura Y 2011 *Philos. Mag. Lett.* **91** 690
[7] Egusa D and Abe E 2012 *Acta Mater.* **60** 166
[8] Lynn K G, MacDonald J R, Boie R A, Feldman L C, Gabbe J D, Robbins M F, Bonderup E and Golovchenko J 1977 *Phys. Rev. Lett.* **38** 241
[9] Asoka-Kumar P, Alatalo M, Ghosh V J, Kruseman A C, Nielsen B and Lynn K G 1996 *Phys. Rev. Lett.* **77** 2097
[10] Nagai Y, Hasegawa M, Tang Z, Hempel A, Yubuta K, Shimamura T, Kawazoe Y, Kawai A and Kano F 2000 *Phys. Rev. B.* **61** 6574
[11] Nagai Y, Tang Z, Hassegawa M, Kanai T and Saneyasu M 2001 *Phys. Rev. B.* **63** 134110
[12] Kelly T F, Miller M K 2007 *Rev. Sci. Instrum.* **78** 031101
[13] Ma S Y, Liu L M and Wang S Q 2013 *J. Mater. Sci.* **48** 1407
[14] Puska M J, Lanki P and Nieminen R M 1989 *J. Phys.: Cond.* Mat. **1** 6081

Journal of Physics: Conference Series **443** (2013) 012030 doi:10.1088/1742-6596/443/1/012030

# Monovacancies and Na substitutional defects in PbTe: theoretical positron annihilation study

**J D Liu[a, b], B Cheng[a, b], L J Zhang[a, b], J Zhang[a, b], H M Weng[a, b] and B J Ye[a, b, 1]**

[a]State Key Laboratory of Particle Detection and Electronics, University of Science and Technology of China, Hefei 230026 China

[b]Department of Modern Physics, University of Science and Technology of China, Jinzhai Road, Hefei, Anhui, 230026, P.R.China

E-mail: bjye@ustc.edu.cn

**Abstract.** The electron structures of cubic PbTe and substitutional defects are investigated using the state of the art pseudo-potential plane wave method, in the framework of the density functional theory within the generalized gradient approximation. The structural dependent positron annihilation in cubic PbTe was calculated and the positron lifetimes of the bulk, the Pb mono-vacancy, the Te mono-vacancy and the Na atom substituting on Pb were calculated respectively. The study provides evidence that positrons can distinguish between mono-vacancies and in situ substitutional Na doping. The positron lifetime values were found to be too similar to be distinguished. But calculations of the positron annihilation momentum distribution spectra did show clear differences between the two types of defect and may be exploited to characterize substitutional defects.

## 1. Introduction

Thermoelectric materials are solid-state energy converters whose combination of thermal, electrical, and semiconducting properties allows them to be used to convert waste heat into electricity or electrical power directly into cooling and heating [1]. To obtain optimized thermoelectric figure of merit (ZT) of a material, a large Seebeck coefficient, high electrical conductivity and low thermal conductivity are simultaneously required. Since the total thermal conductivity includes the two aspects of electronic and lattice-related phonon components, and point defects scatter phonons with a scattering cross section varying as the fourth power of frequency while other interaction processes vary more slowly with frequency, doping has been proposed as an approach to increase the electrical conductivity and decrease the thermal conductivity. PbTe-based compounds exhibit a great figure of merit for thermoelectric generators for the direct conversion of waste heat into electricity in the temperature range 400–800K, with a very low lattice thermal conductivity as well as a large Seebeck coefficient when appropriately p-type and n-type doped, and a good electrical conductivity [2]. In order to optimal control the doping effect on the thermoelectric properties, it is necessary to study in detail the influence of the substitution associated defects behavior in such thermoelectric materials.

[1] To whom any correspondence should be addressed.

Measurements of positron annihilation constitute the reliable experimental tools for studying the electronic structures of condensed media. The annihilation characteristics depend on the type of defect in which the positron is trapped. Such measurements, such as positron lifetime method, could yield information regarding the electron density in the region of electron-positron overlap. It is well known that the experimental measures of total annihilation rate only contain the average electron-positron overlap in the crystal and could not yield information about the details of the overlap. By contrast, the positron lifetime or of data related to the momentum content of the annihilating electron-positron pair in a specific environment could produced accurately basing on the theoretical and computational simulations and more detailed information could be available.

In the present paper we shall focus on the influence of the Na substitution associated defects on positron annihilation in rocksalt structure PbTe, recent experiments have shown that doping Na can result in significant increases to power factor [3]. The electron structures of rocksalt structure PbTe and substitutional defects are investigated using the state of the art pseudo-potential plane wave method, in the framework of the density functional theory within the generalized gradient approximation. And the structural dependent positron lifetimes of the bulk, the mono-vacancy of Pb and Te, and of the Na atom substituting on Pb site were also studied respectively.

## 2. Method

One of the fundamentally important inputs for a clear understanding of the positron annihilation characteristics is the positron state. For the electronic structure calculations we employed the pseudopotential plane wave method (PP-PW), in the framework of the density functional theory within the generalized-gradient approximation (GGA). PbTe adopts the rocksalt structure with 64 atoms in a supercell. The valence electron configurations were $5d^{10}5s^25p^2$, $5s^25p^4$ and $3s^1$ for Pb, Te and Na, respectively. A plane-wave energy cutoff of 400eV and a 4×4×4 grid of Monkhorst-Pack k points were chosen in this study to ensure well convergence of the calculated structure. Then, the calculations for localized positron states at defects in rocksalt structure PbTe are achieved basing on the "conventional" scheme in which the positron does not affect the average electron density around the defects. Once the positron and electron densities are known, the total positron annihilation rate is obtained by

$$\begin{aligned} \lambda &= \pi r_0^2 c\int dr\cdot n_p(r)\cdot n_e(r)\cdot\gamma(n_e) \\ &= \int dr\cdot n_p(r)\cdot\Gamma(n_e) \end{aligned} \tag{1}$$

with $r_0$ denotes the classical electron radius, c denotes the speed of light, $\Gamma(n_e)$ denotes the positron annihilation rate in a homogenous electron gas with density ne and the so-called enhancement factor $\gamma(n_e)$ is a manifestation of electron-positron interaction and it is always a crucial ingredient when calculating the positron lifetime.

The enhancement factor is always a crucial ingredient in the context of calculating the correct positron lifetimes in solids. Local-density parameterizations in which the enhancement is parameterized as a function of the unscreened electron density, $n_e$, at the positron, have been introduced to account for the inhomogeneity of real systems. In this study, the concrete form of enhancement factor parameterized by Boroński and Nieminen (BN model) was chosen which can give reliable results compared with the experiment than other interpolation formula [4]:

$$\Gamma(n_e)=\pi r_0^2 c n_e(1+1.23r_s+0.8295r_s^{1.5}-1.26r_s^2+0.3286r_s^{2.5}+(1-1/\varepsilon)\cdot(r_s^3/6)) \tag{2}$$

where ε is the high-frequency dielectric constant and $r_s$ is the electron density parameter.

## 3. Results and Discussion

The ground-state structural optimization was achieved firstly by a fit to the third-order Birch-Murnaghan equation of state (BMEOS) with the total energy as a function of the cell volume. The optimized values for the lattice parameters of PbTe are a=b=c=6.552Å, which are agreement with experimental results within the acceptable error and other theoretical calculations, as summarized in table 1.

**Table 1.** Theoretical and experimental lattice constants $a_0$, bulk modulus B and its pressure derivatives B' of rocksalt structure PbTe at T=0 and P=0.

| | This work | Other calculations | experiments |
|---|---|---|---|
| $a_0$(Å) | 6.552 | 6.565[a] 6.44[b] | 6.4611(3)[c] |
| B(MBar) | 0.401 | 0.44[a] 0.517[b] | 0.398[d] |
| B' | 4.53 | 3.352[a] 4.52[b] | |

[a] reference 5; [b] reference 6; [c] reference 7; [d] reference 8.

The decomposition of the positron lifetime spectrum is explained by the trapping model which gives the rate equations for the positron annihilating in delocalized states and in localized states. Possibly existing vacancy-type defects can trap positrons, but the results were widely different. The comparison of defect structures yields distinct differences in the electron and positron density distributions as demonstrated in figure 1. Despite the only major internal distinction between the defect-free and its substitution is the fact that the presence of a Na atom with lower atomic number, the positron density distribution difference could not be disregarded.

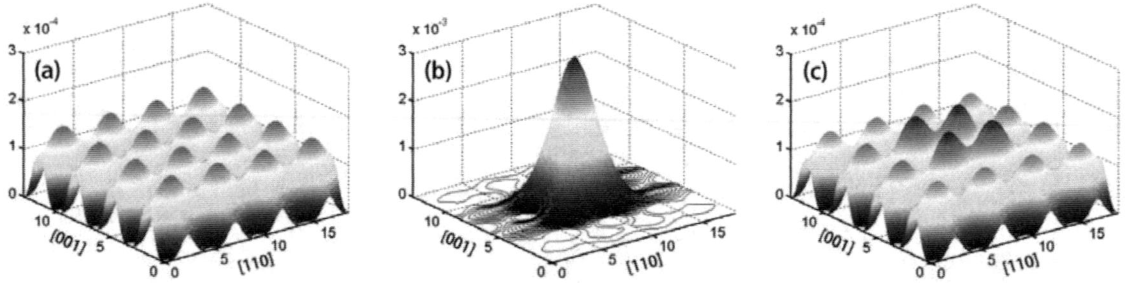


**Figure 1.** Positron density in PbTe [110] direction (Å): (a) defect-free; (b) with mono-vacancy $V_{Pb}$ and (c) Na atom substituting on Pb site.

As for the comparison of theoretical calculations with experimental data, the two lattice parameters are adopted respectively for positron annihilation calculations and the experimental positron lifetimes in rocksalt structure PbTe were also listed in table 2. We can find that the calculated results are in agreement with those previous calculations and experimental values within the allowable limits of error [9, 10]. The positron lifetime of 258 ps, which correspond to positron annihilation at Na atom substituting on Pb site, is slightly higher than that assigned to annihilation at defect-free PbTe. It illustrates that positrons can distinguish mono-vacancies but not the corresponding substituted defects. Furthermore, if the dislocations exist in crystalline PbTe, owing to the similar positron lifetime values, only using positron lifetime technique to identification the defect structures would cause disturbance. Recently, the positron annihilation lifetimes of He-ion-irradiated Fe samples were measured by a pulsed positron beam technique and meet the similar situation and difficulties [11].

**Table 2.** Calculated positron annihilation lifetime in bulk ($\tau_b$), in mono-vacancy Pb ($\tau_{Pb}$) and Te ($\tau_{Te}$) of PbTe, and at Na atom substituting on Pb site ($Na_{Pb}$) respectively

| a0(Å) | 6.552 | 6.462 | |
|---|---|---|---|
| $\tau_b$ (ps) | 252 | 244 | 244[a] |
| $\tau_{Pb}$ (ps) | 335 | 329 | 331[a] |
| $\tau_{Te}$(ps) | 320 | 314 | |
| $\tau_{NaPb}$(ps) | 258 | 246 | |

[a] corresponding experimental results within reference 10.

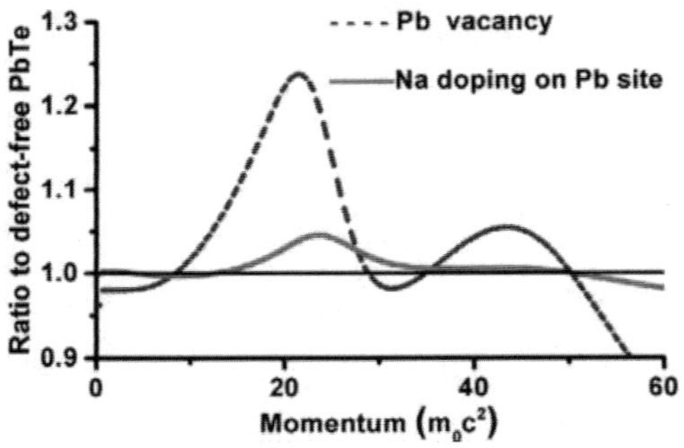


**Figure 2.** Calculated the ratio momentum spectrum of Na atom doping on Pb site in PbTe.

The reconstructed positron annihilation probability density spectra are shown in figure 2. Comparing with defect-free PbTe, the positron annihilation probability density for PbTe with $V_{Pb}$ is varying greatly. The calculated positron binding energy to the $Na_{Pb}$ substitutional defect is 0.011eV lower than it for the defect-free system. As can be seen in figure 1(c), the $Na_{Pb}$ defects can also trap positrons, which may be called shallow positron traps, and compete with vacancies at very low temperatures. The advent of Na atom substituting on Pb site plays an important role in the relative positron annihilation probability density change at low momentum. In contrast to the positron density distributions, we may draw conclusion that the calculated positron annihilation momentum distributions show the remarkable effect of the presence of the substitutional Na impurities.

## 4. Conclusion

The positron lifetimes and positron annihilation momentum distributions in defect-free PbTe and with Na substitution are calculated and the comparison of different defect structures yields distinct differences in the positron probability distributions. The results show that the relative lifetime difference between $Na_{Pb}$ substitutional structure and the defect-free PbTe is small. A comparison of positron annihilation momentum distributions spectra of the two structures demonstrates the strong affection with the existence of different positron traps. And the differences are mainly attributed to the existence of Na atom substituting on Pb site.

## Acknowledgements

This study was partially financially supported by National Natural Science Foundation of China (No. 11105139, 11175171) and the Fundamental Research Funds for the Central Universities.

## References

[1] Bell L E 2008 *Science* **321** 1457
[2] Androulakis J, Todorov I, Chung D, Ballikaya S, Wang G, Uher C and Kanatzidis M 2010 *Phys. Rev. B* **82** 115209
[3] Girard S N, He J, Zhou X, Shoemaker D, Jaworski C M, Uher C, Dravid V P, Heremans J P and Kanatzidis M G 2011 *J. Am. Chem. Soc.* **133** 16588
[4] Boroński E and Nieminen R M 1986 *Phys. Rev. B* **34** 3820
[5] Lach-hab M, Papaconstantopoulos D A and Mehl M J 2002 *J. Phys. Chem. Solids* **63** 833
[6] Wei S and Zunger A 1997 *Phys. Rev. B* **55** 13605
[7] Bouad N, Chapon L, Marin-Ayral R.–M, Bouree-Vigneron F and Tedenac J.-C 2003 *J. Solid State Chem.* **173** 189
[8] Leontyev V G, Ivanova L D, Bente K and Gremenok V F 2012 *Inorganic Materials* **48** 563
[9] Vinokurov A A, Zlomanov V P, Elsayed M and Krause-Rehberg R 2010 *Materials Letters* **64** 661
[10] Dlubek G, Krause R, Mühlberg M and Schenk M 1987 *phys. stat. sol. (a)* **102** K67
[11] Sato K, Kinomura A, Omura T, Xu Q, Yoshiie T, Kasada R, Kimura A and Morishita K 2011 *J. Phys.: Conf.* Ser. **262** 012053

Journal of Physics: Conference Series **443** (2013) 012031 doi:10.1088/1742-6596/443/1/012031

# Vacancy migration process in F82H and Fe-Cr binary alloy using positron annihilation lifetime measurement

**K Sato[1], Q Xu[1], D Hamaguchi[2], SS Huang[1] and T Yoshiie[1]**
[1]Research Reactor Institute, Kyoto University, Kumatori-cho, Sennan-gun, Osaka, 590-0494 Japan
[2]Japan Atomic Energy Agency, Rokkasho-mura, Kamikita-gun, Aomori, 039-3212 Japan

E-mail: ksato@rri.kyoto-u.ac.jp

**Abstract.** Microstructral evolution of electron-irradiated F82H and Fe-8%Cr at 77 K was studied using positron annihilation lifetime measurements. Irradiation-induced vacancies started to migrate at 300 K and 180 K in F82H and Fe-8%Cr, respectively. Solute Cr atoms did not suppress vacancy migration, but they made di-vacancies more stable. Microvoids were not formed by annealing. In F82H, solute atoms acted as trapping site of irradiation-induced defects and annihilation of vacancies and interstitials was facilitated. Pre-existing dislocations and precipitates were also their sinks. These lead to the suppression of microvoids formation. In Fe-8%Cr, small vacancy-type dislocation loops were formed by isochronal annealing test.

## 1. Introduction

Fusion reactor structural materials are irradiated with high dose of neutrons, and a large amount of defects is formed. The formation leads to void swelling and irradiation embrittlement, which are main problems in fusion reactor materials. Reduced activation ferritic/martensitic steel F82H is recognized as attractive candidate materials for neutron interactive structural components of fusion energy systems. Ferritic/martensitic steel has good dimensional stability under high irradiation doses and is suitable for commercial production without a large industrial investment. It is difficult to irradiate them with high dose of neutrons corresponding to actual fusion reactor at the present day. The effects of irradiation on the properties of materials at high doses must be derived from fission neutron irradiation experiments. Kiritani *et al.* reported the comparison of defect structure in many fcc metals irradiated with fission and fusion neutrons [1]. Sato *et al.* also reported the difference of microstructural evolution in vanadium alloy and ferritic/martensitic steel F82H irradiated with fission and fusion neutrons [2]. Lifetime prediction of fusion reactor structural materials is performed by computer simulation, and many physical properties, which are derived from fission neutron irradiation experiments, are necessary for the calculation. But even the vacancy migration energy has not been obtained exactly in F82H. The purpose of this study is to obtain the vacancy migration energy in F82H using positron annihilation lifetime measurement. Fe-Cr binary alloy is also used in order to investigate the effect of added elements except for Cr.

---

[1] To whom any correspondence should be addressed.

## 2. Experimental

F82H sheets were prepared by Japan Atomic Energy Agency (JAEA). Thickness of them was 0.2mm. They were punched to 5 mm in diameter and chemically polished to remove the oxidized area. Composition of F82H is Fe-7.65Cr-2W-0.16Mn-0.16V-0.02Ta-0.11Si-0.09C. Fe-8%Cr alloy was produced by plasma jet melting. It was cold rolled to a thickness of 0.1 mm and 5-mm-diameter discs were punched out. The discs were annealed at 1313 K for 0.5 and then tempered at 1013 K for 2h in vacuum. The electron irradiation was performed by using the electron linac in Research Reactor Institute, Kyoto University under the conditions of 30 MeV, $4\times10^{18}$electrons/cm$^2$, 77 K. Irradiation dose was $1.2\times10^{-3}$ dpa with the displacement threshold energy of 40 eV. Positron annihilation lifetimes were measured using a fast-fast time spectrometer with the $BaF_2$ scintillators and the Hamamatsu photomultiplier tubes H3378. The time resolution of the system was about 190 ps (full width at half maximum). Each spectrum was accumulated to a total of approximately $1.2\times10^6$ counts. Lifetime spectra were analyzed using the PALSfit program [3]. Isochronal annealing experiments were performed at intervals of 20 K and 50 K up to 600 K and above 600 K, respectively. Annealing time was 0.5 h. Measurement temperature of positron annihilation lifetime was liquid nitrogen temperature (77 K) and room temperature (300 K) up to 280 K and above 300 K, respectively.

## 3. Results and discussion

Figure 1 shows the positron annihilation lifetime of electron-irradiated F82H. Lifetime of matrix, single vacancies and di-vacancies in figure is that of pure Fe from reference 4. The long lifetime $\tau_2$ at annealing temperature of less than 280 K was less than lifetime of single vacancies (180 ps [4]). The long lifetime of unirradiated F82H is about 140 ps caused by pre-existing dislocations and precipitates. The long lifetime component contains that of not only single vacancies formed by electron irradiation but also pre-existing defects. After 300 K annealing, the long lifetime increased. The long lifetime of 184ps is shorter than that of di-vacancies (202 ps). A part of vacancies form into di-vacancies and other vacancies which are tightly trapped at solute atoms cannot migrate by annealing from 300 K to 400 K. Vacancy clusters gradually grew up to 420 K. At the same time, the long lifetime intensity (defect concentration) decreased. The reason of suppression of large void formation is that the migrating vacancies find interstitial type defects trapped at solute atoms or sinks (dislocations and interfaces of precipitates) and disappear. The long lifetime decreased to 160 ps by 440 K annealing. By annealing at above 460 K, the long lifetime was about 140 ps and positrons were mainly trapped at dislocation loops and precipitates, which were formed from irradiation-induced defects and existed before irradiation. The decrease of mean lifetime stopped at 600 K.

If all vacancies formed by electron irradiation remain, one vacancy in 833 atoms ($1 / 1.2\times10^{-3}$ dpa) exists. The radius of clusters consisting of 833 atoms is about 6 atomic distances. In order for vacancies to move up to 6 atomic distances in random motion, 36 jumps are needed. When the vacancy jumps is larger than 36 jumps, vacancies can aggregate. From the above results, single vacancies can migrate at 300 K in F82H, and vacancy clusters are formed. Vacancy jump frequency is expressed as $\nu_0\exp(-E_m/kT)$, where $\nu_0$ is the frequency of lattice vibration ($10^{13}$ /s), $E_m$ is the migration energy, $k$ is the Boltzmann constant, and $T$ is absolute temperature. When the vacancy migration energy is 0.87 eV, vacancy jumps exceed 36 jumps at 300 K for 0.5 h. But, because a part of vacancies disappear by interstitial migration during irradiation, the longer-distance migration of vacancies is required for the formation of vacancy clusters. Therefore, the exact vacancy migration energy is expected to be a little lower than 0.87 eV. In order to obtain it, isothermal annealing experiment should be performed.

Figure 2 shows the positron annihilation lifetime of electron-irradiated Fe-8%Cr. Lifetime spectra were not resolved into two components below 120 K. Because the lifetime of 177 ps is almost the same as single vacancy lifetime, almost all positrons were trapped at vacancies. Lifetime spectra could be resolved into two components above 140 K, and the long lifetime increased to about 200 ps by 180 K annealing. This means single vacancy migration and di-vacancies formation. The long lifetime continued to decrease from 260 K to 420 K. By 460 K annealing, it increased once and then started to decrease again. Mean lifetime continued to decrease up to 600 K.

In Fe-8%Cr used in this experiment, α' phase is inevitably formed. There exist nuclei of the phase because of the slow cooling rate after heat treatment. Point defect migration helps the growth of α' phase. Using the method employed to F82H, the vacancy migration energy in Fe-8%Cr was obtained to be 0.52 eV. This value is almost the same as that in pure Fe (0.55 eV [5]). Therefore, solute Cr atoms do not influence the vacancy migration. The decrease of long lifetime intensity from 180 K to 280 K is due to interstitial migration. As binding energy of di-vacancies is 0.2 eV [5], dissociation energy of di-vacancies is 0.75 eV. Di-vacancies dissociate at about 260 K. But di-vacancies remained above 260 K. Although solute Cr atoms did not interact with single vacancies, they made di-vacancies more stable until 300 K. The change in lifetime above 300 K can be explained by the growth of vacancy-type dislocation loops. At annealing temperature of less than 360 K, the long lifetime is longer than that of single vacancies (180 ps) and it is too long compared with the lifetime of vacancy-type dislocation loops. Thus, it is expected that small three-dimensionally-agglomerating vacancy clusters also exist, but we cannot resolve the long lifetime component into two components (three-dimensionally-agglomerated vacancy clusters and dislocation loops). When the vacancy-type dislocation loops grow, the lifetime decreases because the gap between two planes decreases. The decrease of long lifetime from 320 K to 420 K is due to this effect. At 440−500 K, the existence of single vacancies was difficult and three-dimensionally-agglomerating vacancy clusters were not detected by the positron annihilation lifetime measurements. In this temperature range, the short lifetime component $\tau_1$ agreed with that calculated by the two state trapping model, which assumed that positrons are annihilated at matrix and at one kind of defect sites. Therefore, the long lifetime of 160−175 ps at 440−500 K denotes that of small vacancy-type dislocation loops. If we assume that loops are not stable in α' phase at above 440 K, the decrease of long lifetime intensity, i.e., the

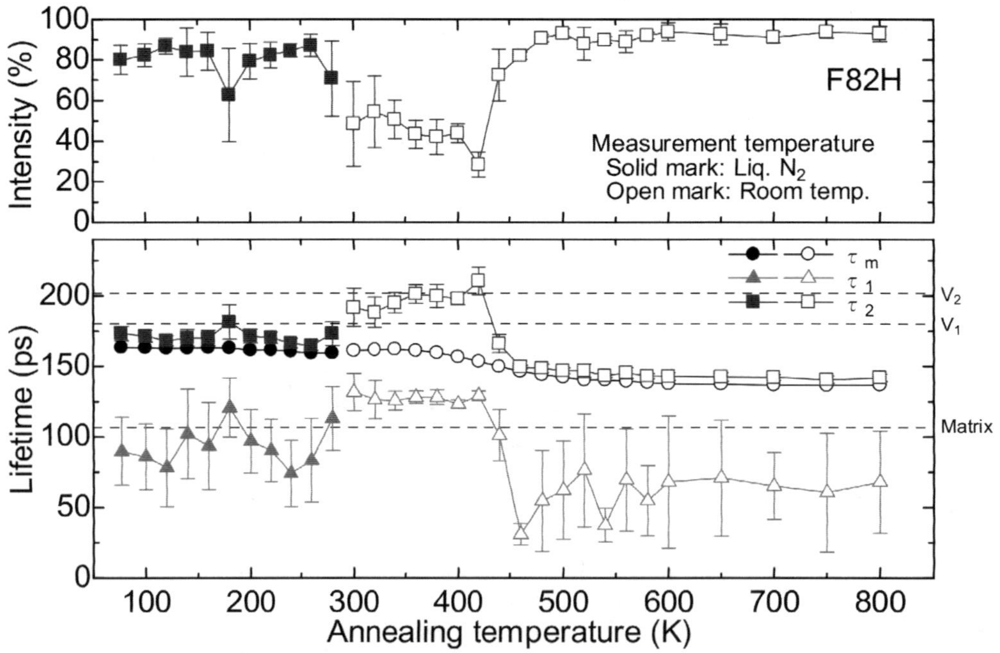


**Figure 1.** Isochronal annealing behavior of positron annihilation lifetime of electron-irradiated F82H. Annealing time was 0.5 h. Circles, triangles and squares denote the mean ($\tau_m$), short ($\tau_1$) and long ($\tau_2$) positron lifetimes, respectively. Solid and open marks denote measurement temperature of liquid nitrogen and room temperature, respectively. Dashed line of Matrix, $V_1$ and $V_2$ denote the calculated positron lifetimes of matrix (106 ps), single vacancies (180 ps) and di-vacancies (202 ps) [4], respectively.

Journal of Physics: Conference Series **443** (2013) 012031 doi:10.1088/1742-6596/443/1/012031

decrease of loop density, is possible to understand. Vacancies generated by evaporation of loops promote the nucleation of small loops and also the growth of α' phase. We have to investigate the phase growth in more detail. In F82H, the formation and growth of phase also should be studied.

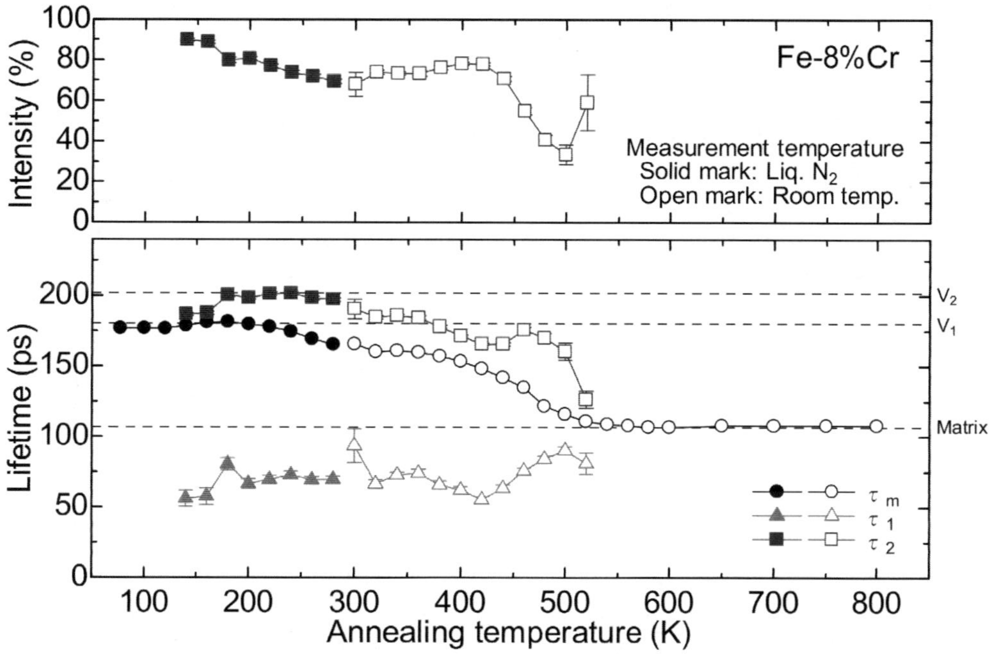


**Figure 2.** Isochronal annealing behavior of positron annihilation lifetime of electron-irradiated Fe-8%Cr. Annealing time was 0.5 h. Circles, triangles and squares denote the mean, short and long positron lifetimes, respectively. Solid and open marks denote measurement temperature of liquid nitrogen and room temperature, respectively. Dashed line of Matrix, $V_1$ and $V_2$ denote the calculated positron lifetimes of matrix (106 ps), single vacancies (180 ps) and di-vacancies (202 ps) [4], respectively.

## 4. Conclusion

Annealing behavior of positron annihilation lifetime of electron-irradiated F82H and Fe-8%Cr was examined. Vacancies started to migrate at 300 K and 180 K in F82H and Fe-8%Cr, respectively. Microvoids were not formed by annealing. In F82H, solute atoms act as trapping site of irradiation-induced defects, and pre-existing dislocations and precipitates also act as sinks. These lead to the suppression of microvoids formation. The effect of added elements except for Cr was quite powerful in F82H. In Fe-8%Cr, the change in long lifetime was related in the growth of vacancy-type dislocation loops.

## References

[1] Kiritani M, Yoshiie T, Kojima S and Satoh Y 1990 *Radiat. Eff. and Def. Sol.* **113** 75
[2] Sato K, Inoue K, Yoshiie T, Xu Q, Wakai E, Kutsukake C and Ochiai K 2009 *J. Nucl. Mater.* **386–388** 203
[3] Olsen J V, Kirkegaard P, Pedersen N J and Eldrup J 2007 *Phys. Stat. Sol.* C **4** 4004
[4] Ohkubo H, Tang Z, Nagai Y, Hasegawa M, Tawara T and Kiritani M 2003 *Mater. Sci. Eng.* A **350** 95
[5] Wollenberger H J 1996 Point Defects *Physical Metallurgy* ed W Cahn and P Haasen (Amsteldam: North-Holland) chapter 18 pp 1621–1721

# Defect investigations of an iron-nickel meteorite

**P Parz[1], M Leitner[1], W Sprengel[1], H Reingruber[2], and W Puff[1]**

[1]Graz University of Technology, Institute of Materials Physics, Petersgasse 16, 8010 Graz, AUSTRIA

[2]Institute for Electron Microscopy, Graz University of Technology, and Center for Electron Microscopy Graz, Steyrergasse 17, A-8010 Graz, Austria

E-mail: p.parz@tugraz.at

**Abstract.** A sample from the Gibeon meteorite, an iron-nickel meteorite with the typical two-phase Widmanstätten structure, was investigated by positron annihilation techniques. A high mean positron lifetime was observed which derives from open volume defects. The meteorite samples were then gradually heated in a high-vacuum furnace and cooled down to room temperature. Upon temperature treatment a decrease of the mean positron lifetime was observed, as well as a gradual evolution of the 2-dimensional Doppler broadening spectra towards pure iron. This leads to the conclusion that the open volume defects are formed during slow cooling in the progress of the formation of the Widmanstätten structure. Upon re-heating these defects start to dissolve and do not re-appear due to fast cooling rates. We therefore attribute the open volume defects to misfit dislocations between the Kamacite (Ni-poor) and Taenite (Ni-rich) phases.

## 1. Introduction

Iron-nickel meteorites are well known for their unique microstructure. Due to their very low cooling rates in space a coarse grained structure, the so-called Widmanstätten pattern is formed [1]. This low cooling rate precludes the reproduction of this unique pattern in laboratory. Therefore, the meteorite sample was expected to be in thermodynamic equilibrium and free of defects. The aim of this study was to verify this condition by positron annihilation.

## 2. Materials and Methods

The meteorite sample with a composition of Ni 7.93 wt.-%, Co 0.41 wt.-%, P 0.04 wt.-%, 91.62 wt.-% Fe and small amounts of carbon [1] was cut into pieces of approximately $8\times8\times3$ mm$^3$ with a low-speed diamond saw. Upon its journey through space the meteorite was subject to very low cooling rates, of approximately 35 K per one million years [2]. The unique cooling rates are the reason for the typical Widmanstätten structure, a two-phase microstructure of Fe-rich Kamacite (ferrite, α-iron) and Ni-rich Taenite (austenite, γ-iron). The meteorite samples were measured at room temperature and subsequent subject to thermal treatment.

Studies of positron lifetime and two-dimensional Doppler broadening (2d-DB) were performed in the same manner as described recently [3]. Summarizing briefly, a conventional fast-fast spectrometer setup with a time resolution of 230 ps full width at half maximum (FWHM) was used for positron lifetime measurements. From the measured positron lifetime spectra ($>5\times10^6$ counts) the mean positron lifetime is derived. Since the focus of this work is on 2d-DB, a detailed lifetime component analysis is not included. The 2d-DB measurements were performed in a coincidence setup of two high-purity Ge detectors with an energy resolution of the two detector system of 0.88 keV (FWHM) related to the 511 keV γ-line. A peak-to-background ratio of $\sim 8\times10^5$ was achieved, which ensured sensitivity to annihilation events with high momentum (core) electrons. Two-dimensional spectra with an energy width of ~1 keV per channel and $(6 \text{ to } 8)\times10^7$ counts were collected, from which the background-free Doppler spectra were obtained. All measurements were performed at room temperature. The 2d-DB spectra were presented and analyzed in the following two ways in order to cross check the reliability of the obtained results.

For a numerical analysis, the 2d-DB spectra were fitted by a weighted linear combination of the spectra measured on pure undeformed and deformed elements following Somoza et al. [4]:

$$S = f_{Fe} \times S_{Fe} + f_{Fe*} \times S_{Fe*} + f_{Ni*} \times S_{Ni*} + f_C \times S_C \qquad (1)$$

with $S_{Fe}$, $S_{Fe*}$, $S_{Ni*}$ and $S_C$ corresponding to the spectra measured on well annealed iron , on deformed iron, on deformed nickel and on pyrolythic carbon, respectively. The weighting factors ($f_x$) characterize the contribution of the pure elements to the fitted spectrum and, therefore, serve as chemical fingerprint. The fitting was performed in a momentum range of $(5 \text{ to } 35)\times10^{-3} m_0c$.

As second method, integral parameters, so-called Shape (S) and Wing (W) parameters are determined. These parameters give information about the changes of the defect concentration (S) and can be correlated to changes of the chemical environment (W). The S parameter is assigned to the area ratio of the centre spectrum $(-2.5 \text{ to } 2.5)\times10^{-3} m_0c$ and the total spectrum, whereas the W parameter is calculated by dividing the counts between electron momenta from $\pm(15 \text{ to } 35)\times10^{-3} m_0c$ by the total measured counts. These values are compared to the $S_0$ and $W_0$ value of the well annealed pure iron spectrum.

Studies on the microstructure were performed on samples, subjected to the same heat treatment, by optical microscopy as well as by scanning electron microscopy using an ESEM QUANTA 600 FEG.

## 3. Results and Discussion

### 3.1. Initial Condition

The meteorite sample exhibits the typical Widmanstätten structure in the cm range visible by the naked eye. Broad Ni-poor Kamacite bars are separated from fields of fine-grained Kamacite and Taenite (Plessite) by narrow bands of Ni-rich Taenite, as shown in figure 1 B. Positron lifetime measurements reveal a relatively high mean positron lifetime of 165ps. Compared to the positron lifetime in defect free iron (110 ps) [5] this shows that open volume defects are present in the meteorite sample. The lifetime of 165 ps is too short for vacancies. Typically vacancies in iron exhibit a positron lifetime of ~175 ps [5]. Due to the low cooling rate no remaining thermal vacancies are expected. This leads to the conclusion, that the positron traps in the meteorite sample are mainly dislocations.

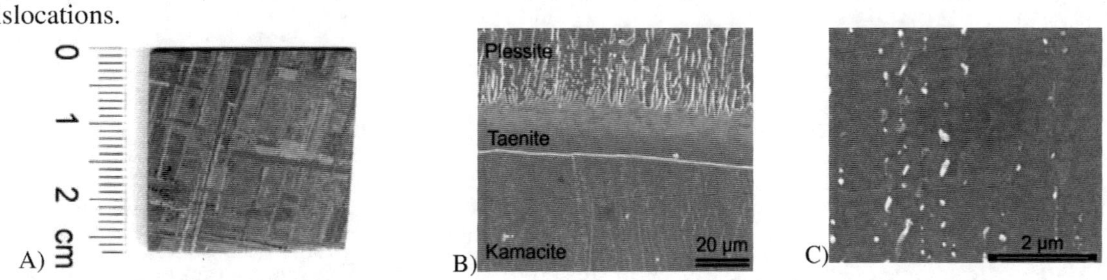


**Figure 1.** A) Optical micrograph of the etched meteorite surface, B) SEM micrograph and C) close up of B, showing precipitates within Kamacite bars.

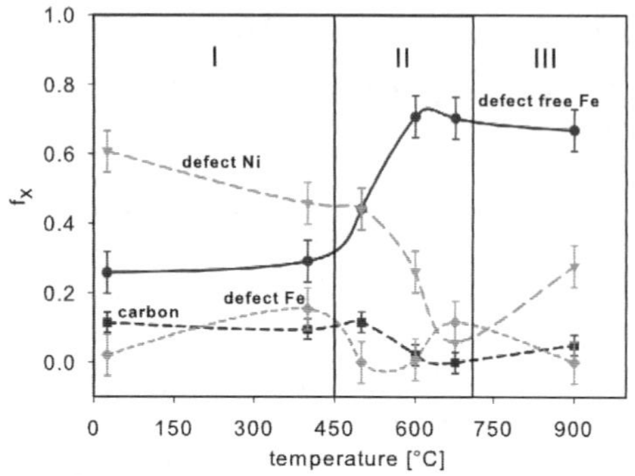


**Figure 2.** Evolution of the Doppler broadening fit parameter $f_x$ (according to eq. 1) as function of annealing temperature.

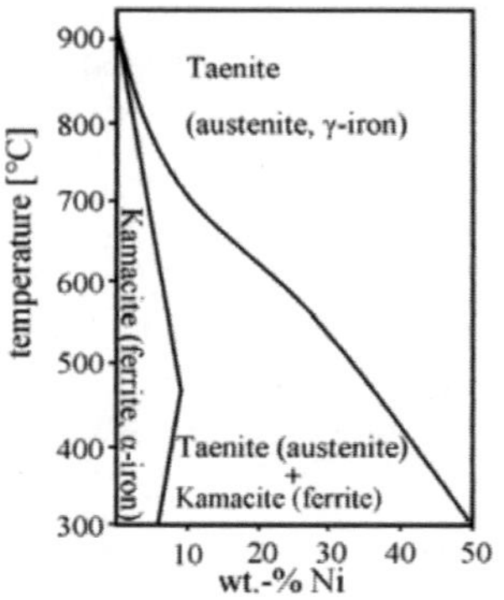


**Figure 3.** Fe-rich section of iron-nickel phase diagram, redrawn after [9].

With 2d-DB experiments and the numerical analysis according to eq. 1 a relatively large nickel signature is observed (Fig. 2). This indicates that misfit dislocations between the Ni-rich Taenite and Ni-poor Kamacite are present. Electron micrographs reveal that within the broad Ni-poor Kamacite bars there are small Ni-rich precipitates present (~200 nm). We, therefore, conclude that at the interface between matrix and precipitates misfit dislocations are present. In fact, the cubic base centered Kamacite (ferrite, α-iron) exhibits a lattice parameter of a = 0.286 nm and the cubic face centered, Ni-rich Taenite (austenite, γ-iron) exhibits a lattice parameter of a = 0.358 nm [7]. Furthermore, according to the phase diagram (Fig. 3) the nickel content of the Kamacite decreases after reaching a maximum at about 460°C and due to the low cooling rates and an extrapolated nickel diffusion coefficient (data from [8]) of $D{\sim}1{\times}10^{-23}$ $m^2s^{-1}$ at 400°C a diffusion of the nickel atoms of $\sim 3x10^{-2}$ μm/a can be estimated. Therefore, Ni-rich precipitates have to be formed during the slow cool down and in fact the misfit dislocations between the Taenite precipitates and the Kamacite matrix are found by positrons and are confirmed by electron microscopy (Fig. 1, C). In addition the numerical analysis of the 2d-DB spectra reveals a minor but detectable carbon signature. This shows that positrons are as well trapped at carbon precipitates within the meteorite sample.

### 3.2. Thermal Treatment

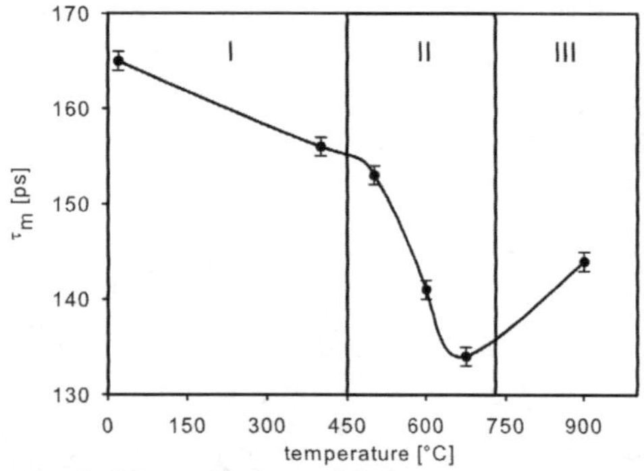


**Figure 4.** Mean positron lifetime $\tau_m$ as function of annealing temperature.

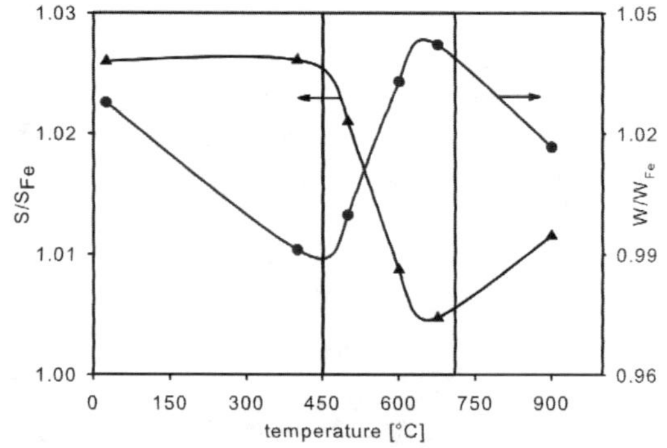


**Figure 5.** S- and W-Parameter as function of annealing temperature.

The meteorite samples were then subjected to subsequent isochronal thermal treatment for one hour at 400°C, 500°C, 600°C, 676°C and 900°C. The evolution of the mean positron lifetime is presented in figure 4. Upon annealing three distinct annealing stages can be observed. First, up to 400°C, a slow decrease of the mean positron lifetime can be distinguished, while the results of the Doppler broadening experiments show no decrease of the S parameter (Fig. 5) upon this temperature step. The W parameter (Fig. 5) on the other hand shows a major decrease. In addition with the numerical analysis of the 2d-DB spectra (Fig. 2) following conclusions can be drawn. Upon this temperature step the chemical environments of the open volume defects change. The nickel signature at defects decreases slightly, while the dislocations sustain. The nickel diffusion length in Fe at this temperature for one hour is approximately 0.2 nm, which is in the range of 1 lattice constant. This change of the chemical environment is as well attributed to lead to the slight decrease of the positron lifetime.

At higher temperatures, stage II, between 500°C and 676°C, the defects start to disappear, which is indicated by the strong decrease of the mean positron lifetime (Fig. 4) while at the same temperatures the S Parameter (Fig. 5) decreases rapidly as well. In addition the chemical sensitive 2d-DB spectra reveal a major increase of the pure iron signature, while the defect nickel signature decreases rapidly and the defect iron signature decreases as well (Fig. 2). The dissolution of the nickel rich precipitates at higher temperatures is confirmed by SEM micrographs. Starting at 600°C the carbon signature of the meteorite sample decreases and disappears at 676°C (Fig. 2). This incident leads to the conclusion, that carbon containing precipitates were dissolved at this temperature and the carbon now occupies interstitial sites and is not detectable by the positrons any more. The dissolution of carbon precipitates in a Fe-Ni meteorite at this temperature has in fact also been reported by Weller and Wegst [10], from internal friction experiments.

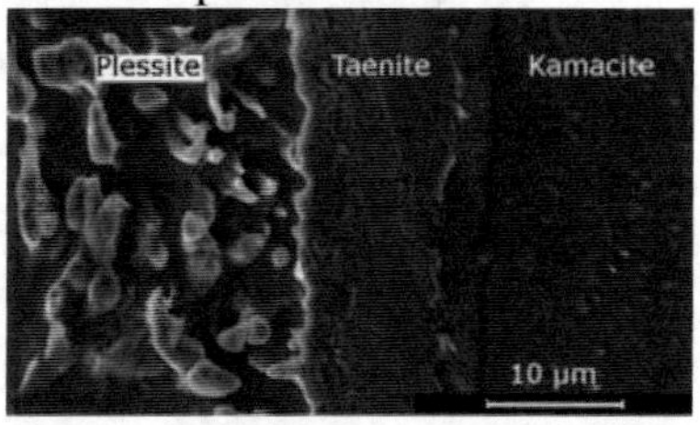


**Figure 6.** SEM micrograph of the Plessite, Taenite, Kamacite phase boundaries after annealing at 900°C.

Stage III, at temperatures as high as 900°C, shows a re-increase of the mean positron lifetime (Fig. 4), as well as the S parameter (Fig. 5) and indicates the formation of new open volume defects at this temperature. This and an increased contribution of the defected nickel signature of the fitting process (Fig. 2) indicate that vacancies are formed [11] and stabilised by impurity atoms. Due to the relatively short annealing times, the coarse Widmanstätten structure is still visible after annealing at 900°C. Furthermore no indications for precipitates within the Kamacite bars can be observed (Fig. 6).

## 4. Conclusion

It was shown, that the meteorite sample is in a stable thermodynamic equilibrium condition with respect to a) thermal vacancies and b) carbon precipitates. However, the structure of the meteorite formed upon the slow cooling exhibits the macroscopic Widmanstätten structure and c) small Ni-rich equilibrium precipitates due to the thermal decomposition of the super saturated Kamacite at low temperatures. These precipitates exhibit misfit dislocations which can be detected by positrons. The positron annihilation techniques give a comprehensive picture of the open volume defects in the meteorite sample.

## References

[1] He Y, Godet S, Jacques P J and Jonas J J, 2006, *Acta Mater.* **54**, 1323
[2] Moren A E and Goldstein J I, 1978, *Earth and Planetary Science Letters* **40**, 151
[3] Puff W, Rabitsch H, Wilde G, Dinda G P and Würschum R, 2007, *J. Appl. Phys.* **101**, 123512
[4] Somoza A, Petkov M P, Lynn K G and Dupasquier A, 2002, *Phys. Rev. B* **65**, 094107
[5] Vehanen A, Hautojärvi P, Johansson J, Yli-Kauppila J and Moser P, 1982, *Phys. Rev. B* **25**, 762
[6] Okrusch M and Matthes S, 2010, *Mineralogie: Eine Einführung in die spezielle Mineralogie*, Petrologie und Lagerstättenkunde (Springer 2010)
[7] Clarke R S Jr. and Scott E R D, 1980, *Am. Mineral.* **65**, 624–630
[8] Cermak J, Lübbehusen M and Mehrer M, 1989, *Z. Metallkde.* **880**, 213
[9] Swartzendruber L J, Itkin V P and Alcock C B, 1993, in: "*Phase Diagrams of Binary Iron Alloys*", Okamoto H (ed.), Materials Information Soc., Materials Park, Ohio
[10] Weller M and Wegst U G K, 2009, *Mat. Sci. And Eng. A* **41**, 521
[11] Shirai Y, Schaefer H E, Seeger A, 1989, in: *Positron Annihilation*, World Scientific Publ. Co., Singapore, p.419

Journal of Physics: Conference Series **443** (2013) 012033 doi:10.1088/1742-6596/443/1/012033

# Comprehensive defect characterization of different iron samples after severe plastic deformation

**Werner Puff, Xiang Zhou, Bernd Oberdorfer, Boris Scherwitzl, Peter Parz, Wolfgang Sprengel and Roland Würschum**

Institute of Materials Physics, Graz University of Technology, Petersgasse 16, A-8010 Graz, Austria

werner.puff@tugraz.at

**Abstract**. Bulk nanostructured materials produced by means of severe plastic deformation such as high-pressure torsion (HPT) show a high strength in combination with good ductility. These materials exhibit a microstructure down to grain sizes of approximately 100 nm. The defects and microstructure produced by HPT are severely affected by the presence of alloying components. Two iron disks with a purity of about 99.8% (ARMCO) and 99.98+% were deformed by HPT conditioning. Slices of the deformed material were then investigated and subjected to thermal treatment. Upon isochronal annealing at increasing temperatures the decrease of the deformation induced defects was monitored.

The defect structure and its annealing behavior are investigated by positron lifetime spectroscopy and Doppler broadening spectroscopy. In addition specimens were analyzed by high precision difference dilatometry, while the evolution of the microstructure is monitored by electron microscopy. In addition, in situ Doppler broadening measurements upon annealing were performed at the high intensity positron beam at FRM II.

The lifetime data in dependence of the annealing temperature are fitted to a model including positron trapping at grain boundaries and vacancy-type defects.

## 1. Introduction

Severe plastic deformation (SPD) is one of the promising methods to produce nanocrystalline bulk metals [1, 2]. The benefit of high pressure torsion (HPT) over other methods like, e.g., equal channel angular pressing (ECAP) is the higher strain leading to finer microstructures [3] down to grain sizes of approximately 100 nm [4]. As reported recently, the defects and microstructure produced by HPT are severely affected by the presence of alloying components [5, 6]. This opens up the question, how minor impurities can affect the defect and microstructure of pure metals.

A main feature of nanocrystalline materials is the high concentration of free volumes and their influence on the physical properties. There are a few methods which are specific to study the free volumes associated with lattice defects directly on an atomic level. One method is positron annihilation which has been applied to probe specifically the size of structural free volumes in a variety of nanocrystalline metals prepared by different methods like cluster-condensation or ball milling. Microstructures of pure metals processed by SPD have been widely investigated during the past two decades [7, 8].

Another method, high-resolution difference dilatometry is able to determine directly and specifically the absolute concentration of non-equilibrium excess volumes even in high concentrations by measuring the irreversible macroscopic length change upon annealing. Time-dependent dilatometry [9] supplements the positron annihilation technique, since it enables to study the kinetics of free volumes and vacancies.

## 2. Experimental

High-purity iron samples (99.98+%) and ARMCO low-purity iron samples (99.8%) were analyzed. One main difference besides others is the lower carbon concentration in pure Fe. Carbon impurities in iron play an important role regarding deformation and structural refinement behavior as well as defect stabilization [10]. The samples, 30 mm in diameter and 10 mm thick, were severely plastically deformed in a high-pressure torsion (HPT) experiment by 10 revolutions applying a pressure of 2.8 GPa. From these deformed disks specimens were cut. The high degree of deformation, given by a van Mises strain of $\varepsilon \sim 36$ was chosen in order to guarantee saturation deformation. According to literature in ARMCO-Fe saturation deformation is reached at an equivalent strain of $\varepsilon = 23$ [11].

For the investigation of the microstructure, scanning electron microscopy (SEM) imaging was performed. For this purpose samples with different annealing states were prepared, namely in the as received state (AR) and after linear heating with 3K/min to different temperatures.

One experimental technique in this paper is differential dilatometry employing a high precision dilatometer (Linseis, L75VD500LT) in vertical setup. This method allows measurements of the irreversible length change due to the annealing of free volumes. With this setup, absolute length changes $\Delta l$ in the order of 90 nm can be measured accurately, which, in combination with specimen length $l$ in the 10 mm range, gives direct access to absolute measurements of the excess volume concentrations of $3\Delta l/l = \Delta V/V$ down to $3\times10^{-5}$.

The positron annihilation experiments, lifetime (LT) as well as Doppler broadening (DB) measurements were performed at room temperature in the standard laboratory technique. For the LT measurements a conventional lifetime spectrometer with a time resolution of 190 ps (FWHM) was chosen. The lifetime spectra were analysed using the PFPOSFIT program [12] taking into account a source correction for the annihilation of positrons in the Al foil and the $^{22}$NaCl source. For a more detailed analysis of the defect characteristic upon annealing an in-situ measurement by means of the high-intensity positron beam at the NEPOMUC facility at the FRM II in Garching was performed [13].

## 3. Results and Discussion

For the investigation of the microstructure, scanning electron microscopy (SEM) imaging was performed. For this purpose samples with different annealing states were prepared. The SEM images were analysed with respect to grain size and grain size distribution. The mean grain sizes are 320 nm after HPT processing in pure iron and 150 nm in ARMCO-Fe. In Fig. 1 the micrographs of the samples after HPT deformation in pure iron and ARMCO-Fe are shown.

Fig.1 : Scanning electron micrographs after HPT deformation with 10 revolutions of pure Fe (left) and ARMCO-Fe (right).

The grain size distribution of undeformed high-purity Fe and ARMCO-Fe is very similar. However, there are substantial differences in the initial grain size and also the grain shapes after HPT deformation between the high and low impurity samples. While the grains of ARMCO-Fe exhibit a strong lamellar shape, those of HP-Fe are virtual spherical. The larger grain size of HP-Fe can be explained with a dynamic recrystallization during the HPT deformation procedure due to the lack of stabilizing impurities [14]. With higher annealing temperatures the grain structure undergoes coarsening losing also the lamellar structure for the low purity samples. Upon annealing at higher temperatures grain sizes of about 5 μm after temperature treatment at ~ 970 K are obtained.

Fig. 2 shows the grain size obtained from the SEM pictures [15] and the mean positron lifetime at different temperatures. The strong correlation between the grain size and $\tau_m$ is evident.

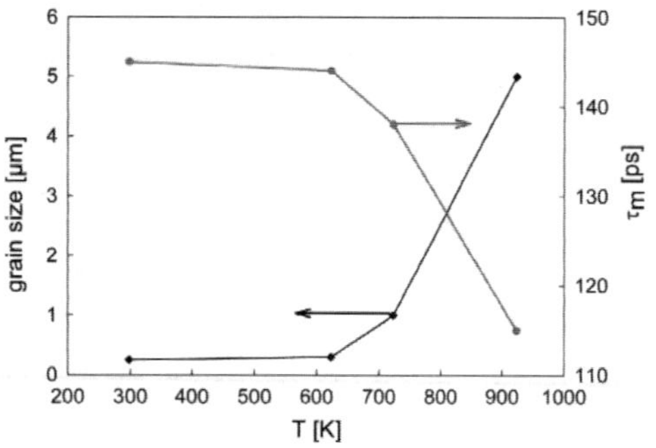


Fig. 2: Grain size and mean positron lifetime at several annealing temperatures for ARMCO-Fe.

For the dilatometric measurements a heating rate of 3K/min was chosen in accordance with the temperature profile of the positron beam measurements (see below). Fig. 3 shows similar annealing characteristics for both samples, however the defect annealing in high-purity Fe starts about 80 K lower and reaches its maximum already at 400 K which is about 100 K less than in ARMCO-Fe. This behaviour can be explained with the different sample purity. Especially the carbon impurities play a crucial role for defect stabilization and recrystallization in iron [10].

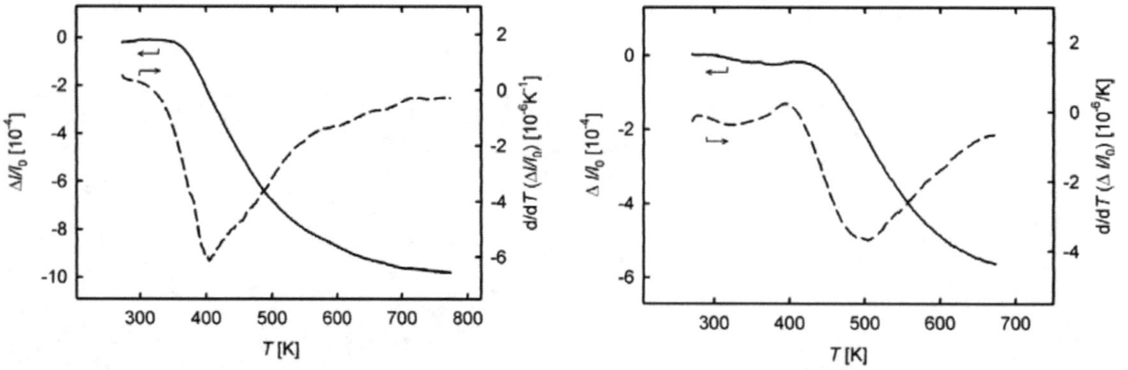


Fig. 3: Dilatometric difference curves of high purity Fe (left). and ARMCO-Fe (right). Also shown are the derivative curves (dashed lines).

Also the different values for the total relative length change are remarkable. The higher total relative length change of nearly $10\times10^{-4}$ in the case of HP-Fe is due to the fact that the overall annealing is shifted to lower temperatures and the maximum annealing temperature applied has even been 100 K higher. The length change results in a relative volume change of about $3\times10^{-3}$. It was discussed already elsewhere [16], that this volume change exceeds the volume change as expected from the loss of relaxed grain boundaries and dislocations upon annealing-induced crystallite growth. The high volume change, therefore, indicates a high amount of excess free volumes either present as vacancy-type defects and vacancy agglomerates within the crystallites or as excess free volume of unrelaxed grain boundaries.

The ARMCO-Fe samples were also subject to an isochronal annealing for 1 hour in steps of 25 °C from RT to 1025 K. After each annealing step the positron lifetime and the Doppler broadening were measured. Up to 600 K the mean positron lifetime, $\tau_m$, stays constant at 144 ps. After annealing at 625 K $\tau_m$ shows a steep decrease from 138 to 118 ps. Above 825 K $\tau_m$ scatters around 115 ps. In this temperature range the lifetime spectra can be separated in two components, yielding a defect component with a lifetime of 186 ps and an intensity of about 21%. The lifetime of 186 ps corresponds quite well to the vacancy lifetime in Fe.

The Doppler S-Parameter after isochronal annealing was measured at room temperature for pure Fe as well as ARMCO Fe. In Fig. 4 the obtained ΔS values in dependence of the annealing temperature for both samples are shown ($\Delta S = S - S_{ann}$). Again a temperature shift is observed.

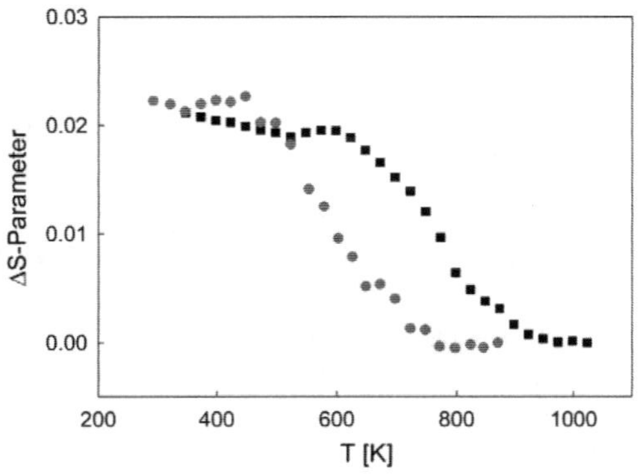


Fig. 4: Difference S-Parameter ΔS as a function of annealing temperature for pure Fe (red dots) and ARMCO-Fe (black squares).

The high number of moderated positrons at the NEPOMUC facility enables studies of fast defect annealing. By choosing the same time profile the results can be combined with dilatometric studies of the annealing of free volumes which are on the same time scale. Fig. 5 shows the difference curve of the S-Parameter of two subsequent measuring runs. Also shown is the resulting relative length change curve. This curve shows a continuous shrinkage from ca. 400 K up to the maximum temperature of 773 K. The annealing characteristics observed by the two complementary techniques of positron annihilation and dilatometry fit together very well. However, a considerable shift of the ΔS-parameter is clearly observable. This shift can be explained with the saturation trapping of positrons at defects for small crystallite sizes [17]. Also shown are the ΔS-Parameter values from isochronal annealing studies. It is obvious that the Doppler data from the two measurements agree quite well.

In a next step the obtained lifetime data for ARMCO-Fe in dependence of isochronal annealing were fitted to a model of competitive positron trapping at point defects within crystallites and diffusion limited positron trapping at grain boundaries [18]. For that purpose values of crystallite diameter in dependence of isochronal annealing were obtained by means of linear interpolation of crystallite sizes measured at selected annealing temperatures (confer Fig. 2).

The experimental data cannot be described by the trapping model when exclusively positron trapping at grain boundaries is assumed, with characteristic values of positron diffusion coefficient, $D_+ = 1\times10^{-4}$ $m^2s^{-1}$, specific grain boundary trapping rates, $\alpha = 600$ $ms^{-1}$, free positron lifetime, $\tau_f = 114$ ps, positron lifetime in grain boundaries $\tau_{GB} = 153$ ps.

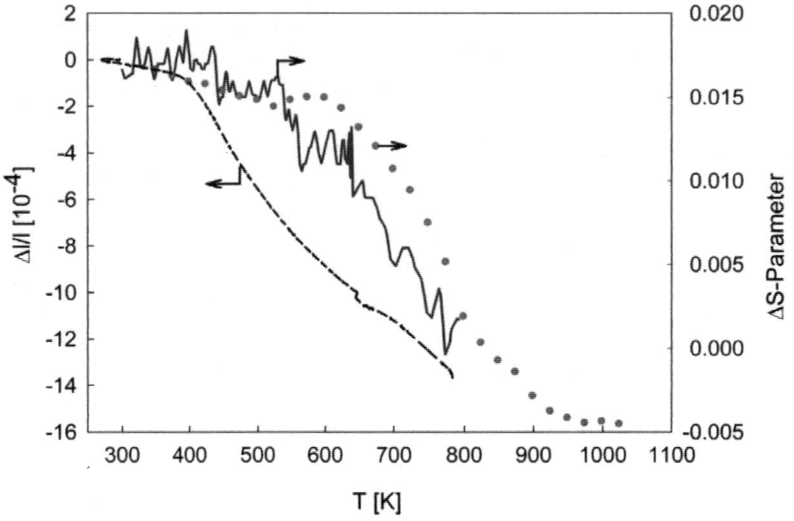


Fig. 5: Doppler broadening ΔS-parameter curve (blue line), dilatometric difference curve (black dashed line) of ARMCO-Fe upon heating with approx. 5K/m and ΔS curve from isochronal annealing (red dots).

The experimental data can only be analysed if in addition to grain boundary trapping vacancy-type positron traps within the crystallites, which anneal in the wake of crystallite growth, are taken into account. The annealing of vacancy-type defects with crystallite growth is simulated with the aid of the well-known theory of Johnson-Mehl-Avrami-Kolmogorov adapted by Henderson for non-isothermal linear heating treatment [19], which has already successfully been applied for describing the recrystallization kinetics in HPT deformed Ni [20]. The pseudo linear heating rate is assumed to be

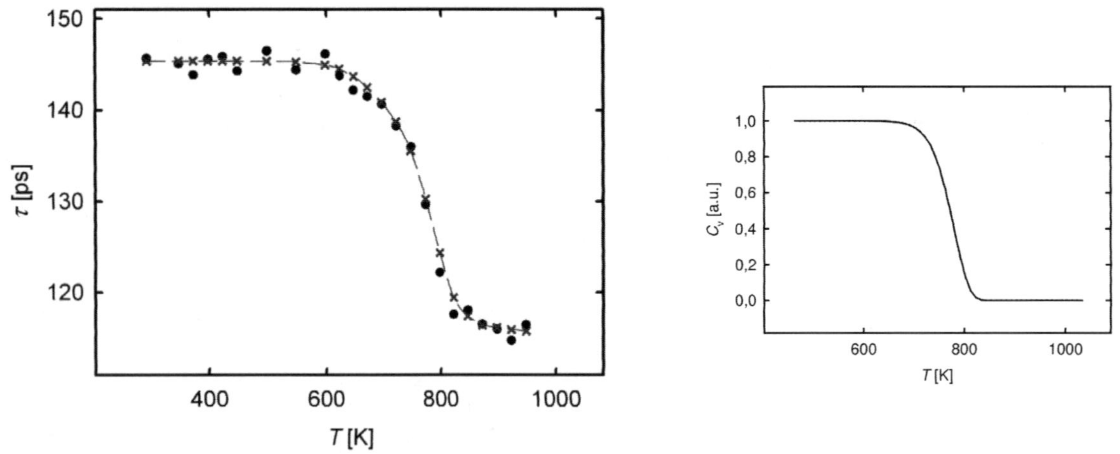


Fig. 6: Fitting of the positron trapping model for grain boundaries and vacancy-type defects to the mean positron lifetime as a function of temperature (left pattern) according to the Henderson model for non-isothermal linear heating treatment (right pattern).

$\delta$ = 0.42 K/min. The dependence of the vacancy concentration on temperature is fitted with the following parameters: Activation energy of $Q$ = 1.8 eV, the pre-exponential frequency $\nu = 1\times10^{8}$ s$^{-1}$, and the Avrami exponent $n$ = 1 (see Fig. 6 right pattern).

With this model of competitive trapping at vacancy-type crystal defects and grain-boundaries, the mean positron lifetime in dependence of annealing temperature can be well described assuming a trap lifetime of 185 ps, an initial concentration of vacancy-type crystal defects $C_{\mathrm{v,\ initial}} = 1.4\times10^{-5}$ and a specific positron trapping rate into vacancies $\mu_{\mathrm{v}} = 3\times10^{14}$ s$^{-1}$.

Vacancy-type crystal defects in iron stable up to temperatures of crystallite growth may be related to carbon impurities. Annealing in the temperature range observed in the present studies was also reported from magnetic measurements of carbon-doped iron after plastic deformation [21].

**Acknowledgment**

The authors would like to thanks R. Pippan for providing us with the HPT deformed samples, C. Kammerhofer for performing the SEM graphs, Ch. Hugenschmidt for support with the positron measurements at NEPOMUC. Financial support by the Austrian Science Fund (FWF) P21009-N20 is appreciated.

**References**

[1] Koch C C 2007 *Nanostructured Materials: Processing, Properties and Applications* (New York: Andrews Appl. Sci. Publ.)
[2] Zehetbauer M, Valiev R (Eds) 2004 *Nanomaterials by Severe Plastic deformation* (Weinheim: Wiley-VCH)
[3] Zhilyaev A P, Kim B K, Szpunar J A, Barò M D and Langdon T G 2005 *Mater. Sci. Eng.* A **391** 377
[4] Pippan P, Wetscher F, Hafok M, Vorhauer A and Sabirov I 2006 *Adv. Eng. Mater.* **8** 11
[5] Parz P, Faller M J, Pippan R, Puff W and Würschum R 2012 *Physics Procedia* **35** 10
[6] Bachmaier A and Pippan R 2011 *Mater. Sci. Eng.* A **528** 7589
[7] Krause-Rehberg, Bondarenko V, Thiele E, Klemm R and Schell N 2005 *Nucl. Instr. Methods Phys. Res.* B **240** 719
[8] Cizek J, Janecek M, Srba O, Kuzel R, Barnovska Z, Prochazka I and Dobatkin S 2011 *Acta Mater.* **59** 2322
[9] Ye F, Sprengel W, Wunderlich R, Fecht H-J and Schaefer H-E 2007 *Proc. Natl. Acad. Sci.* **104** 12962.
[10] Vehanen A, Hautojärvi P, Johannson J, Yli-Kauppila J. and Moser P 1982 *Phys. Rev.* B **25** 762
[11] Wetscher F, Vorhauer A and Pippan R 2005 *Mater. Sci. Eng.* A **410-411** 213
[12] Puff W 1983 *Comp. Phys. Commun.* **30** 359
[13] Oberdorfer B 2012 *Doctoral Thesis,* Technical University Graz
[14] Pippan R, Wetscher F, Hafok M, Vorhauer A and Sabirov I 2006 *Adv. Eng. Mater.* **8** 1046
[15] Hohenwarter A, Kammerhofer C and Pippan R 2010 *J. Mater. Sci.* **45** 4805
[16] Oberdorfer B, Lorenzoni, Unger K, Sprengel W, Zehetbauer M, Pippan R and Würschum R 2010 *Scr. Mater* **63** 452
[17] Oberdorfer B, Steyskal E-M, Sprengel W, Puff W, Pikart P, Hugenschmidt C, Zehetbauer M, Pippan R and Würschum R 2010 *Phys. Rev. Lett.* **105** 146101
[18] Oberdorfer B and Würschum R (2009) *Phys. Rev.* B **79** 184103
[19] Henderson D W 1979 *J. Non-Cryst. Solids* **30** 301
[20] Oberdorfer B, Steyskal E-M, Sprengel W, Pippan R, Zehetbauer M, Puff W and Würschum R 2011 *J. Alloys Compd.* **509S** 5309
[21] Refaat A, Farid Z M and Ghobrial F Z 1994 *Z. Phys.* **94** 227

Journal of Physics: Conference Series **443** (2013) 012034 doi:10.1088/1742-6596/443/1/012034

# Helium implanted RAFM steels studied by positron beam Doppler Broadening and Thermal Desorption Spectroscopy

**I Carvalho,[1,2] H Schut[2], A Fedorov[3], N Luzginova[3], P Desgardin[4] and J Sietsma[5]**

[1]Materials innovation institute (M2i), Mekelweg 2, 2628 CD, Delft, The Netherlands
[2]Delft University of Technology, Faculty of Applied Sciences, Mekelweg 15, 2629 JB, Delft, The Netherlands
[3]Nuclear Research and consultancy Group (NRG), Westerduinweg 3, 1755 LE, Petten, The Netherlands
[4]CEMHTI-CNRS, 3A rue de la Férolerie, 45071 Orléans Cedex, France.
[5]Delft University of Technology, Faculty of Mechanical, Maritime and Materials Engineering, Mekelweg 2, 2628 CD, Delft, The Netherlands

E-mail: i.carvalho@m2i.nl

**Abstract.** Reduced Activation Ferritic/Martensitic steels are being extensively studied because of their foreseen application in fusion and Generation IV fission reactors. To mimic neutron irradiation conditions, Eurofer97 samples were implanted with helium ions at energies of 500 keV and 2 MeV and doses of $5x10^{15}$-$10^{16}$ He /cm$^2$, creating atomic displacements in the range 0.07–0.08 dpa. The implantation induced defects were characterized by positron beam Doppler Broadening (DB) and Thermal Desorption Spectroscopy (TDS). The DB data could be fitted with one or two layers of material, depending on the He implantation energy. The *S* and *W* values obtained for the implanted regions suggest the presence of not only vacancy clusters but also positron traps of the type present in a sub-surface region found on the reference sample. The traps found in the implanted layers are expected to be $He_nV_m$ clusters. For the 2 MeV, $10^{16}$ He/cm$^2$ implanted sample, three temperature regions can be observed in the TDS data. Peaks below 450 K can be ascribed to He released from vacancies in the neighbourhood of the surface, the phase transition is found at 1180 K and the peak at 1350 K is likely caused by the migration of bubbles.

## 1. Introduction

Eurofer97 is a Reduced Activation Ferritic/Martensitic (RAFM) steel used as reference structural material for future fusion reactors. Eurofer97 steel known to possess a high resistance against swelling from gaseous transmutation products (hydrogen, helium) has attractive mechanical properties and shows reduced activation behavior in a fusion neutron spectrum [1]. The effects of neutron irradiation and the accumulation of defects in this material are extensively studied (e.g. [2]) as Eurofer97 will be used for constructing test blanket modules in the ITER fusion reactor [1, 3]. As a consequence of the neutron irradiation, displacement damage will be created together with the production of hydrogen and helium [1, 3]. This leads to changes in the microstructure and ultimately to alterations in the material's mechanical properties. The work described here is part of a larger project in which Thermal Desorption spectroscopy (TDS) and electron microscopy studies will be performed on Eurofer97 neutron irradiated in the High Flux Reactor at NRG. To mimic these neutron irradiations, samples are implanted with 500 keV and 2 MeV $He^+$ ions. The defect structures after implantation and annealing treatments are characterized with the positron beam Doppler Broadening technique (DB). The thermal behavior of the implanted He is studied in parallel by TDS.

## 2. Experiment

The samples used in this study were cut from the European Union Eurofer97 batch, produced by Böhler, Austria, with a nominal composition of Fe-9Cr-1W-0.2V-0.1Ta-0.1C (wt.%). Samples of 12x12x0.5 $mm^3$ were machined so that after the He implantation nine small rectangular cuboids of 2x2x0.5 $mm^3$ (for TDS) could be detached from the larger central piece with dimension of 10x10x5 $mm^3$ (for DB). In order to obtain a tempered martensitic microstructure the samples were annealed after machining in two steps of 30 min at 1253 K and 90 min at 1033 K, both followed by cooling in argon ambient.

Low dose ($5x10^{15}$ and $5x10^{16}$ $He/cm^2$) room temperature helium implantations were performed at CEMHTI-CNRS. For the 500 keV and 2 MeV implantations the beam current was 270-280 nA and 120-212 nA, respectively, with implantation times ranging from 1 up to 15 hours. TRIM calculations for the dose of $1x10^{15}$ $He/cm^2$ show that the defect production (dpa) has a maximum value of 0.08 at a depth of 1 μm for 500 keV and 0.07 dpa at 3 μm for 2 MeV. The maxima in the He distributions are slightly deeper, with a Full Width at Half Maximum of 0.2 μm. For the dose of $1x10^{16}$ $He/cm^2$ similar defects and He distributions are obtained with 10 times higher dpa levels. The positron beam DB measurements were performed up to energies of 25 keV, corresponding to a maximum probing depth of approximately 1 μm. He desorption spectra were obtained in a $10^{-8}$ Torr vacuum, using a linear heating rate of 0.33 K/s from RT up to 1500 K.

## 3. Results and discussion

Figure 1 shows the DB *S* and *W* parameters as a function of the positron implantation energy (bottom axis) and positron mean implantation depth (top axis) for the reference and the four implanted samples. The lines through the data are the fits obtained with the VEPFIT program [4]. In case of the 2 MeV implantation, one layer defining the *S* and *W* parameters and the positron diffusion length was sufficient to obtain a good fit. For the 500 keV implantation two layers (with thicknesses of 200 and 800 nm) plus a layer representing the bulk of the material were needed. This is consistent with the TRIM results which show that for the 2 MeV implantation the defect production is flat over the positron probing range while for the 500 keV implantation the defect profile peaks at the end of the positron range. The decrease of *S* (increase of *W*) for positron implantation energies up to 2 keV is ascribed to non-thermal positrons annihilating at the surface. In the VEPFIT model the *S* and *W* parameters for thermal positrons back-diffused to the surface are found to be rather independent of the implantation conditions. For the reference sample a positron diffusion length ($L_e^+$) of approximately 50 nm is obtained. Due to the He implantation $L_e^+$ has decreased to less than 10 nm, indicative for high positron trapping efficiency.

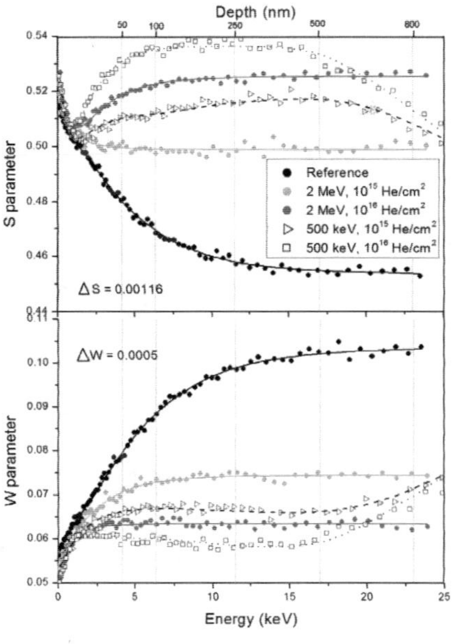


**Figure 1.** *S* and *W* parameter vs. positron implantation energy and depth. Data fitted with VEPFIT.

Figure 2 shows the *S,W* map (for clarity only fitted curves are shown) for all samples with the positron implantation energy as a running parameter starting from 1.5 keV. In this figure the filled symbols (•) represent the characteristic *S,W* values for the reference bulk and surface annihilations. The open symbols (o) are the fitted *S,W* values in the first layer which in case of 2 MeV implantation is the only layer defined. The triangles represent the *S,W* values for the second deeper layer (500 keV only). In this image is also marked the *SW* value of a chromium (Cr) sample annealed at 1200 K for 1 hour.

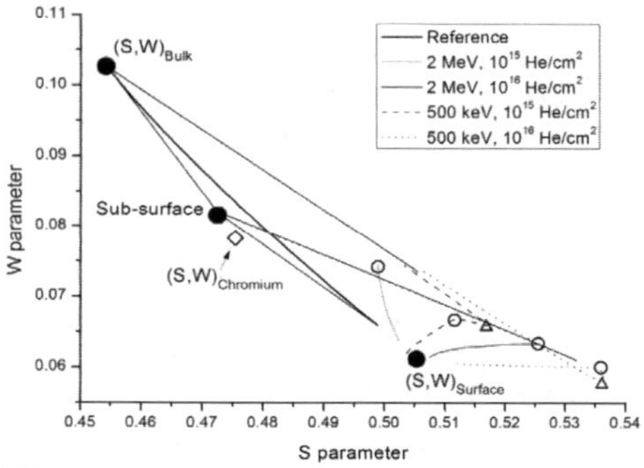


**Figure 2.** *S* and *W* parameter vs. positron implantation energy and depth. Data fitted with VEPFIT.

The *S* and *W* curves for the reference sample shown in figure 1 could be independently fitted using a single layer model. In the *S,W* plot however, a small deviation from the expected straight line connecting the *S,W* point of the surface and bulk is observed. This can be understood by assuming the presence of a near surface layer with *S,W* values different from that of the reference bulk values. With this assumption, the *S,W* analysis program SWAN [5] predicts a layer of 60 nm thickness with characteristic *S,W* values denoted as "sub surface" in figure 2. Note that these values are close to those measured for chromium annealed at 1200 K. This indicates that in the near surface region of the reference sample a small fraction of the implanted positrons annihilates at Cr-rich trapping sites. These may have been formed during tempering treatment to obtain the desired ferritic/martensitic structure. A closer look at the location of all *S,W* points obtained for the He implanted regions shows that they all lie on a single line which extrapolates to the *S,W* point found for the sub-surface region in the reference sample. This means that after the He implantation the positrons annihilate at either vacancy type defects or at defects associated with Cr. The highest S value, which is obtained for the implantation of 500 keV, $10^{16}$ He /cm$^2$, is about 15% higher than the bulk value. In metals this increase is typical for small vacancy clusters [6].

With increasing He dose the *S,W* points shift towards this higher *S* value indicating an enhanced fraction of positrons trapping at the vacancy clusters with the respect to the fraction trapped at vacancies associated with Cr.

In figure 3 a TDS measurement of a sample implanted at 2 MeV, $10^{16}$ He/cm$^2$ and the fitted *S* values (measured at room temperature) after *in situ* annealing for 5 min are shown. From the TDS data it is calculated that the total amount of He desorbed is approximately 10% of the implanted dose.

At temperatures below 450 K two peaks can be observed. In literature, He desorption in this temperature range is attributed to the release of He trapped at a near surface vacancy [8]. At ~ 900 K the rate of desorption of He slightly increases reaching its peak value at ~ 1180 K. The large width of the peak indicates desorption of multiple trapping sites such as simple HeV or larger helium-vacancy clusters, $He_nV_m$ [8]. Retrapping of desorbed helium before reaching the surface can also occur at these temperatures and delays the release of He [8]. The peak at 1180 K is associated with the phase transition, expected to occur at 1160 K, and is likely delayed due to retrapping events. [9]. Finally, a peak is found at ~ 1350 K. It is known that during the annealing carbon-vacancy clusters can form and may combine with $He_nV_m$ to form complex clusters [10]. The thermal stability of these complexes is much

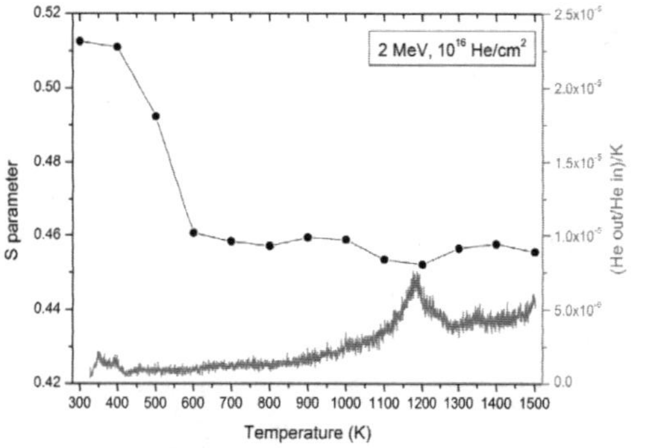


**Figure 3.** *S* parameter and TDS spectra as a function of annealing temperature for the 2 MeV implanted sample with a dose of $10^{16}$ He/cm$^2$.

higher than that of simple HeV and therefore these complexes are expected to desorb at higher temperatures, via dissociation or bubble diffusion mechanism [10].

The decrease of the *S* parameter starting at 400 K can be attributed to the annealing of vacancies, as the first micrometer of Eurofer, when implanted with 2 MeV He ions is expected to have low He content (TRIM). The second step in the *S* curve lies between the 1000 and 1100 K and can be associated with the beginning of the phase transition and the removal of defects by this process. At 1300 K the S-parameter increases slightly after which it finally drops to the near bulk value at 1500 K. This can be understood by the diffusion of He bubbles towards the surface thereby entering the depth region probed by the positrons.

## 4. Conclusions

He implantations have been performed in Eurofer97 steel with energies of 500 keV and 2 MeV and doses of $5 \times 10^{15}$ and $5 \times 10^{16}$ He/cm$^2$. The shape of the *S,W* for the reference sample curve suggests the presence of a 60 nm thick sub-surface layer containing Cr related positron trapping sites. Two types of defects are found in the He implanted samples: vacancy type defects (expected to be $He_nV_m$ clusters) and defects associated with chromium.

In the TDS spectra the phase transition of the steel is observed and the desorption at temperatures above 1300 K can be attributed to migrating He bubbles that are formed during the annealing. In the DB measurements this is seen as an increase in *S* due to the arrival of the He bubbles in the range probed by the positrons.

## Acknowledgements

This research was carried out under project number M74.10.10393 in the framework of the Research Program of the Materials innovation institute M2i (www.m2i.nl).

## References

[1] Lindau R, Moslang A, Rieth M, Klimiankou M, Materna-Morris E, Alamo A, Tavassoli F, Cayron C, Lancha A M, Fernandez P *et al.* 2005 *Proc. 23rd Symp. of Fusion Technology (Venice)* **75-79** 989

[2] Eldrup M, Singh B N and Golubov S I 2010 Final report on investigations of the influence of helium concentration and implantation rate on cavity nucleation and growth during neutron irradiation of Fe and EUROFER 97 *National Laboratory for Sustainable Energy* RISO

[3] van der Schaaf B, Tavassoli F, Fazio C, Rigal E, Diegele E, Lindau R and LeMarois G 2003 *Fusion Eng. and Design* **69** 197

[4] van Veen A, Schut H, Vries J, Hakvoort R A and IJpma M R 1991 Proc. 4th Int. workshop on Slow positron beam techniques for solids and surfaces **218** 171

[5] Fedorov A V, van Veen A and Schut H 2001 *Proc.12th Int. Conf. on Positron Annihilation (Neubiberg)* Materials Science Forum **363-365** 646

[6] Eleveld H and van Veen A 1994 *J Nuclear Materials* **212-215** 1421

[7] Babu S H, Rajaraman R, Amarendra G, Govindaraj R, Lalla N P, Dasgupta A, Bhalerao G and Sundar C S 2011 *Phylosophical Magazine* **92** 2848

[8] Morishita K, Sugano R, Iwakiri H, Yoshida N and Kimura A 2001 Proc. 4th Pacific Rim Int. Conf. on Advanced Materials and Processing (Hawaii) **1-2** 1395

[9] Rieth M, Schirra M, Falkenstein A, Graf P, Heger S, Kempe H, Lindau R and Zimmermman H 2003 *Forschungszentrum Karlsruche GmbH* report number 6911

[10] Oo K Z, Chernov I, Staltsov M, Kalin B, Kalashnikov A, Binyukova S Y, Zaw A K, Ageev V and Nikitina A 2010 *Atomic Energy* **110** 151

Journal of Physics: Conference Series **443** (2013) 012035 doi:10.1088/1742-6596/443/1/012035

# Effect of heterogeneous distributed intermetallic precipitates on accumulation of vacancy-like defects in irradiated Fe-Ni-based alloys studied by positron annihilation

**D.A. Perminov, A.P. Druzhkov and V.L. Arbuzov**

Institute of Metal Physics, Ural Branch of RAS, 18, S. Kovalevskaya str., Ekaterinburg, 620990, Russian Federation

d_perm@rambler.ru

**Abstract**. The effect of heterogeneous distributed intermetallic particles on interaction of point defects with dislocations in cold-worked Fe-Ni-based alloys during electron irradiation was studied by positron annihilation. It was shown that vacancy accumulation occurs in cold-worked alloys during irradiation in spite of high dislocation density. It is caused by the presence of areas with lower dislocation density in alloys. At room temperature, the vacancies are not trapped by dislocations totally because a low their mobility. At elevated temperatures, vacancies may leave dislocations and form VCs. In cold-worked Fe-Ni-Ti alloy the accumulation of defects during irradiation is essentially lowered in comparison with cold worked Fe-Ni alloy. $Ni_3Ti$ precipitates, which are present on dislocations, reduce the efficiency of interaction of dislocations with interstitial atoms and, thereby, enhance the mutual recombination of point defects.

## 4. Introduction

Irradiation-induced swelling is a main limiting factor for using of austenitic stainless steels as structural materials of critical core components of reactors. The swelling is due to a vacancy accumulation occurred under irradiation that is caused by bias factors of various sinks [1]. The preliminary plastic deformation reduces effectively the vacancy accumulation because dislocations, which are formed during deformation, are sinks for vacancies [1-3]. However, a locally inhomogeneous microstructure is formed in austenitic alloys and steels during deformation. Processes of the radiation damage in materials having an inhomogeneous distribution of dislocations are not quite clear. Moreover, an initial dislocation structure substantially changes during irradiation that is caused by dislocations motion (climb and glide). As a result, the effect of deformation drops after irradiation to high doses ($\geq$ 50 dpa).

The dislocation structure may be stabilized by ultrafine intermetallic precipitates which form in aging alloys. These precipitates strongly pin dislocations and thereby block a dislocation motion [4]. Moreover, if they are located on dislocation lines, the precipitates may affect the interaction of point defects with dislocations [1]. However, experimental data about investigations of the effect of intermetallic precipitates on interaction of dislocations with point defects is absent today.

In this paper, we present the results of investigation of the vacancy defects evolution during irradiation of Fe-Ni-based alloys in various initial states (solution annealed, cold-worked) by the positron annihilation spectroscopy. These alloys were used as model materials for fast breeder reactor stainless steels. Defects were induced by electron irradiation, which generated homogeneously distributed Frenkel pairs. The freely migrating point defects (interstitial atoms and vacancies) are only formed at electron irradiation unlike it at the neutron irradiation. The study of interaction of defects with sinks, alloying additions and over features is able in alloys in case of electron irradiation. Experimental results of our research were also published earlier [5].

## 5. Materials and methods

Samples of Fe–36.5 Ni (Fe-Ni) and Fe-36.5 Ni-2.5 Ti (Fe-Ni-Ti) fcc alloys (wt %) were used. After rolling, cutting and electrical polishing, the samples were solution annealed under a $10^{-6}$ Pa vacuum at 1373 K for 1 h and then were cooled quickly (the SA state). Some alloy samples were deformed by rolling to 40% at room temperature. After cold-working, alloys were thermally treated at 650 K for 1 h in order to anneal the vacancy defects formed during deformation (the CW state) [4]. The microstructure of the solution annealed and deformed samples were certified using by transmission electron microscopy (TEM). The samples of the SA alloys had grains ~30 μm in size and the dislocation density of about $10^{11}$ $m^{-2}$. In samples of CW alloys a banded subgrain structure was formed during deformation. The subgrains, which were extended lengthwise, were 50 to 400 nm wide. Dislocations in subgrains formed cellular-network configurations. The mean dislocation density inside subgrains was about $9x10^{14}$ $m^{-2}$ [4].

The samples were irradiated at 300 and 573 K with 5 MeV electrons in a linear accelerator. The maximum electron fluence was $5x10^{22}$ $m^{-2}$, which corresponded to the damaging dose of $5x10^{-4}$ dpa as calculated by the modified Kinchin-Pease model [6].

The angular correlation of annihilation radiation (ACAR), which is a one of positron annihilation methods, is used to investigate of defect structure. ACAR measurements were performed at room temperature. The ACAR method was realized in a one-dimensional ACAR spectrometer providing a resolution of 1 mrad x 160 mrad. A $^{68}$Ge positron source of activity of 400 MBq was used. At least $8x10^5$ coincidence counts were collected in each ACAR spectrum; the peak-to-background ratio was $\sim 10^3$. The ACAR spectra of the alloys were approximated by an inverted parabola and a Gaussian. The approximation quality criterion approached unity [7]. Changes in the shape of the ACAR spectra were characterized by the standard S-parameter which were defined as the ratios of low-regions ($p_z \le 3.5x10^{-3}\, m_0c$) to the total region [8].

## 6. Results and discussion

### 6.1. Irradiation at room temperature

The Figure 1 presents the dependence of the S-parameter on the electron fluence for Fe-Ni and Fe-Ni-Ti alloys in SA and CW states irradiated at 300 K. The detailed description of results was reported in our recent work [5]. We have been estimated from our results a concentration of vacancies presented in alloys after irradiation to various fluencies. It should be noted that high dislocation density (about $9x10^{14}$ $m^{-2}$) is preserved inside subgrains in CW alloys after deformation and annealing [4]. Therefore, we use a Simple Trapping Model (STM) in our estimations. According to three positron state model, the S-parameter is related to the concentration of vacancies presented in samples $C_0$ as [9]:

$$C_0 = \frac{\lambda_f (S - S_f) + \mu_d \rho_d (S - S_d)}{\mu_V (S_j - S)} \qquad (1)$$

where $\lambda_f$ is the positron annihilation rate in the bulk (free) state; $\mu_d$ and $\mu_v$ are the specific positron trapping rates for dislocations and vacancies respectively; $S_f$, $S_d$ and $S_j$ are the S-parameters characteristic of the positron annihilation from the bulk (free), dislocations and vacancy defect-trapped states respectively. The next values of annihilation characteristics, which we obtained in [4,8,9], was used for calculation: $\lambda_f = 9x10^9$ $s^{-1}$, $\mu_d = 1.2\times10^{-4}$ $m^2/s$, $\mu_v = 2.2x10^{15}$ $s^{-1}$, $S_f = 0.526$, $S_d = 0.574$, $S_j = 0.605$. It should be noted that the initial value of S-parameter for Fe-Ni-Ti(CW) alloy is much smaller than $S_d$. As we shown earlier [4], in the Fe-Ni-Ti alloy, the fine intermetallic precipitates of $Ni_3Ti$-type located at dislocations. These precipitates reduce the trapping efficiency of positrons by dislocations. The estimates according trapping model give specific positron trapping rates for dislocation in Fe-Ni-Ti(CW) alloy $\mu_d^{Ti} = (2.1\pm0.3)\times10^{-5}$ m$^2$/s.

As we shown earlier [10], vacancies are mobile in this alloy at room temperature and they form small (3-5 vacancies) three-dimensional clusters (VC). In this case, the $C_0$, which we obtain from Equation 1, correspond to concentration of vacancies bounded in VCs. The results of calculation are shown in Figure 2.

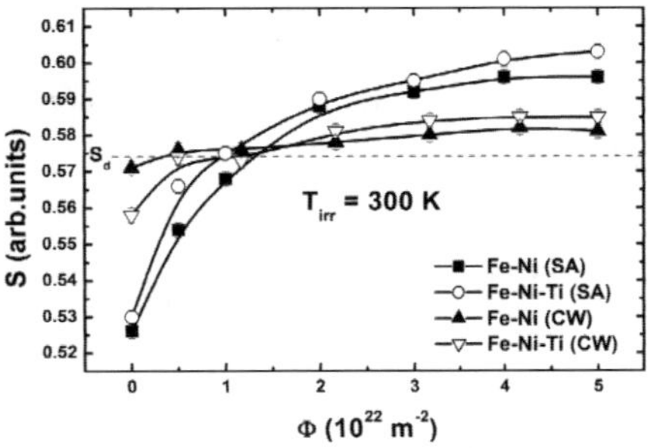


**Figure 1.** S-parameter variation versus irradiation fluence at 300 K for investigated alloys.

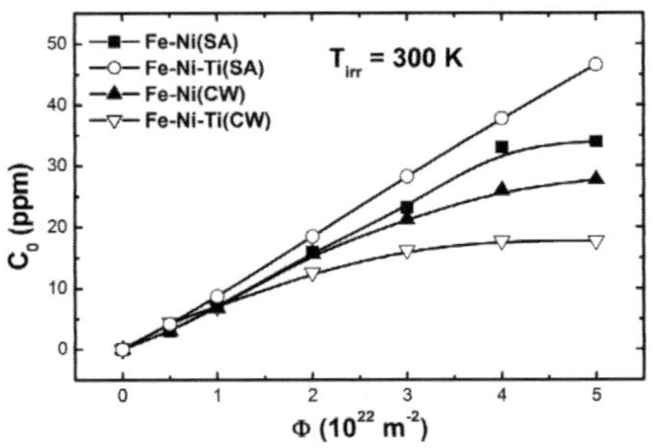


**Figure 2.** The vacancy concentration variation versus irradiation fluence at 300 K for investigated alloys.

As it can be seen from figure, the vacancy accumulation occur in Fe-Ni(SA) during irradiation. In Fe-Ni-Ti(SA), alloy the vacancy accumulation is enhanced in comparison with that in Fe-Ni(SA) alloy. Ti is an over-sized impurity to the Fe-Ni model alloys. Ti atoms interacts with migrating vacancies that result in enhancement of vacancy accumulation [4,5,10].

In Fe-Ni(CW) alloy, the vacancy accumulation is reduced in comparison with that in Fe-Ni(SA) alloy at higher fluencies ($\Phi \geq 2\times10^{22}$ m$^{-2}$). However, the difference between vacancy concentration in SA and CW alloys is very low ($\approx$ 1.5 times) in spite of a high density of sinks presented in cold-worked alloy. As it mentioned above, the dislocations in subgrains formed cellular-network configurations. The dislocation density inside cells is one order of magnitude smaller (about $10^{14}$ m$^{-2}$) than mean density of dislocations. The diffusion length of vacancies is very low at this temperature (a few nanometers) and a deal of vacancies doesn't reach the dislocations [11]. As a result, these vacancies are not absorbed by sinks and form VCs. It should be noted that in fcc metals interstitial atoms have high mobility and they are absorbed totally by dislocations [8].

In Fe-Ni-Ti(CW) alloy, the vacancy accumulation is surprisingly much smaller than that in over alloys. The investigations of annealing of irradiated alloys had shown [4.5] that the Ti-vacancy interaction takes place also in this alloy. Thus, a vacancy accumulation in Fe-Ni-Ti(CW) alloy must be larger than that, at least, in Fe-Ni(CW) alloy. The most probable reason for reduction of vacancy accumulation is enhanced mutual recombination of point defects. It seems that precipitates, which are located at dislocations, reduce the efficiency of interaction of dislocations with interstitial atoms [1]. As a result, some number of interstitials is not absorbed by dislocations and recombines with vacancies.

### 6.2. Irradiation at 573 K

The Figure 3 presents the dependence of the S-parameter on the electron fluence for Fe-Ni and Fe-Ni-Ti alloys in SA and CW states irradiated at 573 K. The detailed description of results was reported in our recent work [5]. The dependence of concentration of vacancies presented in alloys after irradiation on the electron fluence is shown in Figure 4.

As it can be seen from figure, in Fe-Ni(CW) and Fe-Ni-Ti(CW) alloys the vacancy accumulation occurs only at higher fluencies ($\Phi \geq 2\times10^{22}$ m$^{-2}$). However, the vacancy concentration in these alloys at fluence $\Phi = 5\times10^{22}$ m$^{-2}$ is more than that in Fe-Ni(SA) alloy. The vacancy mobility is high at this temperature and vacancies easily reach the dislocations and are trapped by them. However, as it shown in [11], trapped vacancies may leave dislocation core at elevated temperatures. At this, detrapped vacancies form VCs at higher fluencies, when their concentration is sufficiently high.

It should be noted that vacancy accumulation is reduced in Fe-Ni-Ti(CW) alloy in comparison with that in Fe-Ni(CW) alloy. Thus, the precipitates, which are located at dislocations, enhance the mutual recombination of point defects, as it takes place during irradiation at room temperature.

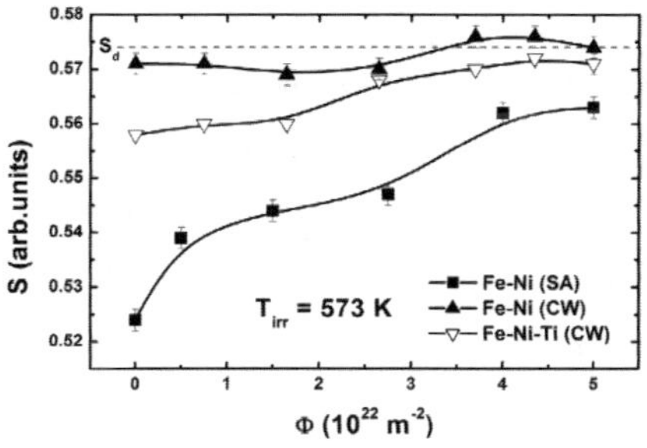


**Figure 3.** S-parameter variation versus irradiation fluence at 573 K for investigated alloys.

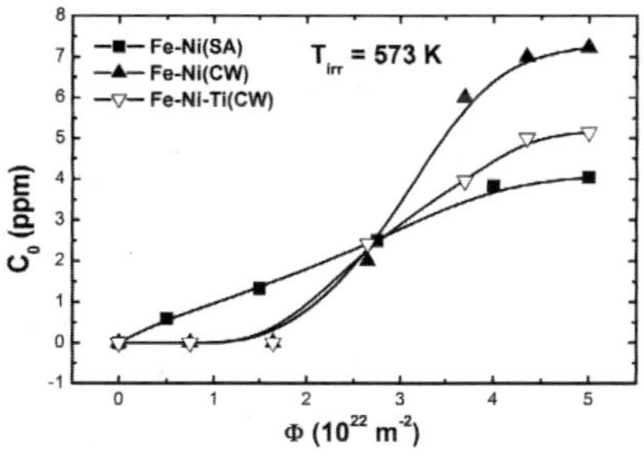


**Figure 4.** The vacancy concentration variation versus irradiation fluence at 573 K for investigated alloys.

## 7. Conclusions

In this paper we present results of ACAR characterization of electron irradiated Fe-Ni and Fe-Ni-Ti alloys with varied initial structure. It was found that vacancy accumulation occurs in cold-worked Fe-Ni alloy during irradiation in spite of high dislocation density. It is caused by the presence of areas with lower dislocation density in alloys. At room temperature, the vacancies are not trapped by dislocations totally because a low their mobility. At elevated temperatures, vacancies may leave dislocations and form VCs. In cold-worked Fe-Ni-Ti alloy the accumulation of defects during irradiation is essentially lowered in comparison with cold worked Fe-Ni alloy. $Ni_3Ti$ precipitates, which are located at dislocations, reduce the efficiency of interaction of dislocations with interstitial atoms and, thereby, enhance the mutual recombination of point defects.

## Acknowledgment

This study was done within RAS Program (Project No. 01.2.006 13394) with partial support of Russian Foundation for Basic Research (Project No 11-03-00018a).

## References

[1] Braislford A D and Bullough R 1978 *J. Nucl. Mater.* **69-70** 434
[2] Johnston W G, Rosolowsky J H, Turkalo A M and Lauritzen T 1973 *J. Nucl. Mater.* **48** 330
[3] Wolfer W G and Glasgow B B 1985 *Acta metall.* **33** 1997
[4] Druzhkov A P and Perminov D A 2010 *Materials Science and Engineering* A **527** 3877
[5] Druzhkov A P, Arbuzov V L and Perminov D A 2005 *J. Nucl. Mater.* **341** 153
[6] Morillo J, de Novion C H and Dural J 1981 *Radiat. Eff.* **55** 67
[7] Druzhkov A P, Perminov D A and Pecherkina N L 2008 *Philos. Mag.* **88** 959
[8] Arbuzov V L, Danilov S E and Druzhkov A P 1997 *Phys. Status Solidi* a **162** 567
[9] Dlubek G, Brummer O and Hensel E 1976 *Phys. Status Solidi* a **34** 737
[10] Arbuzov V L, Druzhkov A P and Danilov S E 2001 *J. Nucl. Mater.* **295** 273
[11] Yoshiie T, Sato K and Xu Q 2009 *Nucl. Instr. and Meth. in Phys. Res. B* **267** 2971

Journal of Physics: Conference Series **443** (2013) 012036 doi:10.1088/1742-6596/443/1/012036

# PLEPS study of ion implanted and annealed Fe-11.62%Cr alloys

**S Sojak[1], V Slugeň[1], W Egger[3], L Ravelli[3], M Petriska[1], S Stanček[1], M Sahul[2], M Skarba[2], P Priputen[2], M Stacho[1], J Veterníková[1], R Hinca[1], V Sabelová[1]**

[1] Slovak University of Technology, Institute of Nuclear and Physical Engineering, Ilkovičova 3, 812 19 Bratislava, Slovakia
[2] Slovak University of Technology, Institute of Materials Science, Paulínska 16, 917 24 Trnava, Slovakia
[3] LRT2, Universitaet der Bundeswehr Muenchen, Werner-Heisenberg-Weg 39, D-85577 Neubiberg, Germany

E-mail: stanislav.sojak@stuba.sk

**Abstract.** Reduced activation ferritic/martensitic (RAFM) steels are one of the candidate structural materials for application in future nuclear facilities. These steels fulfill demands on radiation, thermal, and mechanical resistance during their operational lifetime. Our experiments were focused on study of microstructural changes of binary Fe-Cr alloys after annealing and irradiation, experimentally simulated by ion implantations. Alloys with 11.62% Cr were examined after helium ion implantations at different doses (0.1; 0.3; 0.5 C/cm$^2$). Thermal annealing, motivated by literature and our previous work in the field of reactor steels [1,2], was performed at temperatures of 400, 475, 525 and 600 °C after implantations with aim to study changes of the defect size/amount. The Pulsed Low Energy Positron System (PLEPS) at FRM II reactor in Garching (Munich) was applied for lifetime studies [3]. Damage introduced into the microstructure caused problems with defect interpretations by the PLEPS technique and therefore SEM was applied as well. Positron lifetimes measured at annealing temperatures in range of 400 - 525 °C did not show an expected decrease. Only in the case of temperature of 600 °C the mean lifetime (MLT) decreased significantly and assumptions about the defect size decrease were made according to the literature [2].

## 1. Introduction

Radiation, heat, and mechanical resistance are crucial parameters of structural materials of Nuclear Power Plants (NPP) which limit their planned lifetime. Much higher radiation damage is expected in new generations of nuclear power plants, such as Generation IV and fusion reactors. Therefore, the investigation of new structural materials is, among others, focused on the study of reduced activation ferritic/martensitic (RAFM) steels with good characteristics as lower activation, good resistance to volume swelling, good radiation, and heat resistance (up to 550 °C).

Our work is focused on the study of radiation damage simulated by ion implantations and thermal treatment evaluation of RAFM steels in the form of binary Fe-Cr model alloys. In order to study the microstructure recovery after ion irradiation, we applied an approach for restoration of initial physical and mechanical characteristics of structural materials in the form of thermal annealing, with the goal to decrease the size and amount of accumulated defects. The experimental analysis of material damage at microstructural level was performed by the pulsed low energy positron system (PLEPS) [3] at the high intensity positron source NEPOMUC [4] at the Munich research reactor FRM-II.

## 2. Materials treatment

Detailed chemical composition of studied Fe-Cr alloy can be seen in the table 1. Fabrication processes and treatments of the alloy can be found in [5]. "As-received" material was cut into desired dimensions, ground and polished to mirror-like surface before exposure to helium implantation.

**Table 1.** Chemical composition of studied Fe-Cr alloy (wt%) [5].

| Alloy | Cr* | O* | N* | C* | Mn | P | Ni | Cu | V |
|---|---|---|---|---|---|---|---|---|---|
| Fe-11.62%Cr | 11.62 | 0.031 | 0.024 | 0.028 | 0.03 | 0.05 | 0.09 | 0.01 | 0.002 |

* *measured after heat treatment during fabrication*

Accelerated helium ions were used to obtain cascade collisions in the microstructure of studied material without neutron activation. Helium implantation was performed in two steps with ions energy of 250keV and 100keV, respectively. Implantations at the linear accelerator of the Slovak University of Technology in Bratislava [6] were performed at dose of 0.3 C/cm$^2$ (1.87x10$^{18}$ cm$^{-2}$) corresponding to ~ 55 dpa. The maximum temperature during implantation did not exceed 100 °C.

After PLEPS measurement at different ion implantation doses specimens were thermally treated with the aim to understand the influence of annealing on structure formation and defect behaviour. Thermal treatment was performed at „Universitaet der Bundeswehr" in Neubiberg (Munich, Germany). Specimens were annealed in argon atmosphere (10 kPa) at temperatures of 400, 475, 525 and 600 °C for 2 hours, then gradually cooled down (2 °C/min) and repeatedly measured after each temperature by PLEPS technique.

## 3. Results

Depth profiling of vacancy type defects was performed by PLEPS using positron energies between 2 and 18 keV corresponding to the mean penetration depths 15 – 525 nm. The evaluation of Fe-11.62%Cr measured spectra was performed by PosWin code [7]. The spectra were decomposed into three components.

Mean positron lifetimes (MLT) were calculated from the three partial components for all annealing temperatures. Specimens implanted at a level of 0.3 C/cm$^2$ showed significant decrease of mean positron lifetime at temperature of 600 °C (figure 1).

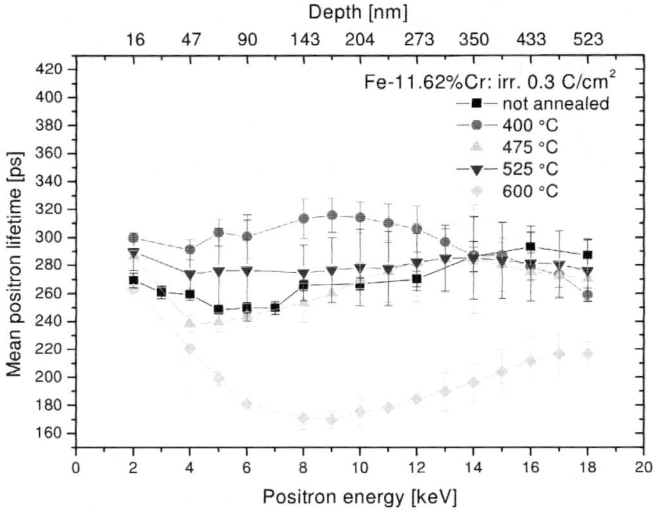


**Figure 1.** Mean positron lifetimes of annealed Fe-11.62%Cr alloys.

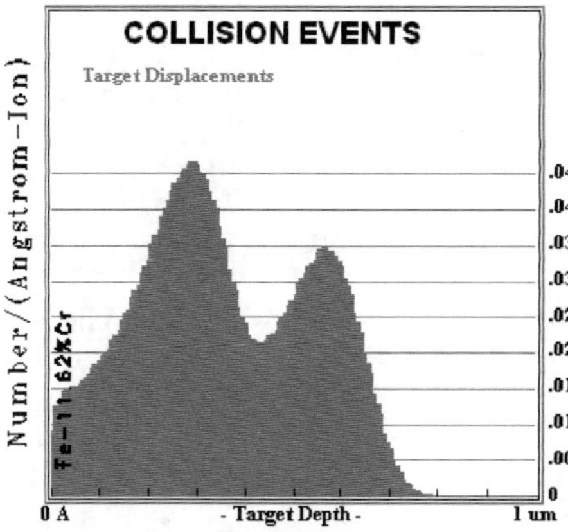


**Figure 2.** SRIM simulation of damage by 100keV (left peak) and 250keV (right peak) He ions.

Analysis based on the MLT at this temperature, at the depth of main damage caused by He ions (figure 2), could be interpreted as extensive decrease of defect size to monovacancies or small vacancy clusters [8, 9]. Components reflecting positron lifetimes in bulk and defects showed a decrease. Small clusters and the bulk with dislocations were recognized. Therefore, taking into account the literature [2], recovery of the microstructure at this temperature can be considered. The growth of MLT from the depth of ~170 nm to the surface could be among others caused by the oxide layer, which was also found by energy dispersive X-ray analysis.

The interpretation of data, obtained by the PLEPS technique at lower annealing temperatures (400, 475, 525 °C), was more difficult due to the behavior of the mean positron lifetimes (figure 1), as no expected differences in trend lines and maximum damage peaks from 100 keV and 250 keV He ions were observed. Therefore, these results were supported by the scanning electron microscopy (SEM) performed at the Institute of Materials Science (Faculty of Materials Science and Technology, STU). A diagonal cut under a 12° angle was performed prior to SEM measurements (figure 3) of an implanted Fe-11.62%Cr specimen. The figure shows major damage in two regions. Considering the results found in SRIM (Stopping and Range of Ions in Matter) simulation, these two regions could be assigned to the maximum damage caused by He ion implantation. The first peak (figure 2) in a depth of ~300 nm is very significant and implies to damage from 100 keV and partially also from 250 keV helium ions. This can be also recognized from SEM figures, in which the area closer to the surface has more significant damage (100 keV He) than the one further from the surface influenced mainly by 250 keV He ions. Therefore, we can assume that MLTs measured at temperatures of 400, 475, 525 °C were influenced by these two heavily damaged regions. Regeneration of the microstructure after annealing at temperature of 600 °C, followed by significant mean positron lifetimes decrease, was also confirmed by preliminary SEM measurement of this structure, where the damaged regions mostly disappeared.

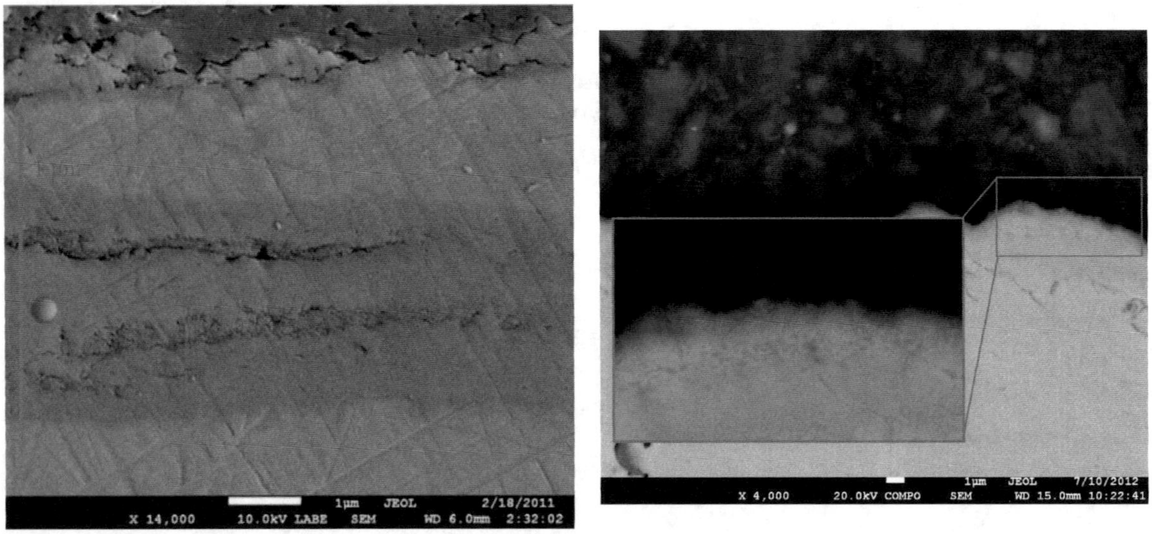


**Figure 3.** Alloy Fe-11.62%Cr implanted at 0.3 C/cm$^2$ level and annealed at 475 °C (left) and 600 °C (right). Annealing at temperatures between 0.3$T_m$-0.4$T_m$ ($T_m$ is melting temperature-in case of RAFM steels it is about 1400-1500 °C depending on the chromium content) can cause void and dislocation structure formation (the dislocation loops are unstable and grow into a dislocation network) and diffusion is sufficient for the formation of precipitates. This corresponds well to results obtained here for an annealing temperature of 475 °C, as ion damaged area was enlarged and voids reached width of tens of nanometers. Heating of the RAFM steels above 0.4$T_m$ led to continuous annealing out of displacement damage, resulting in little change in strength (at these temperatures strength sometimes decreases because irradiation-enhanced diffusion accelerates the normal thermal aging process) [2].

## 4. Conclusions

Investigation of irradiated Fe-11.62%Cr alloys after thermal annealing was performed in this work. Evaluation of the measured data for annealed Fe-11.62%Cr alloys by the PLEPS technique was made considering SEM results for a specimen annealed at 475 °C, which showed major damage in two helium affected regions. Mean lifetimes describing damage of annealed (400-525 °C) specimens were at a level of ~290 ps and in comparison to not annealed specimen, annealing effect was not observed. Only at 600 °C temperature, the mean positron lifetime showed a major decrease. Therefore, considering the literature [2] regarding the temperature influence on reduced activation ferritic/martensitic steels above $0.4T_m$, we could say that such large voids in affected area (<1 μm) were continuously annealed out or at least their size decreased to monovacancies and small vacancy clusters not exceeding size of 5 vacancies. These outcomes were supported also by the preliminary SEM results where the microstructure regeneration is noticeable.

Ion implantation damage produced in specimens was at a very high dpa level and together with thermal annealing and oxide layer on the surface, created complicated system for the final evaluation of measured data. Therefore, even if the positron annihilation spectroscopy is very useful technique for studies of vacancy type defects, we experienced some difficulties with characterization of large defects represented by MLT. Therefore, additional destructive or non-destructive techniques were applied in order to give us a complementary view on the examined materials.

## Acknowledgments

Financial contributions from the grant VEGA 01/0366/12 and 7RP-Euratom/CU are acknowledged. Supports from the Centre of Competence in New Materials, Advanced Technologies and Energy, ITMS: 26240220073 and OP VaV ITMS: 26220120014 are acknowledged (these are promoted by the Reseach and Development Operational programme funded by the European Regional Development Fund). The authors gratefully acknowledge the assistance of the staff of the high-intensity positron source (NEPOMUC) and the PLEPS staff at the Munich research Reactor FRM II.

## References

[1] Slugeň V 2011 *Safety of VVER-440 Reactors : Barriers against Fission Products Release* (London : Springer Verlag London)
[2] Ronald K L and Harries D R 2001 *High-Chromium ferritic and martensitic steels for nuclear applications* (USA: ASTM)
[3] Sperr P, Egger W, Kögel G, Dollinger G, Hugenschmidt Ch, Repper R and Piochacz C 2008 *Appl. Surf. Sci.* **255** 35
[4] Hugenschmidt Ch, Dollinger G, Egger W, Kögel G, Löwe B, Mayer J, Pikart P, Piochacz C, Repper R, Schreckenbach K, Sperr P and Stadlbauer M 2008 *Appl. Surf. Sci.* **255** 29
[5] Matijasevic M and Almazouzi A 2008 *J. Nucl. Mat.* **377** 147
[6] Kovac P, Pavlovic M and Dobrovodsky J 1994 *Nucl. Instr. Meth. B* **85** 749
[7] Kirkegaard P and Eldrup M 1972 *Comput. Phys. Commun.* **3** 240
[8] Troev T, Markowski A, Peneva S and Yoshiie T 2006 *J. Nucl. Mat.* **359** 93
[9] Kršjak V 2008 *Positron annihilation study of advanced nuclear reactor materials* Doctoral thesis (Slovakia: Slovak University of Technology)

# Effects of proton irradiation on Chinese domestic RPV steels by variable energy positron annihilation spectroscopy

**J. Jiang[1], Y C Wu[1], S.J. Wang[1], X. B. Liu[2] and R. S. Wang[2]**

[1] School of Physics and Technology, Hubei Nuclear Solid Physics Key Laboratory, Wuhan University, Wuhan 430072, P. R. China

[2] Center of Plant Life Management, Suzhou Nuclear Power Research Institute, Suzhou 215004, P. R. China

E-mail address: ycwu@whu.edu.cn

**Abstract**. The effects of proton irradiation on Chinese domestic reactor pressure vessel (RPV) A508-3 steels have been investigated by variable energy positron annihilation spectroscopy. The samples were irradiated using 110 keV protons at doses ranging from 0.2 to 2.0 $\times 10^{17}$ $cm^{-2}$ at room temperature. Defect profiles were analyzed by measuring the *S* parameter as a function of incident positron energy from 0.25 to 26keV, corresponding to mean depths of up to ~940nm. It is interesting to note that the *S*-parameter increases rapidly with increasing dose, which implies that matrix damage has a direct relation to dose-dependent effects in microstructural evolution.

## 1. Introduction

Positron annihilation spectroscopy (PAS) has been widely used as a probe for defect studies at the atomic dimension in solids, such as vacancies, vacancy clusters, dislocations, and nanometer-scale voids. The use of variable energy positron beams allows one to probe the depth profile of solids, and is particularly suitable for analysis of surface and near-surface defects [1]. PAS investigations of irradiation-induced defects in reactor pressure vessel (RPV) steels have been reported [2, 3]. Irradiation-induced embrittlement of RPV steels is one of the important degradation phenomena in the long term operation of light water reactors. Embrittlement is considered to be mainly caused by irradiation-induced changes in the microstructure such as the formation of solute nanoclusters, matrix defects (due to irradiation-induced point defect clusters), and dislocation loops [4-6]. However, despite the extensive research, the details of the microstructural changes are not well understood. In the present study, proton irradiation induced defects in Chinese domestic A508-3 steels were investigated by slow positron beam Doppler broadening spectroscopy.

## 2. Experiment

The domestic A508-3 steels were fabricated by China Firstheavy Industries. The forgings of these alloys were heated to 890℃ and kept for 7h, followed by quenching into water, then reheated to 650℃ and kept for 10h, and then cooled in the air. The chemical composition of these steels is shown in Table 1. The samples of $15\times15\times1$ $mm^3$ were cut from as-received materials for PAS measurements. The sample surfaces were carefully mechanically ground on successively finer abrasive papers, and then polished using diamond polishing suspensions in order to remove the effects of surface damage induced by artifacts.

Table 1: Chemical composition of domestic A508-3 steels (wt.%)

| Element | C | Si | Mn | S | P | Cr | Ni | Cu | Mo | V | Total |
|---|---|---|---|---|---|---|---|---|---|---|---|
| A508-3 | 0.167 | 0.193 | 1.35 | 0.002 | 0.005 | 0.086 | 0.738 | 0.027 | 0.481 | 0.007 | 3.056 |

A set of samples were irradiated by protons with the aim to simulate the radiation damage and embrittlement of real RPV steels under neutrons irradiation in nuclear reactors. The irradiation was performed at a terminal chamber of the ECR source 320kV high-voltage platform in Institute of Modern Physics (IMP) in Lanzhou, China. The samples were irradiated using 110 keV protons at doses ranging from 0.20 to 2.0 x$10^{17}$ $cm^{-2}$ at room temperature.

The Doppler broadening parameters *S* and *W*, linked to the annihilation of positrons by low- and high- momentum electrons, respectively, were measured for slow positrons incident with energies E from 0.25 to 26 keV by a high-purity Ge detector (~1.64 keV resolution at 511 keV). Open-volume defects such as vacancies and voids are populated by lower- momentum electrons, and thus have relatively large *S* and small *W* parameters.

## 3. Results and Discussion

Fig.1 shows the *S* and $\Delta S/S$ parameters ($\Delta S/S=(S_{irradiated} - S_{unirradiated})/ S_{unirradiated}$) as a function of incident positron energy (mean implantation depth) in as-received (unirradiated) and irradiated samples. The mean depth of the annihilating positron from the surface is calculated using the following established equation:

$$Z(E)=(4\times10^4/\rho)E^{1.6} \quad (1)$$

Where $Z(E)$ is expressed in units of nanometers, ρ is the density in units of $kg/m^3$ (In the calculation we used the density of pure iron with a value of $7.86\times10^3 kg/m^3$), and $E$ is the incident positron energy in keV. The top axis in figure 1 gives the calculated mean depth from the surface of the annihilated positron, according to equation (1). The *S* parameter for the irradiated sample became much larger than that for the unirradiated ones, which clearly indicates that positrons detect the presence of vacancy-type defects generated during proton irradiation.

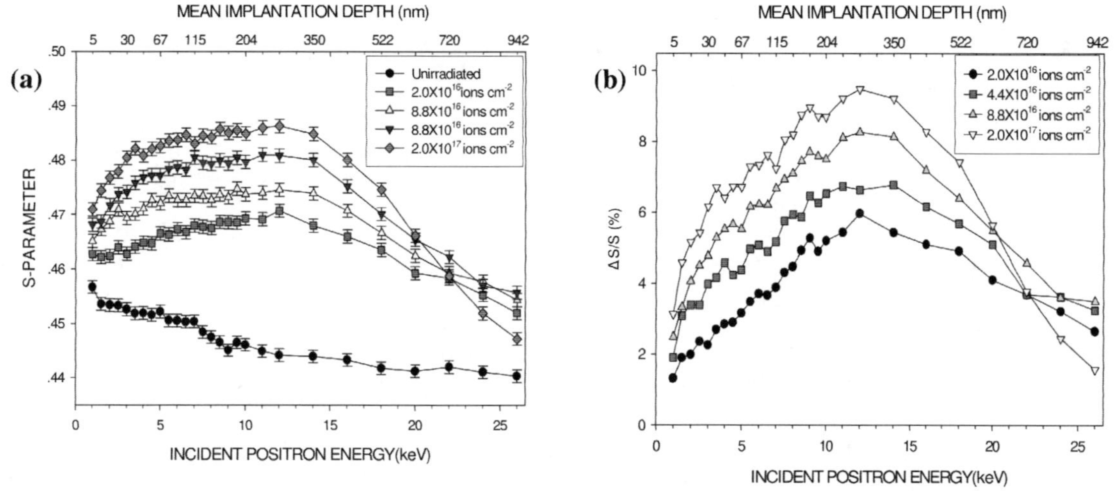


**Figure 1** *S* parameter and $\Delta S/S$ as a function of incident positron energy (mean implantation depth) unirradiated and irradiated steels.

A simple trapping model can be employed. The fraction of positrons annihilating in the bulk ($f_b$), in the surface ($f_s$) and of those trapped and annihilated at defects ($f_d$) can be expressed as follows:

$$S = f_s S_s + S_d f_d + S_b F_b \tag{2}$$

where $S_s$, $S_d$ and $S_b$ are the S parameters corresponding to the surface, the defect and the bulk, respectively, and $f_s$, $f_d$ and $f_b$ are the fractions of positron annihilated in the same three states ($f_s + f_d + f_b = 1$). As can be seen in figure 1(a), *S* for the unirradiated sample decreases from its surface value to its bulk value as positron energy increases, as $f_s$ decreases from 1 to 0. However, for irradiated samples, there exist ranges of positron energy (such as between 3 and 14 keV) over which the essentially constant S value can be directly attributed to damaged region induced by proton irradiation. At higher positron energy, S increases with energy from $S_d$ to $S_b$, as $f_d$ decreases from 1 to 0 and $f_b$ increases from 0 to 1. Implanted and thermalized positrons annihilate either in bulk or at the trapping sites.

When the samples were applied to different dose, the Δ*S/S* parameter increases with the dose, and the increases in the ΔS/S parameter as a function of mean implantation depth appear a peak value at ~12 keV, corresponding to the depth of about 270 nm (see figure 1(b)). Fig.2 shows the proton range distribution and the target displacement profile after proton irradiation with 110 keV protons by SRIM 2006 calculations. The cascades region corresponds to the region where incident ions mainly interact with atoms of the solid via nuclear collisions and finally stop into the lattice. The tracks region is localized between the surface and the nuclear cascades region, and corresponding to the zone where ions slow down mainly by electronic energy loss processes. The other part is non-implanted region. The projected range of hydrogen was at the depth of about 550 nm. Compared to figure 1(b) and figure 2(b), both have a similar peak. The main vacancy source is located in the cascades region, however, the ΔS/S parameters are higher for the tracks region than for the cascades region, i.e. their peaks move to lower positron energy, only half of the projected range of hydrogen. This can be explained by the formation of vacancy-H complex. Ishizaki et al. [7] reported that vacancy–H complexes and vacancy–He complexes were introduced in Fe irradiated with H or He ions. A large number of hydrogen atoms were strongly trapped at vacancies in the cascades region and the vacancy–H complexes formed [8]. As mentioned above, H ions were implanted to depths of about 550 nm; surface-to-peak depths were about 270 nm. Therefore, a large number of H atoms were strongly trapped at vacancies in the cascades region. The H atoms recombined with some vacancy clusters to form vacancy cluster–H complexes. When positrons are trapped at vacancy-type defects coupled with H, the value of *S* is lower than that of vacancy-type defects. Therefore, the *S* parameter of cascades region is lower than that of tracks region due to the suppression of vacancy clustering by vacancy trapping of H near the projected range of H. The number of H atoms in the tracks region is less than that in cascades region. Thus, *S* parameter does not significantly decrease in the tracks region.

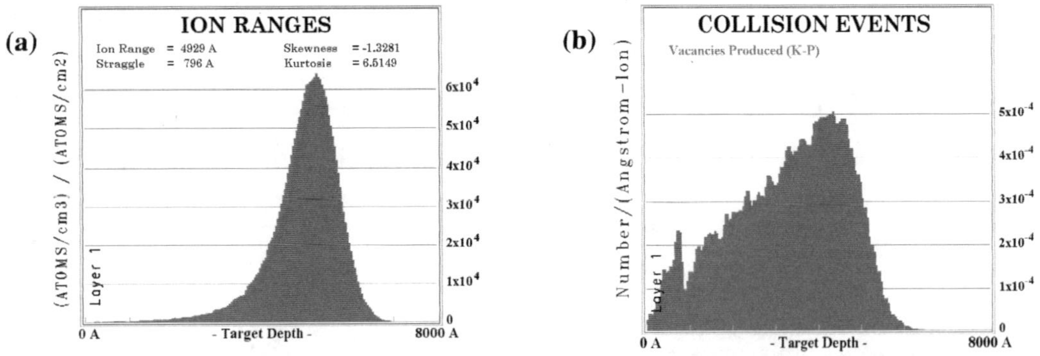


**Figure 2** Results obtained from SRIM 2006 calculations for proton irradiation with 110 keV protons showing: (a) the proton range distribution during irradiation and (b) the target displacement profile after irradiation.

Information about the number of defect types could be revealed by plotting the *S*-parameter as a function of the *W*-parameter (*S*-*W* plots). Fig.3 shows that *S*- and *W*- values for all samples fall on a common locus, which only consists of one straight-line segment. This suggests that only one type of defect (open-volume defect) is present in all measurement both in unirradiated and irradiated steels. Therefore, it is interpreted that an increase in S-parameter in Fig.1 mainly comes from the formation of defects and voids near the surface during proton irradiation. Compared to unirradiated sample, the relative change of $\Delta S/S$ parameter in the most heavily irradiated sample (see Fig.1) exceeds 9%, which means that there exist large-size vacancy clusters or vacancy cluster-H complexes induced due to proton irradiation. Further investigations of annealing experiment at elevated temperature are in progress and will be reported in the future.

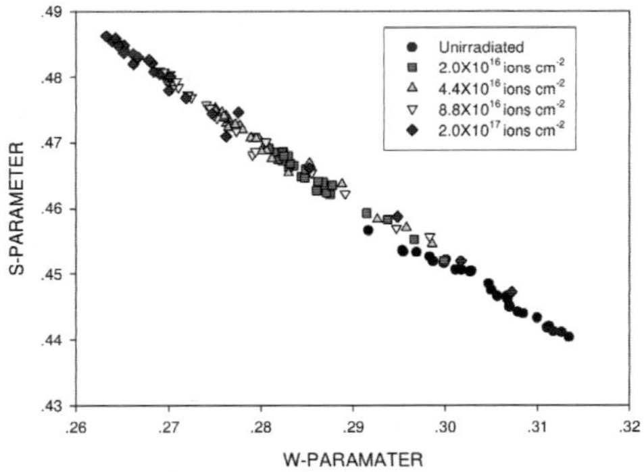


**Figure 3** *S* parameter as a function of the *W* parameter for unirradiated and irradiated steels. (*S* is from figure 2 data)

## 4. Conclusions

Slow positron beam Doppler broadening measurements show that *S* parameter increases evidently with dose after proton irradiation owing to open-volume defects induced by proton irradiation. When positrons are trapped at vacancy-type defects coupled with hydrogen, the value of *S* is lower than that of vacancy-type defects. Therefore, the *S* parameter of cascades region is lower than that of tracks region due to the suppression of vacancy clustering by vacancy trapping of hydrogen near the projected range of hydrogen.

## Acknowledgements

This research is supported by the National Natural Science Foundation of China under grants 10975110, 51071111 and 11175136 and by the National High Technology Research and Development Program of China (863 Program) under grants 2012AA050901, and by the Fundamental Research Funds for the Central Universities, and is also performed under the contract between Suzhou Nuclear Power Research Institute Co., Ltd. and Wuhan University.

## References

[1] Schultz P J and Lynn K G 1988 *Rev. Mod. Phys.* **60** 701
[2] Zeman A, Debarberis L, Slugen V and Acosta B 2006 *Appl. Surf. Sci.* **252** 3290
[3] Slugen V, Kogel G, Sperr P and Triftshauser W 2002 J. Nucl. Mater. **302** 89
[4] Bergner F, Lambrecht M, Ulbricht A and Almazouzi A 2010 *J. Nucl. Mater.* **399** 129
[5] Toyama T, Nagai Y, Tang Z, Hasegawa M, Almazouzi A, van Walle E and Gerard R 2007 *Acta Mater.* **55** 6852
[6] Miller M K and Russell K F 2007 *J. Nucl. Mater.* **371** 145
[7] Ishizaki T, Xu Q, Yoshiie T and Nagata S 2004 *Mater. Trans.* **45** 9
[8] Ju X, Qiu J, Xin Y, Liu S and Wang B.Y 2011 *J. Nucl. Mater.* **411** 20

Journal of Physics: Conference Series **443** (2013) 012038 doi:10.1088/1742-6596/443/1/012038

# Surface states and annihilation characteristics of positrons trapped at the oxidized Cu(100) surface

**N G Fazleev**[1,2], **A H Weiss**[1]
[1] Physics Department, University of Texas at Arlington, Arlington, TX 76019, USA
[2] Physics Department, Kazan Federal University, Kazan 420008, Russian Federation

E-mail: fazleev@uta.edu

**Abstract.** In this work we present the results of theoretical studies of positron surface and bulk states and annihilation probabilities of surface-trapped positrons with relevant core electrons at the oxidized Cu(100) surface under conditions of high oxygen coverage. Oxidation of the Cu(100) surface has been studied by performing an *ab-initio* investigation of the stability and electronic structure of the Cu(100) missing row reconstructed surface at various on-surface and subsurface oxygen coverages ranging from 0.5 to 1.5 monolayers using density functional theory (DFT). All studied structures have been found to be energetically more favorable as compared to structures formed by purely on-surface oxygen adsorption. The observed decrease in the positron work function when oxygen atoms occupy on-surface and subsurface sites has been attributed to a significant charge redistribution within the first two layers, buckling effects within each layer and an interlayer expansion. The computed positron binding energy, positron surface state wave function, and annihilation probabilities of the surface trapped positrons with relevant core electrons demonstrate their sensitivity to oxygen coverage, atomic structure of the topmost layers of surfaces, and charge transfer effects. Theoretical results are compared with experimental data obtained from studies of oxidation of the Cu(100) surface using positron annihilation induced Auger electron spectroscopy (PAES). The results presented provide an explanation for the changes observed in the probability of annihilation of surface trapped positrons with Cu 3p core-level electrons as a function of annealing temperature.

## 1. Introduction

The study of oxidation of single crystal metal surfaces is important in understanding the processes of corrosion and catalysis associated with thin film metal oxides. Copper oxides have also attracted much attention in the area of renewable energy research. The process by which oxide layers are formed on metal surfaces, however, is still not well understood. The structures formed on the oxidized transition metal surfaces vary from simple adlayers of chemisorbed oxygen to more complex structures which result from diffusion of oxygen into subsurface regions. Unlike conventional Auger techniques that average over several atomic layers, positron annihilation induced Auger electron spectroscopy (PAES) has a very high degree of top layer selectivity due to the fact that the positrons are trapped just outside the surface prior to annihilation, hence is ideal to probe the near surface region [1]. Recently this technique has been applied to study the oxidized Cu(100) single crystal [2,3]. Figure 1 shows a comparison of PAES spectra from the clean Cu(100) surface and the oxidized surface. The PAES spectrum from the clean Cu(100) surface is shown in figure 1 (a). Two peaks may be observed at ~60 eV and ~103 eV corresponding to the Cu $M_{2,3}VV$ and Cu $M_1VV$ Auger transitions, respectively. The PAES spectrum obtained from the Cu(100) sample after oxidation is shown in figure 1 (b). These results indicate that the intensity of the Cu $M_{2,3}VV$ Auger peak for the oxidized Cu(100) surface is reduced by ~93% compared to the clean surface. Experimental annihilation probability, $p_{nl}$, of positrons trapped at the oxidized Cu(100) surface with Cu *3p* core-level electrons estimated from the measured intensity of the positron annihilation induced Cu $M_{2,3}VV$ Auger peak shows a significant decrease when compared to its value obtained from the clean Cu(100) surface.

This paper presents the results of theoretical studies of positron surface states and annihilation characteristics of surface trapped positrons with relevant core electrons at the oxidized Cu(100) surface. Such studies are indispensable for the full interpretation of the PAES data. Such interpretation requires an understanding of the localization and spatial extent of the positron surface-state wave function at oxide surfaces, and the surface atomic structure dependence of probabilities of annihilation of surface-trapped positrons with relevant core-level electrons.

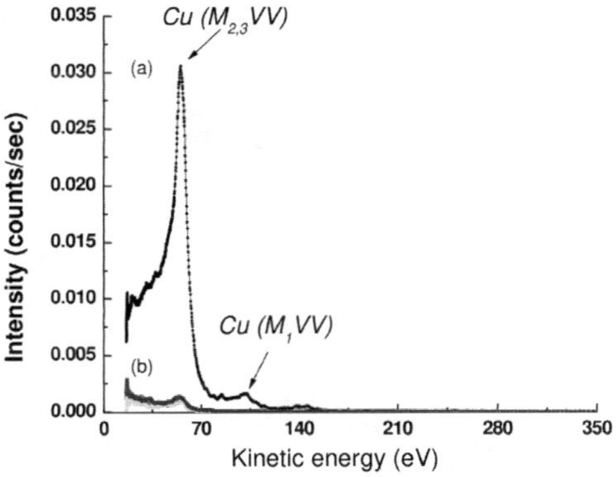


**Figure 1.** PAES spectra of Cu(100) single crystal (a) before and (b) after oxidation.

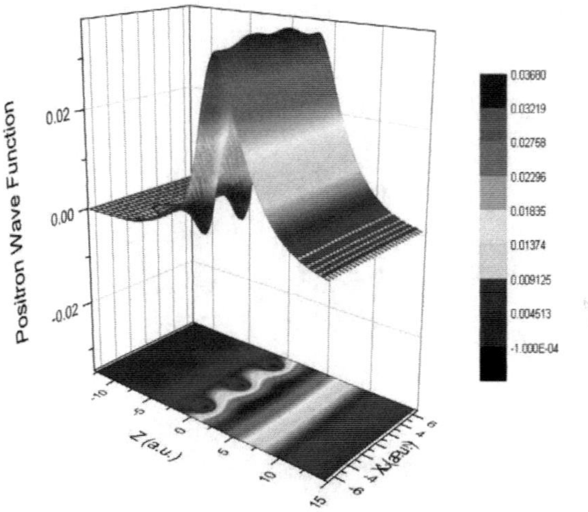


**Figure 2.** Positron surface state wave function at the clean Cu(100) surface.

## 2. Theory

The positron potential $V(\mathbf{r})$ at a semiconductor surface contains an electrostatic Hartree potential $V_H(\mathbf{r})$ and a correlation part $V_{corr}(\mathbf{r})$. $V_H(\mathbf{r})$ is constructed as a superposition of atomic Coulomb potentials. Atomic calculations were performed self-consistently within the local-spin-density approximation [6] using the exchange-correlation functional from Ref. [7]. We followed the method of Weinert and Watson [8] to account for the effects of charge redistribution at the surface and placed the atoms in a "compensating" potential well extending from the atom center out to one Wigner-Seitz radius and then linearly ramping to a value of 0.00 Ry at twice the Wigner-Seitz radius and beyond. The height of the potential well was adjusted for the surface to provide the positron potential at the surface that gives the proper positron work function. The atomic wave functions then provided the overlapping electron densities and corresponding atomic Coulomb potentials via Poisson's equation. In the bulk $V_{corr}(\mathbf{r})$ is described using the local density approximation (LDA) [5, 9]. Outside the surface $V_{corr}(\mathbf{r})$ is described by the image-type potential that is constructed to have the same corrugations as the total electron density [5, 10]. The joining of the image-type potential to the local density correlation potential is done by taking $V_{corr}(\mathbf{r})$ to be the larger of the two at each point outside the surface. The positron was assumed to be in the ground state and delocalized in the XY plane of the surface, and to have a crystal momentum in this plane of $\mathbf{k} = 0$. Positron surface states were obtained by solving Schrödinger's equation numerically with the boundary condition that the positron wave function vanishes sufficiently far into the bulk or the vacuum. Positron annihilation rates, $\lambda_{n,l}$, with specific core electrons, described by quantum numbers $n$ and $l$, and the total annihilation rate, $\lambda$, (and corresponding positron surface state lifetime, $\tau$,) are calculated taking electron-positron correlation effects explicitly into account using the LDA [5, 9]. The probability of annihilation of surface trapped positrons, $p_{n,l}$, with electrons in the core level $n,l$ is obtained dividing $\lambda_{n,l}$ by $\lambda$: $p_{n,l} = \lambda_{n,l} / \lambda$ .

### 3. Results and discussion

Calculations of positron surface states and annihilation probabilities of surface trapped positrons with relevant core electrons, $p_{n,l}$, were performed initially for the adsorption of oxygen at hollow sites of the unreconstructed Cu(100) surface. We considered 0.125, 0.25, 0.50, and 1.00 monolayer (ML) oxygen coverages. Calculations were also performed for the structures constructed by adding oxygen atoms to various surface hollow and subsurface octahedral sites of the 0.5 ML missing row reconstructed phase of the Cu(100) surface with oxygen coverages ranging from 0.5 to 1.5 ML. All surfaces were fully relaxed in calculations. The plots of the calculated positron surface states wave function at the clean and oxidized Cu(100) surface are presented in figures 2-4.

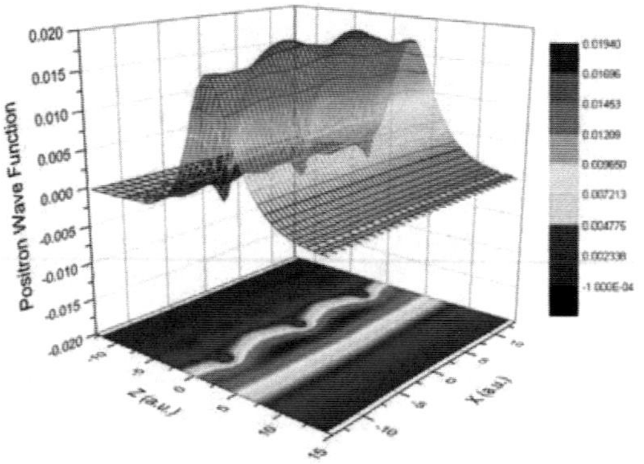


**Figure 3.** Positron surface state wave function at the non-reconstructed Cu(100) surface with 0.5 ML oxygen coverage..

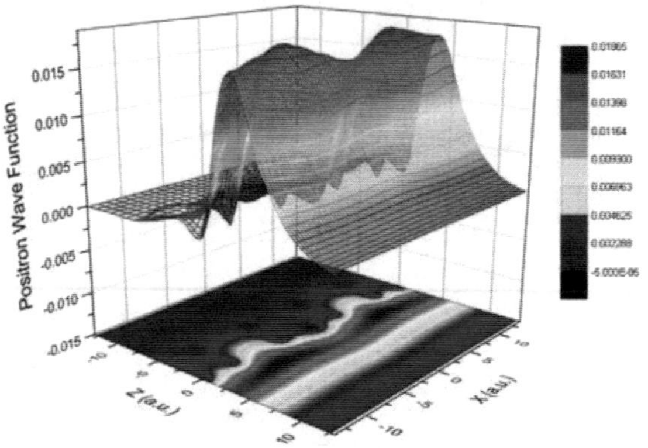


**Figure 4.** Positron surface state wave function at the missing row reconstructed Cu(100) surface with 0.75 ML oxygen coverage.

The computed binding energy, $E_b$, for a positron trapped at the oxidized Cu(100) surface, the positron work function, $\Phi_p$, the surface-state lifetime, $\tau$, and the annihilation probabilities of the surface trapped positron with relevant core electrons, $p_{n,l}$, are given in Tables 1 and 2.

**Table 1.** Theoretical positron binding energy, $E_b$, positron work function, $\Phi_p$, positron surface-state lifetime, $\tau$, annihilation probabilities of the surface trapped positron with relevant core electrons, $p_{n,l}$, computed for the unreconstructed Cu(100) surface with adsorbed oxygen at surface hollow sites.

| System | $E_b$ (eV) (theory) | $\Phi_p$ (eV) (theory) | $\tau$ (ps) surface state (theory) | $p_{n,l}$ (%) surface (theory) | | |
|---|---|---|---|---|---|---|
| | | | | Cu 3s | Cu 2p | Cu 2s+2p |
| Cu(100) | 2.75 | -0.30 | 461 | 1.766 | 6.990 | 8.756 |
| Cu(100) + O (1/8 ML) | 3.14 | -0.61 | 503 | 1.203 | 4.806 | 6.009 |
| Cu(100) + O (1/4 ML) | 3.12 | -1.01 | 495 | 1.043 | 4.172 | 5.215 |
| Cu(100) + O (1/2 ML) | 3.08 | -1.25 | 470 | 0.748 | 2.991 | 3.740 |
| Cu(100) + O (1 ML) | 2.94 | -1.71 | 448 | 0.398 | 1.595 | 1.993 |

It follows from figures 2-4 that the adsorbed oxygen pushes the positron wave function from the top layer of Cu atoms away into the vacuum reducing its overlap with Cu atoms. The calculations show that low coverage oxidation of the Cu(100) surface with the absorbed oxygen up to the total 1ML coverage leads to a decrease in the positron annihilation probabilities with Cu *3s* and *3p* core electrons, and hence to a decrease in the intensity of Cu $M_{2,3}VV$ Auger signal in agreement with the experiment. Performed calculations confirm that $E_b$, $\Phi_p$, $\tau$, and $p_{n,l}$ are sensitive to atomic structure and elemental composition of the topmost layers of the oxidized Cu(100) surface.

The calculations show that no positron bound state exists at the oxidized missing row reconstructed Cu(100) surface under conditions of high coverage oxidation (e.g. when the total on-surface and sub-surface oxygen coverage is 1. 5 ML). Under these conditions the positron annihilation probabilities with Cu *3s* and *3p* core would decrease essentially to zero. Consequently we would expect the intensity of the Cu $M_{2,3}VV$ PAES signal, which is due to the annihilation of the *3p* core to decrease to essentially zero as is the experiment.

**Table 2.** Theoretical positron binding energy, $E_b$, positron work function, $\Phi_p$, annihilation probabilities of the surface trapped positron with relevant core electrons, $p_{n,l}$ , computed for the structures constructed by adding oxygen atoms to various surface hollow and subsurface octahedral sites of the 0.5 ML missing row reconstructed (MR) phase of the Cu(100) surface.

| System | $E_b$ (eV) (theory) | $\Phi_p$ (eV) (theory) | $p_{n,l}$ (%) surface (theory) | | |
|---|---|---|---|---|---|
| | | | Cu 3s | Cu 2p | Cu 2s+2p |
| Cu(100) | 2.75 | -0.30 | 1.77 | 6.99 | 8.76 |
| Cu(100) + O (Total 0.50 ML) | 2.87 | -0.50 | 0.53 | 2.24 | 2.77 |
| Cu(100) + O (Total 0.75 ML) | 3.14 | -0.60 | 0.31 | 1.41 | 1.72 |
| Cu(100) + O (Total 1.00 ML) | 3.12 | -1.05 | 0.23 | 1.17 | 1.40 |
| Cu(100) + O (Total 1.50 ML) | not bound | -1.23 | N/A | N/A | N/A |

## 4. Conclusions

The results of PAES studies of the oxidized Cu(100) single crystal have been analyzed by performing calculations of positron surface and bulk states and positron annihilation characteristics taking into account the charge redistribution at the surface and surface reconstructions. Localization and spatial extent of the positron surface-state wave function, the positron binding energy, work function and annihilation characteristics of the surface trapped positron with relevant core-level electrons have been found to be sensitive to charge transfer, atomic structure and chemical composition of the topmost layers of the surfaces. It has been shown that at low oxygen coverages the annihilation probabilities of surface trapped positrons with Cu *3s* and *3p* core electrons decrease with the increase of the oxygen coverage which is consistent with the observed decrease of the intensity of the positron annihilation induced Cu $M_{2,3}VV$ Auger signal. The calculations show that for high oxygen coverages ( $\geq$1.5 ML) the positron is not bound to the surface. This provides an explanations for the large decrease in the Cu $M_{2,3}VV$ PAES intensity observed in the experiment under conditions of high oxygen coverage on the Cu(100) surface.

**Acknowledgments**
This work was supported in part by the National Science Foundation Grant #DMR-0907679.

**References**

[1] Koymen A R, Lee K H, Yang G and Jensen K O 1993 *Phys. Rev. B* **48** 2020
[2] Fazleev N G, Nadesalingam M P, Maddox W, Mukherjee S, Rajeshwar K and Weiss A H 2010 *Surface Science* **604** 32; Fazleev N G, Maddox W B and Weiss A H 2011 *J. Phys.: Conf. Ser.* **262** 012019.
[3] Nadesalingam M 2007 *PhD Dissertation*, University of Texas at Arlington
[4] Starnes S G 2001 *MS Thesis* University of Texas at Arlington
[5] Fazleev N G, Fry J L and Weiss A H 2004 *Phys. Rev. B* **70** 165309
[6] Gunnarsson O and Lundqvist B I 1976 *Phys. Rev. B* **13** 4274
[7] Ceperley D M and Adler B J 1980 *Phys. Rev. Lett.* **45** 566
[8] Weinert M and Watson R E 1984 *Phys. Rev. B* **29** 3001
[9] Boronski E and Nieminen R M 1986 *Phys. Rev. B* **34** 3820
[10] Nieminen R M and Puska M J 1983 *Phys. Rev. Lett.* **50** 281

Journal of Physics: Conference Series **443** (2013) 012039 doi:10.1088/1742-6596/443/1/012039

# Defects structure characterization of NiMnGa alloys by PALS

**David Merida[1], Jose Angel Garcia[1], Fernando Plazaola[2], Vicente Sanchez-Alarcos[3], Jose Ignacio Pérez-Landazábal[3] and Vicente Recarte[3]**

[1]Fisika Aplikatua II Saila, Euskal Herriko Unibertsitatea UPV/EHU, p.k. 644, 48080 Bilbao, Spain

[2]Elektrizitate eta Elektronika Saila, Euskal Herriko Unibertsitatea UPV/EHU, p.k. 644, 48080 Bilbao, Spain

[3]Departamento de Física, Universidad Pública de Navarra, Campus de Arrosadia, 31006 Pamplona, Spain

E-mail: david.merida@ehu.es

**Abstract**. We have studied the behaviour of defects in off-stoichiometric Ni-Mn-Ga ferromagnetic shape memory alloys by means of positron lifetime spectroscopy. The measurements presented in this work have been performed in six ternary alloys. The studied samples cover a large composition range. Positron experiments have been performed at room temperature after subsequent isochronal annealing at different temperatures up to a maximum temperature of 700ºC. Results show a large variation of the average positron lifetime value with the isochronal annealing temperature in three of these samples, with significant differences between them. In the other three, the response is quite different. The results are discussed in terms of different types of positron trapping defects and their evolution with the annealing temperature. The present work shows a high dependence of recovery behaviour with composition in NiMnGa ferromagnetic shape memory alloys.

## 1. Introduction

Since giant magnetic-field-induced strain (MFIS) was first reported on Ni-Mn-Ga ferromagnetic shape memory alloys (FSMA) by Ullakko et al. [1], a great amount of work has been performed to understand and improve the system in order to be implemented in practical devices [2]. The main drawbacks are the high brittleness, the low values of the martensitic transformation (MT) and the Curie temperatures of Ni-Mn-Ga alloys. These limitations have stimulated the research of new ferromagnetic shape memory alloys having better mechanical properties. Indeed, several studies have been performed to improve the mechanical and thermal properties [3].

Different techniques have been used to analyze diverse properties of Ni-Mn-Ga alloys; however, very little work has been performed to study the role of vacancies in the martensitic transformation [4-5]. Positron annihilation spectroscopy is a very powerful technique to investigate vacancy-type defects in metals [6]. We have used positron annihilation spectroscopy measurements to study the behaviour

[1] To whom any correspondence should be addressed.

of vacancy-type defects in different composition off-stoiquiometry of Ni-Mn-Ga polycrystalline alloys. Positron lifetime experiments have been performed at room temperature after subsequent isochronal annealing up to 700°C.

## 2. Experimental

Polycrystalline ingots of $Ni_2MnGa$ were prepared from high purity elements by arc melting under protective Ar atmosphere. The ingots were homogenized in vacuum quartz ampoules at 1000 °C for 24 hours. Samples for measurements were obtained from discs previously cut from the center of the ingots by slow speed diamond saw. These discs were used for positron lifetime measurements. Subsequent annealing treatments of 30 minutes at 900 °C followed by quenching into ice water were performed on the alloys in a vertical furnace.

For positron lifetime measurements a fast system with a resolution of 240 ps was used and a conventional $^{22}Na$ source on a kapton foil was employed as the positron source. All lifetime spectra were analyzed after subtracting a constant source contribution. The statistic of each spectrum was always better than 1.6 million counts. For the isochronal annealing the same heating/cooling rate of 10 K/min was used. The points in each positron lifetime curve correspond to the average obtained after 5 measurements. For all of the results the error of the measured positron lifetimes is within error bars showed in the figures. B samples exhibit structural modulation; however, A samples do not show modulation [7]. The compositions of the six samples, determined by EDX, are shown in Table 1.

Table 1. Composition of the modulated and non-modulated samples used in the measurements.

| | Ni | Mn | Ga |
|---|---|---|---|
| **A1** | 51.6 | 29.3 | 19.1 |
| **A2** | 55.8 | 19.7 | 24.5 |
| **A3** | 53.9 | 25.4 | 20.7 |
| **B1** | 50.1 | 27.8 | 22.2 |
| **B2** | 48.7 | 26.2 | 25.1 |
| **B3** | 53.5 | 20.8 | 25.7 |

## 3. Results and discussion

Figure 1 shows the behaviour of the average positron lifetime ($\tau_{av}$) as a function of isochronal annealing temperature for sample A1. The first value of $\tau_{av}$ amounts to about 182 ps and corresponds to the sample as quenched from 900°C. Only one positron lifetime component is present in the spectrum indicating that all positrons are annihilating from one positron state, whose lifetime near 182 ps. The second point corresponds to the first annealing after the quench, performed at 200°C. Isochronal annealing performed between 200°C and 350°C do not change within the experimental error the value of $\tau_{av}$ measured after the quench. This lifetime value is typical of positrons annihilating from vacancies in metals [8]. Between 350°C and 450°C $\tau_{av}$ decreases around 20 ps, down to a value near 160 ps, where it remains up to the highest measured isochronal annealing temperature. A second component is needed for a correct fit of the last spectra. The decrease in $\tau_{av}$ indicates the elimination of thermal vacancies retained during the quench. The fact that in the last points there are two components indicates that a total elimination of vacancies is not reached. There can be two reasons to explain this behaviour: 1) the presence of constitutional vacancies, which cannot be eliminated by thermal treatments, 2) the used heat treatment is not the adequate one to obtain the elimination of vacancy type defects up to levels below the sensitivity of the PALS technique.

Figure 2 shows the behaviour of the $\tau_{av}$ as a function of isochronal annealing temperature for sample A2. The first three points remain near 182 ps; however, at 300°C $\tau_{av}$ decreases abruptly to 165 ps. For isochronal annealing temperatures between 300°C and 500°C $\tau_{av}$ increases monotonically up to around 178 ps, where it remains constant up to the highest measured isochronal annealing temperature. There are two significant differences in the behaviour of the previous A samples. The first one is the shift of about 100 ºC on the starting of $\tau_{av}$ decrease from around 180 ps (the shift amounts to about 150 ºC in reaching the minimum of $\tau_{av}$). The second one is the lifetime increase observed in sample A2 for annealing temperatures above 300 ºC. The first difference can be due to different migration energies of the involved vacancy type defects. The $\tau_{av}$ increase observed in the isochronal annealing of sample A2 suggests that new thermal vacancies appear at temperatures above 300 ºC. That is to say, the cooling rate of 10ºC/min is too fast to eliminate the newly created thermal vacancies. The observed significant differences indicate that the migration energies of the vacancy type defects involved in samples A1 and A2 are quite different. It is difficult to imagine such large differences caused by the same vacancy type defect in samples with identical structure but different composition. Therefore, one may conclude that the observed differences are caused by different vacancy type defects in the structure of samples A1 and A2. Moreover, table 1 shows that A1 sample is poor in Ga and sample A2 in Mn, which suggests that Ga vacancies are involved in the observed behaviour of sample A1 and Mn vacancies in the one of sample A2.

Figure 3 shows the behaviour of $\tau_{av}$ as a function of isochronal annealing temperature for sample A3. Up to 300 ºC $\tau_{av}$ remains constant at around 183 ps. It decreases between 300ºC and 450ºC down to around 170 ps, where it remains quite constant. This behaviour is quite similar to that of A1 sample (poor in Ga, too), but with a higher remaining vacancy content.

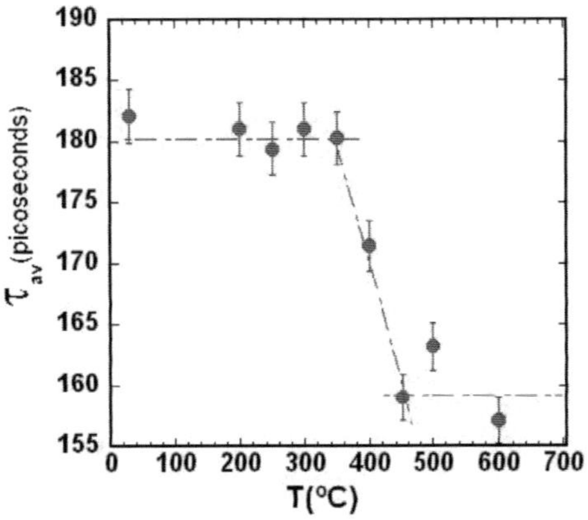


**Figure 1.** $\tau_{av}$ as a function of the isochronal annealing temperature for A1 sample.

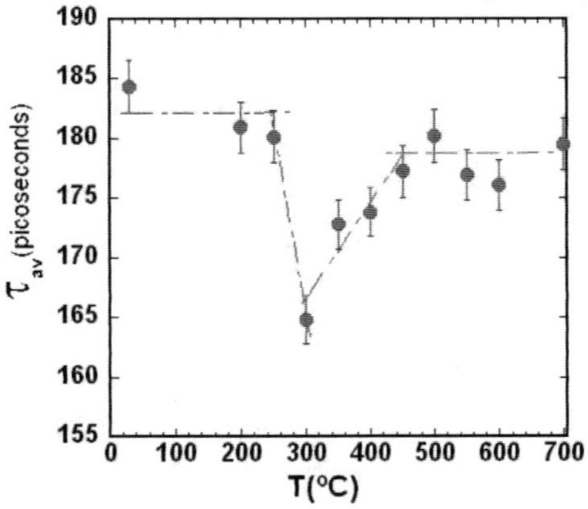


**Figure 2.** $\tau_{av}$ as a function of the isochronal annealing temperature for A2 sample.

Figure 4 shows the behaviour of $\tau_{av}$ as a function of isochronal annealing temperature for B1, B2 and B3 samples. The behaviour of $\tau_{av}$ in these samples is very different to the previous ones. B1 sample shows only a valley between 250 ºC and 450 ºC, but the overall $\tau_{av}$ decrease amounts only to about 3 ps. It is not possible to decompose the spectra in all the studied temperature range and the positron lifetime value of 180 ps, measured below 250 ºC, is about the same value measured above 450 ºC. So, positrons are annihilating in saturation from vacancies below 250 ºC and above 450 ºC. Therefore, the valley observed in the temperature range 250-450 ºC cannot be understood as the elimination of only one type of vacancies.

The value of $\tau_{av}$ measured in B2 sample remains almost constant and in saturation in the studied temperature range. In the case of B3 sample only three points have been measured to confirm the behaviour of the other two samples. In the case of B samples, modulated ones the vacancy content is very high in all the isochronal annealing temperature range.

There exists a clear difference between the results in A and B samples. In A samples, a clear elimination of vacancies is observed, while in the modulated ones, such elimination is not observed at all. This means that the behaviour of vacancies in these two kinds of samples is very different.

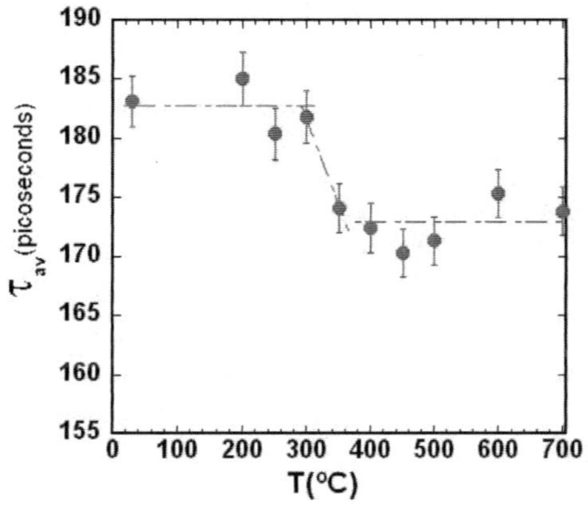


**Figure 3.** $\tau_{av}$ as a function of the isochronal annealing temperature for A3 sample.

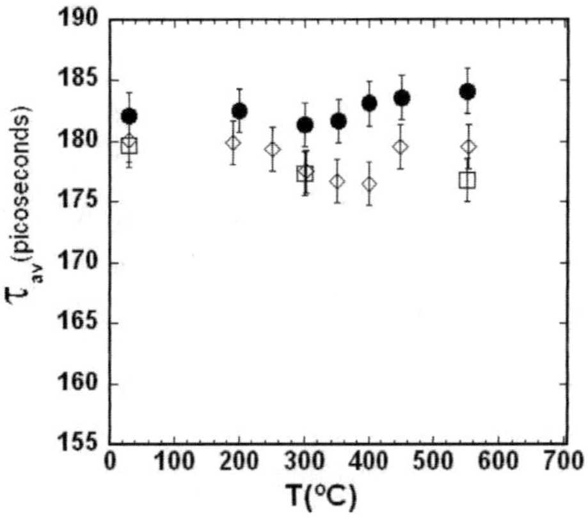


**Figure 4.** $\tau_{av}$ as a function of the isochronal annealing temperature for B1 (open rhombus), B2 (full circle) and B3 samples (open squares).

## 4. Conclusion

The behaviour of vacancy type defects in off-stoichiometry polycrystalline $Ni_2MnGa$ ferromagnetic shape memory alloys has been studied by measuring the effect of the thermal treatment. To analyze the effect of the composition on the isochronal annealing behaviour six different samples, covering a large composition range, were studied. Positron lifetime measurements show a clear and significant difference in the behaviour of positron traps (vacancies) in the two A and B family samples. In B samples the vacancy content remains above the saturation limit in the studied temperature range. However, in the case of A samples there is a clear elimination of vacancies. Within samples of A type a clear difference in the behaviour of the positron traps is measured, too. The obtained results suggest that the type of vacancy present in the sample, after quenching, depends on the composition of the alloy. Indeed, the recovery state can be shifted as much as 150 °C.

## Acknowledgments

To the Basque and the Spanish Governments under Grants IT-443-10 and MAT2012-37923.

## References

[1] Ullakko K, Huang J K, Kantner C, O'Handley R C and Kokorin V V 1996 Appl. *Phys. Lett.* **69** 1966
[2] Mostafa K M, Van Caenegem N, de Baerdemaeker J, Segers D and Houbaert Y 2007 *Physica Status Solidi (c)* **4** 3554
[3] Oikawa K, Ota T, Ohmori T, Tanaka Y, Morito H, Fujita A, Kinuma R, Fukamichi K and Ishida K 2002 *Appl. Phys. Lett* **81** 5201
[4] Merida D, Garcia J A, Apiñaniz E, Plazaola F, Sanchez-Alarcos V, Pérez-Landazábal J I and Recarte V 2010 *Materials Science Forum* **635** 55
[5] Merida D, Garcia J A, Apiñaniz E, Plazaola F, Sanchez-Alarcos V, Pérez-Landazábal J I and Recarte V 2012 *Physics Procedia* **35** 57
[6] Dupasquier A, Kögel G and Somoza A 2004 *Acta Mater* **52** 4707
[7] Brown P J, Crangle J, Kanomata T, et al 2002 *J. Phys.: Condens. Matter* **14** 10159
[8] Campillo Robles J M, Ogando E and Plazaola F 2007 *J. Phys.: Condens. Matter* **19** 176222

Journal of Physics: Conference Series **443** (2013) 012040 doi:10.1088/1742-6596/443/1/012040

# Nano sulfide and oxide semiconductors as promising materials for studies by positron annihilation

**P.M.G. Nambissan**

Saha Institute of Nuclear Physics, 1/AF Bidhannagar, Kolkata 700064, India

E-mail : pmg.nambissan@saha.ac.in

**Abstract.** A number of wide band gap sulfide and oxide semiconducting nanomaterial systems were investigated using the experimental techniques of positron lifetime and coincidence Doppler broadening measurements. The results indicated several features of the nanomaterial systems, which were found strongly related to the presence of vacancy-type defects and their clusters. Quantum confinement effects were displayed in these studies as remarkable changes in the positron lifetimes and the lineshape parameters around the same grain sizes below which characteristic blue shifts were observed in the optical absorption spectra. Considerable enhancement in the band gap and significant rise of the positron lifetimes were found occurring when the particle sizes were reduced to very low sizes. The results of doping or substitutions by other cations in semiconductor nanosystems were also interesting. Variously heat-treated $TiO_2$ nanoparticles were studied recently and change of positron annihilation parameters across the anatase to rutile structural transition are carefully analyzed. Preliminary results of positron annihilation studies on Eu-doped CeO nanoparticles are also presented.

## 1. Introduction

Several of the physical properties of wide band gap sulfide and oxide semiconducting nanomaterial systems are inherently related to the presence of vacancy-type defects and the experimental techniques of positron annihilation spectroscopy are found very useful in their investigations [1-3]. There are a number of changes that may accompany the nanostructure formation of materials like phase transformation, lattice contraction or expansion, inversion (in spinel materials), free volume generation etc. Positron annihilation parameters are highly sensitive to these changes as the processes involve redistribution of the electrons and hence changes in their annihilation characteristics. In this work, we have tried to establish the utility of these techniques for the studies of a number of aspects of nanostructured semiconductors and highlight the role of vacancy-type defects in the modification of material properties.

## 2. Experimental details

The samples used in these studies here are prepared through the chemical route, in order to have better control on the temperature of synthesis and also to ensure better purity of the final products. As a general example, we discuss below about the solvothermal process adopted to get nanocrystalline samples of cadmium sulfide (CdS) of grain sizes varying over a wide range even down to 2 nm. A teflon-lined stainless steel chamber of closed cylindrical shape and 110 ml capacity was used for the purpose. Appropriate amount of $CdSO_4$ or $CdCl_3$ or $Cd(NO_3)_3.4H_2O$ and thiourea ($NH_2CSNH_2$) with molar ratio 1:3 were taken in the chamber and then filled with ethylene glycol ($C_2H_6O_2$) or benzene up to 80 % of its volume. After 30 minutes of stirring, the closed chamber was heated for 12 hours inside a box furnace at the temperature (between 423K and 523K) chosen for the approximate particle size desired to achieve and then cooled down to room temperature. The resulting yellow precipitates were filtered off and washed in ethanol several times to remove the excess traces of reagents. The final products were dried in vacuum at 333K for 6 hrs. The experimental parameters such as the solubility of the Cd sources in the solvent and temperature were varied to get samples of the desired size of CdS nanoparticles. The details of synthesis of the other nanosystems are given in the respective references cited.

The sizes of the nanoparticles were estimated either from x-ray diffraction patterns and/or high resolution transmission electron microscopy images. In carrying out the positron annihilation measurements, a $^{22}Na$ source of approximate strength 400 kBq deposited on a thin (~ 2 mg.cm$^{-2}$) Ni foil is kept immersed in the sample (in powder form) taken in a glass tube and the gamma rays were detected using $BaF_2$-coupled XP2020Q photomultiplier tubes and high pure germanium (HPGe) detectors. The slow-fast gamma-gamma coincidence setup used for positron lifetime measurements had a resolution (FWHM) of 170 ps. The HPGe detectors used for coincidence Doppler broadening (CBD) measurements had energy resolutions 1.27 and 1.33 keV at 511 keV. Typically a million counts were collected under each positron lifetime spectrum and the data were analyzed using the PALSfit program [4]. About 7-8 million coincidence events were generated in each CDB spectrum. The sample and source were maintained under good vacuum conditions to avoid the interference of air and other absorbable gases.

## 3. Results and discussion

The morphology as well as the sizes of the nanocrystallites play crucial roles in governing the annihilation characteristics of positrons in semiconductors. Positrons can also respond differently to nanosystems of different stoichiometric features. This is evident from figure 1 which shows the peak-normalized positron lifetime spectra for undoped and $Mn^{2+}$-doped zinc sulfide (ZnS) nanorod samples. Figure 2 gives an example where the mean positron lifetime $\tau_m$ is shown to continuously increase with decrease in diameter of ZnS hexagonal nanocrystals [5].

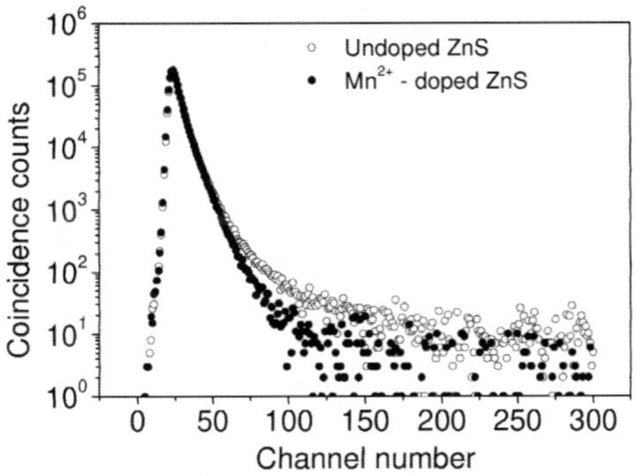


**Figure 1.** Peak-normalized positron lifetime spectra of pure ZnS nanorod sample and after doping with 20% $Mn^{2+}$ ions.

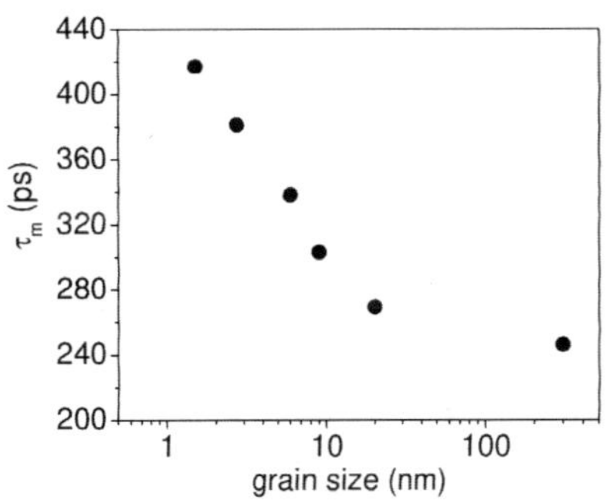


**Figure 2.** The mean positron lifetime versus grain size of ZnS hexagonal nanocrystallites [5].

An interesting example of positron affinity to open-volume defects is further illustrated in figures 3(a) and 3(b) which show the orthopositronium 'pick-off' lifetime $\tau_3$ and its intensity $I_3$ increasing steadily with decreasing grain size in ZnS cubic nanocrystals [5].

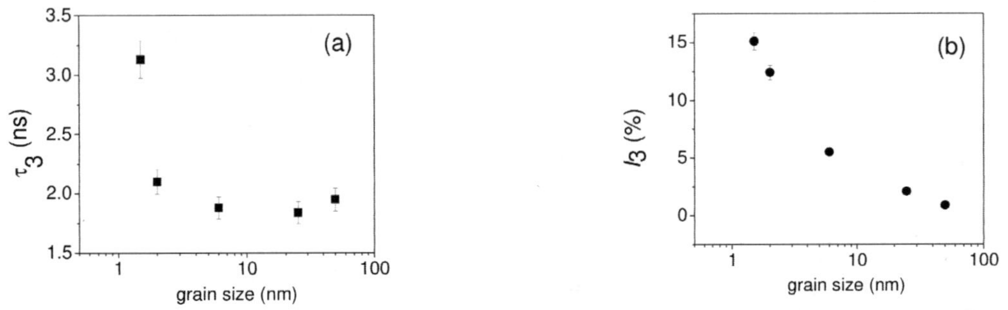


**Figure 3.** (a) Orthopositronium lifetime $\tau_3$ and (b) intensity $I_3$ versus grain size of ZnS cubic nanocrystalline samples [5].

In an experiment conducted to follow the transformation from the hexagonal wurtzite structure to cubic form of nanocrystallites during the substitution by $Mn^{2+}$ ions in ZnS nanorods, as revealed by x-ray diffraction patterns, the positron lifetimes $\tau_1$ and $\tau_2$ and intensity $I_2$ showed distinct features in their variation [6]. The variations of the *S* and *W* parameters and, in particular, the *S* vs. *W* plot shown in figure 4 also showed the transformation due to the sensitivity of positrons to the accompanying defect evolution processes [6].

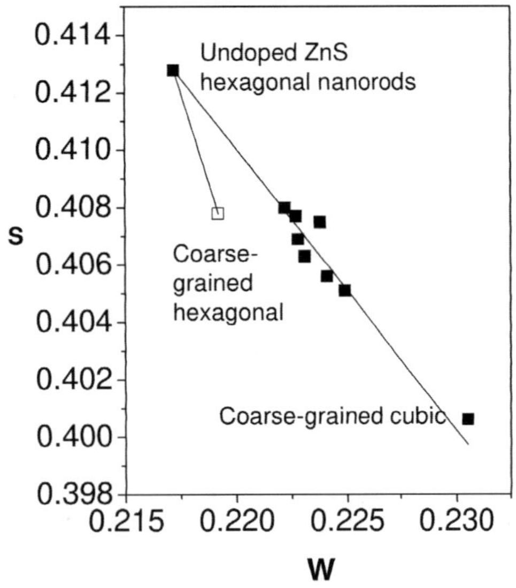


**Figure 4.** *S*-*W* plot of the Mn-doped ZnS nanorod samples [6].

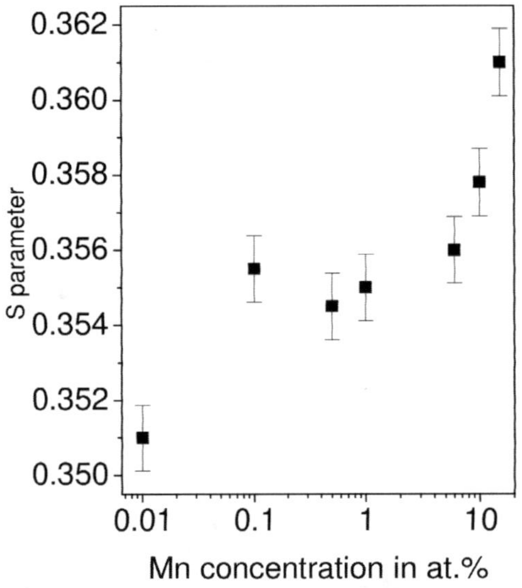


**Figure 5.** *S* parameter versus concentration of Mn doped in ZnS nanocones [9].

A remarkable observation in this study was the collapse of a three-state trapping model situation to a two-state one with the addition of $Mn^{2+}$ ions. The interior of the nanorods became defect-free due to doping [6, 7]. Interestingly, we found no such transformations taking place when identically synthesized ZnS nanoparticles were doped by $Mn^{2+}$ ions [8]. In other words, the morphology of nanosystems has got an important bearing on their structural characteristics.

Positron annihilation studies were also performed on ZnO nanocones doped with $Mn^{2+}$ ions [9]. A significant observation was the continuous decrease of the base diameters of the nanocones from 75 nm to 25 nm when the concentration of $Mn^{2+}$ ions was increased from 0 to 15%. $Mn^{2+}$ ions have larger radius (~ 0.80 Å) compared to that of $Zn^{2+}$ (~ 0.73 Å) and hence the substitution of the latter by the former will result into strain within the nanometer-sized grains. The strain can be gotten rid of by accommodating the $Mn^{2+}$ ions on the surface ionic layer rather than inside the grains. As the concentration of the doped $Mn^{2+}$ ions increase, more surface area is required to contain them and the material takes recourse to the increased surface to volume ratio of smaller nanocones. The *S* parameter showed a continuous increase due to the annihilation of positrons increasingly at the grain surfaces instead of in the interior of the grains (figure 5).

In another experiment, positron lifetime and Doppler broadening measurements were done on two samples of cadmium oxide (CdO), one composed of pure microoctahedrons and another with nanowires growing out of the octahedron surfaces. Three positron lifetimes were obtained, each representing annihilation through different mechanisms and environments. The longest component $\tau_3$ (~1.3-1.4 ns) arising from the annihilation of orthopositronium atoms can be ignored, as its intensity $I_3$ (~0.6%) in both the cases was too small. The other two lifetimes $\tau_1$ and $\tau_2$ were 134 and 323 ps in the case of sample containing only octahedrons and 140 and 342 ps in the other sample. In both the cases, their intensities were the same as 79.1 and 20.3%. As the edge lengths of CdO microoctahedrons were in the range 1.5-3.5 $\mu$m, positron annihilations will be taking place within the octahedrons and not on the octahedron surfaces. For the second sample with nanowires on the surfaces also, the octahedrons were of sizes larger than the thermal diffusion lengths for positrons. But a fraction of the positrons got annihilated on the nanowire surfaces, which are rich in defects because of the highly disordered atomic arrangement. Positrons could get trapped into the defects over there and got an enhanced lifetime.

Quantum confinement effects in semiconductor nanomaterials can be studied by positron annihilation techniques with a high degree of success [11, 12]. For example, both the *S* parameter and the mean positron lifetime $\tau_m$ showed significant increases in the case of cadmium sulfide (CdS) nanocrystalline samples composed of grains whose sizes were below the threshold for observation of the characteristic blue shift in optical absorption studies [11]. Identical observations were obtained in nickel oxide (NiO) nanoparticles synthesized through both solvothermal and sol-gel processes [12]. Positron lifetime $\tau_2$ showed a drastic increase when the grain sizes were below about 7 nm. A more systematic investigation on NiO nanoparticles of different average grain sizes was carried out recently. The optical band gap increased from the bulk value of 3.95 eV to 4.48 eV in nanoparticles of sizes ranging from 15 nm down to 3 nm. The behavior of $\tau_2$ in this case is shown in Figure 6. Detailed analysis is in progress.

The results recently obtained on titanium oxide ($TiO_2$) are partly illustrated figure 7. Details shall be soon published elsewhere. Positron annihilation parameters showed a reversal of variation with annealing temperature due to the transition from the anatase structure to rutile one around 700°C. The shorter lifetime $\tau_1$ in this case was attributed to the bulk lifetime admixed with defects of the mono- and/or divacancy type. The vacancies started annealing above 500°C and the phase transition followed soon after the annealing temperature was raised further high.

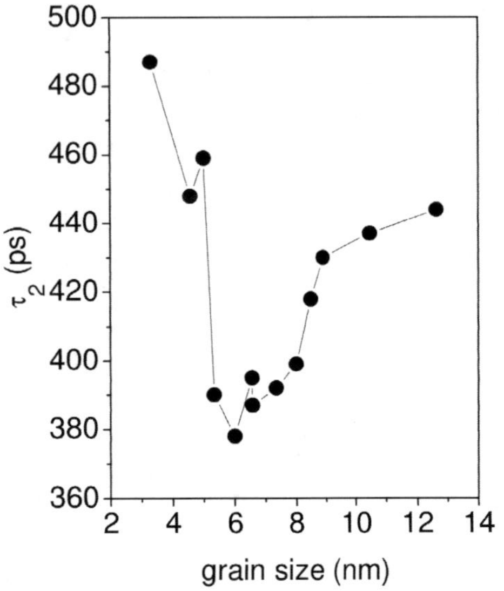


**Figure 6.** The variation of $\tau_2$ versus sizes of NiO nanoparticles.

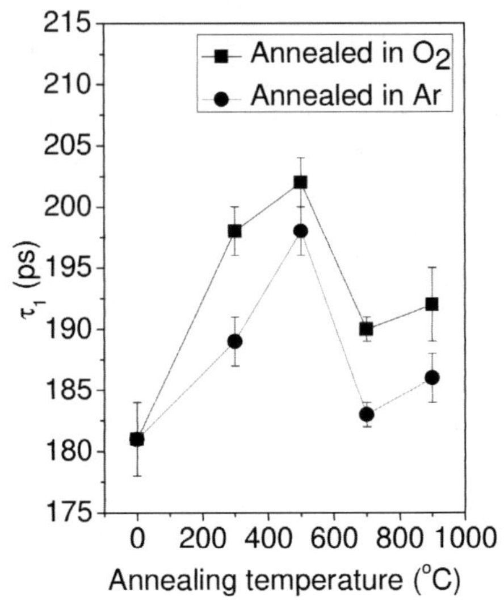


**Figure 7.** The variation of $\tau_1$ with annealing temperature of $TiO_2$ nanomaterial samples.

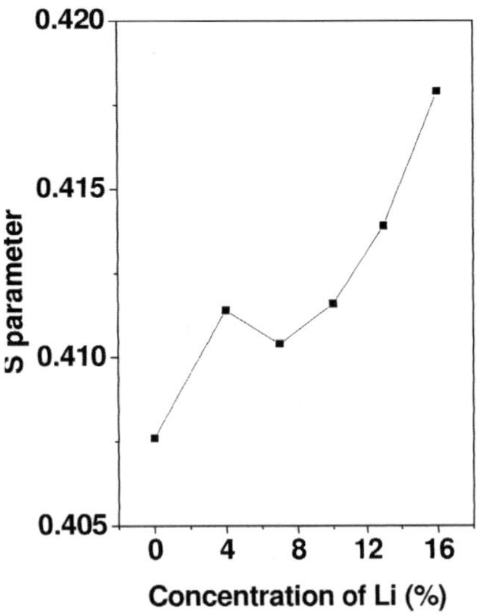


**Figure 8.** *S* parameter versus concentration of $Li^{3+}$ ions in ZnO nanocrystalline samples.

**Table 1.** Positron lifetimes and intensities in Eu-doped CeO nanocrystalline samples.

| [Eu](%) | $\tau_1$(ps) | $\tau_2$(ps) | $\tau_3$(ps) | $I_1$(%) | $I_2$(%) | $I_3$(%) |
|---|---|---|---|---|---|---|
| 0 | 172 | 394 | 2631 | 56.0 | 43.8 | 0.2 |
| 1 | 172 | 403 | 2053 | 56.6 | 43.1 | 0.2 |
| 50 | 173 | 415 | 4642 | 57.0 | 42.7 | 0.3 |

A study of doping ZnO with $Li^{3+}$ ions had been conducted recently. Figure 8 shows the variation of *S* with the doping concentration. The results confirm the interaction of vacancy-type defects present initially in the undoped sample with $Li^{3+}$ ions on doping. The process continues uninterrupted till 16%, the highest doping concentration used in this work. Preliminary results on

Eu-doped CeO nanocrystalline samples are shown in Table 1. The increase in lifetime implies increased positron annihilation on the grain surfaces. Detailed investigations are in progress.

## 4. Summary and conclusions

In semiconductor nanomaterials, the vacancy type defects need to be precisely characterized not only for the knowledge of their size and concentration, but even for their charge states. Positrons are trapped by neutral as well as negatively charged vacancies or their clusters but get strongly repelled by those with a net positive charge. This is crucial when a measured positron lifetime with larger magnitude needs to be attributed to vacancy clusters. Positronium formation, although of small intensities, cannot be ignored totally as the lifetimes and intensities often exhibit systematic variation with grain size, doping concentration, temperature etc. This calls for precise determination of the defect distribution, their location and even orientation. Some of these issues could be addressed successfully with positron annihilation spectroscopy. Although not discussed in detail in this paper, the CDBS experiments are shown to highlight the differing elemental environments around defects in which positrons get trapped at the different stages of variation of the input parameters. These information can be highly useful in studies involving defect-related properties such as magnetism, heat treatment, cold-work, metallurgy etc.

## References

[1] Krause-Rehberg R and Leipner H S 1999 *Positron Annihilation in Semiconductors—Defect Studies* (Springer-Verlag, Berlin), pp. 1–126.
[2] Kar S, Chaudhuri S and Nambissan P M G 2005 *J. Appl. Phys.* **97** 014301
[3] Brauer G, Anwand W, Grambole D, Grenzer J, Skorupa W, Cizek J, Kuriplach J, Prochazka I, Ling C C, So C K, Schulz D and Klimm D 2009 *Phys.Rev. B* **79**, 115212
[4] Olsen J V, Kirkegaard P, Pedersen N J and Eldrup M 2007 *Phys. Stat. Sol. (c)* **4** 4004
[5] Biswas Subhajit, Kar Soumitra, Chaudhuri Subhadra and Nambissan P M G 2006 *J. Chem. Phys.* **125** 164719
[6] Kar Soumitra, Biswas Subhajit, Chaudhuri Subhadra and Nambissan P M G 2007 *Nanotechnology* **18** 225606
[7] Nambissan P M G, Biswas Subhajit, Kar Soumitra and Chaudhuri Subhadra 2007 *Phys. Stat. Sol. (c)* **4** 3889
[8] Biswas Subhajit, Kar Soumitra, Chaudhuri Subhadra and Nambissan P M G 2008 *J. Phys. : Condens. Matter* **20** 235226
[9] Ghoshal Tandra, Kar Soumitra, Biswas Subhajit, De S K and Nambissan P M G 2009 *J. Phys. Chem. C* **113** 3419
[10] Ghoshal Tandra, Biswas Subhajit, Nambissan P M G, Majumdar Gautam and De S K 2009 *Crystal Growth and Design* **9** 1287
[11] Kar Soumitra, Biswas Subhajit, Chaudhuri Subhadra and Nambissan P M G 2005 *Phys. Rev. B* **72** 075338
[12] Das Soumen, Ghoshal Tandra and Nambissan P M G 2009 *Phys. Stat. Solidi C* **6** 2569

## Acknowledgements

The helps received from Mr. Anirban Samanta and Mr. Samudrajit Thapa in the data analysis are gratefully acknowledged. The author also wishes to thank his innumerable collaborators for various inputs that have gone into the experiments and also in the preparation of this manuscript.

Journal of Physics: Conference Series **443** (2013) 012041 doi:10.1088/1742-6596/443/1/012041

# Observation of spatial distribution of vacancy defects in semiconductor by positron microscope and electron beam induced current measurement

**M Maekawa and A Kawasuso**
Advanced Science Research Center, Japan Atomic Energy Agency
Watanuki 1233, Takasaki, Gunma Japan

E-mail: maekawa.masaki@jaea.go.jp

**Abstract.** A complementary study of vacancy defects in Si substrates by using scanning positron microscope (SPM) and electron beam induced current (EBIC) method were demonstrated for the same samples and in the same chamber. Both the S parameter and EBIC contrast were found to be enhanced in the regions containing vacancy defects introduced by ion implantation. That is, the SPM provides a criterion if the spatially resolved carrier recombination centres by the EBIC method are originating from vacancy defects or not.

## 1. Introduction

Electron beam induced current (EBIC) method provides the spatial distribution of defects in semiconductors [1, 2]. This method is based on a phenomenon that electron-hole pairs induced by electron bombardment in a depletion layer recombine at defects and hence the electric current is reduced near the defect region. As the result of persistent improvement, the current EBIC method can provide not only the lateral profile of defects but also the depth profile, the carrier lifetime and energy levels of defects [3, 4].

Scanning positron microscope (SPM) is a powerful tool for observing spatial distribution of vacancy defects near the subsurface region. We have constructed a SPM by using a specially fabricated small source and a solid neon moderator [5-8]. In our system, both the positron and electron beams can be focused on the sample by an objective lens. That is, the simultaneous acquisition of the SPM and EBIC images is possible. From this, one can know if defects visualized as the EBIC contrasts are associated with vacancy defects or not. In this work, we demonstrate a complementary observation of vacancy defects in ion implanted Si by using SPM and EBIC.

## 2. Experimental setup

### *2.1 SPM system*

Figure 1 schematically shows our SPM apparatus. Sodium-22 of 330 MBq was deposited into a 2 mm-diameter hole and sealed by a cap with a Ti window of 5 μm thick. The source is cooled down to 4 K and a solid neon film is grown on the source window as a moderator [9]. The positron beam with the energy of 20 keV is focused onto a sample surface using a commercial scanning electron microscope optics. The positron beam is laterally scanned by moving the sample. The Doppler broadening of annihilation radiation (DBAR) spectra are obtained using a Ge detector. These DBAR spectra are

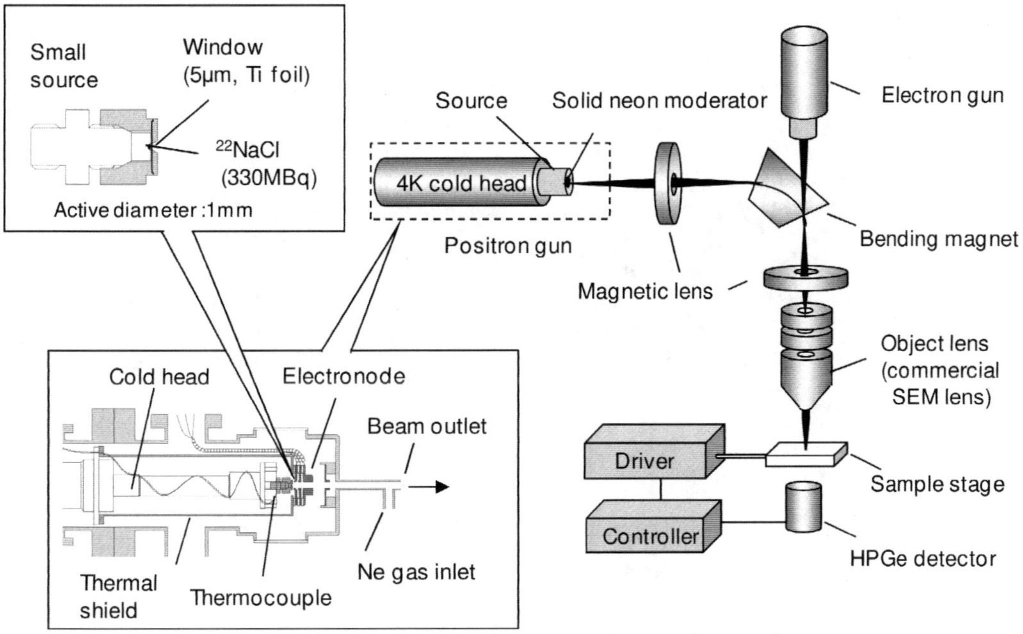


**Figure 1.** Schematics of our scanning positron microscope.

characterized by so-called S parameter, which is defined by the ratio of the intensity at the center of the energy peak of DBAR spectrum. The energy window for S parameter was 510.2 to 511.8 keV. All the S parameters were normalized to those in the Si substrate.

*2.2 EBIC measurement system*

Figure 2 shows the schematics of the EBIC measurement system. A commercial electron gun is installed at the branch of positron microbeam and the electron beam is focused onto the sample. The beam property should be almost the same as that of the positron beam because the same optics is used. Samples are mounted on an insulating holder connecting to the biased electrometer (ADCMT model 8252). Electrons and holes induced by electron bombardment are collected as the EBIC current by an

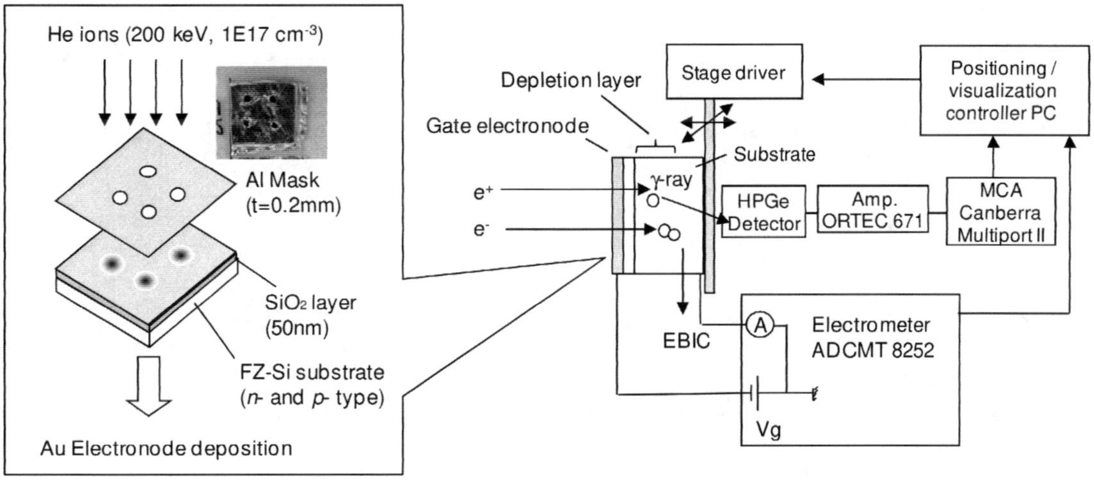


**Figure 2.** Schematics of the sample and the measurement setup.

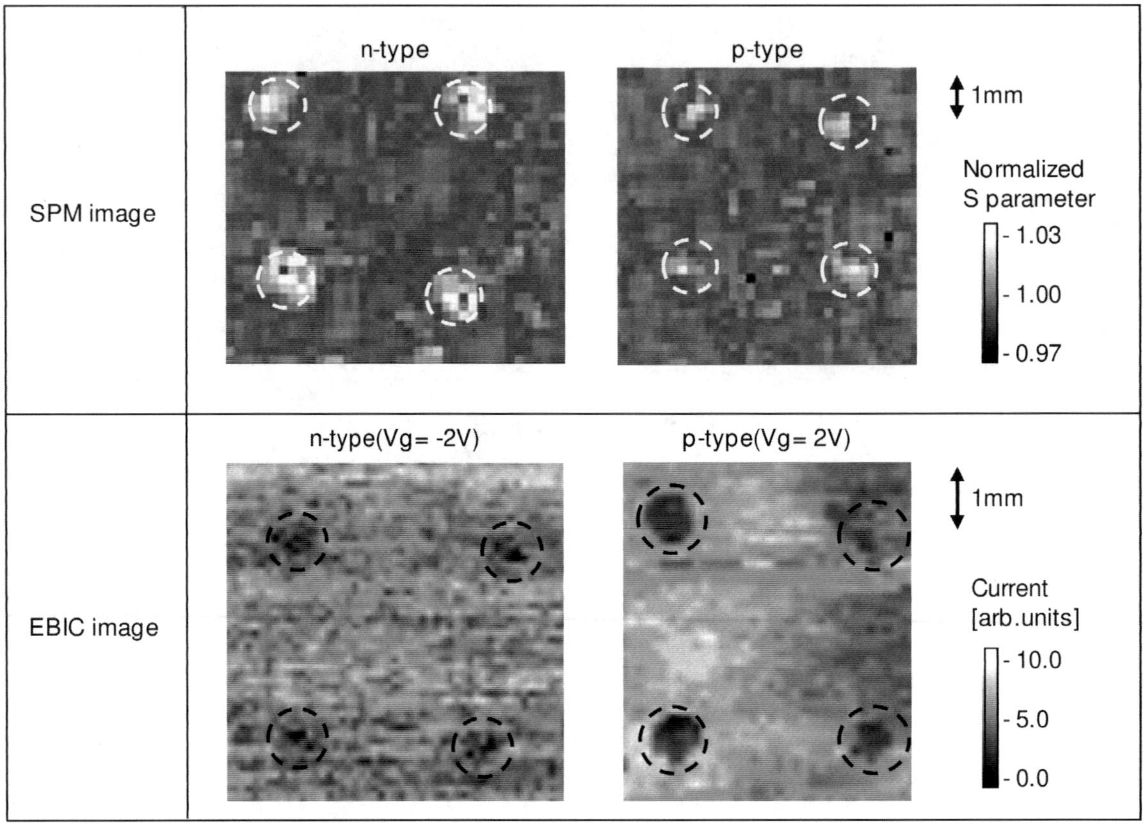


**Figure 3.** Distributions of S parameter and corresponding EBIC images.

internal electric field. If defects exist in the injected area, they mediate the electron-hole recombination. Consequently, by scanning the electron beam in lateral directions, defects in the injected area are visualized as dark contrasts.

*2.3 Sample preparation*

Samples used in this work were *n*- and *p*- type Si crystals (10x10 mm) with $SiO_2$ film of the thickness of 50 nm. Helium ion implantation was carried out through an aluminium mask with four holes as shown in Fig. 2. After implantation, Au gate electrodes and ohmic contacts were fabricated on the front sides and backsides, respectively, of the samples.

## 3. Results and discussion

Figure 3 shows the EBIC and SPM images for the *n*- and *p*-type Si samples. In the ion-implanted regions, dark contrasts are observed when the gate bias, which induces the depletion layer, was applied. This means that implantation-induced defects act as carrier recombination centres. In the ion-implanted region, S parameter also increases. This is attributed to the annihilation of positrons at vacancy defects introduced by the ion implantation. In the *p*-type sample, increase in S parameter is smaller as compared to the *n*-type sample. This may be because the charge state of vacancy defects is more positive in the *p*-type sample.

Figure 4 shows that the DBAR spectrum obtained in the high S parameter region (inside the circle in Fig.3) and several theoretical DBAR spectra of V2 (divacancy), V18 (vacancy cluster), V2+He and V2+2He (complex of divacancy and He atoms) defect structures [10, 11]. The experimental DBAR

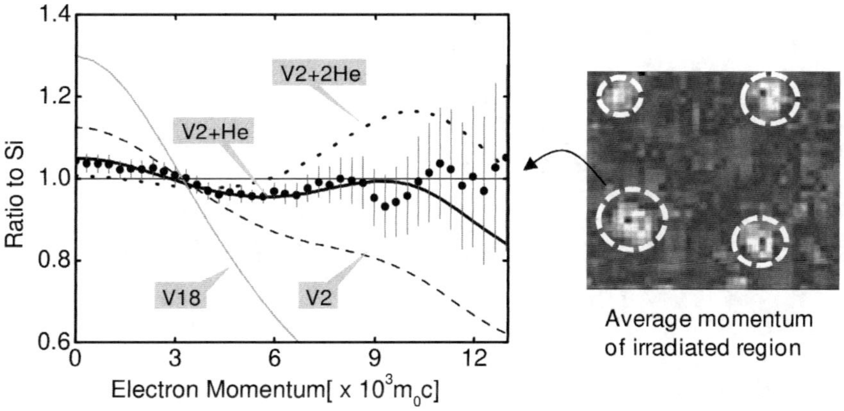


Figure 4. DBAR spectrum in the high S parameter region expressed as a ratio curve to the unimplanted Si. Theoretical curves for several defect structures are also shown.

spectrum is differentiated by the spectrum of the unimplanted silicon. Similarly, the theoretical DBAR spectra are differentiated by the spectrum of the perfect Si. Within these models, the divacancy occupied by one He atoms (V2+He) better reproduces the experimental result.

### 4. Summary
In this study, we demonstrated the simultaneous SPM and EBIC observations for the same samples and in the same chamber. Using the ion-implanted Si, the SPM observation showed that defects observed in the EBIC images contain vacancy defects. The combined SPM and EBIC method, that can provide spatial distribution of electrical active vacancy defects, is expected to be an important evaluation tool in the semiconductor device process.

### References
[1] Leamy H J 1982 J. Appl. Phys. **53** R51.
[2] Wilshaw P. R 1989 Ultramicroscopy **31** 177.
[3] Chen J, Yuan X and Sekiguchi T 2008 Scanning **30** 347-353.
[4] Chen J, Sekiguchi T, Fukata N, Takase M, Chikyo T, Yamabe K, Hasunuma R, Sato M, Nara Y and Yamada K 2008 Appl. Phys. Lett. **92** 262103.
[5] Maekawa M, Kawasuso A 2008 App. Surf. Sci. **255** 39-41.
[6] Maekawa M, Yabuuchi A and Kawasuso A 2010 J. Phys. Conf. Ser. **225** 012033.
[7] Yabuuchi A, Maekawa M and Kawasuso A 2011 J. Nuclear Materials **419** 9–14.
[8] Yabuuchi A, Maekawa M and Kawasuso A 2011 J. Phys. Conf. Ser. **262** 012067.
[9] Mills Jr A P and Gullikson E M 1986 Appl. Phys. Lett. **49** 1121.
[10] Maekawa M and Kawasuso A 2010 J. Phys. Conf. Ser. **225** 012032.
[11] Kawasuso A, Yoshikawa M, Itoh H, Chiba T, Higuchi T, Redmann F and Krause-Rehberg R 2005 Phys. Rev. B **72** 45204.

Journal of Physics: Conference Series **443** (2013) 012042 doi:10.1088/1742-6596/443/1/012042

# Acceptors in undoped GaSb; the role of vacancy defects

**J Kujala, J Slotte and F Tuomisto**

Department of Applied Physics, Aalto University, P.O. Box 11100, FI-00076 AALTO, Finland

E-mail: jiri.kujala@aalto.fi

**Abstract.**
The conventional lifetime setup was used to study Czochralski grown unintentionally *p*- and *n*- type ([*n*]≈ $6 \times 10^{17}\mathrm{cm}^{-3}$) GaSb bulk samples. Several approaches were used to analyze the data. However, it was not possible to successfully analyze the obtained spectrums with the conventional trapping model. From the analyzed data it was derived that the reason for p-type behavior of GaSb was not $V_{\mathrm{Ga}}$. Additionally, the role of gallium vacancy was studied and it's effect to lifetime values are shortly discussed.

## 1. Introduction

Gallium antimonide (GaSb) is an interesting material thanks to its applications in e.g. high speed electronics. GaSb has Zincblende crystal structure and it has a relatively narrow band gap of 0.73 eV. This makes GaSb useful in the fabrication of infrared detectors and sources. GaSb has a melting point of approximately 700 K [1, 2, 3, 4]. The material can be lattice matched with III-V compounds to form a substrate material for various optical communication devices. Furthermore, GaSb has high electron mobility and saturation velocity, which can be utilized in high electron mobility transistors [2]. To some extent, GaSb is a well known compound. There are quite a few studies about it and its structural properties can be found from publicly known sources for e.g. from [1]. However, there are still surprisingly few studies about vacancy properties of said compound. What makes studies on GaSb even more demanding is the fact that GaSb cannot be grown defect free. This makes e.g. the experimental determination of the positron bulk lifetime in GaSb challenging.

## 2. Experimental details

Czochralski grown bulk *n*- type ([*n*]≈ $6 \times 10^{17}\mathrm{cm}^{-3}$) and unintentionally *p*-type GaSb was measured. The standard sandwich setup was used, with a $^{22}$Na $e^+$-source wrapped in a 1.5 $\mu$m thick nickel foil. Aforementioned sample package was placed on a copper holder, which was in thermal contact with a closed cycle helium cryostat for the duration of the measurements. The measurements were performed in vacuum of about $10^{-4}$ Torr at 30-300 K. To confirm the validity of the results, the same measurement set were ran multiple times, in both directions(i.e. from 30 K to 300K and vise versa). The positron annihilations in the source material, in the covering Ni-foil and $e^+$ annihilating as positronium was removed from raw spectra.

Journal of Physics: Conference Series **443** (2013) 012042 doi:10.1088/1742-6596/443/1/012042

## 3. Results

Results for the average positron lifetime ($\tau_{\mathrm{ave}}$) in $n$-type and unintentionally $p$-type GaSb is shown in Fig. 1. For the $p$-type GaSb, a plateau can be seen reaching from 30 K all the way to 160 K. After this plateau, a gradual rise can be seen. In average lifetime, the increase is approximately 10 picoseconds in the temperature range of 30-300 K. For $n$-type bulk GaSb, no similar plateau can be seen. Here, $\tau_{\mathrm{ave}}$ rises gradually from about 255 picoseconds at 30K to 265 picoseconds at 300 K.

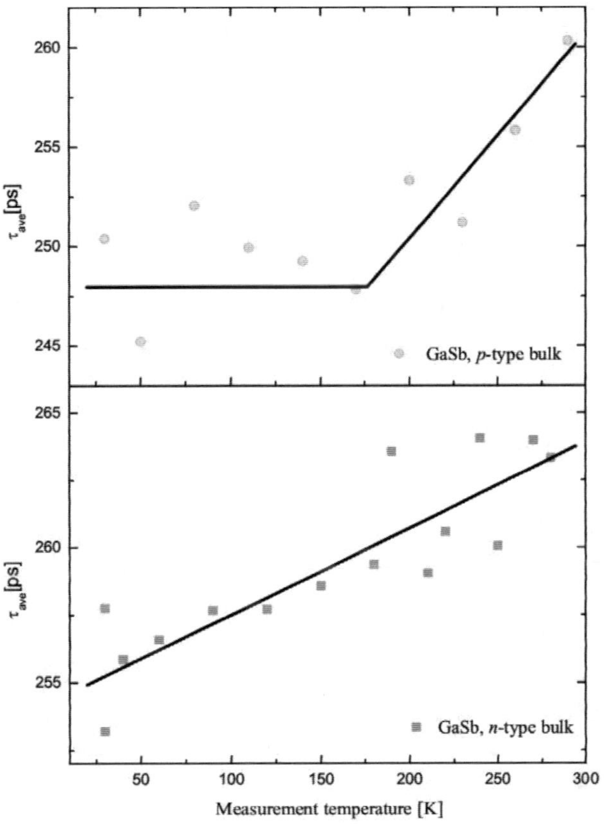


**Figure 1.** Average annihilation lifetimes $\tau_{\mathrm{ave}}$ for $n$- and $p$-type bulk GaSb (The lines are drawn to guide the eyes).

A two component trapping model was used when extracting lifetime data from the spectrum. Fig. 2 shows the resulted graphs for $n$-type and unintentionally $p$-type GaSb.

For unintentionally $p$-type GaSb, it can be seen, that at low temperatures, $\tau_1$ component is around 70 ps and above 170 K, it starts gradually to increase to value of 170 ps at RT. Intensity for this component is quite low; starting from 2-3 % and rising to 7 % at RT. The $\tau_2$ component increases from 260 ps unto 280 ps at room temperature. Its intensity is high; over 90 % and although its value is lower closer to the room temperature, it is still relatively high overall. Just

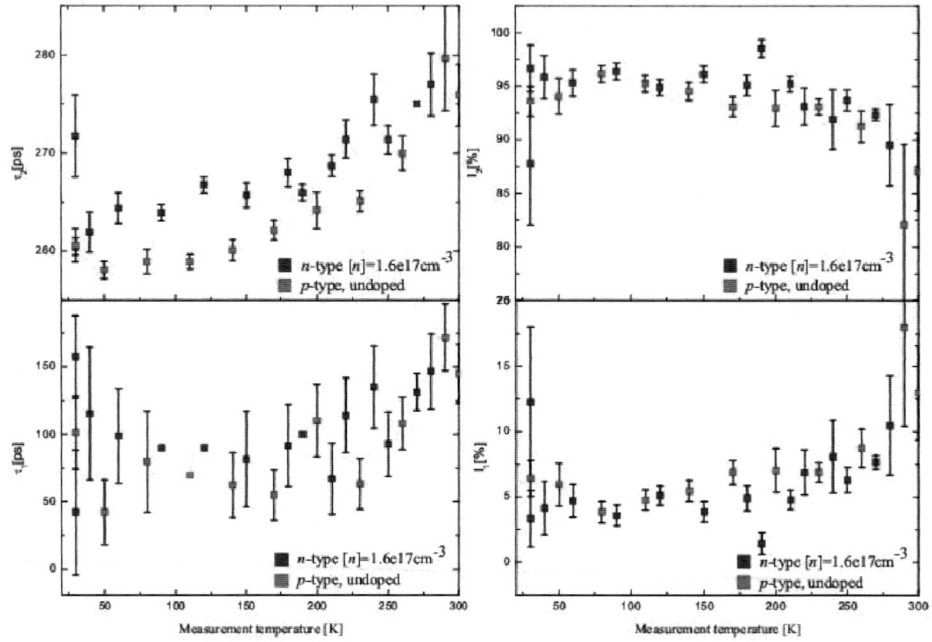


**Figure 2.** Extracted components from positron annihilation data for GaSb

before room temperature,
$I_2$ decreases strongly, being about 80 % at 300 K. For $n$-type GaSb, $\tau_1$ can be seen to stay around 100 ps until 200 K when it slowly starts to rise, all the way to 150 ps. The intensity of $\tau_1$ is around 3 %, until approximately 200 K point, where it increase to 10 %. $\tau_2$ increases first slowly from 260 ps to 265 ps between 30-150 K and then increases faster, being 275 ps in 300 K. The intensity of $\tau_2$ of $n$-type GaSb is quite stable and stays over 95 %, until around 170 K, it starts to decrease, being about 90 % at the room temperature.

## 4. Discussion

The best fit to the spectrum was obtained with a two component trapping model. In the Fig. 1 both curves drop to 250-254 ps, which is assumed to be the bulk lifetime for GaSb. At 30-150 K $\tau_{\text{ave}}$ in the $p$-type GaSb is constant at 257 ps. For $n$-type GaSb, a steady increase of 10 ps can be seen in the average lifetime at 30-300 K. Approximately 5 ps of this increase is can be attributed to thermal expansion. The two-trap analysis in Fig. 2 shows some peculiar behavior for both samples. For both samples the intensity in $\tau_2$ is very high, suggesting saturation trapping. However the temperature dependence of the $\tau_2$ component over the whole temperature interval of the measurement is very strong, i.e. decreases with decreasing temperature over a temperature range of almost 300 K. This could be an indication of a charge transition, as in [3]. As can be seen from typical lifetime spectrum shown in Fig. 3, only one lifetime component is visible.

The average lifetime at 30 K is very close to the assumed bulk lifetime, i.e. the dominant positron trap at 30 K has no open volume. Hence, we do not think that the reduction in lifetime with decreasing temperature is due to a charge transition. Rather, that a negatively charged

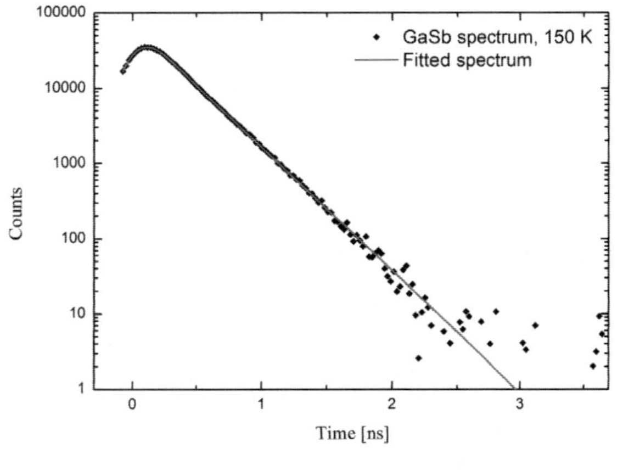


**Figure 3.** Raw spectrum from lifetime measurement at T=150 K

shallow trap (a negatively charged ion) starts to trap positrons with decreasing temperature. This negatively charged ionic defect is most likely the one responsible for the unintentional p-type behavior in as grown GaSb. A possible candidate could be the $Ga_{\mathrm{Sb}}$ antsite. The vacancy type defect that causes the average lifetime to increase with increasing temperature is most likely a Ga vacancy. The fact that the shallow trap starts to compete as a trapping center with the Ga vacancy already at room temperature could suggest that the vacancy is neutral.

## 5. Conclusions

We have used the conventional lifetime setup to study Czochralski grown unintentionally $p$- and $n$- type ([n]$\approx 6 \times 10^{17}\mathrm{cm}^{-3}$) GaSb bulk samples. It was tentatively deduced that a Ga vacancy causes the average lifetime to rise with temperature. Decrease in $\tau_{\mathrm{ave}}$ in low temperatures is possibly due to a negatively charged ion. This trap was assumed to be the $Ga_{\mathrm{Sb}}$ antisite.

## 6. References

[1] Dutta P S and Bhat H L and Kumar V 1997 *J. Appl. Phys.* B **81** 5821
[2] Hakala M, Puska M J and Nieminen R M, 2002 *J. App. Phys.* **91**, 4988
[3] Ling C C, Lui M K, Ma S K, Chen D, Fung S and Beling C D 2004 *Appl. Phys.* **85** 384
[4] Bracht H, Nicols S P, Walukiewicz W, Silveira J P, Briones F and Haller E E 2000 *Nature* **408** 69-72

Journal of Physics: Conference Series **443** (2013) 012043 doi:10.1088/1742-6596/443/1/012043

# In-situ positron lifetime spectroscopy of radiation damage by simultaneous irradiation of slow-positron and ion beams

**A Kinomura, R Suzuki, T Ohdaira, N Oshima, B E O'Rourke and T Nishijima**

National Institute of Advanced Industrial Science and Technology (AIST), 1-1-1 Umezono, Tsukuba, Ibaraki 305-8568, Japan.

E-mail: a.kinomura@aist.go.jp

**Abstract**. An in-situ positron analysis system has been developed for positron annihilation lifetime spectroscopy during ion beam irradiation. An electron linear accelerator was used to produce a slow positron beam for this system. Positron lifetime spectroscopy of $SiO_2$/Si and annealed pure Fe samples was successfully demonstrated during simultaneous irradiation with a 150 keV $Ar^+$ beam. The lifetime spectra changed with increasing ion dose, indicating a decrease in ortho-positronium intensity ($SiO_2$) and the decrease in positron diffusion length (Fe).

## 1. Introduction

Defects induced by neutrons and ions have been important subjects to investigate radiation damage of nuclear materials and to optimize ion implantation processes of semiconductor materials. Radiation induced defects are initially formed as point defects (vacancies and interstitial atoms). Depending on material properties and irradiation temperatures, point defects diffuse, annihilate and form secondary defects [1]. To understand the kinetics of radiation-induced defects, in-situ characterization techniques have been one of the important methods. The most typical in-situ technique is transmission electron microscopy (TEM) combined with ion irradiation [2]. With this technique, the movements of nanometer-sized defects can be observed in the time range from video rates to several hundred seconds (or longer). Because of its benefit, many in-situ TEM systems have been developed.

On the other hand, positron annihilation spectroscopy is recognized as a sensitive probe for atomic-scale open-volume defects below the detection limits of TEM. Because of its sensitivity for vacancy-type defects, in-situ analysis techniques during ion beam irradiation have been developed in recent years. An in-situ positron analysis system using Doppler broadening annihilation radiation (DBAR) with $^{22}$Na-based slow-positron and MeV ion beams was developed by Iwai et al. [3] and successfully applied for various radiation damage studies [3-5]. Positron annihilation lifetime spectroscopy (PALS) can give more direct information (i.e., vacancy size) in comparison with DBAR. Tsuchida et al. reported an in-situ PALS system based on a $\beta^+$-$\gamma$ coincidence method using high-energy positrons from a $^{68}$Ge source [6]. Since ion-beam induced defects are formed near the surface, energy-variable slow-positron beams are preferable to get sufficient depth resolutions. Therefore, a combination of PALS and a slow positron beam was introduced in this study. In particular, an accelerator-based positron source is employed for higher positron intensity.

## 2. Experimental method

An electron beam at about 70 MeV from an electron linear accelerator (LINAC) was converted to a slow-positron beam through pair creation by a Ta converter and a W moderator. A pulse width and a repetition rate of the LINAC were ~1 μs and 10 pps, respectively. A slow positron beam was magnetically guided from the W moderator to a sample chamber. An ion accelerator with a radio-frequency (RF) ion source was used to generate ion beams, where ion species from inert gas sources can be accelerated up to 150 keV. A typical ion beam current was in the range from 1 nA to 1 μA.

Figure 1 shows a schematic view of the sample chamber and the pulse-forming system. The incident angle of the ion beam is 45° off the positron beam direction. A positron spot-size measured by a microchannel plate (MCP) was about 1 cm in diameter. The ion spot size measured by burn marks on Kapton sheets was about 1.5 cm in diameter in the case of 150 keV $Ar^+$. We confirmed that the positron-beam spot was aligned near the center of the ion-beam spot. A sample size was approximately 1.5 cm × 1.5 cm. Ion-beam current was measured by a Faraday cup installed on the sample holder before and after the PALS measurements.

In our previous study, a linear storage section formed a quasi-continuous positron beam to avoid detector saturation [7]. But no such storage device was used in this study. The pulse width of the electron beam was ~1 μs. When an extraction voltage of 5 V was applied to the W moderator, the pulse width increased to ~20 μs during the beam transport from the W moderator to the sample chamber. Each single positron pulse was divided and squeezed into fine pulses by chopper, prebuncher and buncher electrodes [7,8]. Annihilation gamma-rays were detected by a detector with a $BaF_2$ scintillator and a photomultiplier tube. The detector and the chopper signals were recorded by a fast waveform digitizer. The time differences between these two signals were accumulated on a computer to obtain lifetime spectra [9]. The details of the beamline have been described elsewhere [10]. The ion beam was also chopped by deflection-type chopper electrodes. The irradiation timing of ion and positron pulses can be synchronized.

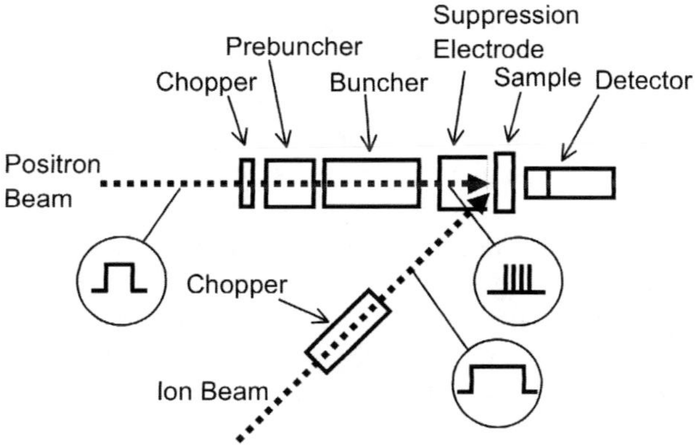


**Figure 1.** Schematic diagram of the sample chamber and the pulse-forming electrodes.

## 3. Results and discussion

A thermally grown silicon-oxide ($SiO_2$) layer with a thickness of 1 μm on Si was firstly measured by the in-situ analysis system with a 3 keV positron beam and a 150 keV $Ar^+$ beam. Note that a high voltage of 3 keV was applied to the sample holder to accelerate positrons. As the high voltage for positrons is also effective for ions, the actual ion-beam energy was 153 keV. The average positron implantation depth and the ion range were calculated to be 0.10 μm and 0.11μm, respectively. All the positrons were implanted into the $SiO_2$ layer and the positron distribution mostly overlaps the ion distribution.

Figure 2 shows the lifetime spectra obtained before the ion irradiation, during the first ion irradiation and after the ion irradiation. Total counts to take one spectrum were 50k - 100k counts and it took 7 - 14 min. The spectrum before irradiation in figure 2 shows the long tail originating from the ortho-positronium (o-Ps) component in $SiO_2$. Following ion irradiation the tail region (i.e., the o-Ps component) decreased because of radiation damage. The measured spectra were decomposed into two lifetime components. The second lifetimes, corresponding to the o-Ps component, were fixed to the initial value. Figure 3 shows the o-Ps intensity as a function of ion dose. The open and closed circles correspond to the spectra with (Beam ON) and without (Beam OFF) the ion irradiation, respectively. As shown in figures 2 and 3, the PALS spectra were successfully measured and the spectrum changes were monitored during the ion irradiation.

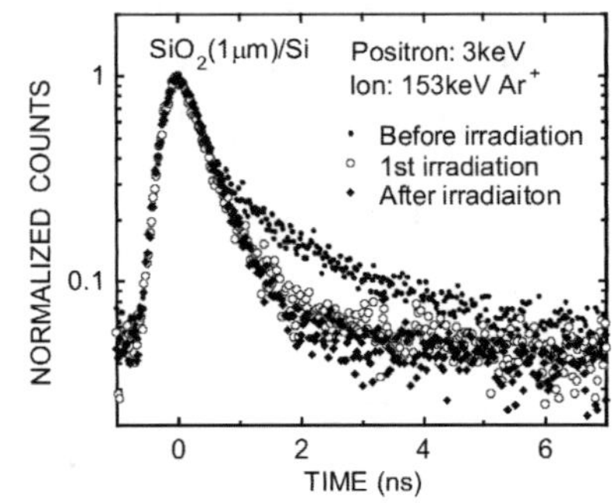


**Figure 2.** Lifetime spectra of a $SiO_2$/Si sample for 153 keV $Ar^+$ irradiation.

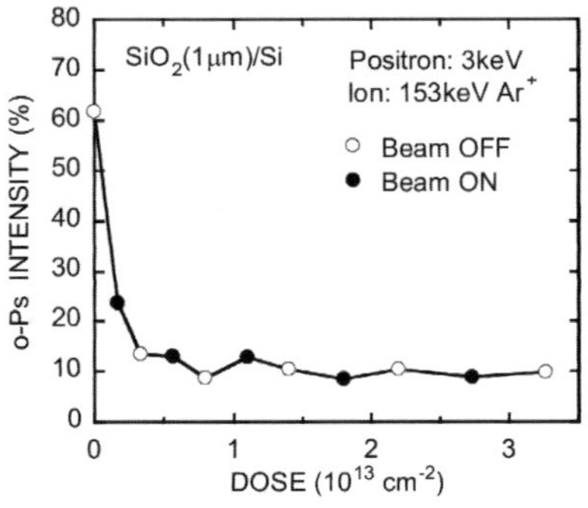


**Figure 3.** o-Ps intensity as a function of ion dose corresponding to figure 2.

A mirror-polished pure Fe (99.994%) sample was also measured by the in-situ analysis system. The sample was annealed at 700 °C for 0.5h in a flowing $H_2$+Ar atmosphere to remove mechanical damage. PALS measurements were performed with a 5 keV positron beam and a 150 keV $Ar^+$ beam (The actual ion energy was 155 keV). The average positron implantation depth and the ion range were calculated to be 0.067 μm and 0.053 μm, respectively. Some of the positrons were implanted slightly deeper than the ion-induced defects but this is expected to have a negligible effect on the analysis.

Figure 4 shows the lifetime spectra obtained before the ion irradiation, during the first ion irradiation and after the ion irradiation. Total counts to take one spectrum were 100k counts and it took 25 min on average. The spectrum changes were also observed in figure 4. If the first lifetime is fixed to the Fe bulk lifetime (0.11 ns), the initial spectrum before the ion irradiation can be decomposed into two lifetime components. The obtained second lifetime was 0.364 ns. This lifetime can be ascribed to the surface oxide layer [11]. Only one lifetime component was identified in the case of the other lifetime spectra. Figure 5 shows the lifetimes as a function of ion dose. The average lifetime was plotted in the case of the initial spectrum. In the same way as figure 3, the open and closed circles correspond to the spectra with (Beam ON) and without (Beam OFF) ion irradiation, respectively. The measured lifetime decreased with increasing irradiation dose. This result can be explained as the reduction of positron diffusion length. The positron diffusion length in the same type of Fe was measured to be 0.15 μm. Before irradiation, a large amount of implanted positrons diffuse back to the surface, reflecting the surface lifetime. With increasing dose, defects induced near the surface reduce the diffusion length and the surface lifetime became less influential. On the other hand, the lifetime of the bulk region increased with increasing defect concentration. The curve in figure 5 was obtained as a mixture of these two effects.

The time resolutions of the spectra in figures 2-5 were in the range of 0.32 – 0.46 ns. The peak background ratios of the spectra were ~20. These values were worse than those of typical positron lifetime spectra. We speculate that these values were caused by (1) insufficient total counts, (2) the data handling method for the waveform digitizer, (3) detector noise, and (4) the response of the scintillator. In terms of (1), as the AIST LINAC can be operated at 100 pps, it is possible to increase total counts and reduce statistical errors. In terms of (2)-(4), by improving the control program and the detector setting, it would be possible to get better time resolutions and peak-background ratios.

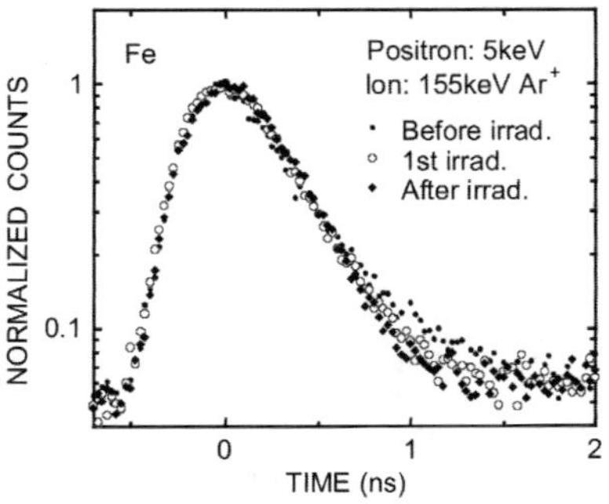


**Figure 4.** Lifetime spectra of an annealed pure Fe sample for 155 keV $Ar^+$ irradiation

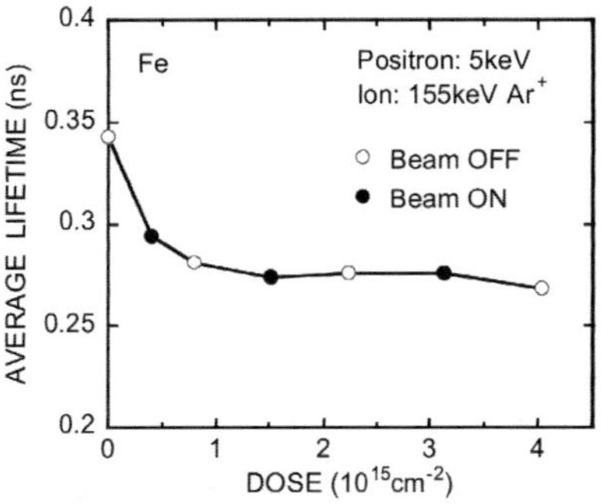


**Figure 5.** Lifetime as a function of ion dose corresponding to figure 4.

## 4. Conclusion
An in-situ analysis system for positron annihilation lifetime spectroscopy during ion beam irradiation has been developed. The slow positron beam was directly generated by a 70 MeV electron beam from a linear accelerator. Preliminary measurements have been successfully performed for a $SiO_2$ layer on Si and annealed pure Fe with 150 keV $Ar^+$ irradiation. The ortho-positronium intensity of $SiO_2$ was observed to decrease with increasing ion dose. The lifetime of the pure Fe decreased with increasing dose, indicating the decrease in positron diffusion length. Both results demonstrated the in-situ monitoring of radiation damage.

## Acknowledgements
This study was financially supported in part by the Budget for Nuclear Research of the Ministry of Education, Culture, Sports, Science and Technology of Japan, based on screening and counselling by the Atomic Energy Commission. It was also supported by JSPS KAKENHI Grant Number 24310080. We would like to thank T. Yoshiie, Q. Xu (Kyoto Univ.), T. Iwai (Univ. of Tokyo), Y. Kobayashi and K. Ito (AIST) for their assistance in the early stage of this study.

## References
[1] Sizmann R 1978 J. Nucl. Mater. **69&70** 386
[2] Birtcher R C, Kirk M A, Furuya K, Lumpkin G R and Ruault M-O 2005 J. Mater. Res. **20** 1654
[3] Iwai T, Ito Y and Koshimizu M 2004 J. Nucl. Mater. **329-333** 963
[4] Tsuchida H, Iwai T, Awano M, Kishida M, Katayama I, Jeong S-C, Ogawa H, Sakamoto N, Komatsu M and Itoh A 2007 J. Phys.: Condens. Matter. **19** 136205
[5] Kinomura A, Suzuki R, Ohdaira T, Oshima N, Ito K, Kobayashi Y and Iwai T 2011 J. Phys. Conf. Ser. **262** 012029
[6] Tsuchida H, Tanaka H, Kasai S, Akiyoshi M and Itoh A Presentation at the Workshop on Clarification of Materials Irradiation Effects and Improvement of Irradiation Techniques, March 10, 2009, Kumatori-cho, Osaka, Japan
[7] Suzuki R, Kobayashi Y, Mikado T, Ohgaki H, Chiwaki M, Yamazaki T and Tomimasu T 1991 Jpn. J. Appl. Phys. **30** L532
[8] Suzuki R, Ohdaira T and Mikado T 2000 Radiat. Phys. and Chem. **58** 603
[9] Kinomura A, Suzuki R, Ohdaira T, Oshima N, O'Rourke B E, Nishijima T and Ogawa H (unpublished)
[10] Kinomura A, Suzuki R, Ohdaira T, Oshima N, O'Rourke B E and Nishijima T 2012 Phys. Procedia **35** 111
[11] Sato K, Kinomura A, Omura T, Xu Q, Yoshiie T, Kasada R, Kimura A and Morishita K 2011 J. Phys. Conf. Ser. **262** 012053

Journal of Physics: Conference Series **443** (2013) 012044 doi:10.1088/1742-6596/443/1/012044

# PALS: A unique probe for the molecular organisation of biopolymer matrices

**M. Roussenova and M. A. Alam**

H. H. Wills Physics Laboratory, University of Bristol, Tyndall Avenue, Bristol BS8 1TL, UK

E-mail: m.roussenova@bristol.ac.uk

**Abstract.** This short review aims to illustrate the versatility of Positron Annihilation Lifetime Spectroscopy (PALS) when utilized for the characterization of biopolymers (e.g.: starch, fractionated maltooligomers, gelatin and cellulose derivatives) commonly used for the formulation of pharmaceutical encapsulants. By showing examples from a number of recent PALS studies, we illustrate that this technique can be used to probe the changes in thermodynamic state and molecular packing for a wide range of biopolymer matrices as a function of temperature, matrix composition and water content. This provides a basis for establishing composition-structure-property relationships for these materials, which would eventually enable the rational control of their macroscopic properties and the design of optimal encapsulating matrices and intelligent drug delivery systems.

## 1. Introduction

Biopolymers are of utmost importance for the pharmaceutical industry where there they are widely used in dense matrices at low water contents for the formulation of pharmaceutical excipients and encapsulation matrices for labile bioactive ingredients (e.g. drugs, vitamins and essential nutrients) [1, 2]. Such pharmaceutical encapsulants are often based on lipids (e.g.: wax, palm fat), proteins (e.g.: gelatin, milk proteins), polysaccharides (e.g: starch) and oligosaccharides (hydrolyzed starch, lactose), as well as cellulose and its derivatives (methylcellulose, hydroxypropylmethyl cellulose). These materials have the ability of forming films and matrices with adjustable morphologies, which have high barrier properties for gases [3] and organic molecules [4]. One of the key issues in the design of optimal encapsulants is the molecular mobility (diffusion/permeation) which governs the barrier properties of these matrices [2]. It is now widely recognised that the molecular mobility is determined by the thermodynamic state [2], as well as the local free volume which exists between molecules in soft matter due to irregular packing, density fluctuations and topological constraints [5, 6]. This free volume consists of a large number of sub-nanometer sized free volume elements ("holes") and a number of theories have been developed to describe phenomena such as self diffusion and the diffusion of guest molecules in terms of the size and size distributions of these free volume elements [6, 7]. The importance of quantifying the free volume in these systems, therefore, cannot be overestimated, since it provides a promising route towards a better understanding of the material properties and the eventual rational design of bio-encapsulants, rather than the current largely trial and error approach practised within the industry.

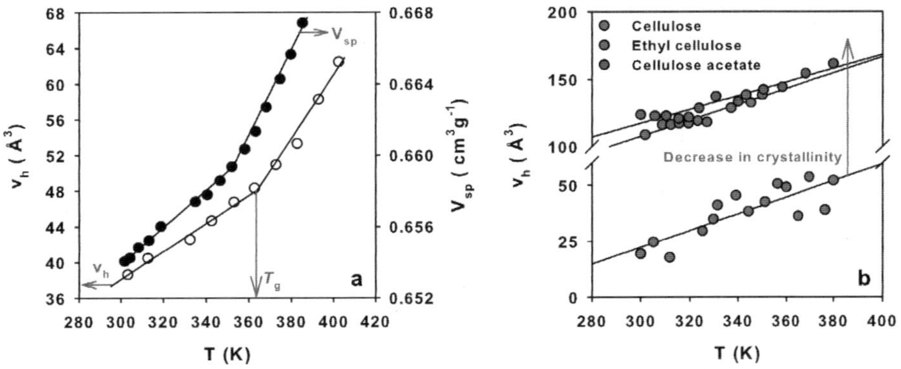


**Figure 1.** **(a)** Temperature dependence of the average molecular hole size, $v_h$, and specific volume, $V_{sp}$, measured for an amorphous maltooligomer matrix equilibrated at a well defined water activity, $a_w = 0.22$ [11][1]. **(b)** Temperature dependence of the average molecular hole size measured for partially crystalline cellulose and cellulose derivative matrices. For both cellulose derivatives the degree of substitution is $\sim 2.5$ [20, 21][2].

Positron annihilation lifetime spectroscopy (PALS) is a unique technique which is capable of measuring the size and size distribution of the free volume elements directly on the sub-nanometre length scale [8, 9]. Over the last decade, this technique has been successfully used to study the phase transitions and changes in molecular packing in a wide range of biopolymers with varying degree of order. Experiments using PALS have shown that the molecular organisation in these materials is strongly influenced by factors such as temperature [10, 11], molecular weight distribution [11, 24], matrix composition (addition of low molecular weight diluents) [12–15] and water content [10, 12, 14]. Furthermore, it was shown that these measurements on the molecular length scale can be successfully correlated with the macroscopic properties of these materials (e.g. density, mechanical strength, permeation properties) [11,13], proving PALS is an invaluable tool for the material characterization of these systems. Here, we illustrate the versatility of this technique by presenting a short review of recent studies on biopolymer systems commonly used for the formulation of pharmaceutical excipients and encapsulants.

## 2. Molecular organisation in biopolmers - Opening it up at the nanolevel

Biopolymers in dense matrices can occur in states of varying degree of order, from perfectly crystalline, to partially ordered and amorphous. In all cases, the molecular packing in these materials is strongly influenced by temperature, as shown in Fig. 1. For amorphous systems, the changes in molecular hole size (and specific volume) with temperature often show two linear branches, as illustrated in Fig. 1a which shows the temperature dependence of the average hole (and specific) volume measured for a completely amorphous maltooligomer matrix. The point of intersection of the two branches can be identified as the glass transition temperature, $T_g$, in agreement with Differential Scanning Calorimetry measurements [11]. For crystalline and partially ordered systems, however, the average molecular hole size generally shows a linear expansion with temperature, as illustrated in Fig. 1b for cellulose and cellulose derivative matrices [20]. It is interesting to note that the formation of cellulose derivatives leads to the marked loss of crystalline structure, accompanied by a significant increase in average molecular hole size, as shown in Fig. 1b [20, 21]. It has been previously shown that the molecular organisation and degree of crystallinity in cellulose based materials is strongly influenced by the type of chemical substitution, as well as the degree of substitution per glucose unit [20, 21]. These two parameters are thus often used to fine-tune the molecular architecture of cellulose derivatives when used for the formulation of pharmaceutical excipients.

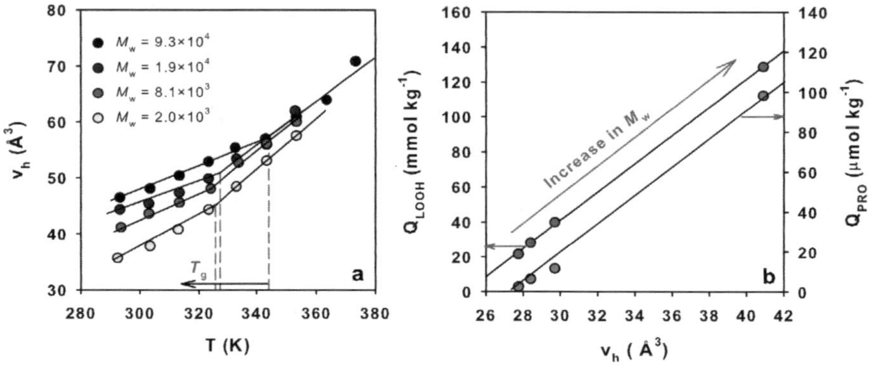


**Figure 2.** (**a**) Average hole volume as a function of temperature for maltooligomer matrices with different molecular weight distributions equilibrated at a well defined water activity, $a_w = 0.54$. [11][3] (**b**) Relation between the average molecular hole size of encapsulating maltooligomer matrices with different molecular weight distributions and the concentration of primary (hydroxiperoxide, $Q_{LOOH}$) and secondary (propanal, $Q_{PRO}$) oxidation products of the encapsulated lipid [23][4].

### *2.1 The effect of molecular weight distribution*

Another method of effectively altering the molecular packing, and hence the physical properties of biopolymer matrices is by changing their molecular weight profile. For example, it has been shown that for relatively low molecular weights both, the molecular packing and $T_g$ are strongly dependent on the average molecular weight of a system, as illustrated in Fig. 2a for a set of maltooligomer matrices [11]. It is clear that the density of the molecular packing increases with decreasing molecular weight and this densification effect is most pronounced in the glassy state, where we measure the largest reductions in $v_h$. Furthermore, this decrease in average molecular hole size has been shown to relate to improvements in the barrier properties of glassy maltooligomer matrices. Drush *et.al* successfully demonstrated that the oxidative stability of a labile lipid encapsulated in glassy fractionated carbohydrate oligomer matrices improved significantly if lower molecular weight fractions were used [23]. A linear relationship was found between the average molecular hole size of the encapsulating matrix and the concentration of primary (hydroxiperoxide) and secondary (propanal) oxidation products of the encapsulated lipid [23], as shown in Fig. 2b. Even small reductions in $v_h$ (achieved by correct molecular weight selection of the encapsulating carbohydrate blend) were shown to significantly improve the oxidative stability of the encapsulated lipid, suggesting that oxygen diffusivity is one of the key determinants of autoxidation of bioactives encapsulated in glassy biopolymer matrices [23].

### *2.2 The effect of low molecular weight additives*

Although the physical properties of biopolymer encapsulants can be controlled by careful selection of their molecular weight distribution, for practical applications these improvements are often achieved more easily by the addition of low molecular weight diluents [2]. Additive selection is normally based on the compatibility with the base biopolymer (in order to prevent phase separation), as well as permanence inside the matrix to prevent changes in material properties due to the time dependent migration of the diluent. The most common additives for the formulation of pharmaceutical encapaulants include polyols (e.g. glycerol, sorbitol), saccharides (e.g. maltose, sucrose and trehalose) and polyethers (e.g. poly(ethylene glycol)) [1]. The effect of these additives on the molecular organisation and physical properties of the matrices are system specific. Some diluents have been shown to act as plasticisers, whilst others act as packing enhancers for the biopolymer matrices. For example, poly(ethylene glycol) has been

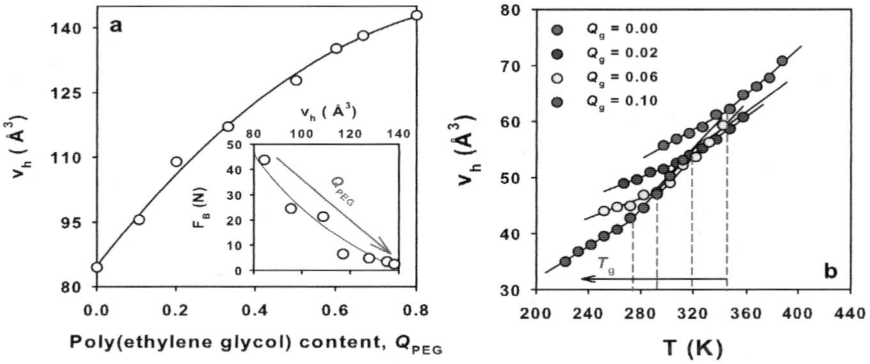


**Figure 3.** **(a)** The plasticising effect of poly(ethylene glycol) on methylcellulose, accompanied by an increase in the average hole size and the strength (*Inset*: decrease in force of breaking, $F_b$) of the matrices [22][5]. **(b)** Temperature dependence of the average molecular hole size for maltooligomer matrices with different compositions (glycerol contents, $Q_g$) equilibrated at a well defined water activity, $a_w = 0.54$. Glycerol is a packing enhancer for the maltooligomer matrices, causing systematic reductions in the average molecular hole size (both, in the glassy and rubbery states), accompanied by a reduction in $T_g$ of the matrices [14][6].

shown to have a plasticising effect on methylcellulose films, as depicted in Fig. 3a, accompanied by an increase in the average hole size and the strength (decrease in force of breaking) of the films [22]. Conversely, low molecular weight additives such as glycerol [14,15] and maltose [12,13] have been shown to act as packing enhancers for a number of biopolymer systems. As illustrated in Fig. 3b, the addition of glycerol to amorphous maltooligomers causes a decrease in the average molecular hole size both, in the glassy and the rubbery states, accompanied by a reduction in $T_g$. In the glassy state, a non-linear decrease in $v_h$ has been observed as a function of increasing glycerol content, indicative of a non-ideal packing behaviour [14]. The rapid initial decrease in $v_h$ may be associated with the reduction in molecular frustration (due to the rapid initial depression of $T_g$) upon the addition of the smaller glycerol molecules to the carbohydrate matrices [14]. Apart from acting as a packing enhancer, glycerol has also been shown to enhance the biostabilisation performance of encapsulating matrices and alter their water vapour sorption behaviour [14].

*2.3 The effect of water*

Indeed, one important property of biopolymers is their ability to absorb water vapour from the atmosphere (i.e. they are highly hygroscopic) [2]. Water is a highly efficient plasticiser for biopolymer matrices and the glass transition temperature of these systems decreases strongly as function of increasing water content [2]. Furthermore, the diffusional mobility of small molecules (e.g. water, gases and volatile organic compounds) in biopolymer matrices has been shown to increase rapidly with increasing water content, both, in the glassy and the rubbery states [3,27]. A number of recent PALS studies have also elucidated that the sorption of water causes complex changes in the molecular organisation of biopolymer matrices [11–15, 19]. To illustrate this point, in Fig. 4, we present examples of the changes in average hole size measured upon the sorption of water vapour for three biopolymer systems. Although in all cases water has a strong plasticising effect on the biopolymer matrices, the observed changes in $v_h$ as probed by Ps are system-specific. In Fig. 4a we show the changes in average hole size as a function of water content for a starch/sucrose blend at various temperatures [17]. The observed changes in $v_h$ with increasing level of hydration appear to be somewhat similar to the changes in $v_h$ as a function of increasing temperature, commonly observed for biopolymer matrices (for example,

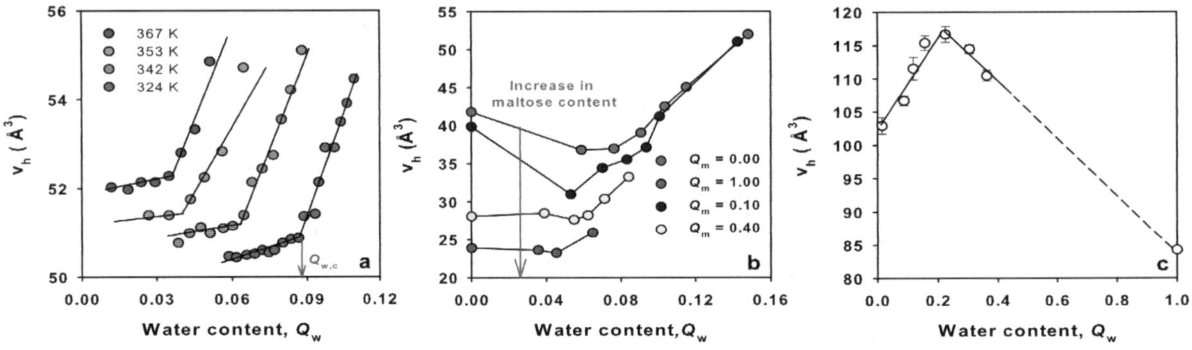


**Figure 4.** **The plasticising effect of water on biopolymer matrices.** The changes in average molecular hole size as a function of increasing water content measured for **(a)** starch/sucrose blends at different temperatures [17][7], **(b)** a series of fractionated maltopolymer/maltose blends [13][8] and **(c)** HPMC [19][9] matrices measured at $T$ = 298 K.

as shown in Fig. 1a). At low water contents we observe a smooth linear increase in $v_h$ up to a critical water content, $Q_{w,c}$, followed by at significantly sharper linear increase in $v_h$ at higher water contents. At the $Q_{w,c}$ the starch/sucrose matrices undergo a transition from the glassy to the rubbery state at a constant temperature, illustrating the superposition of the effects of temperature and water content on the thermodynamic state of the matrices. As expected, the $Q_{w,c}$ values decrease with increasing temperature. Interestingly, this behaviour is not reflected in the changes in molecular packing observed as a function of increasing level of hydration for bidisperse fractionated maltopolymer/maltose matrices, as shown in Fig. 4b [12, 13]. Starting from the anhydrous state, initially a decrease in $v_h$ is measured upon sorption of water. At such low water contents, the water molecules are present in a highly dispersed state and are closely associated with the hydroxyl residues on the carbohydrate chains (confirmed by clustering analysis [11, 14]) so they partially "occupy" the molecular free volume which exists between the carbohydrate chains, causing the observed reduction in $v_h$. The observed decrease in $v_h$ can also result from the reduction in molecular frustration of the system upon the addition of the small, polar water molecules. As the water content increases further ($Q_w \sim 0.05$, but depending on the maltose content, $Q_m$), whilst the matrices are still in the glassy state, we observe an increase in $v_h$ as a function of water content. It has been shown that water molecules have a significant diffusional mobility even in the glassy state [26, 27], thus allowing them to plasticise their local environment, causing an increase in $v_h$ [13]. Upon further hydration ($Q_w \sim 0.10$, but depending on the matrix composition), the matrices become rubbery and the differences in $v_h$ between the various matrix compositions become smaller. Furthermore, clustering analysis has shown that in the rubbery state the preferential homo-coordination between water molecules becomes significant [11, 14], leading to the formation of small clusters of water around the carbohydrate chains. Clustering of water molecules at high levels of hydration has also been observed in other biopolymer systems, such as hydroxypropyl methylcellulose [19] and bovine gelatin [15]. In Fig, 4c we show the changes in average hole size measured for hydroxypropyl methylcellulose (HPMC) as a function of increasing water content. Initially we observe a pronounced increase in $v_h$ as the HPMC matrices become plasticised by water [19]. The average molecular hole volume reaches maximum size at a water content of about $Q_w = 0.23$, after which it begins to decrease steadily upon further sorption of water. The average hole sizes measured for the plasticised HPMC matrices are larger than the average hole size measured for pure bulk water ($v_h = 81.6$ $\AA^3$ at $T = 298$ K). This indicates that the decrease in $v_h$ as a function of increasing water content may be attributed to the formation of small clusters of water in the matrices. This would, in turn, result in the overall *o*-Ps lifetime measured for the sample being a sum of the weighted contributions of an *o*-Ps in the plasticised HPMC matrix and the *o*-Ps in bulk water [19]. Since *o*-Ps probes the local molecular organisation at the sub-nanometer length scale, water clusters of a size less than one nanometer would be probed by *o*-Ps as bulk water.

## 3. Final Remarks

In this short review we have illustrated that PALS is an invaluable tool for probing the molecular organisation/free volume in a wide range of biopolymers commonly used for the formulation of pharmaceutical encapsulants. Factors such as temperature, composition and the amount of water absorbed by the matrices strongly influence their molecular packing and thermodynamic state. Water was shown to act as a strong plasticiser, whereby causing a reduction in $T_g$, generally accompanied by an increase in $v_h$ of the biopolymer matrices, provided clustering of water molecules does not occur. Furthermore, it was shown that the molecular organisation and physical properties of these materials can be improved by careful selection of their molecular weight distribution and through the addition of low molecular weight additives (e.g. glycerol, maltose).

## References

[1] Podczeck F and Jones B E 2007 *Pharmaceutical Capsules* (London: Pharmaceutical Press)
[2] Kasapis S, Norton I T and Ubbink J B 2009 *Modern Biopolymer Science* (Amsterdam: Elsevier)
[3] Schoonman A, Ubbink J, Bisperink Ch, Le Meste M, Karel M 2002 *Biotechnol. Prog.* **18** 139
[4] Gunning Y, Parker R, Ring S 2000 *Carbohydr. Res.* **329** 377
[5] Kusmins C A and Kwei T K 1968 *Diffusion in Polymers* ed Crank J and Park G S (London: Academic)
[6] Vrentas J S and Duda J L 1977 *J. Polym. Sci. Polym. Phys. Ed.* **15** 403
[7] Cohen M H and Grest G S 1979 *Phys. Rev. B* **20** 1077
[8] Schrader D M and Jean Y C (ed) 1988 *Positron and Positronium Chemistry* (Amsterdam: Elsevier)
[9] Jean Y C, Mallon P E and Schrader D M 2003 *Principles and Applications of Positron and Positronium Chemistry* (Singapore: World Scientific)
[10] Kilburn D, Claude J, Mezzenga R, Dlubek G, Alam A and Ubbink J 2004 *J. Phys. Chem. B* **108** 12436
[11] Kilburn D, Claude J, Schweizer T, Alam A and Ubbink J 2005 *Biomacromolecules* **6** 864
[12] Townrow S, Kilburn D, Alam A and Ubbink J 2007 *J. Phys. Chem. B* **111** 12643
[13] Townrow S, Roussenova M, Giardiello, Alam A and Ubbink J 2010 *J. Phys. Chem. B* **114** 1568
[14] Roussenova M, Murith M, Alam A and Ubbink, J 2010 *Biomacromolecules* **11** 3237
[15] Roussenova M, Enrione J, Diaz-Calderon P, Taylor A J, Ubbink J and Alam M A 2012 *New J. Phys.* **14** 035016
[16] Sharma S, Roudaut G, Fabing I and Duplatre G 2010 *Phys. Chem. Chem. Phys.* **12** 14278
[17] Sharma S, Zaydouri A, Roudaut G and Duplatre G 2011 *Phys. Chem. Chem. Phys.* **13** 19338
[18] Roudaut G and Duplatre G 2009 *Phys. Chem. Chem. Phys.* **11** 9556
[19] Trotzig C, Abrahmsen-Alami S and Maurer F H J 2009 *Eur. Polym. J* **45** 2812
[20] Doyle S, Malhotra B, Peacock N and Pethrick R 1984 *Brit. Polym. J* **16** 15
[21] Doyle S and Pethrick R 1987 *J. Appl. Polym. Sci.* **33** 95
[22] Pintye-Hodi K, Regdon G, Eros I, Suvegh K, Marek T, Kery I and Zelko R 2006 *Int. J. Pharm.* **313** 66
[23] Drusch S, Ratzke K, Shaikh M, Serfet Y, Steckel H, Scampicchio M, Voigt I, Schwartz K and Mannino S 2009 *Food Biophys.* **4** 42
[24] Drusch S, Serfet Y, Berger A, Shaikh M, Ratzke K, Zaporojtchenko V and Schwarz K 2012 *Food Hydrocolloids* **27** 332
[25] Liu H, Chaudhary D, Roberts J, Weed R, Sullivan J and Buckman S 2012 *Carbohydr. Polym.* **88** 1172
[26] Akiyama Y, Shibahara Y, Takeda S, Izumi Y, Honda Y, Tagawa S and Nishijima S 2007 *J Polym Sci. B: Polym. Phys.* **45** 2031
[27] Tromp R, Parker R, Ring S 1997 *Carbohydr. Res.* **303** 199

Journal of Physics: Conference Series **443** (2013) 012045 doi:10.1088/1742-6596/443/1/012045

# Positron lifetime study of poly(dimethylsiloxane) based polymer-nanoparticle composites

**D J Hughes[1], M V Roussenova[1], P Beavis[2], A C Swain[2] and M A Alam[1]**

[1] H. H. Wills Physics Laboratory, University of Bristol, Tyndall Avenue, Bristol BS8 1TL, UK.
[2] Atomic Weapons Establishment, Aldermaston, Reading RG7 4PR, UK.

E-mail: D.J.Hughes@bristol.ac.uk

**Abstract.** Positron annihilation lifetime spectroscopy was employed to study the changes in the size of the local free volume elements (holes) of poly(dimethylsiloxane) (PDMS) based nanocomposites incorporating two different types of nanoparticles, *meta*-carborane (*m*-CB) and *iso*-octyl polyhedral oligomeric silsesquioxane (POSS). Materials were prepared with various loading levels of each nanoparticle and the temperature dependence of the *o*-Ps lifetime, and thus the free volume, and its distribution were measured from 100-300 K. It was observed that *m*-CB reduces the size of the local free volume elements while POSS has the opposite effect, in each case the change is only seen in the rubbery state below the so-called "knee temperature" ($T_k \sim 210$ K).

## 1. Introduction

Polymer-nanoparticle composites have received increased interest from the scientific community over the past twenty years or so partly due to their significantly enhanced physical and mechanical properties at much lower filler loading levels compared to traditional composites incorporating micron sized inclusions [1]. The wide range of nano-scale fillers readily available now provides the prospect of designing multifunctional materials optimised for specific applications. The current status is that the mechanism of reinforcement due to these nano-scale inclusions is still poorly understood [2], leading to the lack of accurate prediction of a nanocomposite's properties, thus making the design and production of such materials a somewhat *trial-and-error* process.

This paper reports efforts to study the mechanical reinforcement properties of these nanocomposites via the investigations of the nature of the sub-nanometre free volume elements (holes) that exist between the molecular chains in polymers. It is this free volume that has considerable influence on the molecular mobility and hence mechanical properties of these materials. The size of the free volumes are probed by positron annihilation lifetime spectroscopy (PALS) in a series of polymer-nanoparticle composites incorporating physically dispersed nanoparticles into a cross-linked poly(dimethylsiloxane) (PDMS) network. PALS studies are complemented by thermodynamic characterisation by differential scanning calorimetry (DSC).

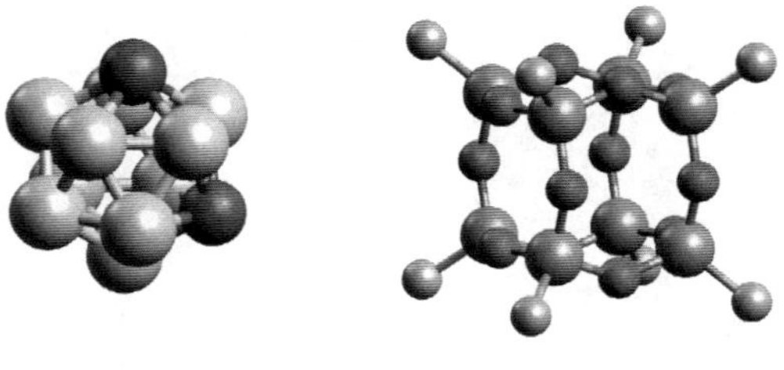

**Figure 1.** Schematics of the structures of *m*-CB (left) and POSS (right). *m*-CB is a pseudo-icosahedral cage consisting of boron (pink), carbon (black) and hydrogen (bonded to each vertex but omitted here for clarity). POSS has a cubic structure consisting of silicon (grey) and oxygen (red), with each silicon vertex bonded to an *iso*-octyl ($C_8H_{17}$) group (blue).

## 2. Materials and methods

### *2.1. Nanoparticles*

Two types of nanoparticles were used in this study, *meta*-carborane (*m*-CB, $C_2B_{10}H_{12}$), known for its excellent thermal and chemical stability, and *iso*-octyl polyhedral oligomeric silsesquioxane (POSS, $(RSi0_{1.5})_n$ where R = *iso*-octyl) (See Figure 1 for schematics of structures). Both nanoparticles were physically dispersed in a cross-linked monomodal PDMS network at various loading levels (expressed as wt.% of total raw material wt.).

### *2.2. Material Synthesis*

Materials were produced using standard tin-catalysed room temperature vulcanisation by the following procedure, only the wt.% of the nanoparticles (*m*-CB and POSS) was varied.
PDMS DMS S31 (5g, Gelest, $M_w = 26,000$, silanol terminated 0.1% OH, uncross-linked) and tetraethoxysilane (TEOS) (0.05 g) were mixed (Siemens SpeedMixer DAC 150 FVZ, 3540 rpm, 1 min.). *m*-CB (0.29 g of *m*-CB / toluene solution, 2.88 M) was added and mixed for 15 s before moulding (circular, diameter 70 mm, depth 2 mm) under pressure until initial curing had taken place (approx. 15 min). The resulting pads were then left to cure at room temperature for 48 hrs followed by final curing at 30 °C for 8 hrs. Samples were prepared containing 1.8, 3.7 and 30 wt.% *m*-CB, 1.24 and 3.05 wt.% POSS, as well as neat PDMS. All samples were elastomeric solids at room temperature.

### *2.3. Positron annihilation lifetime measurements*

Positron annihilation lifetime measurements were performed using a conventional fast-fast coincidence system, with a resolution function (modelled as an exponentially sided Gaussian (ESG)) having a full-width at half maximum (FWHM) of 175 ps. $^{22}$Na sealed between two 8 $\mu$m sheets of Kapton foil was used as the positron source (2.0 MBq). Samples were studied by heating from 100-300 K after being cooled from room temperature at approximately 10 K/min. Each lifetime spectrum contained at least $1.5\times10^6$ counts. Decomposition of the lifetime spectra was performed using LT9.1 [7]. Three sample lifetime contributions and a distribution ($\sigma_3$) on the longest lived component ($\tau_3$) attributable to *o*-Ps pick-off annihilation were found. Lifetime parameters were extracted by fixing the p-Ps lifetime to 125 ps and the ratio of *o*-Ps:*p*-Ps contributions to 3:1 to reduce the number of fitting parameters, supported by initial fitting in which all lifetime parameters were free and theoretical considerations [11].

### *2.4. Thermal characterisation*

Thermal characterisation of the materials was performed using DSC (DSC Q1000, TA instruments) under nitrogen purge. $T_g$ (mid-point) was measured for all samples prepared using a heating rate of 10 K/min from 100-300 K.

## 3. Results and discussion

The temperature dependence of the average *o*-Ps lifetime ($\tau_3$) and its distribution ($\sigma_3$) agree well with those observed previously for uncross-linked PDMS [3, 4, 5] and other low $T_g$ polymers (e.g. poly(isobutylene) and poly(butadiene) [6]). The observed data for the PDMS based materials (presented in both Fig. 2 and 3) can be classified into three distinct regions: 1) $T < T_g$, 2)

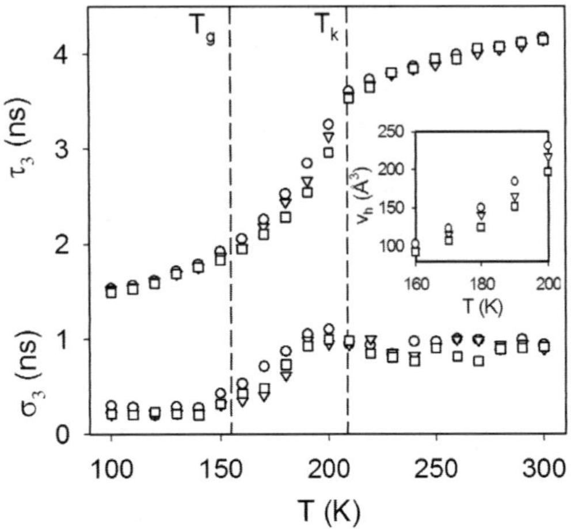


**Figure 2.** Temperature dependence of the *o*-Ps lifetime and dispersion for neat cross-linked PDMS (○ ), PDMS networks containing 1.8 wt.% (▽) and 3.7 wt.% (□) m-CB. The glass transition temperature ($T_g$) and "knee temperature" ($T_k$) are also indicated. Inset: Average hole volume in the rubbery state as calculated using the standard Tao-Eldrup model assuming spherical geometry [11]. Error bars are the same size or smaller than the symbols.

$T_g < T < T_k$ and 3) $T > T_k$, where $T_g$ and $T_k$ denote the glass transition temperature and the so-called "knee temperature" [4] respectively. For $T < T_g$ a small increase in $\tau_3$ with increasing temperature is observed, attributed to the thermal expansion of the free volume due to the anharmonicity of the local molecular vibrations [4]. In the region $T_g < T < T_k$, a change in slope in $\tau_3$ and $\sigma_3$ is observed above $T_g$. The temperature dependence of the *o*-Ps lifetime in this region shows a slight deviation from linearity which can be attributed to the expansion of the free volume cavities by the *o*-Ps zero-point pressure [8]. In the highest temperature region, $T > T_k$, a decrease in slope is seen as a consequence of molecular chain relaxation times in the polymers approaching the time-scale of *o*-Ps lifetimes, leading to formation of the Ps-bubble [8] where $\tau_3$ ceases to mirror free volume size. The same behaviour is seen for $\sigma_3$.

As an aside, it is worth mentioning that at low temperatures the PDMS based materials are semi-crystalline. Our DSC measurements show that cold crystallisation occurs at $T_x$ =195 K, followed by a melting of the crystallites at $T_m$ = 224 K during heating (data not shown). For these PDMS based systems, $T_m$ is very close to $T_k$, which may be attributed to the high flexibility of the siloxane backbone [4].

Figure 2 shows the temperature dependence of $\tau_3$ and $\sigma_3$ for the *m*-CB containing materials. The inclusion of the *m*-CB into the network had no significant effect on the average *o*-Ps lifetime below $T_g$ and above $T_k$. Between $T_g$ and $T_k$ there is a significant decrease in the lifetime and thus the average hole volume as the wt.% of *m*-CB is increased, the effect of this becoming more prominent at higher temperatures. For example, at 200 K the inclusion of 3.7 wt.% *m*-CB results in the average hole volume decreasing by $\sim 35$ Å$^3$ (using the standard Tao-Eldrup model for spherical free volume holes [11]). In spite of this observed change in the average hole volume as a result of including *m*-CB in the PDMS matrix, no measurable change in $T_g$ was seen with PALS for all materials prepared, also confirmed by DSC ($T_{g,DSC} \sim 151$ K, $T_{g,PALS} \sim 154 \pm 4$ K). At 35 wt.% *m*-CB $\tau_3$ shows no significant difference compared to that of neat PDMS over the entire temperature range. Previous studies have shown that in similar systems [9, 10] at high loading levels, physically dispersed nanoparticles in the relatively open structure of PDMS can be highly mobile. This can lead to their agglomeration and/or leeching out of the host matrix leaving behind the neat polymer.

Figure 3 shows the temperature dependence of $\tau_3$ and $\sigma_3$ for the POSS containing materials. As for the *m*-CB, the inclusion of the POSS into the PDMS network resulted in no significant change in the average hole volume below $T_g$ and above $T_k$ as measured by PALS. In the region $T_g < T < T_k$ there is a noticeable increase in the average *o*-Ps lifetime as the loading level of POSS is increased, showing the POSS has the opposite effect of *m*-CB in that it increases the

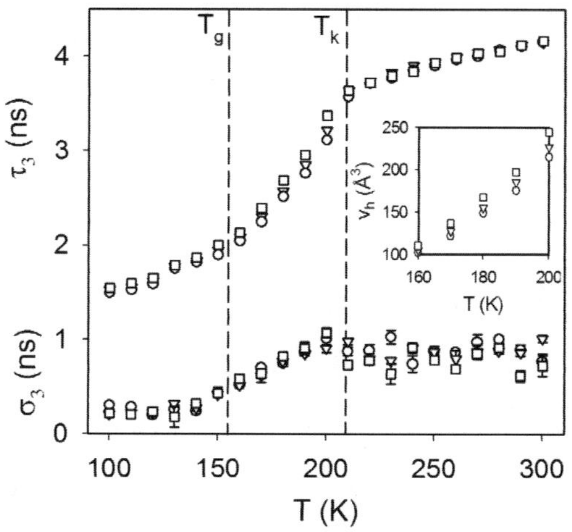


**Figure 3.** Temperature dependence of o-Ps lifetime and dispersion for neat cross-linked PDMS (○), PDMS networks containing 1.24 wt.% (▽) and 3.05 wt.% (□) POSS. The glass transition temperature ($T_g$) and "knee temperature" ($T_k$) are also indicated. Inset: Average hole volume in the rubbery state as calculated using standard Tao-Eldrup model assuming spherical geometry [11]. Errors bars are the same size or smaller than the symbols if not visible.

average hole volume in the rubbery state, as a comparison at 200 K the inclusion of 3.05 wt.% POSS increased the average free volume hole size by $\sim 30$ Å$^3$. There was no significant change in $T_g$ between samples as seen in the PALS analysis and confirmed by DSC.

## 4. Conclusion

We observe that the addition of low levels of physically dispersed *m*-CB into a cross-linked PDMS network reduces the size of the free volume holes in the rubbery state. This would be indicative of reduced molecular/chain mobility resulting in the desired mechanical reinforcement of the material. At higher levels of loading, there are no discernible changes in the free volume compared to pure PDMS matrix indicating either agglomeration of the inclusions within the matrix and/or their diffusing out of the sample. Low loading level inclusion of POSS nanoparticles into the PDMS, on the other hand, increases the free volume sizes suggesting a disruption of the chain packing. Although preliminary, the above findings are interesting pointers towards the usefulness of PALS in correlating free volume changes in polymer nano-composites with its mechanical properties.

## References

[1] Balazs A C, Emrick T and Russell T P 2006 *Science* **314** 1107
[2] Kalfus J and Jancar J 2008 *Composites Science and Technology* **68** 3444
[3] Winberg P, Eldrup M and Maurer F H J 2004 *Polymer* **45** 8253
[4] Dlubek G, De U, Pionteck J, Arutyunov N Y, Edelmann M and Krause-Rehberg R 2005 *Macromol. Chem. Phys.* **206** 827
[5] Dreiss C A, Cosgrove T, Benton N J, Kilburn D, Alam M A, Schmidt R G and Gordon G V 2007 *Polymer* **48** 4419
[6] Zrubcova J, Kristiak J, Pederson W B, Pederson N J and Eldrup M 2001 *Materials Science Forum* **363-5** 359
[7] Kansy J 1996 *Nucl. Instr. Methods Phys. Res.* A **374** 235
[8] Winberg P, Eldrup M and Maurer F H J 2012 *J. Chem. Phys.* **136** (2012) 244902
[9] Lewicki J P, Maxwell R S, Patel M, Herberg J L, Swain A C, Liggat J J and Pethrick R A 2008 *Macromolecules* **41** 9179
[10] Lewicki J, Patel M, Morrell P, Liggat J, Murphy J and Pethrick R 2008 *Sci. Technol. Adv. Mater* **9** 024403
[11] Jean Y C, Mallon P E and Schrader D M 2003 *Principles and Applications of Positron & Positronium Chemistry* (Singapore: World scientific Publishing Co. Pte. Ltd.)

Journal of Physics: Conference Series **443** (2013) 012046 doi:10.1088/1742-6596/443/1/012046

# Study of bicontinuous phase in (TTAB+pentanol)/water/n-octane reverse micellar system using positron lifetime spectroscopy

**K Chandramani Singh[1,*], R Yadav[2] and P H Khani[3]**
[1]Department of Physics, Sri Venkateswara College (University of Delhi), India
[2]Department of Physics, B.R.B College, Samastipur (LNMU), Bihar, India
[3]Department of Physics, Faculty of Science, IHU, Tehran, Iran

E-mail: kongbam@gmail.com

**Abstract.** A phase diagram of (TTAB+pentanol)/water/n-octane has been mapped by using optical method. It exhibits a reverse micellar ($L_2$) phase extending over a wide range of concentrations of the constituents. To investigate the fine structure of the $L_2$ phase, a series of (TTAB+pentanol)/n-octane ternary mixtures having initial concentrations of (TTAB+pentanol) (1:1) in n-octane as 35%, 50% and 65% by weight were prepared. In each of these mixtures, positron lifetime measurements were performed as a function of the concentration of water, using a standard lifetime spectrometer. At water concentrations of 11.8%, 8.5% and 8.4% by weight respectively for the above systems, the o-Ps pick-off lifetime $\tau_3$ shows an oscillatory behaviour while $I_3$ representing the Ps formation exhibits an abrupt change. These changes in the positron annihilation parameters have been explained on the basis of onset of bicontinuity in the microemulsion phase. The positron annihilation technique thus suggests the existence of droplet-like and bicontinuous structures in the $L_2$ phase which is otherwise considered optically to be a single phase as the system remains clear and isotropic throughout this phase. Supporting evidence has been provided by the electrical conductivity measurements performed in these systems. These results are presented in this paper.

## 1. Introduction

The existence of bicontinuous structures in microemulsions has been demonstrated [1-5]. In a water-in-oil (w/o) microemulsion, with increasing water content the system may transform from a droplet like to a bicontinuous structure. Experimental detection of bicontinuity is rather difficult. Scattering techniques, light and x-ray, fail to discriminate between droplet like and bicontinuous structures. Small angle neutron scattering contrast variation technique is also able to provide only some estimate about the mean curvature of the surfactant rich film [6], an indication of bicontinuity. Electrical conductivity measurements have been claimed to demonstrate bicontinuity [7], but they have limited applicability. These measurements can indicate water continuity but fail to establish oil continuity. On the other hand, fluorescence probe method can be employed to establish oil continuity. Recently, positron annihilation spectroscopy has been used in the detection of bicontinuity in w/o microemulsions [8].

In the present work, optical method was used for mapping the w/o microemulsion (or $L_2$) phase in (TTAB+pentanol)/water/n-octane system. The fine structure of the $L_2$ phase was investigated by employing PAS and electrical conductivity measurements.

## 2. Experimental

TTAB (purity > 98%) was procured from Fluka Chemika, Switzerland. Pentanol (purity > 99%) and n-octane (purity > 99%) were obtained from E. Merk, Germany. For making ternary solutions, de-ionised triple-distilled water was used. An optimum proportion of 1:1 by weight of TTAB and pentanol was found suitable for the formation of $L_2$ phase. Accordingly, phase diagram of (TTAB+pentanol)(1:1)/water/n-octane was mapped out by observing the optical activity of the surfactant solution using crossed polaroids. To study the fine structure of $L_2$ phase, three mother solutions were prepared: (i) 20%, (ii) 30%, and (iii) 40% by weight of (TTAB+pentanol)(1:1) in n-octane. To each of these solutions water was added as the third component to yield the required ternary systems. Positron lifetime studies were performed in these (TTAB+pentanol)(1:1)/water/n-octane systems as a function of the concentration of water using a standard lifetime spectrometer, details of which are described elsewhere [9]. The positron source used was $^{22}$Na. In each set of measurements, a total of at least $2\times10^5$ counts were collected. The electrical conductivity measurements were performed using a Delica Impedance Bridge Model 12K.

The positron lifetime spectra recorded were analysed using the software PC-PATFIT developed by Riso National Laboratory, Denmark [10]. To determine the resolution function of the lifetime spectrometer, benzophenone single crystal was grown from a solution of benzophenone in methanol. The resolution function was extracted from the lifetime data using the software package RESOLUTION. The resolution function so obtained was then used to deconvolute the lifetime spectra obtained for a sample. The most plausible fit to a lifetime spectrum was obtained by using the software package POSITRONFIT which fits a given data set to a two-, three-, or four-component multi-exponential fit with several constraints. A three-component fit constraining the shortest component $\tau_1$ to para-positronium (p-Ps) lifetime, 0.125 ns, and the intensities $I_1$ and $I_3$ fixed in the ratio of 1:3 provided the best fit to all lifetime spectra. The longest component $\tau_3$ was attributed to the pick-off annihilation of ortho-positronium (o-Ps) with electrons bound to the molecules of the surrounding medium. The relative intensity component $I_3$ is proportional to the positronium (Ps) formation in the medium. Both these parameters exhibit sensitivity to structural changes occurring in the medium.

## 3. Results and discussion

The phase diagram of (TTAB+pentanol)(1:1)/water/n-octane system in figure 1 shows the $L_2$ phase extending over wide range of concentrations of the constituents.

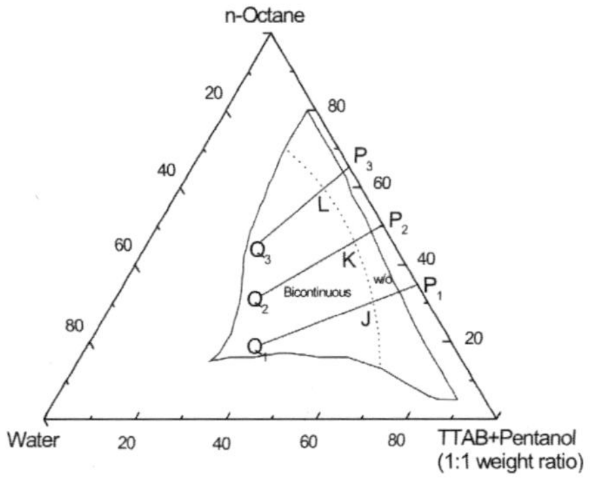


Figure 1. Phase diagram of (TTAB+pentanol) (1:1)/water/n-octane system

The dependence of $\tau_3$ and $I_3$ on water content in (TTAB+pentanol)(1:1)/water/n-octane systems are shown in figures 2a-c. It is observed that up to a certain concentration of water the parameters $\tau_3$ and $I_3$ exhibit monotonous changes. However, when a particular concentration of water is reached, these parameters show either an abrupt or an oscillatory change.

Figure 2a depicts the dependence of $\tau_3$ and $I_3$ on water concentration in a system initially containing 35% of (TTAB+pentanol)(1:1) and 65% n-octane. In this system, the solution is found to exhibit turbidity for water concentration less than ~2%. At water concentration of ~2% the solution turns clear and exhibits isotropic character indicating the formation of $L_2$ phase. Such behaviour exists till ~11.8% of water concentration. The parameters $\tau_3$ and $I_3$ remain almost constant in this region. At water concentration of ~11.8%, $I_3$ value decreases abruptly while $\tau_3$ exhibiting an oscillatory behaviour. Such changes in $\tau_3$ and $I_3$, as also observed in CTAB/water/hexanol system [8], are suggestive of the onset of a structural change in the system. Interestingly, the system continues to be optically clear and isotropic. In the CTAB/water/hexanol system, such behaviour of $\tau_3$ and $I_3$, supported by electrical conductivity measurements, has been interpreted in terms of the onset of bicontinuity in the $L_2$ phase. On further increasing the water content, $\tau_3$ and $I_3$ remain more or less constant till the occurrence of phase separation at a water concentration of ~33%. Optically the system continues to remain clear and isotropic until the phase separation occurs. Abrupt change in $I_3$ and oscillatory change in $\tau_3$ were also observed in other ternary systems investigated (Figures 2a-b).

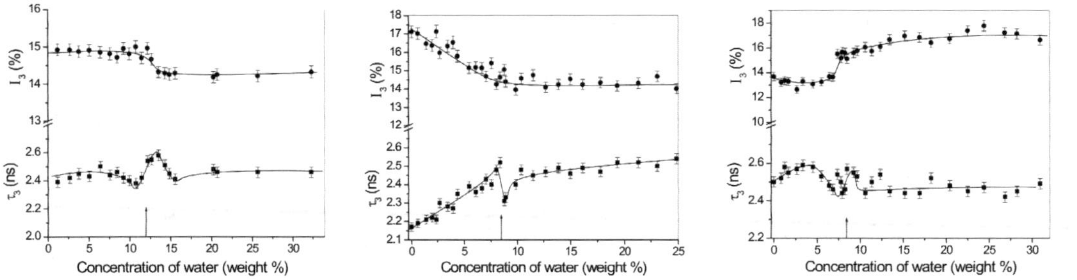


Figure 2 Dependence of $\tau_3$ and $I_3$ on water concentration for (TTAB+pentanol)(1:1)/water/n-octane system initially containing (a) 35% , (b) 50%, and (c) 65%, of (TTAB+pentanol)(1:1).

The corresponding results of the electrical conductivity measurements carried out in these systems are shown in Figures 3a-c. As seen in these figures, after the formation of microemulsion, the specific conductance is found to increase rapidly with increasing water concentration, a phenomenon characteristic of percolation [11-13]. On further increase of water content, the specific conductance plot tends to flatten out. This flattening in specific conductance plot has been attributed to the occurrence of percolation transition or setting in of bicontinuity [7] in the system. The water concentration at which the bicontinuity appears in the system was determined by finding the point at which the change in slope of specific conductance plot occurs.

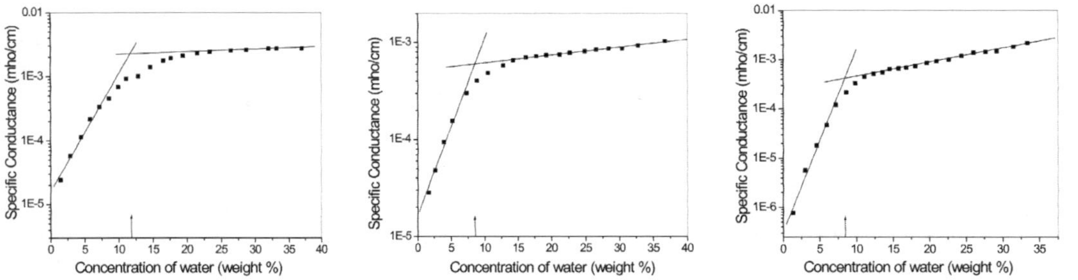


Figure 3 Dependence of specific conductance on water concentration for (TTAB+pentanol)(1:1)/ water/n-octane system initially containing (a) 35%, (b) 50%, and (c) 65%, of (TTAB+pentanol)(1:1).

The electrical conductivity behaviour of w/o microemulsions has been explained on the basis of percolative and effective medium theories [14-16]. However, the mechanism leading to percolation and the charge transport process is still not clear. One approach attributes percolation to the appearance of a "bicontinuous structure"- an open water channel responsible for electrical conduction – the static percolation picture [10,17]. The other approach considers the attractive interaction between water globules as responsible for the formation of percolation clusters; here the charge transport is via hopping of the atomic surfactant on the clusters which also rearrange in time – the dynamic percolation picture [15, 18]. Evidences in support of both the percolation approaches can be found in literature. Frequency dependent electrical conductivity and dielectric measurements carried out by Bhattacharya et al [11] in an AOT/water/decane system support the dynamic percolation model. Eicke et al [19] also used this model to explain the results on AOT/water/isooctane microemulsions.

In the (TTAB + pentanol)(1:1)/water/n-octane system investigated the conductivity in the conducting phase is about $10^{-3}$ mho/cm, which is significantly higher than the corresponding figures for AOT/water/isooctane and AOT/water/decane systems [11,19]. The higher conductivity value suggests that in this system the mechanism of percolation is perhaps the static type i.e., the system undergoes a structural change from droplet-like to bicontinuous structure. Further support to this conjecture is provided by the positron lifetime measurements.

As observed in several other systems [20] an abrupt or oscillatory change in the positron annihilation parameters, $\tau_3$ and $I_3$, is signature of a structural change occurring in the system. The abrupt or oscillatory change exhibited by $\tau_3$ and $I_3$ at a certain water concentration in the above mentioned (TTAB + pentanol)(1:1)/water/n-octane systems is thus suggestive of a structural transformation. The concentration of water at which such a transformation is detected by positron annihilation parameters closely agree with those observed from conductivity measurements. This agreement, therefore, suggests that the changes observed in both the measurements perhaps represent the same change, namely the percolation transition, occurring in the system.

**Summary**

In the present work, w/o microemulsion ($L_2$) phase of (TTAB + pentanol)(1:1)/water/n-octane at 20°C has been mapped out employing optical method. Positron lifetime and electrical conductivity measurements have been employed to investigate the fine structure of this phase. Results obtained from both these measurements agree well and suggest the existence of a fine structure of the $L_2$ phase. Both these techniques support the existence of a phase boundary dividing the $L_2$ phase into droplet-like and bicontinuous structures.

**References**

[1] Chen S J, Evans D F and Ninham B W 1984 *J. Phys. Chem.* **88** 1631
[2] Blum F D, Pickup S, Ninham B W, Chen S J and Evans D F 1985 *J. Phys. Chem.* **89** 711
[3] Evans D F, Mitchell D J and Ninham B W 1986 *J. Phys. Chem.* **90** 2817
[4] Allen M, Evans D F, Mitchell D J and Ninham B W 1987 *J. Phys. Chem.* **91** 2320
[5] Warr G G, Sen R, Evans D F and Trend J E 1988 *J. Phys. Chem.* **92** 774
[6] Auvray L, Cotton J P, Ober R and Taupin C 1984 *J. Phys. Chem.* **88** 4586
[7] Clausse M, Peyrelasse J, Heil J, Boned C and Lagourette B 1981 *Nature* **293** 636
[8] Roy Choudhury S, Yadav R, Maitra A N and Jain P C 1992 *Mat. Sci. Forum* **105-110** 1521
[9] Scriven L E 1976 *Nature* **263** 124
[10] Gennes P G and Taupin C 1982 *J. Phys. Chem.* **86** 2294
[11] Bhattacharya S, Stokes J P, Kim M W, Huang J S 1985 *Phys. Rev. Lett.* **55** 1884
[12] Eicke H F, Hilfiker R and Thomas H 1986 *Chem. Phys. Lett.* **125** 295
[13] van Dijk M A, Casteleijn G, Joosten J G H and Levine Y K 1986 *J. Chem. Phys.* **85** 626
[14] Lagues M, Ober R and Taupin C 1978 *J. Phys. Lett.* **39** L487
[15] Lagues M 1979 *J. Phys. Lett.* **40** L331
[16] Lagourette B, Peyrelasse J, Boned C and Clausse M 1979 *Nature* **281** 60
[17] Talmon Y and Prager S 1978 *J. Chem. Phys.* **69** 2984
[18] Safran S A, Webman I and Grest G S 1985 *Phys. Rev. A* **32** 506
[19] Eicke H F, Borkovec M and Gupta B D 1989 *J. Phys. Chem.* **93** 314
[20] Sharma M, Kaur C, Kumar J, Chandramani Singh K and Jain P C 2001 *J. Phys.: Condens. Matter* **13** 7249

Journal of Physics: Conference Series **443** (2013) 012047 doi:10.1088/1742-6596/443/1/012047

# Positron annihilation lifetime study of interfaces in ternary polymer blends

**D Meghala[1,§] P Ramya[1], T Pasang[1], J M Raj[1], C Ranganathaiah[1,2] and J F Williams[2]**

[1]Department of Studies in Physics, University of Mysore, Manasagangotri, Mysore-570 006, India.

[2]ARC Centre of Excellence, Centre for Antimatter-Matter Studies, School of Physics, University of Western Australia, WA 6009, Australia.

**Abstract:** A new method based on positron lifetime spectroscopy is developed to characterize individual interfaces in ternary polymer blends and hence determine the composition dependent miscibility level. The method owes its origin to the Kirkwood-Risemann-Zimm (KRZ) model for the evaluation of the hydrodynamic interaction parameters ($\alpha_{ij}$) which was used successfully for a binary blend with a single interface. The model was revised for the present work for ternary polymer blends to account for three interfaces. The efficacy of this method is shown for two ternary blends namely poly(styrene-*co*-acrylonitrile)/poly (ethylene-*co*-vinylacetate)/poly(vinyl chloride) (SAN/EVA/PVC) and polycaprolactone /poly(styrene-*co*-acrylonitrile)/poly(vinyl chloride) (PCL/SAN/PVC) at different compositions. An effective hydrodynamic interaction parameter, $\alpha_{eff}$, was introduced to predict the overall miscibility of ternary blends.

## 1. Introduction

The concept used in the development of ternary polymer blends is that a polymer A, which is miscible with two polymers B and C, can make compatible the immiscible binary pair B and C. Miscibility is governed by the interfacial behavior [1] and, generally, the interface is the region formed between two phases. The interface plays an important role with regard to several properties of the blends so formed and is expected to change with composition.

Studies on ternary polymer blends have reported only the overall nature of the blend and in some cases individual binary blends had been constructed and characterized to understand the influence of individual interfaces in ternary blends [2]. However, this information is indirect and the situation concerning three interfaces in a real ternary blend is different and more complicated compared to single individual interface that exists in a binary blend. To understand better the nature of interfaces at the microscopic level we used the novel approach proposed by Schnell and Wolf [3] for polymer/solvent systems in terms of the hydrodynamic interaction parameter ($\alpha$) evaluated from viscosity measurements. This method was modified by Ranganathaiah et al [4] for polymer/polymer binary blends in the solid phase for the characterization of a single interface by invoking the concept that free volume and viscosity are inversely related [5]. The efficacy of this method has been verified for several binary blend systems under varying conditions [6,7] by the determining hydrodynamic interaction parameter from the measured free volume data obtained from positron lifetime spectroscopy. However, the same approach is not directly applicable to ternary blends with three interfaces. So, for three distinct interfaces, three hydrodynamic interaction parameters are to be formulated one for each interface. Hence, in this work, we have investigated two ternary blend systems namely (1) poly(styrene-*co*-acrylonitrile)/poly (ethylene-*co*-vinylacetate)/poly(vinyl chloride) (SAN/EVA/PVC) and (2) polycaprolactone /poly(styrene-*co*-acrylonitrile)/poly(vinyl

§ email: meghala.dinesh@gmail.com

chloride) (PCL/SAN/PVC) for different compositions. The required mathematical formulations for three α's ($\alpha_{12}$, $\alpha_{23}$ and $\alpha_{31}$), one for each interface, have been derived and tested experimentally. From our method the individual α's are derived and used to compute an effective α ($\alpha_{eff}$) using simple additivity rule. The $\alpha_{eff}$ facilitate comparison of the present results with published results.

## 2. Experimental

We have prepared the samples of SAN/EVA/PVC and PCL/SAN/PVC ($T_g$: SAN:106°C, EVA:-20°C PVC:74°C, PCL:-60°C) of different compositions by the conventional solution casting method using a common solvent tetrahydrofuran (THF). To record lifetime spectra of the blends and the respective individual polymers positron annihilation lifetime spectrometer (PALS) with a time resolution of 220 has been used. A 17 μCi $^{22}Na$ positron source on kapton foil backing was used in the measurement of spectra. All the measurements have been done at room temperature in air. The lifetime spectra have been resolved into three lifetime components using computer program PATFIT-88 with proper source and background corrections. Further details can be found from earlier works [6]. The *o*-Ps lifetime $\tau_3$ and intensity $I_3$ are used to obtain the fractional free volume $F_V$ of the blends as well as individual polymers.

## 3. Results & Discussion

Wolf et al [3] introduced two parameters measured from viscosity data; one is the geometric factor (γ) related to the molecular arrangement and to the architecture of the blend, and the other is the hydrodynamic interaction parameter (α) considered as a measure of the excess friction generated at the interface between the blend constituents. The hydrodynamic interaction has its genesis in the KRZ model proposed by Zimm [8] which is an extension of the KSR (Kargin-Slonimsky-Rouse) model proposed by Rouse [9]. According to the KRZ model, 'hydrodynamic interaction' between the polymer and solvent excludes specific interactions but is associated with flow dynamics which influences the visco-elastic behavior of the system either in liquid or solid phase. Further, the friction associated with the visco-elastic flow results in energy dissipation at the interface and is influenced by the composition of the blend. In the case of miscible blends, the polymer chains are brought closer to each other and increase the friction at the interface. Energy dissipation increases and tension at the interface decreases.

The negative sign of the hydrodynamic interaction parameter (α) indicates dissipation of energy. If miscibility is high at certain composition of the blend, α takes large negative values [6]. On the other hand, for immiscible blends, α is close to zero or positive. This is interpreted as the absence of any favorable interaction between the constituent polymers and hence friction at the interface is very less.

For a ternary polymer blend characterized by three interfaces, we define $\varphi_1$, $\varphi_2$, $\varphi_3$ as the volume fractions of component polymers 1, 2 and 3 of the blend and $\alpha_{12}$, $\alpha_{23}$, $\alpha_{31}$ as the hydrodynamic interaction parameters corresponding to interfaces between polymer 1-2, 2-3 and 3-1 respectively, and their relationships as follows.

$$\frac{1}{\Delta F_V} = \delta_1\,[\varphi_1 - \Omega_1(1+\Omega_1\Omega_2)] + \delta_2[\Omega_3(1-\Omega_2^2) - \varphi_3] - \delta_3\Omega_3^2\Omega_1 + \sum_{\substack{i,j=1\\ j=i+1}}^{3}\left[\frac{e^{1/F_{VB}}\rho\Omega_i^2\Omega_j}{1+\gamma_{ij}} + 2\alpha_{ij}(1-\Omega_i)\Omega_i\Omega_j\right] \quad (1)$$

and

$$\frac{1}{\Delta F_V} = \frac{1}{F_{VB}} - \frac{\varphi_1}{F_{V1}} - \frac{\varphi_2}{F_{V2}} - \frac{\varphi_3}{F_{V3}} \quad (2)$$

Here, $F_{VB}$, $F_{V1}$, $F_{V2}$, $F_{V3}$ are the experimentally measured fractional free volumes of the ternary blend, pure polymers (blend constituents) 1, 2 and 3 respectively, $\varphi_1$, $\varphi_2$ and $\varphi_3$ are the volume fractions of the blend components 1, 2 and 3 respectively. Also ρ is density of the blend and $\Omega_1$, $\Omega_2$, $\Omega_3$ are the surface fractions given by,

$$\Omega_1 = \frac{(1+\gamma_{12})(1+\gamma_{31})\varphi_1}{(1+\gamma_{12})(1+\gamma_{31})\varphi_1 + (1+\gamma_{31})\varphi_2 + (1+\gamma_{12})\varphi_3} \quad \Omega_2 = \frac{(1+\gamma_{23})\varphi_2}{(1+\gamma_{12})\varphi_1 + (1+\gamma_{23})\varphi_2 + \varphi_3} \quad \Omega_3 = \frac{\varphi_3}{(1+\gamma_{31})\varphi_1 + (1+\gamma_{23})\varphi_2 + \varphi_3} \tag{3}$$

and

$$\sum_{\substack{i,j=1,\\ j=i+1}}^{3} \delta_{ij} = \frac{1}{F_{Vj}} - \frac{1}{F_{Vi}} \tag{4}$$

$\gamma$ is the geometric factor given by

$$\frac{1}{F_{VB}} = \frac{\varphi_1}{F_{V1}} + \frac{\varphi_2}{F_{V2}} + \frac{\varphi_3}{F_{V3}} + \sum_{\substack{i,j=1\\ j=i+1}}^{3} \left[ \left( \frac{1}{F_{Vj}} - \frac{1}{F_{Vi}} \right) \left( \frac{\gamma_{ij}\,\varphi_j \varphi_i}{1+\gamma_{ij}\varphi_j} \right) \right] \tag{5}$$

In addition to the individual 'α's referring to respective interfaces, we have calculated an effective alpha, $\alpha_{eff}$, for the blend with a view to compare our results with published results. The $\alpha_{eff}$ can be written using the simple additivity rule as,

$$\alpha_{eff} = (\varphi_1 + \varphi_2)\alpha_{12} + (\varphi_2 + \varphi_3)\alpha_{23} + (\varphi_3 + \varphi_1)\alpha_{31} \tag{6}$$

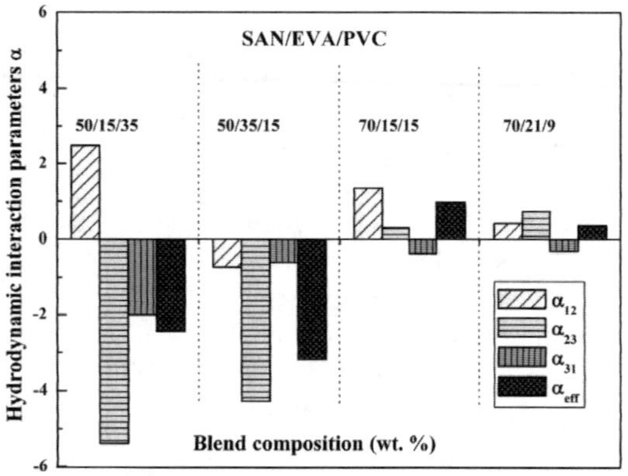


**Figure 1:** Plot of hydrodynamic interaction parameters α as a function of blend composition for the blend SAN/EVA/PVC.

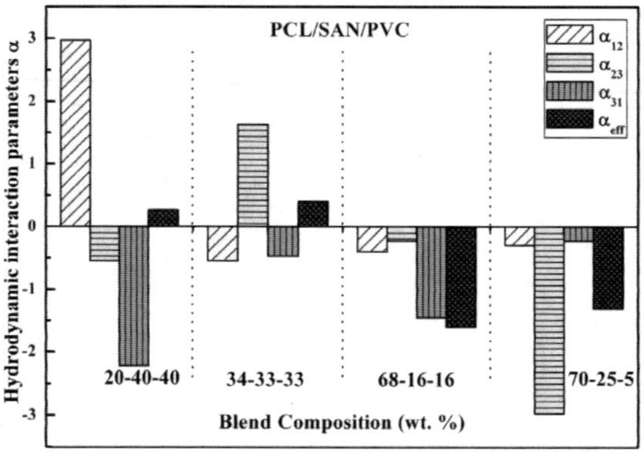


**Figure 2:** Plot of hydrodynamic interaction parameters α as a function of blend composition for the blend PCL/SAN/PVC.

For the ternary blend SAN/EVA/PVC, the $\alpha_{12}$, $\alpha_{23}$, $\alpha_{31}$ denote the hydrodynamic interaction parameters corresponding to respective interfaces between SAN/EVA, EVA/PVC and PVC/SAN evaluated according to equation 1 and shown in figure 1. As seen from figure 1, the interface between SAN/EVA exhibits positive values for the compositions 50/15/35, 70/15/15 and 70/21/9 which suggest a weak interface. An observed small negative value for 50/35/15 suggests little compatibility between the components SAN and EVA. The interface ($\alpha_{23}$) between EVA/PVC indicates negative values which are a signature of a good interface due to excess friction attributable to the physical cross-linking between EVA and PVC phases [2]. The interface ($\alpha_{31}$) formed between PVC/SAN acquires negative values at all the four compositions investigated suggesting that at these compositions, good interfaces are formed. The compatibility between SAN and PVC is due to the repulsive force among the SAN chains which provide sliding pathways for PVC chains. [10]. The influence of these individual interfaces can be seen in the effective α for the blend evaluated using equation 6 and also plotted in figure 1. The $\alpha_{eff}$ is negative for the compositions 50/15/35 (measured $T_g$:66°C) and 50/35/15 (measured $T_g$:45°C) suggesting the blend is miscible at these compositions [2,11] and positive at 70/15/15 (measured $T_g$:44°C and 80°C) and at 70/21/9 (measured $T_g$:24°C and 60°C) indicating that the blend remains phase separated at these compositions [2,11].

The second blend PCL/SAN/PVC can be understood on the same lines as the SAN/EVA/PVC blend. The interfaces between PCL/SAN, SAN/PVC and PVC/PCL are characterized by $\alpha_{12}$, $\alpha_{23}$, $\alpha_{31}$ respectively. The calculations have been done exactly the same way and the results are plotted in figure 2. As observed, the interface $\alpha_{12}$ between PCL/SAN takes on both positive and negative values suggesting for the compositions for which $\alpha$ is negative that corresponding chains are pulled closer and gives rise to the interfacial friction. Interestingly, the negative values of $\alpha$ are observed wherever PCL is the matrix and SAN is the dispersed phase suggesting compatibility between the components. This may be due to the interaction between the carbonyl group of PCL and tertiary hydrogen of SAN, a fact well reported in literature [12].

The second interface is characterized by $\alpha_{23}$ and this takes on a positive value at 34/33/33 composition while the other compositions result in negative values. The negative values of $\alpha$'s can be attributed to the repulsive force among the SAN chains paving way for PVC chains to slide in [10]. Interestingly the third interface in the blend that is between PVC/PCL, produces negative values at all the compositions studied indicating the formation of good interfaces attributable to the interaction between polar oxygen of PCL and proton of PVC [13].

Finally, the $\alpha_{eff}$, plotted in figure 2, shows positive values at 20/40/40 (measured $T_g$:-30°C and 10°C) and 34/33/33 (measured $T_g$:-5°C and 30°C) compositions and thus we infer the blend is immiscible at these compositions. Also the negative values at 68/16/16 (measured $T_g$:-15°C) and 70/25/5 (measured $T_g$:10°C) indicate the blend is miscible at these compositions, in good agreement with the published literature [14].

## 4. Conclusion:

The hydrodynamic interaction parameter ($\alpha$) derived from the measured fractional free volume $F_V$ from positron annihilation lifetime measurements, provides a simple and easy means of characterizing the individual interfaces in ternary blends. Importantly, the composition dependent miscibility level was estimated from this method. Since the information on weak interfaces is readily available, it becomes easy for the designer to choose a proper compatibilization route to stabilize that particular weak interface and so improve the overall properties of the ternary blend. Additionally, $\alpha_{eff}$ from the present method serves as an useful parameter to infer that a given ternary blend is miscible or immiscible.

## References:

[1] Stamm M 2008 *Polymer Surfaces and Interfaces – Characterization, Modification and Applications* (Verlag Berlin Heidelberg: Springer)
[2] Lizymol P P, Sabu Thomas and Jayabalan 1997 *Eur. Polym. J.* **33** 1397
[3] Schnell M and Wolf B A 2000 *J Rheol.* **44** 617 and in 2001 *Polymer* **42** 8599
[4] Jamieson A M, Brian G Olson and Sergei Nazarenko 2010 *Polymer Physics: From Suspensions to Nanocomposites and Beyond* ed. Utracki L A and Jamieson A M, p 491 (Hoboken: John Wiley)
[5] Williams M L, Landel R F and Ferry J D 1955 *J. Am. Chem. Soc.* **77** 3701
[6] Ranganathaiah C and Kumaraswamy G N 2009 *J Appl Polym Sci.* **111** 577 and references there in
[7] Raj J M and Ranganathaiah C 2009 *J Polym Sci Part B: Polym Phys.* **47** 619
[8] Zimm B 1956 *J Chem Phys.* **24** 269
[9] Rouse P J 1953 *J Chem Phys.* **21** 1272
[10] Kim J H, Barlow J W and Paul D R 1989 *Polym Eng Sci.* **29** 581
[11] Lizymol P P and Sabu Thomas 1994 *Thermochimica Acta.* **233** 283
[12] Chiu S C and Smith T G 1984 *J Appl. Polym. Sci.* **29** 1797
[13] Brigitte Albert, Robert Jerome and Philippe Teyssie 1984 *Macromolecules* **17** 2552
[14] Chiarng Huarng J Y H, Kyonsuku Min and James L White 1988 *Polym. Eng. Sci.* **28** 1590

# Interface profile studies in immiscible and partially miscible binary polymer blends from free volume measurement

**P Ramya[1§], D Meghala[1], T Pasang[1], J M Raj[1], C Ranganathaiah[1,2] and J F Williams[2]**

[1]Department of Studies in Physics, University of Mysore, Manasagangotri, Mysore-570006, India.

[2]ARC Centre of Excellence, Centre for Antimatter-Matter Studies, School of Physics, University of Western Australia, WA 6009, Australia

**Abstract.** The diffused interface widths in an immiscible and a partially miscible polymer blend namely Polyvinyl chloride/Ethylene vinyl acetate (PVC/EVA) and Polystyrene/Polymethylmethacrylate (PS/PMMA) are experimentally measured and reported here. A new empirical relation found between hydrodynamic interaction parameter $\alpha$ derived from free volume data and the Flory-Huggins interaction parameter $\chi$ is used to construct density profile across the interface to derive the interface width in above two binary blends.

## 1. Introduction

The combination of two or more chemically and structurally different polymers gives raise to materials with a range of properties, not deliverable by any of the constituents. This is called polymer blending. Blending usually results in immiscible or partially miscible blends. Stabilizing interfaces in such blends is essential for the end use of these materials in several applications. In this context the interface width in such blends needs to be understood. Generally, the interface width is considered as the third phase whose width depends on the inter-diffusion between two component polymers [1,2].
Most of the published literature on interface widths deal with welded interfaces between slabs or thin films of the polymers but very few consider the equilibrium interfaces that exist in an immiscible or demixed blends of different morphology [1]. Further, the interfaces in welded polymer thin films are completely different in morphology compared to engineering or disperse polymer blends produced by chemical or physical blending. In the present work, Positron Annihilation Lifetime Spectroscopy (PALS) is used to measure fractional free volume and hence determine hydrodynamic interaction parameter ($\alpha$) from which the interface widths in two disperse polymer blends are derived by constructing the density profile across the interface.

## 2. Experimental

The blends of Polyvinyl chloride/Ethylene vinyl acetate (PVC/EVA) and Polystyrene/Polymethyl methacrylate (PS/PMMA) of different compositions were prepared by solution casting method using methyl ethyl ketone as the common solvent. Positron lifetime spectra for the blends and pure polymers were recorded using PALS with a time resolution of 220 ps. A 17 $\mu$Ci $^{22}$Na positron source was used

§ email: ramya.dpg1@gmail.com

in the measurement. The acquired positron lifetime spectra were analyzed into three lifetime components with the help of computer programs PATFIT-88 [3] with proper source and background corrections and program CONTIN PALS-2 [4] for free volume distribution analysis. The PATFIT analysis is made to check the correctness of the CONTIN results. The CONTIN analysis provides the annihilation rate probability density function (PDF). The free volume radius PDF and free volume size PDF are transformed by the method of Gregory [4]. Accordingly, free volume radius PDF: f(R) and free volume size PDF: g(V) are expressed as

$$f(R) = 2\Delta R\{\cos[2\pi R/(R + \Delta R)]-1\}\alpha(\lambda)/(R + \Delta R)^2 \quad (1)$$

and $g(V)=f(R)/4\pi R^2$ (2)

where $\Delta R= 1.656$ Å [5]. Further details can be obtained from the earlier publication [5].

## 3. Results and discussion

Figure 1a shows the *o*-Ps lifetime and radius distribution and figure 1b shows the free volume hole size distribution in PS/PMMA with composition 20/80 and PVC/EVA with composition 80/20. The broad distribution in the PVC/EVA blend can be attributed to the formation of interfacial regions between the component polymer chains and hence additional free volume generated [5]. This also indicates a lack of mutual interaction between the component polymers. From figure 1b it can be observed that the curve for PVC/EVA is broader in comparison with PS/PMMA (20/80). The FWHM (full width at half maximum) of the PS/PMMA (20/80) is 30Å$^3$ while that of PVC/EVA (80/20) is 35Å$^3$. The small FWHM of PS/PMMA blend suggests close packing of the constituent polymer chains.

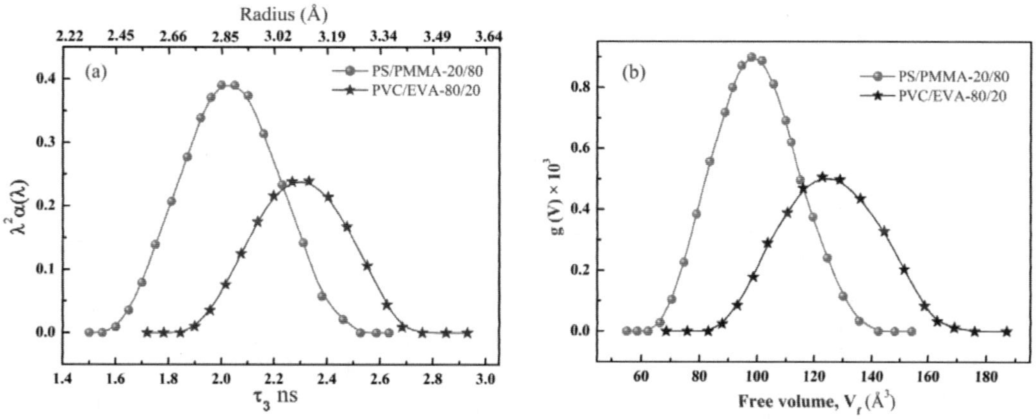


Figure 1. a) *o*-Ps lifetime ($\tau_3$) v/s *o*-Ps probability density function (b) free-volume size ($V_f$) probability density function for PS/PMMA (20/80) and PVC/EVA (80/20) blends.

The hydrodynamic interaction parameter of Schnell and Wolf work [6] for polymer/solvent systems modified by Ranganathaiah and coworkers [7,8] is used here. We derive two parameters namely the geometric factor (γ): which relates to molecular arrangement and architecture of the chains of the blend components and the hydrodynamic interaction parameter (α): which is treated as a measure of excess friction generated at the interface between the blend constituents. For details readers are referred to refs 7 and 8. The efficacy of the method was tested for few binary polymer blends under different conditions and proved to be very successful [8,9]. From the measured free volume results, the hydrodynamic interaction parameter α is evaluated according to the Ranganathaiah method [7-9] using equation (3).

$$\Delta F_V = \left\{\delta\,[\gamma\,(1+\gamma\,\varphi_2\,)^2 - \varphi_1(\,1+\gamma\,)]+2\,\alpha(1+\gamma\,)^2\varphi_2 + e^{\frac{1}{F_V}}\rho\,\varphi_1\right\}^{-1} \frac{(\,1+\gamma\,\varphi_2\,)^3}{\varphi_1\,\varphi_2} \quad (3)$$

where $\Delta F_v = \left[\frac{1}{F_{vb}} - \frac{\varphi_1}{F_{v1}} - \frac{\varphi_2}{F_{v2}}\right]$ and $\delta = \frac{1}{F_{v2}} - \frac{1}{F_{v1}}$ (4)

$F_{V1}$ and $F_{V2}$, are the fractional free volume of the pure polymers, ρ is the density of the blend, $\varphi_1$ and $\varphi_2$ are volume fraction of the blend constituents. The parameter γ is related to fractional free volume [7]

$$F_V = \left[\frac{\varphi_1}{F_{V1}} + \frac{\varphi_2}{F_{V2}} + \delta\left(\frac{\gamma\,\varphi_1\,\varphi_2}{1+\gamma\,\varphi_2}\right)\right]^{-1} \quad (5)$$

To determine interface width, use of the Flory–Huggins interaction parameter χ is made. Chain orientation and entanglements influence the χ parameter similar to the repulsion between unlike monomeric units in addition to other properties of the constituent polymers [1]. Since α is the result of friction at the interface due to mobility of chains is thought to be related to χ. This is used to determine the interface width as both parameters predict miscibility of polymer blends. The χ parameter has been extensively used in the study of interface width based on a mean field theory approach and is well established irrespective of its unrealistic assumptions. We have adopted a similar approach based on α. To meet this requirement a correlation between χ and α parameter is essential and is obtained as follows. A plot of ln α against χ shown in figure 2 is constructed from the α and χ data (from literature) for ten polymer blends which include both partially miscible and immiscible blends.

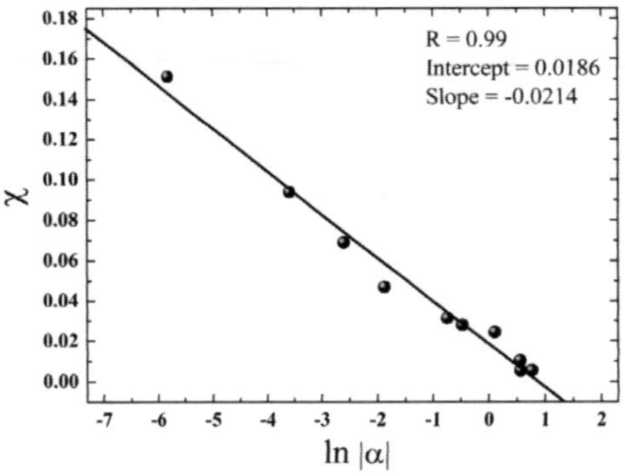


Figure 2 plot of χ against ln |α|

It is clear from figure 2 that for the range of α and χ considered a linear relation fits the data given below:

$$\chi = 0.0186 - 0.0214\ \ln|\alpha| \quad (6)$$

This result is an additional support for the inverse relation between viscosity and free volume. Then the well known Helfand-Tagami theory [10] approach is used to construct a composition density profile across the interface. According to this the composition density profile is given as

$$\rho_i = \rho_{i,0}\, y^2/(1+y^2) \quad (7)$$

where $\rho_i \rightarrow$ density ; i= polymer A or B and $y \equiv \exp\{(6\chi)^{1/2}(x/b)\}$ with x/b designated as the reduced distance across the interface. In the present case, the expression for y is modified to the form

$$y \equiv \exp\{(6H)^{1/2}(x/b)\} \quad (8)$$

where $H = 0.0186 - 0.0214\ \ln|\alpha|$

Using equation 7, composition density profile across the interface of PS/PMMA and PVC/EVA binary blends are constructed and shown in the figure 3a and figure 3b respectively.

In this calculation, the α distribution values obtained from CONTIN analysis are used. From figure 3a and b, the interface is centered at x=0; the positive x region is rich in one polymer while the negative x region is rich with the other. The solid curve is for the first polymer of the blend and the dotted curve is for the second polymer. A tangent is drawn at the point (x/b)=0 and this intersects the x-axis at $(\rho/\rho_0) = 0$ and $(\rho/\rho_0) = 1$ meaning the volume fraction of the constituents corresponds to $\varphi = 0$ and 1 respectively [2,10]. The interface width $\Delta l$ is read between these two points of intersection.

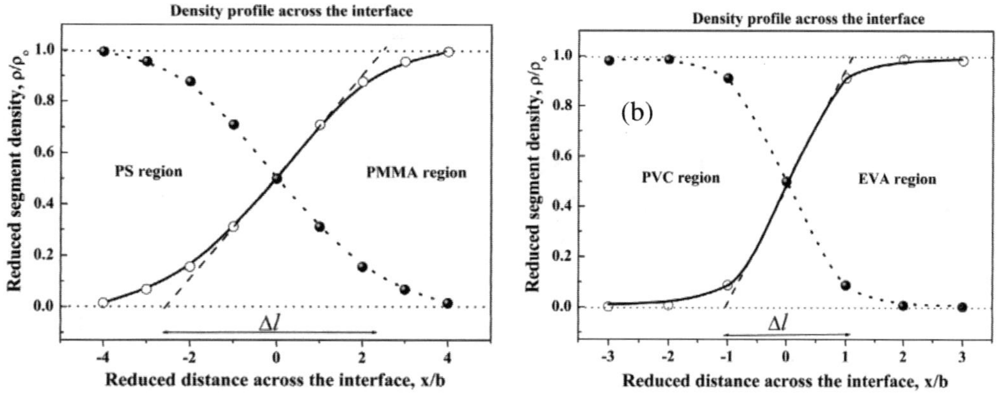


Figure 3. Composition density profile of (a) PS/PMMA-20/80 composition blend (b) PVC/EVA-80/20 composition blend.

The interface width value determined in this way for PS/PMMA blend is 5.04 nm. For this partially miscible blend the possible interaction is between O-$CH_3$ of PMMA and hydrogen of PS [8]. Depending on the strength of this interaction, it is expected that inter-diffusion of PMMA into the PS chains across the interface will be higher and hence wider interface width. The interface width for the immiscible PVC/EVA is determined to be 2.17 nm. An immiscible blend lacks any kind of interaction between the components and so is similar for the PVC/EVA blend. This suggests very little inter-diffusion of PVC into EVA or vice versa and hence a narrow interface width.

## 4. Conclusion

An empirical relation between the hydrodynamic interaction parameter $\alpha$ and the Flory-Huggins interaction parameter $\chi$ has been established for a limited range of $\alpha$ and $\chi$ values. The composition density profile of the constituent polymers across the interface of the blends is constructed for the two polymer blends PVC/EVA and PS/PMMA. From density profiles, the diffused interface width values so obtained are 2.17 and 5.04 nm respectively for the immiscible (PVC/EVA-80/20) and the partially miscible (PS/PMMA-20/80) blend. For other compositions studied, these values are lower than the values reported here.

## References

[1] Stamm M 2008 *Polymer Surfaces and Interfaces - Characterization, Modification and Applications* (Verlag Berlin Heidelberg: Springer)
[2] Ajji A 1998 *Polymer Blends Handbook* ed Utracki L A (London: Kluwer academic publishers) p 295
[3] Kirkegaard P, Pederson N J and Eldrup M 1989 Riso Nat Lab Reports, **M-2740**,
[4] Gregory R B 1993 *CONTIN(PALS-2): a modification of CONTIN (Version2) for the determination of positron annihilation lifetime distributions* (USA: Kent State University)
[5] Liu J, Jean Y C and Yang H 1995 *Macromolecules* **28** 5774
[6] Schnell M and Wolf B A 2000 *J. Rheol.* **44** 617 and in 2001 *Polymer* **42** 8599
[7] Ranganathaiah C and Kumaraswamy G N 2009 *J. Appl. Polym. Sci.* **111** 577
[8] Raj J M and Ranganathaiah C 2009 *J. Polym. Sci. Part B:Polym. Phys.* **47** 619
[9] Jamieson A M, Olson B G and Nazarenko S 2012 *Polymer Physics: From Suspensions to Nanocomposites and Beyond* ed Utracki L A and Jamieson A M (Hoboken: John Wiley)
[10] Helfand E and Tagami Y 1972 *J. Chem. Phys.* **57** 1812

# Investigation of microstructure and conducting mechanism of nanocomposite polymeric electrolytes with rectorite clay by PALS

**S Gao, J Gong, X L Yan, G B Xue, J Zhong and B Wang**
Department of Physics, Key Laboratory of Nuclear Solid State Physics, Wuhan University, Wuhan,430072, China

E-mail: bowang@whu.edu.cn

**Abstract.** Polymeric electrolytes with different modified organic nanoretorite (OREC) content have been prepared. Measurements of the structural transition, the positron annihilation lifetime, free volume and ionic conductivity as a function of the OREC content and the temperature have been performed. According to the variations of the ortho-positronium (o-Ps) lifetimes with temperature, the glass transition temperatures have been determined. A direct relationship between the ionic conductivity and the fractional free volume has been established using based on free volume theory Williams-Landel-Ferry (WLF)equations, implying a free-volume transport mechanism. Our experimental results indicated that the segmental chain motion and ionic migration and diffusion could be explained by the free volume theory. .

## 1. Introduction

As energy storage system and power source of various portable devices, electronic devices, the rechargeable lithium ion batteries have been widely used in our modern society. There has been increasing interest in development of solid polymer electrolytes due to their applications in solid state electrochemical device and particularly in solid rechargeable Lithium batteries. The study on the solid polymer electrolytes polyethylene oxide(PEO) has received more attention since the original work of Wright and Armand in 1973. PEO-based polymer electrolyte is the most attractive host because of its easy fabrication, light weight, low cost, high conductivity above melting temperature. However, high crystalline phase concentration limits the conductivity of PEO-based electrolyte, especially at room temperature. Many attempts have been made to overcome the drawbacks. Recent research has shown that these drawbacks may be particularly overcome by the incorporation of nano-scale fillers, such as nanoclay. The addition of nanoscale fillers not only improve the ionic conductivity, but also enhance the mechanical strength and stability of polymer electrolytes.

The aim of this study is to investigate the effect of nano-clay on the microstructure and conducting behaviour of nanocomposite electrolytes。

## 2. Experimental Section

### *2.1. Materials and sample preparation*

Pristine rectorites, PEO($M_w$=600 000), $LiClO_4{\cdot}3H_2O$ and Hexadecyl trimethyl ammonium bromide (CTAB) as the intercalation reagent to obtain the organophilic rectorite (OREC) were supplied by

Hubei Celebrities Technology Co. (Wuhan, China), Dadi Fine Chemical Co. Ltd, and Shanghai Zhanyun Chemical Co. Ltd (A.R.) respectively. The molar ratio of the ethylene oxide segments over the lithium ions was fixed at [EO]/[Li] =8. PLC0, PLC1, PLC3 and PLC6 are named to denote the nanocomposites with 0, 1, 3 and 6 wt.% OREC content, respectively. PLC is the abbreviation of PEO-$LiClO_4$ Clay. The detailed information on the preparation of OREC nanocomposites can be found elser where[1].

### *2.2. Characterization*

The measurements of small angle X-ray diffraction (SXRD) and wide angle X-ray scattering(WAXS) were completed with a Bruker D8 Advance X-ray Diffractometer using Cu Kα radiation, 1.5406 Å wavelength, at a voltage of 40 KV, under 1°/min scanning rate respectively.

Electrical resistivity was measured by a conventional four-probe method, the detailed information can be found elsewhere[1].

Positron lifetime measurements were carried out using a conventional fast- fast coincident spectrometer. A 15 Ci $^{22}Na$ source was sandwiched between two identical samples, each spectrum contained $1\times10^6$ counts for PATFIT analysis and $4\times10^6$ counts for MELT analysis respectively.

## 3.Results and Discussion

### *3.1. XRD Results.*

The results of small angle x-ray diffraction are shown in figure.1. Using the SXRD results, the intercalated structures can be identified with Bragg's law: with Bragg's law, $\lambda = 2d\sin\theta$, where λ is the wavelength of the X-ray radiation (1.5406 Å), d corresponds to the spacing between the (001) diffraction lattice plane and refers to measured diffraction angle. From this figure, we can see that the interlayer spacing has been extended up to 4.42nm for organic OREC, which is much wider than 2.52 nm of pristine rectorite. On the other hand, for nanocomposites samples with 1wt% and 3wt% clay content, no obvious peaks were observed, which means that the clays were exfoliated.

The WAXS measurements were conducted as shown in figure 2. From this figure, we can see that the intensities of the characteristic peaks of 1wt% and 3wt% clay content samples are lower than that of pure PEO, which means that the small quantity of OREC added into PEO matrix can effectively decrease the crystallinity of PEO matrix[2]. Because polymer intercalation disrupted the normal, three dimensional structure to suppress polymer crystallization[3]. However, the sample with 6% clay content has small effect due to the aggreation of clay

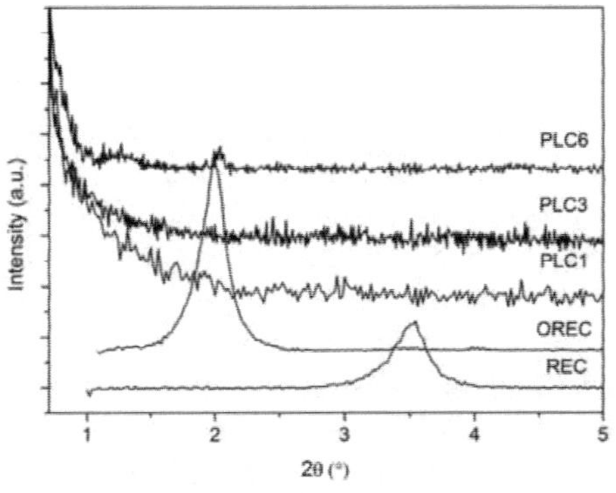


**Figure 1.** SXRD patterns of pristine rectorite (REC), organophilic rectorite (OREC) and PEO/ $Li^+$/ clay composites with different clay contents.

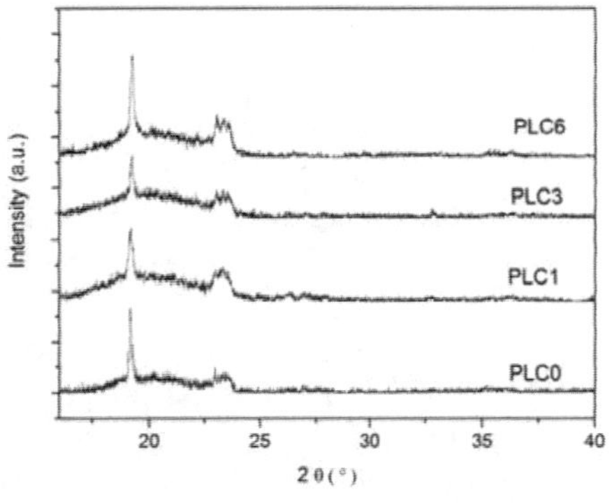


**Figure 2.** WXRD patterns of PEO/ $Li^+$/ clay composites with different clay contents.

*3.2.Free volume distributions*

In order to deeply discern the influence of clay on the free volume distribution, positronium lifetime distributions with temperature and OREC content were measured as shown in figure 3. From figure 3, it is very clear that the values of peaks shift to higher values with increasing temperature due to the free volume thermal expansion. On the other hand, comparing the positronium lifetime distribution between PLC0 and PLC3, an interesting phenomenon was observed that peaks of sample with 3wt% OREC content were spitted into two sub peaks, which suggest that the two different kinds of the free volumes exist. According to the reports, we think that shorter o-Ps lifetime is attributed to the o-Ps annihilation in the free volume at interfacial regions between the polymer matrix and nanoclays and in the interstital free volumes in crystalline region due to the change in the chain packing density induced by nanoclays[3-4]. The longest o-Ps lifetime is attributed to the o-Ps annihilation in the free volume in the amorphous region.

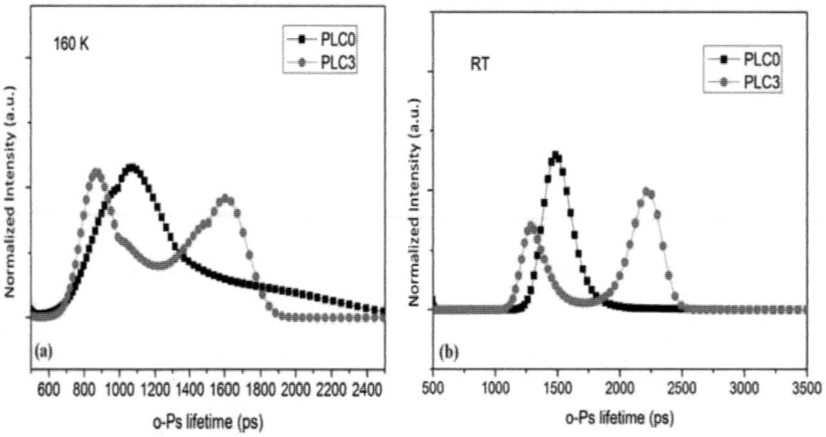


**Figure 3.** Positronium lifetime distribution with temperature((a):160K, (b)room temperature) and clay content.

*3.3.Direct relationship between ionic conductivity and the fractional free volume*

Positronium lifetimes with temperature and clay content were measured as shown in figure 4. Based on the infinite potential spherical well model, we can calculate the free volume V and the relative free volume fraction $f_r(T) = VI_3$ , $I_3$ is the o-Ps intensity. Here, we can neglect the effect of the positronium annihilation in nanoclays because the nanoclay content is very small. From fig.4, we can see that o-Ps lifetimes of 3wt% clay sample are longer than those of pure PEO sample, whereas the lifetimes measured for the 6 wt. % sample are lower than those for the pure PEO. Maybe this results from the dispersion state of clay, which is consist with the WXRD results. In order to determine the glass transition temperature, a least-squares linear fitting program was used to fit the two regions. The temperature corresponding to the intersection of two straight lines is identified as the glass transition temperature Tg. Below Tg, polymer is in the glassy state, the chain motion is frozen, leading to small change in o-Ps lifetime. Above Tg, the matrix is in the rubber state, the motions of the segmental chains and all molecular motions are activated, resulting in marked increase in o-Ps lifetime.

In order to clarify the ionic conductivity mechanism, the WLF equation[5] based on the free volume theory was used to obtain a direct relationship between the ionic conductivity and the fractional free volume fr(T)as follows:

$$\log\frac{\sigma(T)}{\sigma(T_g)} = \frac{C_1(T-T_g)}{C_2+(T-T_g)} \quad (1)$$

$$C_2 = f_r(T_g)/\alpha = T_g - T_0 \quad (2)$$

$$f_r(T) = f_r(T_g) + \alpha(T - T_g) \quad (3)$$

Combining equations(1-3), we obtain

$$\log_{10}\sigma(T) = \log_{10}\sigma(T_g) + C_1\left[1 - \frac{f_r(T_g)}{f_r(T)}\right] \quad (R^2=0.968) \quad (4)$$

Where $C_1$ and $C_2$ are a constant, $\alpha$ is the thermal expansion coefficient of the free volume, $f_r(T_g)$ is the fractional free volume at Tg, $T_0$ is a reference temperature at which the free volume vanishes and the ionic mobility goes to zero. Figure 5 shows a good linear relationship between the ionic conductivity and the fractional free volume, which suggests that polymer chain mobility, ionic migration and diffusion, and conductivity mechanism could be explained by the free volume theory .

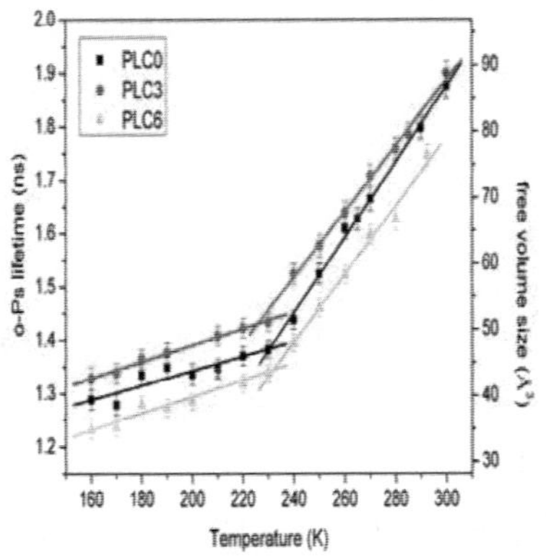


**Figure 4.** Temperature dependence of o-Ps lifetime for PEO-LiClO$_4$-clay composites.

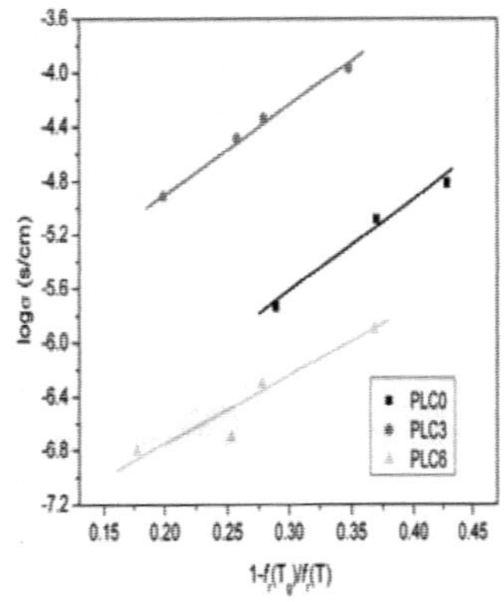


**Figure 5.** logσ versus $1\text{-}f_r(T_g)/f_r(T)$ for PEO-LiClO$_4$-clay composites.

## 4.Conclusion

The free volume and structural transition, and the ionic conductivity have been investigated for PEO/LiClO$_4$/clay nanocomposite electrolytes. Experimental results found that the addition of a small quantity of OREC can effectively effect on the microstructure of PEO. A direct relationship between the ionic conductivity and the fractional free volume has been obtained using WLF equation based on the free volume theory. The conducting mechanics can be explained in terms of free-volume theory.

**Acknowledgments**

This work was financially supported by the National Nature Science Foundation of China(Contract No.11175134).

## 3. References

[1] Gong J, Ph.D.Thesis, May,2012, Investigation of microstructure and physical properties for polymer/ rectorite nanocomposites

[2] Ladewig B P, Knott R B, Hill A J, James D R, White J W, Martin D J, Costa J D, Lu, G Q, 2007, *Chem Mater*, 19, 2372

[3] Bujdak J, Hackett E and Giannelis P, 2000, *Chem.Mater*, 12, 2168

[4] Gong W, Mai, Y Y, Zhou,Y F, Qi N, Wang B and Yan D Y, 2005, *Macromolecules*, 38, 9644

[5] Williams M L, Landel R F and Ferry J D, 1955, J. Am. Chem. Soc. 77, 3701

Journal of Physics: Conference Series **443** (2013) 012050 doi:10.1088/1742-6596/443/1/012050

# Vacancy profile in reverse osmosis membranes studied by positron annihilation lifetime measurements and molecular dynamics simulations

**A. Shimazu[1], H. Goto[2], T. Shintani[1], M. Hirose[1], R. Suzuki[3] and Y. Kobayashi[3]**

[1]Nitto Denko Corporation, 1-1-2, Shimohozumi, Ibaraki, Osaka 567-8680, Japan
E-mail: akira_shimazu@gg.nitto.co.jp
[2]Toyohashi University of Technology, Toyohashi, Aichi 441-8580, Japan
[3]National Institute of Advanced Industrial Science and Technology, Tukuba-shi, Ibaraki 305-8568 and 8565, Japan

**Abstract.** The positron annihilation technique using a slow positron beam can be used for the study of the vacancy profiles in typical reverse osmosis (RO) membranes. In this study, the vacancy profile in the polyamide membrane that exhibits a high permselectivity between ions and water was studied using the positron annihilation technique and molecular dynamics simulations. Ortho-positronium (o-Ps) lifetimes in the surface region of the membranes were evaluated by using a slow positron beam. The diffusion behavior of $Na^+$ and water in the polyamides was simulated by molecular dynamics (MD) methods using the TSUBAME2 supercomputer at the Tokyo Institute of Technology and discussed with the vacancy profile probed by the o-Ps. The results suggested that the large hydration size of $Na^+$ compared to the vacancy size in the polyamides contributes to the increased diffusivity selectivity of water/$Na^+$ that is related to the NaCl desalination performance of the membrane. Both the hydration size of the ions and the vacancy size appeared to be significant parameters to discuss the diffusivity selectivity of water/ions in typical polyamide membranes.

## 1. Introduction

The positron annihilation lifetime (PAL) is a unique probe of molecular size vacancies in certain polymers. The ortho-positronium (o-Ps) lifetime $\tau_3$ for various polymers appears to represent the size of the interstitial space among the molecular chains which are called vacancies in this article. PAL has successfully been applied to glassy polymers for membrane-based separation technology such as a gas separation process [1] and RO process [2]. In addition, MD simulations can be used for the characterization of the vacancy profile in some glassy polymers [3]. However, there have been few studies on the solute diffusion behavior on a molecular level through a typical polyamide as RO membrane. The aim of this study is to discuss the diffusivity selectivity of water/$Na^+$ in a typical polyamide for an RO membrane using o-Ps and MD simulations. The effect of the hydration size of $Na^+$ on the diffusivity selectivity is discussed.

## 2. Experimental

Polyamide composite membranes were synthesized from trimesoyl-chloride (TMC) and 1,3-phenylenediamine (MPD) on a porous polysulfone substrate by an interfacial polycondensation method. Variable-energy PAL measurements were carried out for the polyamide composite membranes using the system at the National Institute of Advanced Industrial Science and Technology (AIST). Details of the system and application have been previously described [4, 5]. The PAL spectra were obtained at a time resolution of 280~300 ps at the full width at half-maximum. The positron incident energy was varied from 0.7keV to 10keV at room temperature. A three-component analysis was used to determine $\tau_3$ of the polyamides. The MD simulation of the polyamides/NaCl aqueous system was carried out using the MD program of the Materials Studio (Accerlys Inc.) using the COMPASS force field [6]. The atomic charges of the models were determined using Density Functional Theory at the B3LYP/6-31G(d) level [7]. A constant number-pressure-temperature method was used for the calculation. A typical molecular model is described in Figure 1.

## 3. Results and Discussion

### *3.1. Vacancies in polyamide thin layer probed by o-Ps*

The typical surface structure observed by SEM of a polyamide composite membrane is shown in Figure 2. About a 400-nm-thick polyamide layer is observed on the porous polysulfone substrate. A positron beam was implanted into the surface of the composite membrane. Figure 3 shows the typical PAL spectra of the composite membrane. The mean implantation depth is determined as a function of the positron energy. For our samples, the PAL of the polyamide layer is measured at the positron energy of 0.7~1keV, corresponding to the mean depth of 17~31 nm, and the PAL of the polysulfone substrate is measured at the positron energy of 10keV, corresponding to the mean depth of approximately 1.2μm. The NaCl rejection of the membrane at room temperature at a pressure of 1.5 MPa with a NaCl 0.15 wt% aqueous solution feed was determined by the following equation.

$$\text{Rejection} = [1-(C_p/C_F)] \times 100\ (\%) \qquad (1)$$

where $C_p$ and $C_F$ are the NaCl concentrations in the permeate and feed side, respectively. Figure 4 shows typical result of the relationships between $\tau_3$ at the positron energy of 0.7keV and the NaCl rejection for the polyamide composite membranes. This result indicates that $\tau_3$ reflects the structural property that is related to the reverse osmosis performance of typical polyamide composite membranes.

According to a simple model, the vacancy is approximated by an infinite height potential well and the electrons available for the pick-off annihilation are assumed to exist as an electron layer on the surface of the well with the thickness ΔR. The o-Ps lifetime $\tau_3$ is then related to the radius R of the vacancies [8].

$$\tau_3 = 0.5\,\{1-R/(R+\Delta R) + \sin[2\pi R/(R+\Delta R)]/2\pi\}^{-1}\ \text{ns} \qquad (2)$$

where ΔR =0.166 nm is known to have an excellent agreement with the experimental data. Using equation (2), radius of vacancy is estimated to be 0.24~0.27nm. Each membrane in Figure 4 exhibits a high NaCl Rejection, therefore, this vacancy size is suggested to contribute to increasing the water/NaCl selectivity.

### *3.2. The effect of the vacancy size on the diffusivity selectivity of water/$Na^+$ in the polyamide.*

In order to consider the effect of the vacancy size on the diffusivity selectivity of water/$Na^+$ in the polyamide, the diffusion behavior of water and $Na^+$ was studied using MD simulations. According to the Einstein equation, the particle diffusion coefficient D is determined by equation (3).

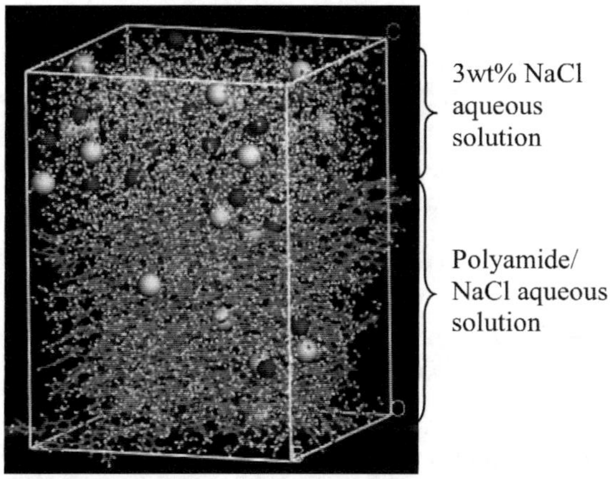


**Figure 1.** Schematics of typical molecular model used for MD simulations. Yellow sphere: $Na^+$, Blue sphere: $Cl^-$, Small molecule consisting of red and white lines: Water.

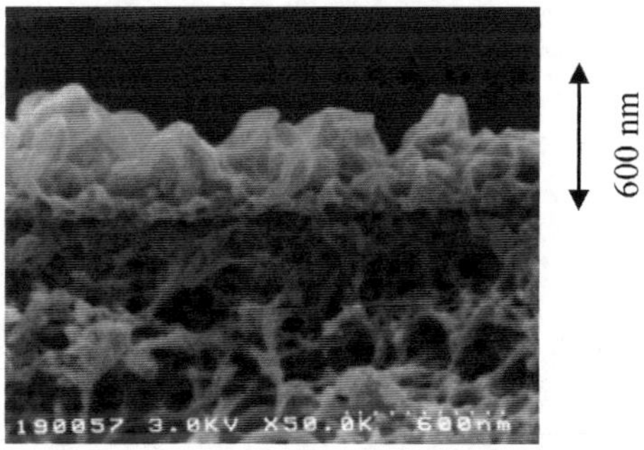


**Figure. 2.** Surface structure of polyamide composite membrane.

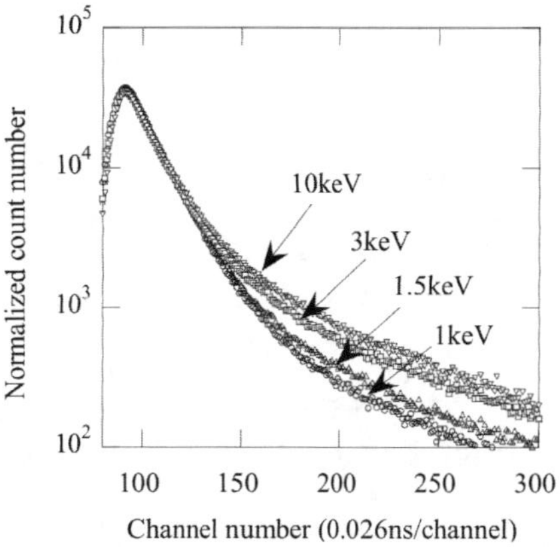


**Figure. 3.** Typical PAL spectra of polyamide composite membrane.

$$D = \frac{d}{dt}\frac{1}{6}\left\langle | r(t) - r(0) |^2 \right\rangle \quad (3)$$

where $<|r(t) - r(0)|^2>$ is the mean square displacement (MSD) of the particle over time t from the start. The MSD of water and $Na^+$ in the surface NaCl aqueous solution on the polyamide layer and in the polyamide layer are demonstrated in Figure 5 and Figure 6, respectively. While the water diffusivity is greater than the $Na^+$ diffusivity in both the surface NaCl aqueous solution and the polyamide, $Na^+$ diffusivity was remarkably low compared to the water diffusivity in the polyamide. The diffusion coefficients of water and $Na^+$ determined by the MD simulations are listed in Table1 with the data measured by the $^1H$ pulsed-field gradient NMR methods [9] in which the NMR measurements were carried out for the water and $Na^+$ in a 3wt% NaCl aqueous solution at 298K. $D_{water}$ and $D_{Na+}$ in the surface NaCl aqueous solution of the molecular model were nearly equal to those in the NaCl aqueous solution measured by $^1H$ pulsed-field gradient NMR methods. This suggests that the simulations reflect the real diffusion behavior of water and $Na^+$. It is noted that the diffusivity selectivity of water/$Na^+$ ($D_{water}/D_{Na+}$) is remarkably high compared to that of the NaCl aqueous solution. According to solution-diffusion model [10], the high $D_{water}$ / $D_{Na+}$ contributes to the increased NaCl rejection of the RO membranes. Figure 7 shows the radius distribution function for $Na^+$ and water $g(r)_{Na+\text{-}water}$ in the polyamide model with the vacancy radius R estimated by the Tao-Eldrup model using the o-Ps lifetimes. Some interesting remarks can be made for the $g(r)_{Na+\text{-}water}$ versus R. $g(r)_{Na+\text{-}water}$ represents the hydration shell size of $Na^+$ in the polyamide. The first peak around 0.22nm and second peak around 0.28nm correspond to the first hydration shell of $Na^+$. The peaks around 0.22nm and 0.28nm are attributed to the distances between $Na^+$ and O of water, and $Na^+$ and H of water, respectively. The peak around 0.48nm corresponds to the second hydration shell of $Na^+$. It is noted that the vacancy size probed by o-Ps is smaller than the first hydration shell size of $Na^+$, which is about 0.35nm. This suggests that the large hydration size of the $Na^+$ compared to the vacancy size contributes to the increased $D_{water}$ / $D_{Na+}$ . Figure 8 illustrates the vacancy site that can be occupied as probed by a hypothetical sphere with a radius of 0.35nm. The yellow sites in the figure represent the surfaces of the region occupied by a hypothetical spherical probe of radius 0.35nm using Voorintholt's method [11]. The calculated yellow sites are apparently few and small. This tendency was consistent with the increased $D_{water}$ / $D_{Na+}$ and corresponding high NaCl rejection due to the sieving effect of the hydrated $Na^+$ in the polyamide.

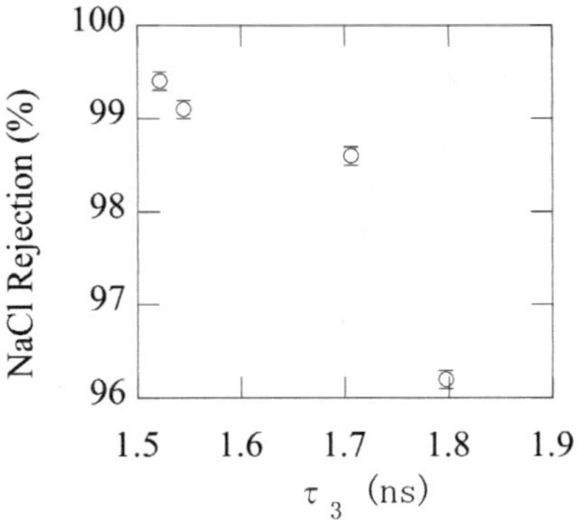


**Figure 4.** Relationships between $\tau_3$ and NaCl rejection for polyamide composite membranes.

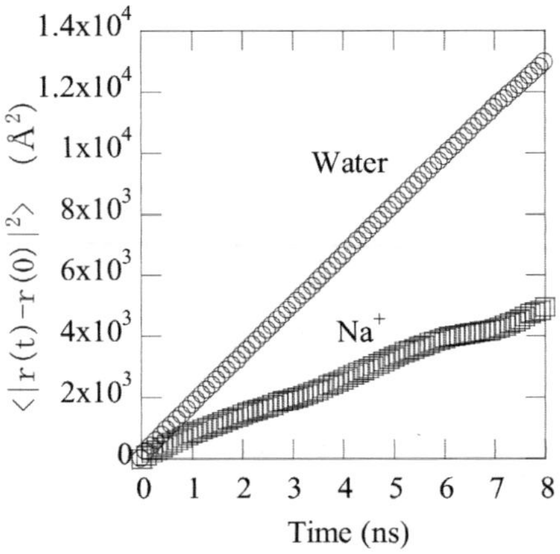


**Figure 5.** Mean square displacement of water and $Na^+$ in suface NaCl aqueous solution described in Figure 1.

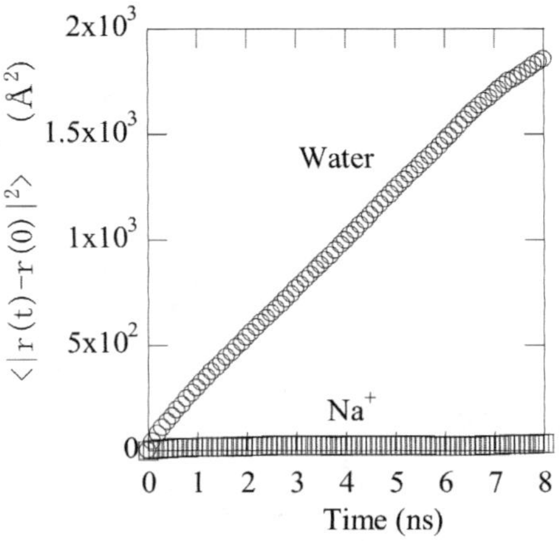


**Figure 6.** Mean square displacement of water and $Na^+$ in polyamide described in Figure 1.

## 4. Conclusions

The effect of the vacancy size on the diffusivity selectivity of water/$Na^+$ in a typical polyamide as RO membrane was discussed using the o-Ps lifetime and MD simulations. A good correlation between the NaCl rejection and o-Ps lifetime was observed. A small vacancy compared to the hydration size of $Na^+$ was considered to contribute to increase the $D_{water}$ / $D_{Na+}$ in the typical polyamide as the RO membrane.

Some of this work has been supported by New Energy and Industrial Technology Development Organization (NEDO), Japan.

**Table 1.** Diffusion coefficient of water and $Na^+$ at 298K.

| Material | $D_{water}$ | $D_{Na+}$ | $D_{water}/D_{Na+}$ |
|---|---|---|---|
| [1]Surface NaCl aqueous solution | 26.7 | 10.2 | 2.61 |
| [2]NaCl aqueous solution$_{exp}$ | 22 | 13 | 1.7 |
| [1]Polyamide | 3.86 | 0.0181 | 213 |

[1]Calculated by molecular dynamics simulation using model described in Figure 1.
[2]Measured by pulsed field gradient NMR method.
D: Diffusion coefficient ($\times 10^{10}$ $m^2/s$)

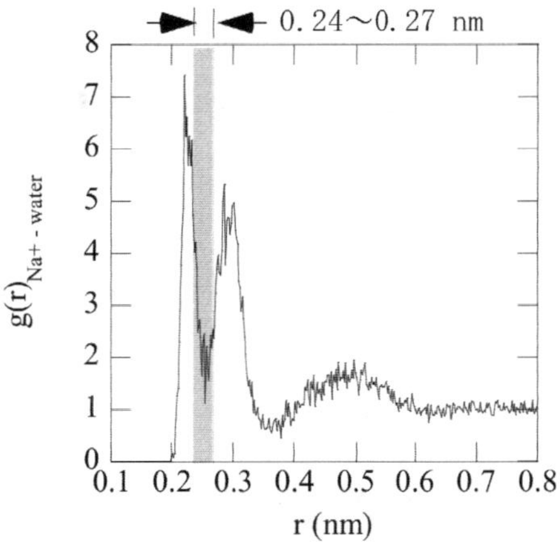


**Figure 7.** Radius distribution function for $Na^+$ and water in polyamide described in Figure 1. 0.24~0.27nm: radius of vacancy estimated using o-Ps lifetime

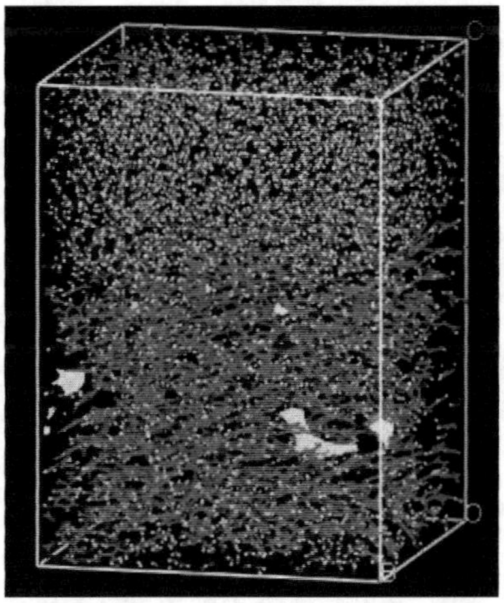

**Figure 8.** Vacancies that can be occupied, probed by a hypothetical sphere with a radius 0.35 nm. Yellow: vacancy site.

## 5. References

[1] Ito Y 1998 Chapter 23 of Structure and Properties of Glassy Polymers (ACS SYMPOSIUM SERIES 710) ed Tant M R and Hill A J (Washington DC: ACS)
[2] Shimazu A, Miyazaki T and Ikeda K 2000 *Radiat. Phys. Chem.* **58** 555
[3] Shimazu A, Miyazaki T, Katayama S and Ito Y 2003 *J. Polym. Sci. Part B Polym. Phys.* **41**, 308
[4] Suzuki R., Ohdaira T and Mikado T 2000 *Radiat. Phys. Chem.* **58** 603
[5] Kobayashi Y, Zheng W, Chang T. B, Hirata K, Suzuki R, Ohdaira T and Ito K 2002 *J. Appl. Phys.* **91**(3)1704
[6] Sun H 1998 *J. Phys. Chem.* B **102** 7338
[7] Frisch M J, Trucks G W, Schlegel H B et al. 2005 Gaussian03 Revision D.02 Gaussian Inc. (Wallingford CT)
[8] Eldrup M, Lightbody D and Sherwood J. N 1981 *Chem. Phys.* **63** 51
[9] Zawodzinski T A, Neeman M, Sillerud L O, Gottesfeld S 1991 *J. Phys. Chem.* **95** 6040
[10] Stern S A and Frisch H L 1981 *Annu. Rev. Mater. Sci.* **11** 523
[11] Voorintholt R, Kosters M T and Vegter G 1989 *J. Mol. Graphics* **7** 243

Journal of Physics: Conference Series **443** (2013) 012051 doi:10.1088/1742-6596/443/1/012051

# Infiltration and Selective Interactions at the Interface in Polymer-Oxide Hybrid Solar Cells

**R Ferragut[1,2†], S Aghion[1,2], F Moia[1,2], M Binda[3], E V Canesi[3], G Lanzani[3] and A Petrozza[3]**

[1] LNESS and Dipartimento di Fisica, Politecnico di Milano, via Anzani 42, 22100 Como, Italy

[2] Istituto Nazionale di Fisica Nucleare, via Celoria 16, 20133 Milano, Italy

[3] Center for Nano Science and Technology @Polimi, Istituto Italiano di Tecnologia, Via Pascoli 70/3, 20133 Milano, Italy

E-mail: rafael.ferragut@polimi.it

**Abstract.** Positron annihilation spectroscopy was used to characterize polymer-based hybrid solar cells formed by poly(3-*hexylthiophene*) (P3HT) finely infiltrated in a porous $TiO_2$ skeleton. A step-change improvement in the device performance is enabled by engineering the hybrid interface by the insertion of a proper molecular interlayer namely 4-*mercaptopyridine* (4-MP). In order to obtain depth-resolved data, positrons were implanted in the sample using a variable-energy positron beam. The characteristics of the partially filled nanoporous structures were evaluated in terms of the depth profile of the positronium yield and the *S*-parameter. A quantitative evaluation of the pore filling in the deep region is given from the analysis of Coincidence Doppler Broadening taken at fixed implantation energy. We note a remarkable difference in terms of the positronium yield when the 4-MP interlayer is introduced, which means a better covering of P3HT on the porous surface.

## 1. Introduction

Hybrid solar cells are characterized by thin films composed by a combination of organic and inorganic materials. This new generation of thin-film is based on a very favourable price/performance ratio and the possibility to manufacture flexible and large-area modules.

We present a polymer-based hybrid photovoltaic cell composed by poly(3-*hexylthiophene*) (P3HT) finely infiltrated in the metal oxide mesoporous structure $TiO_2$. Previous attempts using nanoporous [1] or nanorods [2] structures filled by similar conjugated polymers could not reach power conversion efficiency above 0.5%. Attention has also been paid to the use of possible interlayers whose main role was to control the energy offset at the oxide/polymer interface [3]. By modifying the $TiO_2$ surface through a commercial and inexpensive molecule such as 4-*mercaptopyridine* (4-MP) (see inset in Figure 1), which is optically transparent, we are able to impressively enhance the photocurrent extracted from the nanostructured solar cell, inducing a step change in the overall efficiency. In Figure 1 is shown the current-voltage characteristics for devices with and without 4-MP. When the molecular interlayer is added, the photocurrent (current density when the voltage is zero) is increased, from 1.4 $mA/cm^2$ to 3.7 $mA/cm^2$, together with an enhancement of the $V_{oc}$ (voltage when the current density is

† To whom any correspondence should be addressed.

null) from 0.51 V to 0.60 V. This result proves an improvement in the device performance that is enabled by engineering the hybrid interface by the insertion of a proper molecular interlayer [4].

Positron annihilation spectroscopy (PAS) is an established method for investigating porous oxides [5], polymers [6] and other systems [7, 8]. There are several advantages of PAS over other methods (Grazing incidence X-ray diffraction, TEM and SEM ); it is nondestructive and is not limited in-plane spatial resolution. In recent years, PAS with a variable-energy positron beam has emerged as a powerful tool for the investigation of porous thin films (tens nm up to several μm in thickness). Depth-profiling with PAS has proven to be an ideal way to measure the interconnection length of pores and explore porosity hidden beneath dense layers [5].

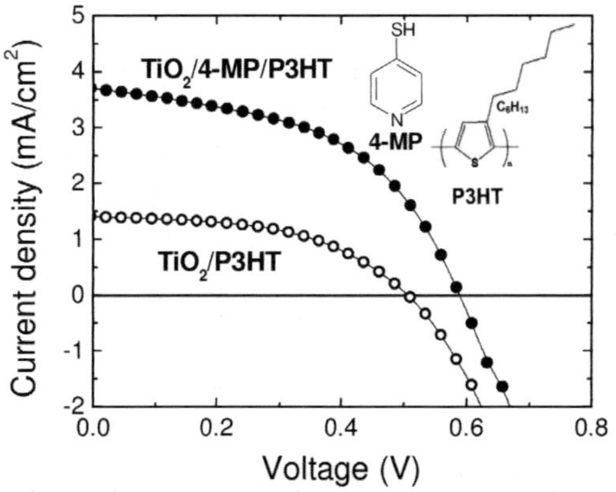


**Figure 1.** Current density/voltage characteristics of the P3HT/$TiO_2$ and P3HT/4-MP/$TiO_2$ devices under illumination with simulated solar light. The P3HT and 4-MP chemical structure are shown in the inset (after Canesi et al. [4]).

In this work, PAS enables a nanometer scale mapping over the hybrid solar cells. We have found out that, contrary to common knowledge, pore filling of the oxide mesoporous structure is not the more important factor for the device performance. The interlayer 4-MP favors a better covering of P3HT on the porous surface. These results are extremely important moving the focus of the investigation to the nature of the polymer/porous oxide hybrid interface.

## 2. Experimental procedure and methodology

The devices were prepared following the procedure reported in Ref. 4. In the case of devices provided with 4-MP interlayer, the porous $TiO_2$ substrates (800 nm in thickness) were immersed in a saturated solution of 4-MP in chlorobenzene for about 20 hours. Immersion was followed by rinsing with pure chlorobenzene. A monolayer of 4-MP is expected to remain at the surface. In order to promote the P3HT infiltration inside the mesoporous $TiO_2$ (with and without 4-MP) the polymer was spin coated onto the substrates at 1000 rpm for 60 s from a 30 mg/ml chlorobenzene solution.

In order to obtain depth-resolved data, positrons were implanted in the sample at various depths using a variable-energy positron beam (from 0.1 to 17 keV) at the *VEPAS* laboratory (LNESS Centre - Politecnico di Milano). The PAS measurements were performed at room temperature at a vacuum level around $10^{-7}$ mbar. Two hyperpure Ge gamma detectors (ORTEC – relative efficiency ~50 % at 1.33 MeV) were used to measure the spectrum of the annihilation radiation. Results of the *S*-parameter, i.e. the fraction of annihilating positron–electron pairs with momenta $p_L \leq 0.456$ atomic units (energy range 511±0.85 keV), are presented. The tails of the two-gamma peaks at high Doppler shift, which come from annihilation with inner atomic shells, are more sensitive than the *S*-parameter to the average chemical composition in the annihilation region and are suitable for quantitative compositional analyses. However, the low counting rate in the high momentum region requires background suppression by adopting the coincidence Doppler broadening technique (CDB), i.e. to take both annihilation gammas in coincidence. CDB spectra with about $2\times10^7$ total counts were measured with a peak/background ratio of $10^5$-$10^6$. The momentum resolution (FWHM) was $3.7\times10^{-3}$ $m_0c$. The experimental procedure and the data analysis used in this study are similar to those described in Refs. 9 and 10.

An additional PAS parameter, sensitive to the presence of free volume and to the extension of inner pore surface, is the positronium (*Ps*) fraction. *Ps* fraction measurements were performed using the well-known "3γ method" [11, 12].

## 3. Results and discussion

Positron annihilation spectroscopy was used to monitor the effect of 4-MP on the penetration of P3HT in the porous $TiO_2$. The $S$ and $Ps$ yield profiles are reported in Fig. 2. Figure 2.a displays an evolution from the high $S$-value characteristic of P3HT ($S \approx 0.57$) toward lower values expected for porous $TiO_2$ with or without 4-MP (full and open stars, respectively).

The zero of the depth scale is fixed at the surface of the porous oxides. Grey and black ticks are presented in the upper frame of Fig. 2.b (positron implantation energy), that correspond to the devices without and with a capping layer of P3HT, respectively. The positron mean implantation depth profile of the oxides infiltrated with P3HT was calculated using the estimation of the polymer concentration infiltrated in the oxide taking into account the density of each layer (see Ref. 4). The curves through the experimental points are the result of a fitting procedure (VEPFIT [13]) which identifies four layers in the explored structure: a solid P3HT cap, a thin porous layer heavily infiltrated with P3HT, a second thick porous layer containing less P3HT and the glass substrate. Table 1 shows the most relevant parameters of the fit. From the results of the first and the second infiltrated layers we can conclude that there is a P3HT concentration gradient within the oxide in both samples, with and without the interlayer; however, in presence of 4-MP the degree of P3HT infiltration is reduced in the first layer and enhanced in the deeper region. A quantitative evaluation of the deep region is given from the analysis of CDB taken at fixed energy (5 keV, see Ref. [4]). CDB is sensitive to the average chemical composition in the annihilation region [9, 10, 14].

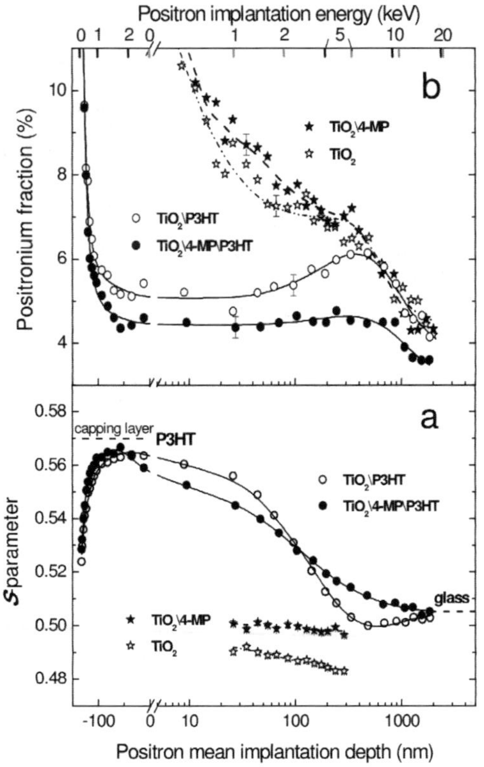


**Figure 2.** S-parameter (a) and Ps yield (b) profiles for P3HT embedded in porous $TiO_2$ without and with 4-MP together with the reference values for P3HT and for $TiO_2$ with and without 4-MP. The continue lines through the data are obtained using VEPFIT. The lower scale before the break at zero is linear.

**Table 1.** Results obtained in the heterostructure after the polymer implantation from the positron lineshape profiles (Fig. 2.a) using VEPFIT. Thickness $Z_i$, positron diffusion length $L_i$ and $S_i$-parameter. Fixed parameters are marked with the letter F.

| $TiO_2$/ | $Z_{CL}$ (nm) | $S_{CL}$ | $L_{CL}$ (nm) | $Z_{1st}$ (nm) | $S_{1st}$ | $L_{1st}$ (nm) | $Z_{2nd}$ (nm) | $S_{2nd}$ | $L_{2nd}$ (nm) | $S_{glass}$ | $L_{glass}$ (nm) |
|---|---|---|---|---|---|---|---|---|---|---|---|
| P3HT | 130(20) | 0.568(3) | 20(5) | 70(10) | 0.564(4) | 15(3) | 730F | 0.486(3) | 60(10) | 0.505(3) | 40F |
| 4-MP/P3HT | 135(20) | 0.569(3) | 17(5) | 43(10) | 0.542(5) | 10(3) | 757F | 0.510(4) | 60(10) | 0.505(3) | 40F |

In Fig. 3 the CDB spectra are presented in terms of relative change respect to the reference spectrum of the porous $TiO_2$. The curves for $TiO_2$/P3HT and $TiO_2$/4-MP/P3HT are fitted using synthetic spectra formed as linear combinations

$$\rho = w\,\rho_{P3HT} + (1-w)\,\rho^{*}_{TiO_2}\,, \qquad (1)$$

of the spectra taken separately for P3HT ($\rho_{P3HT}$) and for $TiO_2$ or for $TiO_2$/4-MP ($\rho^{*}_{TiO_2}$), depending if the oxide contains or does not contain 4-MP. The physical meaning of the weight ($w$) in the linear combination is the fraction of positron annihilating in a P3HT environment and therefore is a good index of infiltration. The weights $w$ obtained from the linear combination procedure are $0.37 \pm 0.03$ and $0.44 \pm 0.03$ respectively for $TiO_2$/P3HT and for $TiO_2$/4-MP/P3HT. It can be concluded that the presence of 4-MP increases P3HT infiltration at the selected depth, although the effect is barely beyond the error margin. The $w$ weights are univocally related to the pore filling factor [4]. The limited difference observed in terms of $w$ indicates that the presence of the 4-MP does not affect sizably the polymer filling factor suggesting that the pores filling by itself is not the limiting factor controlling the device performances shown in Fig. 1.

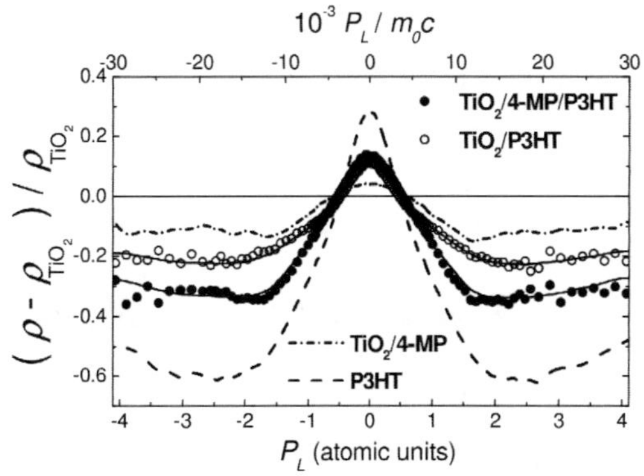


**Figure 3.** CDB spectra of $TiO_2$/P3HT and $TiO_2$/4-MP/P3HT taken at a fixed implantation energy (5 keV). The dashed lines represent P3HT and $TiO_2$/4-MP. The continues lines are calculated using a linear combination of the reference spectra (eq. 1) [4].

On the other hand, remarkable differences in terms of the positronium fraction can be observed when the 4-MP interlayer is introduced. In fact, the *Ps* depth profile of Fig. 2.b drops drastically when $TiO_2$ is infiltrated with P3HT, specially in presence of 4-MP (a monolayer of 4-MP in bare $TiO_2$ does not reduce the *Ps* yield). In particular, the bump in the $TiO_2$/P3HT sample (open circles) which appears in the deeper region (200nm-800nm) is decreased in presence of 4-MP. *Ps* yield reduction can be the consequence of: *i)* a better pores filling that favours 2-γ annihilation and/or *ii)* a larger P3HT covering of the free inner titania surface where *Ps* is formed. In both cases the polymer induces *Ps* inhibition and ortho-*Ps* quenching [15]. The limited difference in the filling factors measured in the deeper region by CDB in samples with and without 4-MP cannot explain the large reduction in *Ps* yield. Moreover, if we look at the shallowest region, the *S*-parameter shows a lower concentration of the P3HT phase in presence of the 4-MP monolayer, while the *Ps* yield is again reduced. Therefore, we conclude that the *Ps* fraction reduction is due to a reduction of free $TiO_2$ surface, which means a better covering of P3HT on the porous surface. The presented information, inaccessible by optical and electron microscopy techniques previously used, coincides with an improvement of charge generation and device photocurrent. The interpretation is in agreement with X-ray photoemission spectroscopy and with atomistic simulations results [4]. This study indicates that tailoring the interaction between the polymer and metal oxide is the critical area to focus attention to improve these hybrid solar cells.

## References

[1] Coakley K M and McGehee M D 2003 *Appl. Phys. Lett.* **83** 3380
[2] Mor G K et al. 2009 *Nano Lett.* **9** 4250
[3] Goh C, Scully S R and McGehee M D 2007 *J. Appl. Phys.* **101** 114503
[4] Canesi E V et al. 2012 *Energy Environ. Sci.* **5**, 9068
[5] Gidley D W, Peng H G and Vallery R S 2006 *Annu. Rev. Mater. Res.* **36** 49
[6] Mallon P E 2003 *Principles and Applications of Positron and Positronium Chemistry*, ed Y C Jean, P E Mallon and D M Schrader (Singapore: World Scientific) p 253
[7] Somoza A, Dupasquier A and Ferragut R 2009 *phys. stat. sol.* (c) **6** 2295
[8] Krause-Rehberg R and Leipner H S 1999 *Positron Annihilation in Semiconductors. Solid-State Sciences*, Vol. 127 (Berlin: Springer)
[9] Dupasquier A, Ferragut R, Iglesias M M, Massazza M, Riontino G, Mengucci P, Barucca G, Macchi C E and Somoza A 2007 *Phil. Mag.* **87** 3297
[10] Ferragut R 2012 *Physica* B **407** 2676
[11] Mills A P Jr 1978 *Phys. Rev. Lett.* **41** 1828
[12] Ferragut R et al. 2010 *J. Phys.: Conf. Ser.* **225** 012007
[13] Van Veen A, Schut H, de Vries J, Hakvoort R A and IJpma M R 1991 *AIP Conference Proceedings* **218** 171
[14] Asoka-Kumar P et al. 1996 *Phys. Rev. Lett.* **77** 2097
[15] Goldanskii V I 1967 in *Positron Annihilation*, ed L O Roelling and A T Stewart (New York: Academic Press) p 183

# Degradation process of fuel cell membrane observed by positron

**Y Honda[1*], Y Aoyagi[1], S Tojo[1], G Watanabe[2], Y Akiyama[2], S Nishijima[2]**

[1]Institute of Scientific and Industrial Research, Osaka University, 8-1 Mihogaoka, Ibaraki, Osaka 567-0047, Japan
[2]Graduate school of Engineering, Osaka University, 2-1 Yamadaoka, Suita, Osaka 565-0871, Japan

Email: honda@sanken.osaka-u.ac.jp

**Abstract**. To investigate degradation process due to radicals in fuel cell membrane by means of positron annihilation spectroscopy, three kinds of radicals, $HO^•$, $H^•$ and $O_2^{•-}$ are produced through water radiolysis. The results show that the cluster structure and proton conductivity was greatly affected by reductive radicals. This is because the oxidative radical is responsible for the dissociation of sulfonic group, whereas the reductive radical breaks down the cluster in the membrane and disrupts proton conduction, which is consistent with solution analysis.

## 1. Introduction

Elucidation of degradation process of polymer electrolyte membrane (PEM) for fuel cell is very important to develop a better membrane. In practical fuel cell operation, generation of radicals such as hydroxyl radical ($HO^•$), hydrogen radical ($H^•$) and peroxy radical ($HOO^•$) is reported and these are regarded as the sources of degradation. [1] We have investigated the influence of radical species such as $HO^•$, $H^•$, superoxide anion ($O_2^{•-}$) on Nafion®-117. [2] Nafion produced by DuPont is known as a most representative perfluorinated sulfonic acid membrane and is well investigated. [3] This membrane has polytetrafluoroethylene (PTFE) backbone and side chain terminated with sulfonic group through two ether bonds. Aggregation of hydrophilic sites, sulfonic groups, forms clusters and such clusters are connected each other with channels where hydrated proton can be transferred, which is known as a cluster-network model and is most widely referenced, though other models have been proposed. [3] Ortho-positronium (*o*-Ps) as well as positron has been thought to be a sensitive probe to find structural change in soft materials. To evaluate a free volume with positron, it should be taken into account that positron may annihilate in a special place due to local electric field relevant to functional group, followed by localization of *o*-Ps, and the deduced free volume size might not reflect overall property of the material. However, such character would be attractive in turn, if such a local place is closely relating to the function of material. Indeed, based on the cluster-network model, which would be valid for a water volume fraction less than 0.5, and taking into account that the time-dependent S-parameters for PTFE and Nafion®-117, the lifetime of *o*-Ps in swelling process and the evaluated space size based on the lifetime of *o*-Ps, it could be found that *o*-Ps stays in the space made up of side chains and hydrophobic backbone, and the change of its lifetime reflects the change of cluster structure. [4] In the followings, the influences of three kinds of radicals on two kinds of PEMs are shown by using positron annihilation spectroscopy.

## 2. Experimetal setup

Two kinds of perfluorinated membranes, Nafion®-117 and Aquivion® H87-10 produced by Solvay Solexis, were adopted as PEM (Figure 1). Chemical form of Aquivion® is similar to that of Nafion® but has shorter side chain, and it also has clusters. [5] Both kinds of membranes were exposed to $HO^{\bullet}$, $H^{\bullet}$ and $O_2^{\bullet -}$ radicals which were produced through water radiolysis due to γ-irradiation. [2] Absorbed dose was ranged from 0.1 kGy to over 1 MGy. The number of generated radicals depends on absorbed dose and it can be evaluated based on G-value. In case of 1 MGy, the concentration of $HO^{\bullet}$ is evaluated to be about 0.5 mol/liter. Production rate of these radicals is $HO^{\bullet} : H^{\bullet} : O_2^{\bullet -} \sim 2 : 1.2 : 1$ for the same dose. Nafion® was stacked in atmosphere with 6 sheets and Aquivion® was with 10 sheets as the samples. All the membranes were pre-treated with sulfuric acid and hydrogen peroxide solution to remove impurities. [2] We regarded proton conductivity as an index of degradation and this was measured by using 4-probe AC impedance meter. Na-22 of 1 MBq was used as a positron source and sandwiched with the samples.

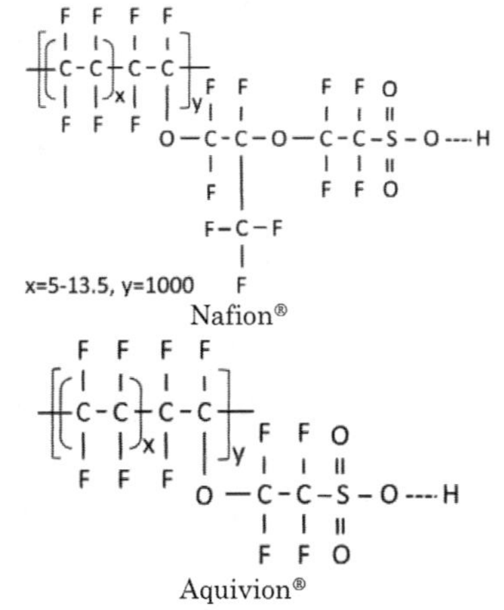


Figure 1. Chemical form of Nafion® and Aquivion®.

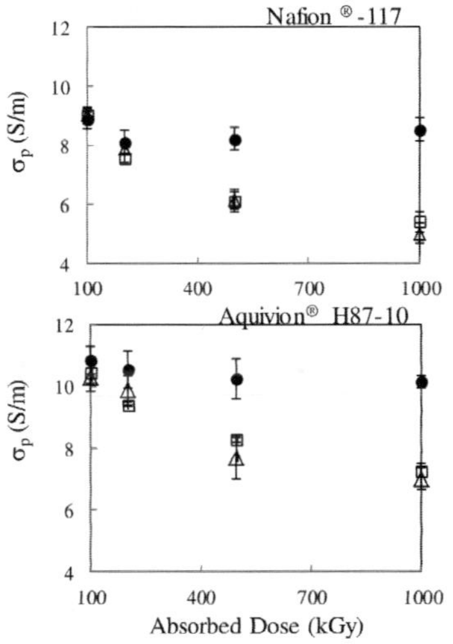


Figure 2. Proton conductivity vs. absorbed dose. ●: $HO^{\bullet}$, □: $H^{\bullet}$, Δ: $O_2^{\bullet -}$.

## 3. Results and Discussion

### 3.1. Proton conductivity

Dependencies of proton conductivities on absorbed dose for Nafion®-117 and Aquivion® H87-10 are shown in Figure 2. Up to 100 kGy, no significant change was observed for both kinds of membranes. Above 100 kGy, the proton conductivity, $\sigma_p$, was decreased for the samples exposed to $H^{\bullet}$ and $O_2^{\bullet -}$, whereas $\sigma_p$ of the sample exposed to $HO^{\bullet}$ did not decrease significantly. Taking into account of the number of $HO^{\bullet}$ being almost twice that of other radicals, it is found that the reductive radicals such as $H^{\bullet}$ and $O_2^{\bullet -}$ caused more serious damage on the membrane rather than oxidative radical, $HO^{\bullet}$.

### 3.2. Positron annihilation lifetime spectroscopy

Figure 3 shows the dose-dependencies of lifetime and relative intensity of *o*-Ps for Nafion®-117 exposed to different radicals. There were no significant changes in the lifetime of *o*-Ps below 100 kGy, but it decreased drastically for the samples exposed to $H^{\bullet}$ and $O_2^{\bullet -}$, whereas for the sample exposed to $HO^{\bullet}$ it was only scattered. As the lifetime of *o*-Ps reflects the size of cluster [4], these results show that the cluster structure was not influenced significantly by $HO^{\bullet}$, but damaged heavily by the reductive radicals above 100 kGy. The dose-dependencies of the relative intensity of *o*-Ps for the samples exposed to reductive radicals were very similar to those of lifetime. The relative intensity of *o*-Ps in the sample exposed to $HO^{\bullet}$ started decreasing from a lower dose region but the decreasing rate with respect to dose was less. These results show that the damage due to radical reactions clearly depends on radical species. $HO^{\bullet}$ does not reduce the cluster size but decreases the number of the space *o*-Ps can stay or deprives positron of chance to form *o*-Ps, whereas other radicals reduce the cluster size and decrease the number of the space *o*-Ps can stay. Along with the solution analysis [2], it can be concluded that $HO^{\bullet}$ brings about the dissociation of side chain and the cluster structure remains without serious damage. On the other hand, other radicals are thought to cause destruction of cluster above 100 kGy. As the behaviors of proton

conductivity and the lifetime of *o*-Ps with respect to dose are quite similar, the serious damage is thought to be brought by the disruption of proton conduction channel due to destruction of cluster.
Figure 4 shows the results for Aquivion®. The lifetime of *o*-Ps for Aquivion® is a little longer than that for Nafion® in spite of the shorter side chain, which is probably due to the longer interval of side chain than that of Nafion®. With respect to the diameter of the space where *o*-Ps stayed, it was reduced roughly from 0.7 nm, which is equivalent length between sulfonic group and backbone, to 0.65 nm for Nafion® exposed to reductive radicals. Corresponding dropping of the diameter for Aquivion® was from 0.72 nm to 0.64 nm. Both behaviors of lifetime and relative intensity of *o*-Ps showed similar dose-dependencies to those of Nafion®, but there were slight differences. Decrease of the relative intensity of *o*-Ps at a higher dose range in Aquivion® exposed to $HO^{\bullet}$ was less compared to that in Nafion®. Aquivion® was also more affected by the reductive radicals than the oxidative radical and the structural change caused by the reductive radicals seemed to progress from a lower dose range compared to Nafion®. However there was no remarkable difference in proton conductivity. In case of Aquivion®, it seems that successive structural change takes place so as not to lose proton conduction ability.

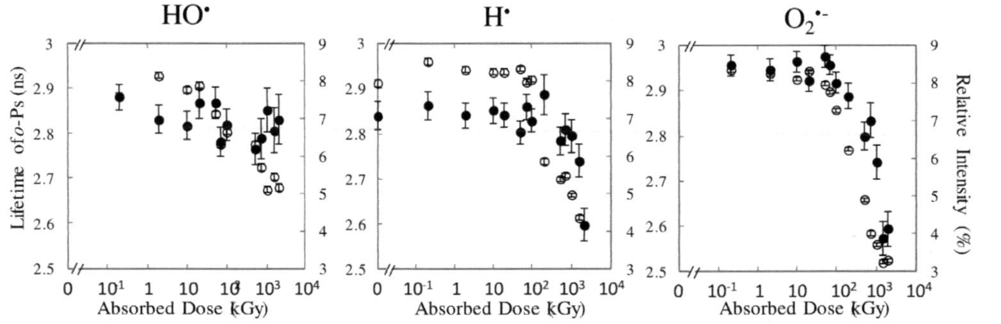


Figure 3. Lifetime and relative intensity of *o*-Ps for Nafion®-117 exposed to different radicals. Solid and open circles represent lifetime and relative intensity, respectively.

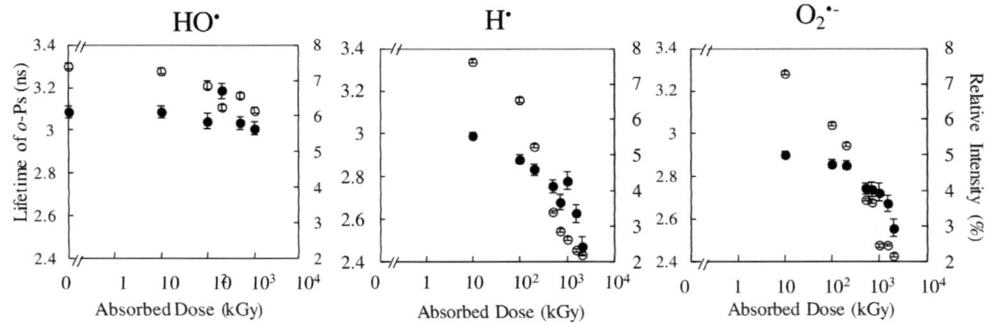


Figure 4. Lifetime and relative intensity of *o*-Ps for Aquivion® H87-10 exposed to different radicals. Solid and open circles represent lifetime and relative intensity, respectively.

### 3.3. Energy spectrum analysis

As the fraction of *o*-Ps is less than 10 %, the momentum of free annihilation electron is dominantly reflected on S-parameter. As the location of *o*-Ps is limited to the space formed with backbone and side chains, major free annihilation is probably taking place near there, i.e. around sulfonic group. Figure 5 shows the ratios of the Doppler broadened spectra of PTFE (Aldrich; powder 100 micron) and Aquivion® to Nafion® obtained by coincident Doppler broadening measurements. Each spectrum was normalized by the area of higher energy side of photopeak. The shapes of energy spectrum of Nafion® and Aquivion® were almost same, whereas annihilation with a larger momentum electron was more likely to take place in PTFE, indicating that the dominant free annihilation in perfluorinated sulfonic acid membrane is taking place near sulfonic group, probably oxygen and hydrogen bond, since some water molecules are bonded to sulfonic group through hydrogen bond even in dry state. Thus sulfonic group is an attractive site for positron. Figure 6 shows dose-dependencies of S-parameter for Nafion® exposed to different kinds of radicals. The S-parameter of the sample exposed to HO• was slightly decreased with increase of dose, whereas it remarkably increased for the sample exposed to H• in a higher dose region. Decreasing of S-parameter would be attributed to either the increase of annihilation with energetic electrons belonging to C-F bond or the decrease of annihilation with electrons relevant to hydrogen bond. Such change is thought to be brought by the lost of sulfonic group due to HO•. Increasing of S-parameter can be understood in such a way that the destruction of cluster leads to the dispersion of sulfonic groups along with associated water molecules, and eventually free annihilation with a lower energy electron relevant to hydrogen bond would increase. In the sample exposed to $O_2^{\bullet -}$, a similar behavior to the samples exposed to H• was expected but there was no significant dependency on absorbed dose. The radicals produced at the instance of scission would react with abundant oxygen or oxygen radical and eventually the scission site would be terminated with an oxygen, leading to the suppression of annihilation with the electrons belonging to hydrogen bond and C-F bond. In γ-irradiated Nafion® in dry state, there was a reverse correlation between proton conductivity and S-parameter, and cross-linking was thought to happen between neighboring sulfonic groups. [4] In dry state, HO• radicals would be generated from bound water to sulfonic group. However there was no such correlation between S-parameter and proton conductivity for the sample exposed to HO• in water. This indicates that water greatly influences on the degradation process caused by radicals.

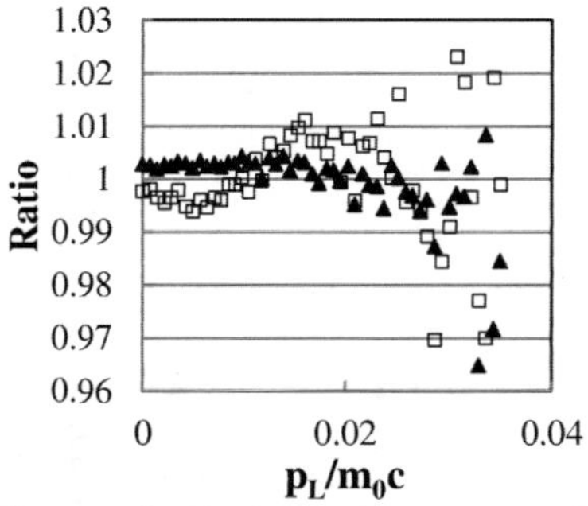


Figure 5. Ratios of spectra of PTFE (□) and Aquivion® (▲) to Nafion®.

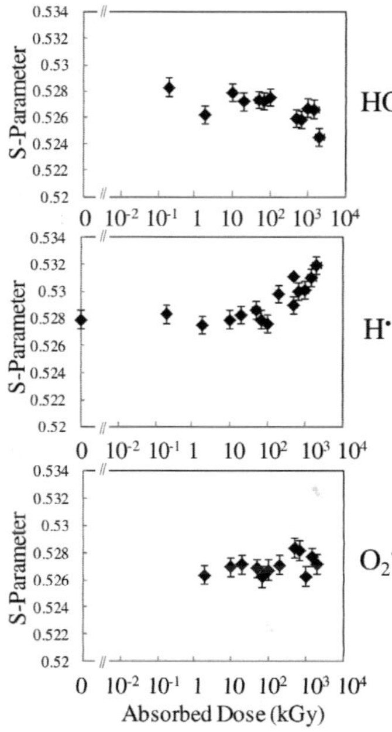


Figure 6. Dose-dependencies of S-parameter for different radicals.

## 4. Conclusion

Degradation process of PEM caused by different kinds of radicals was investigated by positron annihilation spectroscopy. The catastrophic damage was brought by reductive radicals, i.e. destruction of cluster structure was more severe rather than loss of sulfonic group in cluster for proton conductivity. Such degradation processes due to structural change were able to be drawn by positron annihilation spectroscopy.

**Acknowledgement**
This work was supported by NEDO.

## References

[1] Danilczuk M, Coms F D and Schlick S 2009 *J. Phys. Chem.* B **113** 8031
[2] Uegaki R, Akiyama Y, Tojo S, Honda Y and Nishijima S 2011 *J. Power Sources* **196** 9856
[3] Mauritz K A and Moore R B 2004 *Chem. Rev.* **104** 4535
[4] Honda Y, Aoyagi Y, Tojo S, Akiyama Y and Nishijima S 2013 *Material Sci. Forum* **733** 53
[5] Halim J, Buchi F N, Haas O, Stamm M and Scherer G G 1993 *Electrochimica Acta* **39** 1303

Journal of Physics: Conference Series **443** (2013) 012053 doi:10.1088/1742-6596/443/1/012053

# Free volumes studies in Thymoquinone and Carvone β-cyclodextrin nanoparticles by positron annihilation lifetime spectroscopy

**M F Ferreira Marques[1,2], P M Gordo[2] and A Moreira da Silva[3]**
[1]Department of Chemical and Biological Engineering, ISEC-IPC, R. Pedro Nunes, Quinta da Nora, 3030-199 Coimbra, Portugal
[2]CEMUC®, Department of Physics, Coimbra University, R. Larga, 3004-516 Coimbra, Portugal
[3]Department of Food Science and Technology, ESAC-IPC, Bencanta, 3040-316 Coimbra, Portugal

E-mail: fatima@coimbra.lip.pt

**Abstract.** Positron annihilation lifetime spectroscopy is used to study free volume in β-cyclodextrin with the encapsulation of thymoquinone and S-carvone, in samples covering the guest to host fraction range from 1:0.1 to 1:1. The results clearly indicate the presence of long lifetime components related to Ps-formation. Although the behavior of the two guests is different, in both cases the results indicate the formation of 1:1 cyclodextrin inclusion compounds. Data show that the addition of carvone to β-cyclodextrin results in a decrease of the *o*-Ps lifetime corresponding to a reduction of the average radius of cavities from 2.41 Å to 2.29 Å, whereas the addition of thymoquine decreases the radius from 2.57 Å to 2.35 Å. In turn, the intensity varied from 20.55 to 19.20% and from 20.83 to 0.41%, respectively.

## 1. Introduction

β-cyclodextrin (β-CD) is a cyclic oligosaccharide consisting of seven glucopyranose units linked by α-1,4 glycosidic bonds in normal chair conformations. The steric arrangement of glucose units in the CD molecule results in the shape of a hollow truncated cone with a hydrophilic outer surface (which makes CD water soluble) and a hydrophobic internal cavity. This enables CD to form inclusion complexes with various hydrophobic guest molecules [1]. Due to their extensive applicability in different fields (food and pharmaceutical industry, agriculture, etc.) CDs, capable of hosting a wide variety of molecules, are in the forefront of structural investigations [1,2,3].

The present work aims at studying by positron annihilation lifetime (PALS) the encapsulation in β-CD of two different molecules: thymoquinone (ThQ), the main constituent of *Nigella sativa* seeds, traditionally employed in folk medicine due to various pharmacological effects, and (4S)-(+)-Carvone (S-Crv), present in the essential oil of caraway seeds *Carumcarvi L.*, that acts as a sprouting inhibitor agent for potato tubers in storage. PALS became a powerful technique for the investigation of microstructures, with applicability in the determination of free volume size and volume fractions [4,5], as it allows one to obtain both the lifetimes ($\tau_i$) and the relative abundances (*i.e.*, the intensities, $I_i$) of the various annihilation channels. In order of increasing lifetimes, one normally finds the singlet (*p*-Ps, i=1), free positrons (i=2) and triplet Ps state (*o*-Ps, i=3). In samples where no positronium (Ps) formation takes place, positron trapping provides the annihilation channel which could sense structural imperfections. In turn, in media facilitating Ps-production the trapping of Ps reveals characteristic parameters of internal voids and cavities. The *o*-Ps channel constitutes the most useful probe, since the associated lifetimes are long and distinct. The primary annihilation mechanism of this long-lived *o*-Ps is the "pick-off" process, in which the positron in *o*-Ps annihilates with one of the surrounding spin opposite electrons, which obviously depends on the local physicochemical properties. The lifetime and intensity of this component can be translated into hole size (radius) and density of holes, respectively, as a probe to characterize various properties of matter [5].

## 2. Experimental

The β-CD, kindly offered by Wacker-Chemie, Munich (Germany), was recrystallized by cooling concentrated aqueous solutions in a Dewar flask from *ca.* 80ºC to room temperature. The S-carvone (2-methyl-5-(1-methylethenyl)-2-cyclohexen-1-one or (4S)-(+)-Carvone) was purchased from Aldrich Chemical Co, and the Thymoquinone (2-isopropyl-5-methylbenzo-1,4-quinone) from Sigma-Aldrich. The preparation of the inclusions of S-carvone and of thymoquinone in β-CD followed previous description [6]. Starting with those powders, samples were pressed under ~70 bar into disks of 10 mm diameter and about 0.5 mm thickness. The positron $^{22}Na$ source (*ca.* $7\times10^4$ Bq, closed between Kapton® foils), was sandwiched between two identical sample disks. As the thickness of the specimens was not sufficient to stop all the positrons, "backing" of SS316 stain steel was used and the set finally wrapped in Al-foil. All measurements were performed at 298 K with the sample assemblies placed into an evacuated stainless steel tube.

The lifetime (LT) spectra were recorded in a fast-fast coincidence PALS setup (featuring Pilot-U scintillators and XP 2020 photomultipliers) with a time resolution of 270 ps. Each sample was counted three times. The lifetime spectra had a total number of *ca.* $2.3\times10^6$ integral counts and were evaluated through the LT (version 9) code [7].

## 3. Results and discussion

All spectra measured are well fitted with three lifetime components, the longest one ($\tau_3$ and $I_3$), being associated with pick-off annihilation of *o*-Ps in the CDs free volume.

For an *o*-Ps confined in an infinitely deep spherical potential well there is a correlation between the free volume hole radius, R, and the long-lived lifetime, according the semi-empirical model of Tao-Eldrup [8,9] that establishes the correlation between free-volume hole radius, R (in Å), and *o*-Ps pick-off lifetime, $\tau_3$,

$$\tau_3 = 0.5\left[1-\frac{R}{R_0}+\frac{1}{2\pi}\sin\left(\frac{2\pi R}{R_0}\right)\right]^{-1} \quad \text{(in ns)} \tag{1}$$

where $R_0 = R+\Delta R$ and $\Delta R = 1.656$ Å is the thickness of a homogeneous layer for the pick-off annihilation [10]. The average radius R of the free volume holes in polymers and the average volume of the holes, taken as $V_f = (4/3)\pi R^3$, can be calculated from the measured *o*-Ps lifetime. Despite of its principal limitation which cannot be circumvented (the apparent free volume integrates only the sites seen by the *o*-Ps [11]), the model has been successfully used in PALS studies on a wide class of substances including polymers, membranes [11,12] and cyclodextrins [6,13]. Although equation (1) was derived for spherical holes, which is not the case of CDs, the value of R extracted from $\tau_3$ can be considered as an "average" radius and it is most often further translated into an "average" free volume. Given the chemical interaction of *o*-Ps with the surrounding medium, it must not be expected that the assessed volumes coincide accurately with those determined by other techniques.

Figure 1 shows the lifetimes ($\tau_3$) and intensities ($I_3$) observed in samples of complexes of S-Crv and of ThQ in β-CD, also including pure β-CD as a reference. The radii obtained with equation 1 are also plotted. Previous results [6] for pure β-CD are in agreement with the present one and with those reported in literature [14].

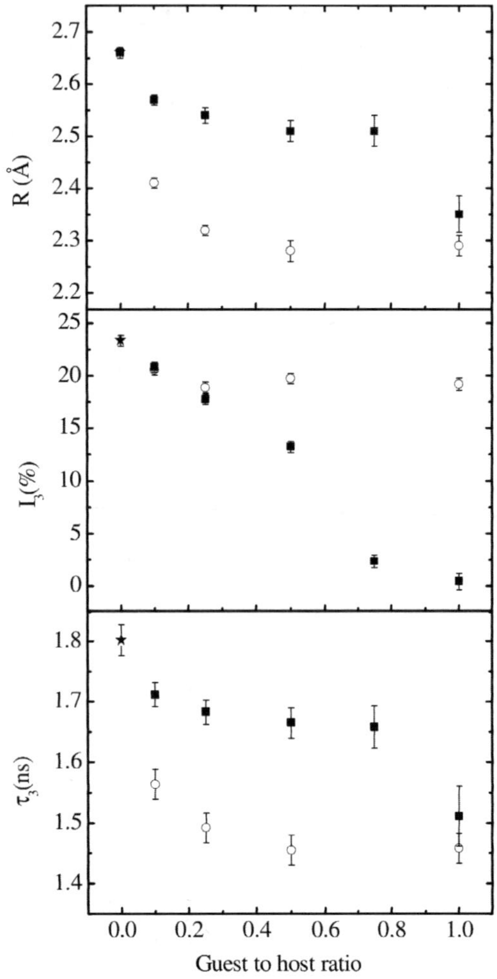


**Figure 1.** PALS parameter and calculated average radii of *o*-Ps trapping sites for the different samples studied:

- ★ pure β-cyclodextrin (β-CD)
- ■ β-CD with ThQ
- ○ β-CD with S-Crv

For the inclusion of S-Crv in β-CD, the value of $\tau_3$ (and thus R) decrease until 0.25 guest to host ratio and then remains practically constant, the resulting free volume holes varying from 78.8 (pure β-CD) to 52.9 Å$^3$. $I_3$, in turn, shows an initial slight decrease remaining practically constant above 0.25 guest to host fraction.

For the ThQ inclusion compound, the value of lifetime $\tau_3$ remains roughly constant within the error bars with increasing guest concentration, except for the 1:1 guest to host ratio. The computed average free volumes vary from 78.8 Å$^3$ (pure β-CD) to 66.4 Å$^3$ for 0.75 and then to just 27.7 Å$^3$ for 1:1 ratio. However the intensity $I_3$ for this 1:1 ratio is just 0.41%, and such a low value is actually compatible with no Ps formation. Parameter $I_3$, related with both the density of free volume elements and the chemical environment that determines the probability of *o*-Ps formation, shows a continuous decrease with increasing guest to host ratio. The behavior of $I_3$ agrees well with the reduction of free volume on the hollow truncated cone of the β-CD due to the inclusion of the guest molecules, implying that the holes become fully occupied, thus pointing to a 1:1 stoichiometry.

Furthermore, this latter observation is in agreement with results of Differential Scanning Calorimetric (DSC) and Fourier Transform Infrared Spectroscopy (FTIR) measurements [15]. DSC provides evidence for the inclusion of the guest, revealing the absence of the endothermic peak assigned to ThQ melting for the highest guest to host fraction. In turn, the FTIR spectrum for 1:1 guest to host ratio with ThQ is very similar to the pure β-CD, implying that ThQ would be included and not interposed between β-CD molecules [15].

The behavior of $I_3$ with increasing guest content is similar for both S-Crv and ThQ up till 0.25 guest to host ratio. However, whereas above this concentration the inclusion of S-Crv does not cause changes in $I_3$, for the ThQ case a continuous decrease is observed and the smallest values are reached.

The product of $\tau_3$ and $I_3$ is an interesting quantity, since it can reflect the free volume of the sample [16]. In the present case such quantity reveals a different behavior for the two guests: whereas for ThQ it shows a linear decrease, for S-Crv it presents a linear decrease until 0.25 guest to host ratio (similar to the thymoquinone) and for higher guest fractions it remains practically constant. As stated above, the different behavior observed for the two guests may be associated with distinct chemical interaction between β-CD and guests with diverse molecular structures.

## 4. Conclusions

The present study of β-cyclodextrin and their complexes by PALS provides a strong assessment of way the free volume is used, proving that the technique is adequate for gathering micro structural information in such systems.

Although the difference between the inclusion of ThQ and S-Crv was made evident, in order to clarify the mechanisms involved, further systematic studies must be made. Particular attention should also be devoted to clarifying the evolution with the time elapsed since encapsulation, an aspect of utmost practical importance. For the characterization of S-Crv inclusion in the β-CD, DSC and FTIR have already proven to be important. However, in order to achieve a better understanding of the encapsulation mechanisms, the use of other complementary techniques like Raman, NMR and Doppler broadening is foreseen.

## References

[1] Martin Del Valle E M 2004 *Process Biochemistry* **39** 1033
[2] Szente L and Szejtli J 2004 *Trends in Food Science & Technology* **15** 137
[3] Astray G, Gonzalez-Barreiro C, Mejuto J C, Rial-Otero R and Simal-Gándara J 2009 *Food Hydrocolloids* **23** 1631
[4] Kilburn D, Dlubek G, Pionteck J and Alam M A 2006 *Polymer* **47** 7774
[5] He C, Suzuki R, Ohdaira T, Oshima N, Kinomura A, Muramatsu M and Kobayash Y 2007 *Chem. Phys.* **331** 213
[6] Ferreira Marques M F, Moreira da Silva A M G, Gordo P M and Kajcsos Zs 2011 *Material Science Forum* **666** 99
[7] Kansy J 1996 *Nucl. Instrum. Methods Phys. Res.* A **374** 235
[8] Tao S J 1972 *J. Chem. Phys.* **56** 5499
[9] Eldrup M, Lightbody D and Sherwood J N 1981 *Chem. Phys.* **63** 51
[10] Nakanishi H, Wang S J and Jean Y C 1987 *Positron Annihilation Studies of Fluids*, edited by S.C. Sharma, World Scientific, Singapore 292
[11] Maurer F H J and Schmid M 2000 *Radiat. Phys. Chem.* **58** 509
[12] Ferreira Marques M F, Gordo P M, Lopes Gil C, de Lima A P, Queiroz D P, de Pinho M N and Kajcsos Zs 2007 *Radiat. Phys. and Chem.* **76** 129
[13] Roussenova M, Murith M, Alam A and Ubbink J 2010 *Biomacromolecules* **11** 3237
[14] Itoh Y, Shimazu A, Sadzuka Y, Sonobe T and Itai S 2008 *Int. Journal of Pharmaceutics* **358** 91
[15] Cardoso T, Galhano C I C, Ferreira Marques M F and Moreira da Silva A 2012 *Spectroscopy: An International Journal* **27** 329
[16] Dlubek G, Buchhold R. Hübner Ch and Nakldal A 2010 *Macromolecules* **32** 2348

# Positron annihilation lifetime spectroscopy of mechanically milled protein fibre powders and their free volume aspects

**K Patil[1], S Sellaiyan[2,3,4,§], R Rajkhowa[1], T Tsuzuki[1], T Lin[1], S V Smith[5], X Wang[1] and A Uedono[6]**

[1]Institute for Frontier Materials, Deakin University, Geelong, VIC 3217, Australia
[2]Center of Excellence in Antimatter Matter Studies, Australian Nuclear Science and Technology Organisation, Kirrawee, 2232, Australia
[3]Center of Excellence in Antimatter Matter Studies, Research School of Physics and Engineering, Australian National University, Canberra, 0200, Australia
[4]University of Tsukuba Tandem Accelerator Complex, University of Tsukuba, Tennodai 1-1-1, Ibaraki 305-8577, Japan
[5]Collider- Accelerator Department, Brookhaven National Laboratory, Upton, NY 11973, USA
[6]Division of Applied Physics, Faculty of Pure and Applied Science, University of Tsukuba, Tsukuba, Ibaraki 305-8573, Japan

E-mail: selva2d@yahoo.co.in

**Abstract.** The present study reports the fabrication of ultra-fine powders from animal protein fibres such as cashmere guard hair, merino wool and eri silk along with their free volume aspects. The respectively mechanically cleaned, scoured and degummed cashmere guard hair, wool and silk fibres were converted into dry powders by a process sequence: Chopping, Attritor Milling, and Spray Drying. The fabricated protein fibre powders were characterised by scanning electron microscope, particle size distribution and positron annihilation lifetime spectroscopy (PALS). The PALS results indicated that the average free volume size in protein fibres increased on their wet mechanical milling with a decrease in the corresponding intensities leading to a resultant decrease in their fractional free volumes.

## 1. Introduction

Natural fibres derived from animal sources are proteinaceous in nature. Their molecules consist of a long polypeptide chain constructed from the eighteen commonly encountered amino acids. Due to the diverse chemical nature of these amino acids, different nucleophilic groups are available on these fibres such as amino (-$NH_2$), hydroxyl (-OH) or thiol (-SH) groups which are present in the amino side-chains of the polypeptide [1]. These functional groups can be used as active sites for binding of molecules targeting specific applications [2]. To explore the functional properties of animal protein fibres, there has been growing interest in converting protein fibres such as cashmere guard hair, wool and silk into fine powders [3,4]. Apart from the various possible applications [5], animal protein fibre

[§] Correspondence to Selvakumar Sellaiyan, University of Tsukuba Tandem Accelerator Complex, University of Tsukuba, Tennodai 1-1-1, Ibaraki 305-8577, Japan.

powders have demonstrated to improve absorption kinetics towards organic [6] and inorganic [7] molecules due to their increased specific surface area and reactivity.

The absorption kinetics is mainly governed by the free volumes, also known as cavities, voids, holes and pores, present in the sorbent, besides its specific surface area and reactivity. These free volumes can act as a transport system for the molecular diffusion into the bulk of the sorbent material. Therefore, the mean free volume size, its distribution and the fractional free volume (i.e. total free volume as a percentage of the complete volume of the material) in the sorbent are important factors which determine the absorption capacity of the sorbent for the given molecules [8]. Due to the highly realized potential of animal protein fibre powders in the space of molecular absorption, this study reports on the free volume aspects of certain animal protein fibres such as cashmere guard hair, merino wool and eri silk fibre and the effect of their mechanical milling on the same using positron lifetime spectroscopy (PALS).

## 2. Experimental

Cashmere guard hair (CGH) supplied by *M/s Cashmere Connections Pty Limited, Australia*, scoured merino wool fibres (MWF) procured from *M/s Australian Country Spinners, Australia* and eri silk (ES) in the form of cocoons from north east India were used for their respective CGHP, MWFP and ESP powder preparations. MWF were used as received, while CGH and ES cocoons were cleaned prior to milling. CGH fibres were cleaned on a *Mesdan* sample carding machine with three passes, followed by manual removal of any adhering traces of foreign matters. Whereas, ES cocoons were degummed for 20 min in a *Thies* laboratory dyeing machine using laboratory grade 2 g/L $Na_2CO_3$ and 0.6 g/L sodium dodecyl sulphate at 100 °C with a material-to-liquor ratio of 1:25 (kg/L). The cocoons were subsequently washed thoroughly with warm distilled water followed by copious amounts of cold distilled water. The degummed ES cocoons were finally dried at 60 °C overnight. The cleaned CGH, merino wool fibres and degummed ES cocoons were converted into dry powders by following the previously developed powder fabrication sequence in our research group [3].

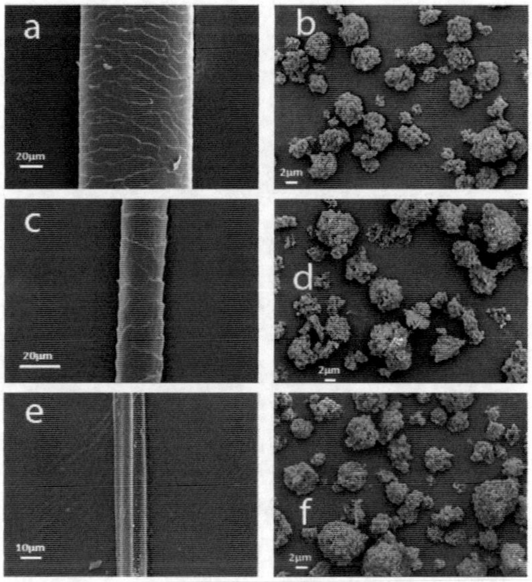

**Figure 1**. SEM image of a. CGH, b. CGHP, c. MWF, d. MWFP, e. ES and f. ESP

The morphologies of the milled powder samples and corresponding fibres were observed under a Scanning Electron Microscope, SEM (*Zeiss Supra 55VP*) at 5-10 kV accelerated voltage and 8-9 mm working distance. The samples were coated with 15 nm gold layer on a sputter coater (*Bal-Tec sputter Coater SCD 050)* before imaging. All powder samples' particle size distribution was measured by using the laser diffraction technique on *Mastersizer 2000* from *Malvern Instruments, UK*. The powder particles were dispersed in *propan-2-ol* (from *Sigma-Aldrich*) during the measurement. The particle size measurement was repeated three times and the volume based particle size distribution was reported along with the mean particle size (d(0.5)). As there was no significant variation among the measurements, no error bars are plotted while reporting the results. For each of the studied protein fibre types and their respective powder samples, PALS measurements were performed at room temperature. The lifetime spectra of positrons were measured using a conventional lifetime system with digital oscilloscope (LeCroy Wavepro). In order to determine a number of lifetime components, deconvolution was performed on the lifetime spectra. The observed spectra were analysed with a time resolution of about 170 ps by using the RESOLUTION computer program [9]. The free volume radius (R) was calculated as per the simple relation (equation 1), according to Nakanishi et al [10].

$$\tau_3 = \frac{1}{2}\left[1 - \frac{R}{R_0} + \frac{1}{2\pi}\sin\left(\frac{2\pi R}{R_0}\right)\right]^{-1} \quad (1)$$

where, $\tau_3$ (o-Ps lifetime) and $R$ (free volume radius) are expressed in ns and nm, respectively and $R_0 = R + \Delta R$, where $\Delta R$ is a constant for an electron layer thickness (0.166 nm). The average free volume size ($V_f$) was then simply evaluated as $V_f = (4/3)\pi R^3$, whereas the fractional free volume ($F_v$) was estimated as $F_v = CV_fI_3$ [11] and $C$ may be assigned a value of 1 due to absence of variation of free volume data with temperature.

## 3. Results and Discussion

The surface morphology of CGH fibre is very much similar to MWF with near cylindrical shape and surface scales pointing towards the fibre tip (figure 1a and 1c), whereas ES fibres have well known smooth and trilobal morphology (figure 1e). CGH fibres are distinctly large in diameter, ~ 80 μm as compared with MWF and ES fibres which were typically around 20 μm. Although each protein fibre type had distinct surface morphology and structure, the spray dried particles from their 6 hr attritor milled slurry appeared very similar in shape and size. The particles were nonspherical, globular in shape with a mushroom like structure (figure 1b, 1d and 1f). The volume based mean particle sizes (d(0.5)) were 4.63 μm, 6.05 μm and 5.85 μm for CGHP, MWFP and ESP respectively (figure 2). Figure 3 shows PAL spectra of CGH. The spectra were similar for all fibre and powder samples. We observed that three lifetime components give the best $\chi^2$ and reasonable standard deviation. PAL component, $\tau_1$, is the lifetime of para-positrons (p-Ps) and $\tau_2$ is related to free annihilations. The longest lifetime component ($\tau_3$) is attributed to the pick-off annihilation of ortho-positrons (o-Ps) in the free volume sites present in the amorphous regions of the sample [11]. Thus, the $\tau_3$ component of the PAL spectra and its corresponding intensity ($I_3$) keeps relevance from the molecular diffusion point of view. Even though the PAL spectra were similar for all the studied fibre and powder samples (figure 3), slight differences in o-Ps ($\tau_3$) and their corresponding intensities ($I_3$) were observed (table 1). From the o-Ps results, mean free volume radius ($R$), $V_f$ and $F_v$ were calculated and listed in table 1. Among the studied animal protein fibre samples, ES has the highest $F_v$ (17.9 %), whereas the CGH and MFW have similar $F_v$ (~15 %). Although relatively small, but very similar values for silk, wool and few other hair fibres have also been reported elsewhere [12]. Most importantly, table 1 also depicts that the $F_v$ for CGH, MWF and ES has decreased on milling by 3.56 %, 19.68% and 11.61 % respectively.

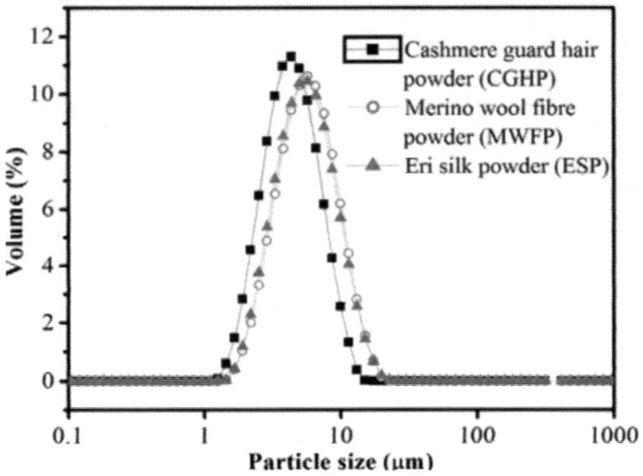


**Figure 2.** Particle size distributions of mechanically milled protein fibre powders

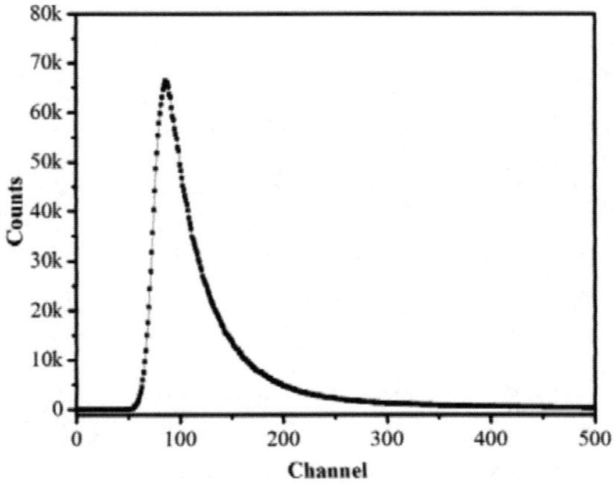


**Figure 3.** PAL spectra of cashmere guard hair (CGH)

On milling, the intensity of o-Ps lifetime component $\tau_3$ was decreased for all the animal protein fibre powders (table 1). It is likely that the wet conditions during the attritor milling of the protein fibres significantly decreased their glass transition temperature [13] and made their respective molecules in the amorphous region more labile at the milling temperature. Under the influence of compressive forces of milling media on the labile protein fibre molecules, the $V_f$ in the amorphous region would have undergone coalescence phenomenon, resulting in an increase in the $V_f$ and a decrease in its intensity. From the molecular absorption application point of view, although the ultrafine protein fibre powder particles would provide improved surface area for immediate molecular adsorption, the decreased $F_v$ is likely to retard the time dependant molecular diffusion phenomenon.

Journal of Physics: Conference Series **443** (2013) 012054 doi:10.1088/1742-6596/443/1/012054

Similar results have been realized by us [12] where $Zn^{2+}$ and $Cr^{6+}$ have been instantaneously adsorbed onto the CGHP with negligible diffusion controlled, time dependent absorption of the species.

**Table 1.** Analysed PAL spectra data for animal protein fibres and their mechanically milled powder

| Sample | | o-Ps Lifetime (ns) | Intensity (%) | Average free volume radius (Å) | Average free volume size (Å$^3$) | Fractional free Volume |
|---|---|---|---|---|---|---|
| | | $\tau_3$ | $I_3$ | $R$ | $V_f$ | $Fv$ |
| Cashmere guard hair | Fibre | 1.79 | 19.5 | 2.65 | 77.8 | 15.17 |
| | Powder | 1.79 | 18.6 | 2.66 | 78.7 | 14.63 |
| Merino wool | Fibre | 1.81 | 19.3 | 2.67 | 79.5 | 15.34 |
| | Powder | 2.00 | 12.6 | 2.86 | 97.8 | 12.32 |
| Eri silk | Fibre | 1.85 | 21.3 | 2.72 | 84.1 | 17.91 |
| | Powder | 1.81 | 19.7 | 2.68 | 80.4 | 15.83 |

Thus, protein fibre powders fabricated by the present wet attritor milling route are likely to be more suitable in applications where ultrafine materials, accompanied with biocompatibility and biodegradability, are required. However, molecular absorption trajectory of the potential protein fibre powder application could be more efficiently explored when the free volume aspects of the parent protein fibres get retained or improvised during their fabrication.

## 4. Conclusion

The wet mechanical milling of viscoelastic protein fibres promote coalescences of the free volumes present in the fibre structure, leading to the increased average free volume size, along with decrease in the corresponding relevant intensity. Consequently, the fractional free volume in the protein fibre powder gets decreased and is likely to affect their performance in molecular absorption applications. For better molecular absorption capacities of the mechanically milled protein fibre powders, free volume aspects of the parent protein fibres need to be either retained or improved by exploring different powder fabrication process parameters or routes.

## Acknowledgement

Financial support from Australian Research Council (DP 1094979) is acknowledged.

## References

[1] Christie R M 2001 *Colour Chemistry* (Cambridge: RSC Publishing)
[2] Zhao T and Sun G 2007 *J. Appl. Polym. Sci.* **103** 482
[3] Rajkhowa R, Wang L, Kanwar J and Wang X 2009 *Powder Technol.* **191** 155
[4] Patil K, Rajkhowa R, Dai X J, Tsuzuki T, Lin T and Wang X 2012 *Powder Technol.* **219** 179
[5] Huang J, Liu X, Li W and Xu W 2012 *J. Thermoplast. Compos. Mater.* **25** 75
[6] Wen G, Cookson P G, Liu X and Wang X G 2010 *J. Appl. Polym. Sci.* **116** 2216
[7] Patil K, Smith S V, Rajkhowa R, Tsuzuki T, Wang X and Lin T 2012 *Powder Technol.* **218** 162
[8] Ehrhardt C and Kim K 2008 *Drug Absorption Studies: In Situ, in Vitro and in Silico Models* (New York: Springer)
[9] Saito H and Hyodo T 2003 *Radiat. Phys. Chem.* **68** 431
[10] Nakanishi H, Wang S and Jean Y C 1987 *Int. Symp. on Positron annihilation studies of fluids (Texas)* ed S C Sharma (Singapore: World Sceintific) pp 292-8
[11] Jean Y C 1990 *Microchem. J.* **42** 72
[12] Chandrashekara M N and Ranganathaiah C 2011 *AIP Conf. Proc. (Manipal)* vol 1349/1 (New York: AIP Publishing) pp 228-9
[13] Rowland S P 1980 *Water in polymers* (Washington: ACS Publishing)

# Positron annihilation lifetime spectroscopy study on epoxy moisture absorption

**P M Gordo[1], J M M Luis[1], M F Ferreira Marques[1,2], M Duarte Naia[3] and J J M Morais[4]**
[1] CEMUC®, Department of Physics, Coimbra University, R. Larga,3004-516 Coimbra, Portugal
[2] Department of Chemical and Biological Engineering, ISEC-IPC, R. Pedro Nunes, Quinta da Nora, 3030-199 Coimbra, Portugal
[3] CEMUC®, Department of Physics, ECT-UTAD, PO Box 1013, 5001-801 Vila Real, Portugal
[4] CITAB, Department of Engineering, ECT-UTAD, PO Box 1013, 5001-801 Vila Real, Portugal

E-mail: pgordo@ci.uc.pt

**Abstract.** Positron annihilation lifetime spectroscopy was used to study the effect of water uptake on the free volume of a commercial epoxy exposed to atmospheres with relative humidity ranging from 30.5% to 95.8%. The absorption process was also analyzed by gravimetric measurements. When exposed to humidity the mean hole volume remains constant and the number of holes not occupied by water molecules decrease with increasing of both the exposure time and the relative humidity as indicated by the constancy of $\tau_3$ and the decrease of $I_3$.

## 1. Introduction

Epoxy systems are polymers heavily cross-linked in which the cross-linking is derived from reactions of the epoxy group and thus it is a rigid and strong material. They have in general excellent adhesion to many surfaces, superior thermal resistance, a relative low dielectric constant, and ease of processing [1]. Owing to their superior mechanical and dielectric properties, coupled with excellent thermal stability and great chemical resistance, epoxy systems has been widely used in many applications as aircraft components, insulating layers and encapsulants in microelectronics industry, aerospace structural composites and many more [1, 2]. A major drawback of some epoxy systems, like the one used in this work, is their hygroscopic nature which results in the absorption of few percent moisture by weight under ordinary environmental conditions. The absorbed water plasticize the epoxy system, induces differential swelling stresses and generally degrades the physical properties of the epoxy. Some of the effects are a decrease of the glass-transition temperature ($T_g$), i.e. a decreased thermal stability [3]; a decrease in yield strength, a change of the yield/deformation mechanisms, or both [4]; the creation of micro-cracks or crazes [3, 5]; and chain scission through hydrolysis reactions [5]. Due to these factors it is important to control or inhibit moisture absorption to prevent these undesirable effects. Several studies have been undertaken to investigate the process of moisture absorption in epoxies, but the exact mechanisms that govern humidity-induced water uptake and diffusion of water molecules within the material are still unclear. It is generally believed that moisture absorption is closely related to the free volume properties of the epoxy [2]. Positron Annihilation Lifetime Spectroscopy (PALS) is a well-established and very sensitive technique for probing sub-nanometer-sized local free volumes between molecular chains in polymeric structures. In the present work, we use PALS to investigate the effect of water absorption process on the free volume in one commercial epoxy submitted different relative humidity environments.

## 2. Experimental

The samples used in this study were made from one commercial epoxy resin known as SR1500 and one hardener known as SD2505 both produced by Sicomin Composites Company (France). The formulation bases of the epoxy resin SR1500 are bisphenol A and bisphenol F and this epoxy resin has been modified to be crystallization free and have low toxicity. The epoxy resin when mixed with the hardener, upon polymerization, forms the epoxy. The samples used were prepared from one continuous layer of approximately 2 mm of thickness.

The drying of the epoxy samples was performed by placing the samples in a sealed housing containing silica gel at 50ºC for several months (typically for 2 months).

The different relative humidity (RH) atmospheres were controlled between 0% and 95.8% and the temperature was constant at 50ºC except during the positron measurements. To maintain a precisely controlled RH atmosphere over long periods of time, saturated salt solutions were used. The atmosphere above these solutions had RH of 30.5% ($MgCl_2/H_2O$), 74.4% ($NaCl/H_2O$) and 95.8% ($K_2SO_4/H_2O$), respectively.

To determine the humidity-induced mass uptake in the epoxy samples submitted to different RH environments a precision balance (Perkin-Elmer microbalance, accuracy 0.01 mg) setup was used.

The positron lifetime measurements were carried out at room temperature (24 ºC) using a conventional fast-fast coincidence PALS setup (featuring Pilot-U scintillators and XP2020 photomultipliers) with a time resolution of 260 ps (FWHM). For each experiment, the positron $^{22}Na$ source (*ca.* ~30μCi, closed between Kapton® foils) was sandwiched by two identical samples. The lifetime spectra had a total number of *ca.* $2x10^6$ integral counts and were evaluated using the LT (version 9) code [6].

## 3. Results and discussion

It is generally accepted that the behavior of water absorption in the epoxy is well described by Fick's second law [7],

$$\frac{M}{M_{max}} = 1 - \sum_{n=0}^{\infty} \frac{8}{(2n+1)^2\pi^2} exp\left[\frac{-D(2n+1)^2\pi^2 t}{4L^2}\right] \quad (1)$$

where $M$ is the water absorption at time $t$, $M_{max}$ is the equilibrium water absorption, $L$ the sample thickness and $D$ is the diffusion coefficient. For values of $M/M_{max}$ lower than 0.5, equation 1 takes the simplest form,

$$\frac{M}{M_{max}} = \frac{4}{L}\sqrt{\frac{Dt}{\pi}} \quad (2)$$

The gravimetric results obtained for the relative mass uptake of the samples as a function of the square root of the exposure time for the 3 different RH environments, 30.5%, 74.4% and 95.8%, respectively, are shown in figure 1. The material shows increasing mass uptake from RH=30.5% to 95.8%. Higher RH environment values (74.4 and 95.8%) correspond to higher values of the equilibrium water absorption, achieving *ca.* 1.5% and 3% of relative mass percentage. For the samples submitted to higher RH environments, the water uptake is linearly proportional to $\sqrt{t}$ at the initial stage; for the lowest RH value it is difficult to see clearly this proportionality due to the small amount of uptaken water (less than 0.5%) for the 30.5% RH ambient. It can be also observed that the beginning of the equilibrium water absorption is achieved at approximately the same humidity exposure time ($\sqrt{t}$~20 $h^{1/2}$) and this is an indication that the diffusion coefficient, $D$, is the same for all samples.

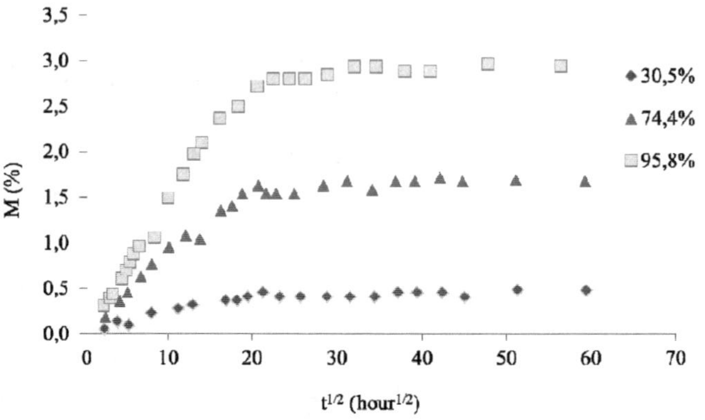


**Figure 1.** Relative mass uptake as a function of the square root of exposure time of dry samples placed in different RH environments.

In the systematic LT analysis it was observed that all spectra were well fitted with three components: two short-lifetime components ($\tau_1$, $I_1$) and ($\tau_2$, $I_2$) are normally assigned to self-annihilation of *p*-Ps and free/trapped positron annihilation, respectively; the longest lifetime component ($\tau_3$, $I_3$) is due to the annihilation of *o*-Ps. $\tau_3$ is considered to be related to the size of intermolecular-space holes, and a semi-empirical relationship between $\tau_3$ and the size of holes for Ps localized on spherical holes [8] has been used to estimate the volume created among polymer chains. $I_3$ is often treated as a measure of the density of holes [9].

In our analysis the value of $\tau_1$ was varying systematically around 155 ps and it was fixed to that value in order to reduce the artificial effects. Assuming that the pick-off annihilation is the only *o*-Ps quenching process the ratio between $I_1$ and $I_3$ is 1/3 and in the analysis the value of the intensity $I_1$ was forced to be equal to $I_3/3$.

The second lifetime component $\tau_2$ was a free parameter in the fitting process and it had a regular value around 360 ps.

Since only the long-lived *o*-Ps lifetime could be related to the properties of the free volume size in the samples, the following discussion are mainly based on the annihilation characteristics of $\tau_3$ and $I_3$.

Assuming that the holes are spherical, the relationship between $\tau_3$ and the radius ($R$) is described as [8,10,11]

$$\tau_3 = 0,5\left[1 - \frac{R}{R+\Delta R} + \frac{1}{2\pi} sin\left(\frac{2\pi R}{R+\Delta R}\right)\right]^{-1} \quad (ns) \tag{3}$$

where $\Delta R$=0.166 nm is the thickness of a homogeneous layer for the pick-off annihilation and the radius of the finite spherical potential is $R+\Delta R$.

The free volume fraction, $F_v$, is directly related to $I_3$ and $\tau_3$ through the expression [8],

$$F_v = C\frac{4}{3}\pi R^3 I_3 \tag{4}$$

where $C$ is an empirical scaling constant that reflects the probability of *o*-Ps formation.

Figure 2 represents the values of $\tau_3$ and $I_3$ as a function of the square root of the exposure time of the dry samples in different relative humidity atmospheres. It can be seen that the value of $\tau_3$ is independent of the amount of absorbed water of each sample even in the non-saturation regimen ($\sqrt{t}$<20 $h^{1/2}$). On the other hand, the values of $I_3$ decrease monotonically with the exposure time and become constant approximately for $\sqrt{t}$>20 $h^{1/2}$ when the water uptake saturation regimen is achieved. The decrease in $I_3$ is more pronounced for samples exposed to atmosphere with higher relative humidity revealing that the water molecules are occupying the free holes in the epoxy. It is well known that the inclusion of water in polymers has an effect of reducing positronium formation process and consequently in a reduction of $I_3$. A linear relation between the relative variation of $I_3$ and the

relative variation of the uptake water mass seems to exist and this behavior maybe suggest that the water absorption is determined mainly by the free volume fraction of the material, but we have only 3 different RH atmospheres and more data are needed to confirm this sentence.
The constant value of $\tau_3$ is 1.76±0.02 ns and, according to equation 3, corresponds to "spherical" holes with an average radius of 2.63±0.02 Å, which is comparable to the kinetic diameter of a water molecule, 3 Å. The constancy in $\tau_3$ and the behaviour of $I_3$ with the RH atmosphere reveal that each occupied hole accommodates only one water molecule and in the occupied hole there is no more Ps formation.

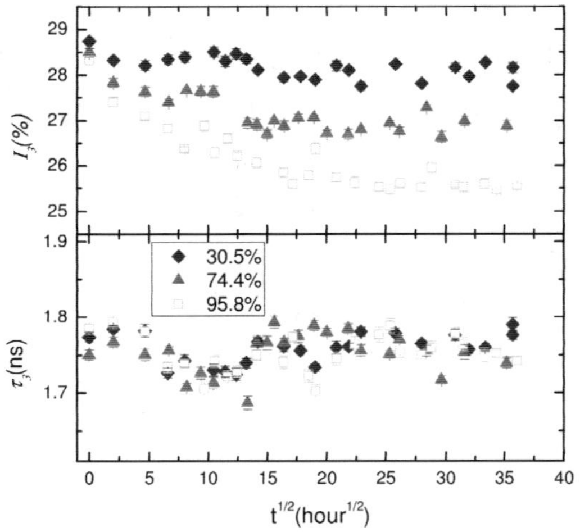


**Figure 2.** *o*-Ps lifetime and intensity, $\tau_3$ and $I_3$, as a function of the square root of exposure time of initially dry samples in different humidity atmospheres.

## 4. Conclusion

The positron annihilation lifetime technique has been applied to study water absorption in a commercial epoxy exposed to atmospheres of controlled relative humidity. The epoxy shows increasing mass uptake from 30.5% to 95.8%, with a maximum mass uptake of approximately 3% at RH of 95.8%. PAL measurements indicate a decrease in intensity $I_3$, which is strongly dependent on the atmosphere exposure time and on the relative humidity. The value of $\tau_3$ is constant and independent of RH and the exposure time revealing that in the hole occupied with water there is no Ps formation.

## References

[1] Soles C L, Chang F T, Gidley D W and Yee A F 2000 *J. Polym. Sci. B: Phys.* **38** 776
[2] Dlubek G, Buchhold R, Hübner Ch and Nakladal A 1999 *Macromolecules* **32** 2348
[3] Browning C E 1978 *Polymer Engineering & Science* **18** 16
[4] Mostovoy S and Ripling E J, 1971 *J. Appl. Polym. Sci.* **15** 641
[5] Morgan R J 1979 *J. Appl. Polym. Sci.* **23** 2711
[6] Kansy J 1996 *Nucl. Instrum. Methods Phys. Res.* A **374** 235
[7] Barrie J A 1968 *Diffusion in Polymers* ed Crank J, Park GS (London Academic Press) Chapter 1
[8] H. Nakanishi and Y.C. Jean 1988 Studies in physical and theoretical chemistry *Positron and Positronium Chemistry* ed Schrader D M, Jean YC (Amsterdam: Elsevier Sci. Publ.) pp159-192
[9] Okamoto K, Tanaka K, Katsube M, Sueoka O and Yto Y 1993 *Radiat. Phys. Chem.* **41** 61
[10] Tao S J 1972 *J. Chem. Phys.* **56** 5499
[11] Eldrup M, Lightbody D and Sherwood J N 1981 *Chem. Phys.* **63** 5

Journal of Physics: Conference Series **443** (2013) 012056 doi:10.1088/1742-6596/443/1/012056

# A nanostructural investigation of the effect of glycerol on the molecular packing, thermodynamic state and water sorption behaviour of glassy gelatin films

**P Díaz-Calderón[1], M Roussenova[2], J Enrione[3*] and M A Alam[2]**

1 Department of Food Science and Technology, Universidad de Santiago de Chile, Av. Ecuador 3343, Estación Central, Santiago, Chile.
2 H. H. Wills Physics Laboratory, Tyndall Avenue, Bristol, BS8 1TL, United Kingdom.
3 Facultad de Medicina, Escuela Administración de Servicios, Universidad de los Andes, San Carlos de Apoquindo 2200, Las Condes, Santiago, Chile.

*Corresponding author email*: jenrione@uandes.cl

**Abstract**. This work explores the effect of glycerol, a low molecular weight polyol, on the molecular packing, thermodynamic state and water sorption of low water content gelatin films. For this purpose, bovine gelatin films with different glycerol contents (0-10% wt.) were equilibrated at a range of relative humidities (RH=11-44%, $T$=298K). Our PALS measurements show that over the concentration range studied, glycerol acts as a packing enhancer (causing a non-linear decrease in the average molecular hole size), whilst reducing the glass transition temperature of the gelatin films. Glycerol also alters the water sorption behaviour of the glassy gelatin films, (whereby reducing the amount of water absorbed at well defined relative humidities), highlighting the importance of molecular packing for the sorption of water vapour in the glassy state.

## 1. Introduction

Gelatin is a hydrocolloid widely used in the food, pharmaceutical and biomedical industries [1, 2]. It is obtained from collagen, the principal component of fibrous connective tissue by hydrolytic degradation using either acids or alkali [3]. Gelatin is one of the most versatile biopolymers and its traditional uses are based on its well known gelling capability and its properties as a viscosity enhancer [2]. Gelatin also has the ability of forming strong, clear films which are readily soluble in water, making it an ideal raw material for the manufacture of hard and soft pharmaceutical capsules, as well as microcapsules for the food industry [4]. Such capsules are commonly used to encapsulate labile bioactive compounds (e.g. drugs, probiotics, flavors, fatty acids, etc), protect them from undesirable external influences (such as oxygen and water) and achieve their targeted release when required [5]. Unfortunately, low water content glassy gelatin capsules have the disadvantage of high brittleness which may lead to their undesired failure during handling and storage, allowing the loss of the bioactive or the migration of surrounding molecules (e.g. oxygen and water) [6]. Furthermore, gelatin is highly hygroscopic (it can readily absorb water vapor from the atmosphere [7]) which can cause significant modifications in the barrier properties of gelatin capsules, leading to the oxidation or premature release of the encapsulated bioactive compound. In order to overcome these problems,

Journal of Physics: Conference Series **443** (2013) 012056 doi:10.1088/1742-6596/443/1/012056

gelatin films are commonly modified by the inclusion of small amounts of low molecular weight polyols [6] such as glycerol. Glycerol has been widely used in the pharmaceutical and food industries to alter the mechanical, water sorption and permeation (towards gases and water vapour) properties of biopolymers in order to design optimal encapsulating matrices [7, 8].

One of the main issues in the design of encapsulating gelatin matrices with optimal barrier properties is the molecular mobility through the matrix, which in the glassy state is governed by the density of the molecular packing of the gelatin oligomer chains [4]. In amorphous and partially ordered soft matter, the irregular arrangement of the constituent molecules results in a certain amount of local free volume. This local free volume consists of a large number of sub-nanometer sized free volume elements and it plays an important role in phenomena such as self-diffusion, the diffusion of guest molecules and the glass transition [9, 10].

The objective of this work is to assess the effect of glycerol on the molecular packing, thermodynamic state and water vapour sorption properties of low water content gelatin films by using Positron Annihilation Lifetime Spectroscopy (PALS) in conjunction with a number of complementary techniques.

## 2. Materials and Methods

Films of bovine gelatin ($M_w$=2.84 x $10^5$ Da, Rousselot, Brazil) and glycerol (purity > 99.99%, Merck, Germany) were prepared by cold casting from suspension. A suspension of 7% w/v gelatin was prepared with various amounts of glycerol, $Q_g$, expressed as mass fractions on an anhydrous basis $Q_g = m_g/(m_{gel} + m_g)$ (where $m_g$ and $m_{gel}$ denote the mass of glycerol and gelatin, respectively). Films with compositions of $Q_g$ = 0.0, 0.02, 0.06 and 0.10 were measured in this study. The films were equilibrated at different relative humidities (RH=11-44%) at $T$ = 298K in desiccators containing saturated salt solutions [11]. The sorption of water was followed gravimetrically until equilibrium was achieved (generally within 25 days). Further details of the method used to determine the water content of the films are given in Ref.[12]. The glass transition temperatures, $T_g$, of the films were measured by Differential Scanning Calorimetry (Diamond DSC, Perkin Elmer, USA) performing two temperature scans between 233-450K at a heating rate of 10K/min. The changes in $T_g$ as a function of glycerol content were modeled using the semi-empirical Gordon-Taylor equation [13]. See reference [12] for further details. Positron Annihilation Lifetime Spectroscopy (PALS) experiments were performed

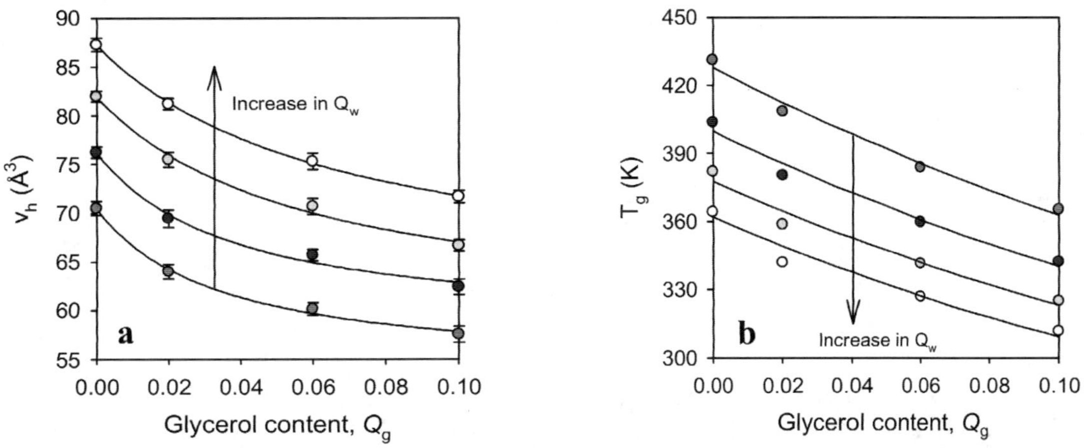


**Figure 1.** Average hole volume, $v_h$, ($T$=298 K) **(a)** and glass transition temperature, $T_g$, **(b)** as a function of increasing glycerol content for glassy gelatin films with a well defined water contents: $Q_w$ = 0.02 (pink series), $Q_w$ = 0.04 (purple series), $Q_w$ = 0.06 (cyan series) and $Q_w$ = 0.08 (white series). The data presented in this figure were derived by interpolation of the data series presented in Figures 6a and 5, respectively in Ref. [12].

using a fast-fast coincidence system. $^{22}$Na was used as a positron source which was prepared by depositing aqueous $^{22}$NaCl between two 7.5 μm sheets of Kapton foil. Spectra were collected over a period of two hours to generate at least 2.5 million events per spectrum and spectra were decomposed by three components using the routine Life Time (version 9.1) [14]. Specific details of the spectra analysis can be found in references [12] and [15].

## 3. Results and Discussion

The effect of glycerol on the molecular packing of the glassy gelatin films is presented in Figure 1a. It can be seen that over the concentration range studied, glycerol acts as a packing enhancer for the glassy gelatin films. For films with well defined water contents, glycerol causes a non-linear decrease in the average molecular hole size which is indicative of non-ideal packing behavior in the glassy state [15-17]. This effect has been previously observed in glassy maltooligomer matrices upon the addition of low molecular weight diluents, such as maltose and glycerol [15, 17]. The initial rapid decrease in $v_h$ may be related to the reduction in the molecular frustration of the system, accompanied by a sharp reduction in the glass transition temperature, $T_g$, of the gelatin films, as is shown in Figure 1b. For films with well defined water contents, we observe a non-linear reduction in $T_g$ as a function of increasing glycerol content, which was fitted well to the semi-empirical Gordon-Taylor equation [13] (yielding a Gordon-Taylor coefficient for glycerol, $k_g = 3.6 \pm 0.3$, in agreement with previous studies [17, 18]). Figure 1 (a and b) also illustrates the plasticizing effect of water on the gelatin films, causing an increase in the average molecular hole size accompanied by a reduction in $T_g$. Water has been shown to be a strong plasticizer for biopolymer matrices due to its small molecular size and its propensity to form hydrogen bonds (see Refs. [15-17] for a more detailed discussion).

The presence of glycerol in the glassy gelatin films also has an effect on their water sorption behavior as illustrated in Figure 2. It can be seen that at all relative humidities, the presence of glycerol reduces the amount of water absorbed by the glassy gelatin films. For example, at RH=11% the water content

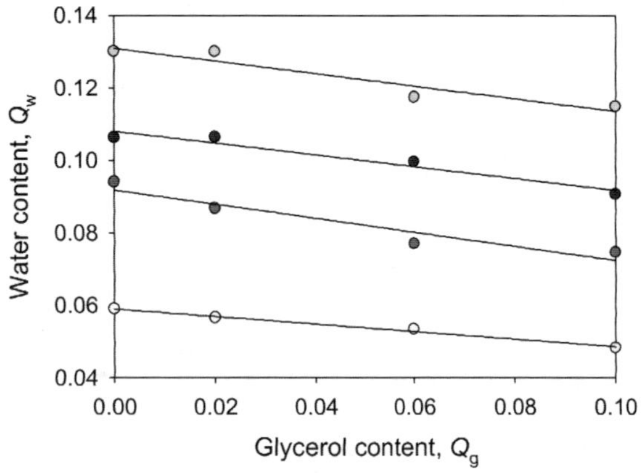


**Figure 2.** Water content as a function of increasing glycerol content for glassy gelatin films at well defined relative humidities at $T$=298K: RH=11% (yellow series), RH=22% (red series), RH=33% (blue series) and RH=44% (green series). All linear regression coefficients are $R^2$>0.95 and the errors in $Q_w$ are the same size as, or smaller than the symbols shown.

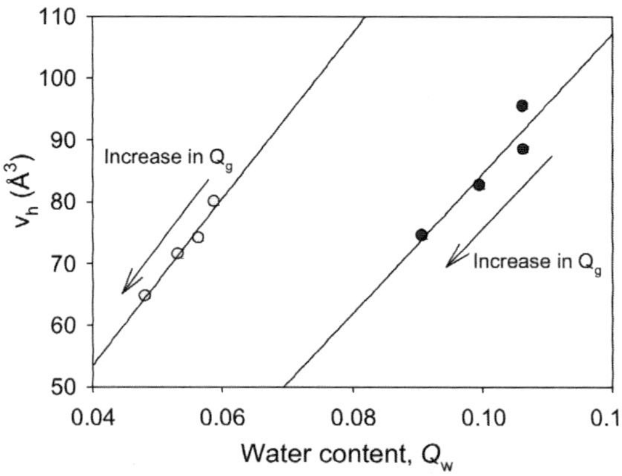


**Figure 3.** Relation between average molecular hole size and the amount of water absorbed by gelatin films ($T$=298K) with different compositions equilibrated at well defined relative humidities, RH=11% (yellow series), RH=33% (blue series). All linear regression coefficients are $R^2$>0.94 the errors in $v_h$ are the same size as, or smaller than the symbols shown.

$Q_w$=0.059 for the pure gelatin films and $Q_w$=0.048 for gelatin films containing 10 wt. % glycerol (i.e. a 19% reduction in water content). In fact, there is a positive linear relationship between the average molecular hole size and the amount of water absorbed by films with different compositions at well defined relative humidities (Figure 3), further illustrating the effect of glycerol as a packing enhancer. This effect has been previously reported for maltooligomer-maltose [15] and maltooligomer-glycerol [17] systems and it highlights the importance of molecular packing for the sorption of water by biopolymers in the glassy state. It is worth mentioning that one would expect to see the opposite effect for gelatin films in the rubbery state, i.e. an increase in the amount of water absorbed by the films at a well defined relative humidity as a function of increasing glycerol content [17, 18]. This is due to the fact that in the rubbery state, water sorption is related to the entropy of mixing of the constituent molecules [18].

## 4. Final Remarks

Our study shows that the molecular organisation, thermodynamic state and water sorption properties of low water content gelatin films can be modified by the addition of low molecular weight polyols such as glycerol. Glycerol acts as a packing enhancer, causing a reduction in the average molecular hole size of the gelatin films, whilst reducing their glass transition temperature and ability to absorb water in the glassy state. By establishing composition-structure relationships for this biopolymer we present a new step towards enabling the rational control of its material properties leading to the eventual design of optimal pharmaceutical encapsulants.

## Acknowledgements

The authors acknowledge financial support from FONDECYT (Chile) Nº1110607 and CONICYT (Chile) PhD Grant Nº21110898 and N°75120091.

## References

[1] Sobral P J A and Habitante A M Q B 2001 *Food Hydrocolloid.* **15** 377
[2] Karim A A and Bhat R 2008 *Trends Food Sci. Tech.* **19** 644
[3] Yakimets I, Wellner N, Smith A C, Wilson R H, Farhat I and Mitchell J 2005 *Polymer* **46** 1257
[4] Ubbink J 2009 *Modern Biopolymer Science, Bridging the Divide between Fundamentals Treatise and Industrial Application* ed S Kasapis, I Norton and J Ubbink (London: Academic Press) p 277
[5] Ubbink J and Kruger J 2006 *Trends Food Sci. Tech.* **17** 244
[6] Vanin F M, Sobral P J, Menegalli F C, Carvalho R A and Habitante A M 2005 *Food Hydrocolloid.* **19** 899
[7] Coppola M, Djabourov M and Ferrand M 2008 *Macromol. Symp.* **273** 56
[8] Thomazine M, Carvalho R A and Sobral P J 2005 *J. Food Sci.* **70** E172
[9] Vrentas J S and Duda J L 1977 *J. Polym. Sci.* **15** 403
[10] Cohen M H and Grest G S 1979 *Phys. Rev. B* **20** 1077
[11] Greenspan L 1977 *J. Res. NBS. A Phys. Ch.* **81** 89
[12] Roussenova M, Enrione J, Díaz-Calderón P, Ubbink J, Taylor A J and Alam M A 2012 *New J. Phys.* **14** 035016
[13] Gordon M and Taylor J 1952 *J. Appl. Chem.* **2** 493
[14] Kansy, J 1996 *Nucl. Instrum. Methods* ***374***, 235
[15] Townrow S, Roussenova M, Giardello M-I, Alam M A and Ubbink J 2010 *J. Phys. Chem. B* **114** 1568
[16] Townrow S, Kilburn D, Alam M A and Ubbink J 2007 *J. Phys. Chem. B* **111** 12643
[17] Roussenova M, Murith M, Alam M A and Ubbink J 2010 *Biomacromolecules* **11** 3237
[18] Ubbink J, Giardiello M-I and Limbach H-J 2007 *Biomacromolecules* **8** 2862

# Account of the intratrack radiolytic processes for interpretation of the AMOC spectrum of liquid water

**D S Zvezhinskiy[1], M Butterling[2], A Wagner[2], R Krause-Rehberg[3] and S V Stepanov[1]**

[1] Institute for Theoretical and Experimental Physics, B. Cheremushkinskaya, 25, Moscow, 117218, Russia

[2] Institute of Radiation Physics, Helmholtz-Zentrum, Dresden-Rossendorf, P.O. Box 510119, Dresden, 01314, Germany

[3] Martin-Luther-University Halle-Wittenberg, Dept. of Physics, Halle, 06099, Germany

E-mail: zmitja@yandex.ru

**Abstract.** Recent development of the Gamma-induced Positron Spectroscopy (GiPS) setup significantly extends applicability of the Age-Momentum Correlation technique (AMOC) for studies of the bulk samples. It also provides many advantages comparing with conventional positron annihilation experiments in liquids, such as extremely low annihilation fraction in vessel walls, absence of a positron source and positron annihilations in it. We have developed a new approach for processing and interpretation of the AMOC-GiPS data based on the diffusion recombination model of the intratrack radiolytic processes. This approach is verified in case of liquid water, which is considered as a reference medium in the positron and positronium chemistry.

## 1. Introduction

The important step towards further development of AMOC was made by adopting an electron LINAC to produce positrons inside an investigated sample with the help of bremsstrahlung [1]. This experimental method is known as Gamma-induced Positron Spectroscopy (GiPS). The absence of a conventional positron source makes possible to produce positrons over the whole volume of an investigated medium. This might be potentially useful for easier treatment of biological and chemical samples of solid and liquid state.

Before starting various physico-chemical investigations on GiPS setup, we have decided to measure AMOC spectrum of liquid water, which is usually considered as a reference medium for positron studies. Available experimental data for water [2, 3] may provide good quantitative comparison of lifetime and Doppler parameters obtained by novel AMOC-GiPS setup.

## 2. Theoretical model for AMOC spectrum

Importance of taking into account the radiolytic processes for interpretation of annihilation lifetime spectra was demonstrated in [4]. In the present work we generalize this model to apply for AMOC data. We have attributed some energy distribution for annihilation gamma-quanta produced from different positron states participating in intratrack radiolytic reactions. Further details concerning the diffusion-recombination model can be found in [5, 4].

Journal of Physics: Conference Series **443** (2013) 012057 doi:10.1088/1742-6596/443/1/012057

Energy distributions of annihilation radiation, which correspond to certain positron states, are usually described by the Gaussian functions centered at 511 keV energy [6].

In present work we use the same approximation except for the shape of the "narrow" (p-Ps) component, which is calculated according to the so-called finite well Ps bubble model. This model yields positronium annihilation rates as well [7]. Being linked to the energy dimension, it allows to reduce overall number of the adjustable parameters. The lifetime and energy shape of p-Ps AMOC relief depend on the radius $R$ of the Ps bubble state and depth $U$ of the potential well (or on electron penetration depth $\delta$) [8].

Shape of the Doppler spectrum, that is a distribution of annihilation gammas on their energy, is given by the following expression:

$$C(\Delta) \sim \int \rho(\mathbf{p}) \cdot \delta\left(\Delta - \frac{cp}{2}\cos\theta\right) \frac{d^3p}{(2\pi\hbar)^3}. \quad (1)$$

Here $\rho(\mathbf{p})$ is the momentum distribution of the positron-electron pair, $\Delta$ is the difference between the energy of annihilating photon and $mc^2$ =511 keV and $\theta$ is the angle between $\mathbf{p}$ and direction to the energy detector. In the spherically symmetrical case $\rho(\mathbf{p}) = \rho(p)$ after integration over $\theta$ Eq.(1) is reduced to a simpler form:

$$C(\Delta) \sim \int_{2|\Delta|/c}^{\infty} p\rho(p)dp. \quad (2)$$

Expression for $\rho(k) = \rho(p/\hbar)$ can be calculated by taking Fourier transformation from the wave-function of the point-like Ps, which is given by the finite well Ps bubble model (e.g. see [8] for the description of calculation procedure):

$$\rho(k) \sim \left[\frac{1}{k_U^2 - k^2}\left(\sin k_U R_U \cos kR_U - \frac{k_U}{k}\cos k_U R_U \sin kR_U\right) + \frac{\sin k_U R_U}{k^2+\kappa^2}\left(\cos kR_U + \frac{\kappa}{k}\sin kR_U\right)\right]^2. \quad (3)$$

Here $\kappa = -k_U/\tan(k_U R_U)$, $k_U = \frac{\sqrt{2E_{\rm Ps}/Ry}}{a_0}$, $a_0$ is the Bohr radius (=0.53 Å), Ry=13.6 eV, $E_{\rm Ps}$ is the energy of Ps inside the finite well with radius $R_U$ and depth $U$.

Substitution of Eq.(3) into Eq.(2) gives theoretical shape of the Doppler line for p-Ps. For fitting of the experimental data it must be convolved with the energy resolution function.

### *2.1. Background events*

The model of background does not contain the Compton plateau, which is usually produced by annihilation gammas below 511 keV. This effect is significantly reduced with BGO escape-supression shields attached to HP-Ge energy detectors [9]. Though the Compton scattered gammas are observed, they result in negligible difference between low- and high-energy tails of spectrum.

Let us review the reason for appearance of background events in GiPS-AMOC experiment (see Fig. 1). Prompt gammas give birth to the positron. Though energy of bremsstrahlung, which enters the sample chamber, has mainly an order of some MeV, its scattered fraction is able to trigger energy detector by releasing the energy of ~511 keV in it. On the other hand, the positron (and positronium) appeared in the sample produces annihilation gammas, one of which may trigger the time detector. Such a mixture of true (annihilation) and false (caused by scattered bremsstrahlung) events recorded by age and energy detectors correspondingly gives rise to the relief of background events

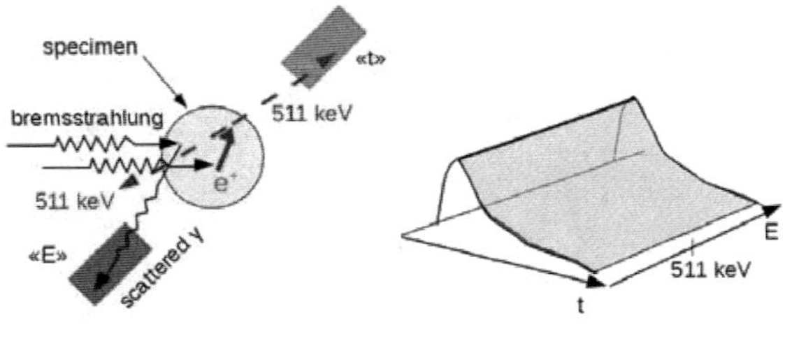


**Figure 1.** Appearance of background events

shown in Fig. 1. Thus the projection of background component on the time axis is similar to the lifetime spectrum of the investigated medium and its energy projection is constant.

We have taken into account an appearance of the "reversed" events, which come from detection true annihilation events (one of the annihilation gammas is detected by the energy detector, while scattered bremsstrahlung triggers the time detector). However, overall contribution of these events to the observed spectrum, turned out to be quite small.

### *2.2. Time and energy resolution function*

The time resolution function is important for the spectrum processing. Usually it is fixed from independent experiments done with the reference samples such as annealed Fe, monocrystalline Si, kapton etc.

In GiPS, parameters of the temporal resolution function may depend on the investigated medium along with the sample dimensions. Distinct spatial distribution of annihilation events, which depends on positron birth rate in a certain medium, leads to variation of the parameters of the resolution function (e.g. its FWHM varies from 180 ps in Fe sample to 215 ps in water). Therefore, dispersion of the resolution function is accepted as ajustable parameter in our model.

The energy resolution function was measured with a help of $^{137}$Cs and $^{60}$Co isotopes. Then it was extrapolated to 511 keV energy region. We fixed its FWHM on 1.52 keV value.

## 3. Experimental data processing

### *3.1. Gauss-exponential decomposition*

The AMOC spectrum of twice distilled water was measured at room temperature. Water was poured to the kapton tube. The oxygen removal was made by bubbling nitrogen through the sample before the measurement. The spectrum contains $\sim 4.5\times 10^6$ events, which required a week of measurement session. Energy range of annihilation radiation acquired to the spectrum is from 501 to 520 keV.

We have used the ROOT framework [11] to fit theoretical models to the experimental data. The fitting strategy was the following.

To compare the data with the previous experimental data [3, 2], we have decomposed it into a set of the simplest gauss-exponential functions $h_i(E,t) = g_i(E) \cdot f_i(t)$, where $g_i \sim \exp(-\frac{(E-511\text{ keV})^2}{2\sigma_i^2})$ describes the energy distributions of the annihilation gammas. Age distribution functions, $f_i$, were taken in the form of exponents: $f_i(t) \sim \exp(-t/\tau_i)$.

Data decomposition was made into three $h_i$ functions (p-Ps, positron and o-Ps states) and the background contribution by means of maximization of the likelihood function. We have obtained that "narrow" energy data interval used for fitting yields smaller dispersion of the p-Ps component (Tab. 1). Anyway the results agree well with the literature data [2, 3].

**Table 1.** Fitting of the water GiPS-AMOC spectrum with a help of the "gauss-exponential" model with the fixed $I_3/I_1 = 3/1$ ratio.

| | Full E range | | | E= 506.5...515.5 keV | | |
|---|---|---|---|---|---|---|
| | Intensity | $\tau$, ns | $\sigma$, keV | Intensity | $\tau$, ns | $\sigma$, keV |
| o-Ps | 0.260(1) | 1.767(3) | 1.053(2) | 0.257(1) | 1.788(3) | 1.034(2) |
| $e^+$ | 0.653 | 0.357(1) | 1.146(1) | 0.657 | 0.368(1) | 1.097(1) |
| p-Ps | 0.087 | 0.187(2) | 0.209(10) | 0.086 | 0.178(2) | 0.346(10) |

The reason of such a behavior may be more complicated shape of the energy background projection (it could be not a flat one, but of parabolic or gaussian shape), which is neglected in

**Table 2.** Parameters of the Ps bubble and the diffusion-recombination model obtained from the experimental data fitting in the narrow energy range (506.5...515.5 keV). Uncertainties were estimated as the differences between adjustable parameters obtained for the narrow and full energy ranges

| Par. | R, Å | $R_{ox}=R_{opc}$, Å | $\sigma_{e+}$, keV | $\sigma_{o-Ps}$, keV | $\tau_{e+}$, ns | $\delta$, Å | $\eta$ |
|---|---|---|---|---|---|---|---|
| Value | 3.4(2) | 5.8(6) | 1.20(1) | 1.04(12) | 0.31(2) | 0.85(1) | 0.74(20) |

the present fitting.

### *3.2. Diffusion-recombination model*
To reduce a number of free parameters at fitting the data with more complicated theoretical model, we have fixed parameters of the time resolution function and the background, which were obtained from the gauss-exponential decomposition. Table 2 summarizes the obtained results.

The set of parameters yielded by the diffusion-recombination model agree well with the previously reported values [4]. The value of $\delta$, also called as "electron penetration depth", is lesser than reported previously from conventional lifetime spectra [4], which means higher value of corresponding potential well depth (U$\sim$6 eV). The relative contact density parameter, $\eta$, also agrees well with the results known from the magnetic quenching experiment [12], though its estimated variability implies that it should be fixed. Both energy and time resolution functions were fixed to that obtained from the gauss-exponential decomposition. In the fitting we also assumed that the oxidation reaction radius $R_{ox}$ equals to the ortho-para conversion reaction radius $R_{opc}$. Within the present model this radius should be considered as an averaged reaction characteristic for Ps bubble state over all oxidizers (OH radical, $H_3O^+$) and ortho-para convertors (OH, H, radical-cation $H_2O^+$, hydrated electron).

## 4. Summary
Annihilation spectrum measured in water by means of GiPS-AMOC setup is consistent with previously reported results. Its convetional (gauss-exponential) decomposition does not show any deviations from the known values.

Our attempt to describe this spectrum with a help of the diffusion-recombination model shows that the model is also consistent with the measured spectrum and parameters of the model are in agreement with previously reported values [4].

## References
[1] Butterling M et al. 2011 *NIM B* **269** p 2623
[2] Mogensen O E 1995 *Positron annihilation in chemistry* (Springer-Verlag)
[3] Castellaz P et al. 2002 *J. Nucl. Radiochem. Sci.* **3 (2)** p R1
[4] Stepanov S V et al. 2011 *Mat. Sci. Forum* **666** p 109
[5] Stepanov S V and Byakov V M 2005 *High Energy Chemistry* **39 (3)** p 131; *High Energy Chemistry* **39 (5)** p 282
[6] Castellaz P et al. 1996 *J.Radioanal.Nucl.Chem.* **210(2)** p 457
[7] Byakov V M and Stepanov S V 2000 *Rad.Phys.Chem.* **58** p 687
[8] Stewart A T, Briscoe C V and Steinbacher J J 1990 *Can. J. Phys* **68** p 1362
[9] Schwengner R et al. 2005 *NIM A* **555 (1-2)** p 211
[10] http://kinetics.nist.gov/solution/
[11] Brun R, Rademakers F 1997 *Proc. AIHENP'96 Workshop* **389** p 81, `http://root.cern.ch`
[12] Bisi A, Consolati G, Gambarini G and Zappa L 1985 *Nuovo Cimento* **5D** p 358

Journal of Physics: Conference Series **443** (2013) 012058 doi:10.1088/1742-6596/443/1/012058

# Observations of the growth and sublimation of water ice films by variable-energy positron annihilation spectroscopy

**S. Townrow[1], P.G. Coleman[1], Y.C. Wu[2], J. Jiang[2] and S.J. Wang[2]**

[1] Department of Physics, University of Bath, Bath BA2 7AY, UK
[2] School of Physics and Technology, Hubei Nuclear Solid Physics Key Laboratory, Wuhan University, Wuhan 430072, P.R.China

E-mail: s.townrow@bath.ac.uk

**Abstract**. The near-surface structure of thin vapor-deposited water ice films during formation and sublimation in vacuum has been probed using in-situ positron annihilation spectroscopy. While amorphous solid water films prepared at 120K are highly dense with low porosity micro- and interconnected meso-pores have been observed in the top few tens of nm of ice films prepared at temperature of 170 - 180K as the films are forming and during sublimation. Responses to these near-surface pores are much more pronounced in films grown at 170 - 180K than in those grown at lower temperatures.

## 1. Introduction

The sublimation of ice is a fundamental phenomenon with relevance to the understanding of many astrophysical phenomena such as the outgassing of volatiles from cometary systems [1]. Sack and Baragiola studied the dynamics of sublimation of water ice grown under different conditions, quoting relevance not only to cometary activity but also to the evolution of the surfaces of particles in Saturnian rings, satellites in the outer solar system and interstellar grains [2]. Recently Gundlach et al. [3] made laboratory measurements of the sublimation of hexagonal water ice and diffusion through a dust layer to allow more realistic modelling of cometary activity.

With the aim of providing new information on the near-surface structure of water ice films during sublimation in vacuum conditions, a study has been carried using in-situ variable-energy positron annihilation spectroscopy (VEPAS). This method has recently been shown to provide unique depth-dependent information on the phase of thin ice films [4] extended to the measurement of the probability of positronium (Ps) formation and decay into three gamma photons, which yields direct information on the formation and collapse of micro- and meso-pores (diameter > 2nm) [4,5].

Positrons of energies $E$ between 0.25 and 30 keV (± 1eV) were implanted into water ice films and the annihilation parameters $S$ and $F$ obtained from gamma ray energy spectra recorded by a Ge detector. $S$ is the fraction of the annihilation line between two pre-set energies symmetrically on either side of 511 keV – this is a measure of the extent of the Doppler broadening of the line, being particularly sensitive to annihilation with low-momentum electrons, and thus to the phase of the ice and/or the presence of vacancy clusters or small closed pores [4]. $S$, which also reflects the formation and decay of para-positronium (p-Ps) because its two-gamma decay contributes a narrow component to the annihilation line, is normalized to the bulk value for the copper substrate. $F$, determined from the ratio of annihilation events in the gamma energy range 475-505 keV to those in the 511 keV annihilation line, is a measure of the ratio of 3γ:2γ annihilations, and thus of the fraction of implanted positrons which are annihilated as triplet, or ortho-positronium (o-Ps), which decays in vacuum overwhelmingly via 3γ emission. $F = 1$ refers to the maximum o-Ps fraction observed in the current measurements, and $F = 0$ to zero o-Ps formation (from data for the copper substrate at $E = 30$ keV). $F$ – here a relative rather than absolute value for the o-Ps fraction - is an excellent measure of the size and concentration of large pores in materials, especially of their interconnectedness [5-7].

## 2. Experimental procedure

Water ice films were grown to thicknesses of ~ 10 μm by bleeding pure water vapor through a needle valve into a chamber evacuated to 1.3 μPa and condensing on to a copper finger cooled by a closed-cycle He refrigerator to pre-set temperatures in the range 50 – 180K. At 50K amorphous solid water (ASW) is formed; at 120K a compact film results, with very few pores; and at 150K or above crystalline films are grown. At the highest temperatures the growth rate has to exceed that for sublimation. For example, at 170K $S$ and $F$ were measured as a function of time at $E = 1$ and 30keV, for different background vapor pressures $P$ (i.e. varying growth rates). For $P = 20$ μPa, no growth was observed after 10h; for $P = 130$ and 1300 μPa, a 10 μm film was grown in ~ 7 and 0.6h, respectively.

Data for individual points were typically collected over 1000 seconds. For temperatures up to 150K the ice is effectively steady state on these timescales, though this is not valid for higher temperatures.

## 3. Results

For measurement temperatures at or above 170K a 'peak' appears in $S(E)$ and an enhanced peak in $F(E)$ at $E$ ~ 1.5keV (equivalent to a mean positron implantation depth of ~80nm in a solid of unit density), as shown in Fig.1. The behavior of $S$ and $F$ at $E$ < 1.5 keV is probably a result of incomplete

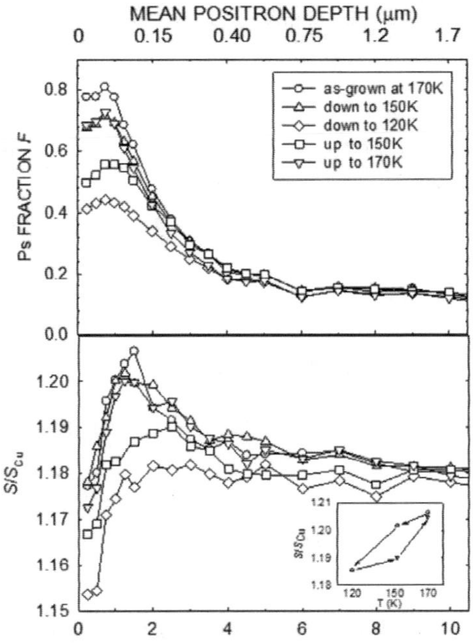


**Figure 1.** The Ps fraction $F(E)$ and $S(E)$ for ice film grown at 170K, measured at 170, 150 and 120K. The inset shows the hysteresis in the peak $S$ value as the temperature is cycled.

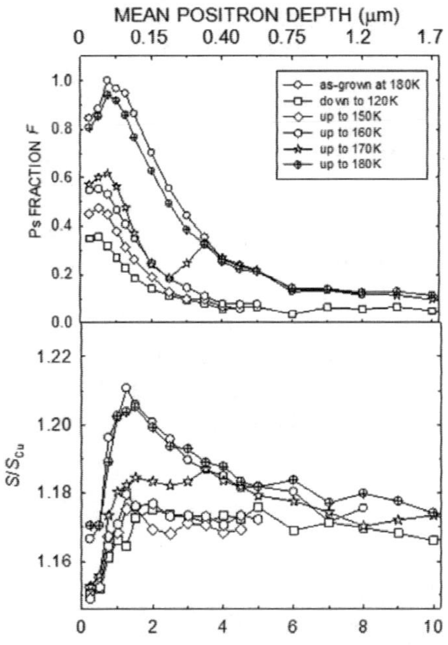


**Figure 2.** $F(E)$ and $S(E)$ for ice film grown at 180K, measured at 180, 170, 160, 150 and 120K. Note the jump from low to high parameter during the second 170K run.

positron thermalization; thus, the increases in *S* and *F* should be considered to be a result of Ps escape into the vacuum from depths of up to a few hundred nm. As a greater proportion of the o-Ps escapes to vacuum and is lost, the increase in *S* results from an increased contribution from p-Ps decay, verified by fitting the annihilation line as a combination of Gaussian curves and observing changes in the intensity of the narrowest (p-Ps) component.

Fig. 1 shows *S* and *F* for *E* up to 10 keV for a film grown at 170K; the parameter values remain effectively constant at higher values of *E* and suggest that the film is largely crystalline [4]. The high values of *S* and *F* at ~1.5 keV suggest the presence of interconnected pores leading to the surface of the ice, which is subliming during the measurement. The reversibility of this effect as the measurement temperature was cycled down to 120K and back to 170K, is shown in the inset in Fig.1, demonstrating hysteresis. Whereas it takes only timescales on the order of tens of minutes for the enhanced surface response to appear in a film grown at 170K, it takes two orders of magnitude longer for the response to disappear when the temperature is dropped to 120K.

Fig. 2 shows data similar to those in Fig.1, but for a film grown at 180K. Here the differences between measurements is even more marked than in Fig.1; and the rapid change can be directly seen after the first eight measurements at 170K (as the temperature is being increased from 120 to 180K).

Fig. 3 shows the much smaller near-surface change associated with a 'compact' film grown at 120K [8] measured at 120 to 170K. No significant effect is seen until the temperature reaches 170K and the film is subliming, when a modest increase in *S* is seen, probably associated with the formation of closed pores near to the surface. Similar results were obtained for films grown at 50 K (amorphous) and 150 K (crystalline).

Fig. 4 shows *S* and *F* measured as a function of time at 170K, both near the surface ($E$ = 1keV) and in the bulk ($E$ = 10keV) of films grown at 50K. *S* increases steadily throughout the film until, after about 580 mins, the film has completely sublimed and *S* drops to the substrate value. The sublimation rate at 170K has been measured to be ~ 12 nm min$^{-1}$. Equivalent measurements of *F* exhibit a large absolute increase near the surface compared to the bulk. Larger changes are observed at 180K, at which the sublimation rate is ~ 80 nm min$^{-1}$.

## 4. Discussion

We propose a model in which ice grown at high (eg 170-180K) temperatures is characterized by

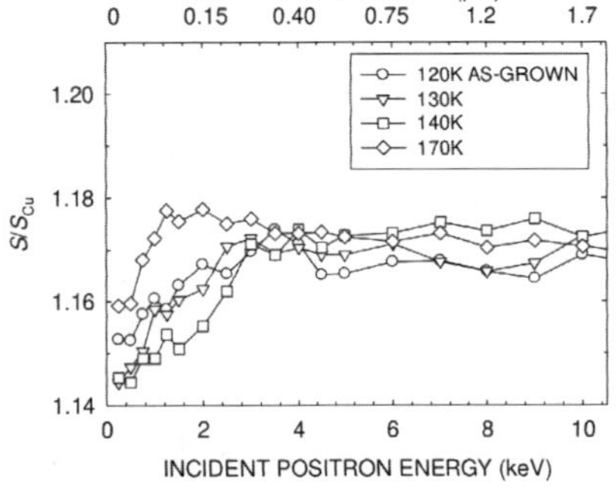


**Figure 3.** *S*(*E*) for ice film grown at 120K, measured at 120, 130, 140 and 170K.

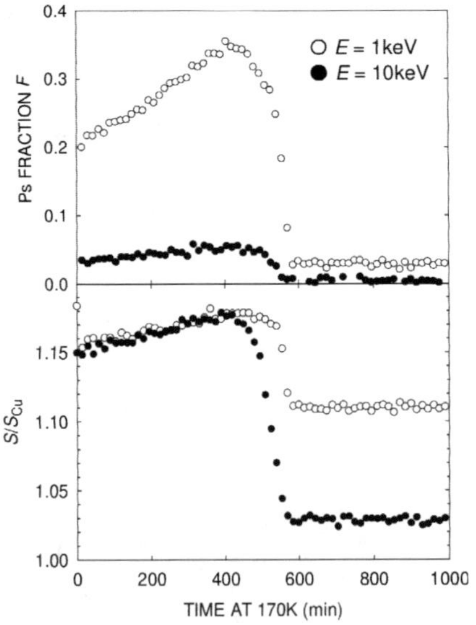


**Figure 4.** *F*(1,10) and *S*(1,10) for ice films grown at 50K, measured at 170 K as a function of time.

many open, interconnected pores via which o-Ps can escape into the vacuum and decay via 3-gamma emission. As the temperature is decreased to 120 K, water vapor freezes on the topmost surface of the film, closing off the open channels and preventing o-Ps escape. The associated hysteresis (Fig. 1) reflects the time required for the 'ice cap' to form (~ 10h at 120K) and sublime on re-heating. Molecular reorganization at the surface at 120K [9] may also help to form a stable morphology in the near-surface region, which may induce pore closure/collapse.

$F(E)$ in Figs. 1 and 2 were fitted using the standard code VEPFIT [10] to yield effective Ps diffusion lengths $L_{Ps}$. For films grown at 170K $L_{Ps}$ increases reversibly from ~100 to ~150nm when the measurement temperature is decreased from 170 to 120K, whereas for films grown at 180K $L_{Ps}$ *falls* from ~ 140 to ~ 50nm at temperatures of 180 and 120K. For fast-grown 170K-grown films the bulk $F$ value is always ~ 14%, but for 180K-grown films drops from ~10 to 4% at 120K. In the bulk of the 170K-grown film closed pores hamper Ps diffusion [see 11] - and their thermal contraction at 120K leads to an increase in $L_{Ps}$ without significantly changing $F$. In the 180K-grown film a larger concentration of inter-connected pores allows Ps to diffuse readily, but on cooling to 120K these shrink and become more isolated, and $L_{Ps}$ falls. Thus the results suggest that the degree of pore interconnectedness in bulk films increases with growth temperature and rate.

The increase of $S$ and $F$ with time at 170K shown in Fig. 4 can be explained in two ways. The first involves the development of a pore structure with time, in which ice films start as 'compact', and transit slowly to porous. This change strongly depends on sublimation rate (which is 7 times greater at 180 than at 170K). The second model is based on crystallization of the whole film; restrained amorphous ice ($I_{ar}$) can coexist with cubic crystalline ice ($I_c$) for at least 16h at 178K [12]. Here crystallization of $I_{ar}$ would take more than 10h at 170K. Both crystallization and pore development could be occurring coincidentally, $S$ being more sensitive to the former and $F$ to the latter.

In summary, the VEPAS results indicate the formation of interconnected pores in ice films grown while subliming which provide channels to the vacuum above the surface through which o-Ps can escape and decay. These escape channels are closed on reducing film temperature, and reappear on re-heating. At higher growth temperatures and rates the relative concentrations of open to closed pores in the bulk films increases. Future work will focus on the energetics of these processes.

## Acknowledgments

The authors are grateful to the China Scholarship Council for providing financial support for YCW during the performance of this research. This work was supported by EPSRC, UK, under grant EP/I016767/1, and by the National Natural Science Foundation of China under Grants No. 10975110, 51071111 and 11175136.

## References

[1] Jenniskens P and Blake DF 1996 *Astrophysical J.* **473** 1104
[2] Sack NJ and Baragiola RA 1993 *Phys. Rev. B* **48** 9973 and references therein
[3] Gundlach B, Skorov YV and Blum J 2011 *Icarus* **213** 710
[4] Wu YC, Kallis A, Jiang J and Coleman PG 2010 *Phys. Rev. Lett.* **105** 066103
[5] Wu YC, Jiang J, Wang SJ, Kallis A and Coleman PG 2011 *Phys. Rev. B* **84** 064123
[6] Gidley DW, Frieze WE, Dull TL, Yee AF, Ryan ET and Ho HM 1999 *Phys. Rev. B* **60** R5157
[7] Weber MH and Lynn KG 2003, in *Principles and Applications of Positron and Positronium Chemistry*, eds. Jean YC, Mallon PE and Schrader DM (World Scientific: Singapore) 167-210
[8] Fillion J-H, Amiaud L, Congiu E, Dulieu F, Momeni A and Lemaire J-L 2009 *Phys. Chem. Chem. Phys.* **11** 4396
[9] Watkins M, Pan D, Wang EG, Michaelides A, VandeVondele J and Slater B 2011 *Nature Mat.* **10** 794
[10] van Veen A, Schut H, de Vries J, Hakvoort RA and IJpma MR 1990 *AIP Conf. Proc.* **218** 171
[11] Eldrup M, Mogensen O and Trumpy G 1972 *J. Chem. Phys.* **57** 495
[12] Jenniskens P and Blake DF 1994 *Science* **265** 753

Journal of Physics: Conference Series **443** (2013) 012059 doi:10.1088/1742-6596/443/1/012059

# PALS and DSC study of nanopores partially filled by hexadecane

**O Šauša[1], E Illeková[1], J Krištiak[†1], D Berek[2], and E Macová[2]**

[1]Institute of Physics, Slovak Academy of Sciences, Dúbravská cesta 9, 845 11 Bratislava, Slovakia

[2]Polymer Institute, Slovak Academy of Sciences, Dúbravská cesta 9, 845 41 Bratislava, Slovakia

E-mail: ondrej.sausa@savba.sk

**Abstract**. The controlled porosity glasses (CPG) filled with various amount of hexadecane (HXD) in nanopores were studied both by the positron annihilation lifetime spectroscopy (PALS) and the differential scanning calorimetry (DSC) methods. Two types of CPG matrices were used with average pore sizes 12.6 and 22.2 nm. The PALS measurements showed, that when the process of large pores filling by HXD has started, the long o-Ps lifetime went down to HXD o-Ps lifetime about 3ns [1]. DSC measurements at partially filled nanopores showed always two crystallization peaks [2]. Their positions depended on average pore size of matrix. Third crystallization peak was identified in overfilled samples (only short o-Ps lifetimes were present) and their position in temperature scale was the same as for the bulk HXD peak. The latter peak was independent of the average pore size of matrices. This fact confirms the assumption that processes studied by PALS with the samples that contained smaller amount of HXD in CPG occured inside of nanopores of the matrix.

## 1. Introduction

In our previous PALS study of confined molecules of hexadecane (also called cetane) in fully filled silica gel pores we found some basic differences between dynamics of bulk and confined hexadecane in wide temperature range from 15 K to 300 K [1]. While the free-volume hole size decreased sharply in bulk HXD at 292 K and then no change for low temperature region has appeared the slow change was observed of the free-volume hole size with decreasing temperature was observed in case of confined HXD. Similar differences were reported in [3]. DSC measurements indicated two crystallization peaks (one for confined state with the position depending on the pore size and one for the bulk state with a stable position in temperature range) [1]. Presence of two crystallization peaks [2] for the confined hexadecane in CPG samples evokes the questions, what induces these different states, where they are localized, etc. The second interesting point is process of successive filling (or release) of materials to (from) nanopores utilized for instance at drug release. The aim of this work was to look for some relationships between quantity of hexadecane in partially filled nanopores of CPG samples with different pore size and positron lifetime or DSC parameters.

## 2. Experimental

Samples of liquid n-hexadecane, with the chemical formula $CH_3(CH_2)_{14}CH_3$, was obtained from

Lachema, Czechia. Melting point of hexadecane is 292±0.1 K [2], molar mass is 226.45 g mol$^{-1}$, density is 0.773 g mL$^{-1}$ and length of molecule is app. 2.2 nm. Controlled porosity glasses CPG-10 with 12.6 nm and CPG-10 with 22.2 nm mean pore diameter, respectively were obtained from BDH Chemicals Ltd., England. Porous glass was filled by a defined amount of HXD at ambient temperature. The actual amount of HXD was controlled by weight with precision of ±0.01 wt.%. The quantity of hexadecane in CPG is characterized by the coefficient of filling k defined as $k=m_{HXD}/(m_{CPG}+m_{HXD})$, where $m_{HXD}$ and $m_{CPG}$ is mass of HXD and CPG in sample, respectively.

The powdered samples were located in two aluminium containers, which were placed on both sides of the Na-22 positron source (Kapton foil, cca 1.7 MBq activity). Positron annihilation lifetime spectra were measured by standard fast-fast coincidence method using plastic scintillators and XP2020 photomultipliers. The time resolution of system was ~320 ps. This was assessed by the defect-free Al sample, employing three Gaussian approximations of resolution function. PATFIT-88/POSITRONFIT [4] code was used for the analysis of spectra. Source correction was taken into account.

PALS measurements were performed in vacuum 2 Pa (continuous pumping) and at temperature 300 K. Time of each measurement was about 24 hours. The escape of hexadecane from the sample was less than 2 wt. %. This was confirmed by the weight of samples before and after measurements.

Differential scanning calorimetry (DSC) is the principal thermo-analytical method which, by measuring the enthalpy evolved or absorbed by the investigated sample, indicates the change of thermodynamic state of sample, namely the onset ($T_{x,s}$) and minimum ($T_{p,s}$) peak temperatures and heat of the solidification ($\Delta H_s$) or of the melting of hexadecane ($T_{x,m}$, $T_{p,m}$, $\Delta H_m$). Power-compensation DSC8500 (Perkin-Elmer) with automatic intracooler (203 K), encapsulated samples of 15 – 50 mg and dynamic argon atmosphere were used. The scanning rate was ± 2 K.min$^{-1}$.

## 3. Results and discussion

The five-component analysis of lifetime spectra was performed for a better differentiation of o-Ps lifetime. Two shorter o-Ps lifetimes were near to value of bulk hexadecane about 3ns [1]. The PALS measurements have indicated that the large pores alter their apparent size above k ~ 0.2 for (Figs. 1 and 2). This corresponds to about 40% and 30% pore saturation for CPG 12.6 and CPG 22.2, respectively - as estimated from the pore volume given by producer and the weight of HXD added.

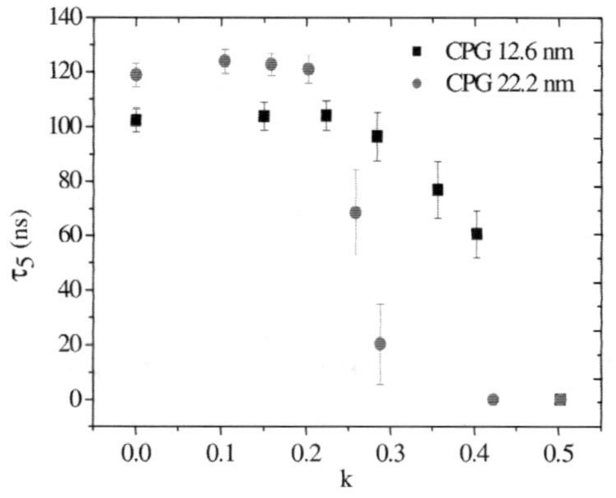


**Figure 1.** Dependence of lifetime $\tau_5$ on the coefficient of filling k.

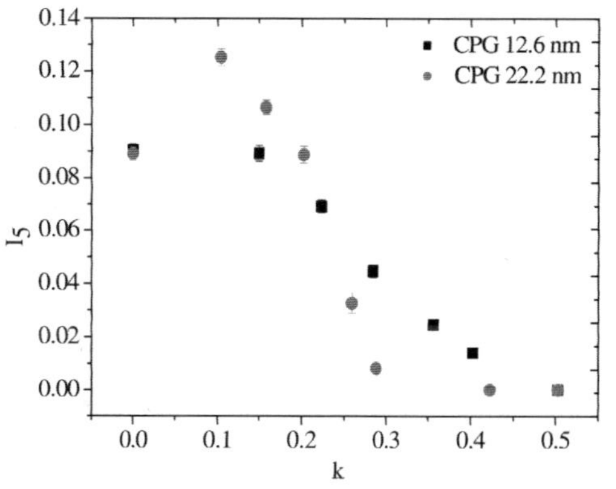


**Figure 2.** Dependence of relative intensity $I_5$ on the coefficient of filling k.

The CPG 22.2 nm is fully filled above k=0.3, while CPG 12.6 nm with greater surface of pores is fully filled at approx. k=0.5. Presence of two shorter o-Ps lifetimes (Fig.3,4) in two distinct matrices (1.4 ns, at 8% and 4.5 ns, at 0.4% for CPG 22.2 nm, as well as 1.7 ns, at 4.9% and 7.4 ns, at 1.5% for CPG 12.6 nm) caused some difficulties in the interpretation of PALS spectra in the ns region. These lifetimes correspond to the free volume of matrix skeleton similar to [5], and are overlapped with the o-Ps lifetime of HXD for filled samples. As evident from Fig. 5, the increase of HXD amount in CPG is accompanied with the growth of intensity of the sum of $I_3$ and $I_4$.

At small k values, the HXD molecules may be situated as continuous or discontinuous thin film on the inner pore walls and they do not affect value of $\tau_5$. Shorter o-Ps lifetimes connected with the free volumes of disordered molecules HXD in the film or free volumes between HXD molecules and CPG walls are larger then lifetime of HXD in full-filled pore. It is in accordance with some ideas of work [5], about the behaviour of o-Ps lifetimes at adsorption and desorption of n-heptane on porous silica gel at pressure measurements.

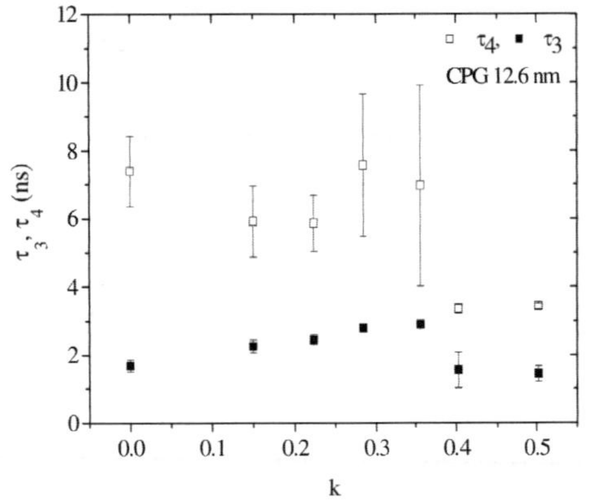


**Figure 3.** Lifetime $\tau_3$ and $\tau_4$ for different coefficient of filling k. CPG 12.6 nm.

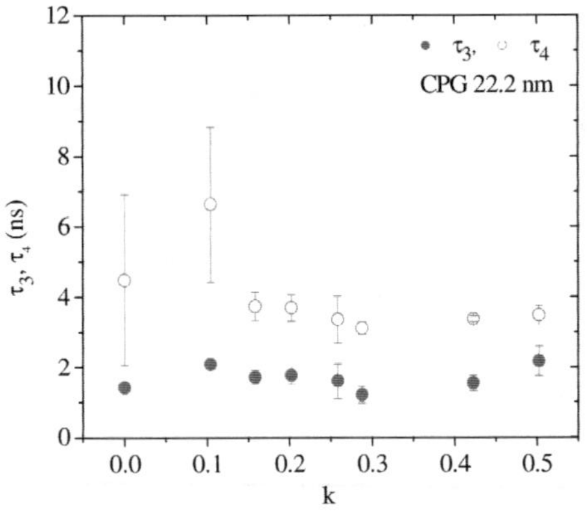


**Figure 4.** Lifetime $\tau_3$ and $\tau_4$ for different coefficient of filling k. CPG 22.2 nm.

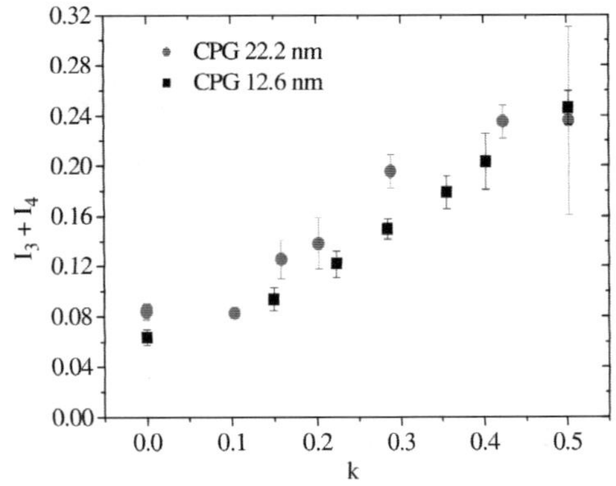


**Figure 5.** Sum of $I_3$ and $I_4$ for both CPG for different k.

Three independent phase transition peaks were observed in each DSC measuring cycle. One crystallization exotherm Rb of the bulk HXD was observed only in the overfilled samples. Its position on the temperature scale is the same as for the bulk sample. Two crystallization sub-peaks Rc1 and Rc2 belong to the confined HXD in the case of continuous cooling (Fig. 6). Temperatures of both Rc1 and Rc2 transitions decrease with the decreasing pore diameter of CPG. Similar situation appeared with other samples exhibiting different pore size [6]. Transition temperature is a function of $1/R_p$ ($R_p$ – radius of matrix pores, Gibbs-Thompson rule). The onset temperature of both transitions Rc1 and Rc2 is independent of k.

Influence of extent of pore-filling k on the total enthalpy of solidification of hexadecane confined in CPG is manifested in Fig. 7. The peak area for bulk hexadecane [2] is larger than areas of all peaks for the HXD confined in CPG. This difference is expressed as $|\Delta H|$ in %. $|\Delta H|$ is calculated as the difference of $\Delta H$ between -219.5 J $g^{-1}$ (measured for the bulk HXD sample) and the sum of all three solidification heats, namely of Rb, Rc1 and Rc2 for the confined systems. This apparently missing HXD might indicate that not entire HXD present in pores takes part in the crystallization process. This deficit decreases with increasing k.

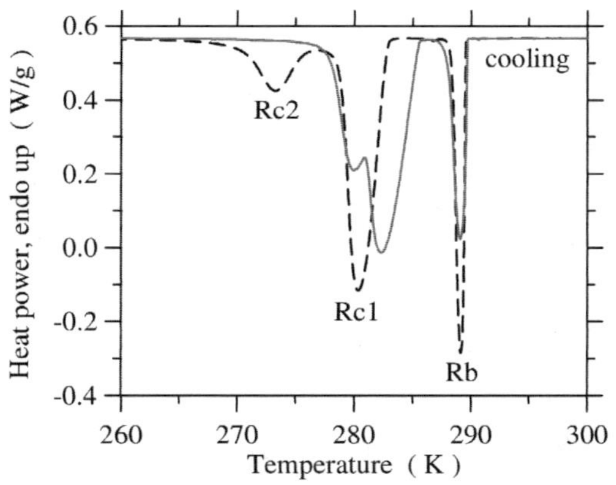


**Figure 6.** DSC cooling curves for hexadecane confined in CPG (diameter of pores 12.6 nm - - - and 22.2 nm ___ ). The scanning rates were ± 2 K min$^{-1}$.

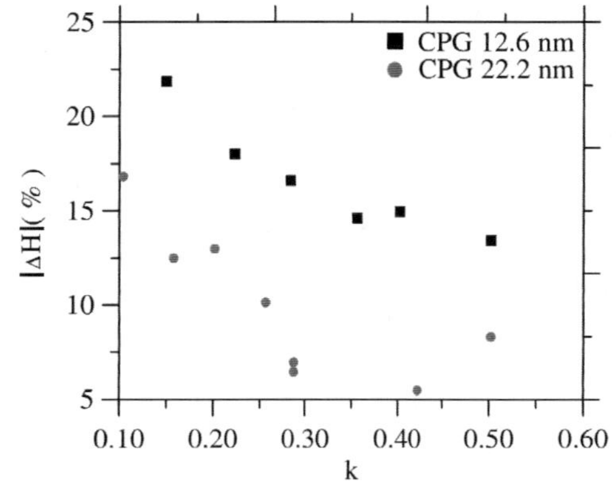


**Figure 7.** Difference |ΔH| between bulk hexadecane and total enthalpy of all DSC crystallization peaks in CPG sample for different k.

## 4. Conclusions

The PALS measurements showed that hexadecane filled up pores. The process of the filling of the large pores has started approx. at k ~ 0.25 and 0.30 for CPG 22.2 nm and 12.6 nm, respectively. Long o-Ps lifetime decreases at large filling coefficient k to the value 3.4 ns, close to the typical HXD o-Ps lifetime about 3 ns. In comparison with bulk HXD, the arrangement of HXD molecules in fully filled pores is partially disordered. This is similar to silica gels [1]. At small k values, the HXD molecules may be situated as thin film on the inner pore walls. They do not affect value of o-Ps lifetime $\tau_5$. Due to strong capillary forces, their location on the outer surface of CPG particles is improbable.

Two crystallization sub-peaks Rc1 and Rc2 are observed by the DSC measurements in the process of cooling for hexadecane confined in CPG. Their onset temperature depends on 1/Rp and is independent of k. The absolute value of the enthalpy of both peaks increases with the increasing filling coefficient k. For confined hexadecane, the absolute value of the total solidification enthalpy is always lower than the solidification enthalpy for bulk hexadecane and this difference decreases with increase of k.

## Acknowledgements

The authors wish to thank the Agency of the Ministry of Education of the Slovak Republic for the Structural Funds of the EU (CEKOMAT II, ITMS 26240120020, 80%), and for support to the grant agencies VEGA Slovakia, projects no. 2/0099/10, 2/0111/11, and 2/0001/12, as well as APVV, projects 0647-10, 0109-10, and 0492-11.

## References

[1] Iskrová M, Majerník V, Illeková E, Šauša O, Berek D and Krištiak J 2009 *Materials Science Forum* **607** 235
[2] Illeková E, Miškovičová M, Šauša O and Berek D 2012 *J. Therm. Anal. Calorim.* **108** 497
[3] Kilburn D, Sokol P E, Sakai V G and Alam M A 2008 *Appl.Phys. Lett.* **92** 033109
[4] Kirkegaard P, Pedersen N J and Eldrup M 1989 *PATFIT-88*, Riso Report – M2740
[5] Zaleski R, Dolecki W, Kierys A, Goworek J 2012 *Microporous Mesoporous Mater* **154** 142
[6] Illeková E, Krištiak J, Macová E, Maťko I and Šauša O 2012 *J. Therm. Anal.* submitted

16th International Conference on Positron Annihilation (ICPA-16) IOP Publishing
Journal of Physics: Conference Series **443** (2013) 012060 doi:10.1088/1742-6596/443/1/012060

# Positronium bubble formation in room temperature ionic liquids

**T. Hirade**[1]
*Nuclear Science and Engineering Directorate, Japan Atomic Energy Agency,*
Tokai, 319-1195 Japan
*Graduate School of Science and Engineering, Ibaraki University,*
4-12-1 Narusawa, Hitachi, Ibaraki, 316-8511 Japan

E-mail:hirade.tetsuya@jaea.go.jp

**T. Oka**
*Research Institute for Science and Engineering, Waseda University*
3-4-1, Shinjuku, Tokyo, 169-8555 Japan

E-mail: kotobuki@kurenai.waseda.jp

**Abstract.** Positron annihilation age-momentum correlation (AMOC) measurements were performed for a room temperature ionic liquid (IL) to investigate positronium (Ps) bubble formation process. The Ps just after the formation must be squeezed Ps that has larger probability of annihilation with core electrons in liquids. However, it has been believed that the bubble formation in liquids is very fast and difficult to observe the squeezed Ps experimentally. When the bubble formation is slow, it can give broader energy distribution of annihilation gamma-rays at young age region. There have been several experimental results showing young age broadening and some of them are interpreted as the delayed Ps formation. However, if the bubble formation is slow, it also can be a reason of the young age broadening. If the squeezed Ps and the free positron give different momentum distributions by annihilation with core electrons on different atoms, it is possible to detect the annihilation from squeezed Ps at the young age, i.e. the bubble formation can be observed. We found larger high momentum distribution at young age in IL and it might be caused by the delayed bubble formation in IL. Positron annihilation methods can be a tool to investigate the IL molecular dynamics at the time range of pico-nano second and in the scale of sub-nano to nano meter.

## 1. Introduction

Positron can form positronium (Ps, a bond state of an electron and a positron) with an electron. Ps formation in insulating materials, such as liquids or solids, is well explained by the spur reaction model.[1] Injected positrons thermalize at the terminal spur where many cations and excess electrons formed. Positrons have a chance of Ps formation with one of excess electrons before electron and positron localization, usually. There are two different Ps states because of spins of the electron and the positron. Para-Ps has anti-parallel spins and ortho-Ps has parallel spins. The mass of the electron and

[1] To whom any correspondence should be addressed.

Journal of Physics: Conference Series **443** (2013) 012060 doi:10.1088/1742-6596/443/1/012060

the positron disappears and the energy appears as gamma rays instead. The allowed number of annihilation gamma rays is even for para-Ps and odd for ortho-Ps. One gamma ray emitting annihilation is not allowed because of the low of conservation of momentum. Therefore para-Ps gives the fastest intrinsic-annihilation by emitting two gamma rays with lifetime of 125ps, and the ortho-Ps intrinsic-annihilation by emitting three gamma rays has the lifetime of 142ns in vacuum. In condensed matter, the positron in the Ps can overlap with electrons on surrounding molecules and pair annihilation with those electrons is also possible. This annihilation process is called □pick-off annihilation□ Almost all of pick-off annihilation gives two gamma rays. The ortho-Ps lifetime in condensed matter is about 1 ~ 10 ns mainly by the pick-off annihilation. Some of the positrons do not form Ps and annihilate with electrons on the surrounding molecules. The lifetime of these positrons is about 400ps in condensed matter.

Room temperature ionic liquids are liquids consisting of ions. Negligible vapour pressure and other interesting properties are considered to be caused by the ionic nature. There are many considerable applications of ionic liquids. One of the reasons is their environmentally friendly property.

It is reported that thermalization distance of electrons that has sub-excitation energy is large and that electrons before full solvation in ionic liquids can diffuse [2]. Ps formation is a good tool to investigate the electrons before full solvation. If these electrons can diffuse for long time, Ps formation is quite possible even at the later positron age such as 10~100ps and hence the Positron annihilation age-momentum correlation (AMOC) experimental results measured for IL was explained by the delayed Ps formation [3].

There were just the arguments for the change of the shape of the annihilation line measured by Ge semiconductor detector by use of time dependent *S*-parameter in [3]. Larger *S* means the sharper line shape. There is the other parameter, *W*-parameter, i.e. wing parameter. It gives the information of annihilation with core electrons. Hence for the change between two positron annihilation processes, the sets of *S* and *W* appear on one straight line on the *S*-*W* plane.

Here, we are discussing the change of positron annihilation processes by plotting on the *S*-*W* plane and discussing the possibility of the delayed Ps bubble formation.

## 2. Experimental

N,N,N-trimethyl-N-propylammonium bis (trifluoromethanesulfonyl)imide (TMPA-TFSI) (Kanto Chemical Co.) was used without any purification in this study. It has the viscosity of 72mPas and the melting temperature of 19C. The oxygen gas in the sample was removed by nitrogen gas bubbling for about 30 min.

AMOC was performed by use of a triple coincidence method by use of a digital oscilloscope (Wavepro7100A, LeCroy) [4], scintillation detectors, i.e. truncated-cone-shaped $BaF_2$ scintillators and Hamamatsu H3378 photomultiplier tubes, and LO-AX (SEIKO EG&G) type semiconductor detector. Positron annihilation lifetime (PAL) was performed by the same scintillation detectors for AMOC by just removing the semiconductor detector. $^{22}Na$ was used as a positron source. About 50kBq $^{22}Na$ was sealed with two kapton films whose thickness is about 1.1mg/cm$^2$. The time resolution of PAL and AMOC was about 250ps (fwhm). The counting rate was about 1cps for AMOC. The total count for every spectrum was about 1 million. The PAL spectra analysis was performed by the PALSfit programme. [5]

Every *S* and *W* parameter was obtained by fraction of the counts appeared at 0~4x10$^{-3}$$mc$ and 15~30 x10$^{-3}$$mc$ from the centre to the total counts for the annihilation peak. Here, $m$ is the mass of electron, $c$ is the velocity of the light.

## 3. Results and Discussion

*S* and *W* parameters at the young positron ages from about -150ps in fused quartz and water were indicated in Fig.1. There are straight lines indicated by arrows up to about 1ns. The time interval between plots is 150ps. The straight lines indicated by arrows show the change from the p-Ps intrinsic annihilation to the free positron annihilation. In the case of the intrinsic-annihilation of para-Ps, momentums of the electron and the positron are cancelled out with each other. Hence, only the

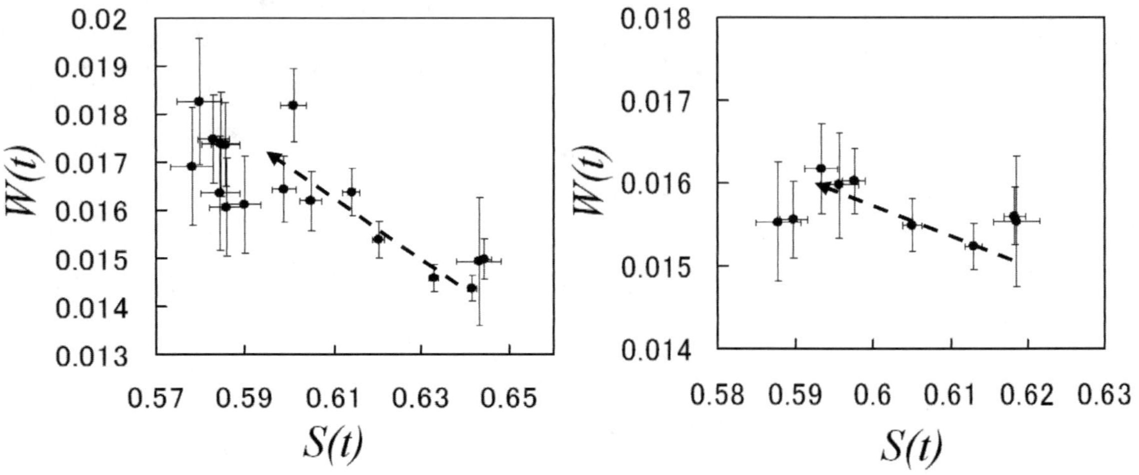


**Figure 1.** The change of *S* and *W* at young positron age in fused quartz (left) and water (right). The time interval between points were 150ps.

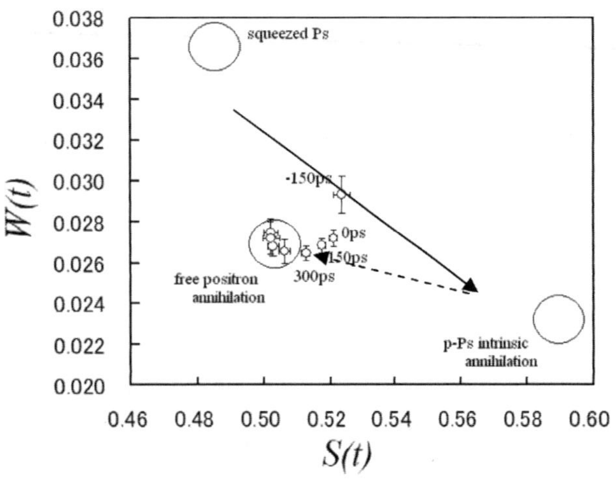


**Figure 2.** Positron age dependence of the *S* and *W* in TMPA-TFSI from -150ps to 900ps. Large circles of *S*-*W* area for squeezed Ps, p-Ps intrinsic annihilation and free positron annihilation are indicated for easy understanding. Arrows indicate the possible change between the states. The arrow by solid line indicates the bubble formation that occurs in very short time and hence only the arrow by dashed line is observable for usual liquids. The change of the measured values suggests that the changes indicated by the arrows occur simultaneously.

annihilation gamma-rays from para-Ps have a narrow energy distribution, i.e. large *S* and small *W*. When you observe positron age dependent shape of annihilation gamma-rays, you can find narrower distribution of annihilation gamma-rays at younger age region where para-Ps intrinsic-annihilation is favourable.

*S* and *W* parameters at the positron ages from -150ps to 900ps in IL, TMPA-TFSI, were indicated in Fig.2. For TMPA-TFSI, smaller *S*, i.e. the broader distribution was found at young age region, and interpreted as the delayed Ps formation [3]. In the case of the delayed Ps formation, it is expected to have longer lifetime of free excess electrons, i.e. they can diffuse longer time. Moreover positrons also need to diffuse longer time. It means that there is more fraction of the annihilation from free positrons and probably freely mobile positrons at young age region. Therefore the annihilation at young age should give *S* and *W* on or close to the arrow in dashed line in Fig.2. Therefore, the larger *W* at young age in Fig.2 is difficult to explain with the delayed Ps formation.

The possible explanation for the *W* at young age is, probably, delayed Ps bubble formation. If the bubble formation is slow, Ps is squeezed just after Ps formation. If the electron and the positron in squeezed Ps prefer to be on the ions like in $CS_2$ at low temperatures [6], *W* should be similar to the *W* at the free positron state. However, *W* is larger and hence the positron and the electron do not separated even though Ps is squeezed and Ps is distributed on surrounding molecules or ions, i.e. it is probably similar to the delocalized Ps in ionic crystals. The negatively charged nitrogen atom in the TFSI anion attract positrons, which might be a suitable site for the free positron. TFSI anion contains fluorine, oxygen, surfer atoms and hence the squeezed Ps can give larger *W* parameter than that for the free positron state. Therefore the change on the *S*-*W* plane during the bubble formation might be like the arrow in solid line as indicated in Fig.2. Indeed, the real p-Ps intrinsic annihilation can probably have larger *S* and smaller *W*. The squeezed Ps also can have larger *W* and smaller *S*. The circles are just indicated for easy understanding.

The annihilation rate of the squeezed Ps is smaller than that of the p-Ps intrinsic annihilation, therefore the shoulder like shape on the lifetime spectra is expected and it can not be fitted by the exponentially decaying components. If the *S* and *W* change in Fig.2 indicates the delayed bubble formation, it might be possible to see the effect on the lifetime spectra. Indeed, the shortest lifetime is much longer than the one obtained in usual materials [3]. As mentioned above, it is impossible to fit the lifetime spectra of TMPA-TFSI with the exponentially decaying components and the difference between measured curves and fitted curves can make the shoulder visible. The differences for 47.5C and 22.5C are plotted in Fig.3. The shoulders on the lifetime spectra can be seen as the peaks indicated in Fig.3. The lifetime spectra of IL are very difficult to fit and then three fixed lifetimes, 0.2959ns, 0.4000ns, 1.0000ns and one free lifetime for the longest lifetime component were used. Intensities were free. The peaks indicate just the tendencies but it is larger and shifts to the older age at the lower temperature, 22.5C. Probably it indicates that the bubble formation is slower at lower temperatures.

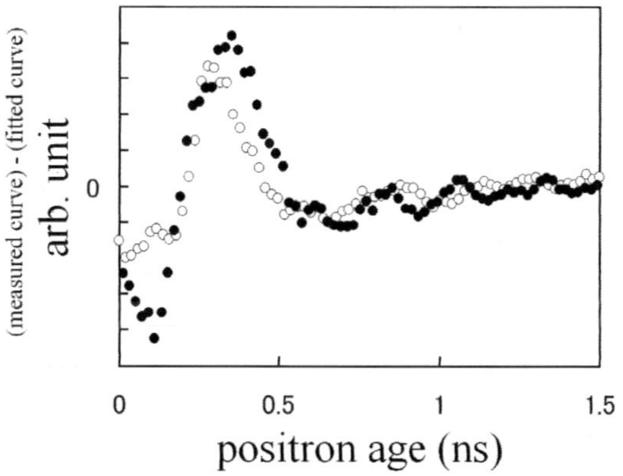


**Figure 3.** Differences between measured curves and fitted curves by use of fitting programme with exponentially decaying components, i.e. PALsfit[5] for TMPA-TFSI. Open circles were obtained at 47.5C and closed circles were at 22.5C. The values were standardised by the total count of every spectrum.

## 4. Conclusion

Large *W* was found at young age in IL, TMPA-TFSI, by the AMOC measurement and probably the delayed Ps bubble formation is only the possible explanation now. The shoulder like shape on the lifetime spectra was expected and it was confirmed by the peaks obtained by subtracting the fitted curve by PALSfit[5] from the measured curve, as indicated in Fig.3. The peaks at 22.5C and 47.5C indicate that the bubble formation is slower at lower temperatures.

## 5. Acknowledgement

This research was partially supported by a Ministry of Education, Culture, Sports, Science and Technology Grant-in-Aid for Scientific Research (C), 23600011, 2011-2013.

## 6. References

[1] Mogensen O E 1974 *J. Chem. Phys.* **60** (3) 998
[2] Katoh R, Yoshida Y, Katsumura Y and Takahashi K 2007 *J. Phys. Chem. B* **111** (18) 4770
[3] Hirade T 2009 *Materials Science Forum* **607** 232
[4] Inoue K, Nagai Y, Tang Z, Toyama T, Hosoda Y, Tsuto A and Hasegawa M 2011 *Physical Review B* **83** 115459
[5] Kirkegaard P, Olsen J V, Eldrup M and Pedersen N J 2009 PALSfit Risø-R-1652(EN)
[6] Hirade T 1996 *Bull. Chem. Soc. Jpn.* **69** 2153

Journal of Physics: Conference Series **443** (2013) 012061 doi:10.1088/1742-6596/443/1/012061

# Study of Positronium formation in nano-channelled silicon as a function of sample temperature

S Mariazzi[1,3], L DiNoto[1], L Ravelli[2], W Egger[2], R S Brusa[1]
[1]Dipartimento di Fisica, Università di Trento and INFN, Gruppo collegato di Trento, Via Sommarive 14, I-38123 Povo, Trento, Italy
[2]Institut für Angewandte Physik und Messtechnik, Universität der Bunderswehr München, 85577 Neubiberg, Germany

E-mail: mariazzi@science.unitn.it

**Abstract.** Oxidized nanochannel in silicon have been demonstrated to be suitable for positronium (Ps) formation and cooling also at low sample temperature. To investigate the Ps yield and to clarify the Ps formation mechanism we studied, by Positron Annihilation Spectroscopy (PAS), nanochanneled Si p-type samples in the 150-430 K temperature range. Ps yield was found to be constant in the 150-300 K temperature range, then it increases up to ~50% of its value from 350-400 K. This effect is associated to a decrease of the fraction of positrons annihilating in Si and in the $SiO_2$ layer on the nanochannels surface. This finding is compatible with the thermal decrease of the positive charge distribution at the $Si/SiO_2$ interface limiting $e^+$ reaching the $SiO_2$ layer and to a charge rearrangement at the $SiO_2$ surfaces.

## 1. Introduction

Production of large amounts of Positronium (Ps) is required for many advanced experiments in fundamental physics. Confinement of ~$10^{17}$ Ps/cm$^3$ in nanometric cavities at temperature T<30 K would allow one to achieve the Ps Bose Einstein Condensation (Ps-BEC) [1]. Ps-BEC would open the route for linked experiments as making a plane wave monoenergetic Ps beam and a Ps annihilation $\gamma$ ray laser [2]. The production of Ps beams at T < 160 K are required for high precision spectroscopy studies [3] and anti-hydrogen formation by charge exchange process with antiprotons [4, 5].
In recent years, porous [6] and channeled [7] insulating materials have attracted much attention concerning Ps production. These kind of materials are promising for two reasons: i) they efficiently produce Ps also at low sample temperatures and ii) provide a cooling mechanism [8–10]. Ps formed in the bulk is emitted with an energy of the order of few eV [11] into pores, then it cools down reducing its kinetic energy by collisions with the walls of the pores reaching the sample temperature [8]. Indeed when a network of connected open pores is present, Ps can escape into the vacuum after a hopping diffusion. For a review about Ps formation and cooling mechanism see Ref. [12].
Ps formation, cooling and emission into the vacuum from ordered nanochannels etched into a Si substrate and subsequently oxidized have been recently investigated [7, 8]. It was found that up to

[3] To whom any correspondence should be addressed.

27% of implanted $e^+$ at 7 keV escapes into vacuum as Ps and a fraction of this Ps is emitted with thermal energy in the 150-300 K sample temperature range [8]. The high Ps amount was inferred to be due to the high probability of thermalized $e^+$ in Si to diffuse to the Si/$SiO_2$ interface due to the small distance among nanochannels (few nanometers) with respect to the diffusion length (more than 200 nm). The $e^+$ then pass into the $SiO_2$ because this is energetically favored [13]. Here Ps can be formed and emitted into the nanochannels.
In this work we present a study of Ps formation mechanism in nano-channelled Si by investigating the Ps yield in the 150-430 K temperature range by using 2γ-3γ annihilation ratio of Ps (3γ-PAS) and Positron Annihilation Lifetime Spectroscopy (PALS).

## 2. Experiment

Two Si samples with oxidized nanochannels were synthesized by electrochemical etching according to the procedure described in Ref. 7 and 8. p-type Si (100) with resistivity 0.15–0.21 Ω cm was etched in an aqueous solution at 48 % of HF with added ethanol in volume ratio of 1:3=HF:ethanol. The etching current and time were I=10 mA and t=13 min, respectively. The samples were subsequently oxidized in air at 100°C for 2h and then subjected to 2 cycles of re-etching in HF for 60 s with a successive 2h annealing in air at 100°C in order to tune the nanochannels size between 10 and 18 nm [7]. One sample was capped with ~30 nm of $TiO_2$ deposited by RF reactive sputtering (13.56 MHz) from a pure $TiO_2$ target, at a pressure of 6.6 Pa. The sputtering gas was Ar (30 sccm). The sample was mounted on a rotating support and not heated during deposition. The cathode was maintained at -850 V dc self-bias, corresponding to an effective load power of approximately 84 W.
o-Ps 3γ annihilations were measured in the capped and un-capped samples by 3γ-PAS with the Trento variable slow positron beam [14]. As a first approximation, the mean $e^+$ implantation depth $z$ can be related to the $e^+$ implantation energy $E$ through the equation $z = 40/\rho E^{1.6}$ with $z$ in nanometers when the material density $\rho$ and $E$ are expressed in g/cm$^3$ and keV, respectively [15]. The $\rho$ of the present samples was estimated to be 1.9 g/cm$^3$ [7]. The annihilation γ rays were detected by two high purity germanium detectors (HpGe) with 45% efficiency and 1.4 keV energy resolution at 511 keV [16]. The distribution of γ-ray energy ($E_\gamma$) was subdivided in the 511 keV peak area ($P$), attributable prevalently to 2γ annihilation ($|511 - E_\gamma| \le 4.25$ keV) and in the valley area ($V$), due to o-Ps 3γ annihilation ($410 \le E_\gamma \le 500$ keV). The 3-2γ ratio, $R(E)=V(E)/P(E)$ parameter was calibrated by measuring the Ps formation in a Ge crystal as a function of T [17, 18]. The fraction of $e^+$ annihilating as o-Ps is defined as $F_{3\gamma}(E) = \frac{3}{4}\left[1 + \frac{P_1(R_1 - R(E))}{P_0(R(E) - R_0)}\right]^{-1}$ where $R_1$, $R_0$, $P_0$, $P_1$ are the $R$ and $P$ values corresponding to 100% and 0% o-Ps formation respectively [19].
Information about the $e^+$ and Ps pick-off annihilations were obtained by PALS measurements carried out on the capped sample. PALS was done with apparatus PLEPS (Pulsed Low Energy Positron Beam) [20] at NEPOMUC (Neutron induced POsitron source MUniCh) [21]. The detector plus pulsing system time resolution was 330 ps. Both 3γ-PAS and PALS measurements were carried out ~~by~~ while varying T from 150 to 430 K.

## 3. Results and Discussion

The $F_{3\gamma}(E)$ o-Ps fraction was measured at $E$= 2, 3.3, 5 and 7 keV in the un-capped sample and at 3.3, 5 and 8 keV in the capped one (Fig.1). $E$ was chosen to be higher in the capped sample in order to implant $e^+$ deeper than the $TiO_2$ layer (ρ ~ 4 g/cm$^3$) of ~30 nm, corresponding to $E$~2 keV. In both samples, $F_{3\gamma}$ is roughly constant between 150 and 300 K at each $E$. Then it increases reaching a maximum value at T=350-400 K. At higher T, $F_{3\gamma}$ remains approximately constant at the reached value. In the un-capped sample, the decrease of the absolute $F_{3\gamma}$ values, at a given T increasing $E$, is mainly due to o-Ps pick-off annihilations. In nanochannels the o-Ps diffusion length is very large, of the order of 1400 nm [7], thus o-Ps formed by $e^+$ implanted with $E$=2 keV have a high probability to reach the vacuum and to annihilate via 3γ. Instead, o-Ps formed deeper in the sample undergoes to many interactions with the nanochannels inner surface and the pick-off probability increases. In the capped sample, as o-Ps formed at each depth is expected to overtake roughly the same number of interactions with the channel surface, the pick-off probability should be the same. The observed $F_{3\gamma}$ decrease from $E$= 3.3 keV to $E$= 8 keV is due to a lower o-Ps formation at high $E$ as an increasing fraction of $e^+$ annihilate in the bulk silicon underlying the nanochannels.

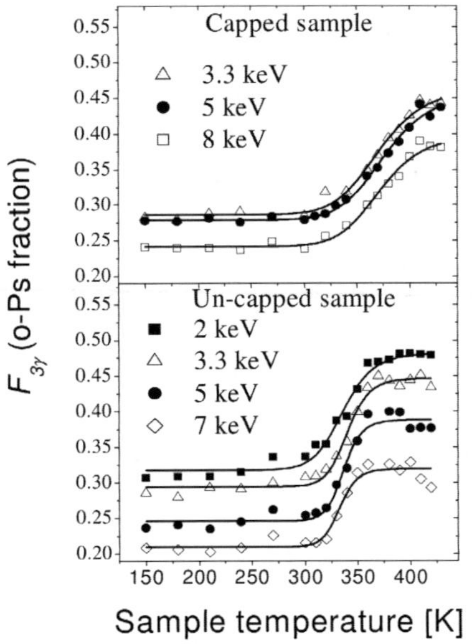


**Figure 1.** $F_{3\gamma}$ (o-Ps fraction) vs. sample temperature measured at different positron implantation energy in capped (upper panel) and un-capped sample (lower panel). The continuous lines are best fit to the data (see text)

Both in capped and un-capped samples, the maximum values of $F_{3\gamma}$ at T > 350 K are ~50% higher than the values measured in the same sample and at the same $E$ at T < 300 K. This increase is sensitively higher than the increment due to the thermal activated detrapping of o-Ps from the silica surface that is expected to be less than 10% in the investigated temperature range [22]. Moreover, fitting the data of the capped and un-capped samples with the equation of thermal activated detrapping $F_{3\gamma}(T)=A\left(1+Be^{-\frac{E_b}{k_bT}}\right)/\left(1+Ce^{-\frac{E_b}{k_bT}}\right)$(where $k_b$ is the Boltzmann constant and the parameters $A$, $B$ and $C$ are defined as $F_{3\gamma}(T=0)=A$ and $F_{3\gamma}(T=\infty)=\frac{A(1+B)}{1+C}$) an activation energy $E_b$= 0.7 ± 0.3 eV, up to eight times higher than the $E_b$ values of 0.11-0.16 eV estimated on a single crystal $SiO_2$ surface [23] and in porous silica [22], is found. These findings point out that the observed $F_{3\gamma}$ increase can not be entirely attributed to thermal Ps detrapping.

To understand the origin of the observed Ps increase, PALS measurements were carried out on the capped sample. In Fig.2 the three achieved lifetimes are reported as a function of *E*. Beyond the $TiO_2$ capping layer, $E > 3$ keV, the lifetimes are constant: $\tau_1$, due to an average of *p*-Ps self-annihilation and $e^+$ free annihilations in Si, results ≅ 200 ps [24], $\tau_2$, coming from the annihilation of $e^+$ trapped in the bulk silica layer at the nanochannels surface, is ≅ 420 ps [25] and $\tau_3$, due to o-Ps pick-off annihilation, is ≅ 48 ns. At $E < 3$ keV $\tau_3$ is due to Ps annihilating in open volume in the $TiO_2$ capping layer while $\tau_1$, that reaches ≅ 250 ps, is compatible with $e^+$ lifetime in amorphous $TiO_2$ [26].

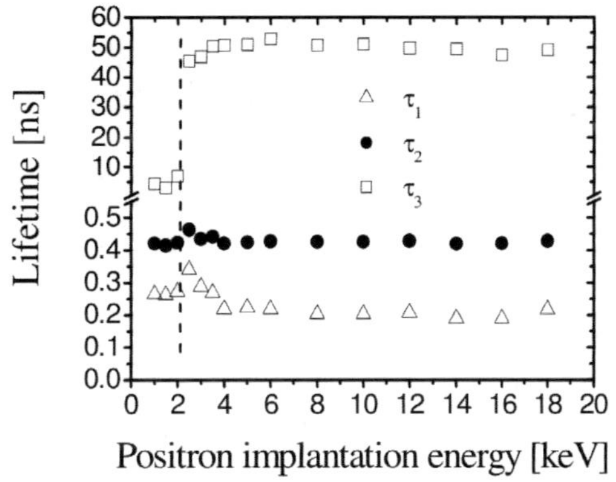


**Figure 2.** Positron lifetimes vs. positron implantation energy in the capped sample. Errors are inside the symbols. The vertical dashed line marks the border of the $TiO_2$ capping layer.

The lifetimes in the nanochannels measured as a function of T at *E*=5 keV (Fig.3), result approximately constant in the considered temperature range. Only a very small decrease of $\tau_3$ from ≅ 48 ns to ≅ 46 ns is observed at T > 350 K.

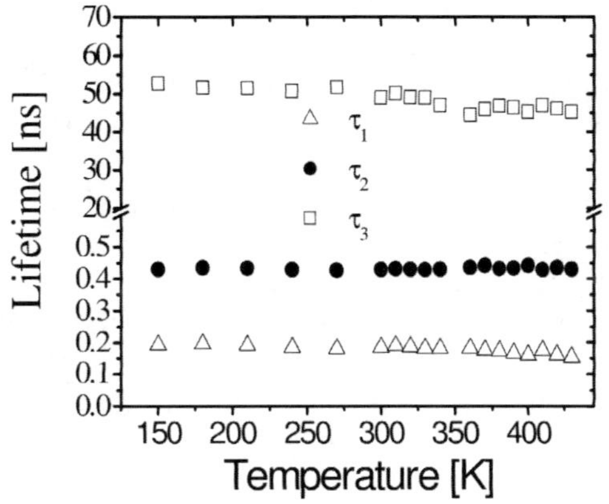


**Figure 3.** Capped sample: lifetimes vs. sample temperature at *E* = 5 keV Errors are inside the symbols.

The intensities $I_i$, corresponding to lifetimes of Fig.3, have been normalized to take into account the o-Ps fraction annihilating via 3γ and not detected by PALS. The normalized intensities to the total number of implanted $e^+$, $I_i^* = I_i\,[1 - F_{3\gamma}(E = 5keV)]$ with *i* from 1 to 3, are reported in Fig.4.

$I_1^* \cong 20\%$, $I_2^* \cong 40\%$ and $I_3^* \cong 15\%$ as the total fraction $F_T = F_{3\gamma}(E = 5keV) * 100 + I_3^*$ of formed o-Ps are constant below T = 300 K (see Fig. 4). Above 300 K it is observed : i) an increase of $F_T$ mainly due to the increase of $F_{3\gamma}$, ii) a strong decrease of $I_2^*$ pointing out the decrease of $e^+$ annihilation in $SiO_2$, iii) a small decrease of $I_1^*$ and $I_3^*$ , the last due to the decrease in o-Ps pick-off. $F_T$ was fitted by using the equation of the thermal activated detrapping (see above) in order to find a more accurate value of $E_b$ not altered by pick-off (Fig.4). The $E_b$ pointed out by this fitting procedure is 0.64 ± 0.15 eV.

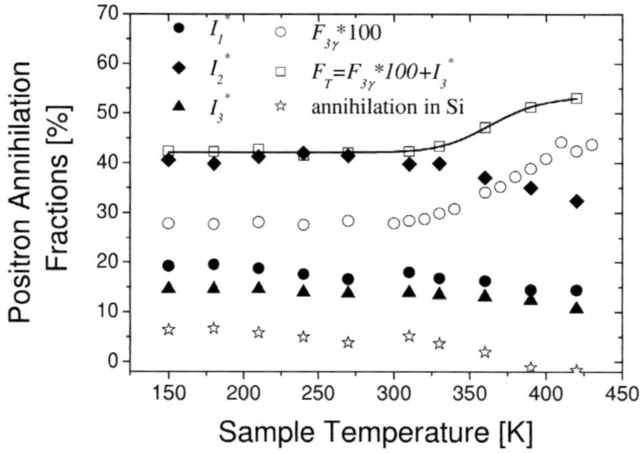


**Figure 4.** $e^+$ and Ps annihilation fractions vs. sample temperature. The fractions have been estimated according to the procedure described in the text for the capped sample at $E$ = 5 keV. The continuous line is the best fit of the total o-Ps fraction formation (see text).

The fraction of $e^+$ annihilating in Si can be estimated as $I_1^* - 1/3 * F_T$ where $1/3 * F_T$ is the p-Ps fraction taking into account the p-Ps:o-Ps ratio 1:3. As appreciable by Fig. 4, this fraction is around 5-7 % at T < 300 K and it approaches to zero at higher T.

In light of the data presented in Fig. 4, we can conclude that the observed increase of Ps is related to: (a) a decrease of $e^+$ annihilating in $SiO_2$ and (b) an increment of $e^+$ that reach $SiO_2$ layer coming from Si. Both this effects could be ascribed to a modification of the charged defects at the Si/$SiO_2$ interface and in the $SiO_2$ layer where it changes the electronic structure at the walls of the nanochannnels. At room temperature the p-type Si /$SiO_2$ interface is positively charged due to donor's levels above the Fermi level and it limits $e^+$ reaching the surface [27]. The effect of the temperature appears to be that of lowering the positive charge promoting more positrons reaching $SiO_2$. Positrons in the thin $SiO_2$ layer and/or at the $SiO_2$ surface must be also favored to form Ps. This could occur with the formation of thermally generated negatively charged dangling bonds [28].

## 4. Conclusion

In the present paper, we have studied the o-Ps formation as a sample temperature (range 150-430 K) performed by 3γ-PAS and PALS on Si samples with oxidized nanochannels synthesized by electrochemical etching. The measurements point to an increase of Ps formation between 300 and 400 K. This increase of Ps yield is associated to a decrease of the fraction of $e^+$ annihilating in silica and the disappearing of $e^+$ annihilation in the Si substrate. These combined effects has been interpreted as a thermal charge rearrangement at the Si/$SiO_2$ interface and at the $SiO_2$ nanochannels surface at high T that contrasts the positive electric field present in Si p-type. Future measurements in similar samples etched in n-type Si will be clarifying. In fact, if the above hypothesis is correct, a higher Ps formation at $T \leq 300$ K would be expected in n-type Si where at room temperature $e^+$ diffusion towards the $Si/SiO_2$ interface is favored [27]. As a consequence, the Ps increase at $T > 300$ K would be less pronounced.

**Acknowledgement**

We thank Laidani N and Luciu I of *Fondazione Bruno Kessler* for the $TiO_2$ capping deposition. Beam time at FRMII is gratefully acknowledged.

[1] Platzman P M and Mills Jr. A P 1994 *Phys. Rev. B* 49 454
[2] Cassidy D B and Mills Jr. A P 2007 *Phys. Stat. Sol. (c)* 10 3419
[3] Castelli F, Boscolo I, Cialdi S, Giammarchi M G and Comparat D 2008 *Phys. Rev. A* 78 052512
[4] Charlton M, *Phys. Lett. A* 1990 143 143
[5] Kellerbauer A *et al.,* 2008 *Nucl. Instrum. Methods Phys. Res. B* 266 351
[6] Liszkay L *et al.,* 2008 *Appl. Phys. Lett.* 9 063114
[7] Mariazzi S, Bettotti P, Larcheri S, Toniutti L and Brusa R S 2010 *Phys. Rev. B* 81 235418
[8] Mariazzi S, Bettotti P and Brusa R S 2010 *Phys. Rev. Lett.* 104 243401
[9] Cassidy D B *et al.,* 2010 *Phys. Rev. A* 81 012715
[10] Crivelli P, Gendotti U, Rubbia A, Liszkay L, Patrice Perez and Corbel C 2010 *Phys. Rev. A* 81 052703
[11] Nagashima Y *et al.,* 1998 *Phys. Rev. B* 58 12676
[12] Brusa R S and Dupasquier A in *Physics with many Positrons*, ed. by Brusa R S, Dupasquier A and Mills A.P jr. (North Holland, Amsterdam, 2010) p 245.
[13] Brauer G, Anwand W, Skorupa W, Revesz A G and Kuriplach J 2002 *Phys. Rev. B* 66 195331
[14] Zecca A, Bettonte M, Paridaens J, Karwasz G P and Brusa R S 1998 Meas. Sci. Technol. 9 409
[15] Schultz P J and Lynn K G 1988 *Rev. Mod. Phys.* 60, 701
[16] Macchi C *et al.,* 2006 *Phys. Rev. B* 74 174120
[17] Brusa R S, Macchi C, Mariazzi S and Karwasz G P 2005 *Acta Physica Polonica A* 107 702
[18] Mills A P Jr. 1978 *Phys. Rev. Lett.* 41 1828
[19] Eldrup M, Vehanen A, Schultz P J and Lynn K G 1985 *Phys. Rev. B* 32 7048
[20] Egger W in Physics with many Positrons, ed. by Brusa R S, Dupasquier A and Mills A P jr. (North Holland, Amsterdam, 2010) p 419.
[21] Hugenschmidt C in *Physics with many Positrons*, ed. by Brusa R S, Dupasquier A and Mills A P jr. (North Holland, Amsterdam, 2010) p 399.
[22] Cassidy D B and Mills A P Jr. 2008 *Phys. Rev. Lett.* 100 013401 [19] Brusa R S *et al.,* 1994 *Phys. Rev. B* 49 7271
[23] Sferlazzo P, Berko S and Canter K F 1985 *Phys. Rev. B* 32 6067
[24] Brusa R S *et al.,* G 2005 *Phys. Rev. B* 71 245320
[25] Mazzoldi P, Mattei G, Ravelli L, Egger W, Mariazzi S and Brusa R S 2009 *J. Phys. D: Appl. Phys.* 42 115418
[26] Murakami H, Onizuka N, Sasaki J and Thomghai N 1998 *J. Mat. Sci.* 33 5811
[27] Brusa R S *et al.,* 1994 *Phys. Rev. B* 49 7271
[28] Helms C R and Poindexter E H 1994 *Rep. Prog. Phys.* 57 791

# Ortho-positronium localization in pores of Vycor glass at low temperature

**Radosław Zaleski**

Maria Curie-Skłodowska University, Institute of Physics, Pl. M. Curie-Skłodowskiej 1, 20-031 Lublin, Poland

E-mail: radek@zaleski.umcs.pl

**Abstract**. An explanation of a discrepancy between the Extended Tao-Eldrup model and experimental lifetimes observed in some porous materials at low temperatures is presented. The discrepancy is expected in pores with a varying thickness. In such pores thermalized ortho-positronium (o-Ps) could locate in the widest parts, between potential barriers formed by a higher ground energy levels in the narrower parts of the pore. Such an effect can occur, if the energy, which o-Ps can obtain from thermal excitation, is not sufficient enough to pass the energy barrier (i.e., at low temperatures). As a consequence, the lifetime of a localized o-Ps is characteristic for the widest part of the pore only, and not to the averaged width, which can be observed when o-Ps has the ability to move freely along a pore. The problem is discussed basing on the temperature dependence of the results for Vycor glass obtained by o-Ps annihilation lifetime spectroscopy. Additionally to the mean lifetimes, the width of the lifetime distribution in the component is taken into account in the discussion.

## 1. Introduction

Positronium (Ps) is an exotic atom, similar to hydrogen, where a positron plays the role of a proton. This antiparticle-particle system is unstable due to intrinsic (with its own electron) or pick-off (with an electron from a surroundings) annihilation [1]. This second process concerns mostly long-lived ortho-positronium (i.e., state of Ps, where a positron and an electron have parallel spines) and it makes o-Ps interesting for material studies. However, in order to correctly interpret the data obtained from o-Ps used as a probe, its behavior in the material has to be known as well as it is possible.

A relatively new method of material study is positron porosimetry based on positron annihilation lifetime spectroscopy (PALS) [2-6]. This method is dedicated to the reconstruction of pore size distributions from the PALS spectra. The extended Tao-Eldrup model [7] is used to transform o-Ps lifetimes to pore sizes. An additional calculation scheme based on a simple model [8,9] is used for the transformation of o-Ps component's intensities to the volume of the pores of a particular size. The main advantage of this technique is the possibility to perform measurements in various conditions (e.g., a wide temperature range [10-13], various mechanical pressures [14] or in the presence of a gas [15-17]). Nevertheless, positron porosimetry still requires testing until it can be established as commonly used technique for the characterization of porous material.

The most disturbing discrepancy from the predictions of the models used in positron porosimetry is the elongation of the o-Ps lifetime at low temperatures [11,12,18]. This effect is not common for every porous material (e.g., it is not observed for porous polymers [9], polymer-silica composites [19] and is also negligible for most of the ordered porous silica materials [20]). It suggests that it is not a problem with the model assumptions, but it is the result of a specific o-Ps behaviour caused by the properties of the materials. Among the possible sources of uncommon o-Ps lifetime are chemical bounds of positronium (which need to be verified in the future) or physical effects influencing the observed lifetimes.

One of the physical effects, which can distort the expected o-Ps lifetime is positronium migration observed, particularly, in relatively large free volumes, such as mesopores [21-24]. However, o-Ps migration, besides distorting positron porosimetry results, can also be beneficial, if it is fully understood and described. The migration of o-Ps from open pores makes positron porosimetry results complementary to standard sorption techniques [6], which are sensitive only to open pores. Moreover, in experiments like AEGIS or GBAR production of positronium dedicated for antihydrogen creation relies on the migration of o-Ps outside the open mesopores [25,26].

An explanation of the unexpected low temperature o-Ps lifetime elongation, using assumption that this effect is connected with the migration of o-Ps, will be presented in this paper.

## 2. Experimental

### 2.1. Experimental methods

The conventional fast-slow delayed coincidence spectrometer was used for the PALS measurements. Its detectors were equipped with $BaF_2$ scintillators. The energy windows covered both photopeak and energies of about 150-200 keV below Compton edge in order to obtain a high overall coincinence rate and, first of all, highest possible contribution of three gamma annihilation events in the spectrum. The positron source, made of $^{22}$Na (0.4 MBq) enclosed in a Kapton envelope, was surrounded by two layers of the sample. A sample holder was placed on the copper rod, which could be cooled by liquid nitrogen when necessary. The main temperature regulator was a heater placed inside the rod, just below the sample holder. The current flowing through the heater was controlled by a Shimaden FP 21 unit, which allowed us to obtain temperature stability ±0.1 K. A vacuum below 0.1 Pa was maintained in the sample chamber by a turbo-molecular pump. It allowed the preventing of both the o-Ps to p-Ps conversion in the presence of oxygen and the contamination of the pore surface.

The total count number in each spectrum slightly exceeded $10^7$. The lifetime spectra were initially analysed numerically by Bayesian methods using the MELT routine [27], which allowed us to find the total number of components in the spectrum as well as the lifetime distributions. Thereafter, fitting of the coefficients of a spectrum model (assumed basing on the MELT results) was performed by the LT programme [28]. This second type of analyses diminished the scatter of the results. The time resolution function was approximated by two Gaussian functions giving together FWHM ~ 280 ps. Except the o-Ps components described later, para-positronium $\tau_1$ = 120-140 ps and free positrons $\tau_2$ = 400-600 ps components were found.

Independent from PALS, a material characterization was performed by measuring the nitrogen adsorption/desorption isotherms at 77 K with the use of AUTOSORB-1CMS adsorption analyser (Quantachrome Instruments). The specific surface area ($S_{BET}$) of the sample was found using the BET method in the relative pressure $p/p_0$ range of 0.05 to 0.3. The total pore volume ($V$) was estimated from the amount of nitrogen adsorbed at the relative pressure of 0.99. The mesopore diameter at the maximum of the distribution was found using the Barrett-Joyner-Halenda method ($D_{BJH}$) and alternatively using the Density Functional Theory equilibrium model for cylindrical pore geometry ($D_{DFT}$).

### 2.2. Material

The initial Vycor glass was prepared by melting $NaNO_3$, $Na_2B_4O_7$ and $SiO_2$ (from Brazilian quartz) at 1725 K. The homogenous mixture was then rapidly cooled in water resulting in solid material, which was then grinded to a fine powder. The glass obtained was composed of 7% $Na_2O$, 23% $B_2O_3$ and 70% $SiO_2$. Afterwards, the material was heated at 770 K in order to demix its components (liquation process). The structure of pores was created by leaching the cooled glass in an acidic bath, which removed alkali-borate phase from the material leaving empty spaces (i.e. mesopores) surrounded by walls made mostly of $SiO_2$ [29].

Characterization of the obtained Vycor glass by the nitrogen sorption method showed that its specific surface area $S_{BET}$ = 297 m$^2$/g and total pore volume $V$ = 0.207 cm$^3$/g. Depending on the method used for the determination of the most probable pore size $D_{BJH}$ = 2.3 nm or $D_{DFT}$ = 3.1 nm.

## 3. Results and discussion

Numerical analysis of the positron lifetime spectra of the Vycor glass revealed three o-Ps components. The origin of the shortest-lived one ($\tau_3$ = 1.7-2.2 ns, $I_3$ = 4-7%) are undoubtedly free volumes in the glass structure, between the atoms of amorphous $SiO_2$. Likewise, the presence of the medium-lived component with a lifetime scattered around 14 ns and a low intensity ($I_4$ = 1-3%) is probably caused by the existence of micropores (e.g., microcracks) in the material. The most interesting, from the point of view of the present study, is the long-lived component reflecting the annihilation of o-Ps in the mesopores produced by leaching.

Above 220 K, the long-lived component's lifetime agree very well with predictions of the ETE model (Fig.1), which were calculated assuming a cylindrical pore shape of diameter $D$ = 2.8 nm and the value of the empirical parameter (required by the model) $\Delta$ = 0.193 nm [30]. The deviation from the curve predicted by the model is visible below 220 K. Instead of diminishing changes of the lifetime with decreasing temperature, it increases very fast, reaching a value 30% larger than expected at the lowest achieved temperature of 106 K. This value of lifetime according to the ETE model corresponds to $D$ = 3.4 nm, a diameter 20% larger than the one calculated at temperatures above 220 K. A change of the material structure leading to such growth of the pores is highly improbable, and it was not observed in the results of the nitrogen sorption method at an even lower temperature of 77K. Therefore, the effect should be ascribed to special o-Ps behavior in the Vycor glass.

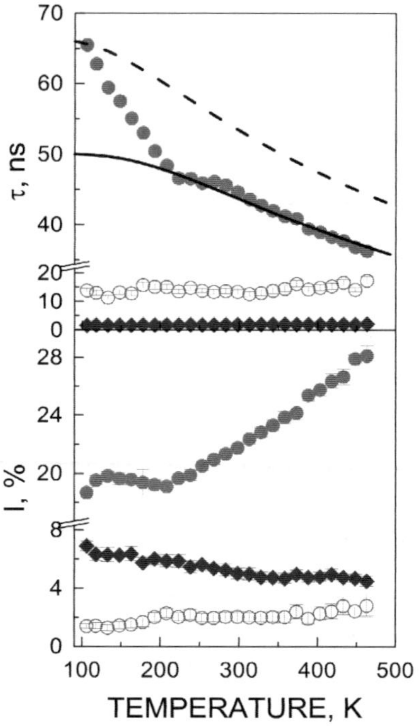


**Figure 1.** Temperature dependence of the o-Ps components lifetime and intensity. Lines represent predictions of the ETE model for cylindrical pores with diameter 2.8 nm (solid) and 3.4 nm (dashed).

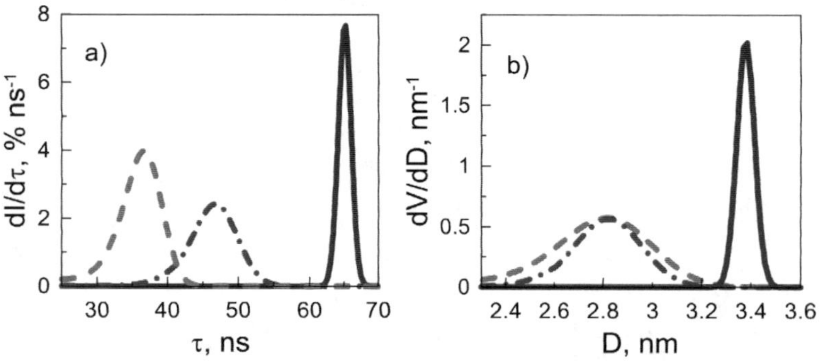


**Figure 2.** The distribution of lifetimes (a) in the longest-lived component obtained by MELT and the relative distribution of pore sizes (b) calculated using the ETE model form the presented distribution of lifetimes at 106 K (dashed line), 223 K (dashed-dotted line) and 463 K (solid line).

The intensity of the longest-lived component increases with temperature from 19% to 28% allowing us to determine the lifetime of this component with a sufficient accuracy. The increase of the intensity reflects the changes of the positron mobility, which rises with temperature, causing a higher probability of a positron encountering a mesopore, which is then followed by o-Ps creation [31]. The deviation from this dependence is observed below 220 K, in the same temperature range where the lifetime disagrees with the ETE model. However, almost constant intensity in this region is most likely a consequence of different detection efficiencies of three-gamma and two-gamma annihilation. Much longer lifetime, observed at low temperature, is accompanied with increased ratio of three-gamma to two-gamma annihilation. Wide energy window in the stop detector results in more efficient registration of three-gamma events comparing to two-gamma ones. Therefore an intensity is dependent not strictly on number of o-Ps formed in a particular free volume, but also, to a lesser extent, on their lifetime. This is probably a reason of the intensity distortion at low temperatures, where large changes of lifetime are detected. Elongation of the lifetime with temperature decrease causes artificial increase of the intensity, which seems to be constant instead of the continuation of its decrease.

A more detailed investigation of the longest-lived component was performed using Bayesian methods, which give a quasi-continuous intensity distribution over a dense lifetime grid (Fig.2a). The distributions in the spectra above 220 K are both quite wide (FWHM $\approx$ 7 ns), while a 2-3 times narrower peak (FWHM $\approx$ 2 ns) is observed at 106 K. Positron porosimetry calculation schemes [9] allow the reconstruction of a pore size distribution at chosen temperatures (Fig. 2b). As it was found before, the pore size at the maximum of the distribution is in very good agreement with each other for temperatures above 220 K, while at low temperatures the maximum is definitely shifted towards larger pore size. Moreover, the width of a pore size distribution is similar for larger temperatures. The difference between the width of the lifetime distribution obtained at 203 K and 463 K is most likely cause by an inaccuracy of the MELT analysis rather than a result of a physical effect. However, the width of the distribution is undoubtedly much narrower for 106 K. Moreover, it is significant that low temperature distribution is located within the tail of the distributions found for temperatures above 220 K. It would suggest, that o-Ps annihilates only in the largest pores at low temperature, while it can annihilate in pores of various sizes at higher temperatures.

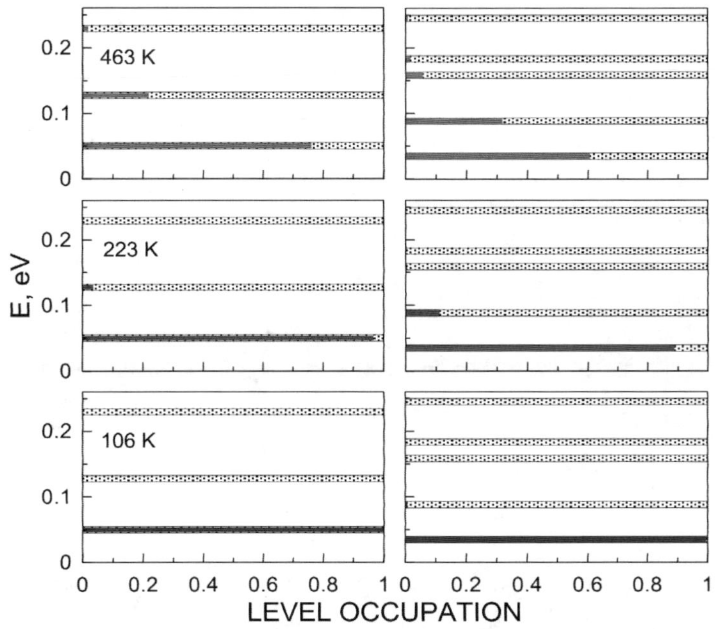


**Figure 3.** Several lowest energy levels of a particle (e.g., positronium) in a rectangular potential well of depth 3 eV (dotted lines). A cylindrical shape of the potential well with a diameter of 2.8 nm (left) and 3.4 nm (right) is assumed. Filled parts of the lines show the probability of energy levels occupation by the thermalized particle at 106 K (down), 223 K (middle) and 463 K (top).

In order to find the origin of such an effect, the occupation of energy levels by positronium at various temperatures trapped inside the cylindrical pores of various size was investigated (Fig.3). A square potential well with depth 3eV [32,33] was assumed. It is worth noticing, that the depth of the well is not a very important parameter, because the difference between the energy levels in a finite well of 3 eV depth and an infinite one is almost constant and less than 8% for the presented energy levels. Two diameters of the well were investigated: 2.8 nm, the mean size found by PALS, which is close to an averaged pore size from nitrogen sorption, and 3.4 nm, the size of pores found at 106 K. The difference between energy levels for these both pore sizes is easy to observe. The ground energy level in the smaller pore is equal to about 70% of the energy in the larger one. Differences between the energies of the excited states are even larger. The probability of energy levels occupation by thermalized positronium at various temperatures can be determined applying the Boltzmann distribution to the energy level structure, in the same way as it was done in the ETE model. Results presented in Fig.3 show, that even at 463 K only several energy levels can be populated by positronium with a significant probability. Only the two lowest energy levels can be populated at 223 K, and practically only the ground level is populated at 106 K. This quantum property of positronium trapped in the well can help in explaining the discrepancy between the ETE model and the experimental results at low temperature.

The wide distribution of pore sizes, like the one observed in the Vycor sample, can be interpreted in two ways for the cylindrical pores: pore width is different in each pore, but uniform along a pore, or pore width varies along a pore (Fig.4). The second case is much more probable for Vycor glass, where pores created during the leeching form an interconnected, disordered structure. The energy levels of positronium trapped in such pores change fluently along a pore between energy level configurations similar to the ones presented in Fig.3. As long as positronium at high temperatures can occupy excited energy levels, a transfer from wider pores to the narrower ones requires gaining some thermal energy form the wall if o-Ps stays on the same level or losing some energy, if o-Ps changes to a lower level. Since it was proven that during its lifetime o-Ps changes the occupied level many times [7], migration of o-Ps along the pore is almost uninterrupted. Therefore, its lifetime is averaged over lifetimes characteristic for the pores with random sizes, which Ps passes during its lifetime. In contrary, o-Ps almost certainly occupies the ground level at low temperatures. There is no thermal energy which can be acquired by o-Ps when it encounters smaller a pore fragment with a higher energy ground level. For that reason o-Ps becomes trapped in the widest part of the pore (Fig.4) and annihilates with a lifetime characteristic for this part of the pore only. According to the explanation presented above, the systematic increase of the longest-lived component lifetime with decreasing temperature observed in Fig.1, would be a result of successive fractions of the narrowest pores getting inaccessible for o-Ps. In other words, it can reflect a diminishing probability of finding o-Ps inside narrower pores, as its movement from the wider part of a pore to the narrower ones is less and less probable comparing to the movement in the opposite direction when temperature decreases.

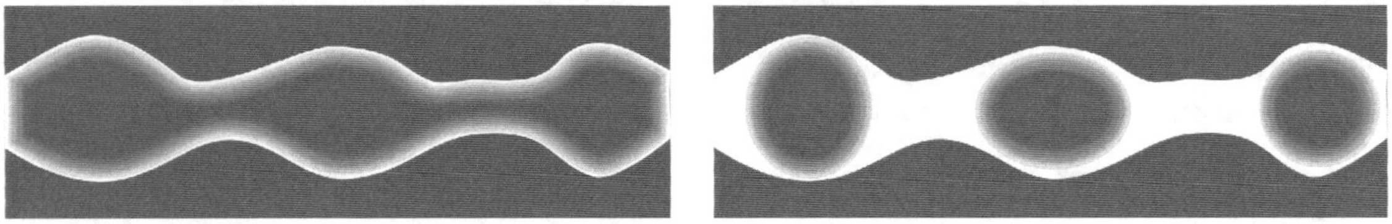

**Figure 4.** A schematic representation of the probability of finding o-Ps in the pore of Vycor glass at high (left) and low (right) temperatures.

## 4. Conclusions

Elongation of the o-Ps lifetime at low temperatures can be attributed to o-Ps localization in the wider parts of the pores. This effect should be observed in the pores which have varying, highly non-uniform width. The temperature below which this effect occurs, should depend on the narrowest pore size (i.e. on both mean pore size and width of pore size distribution), because the distance between the energy levels depend on the pore size. The temperature, below which deviation is observed, should be lower for larger pores or for narrower pore size distribution. Observation of these relations in various porous materials can verify the presented model.

It is highly unlikely that this effect explains all deviations from the ETE model. Nevertheless, it should be taken into account during the characterization of porous solids. It can provide, or at least verify, some information about material structure, which are the width of pore size distribution and the uniformity of single pore thickness. The property of pores with varying thickness, which consist in letting through only positronium with energy above some threshold could be useful for some applications where energy of positronium emitted from pores is important (e.g. antihydrogen production). A porous material with tailored pore outlets can provide o-Ps with a narrow energy distribution.

**Acknowledgment**
Author thanks Bożena Jasińska for providing the Vycor glass sample.

**References**

[1] Goworek T, Jasinska B, Wawryszczuk J, Zaleski R, Suzuki T 2002 *Chem. Phys.* **280** 295
[2] He C, Suzuki R, Ohdaira T, Oshima N, Kinomura A, Muramatsu M, Kobayashi Y 2007 *Chem. Phys.* **331** 213
[3] Kullmann J, Enke D, Thraenert S, Krause-Rehberg R, Inayat A 2010 *Colloids and Surfaces A: Physicochemical and Engineering Aspects* **357** 17
[4] Zaleski R, Stefaniak W, Maciejewska M, Goworek J 2010 *J. Colloid Interface Sci.* **343** 134
[5] Zaleski R, Kierys A, Grochowicz M, Dziadosz M, Goworek J 2011 *J. Colloid Interface Sci.* **358** 268
[6] Zaleski R, Kierys A, Dziadosz M, Goworek J, Halasz I 2012 *RSC Advances* **2** 3729
[7] Goworek T, Ciesielski K, Jasinska B, Wawryszczuk J 1998 *Chem. Phys.* **230** 305
[8] Gidley DW, Frieze WE, Dull TL, Sun J, Yee AF, Nguyen CV, Yoon DY 2000 *Appl. Phys. Lett.* **76** 1282
[9] Zaleski R, Goworek J, Maciejewska M 2009 *physica status solidi (c)* **6** 2445
[10] Zaleski R, Wawryszczuk J, Goworek J, Borówka A, Goworek T 2003 *J. Colloid Interface Sci.* **262** 466
[11] Fischer CG, Weber MH, Wang CL, McNeil SP, Lynn KG 2005 *Phys. Rev. B: Condens. Matter Mater. Phys.* **71** 180102
[12] Thraenert S, Hassan EM, Enke D, Fuerst D, Krause-Rehberg R 2007 *physica status solidi (c)* **4** 3819
[13] Kunishige S, Koshimizu M, Asai K 2009 *Radiat. Phys. Chem.* **78** 1088
[14] Zaleski R, Goworek J, Borówka A, Kierys A, Wiśniewski M 2009 in eds Seaton N, Reinoso FR, Llewellyn P, Kaskel S *Characterisation of Porous Solids VIII* (RSC Publishing: Cambridge), pp 400
[15] Goldanskii V, Mokrushin A, Tatur A, Shantarovich V 1975 *Appl. Phys. A: Mater. Sci. Process.* **5** 379
[16] Wilkinson NJ, Alam MA, Clayton JM, Evans R, Fretwell HM, Usmar SG 1992 *Phys. Rev. Lett.* **69** 3535
[17] Zaleski R, Dolecki W, Kierys A, Goworek J 2012 *Microporous Mesoporous Mater.* **154** 142
[18] Sniegocka M, Jasinska B, Goworek T, Zaleski R 2006 *Chem. Phys. Lett.* **430** 351
[19] Gorgol M, Tydda M, Kierys A, Zaleski R 2012 *Microporous Mesoporous Mater.* **163** 276
[20] Zaleski R, Goworek J, Kierys A 2007 *physica status solidi (c)* **4** 3814
[21] Shantarovich V, Berdonosov S, Znamenskaya I, Melikhov I, Gustov V, Kevdina I 2006 *Radiochemistry* **48** 505
[22] Zaleski R, Wawryszczuk J 2008 *Acta Phys. Polon. A* **113** 1543
[23] Zaleski R, Sokół M 2011 *Mater. Sci. Forum* **666** 123
[24] Zaleski R, Błażewicz A, Kierys A *Nukleonika* (submitted)

[25] Ferragut R, Calloni A, Dupasquier A, Consolati G, Quasso F, Giammarchi MG, Trezzi D, Egger W, Ravelli L, Petkov MP, Jones SM, Wang B, Yaghi OM, Jasinska B, Chiodini N, Paleari A 2010 *Journal of Physics: Conference Series* **225** 012007

[26] Liszkay L, Guillemot F, Corbel C, Boilot J-P, Gacoin T, Barthel E, Pérez P, Barthe M-F, Desgardin P, Crivelli P, Gendotti U, Rubbia A 2012 *New Journal of Physics* **14** 065009

[27] Shukla A, Peter M, Hoffmann L 1993 *Nucl. Instrum. Methods Phys. Res., Sect. A* **335** 310

[28] Kansy J 1996 *Nucl. Instrum. Methods Phys. Res., Sect. A* **374** 235

[29] Jasińska B, Dawidowicz AL, Goworek T, Radkiewicz S 2000 *Phys. Chem. Chem. Phys.* **2** 3269

[30] Thränert S, Enke D, Dlubek G, Krause-Rehberg R 2009 *Mater. Sci. Forum* **607** 169

[31] Stepanov SV, Byakov VM 2002 *J. Chem. Phys.* **116** 6178

[32] Nagashima Y, Morinaka Y, Kurihara T, Nagai Y, Hyodo T, Shidara T, Nakahara K 1998 *Phys. Rev. B: Condens. Matter Mater. Phys.* **58** 12676

[33] Tanaka HKM, Yamauchi Y, Kurihara T, Sakka Y, Kuroda K, Mills AP 2008 *Advanced Materials* **20** 4728

Journal of Physics: Conference Series **443** (2013) 012063 doi:10.1088/1742-6596/443/1/012063

# Positron Annihilation Studies of Mesoporous Silica MCM-41

**J F Williams[1], P Guagliardo[1], K Sudarshan[1], C Ranganathaiah[1], G Koutsantonis[2], N Hondow[2] and S Samarin[1]**

[1]Centre for Antimatter-Matter Studies, School of Physics, The University of Western Australia, 35 Stirling Highway, Crawley, WA 6009, Australia
[2]School of Chemistry and Biochemistry, The University of Western Australia, 35 Stirling Highway, Crawley, WA 6009, Australia

E-mail: paul.guagliardo@uwa.edu.au

**Abstract**. Positron annihilation has been used to study the mesoporous silica MCM-41. Lifetime spectra of evacuated MCM-41 indicate a significant contribution from 3$\gamma$ annihilation events with $\tau_4 = 116$ ns and $I_4 = 24.5$ %. This is supported by measurements of the full energy distribution, where MCM-41 shows enhanced counts in the low energy region (below 511 keV) relative to a pure 2$\gamma$ sample. MCM-41 was also studied under air and oxygen atmospheres. The presence of atmosphere has a significant effect on both the lifetime and Doppler patterns, with both the lifetime data ($\tau_4$ and $I_4$) and the 3$\gamma$-fraction decreasing with increasing oxygen concentration. This is indicative of paramagnetic quenching of o-Ps by oxygen.

## 1. Introduction

Positron annihilation has been used to study MCM-41, a mesoporous silica composed of an array of non-intersecting hexagonal channels or pores of uniform size. This material has been extensively investigated for applications in catalysis due to its high surface area and structured array of pores [1]. In the following, we have utilised positron lifetime spectroscopy and Doppler broadening spectroscopy (registering the full energy distribution of annihilation radiation) to study MCM-41. We have compared the pore size determination from lifetime measurements with x-ray diffraction (XRD) data and have looked at the effects of atmosphere (vacuum, air and oxygen) on positron lifetime and Doppler broadening results.

## 2. Experimental

X-ray diffraction (XRD) studies indicated that the MCM-41 powders used had an ordered pore structure surrounded by amorphous pore walls with a uniform diameter of approximately 29 Å [2].

Lifetime spectra were recorded with $BaF_2$ detectors angled at 60°, and approximately 20 μCi of $^{22}$NaCl sealed in Kapton foil was used as the positron source. The time dispersion of the lifetime spectrometer was 119 ps per channel (1 μs TAC range), and the time resolution was approximately 380 ps owing to the large stop windows employed, extending into the Compton region so as to register 3$\gamma$-annihilation events. Each spectrum contained at least 4 million counts and good reproducibility of the spectra was found upon reprocessing of the samples. Data analysis was carried out using PAScual version 1.30 [3]. Doppler broadening measurements were recorded simultaneously, with the HPGe detector sufficiently removed from the source; the energy resolution of the detector was approximately 1.5 keV at 511 keV.

The samples were heated at 150 °C under vacuum for ~ 48 hours prior to measurement in order to remove free water. After this time they were briefly exposed to air (for less than 10 minutes) while packed with the source and then transferred to the main vacuum system for positron annihilation measurements. Experiments were performed at a base pressure of $4 \times 10^{-1}$ Torr. The vacuum chamber was equipped with a gas handling system to admit air (atmospheric gas) and oxygen (ultrahigh purity).

## 3. Results and Analysis

Figure 1 shows annihilation gamma-ray spectra measured using a HPGe detector for samples of MCM-41 (evacuated) and copper. The spectra have been normalised to the area of the 1.274 MeV peak. The MCM-41 spectrum shows enhanced counts to the left of the 511 keV photo peak relative to copper, and to the right of the photo peak the counts coincide. The reduced photo peak intensity and enhanced low energy background for MCM-41 suggests a significant contribution from o-Ps self-annihilation into 3γ-quanta, as for 3γ -annihilation, the energy distribution of the individual quanta is continuous and increases monotonically up to 511 keV [4]. The fraction of 3γ events ($f_{3\gamma}$) can be quantified by comparing the relative integral counts in the 511 keV peak for a 3γ-sample ($A_{2\gamma}$) with that of a pure 2γ–sample ($A^0_{2\gamma}$) (typically a metal sample to exclude Ps formation) [5]. The relative number of positrons annihilating in the 3γ channel can be expressed as

$$f_{3\gamma} = 1 - A_{2\gamma}/A^0_{2\gamma}$$

This measure gives a value of 23.5 % for MCM-41.

Lifetime measurements were also performed on MCM-41 and similarly indicate a significant 3γ contribution. The spectrum could be decomposed in terms of four lifetime components. The shorter components $\tau_1$ and $\tau_2$ are associated with p-Ps and free positron annihilation. The total intensity of these components comprises 74.9% of the spectrum. The longer components of $\tau_3$ = 7.1 ns and $\tau_4$ = 116 ns, with respective intensities of 0.9% and 24.3%, correspond to o-Ps annihilation. These results differ somewhat from MCM-41 studies described in the literature [6, 7]; however, there were significant differences in the sample synthesis as well as experimental parameters, such as the vacuum level.

The shorter of these, $\tau_3$, could be attributed to o-Ps pick-off annihilation in nano regions within the silica matrix; the rectangular Tao-Eldrup (RTE) model [8] attributes a spherical pore diameter of 12 Å to this lifetime value. The long lifetime $\tau_4$ and its high relative abundance suggest a high proportion of 3γ events. When $\tau_4$ = 116 ns is converted to a pore size using the RTE model, the values obtained are significantly larger than the cavity size determined by XRD measurements (adopting a square channel geometry, a side length of 169 Å is obtained). This could indicate that annihilations occur predominantly in the free space between powder particles rather than in the inner space of the cavities. It is difficult to distinguish between these two possible contributions for the very long lifetime, and $\tau_4$ could be the average of positronium annihilation in the channels and in the outer space.

The influence of atmosphere was also investigated. Lifetime spectra are shown in figure 2 for MCM-41 recorded under vacuum and in air and oxygen atmospheres (≈ 1 atm). It can be seen that the presence of atmosphere leads to drastic changes in the lifetime spectra. For each case the spectrum could be analysed in terms of four components (see table 1). The long lifetime component $\tau_4$ decreases by a factor 1.7 in the presence of air and by a factor of 3.4 in the presence of oxygen, relative to the vacuum value. The intensity of this component $I_4$ also decreases with increasing oxygen concentration, with the weighting shifting mostly to $I_1$, associated partly with p-Ps. This is indicative of paramagnetic quenching by oxygen [9]. The adsorption of gas molecules can also lead to a reduction in the apparent pore size; furthermore, the alteration of the chemical nature of the pore surface due to gas adsorption can change the probability of pick-off annihilation in each collision. The positronium quenching rate constant of oxygen is reported to be 0.023 ± 0.003 $ns^{-1}$ $atm^{-1}$ [10, 11]. Taking the partial pressure of oxygen in air to be 0.2 atm and the same quenching constant to be valid for oxygen present in the mesopores, the rate constant in air would be 0.0046 $ns^{-1}$ $atm^{-1}$. If we calculate the quenching rates from the measured lifetimes we obtain 0.0059 $ns^{-1}$ $atm^{-1}$ for air and 0.020 $ns^{-1}$ $atm^{-1}$ for oxygen. The value for oxygen is within the uncertainty of the literature result; however the rate for air is higher than the expected value.

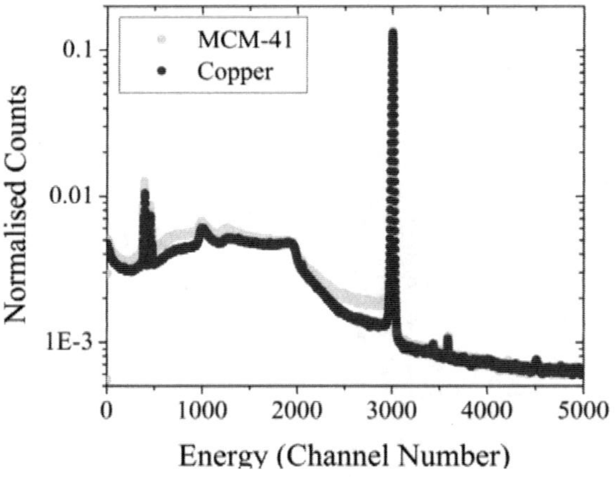


**Figure 1.** Energy spectra of MCM-41 and Copper recorded with a HPGe detector.

However, the enhancement of the annihilation rate caused by air has shown a dependence on pore size [11]; this is likely due to the absorption of air molecules on the pore walls, and for the case of small pores the number of surface adsorbed molecules has been shown to exceed the number of free molecules in the volume of the pore [12]. This dependence is depicted in the right panel of figure 2 where the quantity $\lambda_{air}$-$\lambda_{vac}$ is plotted for polymeric and silica samples containing spherical pores of differing sizes; a decrease in $\lambda_{air}$-$\lambda_{vac}$ is seen with increasing pore radius. The value of $\lambda_{air}$-$\lambda_{vac}$ for MCM-41 is also plotted as a green triangle in figure 2 and is somewhat below the trend depicted by this data. This could suggest a differing trend for samples containing regular channels; however the result of He et al. [6] for MCM-41 coincides with the data.

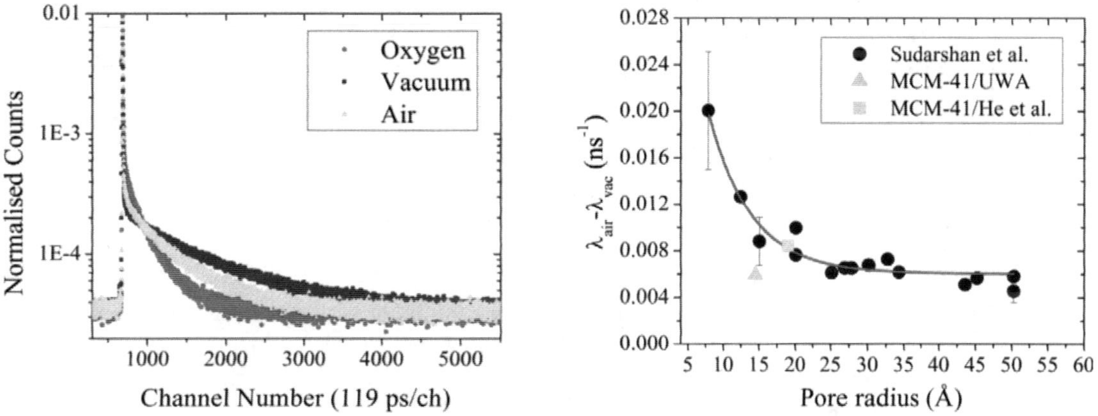


**Figure 2.** Left: Lifetime spectra of MCM-41 recorded under vacuum and in oxygen and air atmospheres (the peaks extends to about 0.1); right: $\lambda_{air}$-$\lambda_{vac}$ vs. pore size; the blue dots are the results of Sudarshan et al. [11], the cyan square is data from He et al. [6], and the green triangle is the MCM-41 data from this study.

Annihilation gamma-ray energy spectra for the three atmospheres were also recorded and the intensity in the Compton region was found to decrease with increasing oxygen concentration, indicating a decrease in the 3γ contribution due to paramagnetic quenching by oxygen. This is evidenced by the 3γ fraction (given in table 1) which decreases with increasing oxygen concentration, signalling the

Journal of Physics: Conference Series **443** (2013) 012063 doi:10.1088/1742-6596/443/1/012063

conversion of o-Ps to p-Ps. Nitrogen was also studied and gave comparable results to those recorded under vacuum with only a slightly reduced 3γ fraction at 20.49%. This is consistent with literature results which show that nitrogen is not a strong positronium quencher.

**Table 1.** PALS fitting results for MCM-41 recorded in vacuum and in air and oxygen atmospheres; $f_{3\gamma}$ calculated from energy spectra

| | Vacuum | Air | Oxygen |
|---|---|---|---|
| $\tau_1$ (ps) | 139 (2) | 129 (2) | 132 (2) |
| $\tau_2$ (ps) | 443 (2) | 446 (2) | 445 (2) |
| $\tau_3$ (ns) | 7.3 (0.2) | 8.4 (0.2) | 7.3 (0.2) |
| $\tau_4$ (ns) | 116 (0.3) | 68.7 (0.3) | 34.5 (0.2) |
| $I_1$ (%) | 42.7 (0.2) | 46.9 (0.3) | 49.1 (0.3) |
| $I_2$ (%) | 31.7 (0.2) | 33.7 (0.3) | 32.9 (0.4) |
| $I_3$ (%) | 1.0 (0.1) | 1.9 (0.1) | 3.4 (0.1) |
| $I_4$ (%) | 24.5 (0.1) | 17.3 (0.1) | 14.5 (0.1) |
| $\chi^2$ | 1.085 | 1.056 | 1.082 |
| $f_{3\gamma}$ (%) | 20.5 (0.1) | 10.3 (0.1) | 4.2 (0.1) |

## 4. Conclusions

Lifetime spectra of evacuated MCM-41 indicated a significant contribution from 3γ annihilation events with lifetimes exceeding 100 ns with high intensity. This is supported by measurements of the full energy distribution, where MCM-41 shows enhanced counts in the low energy region relative to a copper sample. The presence of atmosphere was found to drastically change the positron lifetime and Doppler patterns due to paramagnetic quenching by oxygen molecules.

## References

[1] Taguchi A and Schüth F 2005 *Micro. and Meso. Mater.* **77** 1

[2] Hondow N S, Koutsantonis G A, Nealon G L and Saunders M 2012 *Micro. and Meso. Mater.* **151** 264

[3] Pascual-Izarra C, Dong A W, Pas S J, Hill A J, Boyd B J and Druumond C J 2009 *Nucl. Instrum. Meth. Phys. Res. A* **603** 456

[4] Ore A and Powell J L 1949 *Phys. Rev.* **75** 1696

[5] Liszkay L, Kajcsos Z, Duplâtre G, Lázár K, Pál-Borbély G and Beyer H K 2001 *Mater. Sci. For.* **363** 377

[6] He Y J, Zhang H Y, Chen Y B, Wang H Y and Horiuchi T 2001 *J. Phys.: Conden. Mat.* **13** 2467

[7] Zaleski R, Wawryszczuk J and Goworek T 2007 *Rad. Phys. Chem.* **76** 243

[8] Dull T L, Frieze W E, Gidley D W, Sun J N and Yee A F 2001 *J. Phys. Chem. B* **105** 4657

[9] Klobuchar R L and Karol P J 1980 *J. Phys. Chem.* **84** 483

[10] Celitans G J, Tao S J and Green J H 1964 *Proc. Phys. Soc.* **83** 833

[11] Sudarshan K, Dutta D, Sharma S K, Goswami A and Pujari P K 2007 *J. Phys.: Conden. Mat.* **19** 386204

[12] Zaleski R and Sokół M 2011 *Mater. Sci. Forum* **666** 123

# Production and thermalization of positronium in homogeneous porous silica

**S Aghion[1,2], R Ferragut[1,2], F Moia[1,2], M P Petkov[3] and S M Jones[3]**

[1] LNESS and Dipartimento di Fisica, Politecnico di Milano, via Anzani 42, 22100 Como, Italy
[2] Istituto Nazionale di Fisica Nucleare, via Celoria 16, 20133 Milano, Italy
[3] Jet Propulsion Laboratory - NASA, California Institute of Technology, 4800 Oak Grove Drive, Pasadena, CA 91109, U.S.A.

E-mail: stefano.aghion@mail.polimi.it

**Abstract.** Positronium yield using the "3γ method" and lifetime measurements were performed at cryogenic and room temperature by means of a variable energy positron beam in homogeneous porous silica (Aerogel). An estimation of the positronium (*Ps*) mean diffusion length was obtained by measuring capped samples. An efficient formation of cooled *Ps* atoms is a requisite for the production of antihydrogen, with the aim of a direct measurement of the Earth gravitational acceleration *g* of antimatter, which is the primary scientific goal of AEgIS (Antimatter Experiment: gravity, Interferometry, Spectroscopy; CERN, Geneva). Porous materials are necessary to obtain a high *Ps* yield as well as to thermalize *Ps*. Our results indicate a high *Ps* production, long survival time and diffusion length in Aerogel samples. It will be shown that positronium yield, lifetime and diffusion length are independent on temperature and on the effect of gas adsorption at low temperature. The results indicate that Aerogel is a good candidate for an efficient formation of cold *Ps* for the AEgIS experiment.

## 1. Introduction

Some fundamental questions of modern physics relevant to unification of gravity with the other fundamental interactions, models involving vector and scalar gravitons, matter–antimatter symmetry can be enlightened via experiments with antimatter [1]. The AEgIS experiment is carried out at the CERN Antiproton Decelerator to directly measure the gravitational acceleration *g* of a beam of cold antihydrogen [2, 3]. Gravitational interaction between matter and antimatter has never been tested.

An efficient formation of cooled positronium (*Ps*) atoms is a requisite for the production of antihydrogen ($\overline{H}$) in the AEgIS experiment. Porous materials are necessary to produce a high yield of *Ps* atoms as well as to cool *Ps* through collisions with the inner walls of the pores. The collisions between *Ps* and the internal surface of the pores involve a weak coupling to phonons or other surface modes. The velocity distribution of the *Ps* atoms coming out of the target should be of the order of $10^4$ m/s to allow *Ps* laser excitation to a Rydberg state ($Ps^*$) [4] and for efficient $\overline{H}$ formation, which requires that the relative velocity of antiprotons ($\overline{p}$) and $Ps^*$ must be not higher than the classical orbital velocity of the positron in the Rydberg *Ps* atom.

The formation of cold Rydberg antihydrogen $\overline{H}^*$ will be possible by means of the charge exchange reaction (more details in Refs. 2 and 3). The AEgIS experiment will be performed at cryogenic temperatures (~100 mK). A negative aspect of the cryogenic environment is the formation of ice at the surface of porous materials, as well in high vacuum condition. This effect could avoid the *Ps* escape in the free space outside the sample.

The Aerogels tested in the present work were characterized at room and cryogenic temperatures. Since the experimental evidences show a high *Ps* production, mobility and long survival time independent on the temperature, it is possible to conclude that the studied Aerogels are good candidates for the AEgIS experiment.

## 2. Experimental procedure and methodology

Silica aerogel samples were produced using a procedure developed for the particle capture collectors of the NASA Stardust project [5]. The first step for preparing ultra-low density aerogels involves the formation of a silica sol. The sol was prepared from tetraethyl ortho silicate (TEOS) in a way similar to that developed by Tillotson and Hrubesh [6], where the hydrolysis and condensation of a silicon alkoxide forms a suspension of micro-particles. The aerogel precursor mixture was prepared by combining acetonitrile, sol, water, methyl triethoxy silane and ammonium hydroxide. The density of the final aerogel is controlled by varying the ratio of the acetonitrile to that of the sol. The methyl triethoxy silane was added to make the aerogel hydrophobic. The resulting wet gel was dried in an autoclave, where the temperature was ramped up to 568 K at approximately 36 K/h, after pressurizing the system to 54 bar with argon. The vessel was then depressurized at a rate of 2.25 bar/h. Once the temperature had reached ambient pressure, the system was opened and allowed to cool before removing the aerogels. This process of "supercritical solvent extraction" allows the solvent to be removed while the silica network remains in a highly expanded state in the form of a three dimensional array of filaments. The aerogel samples have densities of 85 and 150 mg $cm^{-3}$. The sample with lower density was partially dried at room temperature in air before supercritical solvent extraction. Some samples were capped with an Al layer of about 110-120 nm. The Al layers were deposited by molecular beam epitaxy.

Both the lifetime and *Ps* formation measurements were performed, in high vacuum conditions (about $10^{-8}$ mbar), at two different temperatures: cryogenic temperature (18 K) and room temperature (295 K). The measurements of *Ps* formation yield were made by means of a monoenergetic positron beam, with energy up to 18 keV, using the well-known "3$\gamma$ method" [7-9] and detecting the $\gamma$ rays with a high purity germanium detector. Lifetime measurements were performed with a fast-fast timing coincidence system with a time resolution (FWHM) of 270 ps. A 15 mCi source of $^{22}$NaCl deposited between two thin Kapton foil (7.5 μm) was sandwiched between two identical porous samples. The thickness of the samples, of about 1.5 cm, was sufficient to assert that the positrons annihilate only into the samples. The samples were covered with an Al foil to ensure good thermal contact between the cold finger and the sample. Each spectrum contained about $2\times10^6$ counts and it was analysed with the POSITROFIT program [10] in four discrete components.

## 3. Results and discussion

Figure 1.a shows the *Ps* yield as a function of the positron implantation energy in Aerogel 85 and 150 mg/$cm^3$ measured at room temperature (full symbols). *Ps* yield is very high for any implantation energy. The more promising Aerogel (85 mg/$cm^3$) was also tested at low temperature (18 K, open symbols). The results indicate that *Ps* yield does not depend on temperature for any implantation energy. A similar result was recently found in silica films [11] and in MCM-41 [12].

Water absorption by porous $SiO_2$ was ascertained [13] even in ultra-high vacuum conditions, where at low temperatures (below 150 K) ice is formed [14]. The results presented in Fig. 1.a show that the effect of water adsorption at low temperature in the studied porous silica does not influence the *o-Ps* yield. The main reason is attributed to the hydrophobicity of the Aerogels presented here [15].

In order to obtain the positronium diffusion length the mesoporous samples were capped with an Al film to prevent *o-Ps* escape into the vacuum. The obtained results in Aerogel 85 and 150 mg/$cm^3$ at room temperature are shown in Fig. 1.b (full symbols). The Aerogel 85 was also measured at 18 K (open symbols) and the results show that the distribution does not depend on temperature. Only at low energy, where a fraction of the positrons form *Ps* at the Al surface, there is a small difference between the results obtained at room temperature and at 18 K due to the formation of ice at the Al surface. *Ps* formation is almost zero for implantation energies of about 2 keV; indeed, all the incident positrons annihilate in the metallic layer where no *Ps* formation is expected. The results of the *Ps* fraction as a function of the positron incident energy obtained in a capped porous sample were analyzed by means of a semi-linear fitting procedure using the VEPFIT program [16]. This program extracts relevant parameters from positron measurements on layered structures. In the present case our attention was focused on the diffusion *Ps* length inside the porous samples. The experimental data were fitted using a model of layers comprised of: surface, an Al layer, an interface and a semi-infinite layer of the porous material. The interface between Al and the porous material used in the analysis model takes into account a possible contamination effect of few nanometers ($z_{int}$) on the porous material with Al during deposition. The transport of positrons after implantation and thermalization in a solid is approximated by the classical diffusion theory. The depth dependent implantation rate of thermalized positrons is obtained by adopting the implantation-profiles of the so-called Makhov distribution [17]. According to the fitting results the positron diffusion length after thermalization, of about 20 nm, is much lower that the *Ps* diffusion length. In this condition it is possible to assume that the diffusion equation holds only for *Ps*. The adopted model is based on the assumption that the diffusion coefficient does not depend on the *o-Ps* temperature during its cooling. The continuous and dashed lines in Fig. 1.b correspond to the best fit values of the *o-Ps* diffusion length. The most relevant parameters of the fitting procedure are shown in Table 1. These high diffusion lengths are consistent with a high *o-Ps* emission outside the material also when positrons are implanted at relatively high energies.

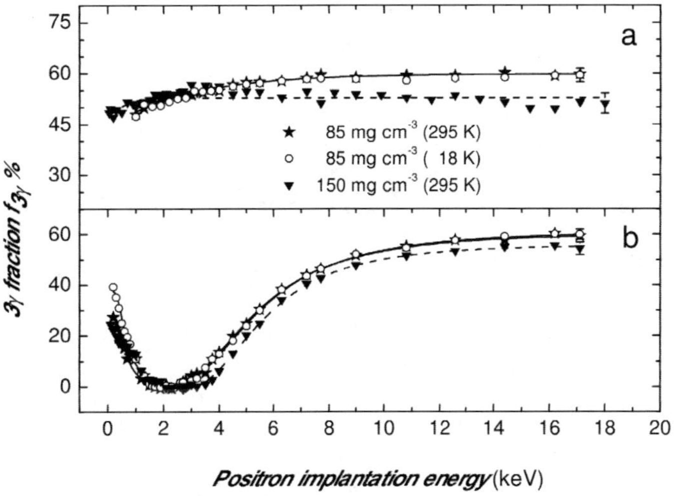


**Figure 1.** *Ps* fraction $f_{3\gamma}$ as a function of the positron implantation energy in Aerogel 85 and 150 mg cm$^{-3}$ at 18 K and 295 K. In panel b the samples were capped with Al to avoid the positronium escape. Error bars are shown for one point only in each evolution.

**Table 1.** Lifetime $\tau_4$, mean free path $l$, thicknesses of the Al capping layer $z_{Al}$ and the interface $z_{int}$ and *o-Ps* diffusion length $L_{Ps}$ in Aerogel 85 and 150 mg cm$^{-3}$ samples. The value indicated with F is a fixed parameter.

| Aerogel | T(K) | $\tau_4$ (ns) | $l$ (nm) | $z_{Al}$ (nm) | $z_{int}$ (nm) | $L_{Ps}$ (μm) |
|---|---|---|---|---|---|---|
| 85 | 300 | 129(3) | 23(2) | 108(6) | 15F | 2.30(0.25) |
| 85 | 18 | 128(3) | 22(2) | 108(6) | 15F | 2.25(0.25) |
| 150 | 300 | 122(3) | 14(2) | 120(5) | 15F | 1.90(0.20) |

Positron lifetime measurements were performed in vacuum in the studied materials at room temperature and also at 18 K in Aerogel 85 mg/cm$^3$. The best fits of the spectra were obtained with four discrete lifetime components. The first shortest is ascribed to *p-Ps* as well as to free positrons which annihilate in the bulk $SiO_2$ ($\tau_1$= 0.2(1)). The second component is ascribed to structural defects ($\tau_2$= 0.6(1) ns, with ($I_1$ + $I_2$)≈ 80%). The third component probably comes from o-Ps annihilating in structural defects ($\tau_3$= 3(1) ns, $I_3$≈ 0.80(5)%). The fourth lifetime component $\tau_4$ reported in Table 1 is the most interesting one, since it can be attributed to *o-Ps* annihilations into the pores. The results indicate the presence of a long *o-Ps* lifetime that is independent on the sample temperature (within the experimental uncertainty). The intensity of this component is about 20% in all cases; because this intensity depends also on the energy window of the stop channel in the lifetime spectrometer, only a fraction of *o-Ps* is revealed with a lifetime measurement.

It is known that the lifetime of *o-Ps* confined into closed pores depends on temperature according to the extended Tao-Eldrup (TE) model [18, 19]. But in silica when the porosity is very high (93-96%) and the pores are large and interconnected, as in the aerogels presented here, there are evidences that the lifetime is independent on temperature, i.e. there is no decrease in the pick-off rate at low temperature compared to room temperature [20-23]. In this case *Ps* is partially confined into the pores and it diffuses long distances during cooling as has been discussed before. The value of the fourth lifetime component allows to estimate the *o-Ps* mean free path $l$ [24, 25] between collisions with the pore walls (see Table 1).

The results obtained in Aerogels indicate that these materials are good converters for the AEgIS experiment under the following aspects: high *o-Ps* yield, long survival time of *o-Ps* and very high *o-Ps* mobility independently on temperature and on the effect of gas adsorption at low temperature. These characteristics indicate that *o-Ps* cooling in Aerogel is feasible. In fact, the literature tells that in this homogeneous porous materials *Ps* thermalization and escape in vacuum were ascertained [26-28]. In particular, Mills et al. [27] have observed 2% and 8% of *Ps* thermalized (with respect to the implanted positrons) into the vacuum at 4.2 and 77 K, respectively, in an homogeneous porous silica with a mass density (180 mg cm$^{-3}$) similar to the materials studied here.

## References

[1] Hughes R J 1993 *Nucl. Phys.* A **558** 605
[2] Kellerbauer A et al. (AEgIS collaboration) 2008 *Nucl. Instr. & Methods* B **266** 351
[3] Ferragut R et al. (AEgIS collaboration) 2011 *Canadian J. Physics* **89** 17
[4] Castelli F, Boscolo I, Cialdi S, Giammarchi M G and Comparat D 2008 *Phys. Rev.* A **78** 052512
[5] Brownlee D et al. 2006 *Science* **314** 1711
[6] Tillotson T M and Hrubesh L W 1992 *J Non-Cryst Solids* **145** 44
[7] Mills A P Jr 1978 *Phys. Rev. Lett.* **41** 1828
[8] Ferragut R et al. 2010 *J. Phys.: Conf. Ser.* **225** 012007
[9] Canesi E V et al. 2012 *Energy Environ. Sci.* **5** 9068
[10] Kirkegaard P and Eldrup M 1972 *Comput. Phys. Commun.* **3** 240
[11] Crivelli P et al 2010 *Phys. Rev.* A **81** 052703
[12] Ferragut R et al. 2011 *J. Phys.: Conf. Ser.* **262** 012020
[13] Bisi O, Ossicini S and Pavesi L 2000 *Surf. Sci. Rep.* **38** 1
[14] Sneh O, Cameron M A and George S M 1996 *Surf. Sci.* **364** 61
[15] Fesmire J E 2006 *Cryogenics* **46** 111
[16] Van Veen A, Schut H, de Vries J, Hakvoort R A and IJpma M R 1991 *AIP Conference Proceedings* **218** 171
[17] Nieminen R M and Oliva J 1980 *Phys. Rev.* B **22** 2226
[18] Goworek T, Ciesielski K, Jasinska B and Wawryszczuk J 1998 *Chem. Phys.* **230** 305
[19] Zaleski R 2002 *Positronium lifetime vs. temperature and free volume size TABLES – Pick-off model calculations*. Institute of Physics, Maria Curie-Sklodowska University, Lublin, Poland
[20] Saito H, Nagashima Y and Hyodo T 1995 *Phys. Rev.* B **52** R689
[21] Thraenert S et al. 2007 *Phys. Stat. Sol.* (c) **4** 3819
[22] Šauša O et al. 2008 *Acta Phys. Pol.* A **113** 1517
[23] Kiefl R F and Harshman D R 1983 *Phys. Lett.* A, **98,** 447
[24] Dull T L, Frieze W E, Gidley D W, Sun J N and Yee A F 2001 *J. Phys. Chem.* B **105** 4657
[25] He C et al. 2007 *Phys. Rev.* B **75** 190454
[26] Nagashima Y et al 1995 *Phys. Rev.* A **52** 258
[27] Mills A P Jr, Shaw E D, Chichester R J and Zuckerman D M 1989 *Phys. Rev.* B **40** 2045
[28] Tanaka H K M, Kurihara T and Mills A P Jr 2006 *J. Phys.: Condens. Matter.* **18** 8581

Journal of Physics: Conference Series **443** (2013) 012065 doi:10.1088/1742-6596/443/1/012065

# Positron annihilation characteristics in multi-wall carbon nanotubes with different average diameters

**L A Tuyen[1], Zs Kajcsos[2], K Lázár[3], T D Tap[4], D D Khiem[1], P T Phuc[1]**

[1]Center for Nuclear Techniques-Vietnam Atomic Energy Institute, 217 Nguyentrai Street, District 1, Hochiminh City, Vietnam
[2]Institute for Particle and Nuclear Physics-Wigner Research Center for Physics, Hungarian Academy of Sciences, P.O.B. 49, H-1525 Budapest 114, Hungary (deceased)
[3]Centre for Energy Research, Institute of Isotopes, Hungarian Academy of Sciences, P.O.B 49, H-1525, Budapest 114, Hungary
[4]Department of Nuclear Physics-University of Natural Science, 227 Nguyenvancu Street, District 5, Hochiminh City, Vietnam

E-mail: tuyenluuanh@gmail.com

**Abtract.** Positron lifetime spectroscopy was used to study multi-wall carbon nanotubes. The measurements were performed in vacuum on the samples having different average diameters. The positron lifetime values depend on the nanotube diameter. The results also show an influence of the nanotube diameter on the positron annihilation intensity on the nanotube surface. The change in the annihilation probability is described and interpreted by the modified diffusion model introducing the positron escape rate from the nanotubes to their external surface.

## 1. Introduction

Soon after their discovery, multi-wall carbon nanotubes (MWNTs) became promising high technology materials with a wide variety of possible applications. The influence of synthesis conditions on the quality and physical properties of MWNTs was studied by various methods [1, 2, 3, 4, 5]. Recently, positron techniques have been considered to be useful methods for studying nanomaterials [4, 6, 7, 8, 9, 10, 11, 12, 13, 14]. The unique sensitivity of the positron techniques for probing properties and special imperfection structures of material is due to positron trapping in the low-electron density regions (such as surfaces, interstitial sites, free volumes, and voids) of real solids [15]. Early studies reported on different values of positron lifetime and annihilation probability on the surface of the nanotubes [4, 5, 7, 8, 9, 10]. Our previous study also showed that the presence of amorphous structure, the kind and concentration of defects as well as the presence of possible catalytic agents can affect the positron annihilation data. It was also indicated that the positron lifetime on the external surface of the nanotubes could be related to the external surface area of the nanotube, which depends on the average nanotube diameter [6].

In the present work lifetime spectroscopy (LT) is used to study positron annihilation characteristics of high quality MWNT samples produced by Thermal Chemical Vapor Deposition (CVD) technique with different average diameters [2, 11]. The samples were carefully selected to avoid the influence of the amorphous fraction, of the defects as well as of the catalytic agents in the experimental data. Besides, a modified positron diffusion model for the nanomaterial was used to include the positron escape rate from the nanotubes in the positron annihilation probabilities.

## 2. Experimental

Samples of multiwall carbon nanotubes with average diameters (*d*) in the range of 50-100 nm were investigated by transmission electron and atomic force microscopy [2, 6, 11]. The high quality samples were selected for LT measurements which will be referred to as M1 (d≈50 nm), M2 (d≈60 nm), M3 (d≈75 nm), M4 (d≈80 nm), M5 (d≈90 nm) and M6 (d≈100 nm). The powder samples were densely packed and pressed by c. 5 MPa into pellets. The $^{22}$Na source (of c. 20 MBq activity kept between thin Al foils) was inserted between two identical pellets and this sandwich was then covered with a thin Al foil. LT measurements were carried out with the sample in vacuum using a conventional fast-fast spectroscopy with a time resolution of 210 ps. The spectra were recorded in a Microfast 16k PC MCA with integral counts of $2\text{x}10^6$. Data evaluation was carried out using the LT v. 9 program applying a source correction of 10% and accounting for annihilation in the various aluminium foils.

## 3. Results and discussion

In our previous studies [6, 11], the LT spectra of MWNTs samples were decomposed into three lifetime components. The short lifetime component $\tau_1$ (≈215 ps as case of positron annihilation in graphite [9, 14]) was assigned to originate from positron annihilation in the nanotubes, the characteristic lifetime $\tau_2$ to the annihilation on the external surface of the nanotubes, and the longer lifetime component $\tau_3$ (≈1 ns) was related to the formation of positronium in defects and/or free volumes [6, 11]. A similar result was also indicated by Kun et al [9]. However, for the high quality samples of this study, the long lifetime component $\tau_3$ was not observed in any of the LT spectra. It seems that positronium formation is negligible in the high quality samples. The positron lifetime spectra are shown in figure 1.

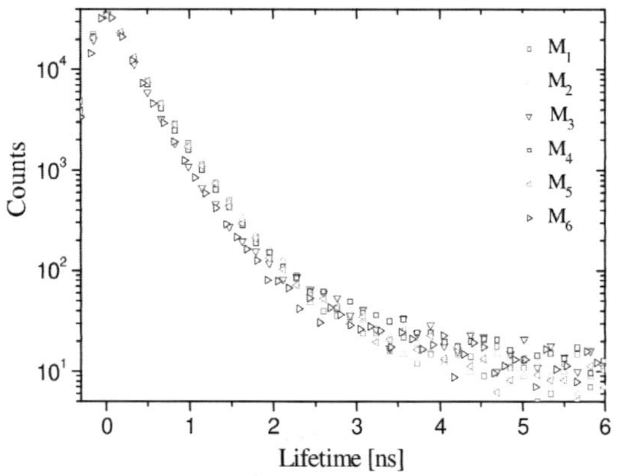


**Figure 1.** Positron lifetime spectra of the samples of various diameters

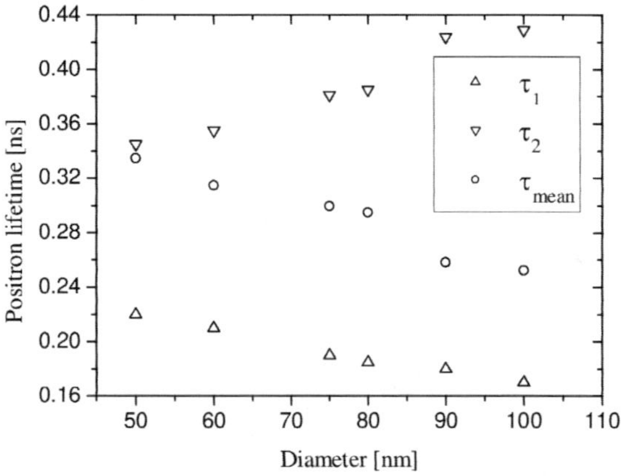


**Figure 2.** Positron lifetimes versus the average diameter of the samples

Figure 2 shows the correlation between the positron lifetimes and the average nanotube diameter. The short lifetime and the mean lifetime component ($\tau_1$ and $\tau_{mean}= (I_1.\tau_1+I_2.\tau_2)/(I_1+I_2)$, respectively,) of the positrons are decreasing upon increasing average nanotube diameter. Similarly, the mean positron lifetime was found to decrease with increasing particle size in ZnO [16] and silver [17] nanoparticles. Conversely, the characteristic lifetime ($\tau_2$) assigned to positron annihilation on the external surface of the nanotubes increases with increasing average nanotube diameter. The increase of the diameter of the nanotubes leads to an increase of the external surface area as well as the interstitial region volume among the nanotubes. Hence, the increase of $\tau_2$ can be interpreted by assuming that the larger the diameter of the nanotube, the larger the chance for the positron to get attached to the external surface so the longer the resulting lifetime will become [6, 11]. This result may also explain many different values of the characteristic positron lifetime in the early experiments performed by different authors [4, 5, 7, 8, 9, 10] and also reflects some conditions for MWNT synthesis with different average diameters (not shown here).

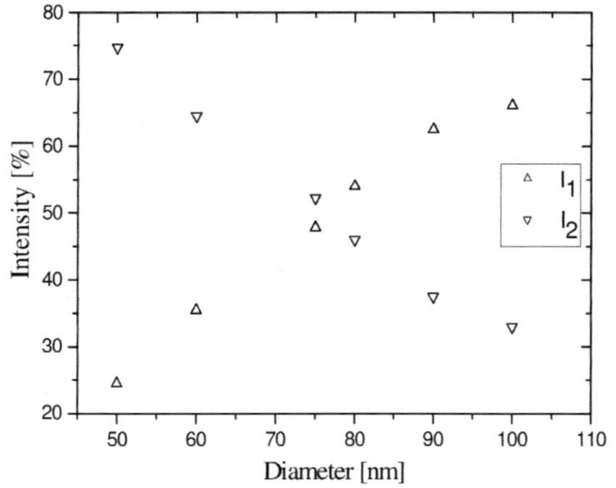


**Figure 3.** Positron annihilation intensities of samples with different average diameters.

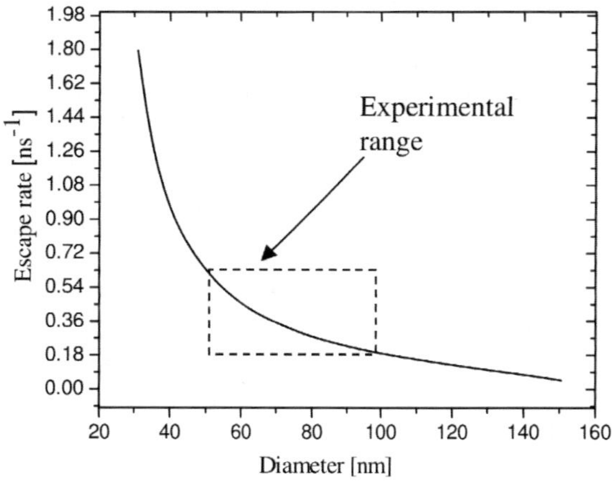


**Figure 4.** The escape rate versus the nanotube average diameter

The result of the LT measurements indicate that the intensity of the positron lifetime component associated with the external surfaces is decreasing as a function of average nanotube diameter (figure 3). Actually, for the nanomaterial or the fine-grain powder, the surface to volume ratio is inversely proportional to the diameter but proportional to the escape probability of positron to the surface [15, 18]. This suggests that the escape probability of the positrons decreases with increasing nanoparticle diameter. To account for the latter, positron diffusion theory developed for nanomaterials was considered [13, 15]. We suppose that the diffusion length of the thermalized positrons in the MWNTs is approximately equal to that in the graphite material. Indeed, in MWNTs the distance between the coaxially rolled graphene sheets is approximately equal the distance between the graphene sheets in graphite (≈ 0.34 nm) [1, 5]. As a result, the thermalized positron diffusion length in MWNTs can be calculated by $L \approx \sqrt{\tau_f . D} = 102$ nm, where, $\tau_f = 215$ ps and $D = 0.49 \times 10^{-4} m^2 s^{-1}$ are the bulk positron lifetime and the positron diffusion coefficient in the graphite material, respectively [9, 14]. Therefore, if the radius (half of the diameter) of the nanotube is below 100 nm, the thermalized positrons diffusing in the nanotube can escape to the external surface before annihilation [12]. The positron annihilation intensity on the external surface of the nanotube can be approximated as

$$I_2 \approx \frac{\alpha D}{\alpha D + d^2(\lambda_1 - \lambda_2)} \tag{1}$$

and the positron escape rate $k$ from the nanotube can be calculated using

$$k \approx \alpha D / d^2 \tag{2}$$

where α, $d$, $I_2$ are the shape and construction parameter of the nanotube, the average nanotube diameter and the annihilation intensity on the external surface of the nanotube, respectively. Additionally, $\lambda_1$ and $\lambda_2$ are the positron annihilation rates in and on the external surface of the nanotube. The value of α ≈ 0.034 determined by experimental data fitting of equation (1) was used for calculating the escape rate $k$ versus the average nanotube diameter in equation (2). The results show the influence of the the average nanotube diameter on the positron escape rate to the external surface of the nanotubes (figure 4). In the average nanotube diameter range below 30 nm (the small diameter range), the positron escape rate is very large, consequently the positron annihilation mainly occurs on the external surface of nanotubes. This is confirmed by the single lifetime component found in the small diameter MWNTs samples by the early studies [4, 5, 7]. Conversely, in the nanotube diameter range above 100 nm the escape rate is about one order of magnitude smaller than for the small diameter range, and the positrons predominantly annihilate in the nanotubes. The MWNT diameters in this study cover the range of 50-100 nm, therefore, the positrons annihilate both in and on the external surface of the nanotubes.

## 4. Conclusions

In this study, the correlation between the positron lifetimes with the average nanotube diameter was established for high quality multi-wall carbon nanotube samples produced by the Thermal CVD technique. The short and mean lifetime components decrease as the average nanotube diameter increases. The results also indicate the dependence on nanotube diameter of the probability of positron annihilation on the external surfaces of the nanotubes. A modified diffusion model was applied to the multi-wall nanotubes and an escape rate *k* was introduced for the positrons from the nanotubes to the surface to properly account for the positron annihilation probability in and on the external surface of the nanotubes. The present study may contribute to a better understanding of positron annihilation in MWNTs and the obtained results will be applied to better control the synthesis process of nanotubes.

**Acknowledgement**

This study was supported by the inter-governmenatal scientific cooperation project TÉT 10-1-2011-0671 between Vietnam and Hungary, for which the authors express their sincere gratitude. Helpful discussions with Dr. L. Bottyán, Wigner Research Centre for Physics, Budapest are gratefully acknowledged.

**References**

[1] Gerard L J, Subramoney S, Ruoff R S, Berber S and Tomanek D 2002 *Carbon* **40** 1123
[2] Thi P T, Han G H, Lee I H, Thanh L B and Chien D M 2007 *Proc. Int. Workshop on Nanotechnology and Application* (Vungtau-Vietnam) p 767
[3] Ishibashi S 2002 *J. Phys.: Condens. Matter* **14** 9753
[4] Ito Y and Suzuki T 1999 *Phys. Rev. B* **60** 15636
[5] Ohdaira T, Suzuki R, Kobayashi Y, Akahane T and Dai L 2002 *Applied Surface Science* **194** 291
[6] Luu A T, Kajcsos Z, Thanh N D, Dung T Q, Nhon M V, Lázár K, Havancsák K, Huhn G, Horváth Z E, Tap T D, Son L T and Phuc P T 2009 *Phys. Status Solidi C* **6** 2578
[7] Schaefer H E, Forter M and Wurschum R 1992 *Phys. Rev B* **45** 12164
[8] Chakrabarti K, Nambissan P M G, Mukherjee C D, Bardhan K K, Kim C and Yang K S 2006 *Carbon* **44** 948
[9] Kun M X, Ong E and Ji H Y 1999 *Acta Phys. Sin.* **8** 783
[10] Kristiak J, Krstikova K and Sausa O 1994 *Phys. Rev. B* **50** 2792
[11] Tuyen L A, Kajcsos Z, Thanh N D, Horváth Z E, Tap T D and Phuc P T 2011 *Proc. Na. Conf. on Nuclear Science and Technology* (Ninhthuan-Vietnam) (Hanoi: Science and Technique Publication House) p 809
[12] Murakami H and Sano M 2002 *J. Phys. Soc. Jpn.* **71** 125
[13] Varisov A Z, Grafutin V I, Zaluzhnyi A G, Ilyukhina O V, Myasishcheva G, Prokopev E P, Timoshenkov S P and Funtikov Y V 2008 *J. Surface Investigation, X-Ray, Synchrotron and Neutron Techniques* **2** 959
[14] Brusa R S, Somoza A, Huck H, Tiengo N, Karwasz G P, Zecca A, Reinoso M and Halac E B 1999 *Applied Surface Science* **150** 202
[15] Brandt W 1974 *Appl. Phys.* **5** 1
[16] Sharmaa S K, Pujari P K, Sudarshan K, Dutta D, Mahapatra M, Godbole S V, Jayakumar O D and Tyagi A K 2009 *Solid State Communications* **149** 550
[17] Mukherjee M, Chakravorty D and Nambissan P M G 1998 *Phys. Rev. B* **57** 848
[18] Paulin R, Ripon R and Brandt W 1973 *Phys. Rev. Letters* **31** 1214

# Tuning porosity of silica films by using various surfactants and changing their loading: A study of positron annihilation Doppler broadening based on a slow positron beam

**Bangyun Xiong, Wenfeng Mao, Chunqing He**[1]
Key Laboratory of Nuclear Solid State Physics Hubei Province, School of Physics and Technology, Wuhan University, Wuhan 430072, China

E-mail address: hecq@whu.edu.cn

**Abstract**. Porous silica films were synthesized via a sol-gel method using a nonionic amphiphilic triblock copolymer (F127) and a cationic surfactant (CTAB) as the structural templates with varying weight ratio. Positron annihilation Doppler broadening spectroscopy based on a slow positron beam was used to study the prepared silica films. For the porous silica films, the S parameter increased gradually with increasing the surfactant loading, which showed that higher porosity was introduced in the silica films with more porogen amount.

## 1. Introduction

Porous silica films have attracted considerable attention for their potential applications, such as adsorbents, catalysts and catalyst supports, low-k dielectric materials, chemical sensors, host-guest chemistry, antireflective coatings and membrane separations. However, tailoring of the porous structures is indispensable for these practical applications which benefit from highly ordered structures and high porosity. Since the synthesis of silicate mesoporous molecular sieves designated as M41S [1], nonionic amphiphilic block copolymers [2-4] have been subsequently used to direct the formation of a family of mesostructured materials with tunable mesoporous structures and various morphologies. Ionic surfactants [5-7] have also been devoted to design porous structures. Such surfactant-templated porous materials rely on a successful application of sol-gel chemistry and self-assembly between the organic structure-directing agents and the inorganic precursor species. In this method, selection of surfactant is of most importance for determining the porous structure.

In this study, porous silica films were tuned by two kinds of surfactants as pore generators with various loading. The silica films were studied by Doppler broadening spectroscopy of positron annihilation based on a slow positron beam. The porous structure of the silica films was found to be significantly influenced by the additive of surfactant template and preparation conditions.

## 2. Experimental

### *2.1. Sample preparation*

Tetraethoxysilane (TEOS) was used as the silica source and a nonionic amphiphilic triblock copolymer ($EO_{106}PO_{70}EO_{106}$, Pluronic F127) and a cationic surfactant hexadecyltrimethylammonium bromide (CTAB: $CH_3(CH_2)_{15}N^+(CH_3)_3Br^-$) were used as templating agents, respectively. The

[1] To whom any correspondence should be addressed.

surfactant templates were introduced with various weight ratios, i.e. $W_{template}/(W_{template} + W_{TEOS})$, which were assigned as 0, 5, 10, 15, 20 and 25 wt%, respectively. The silica thin films were prepared on polished silicon wafers by dip-coating using precursor sols of TEOS, surfactant, anhydrous ethanol (EtOH), distilled water and hydrochloric acid (HCl) (the mole ratios: 1 TEOS: x surfactant: 22 EtOH: 8 $H_2O$: 0.002 HCl; x was determined by the surfactant weight ratio). There was a thin layer of non-porous $SiO_2$ (~ several nm) on the as-grown silicon substrate. The as-deposited films were elevated at a heating rate of 1 °C/min and subsequently calcined at 450 °C in air for 5 hours in order to remove the organic species. To study the evolution of the pore structure, the same as-deposited silica film templated by CTAB with the weight ratio of 15 wt% was treated successively at 200, 300, 350 and 400 °C in air for 2 hours, respectively.

### 2.2. *Characterization of silica films*

Positron annihilation Doppler broadening spectroscopy based on a slow positron beam was used to characterize the synthesized silica films. The Doppler broadening spectra were measured with a high-purity Ge detector as a function of the incident positron energy E by using a monoenergetic positron beam at Wuhan University. A spectrum with about $10^6$ counts was collected for each incident positron energy. The low-momentum part was characterized by the S parameter, defined as the fraction of the annihilation line over the energy range of 511 keV ± $|\Delta E_\gamma|$ (where $\Delta E_\gamma \sim 0.76$ keV) around the central region. The S parameters were used to evaluate the porosity of the mesoporous films, since S parameters in films strongly depend on pore size and/or pore density, which govern positronium formation and annihilation in them.

## 3. Results and discussion

### 3.1. *Positron annihilation characteristics in silica films*

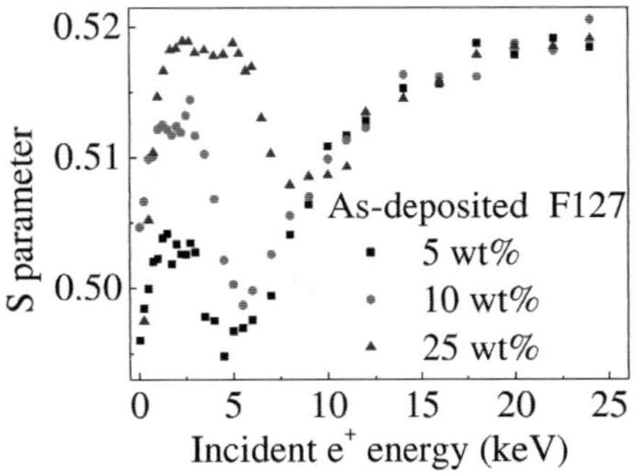


**Figure 1.** S-E curves of the as-deposited silica films with various loading.

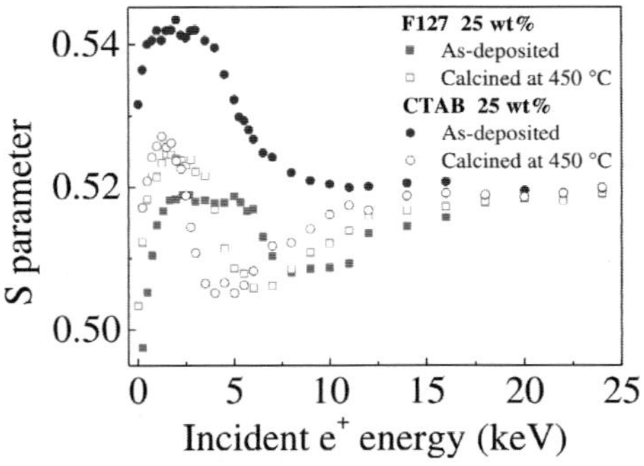


**Figure 2.** S-E curves of different surfactant-silica films before and after calcination.

Figure 1 showed the S-E curves of the as-deposited silica films using F127 as the structure-directing agent. For a typical S-E curve, for instance, in as-deposited film with 25 wt% F127, the S parameter increased sharply with increasing positron energy from 0 keV to around 2 keV, from which to ca. 5 keV it almost kept at a constant value resulted from positron annihilation in the as-deposited film. With increasing positron energy from about 5 to 10 keV, the S parameter showed a dip, which was attributed to positron annihilation in the nonporous $SiO_2$ thin layer on the silicon substrate. Finally, with implanting positron deeper the S parameter gradually increased and reached a plateau. Considering the positron implantation profile in condensed matters [8, 9], three regions were revealed by the S-E curves for the as-prepared films on silicon wafers: The first region was the surfactant-silica film; the second one was the nonporous $SiO_2$ thin layer on silicon substrate, and the third region was the silicon itself. Keep in mind that S parameters in the films on Si substrates were closely associated with physiochemical characteristics of the films. The S parameters of the as-deposited films were found to increase gradually with loading more additive from 5 to 25 wt%. For the as-deposited films with the same weight ratio, much higher S parameter was found in the CTAB-silica film than that in the F127-silica film (Figure 2), due to the different characteristics of positron annihilation in the two kinds of surfactant-silica nanocomposites. Several factors could be taken into consideration to interpret the different S parameters in silica films: film chemical composition [10, 11], Ps formation probability and porosity of the film. In this work, different chemical compositions in CTAB-silica and F127-silica were responsible for the large difference in S parameters of the as-deposited films.

After being calcined at 450 °C for 5 hours, the S parameter of CTAB-templated film decreased significantly, meanwhile, a distinct increase was observed for F127-templated film due to the removal of the additives. Similar S values in both calcined films were due to Ps annihilation in porous structures [12, 13] resulted from removal of the surfactants with the same loading. Because of thinning of the porous silica film due to condensation of inorganic species, and formation of mesopores during calcination, the dip of S-E curves shifted to the lower energy region, for instance, from ac. 9 to 6 keV for the F127-templated silica film.

### *3.2. Evolution of porous structure depending on treatment temperature*

Figure 3 showed the S-E curves of the same silica film treated successively at different temperatures. The S parameter of as-deposited silica film increased gradually and approached a plateau ac. 0.54 in the energy range of 1 to 4 keV. After being heated at 200 °C for 2 hours, the S parameter distinctly increased to ac. 0.55, which was attributed to the removal of water and ethanol in surfactant-silica film during the drying process. Little change in the width of the plateau could be found after the treatment at 200 °C, indicating that the template was hard to decompose at about 200 °C. However, the S parameter decreased remarkably to around 0.53 by successive heating at higher temperatures due to the removal of CTAB template, meanwhile, significant reduction in the width of the plateau was observed upon successive treatment at higher temperature up to 300 °C due to the condensation of inorganic species and the formation of porous framework. No changes were found for the S parameters in the porous silica film treated from 300 °C to 400 °C, revealing good thermal stability of the porous silica film.

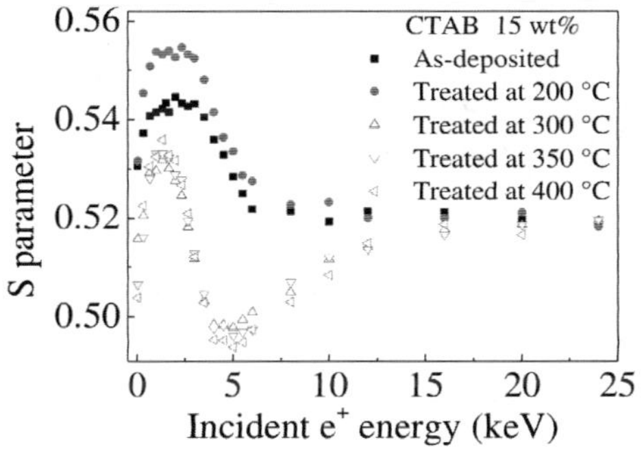


**Figure 3.** S-E curves of the same silica film treated successively at different temperatures.

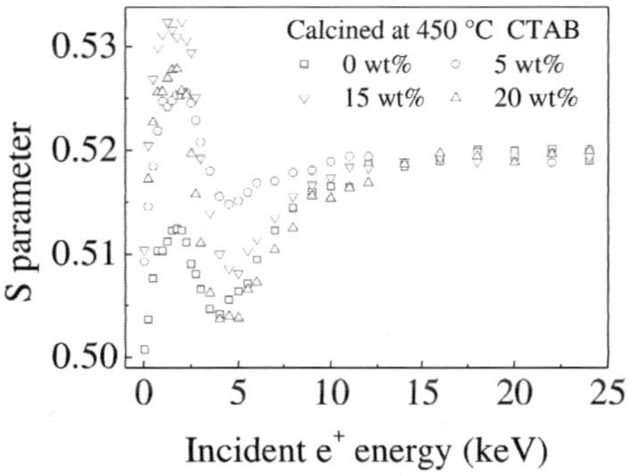


**Figure 4.** S-E curves of porous silica films with increasing loading of CTAB.

### *3.3. Evolution of pores depending on porogen loading*

Figure 4 showed S-E curves of the silica films, which were prepared by adding CTAB with various weight ratios and calcined at 450 °C for 5 hours. The S parameter of silica film was low (around 0.51) when no surfactant was loaded. A sharp increase of S parameter was found in the calcined silica film using 5 wt% CTAB as the template, revealing that pores were introduced by the addition of surfactant. For the porous silica films, the S parameter was found to increase gradually with increasing porogen loading from 5 wt% up to 15 wt%, which showed that higher porosity was introduced in the silica films prepared with more porogen amount. Similar results of pore evolution depending on the porogen content were reported in a previous paper, in which a triblock polymer Pluronic F38 (BASF surfactant) was used as a pore generator [14]. In this study, the S parameter decreased distinctly with further increasing the porogen loading up to ca. 20 wt% perhaps because of formation of ordered mesostructures [6] and their contraction [15] upon calcination.

## 4. Conclusion

In conclusion, porous silica films were prepared via a sol-gel synthesis by two kinds of surfactants as the templating agents: a nonionic copolymer F127 and a cationic surfactant CTAB. Positron annihilation Doppler broadening spectroscopy based on a slow positron beam was used to study the synthesized silica films. Results indicated that the mesostructures of the silica films were successfully tuned by varying the additive and loading. The S parameter of porous silica films increased gradually with increasing the porogen loading, which showed that higher porosity was introduced with more loading. Furthermore, the temperature dependence of evolution of the mesostructures was also studied. After being treated at 200 °C for 2 hours, the S parameter of the CTAB-silica nanocomposite films increased distinctly, however, it decreased remarkably due to the removal of the surfactant template when successively treated at higher temperatures.

## Acknowlegements

This work was supported in part by National Natural Science Foundation of China (NSFC) under Grants 10975108 & 10974149, Research Fund for the Doctoral Program of Higher Education of China (RFDP) under Grant 20090141120069, Wuhan University Research Fund (Grant 1081004) and the Fundamental Research Funds for the Central Universities (Grant 2012202020218).

## References

[1] Beck J S et al. 1992 *J. Am. Chem. Soc.* **114** 1083
[2] Yang P D, Zhao D Y, Margolese D I, Chmelka B F and Stucky G D 1998 *Nature* **396** 152
[3] Zhao D Y, Feng J L, Huo Q S, Melosh N, Fredrickson G H, Chmelka B F and Stucky G D 1998 *Science* **279** 548
[4] Tanaka S, Nishiyama N, Oku Y, Egashira Y and Ueyama K 2004 *J. Am. Chem. Soc.* **126** 4854
[5] Ogawa M 1996 *Chem. Commun.* 1149
[6] Honma I, Zhou H S, Kundu D and Endo A 2000 *Adv. Mater.* **12** 1529
[7] Ma C Z, Han L, Jiang Z, Z Huang H H, Feng J, Yao Y and Che S N 2011 *Chem. Mater.* **23** 3583
[8] Ritley K A, Lynn K G, Ghosh V J, Welch D O and McKeown M 1993 *J. Appl. Phys.* **74** 3479
[9] Petkov M P, Weber M H, Lynn K G, Rodbell K P and Cohen S A 1999 *J. Appl. Phys.* **86** 3104
[10] Ito K, Kobayashi Y and Nanasawa A 2003 *Appl. Phys. Lett.* **82** 654
[11] Ito K and Kobayashi Y 2005 *Acta Phys. Pol.*, A **107** 717
[12] Brusa R S, Spagolla M, Karwasz G P, Zecca A, Ottaviani G, Corni F, Bacchetta M and Carollo E 2004 *J. Appl. Phys.* **95** 2348
[13] Liszkay L et al. 2012 *New J. Phys.* **14** 065009
[14] He C Q, Oka T, Kobayashi Y, Oshima N, Ohdaira T, Kinomura A and Suzuki R 2008 *Appl. Surf. Sci.* **255** 183
[15] He C Q, Suzuki R, Ohdaira T, Oshima N, Kinamura A, Muramatsu M and Kobayashi Y 2007 *Chem. Phys.* **331** 213

Journal of Physics: Conference Series **443** (2013) 012067 doi:10.1088/1742-6596/443/1/012067

# Material characterization for advanced Si LSI process technology by means of positron annihilation

**A Uedono[1], N Oshima[2], T Ohdaira[2], R Suzuki[2] and S Ishibashi[3]**

[1] Division of Applied Physics, Faculty of Pure and Applied Science, University of Tsukuba, Tsukuba, Ibaraki 305-8573, Japan
[2] Research Institute of Instrumentation Frontier, National Institute of Advanced Industrial Science and Technology (AIST), Tsukuba, Ibaraki 305-8568, Japan
[3] Nanosystem Research Institute (NRI) "RICS", National Institute of Advanced Industrial Science and Technology (AIST), Tsukuba, Ibaraki 305-8568, Japan

E-mail: uedono.akira.gb@u.tsukuba.ac.jp

**Abstract.** Vacancy-type defects in gas cluster ion implanted Si and electroless deposited Cu were studied by monoenergetic positron beams. For Ar gas cluster ion implanted Si, we found that the vacancy-rich region was localized at a depth of 0-13 nm. Two different defect species were found to coexist in the damaged region, and they were identified as divacancy-type defects and vacancy clusters filled with Ar. For electroless deposited Cu films, the major defect species were identified as vacancy complexes ($V_3$-$V_4$) and larger vacancy clusters ($\sim V_{10}$). Annealing behaviours of the defects and the relation between the defects and impurities were also discussed. We have demonstrated the efficacy of positron annihilation to aid in the optimization of process parameters for advanced Si LSI processes.

## 1. Introduction

Developments of Si-technology based industries and the qualities of their products mainly depend on the ability to decrease the size of integrated circuits [1]. The first transistor was made by eyes and fingers in 1947, but a third dimension 22-nm high-*k*/metal gate transistor was now shown to be a production level. Modern semiconductor devices are fabricated though numerous processing steps. They are separated into two major categories; front-end-of-line (FEOL) and back-end-of-line (BEOL). FEOL processing refers to the formation process of transistors such as junction formation, insulator/metal-gate fabrication, etc. BEOL refers to the process connects those transistors using metal wires isolated by dielectric layers. Significant efforts to introduce new materials have been made in order to achieve "scaling". However, one often faces the difficulties in the control of materials properties which caused by atomic scale disorders. We have shown that the positron annihilation technique is a powerful characterization tool for modern Si-technology related materials, and can contribute developments of both FEOL and BEOL processing [2-10]. In the present paper, as typical applications of this technique to those processing technologies, we report studies of defects in ultra-shallow region of gas cluster ion implanted Si and in electroless deposited Cu for composite barrier/seed layers of Cu interconnects.

The gas cluster ion beam (GCIB) is based on electrostatic acceleration of charged cluster ions with a few hundred to thousands of atoms or molecules. This technique and related phenomena have been extensively studied because of its unique characteristics [11]. It has been suggested that GCIBs have great potential as a tool for ultra-shallow implantation, high yield sputtering, surface smoothing, ion-assisted deposition/modification, and time-of-flight secondary ion mass spectrometry. When cluster ions are implanted into solids, high density electric excitation states of the targets and the projectile are created. The defect-introduction mechanism and resultant damage by GCIBs are therefore thought to differ from those for conventional ion implantation. Cu interconnects and related technologies have been extensively developed to meet requirements of high-performance metal lines used in modern ULSI devices. The dual damascene scheme using electroplated Cu is the mainstream technology [1]. Recently, composite barrier/seed layers using physical vapor deposition and electroless Cu plating are proposed to obtain an efficient bottom-up fill during Cu electroplating [12]. The integration of an electroless Cu process into the Cu metallization process, however, requires detailed knowledge about the behavior of Cu atoms and impurities which may affect blistering and post-anneal void formation.

## 2. Experiments

The samples investigated were Ar and B cluster ion implanted Si fabricated using a GCIB developed by TEL Epion Inc. [13]. Neutral clusters were generated by adiabatic expansion of Ar and $B_2H_6$/He gas into a vacuum chamber, and they were ionized by electron bombardment. Ionized clusters with a mean size of $2\times10^3$ atoms were implanted with energy of 20 - 60 keV to Cz-Si. The cluster ion doses were $1\times10^{14}$ cm$^{-2}$ and $2.5\times10^{13}$ cm$^{-2}$ for Ar and B gas clusters, respectively.

ELD-Cu films with high and low residual impurity concentrations were deposited with a novel electrolyte using a metal ion complex as a reducing agent [14] to eliminate hydrogen evolution during the electroless plating process (hereafter, these are referred as ELD-Cu(H) and ELD-Cu(L)). The thickness of ELD-Cu(H) and ELD-Cu(L) were 1.6 and 2.3 μm, respectively. For ELD-Cu(H), the major impurity species are nitrogen ($10^{21}$ cm$^{-3}$), carbon ($10^{20}$ cm$^{-3}$), oxygen ($10^{19}$ cm$^{-3}$), and sulfur ($10^{18}$ cm$^{-3}$). For ELD-Cu(L), these concentrations are an order of magnitude lower. The mean grain size is calculated to be 150 and 540 nm for ELD-Cu(H) and ELD-Cu(L), respectively. Isochronal annealing is performed up to 750°C in vacuum (60 min).

With a monoenergetic positron beam, the Doppler broadening spectra of the annihilation radiation were measured with a Ge detector as a function of the incident positron energy *E*. The low-momentum part was characterized by the *S* parameter, defined as the number of annihilation events over the energy range of 511 keV$\pm\Delta E_\gamma$ (where $\Delta E_\gamma$=0.76 keV) around the center of the peak. The relationship between *S* and *E* was analyzed to obtain the depth distribution of *S* [15]. We also used a coincidence-detection system, and the obtained spectra were compared with those theoretically calculated using QMAS (Quantum MAterials Simulator) code based on the projector augmented-wave (PAW) method [16]. The (*S*,*W*) value was used to identify the defects, where the *W* value calculated from the tail of the Doppler broadening spectrum in the range of 3.4 keV$\leq|\Delta E_\gamma|\leq$6.8 keV. The lifetime spectra of positrons were also measured using a pulsed monoenergetic positron beam, and analyzed using RESOLUTION computer program [17].

## 3. Results and discussion

### *3.1. Vacancy-type defects in gas cluster ion implanted Si*

Figure 1 shows the *S* values of Ar and B gas cluster implanted Si as a function of incident positron energy *E*. For Ar gas cluster implanted samples, the *S* value at low *E* (1 keV) was larger than the *S* value for defect-free Si (at *E*≅20 keV), suggesting the annihilation of positrons trapped by vacancy-type defects introduced by the implantation. The solid curves are fits to the experimental data, and the derived depth distributions of the *S* values are shown in Fig. 2. For all samples, the damage depth was 10-13 nm. The projected range of monomer $Ar^+$ with energy of 10 keV is about 15 nm, which is close to the damaged area shown in Fig. 2. In the present experiment, however, the average energy of each monomer ion is an order of 10 eV. A cluster ion interacts with the surface almost simultaneously, and deposits its energy into a very small volume of the target material. The impact area of gas cluster ion implantation thus experiences extremely high temperature and pressure/temperature transients, which cause non-linear implantation effects. The observed extremely deep damaged region can be attributed to such properties of cluster ion implantation.

Lifetime spectra of positrons for Ar gas cluster implanted Si were measured at $E$=1 keV. The measured spectra were decomposed into two components, the short- and long-lived positron lifetimes, $\tau_1$ and $\tau_2$, and the intensities for the long-lived component, $I_2$, are shown in Fig. 3. No large change in the lifetimes was observed for the samples with different Ar gas cluster acceleration energies. The value of $\tau_1$ is close to the lifetime of positrons trapped by a divacancy $V_2$, which is typical for vacancy-type defects introduced by conventional ion implantation. In the present experiments, because the damaged region was expected to be amorphous [11], the defects detected by the first component are considered to be the open spaces which have an open space close to that of $V_2$. The defects detected by the second component can be identified as large vacancy-clusters, where the open space of such defects could be larger than that of $V_{10}$ [18]. The value of $I_2$ increased with increasing gas cluster acceleration energy, suggesting an increase in the concentration of the vacancy clusters, and this was an origin of the increase in the $S$ value observed for those samples.

Figure 4 shows the $S$–$W$ relationships for gas cluster ion implanted samples measured using the

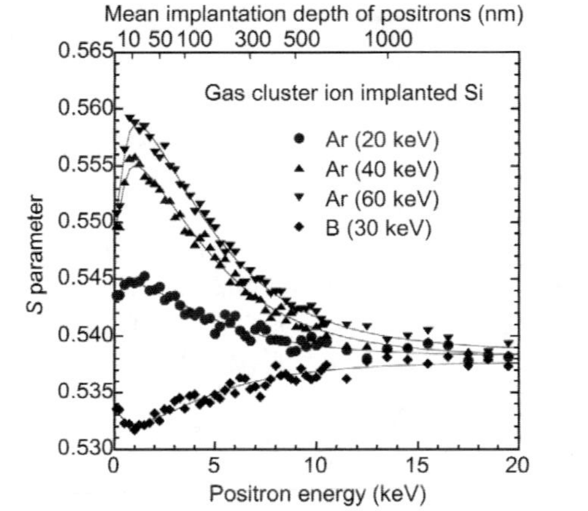


**Figure 1.** $S$ parameters as a function of incident positron energy $E$ for gas cluster ion implanted Si. The solid curves are fits to the experimental data.

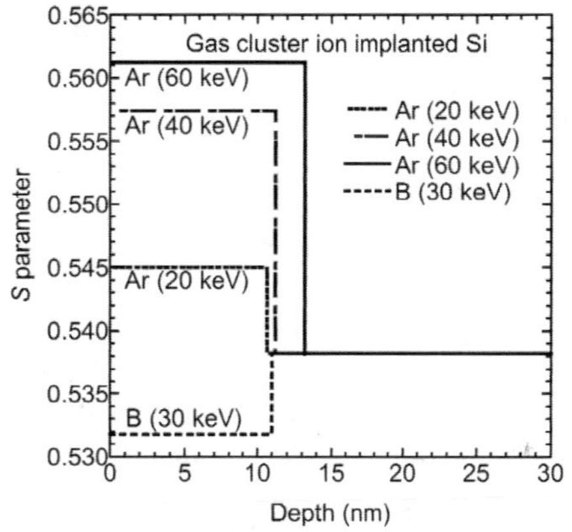


**Figure 2.** Derived depth distributions of $S$ from the $S$–$E$ curves shown in Fig. 1.

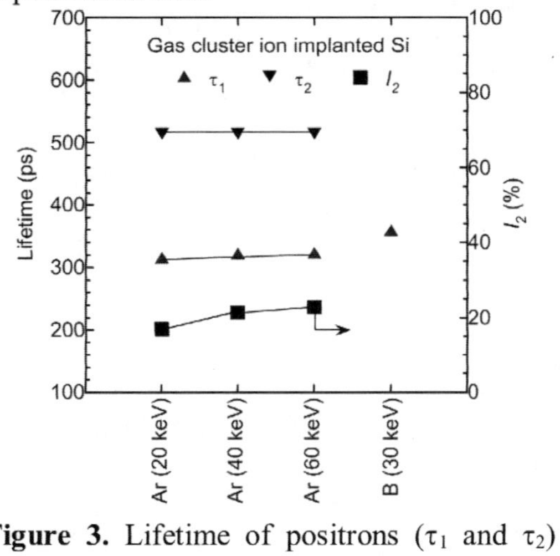


**Figure 3.** Lifetime of positrons ($\tau_1$ and $\tau_2$) and the second intensity ($I_2$) for gas cluster ion implanted Si.

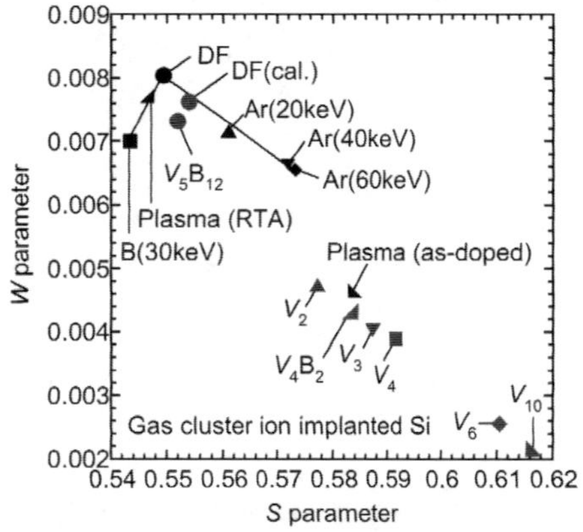


**Figure 4.** $S$–$W$ relationships for gas cluster ion implanted Si, and the ($S,W$) values calculated by the PAW method. The ($S,W$) values for the damaged region for plasma immersion B-implanted Si before and after spike RTA were also shown [5].

coincidence-detection system at $E$= 1 keV. The $(S,W)$ values obtained using the PAW method are also shown, where calculations were made for the annihilation of positrons in the delocalized state (DF: defect free) and typical vacancies such as $V_2$-$V_{10}$. The obtained $S$ (and $W$) values for Ar gas cluster implanted Si were smaller (or larger) than those for $V_2$. This means that the momentum distribution of electrons in the vacancy-type defects introduced by Ar gas cluster implantation was modified from that of 'pure' vacancies. Maekawa and Kawasuso [18] studied high-dose He-implanted Si using positron annihilation and the first-principles calculation. They reported that the trapping of positrons by He-storing microvoids enhanced the annihilation rate of high-momentum electrons. In the present experiment, therefore, a candidate for the modification is the trapping of positrons by Ar-storing vacancy-clusters (micro gas bubbles). In general, inert gas storing defects are clearly observed after post-implantation annealing. Simpson *et al.* [19] reported that the annealing temperature ramp-rate had a significant impact on residual defects for He-implanted Si. They found that the $S$ value increased with an increasing ramp-rate, suggesting that the formation of He-related defects was accelerated by increasing the ramp-rate. In the present experiment, Ar-storing clusters (the second component: $\tau_2$=514 ps) were observed without annealing. Thus, the formation of micro gas bubbles can be attributed to the extremely high temperature and its rapid gradient in impact regions of Ar gas clusters.

As shown in Figs. 1 and 2, for B gas cluster implanted Si, the $S$ value in the damaged region was smaller than that for defect free Si. In Fig. 3, the lifetime of positrons (360±1 ps) for this sample was longer than the defect-free lifetime. In Fig. 4, the $(S,W)$ values for the damaged region for plasma immersion B-implanted Si before and after spike rapid thermal annealing (RTA, 1075°C) were shown [5]. For the as-implanted sample, the $(S,W)$ value was close to the calculated value for $V_4B_2$ (four Si atoms were removed and two B atoms were inserted into substitutional sites), suggesting that major defect species is such a vacancy-boron complex. After RTA, the $S$ ($W$) value was smaller (larger) than the value for defect-free Si, and the relationship between the $(S,W)$ values for defect-free Si and the damaged region was close to that for calculated defect-free Si and an icosahedral B cluster [20] ($V_5B_{12}$), where the damaged region was crystal after RTA and the calculation of $V_5B_{12}$ was done for the crystalline structure. For B gas cluster ion implanted Si, the observed $(S,W)$ value was apart from the calculated value for $V_5B_{12}$, but lie along the line connecting the values for defect-free Si and for the plasma immersion B-implanted Si after RTA. The difference between the $(S,W)$ value for B gas cluster and plasma immersion B-implantations could be due to an amorphousization of Si matrix by gas cluster implantation, and a resultant enhancement of the effect of B clusters on the $(S,W)$ value.

### *3.2. Agglomeration and dissociation of vacancies in ELD-Cu*

Figure 5 shows the $S$–$E$ curves for ELD-Cu(H) before and after annealing. Since the $S$ value for the defect-free Cu was measured to be 0.4160 [10], the observed large $S$ value for ELD-Cu is due to the annihilation of positrons trapped by vacancy-type defects. The depth distributions of $S$ were obtained from the $S$–$E$ curves and the results are shown in Fig. 6. For the as-deposited sample, the $S$ value starts to increase at <800 nm. Thus, the vacancy-type defects introduced after the deposition diffuse toward the surface and agglomerate in the subsurface region. After annealing at 170°C, the $S$ value at $E$=2 keV increases. This corresponds to defect agglomeration in the region of <50 nm. As shown in Fig. 6, SIMS measurements show that the sulphur concentration increases near the surface for the sample annealed at 170°C. Vacancies and sulphur both tend to migrate to the surface. Their interaction and the resultant formation of vacancy-impurity complexes determine the annealing behavior of vacancy-type defects. For the sample annealed above 300°C, such vacancy agglomerates are annealed out. At 740°C, the $S$ value below 1200 nm is higher than that in the shallower region, suggesting that vacancy-type defects exist near the Cu/barrier-metal interface even at this temperature and that the Cu/ barrier-metal interface does not act as a sink for vacancies.

Figure 7 shows $S$ at $E$=18 keV for ELD-Cu(H), ELD-Cu(L) and electroplated Cu with high and low residual impurity concentrations [E-Cu(H) and E-Cu(L)] [21]. For ELD-Cu(L) before annealing, the $S$ value is smaller than that for ELD-Cu(H), suggesting that the mean size (and/or concentration) of vacancy-type defects increases with increasing residual impurity concentration. This is attributed to an interaction between vacancies and impurities, and the resultant formation of vacancy-impurity complexes. For ELD-Cu(L) and E-Cu(H), the observed increase of $S$ in the temperature range between 200-400°C corresponds to the agglomeration of vacancy-type defects, which is close to the annealing stage of vacancy-type defects in light-particle irradiated Cu (stage V) [22]. Since the residual impurity concentration in ELD-Cu is higher than that in E-Cu, the annealing of defects requires higher temperature for Cu films with high impurity concentration.

Figure 8 shows the results of positron lifetime measurements on ELD-Cu(H) as a function of annealing temperature. It was found that $\tau_2$ to be almost constant (350 ps), but $\tau_1$ decreases from 250 ps to 120 ps with increasing annealing temperature. Using these results, the defect species detected by the first and second component can be identified as vacancy complexes ($V_2$-$V_4$) and larger vacancy clusters (such as $V_{10}$, ref. 23), respectively. The intensity corresponding to the vacancy clusters ($I_2$) decreases after 750°C annealing, corresponding to the decrease in $S$ in this temperature range. Although the size of vacancy clusters is stable in the temperature range below 750°C annealing, that of the vacancy complexes ($V_2$-$V_4$) decreases with increasing temperature. Because the 'pure' vacancies with the size of $V_2$-$V_4$ are unstable at 600-800°C, they must couple with residual impurity to make stable complexes.

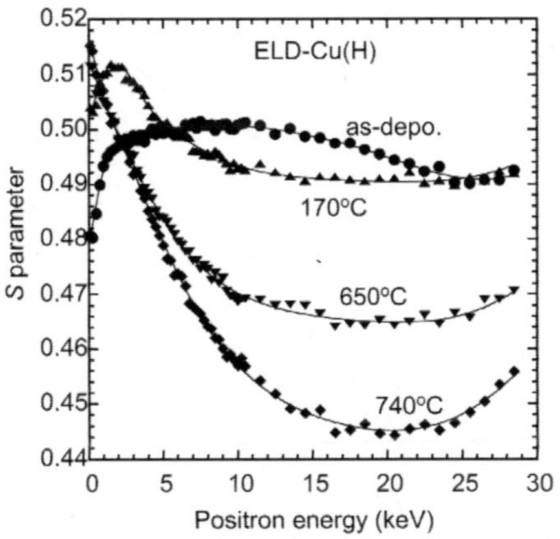


**Figure 5.** $S$ as a function of $E$ for the ELD-Cu(H) before and after annealing.

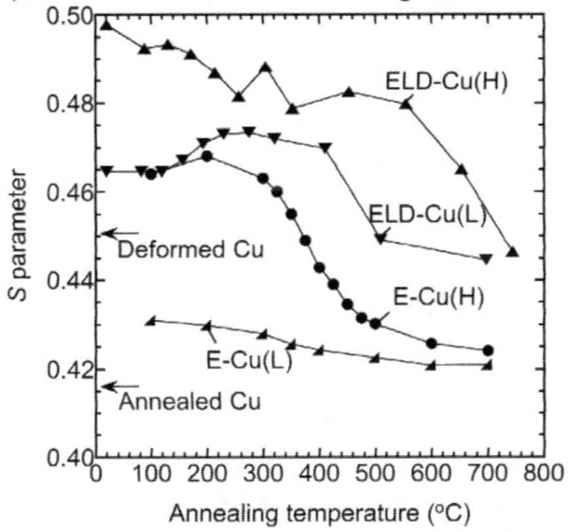


**Figure 7.** Annealing behaviours of $S$ for ELD-Cu(H) and ELD-Cu(L). The results for electroplated Cu with high and low residual impurity concentrations are also shown [E-Cu(H) and E-Cu(L)].

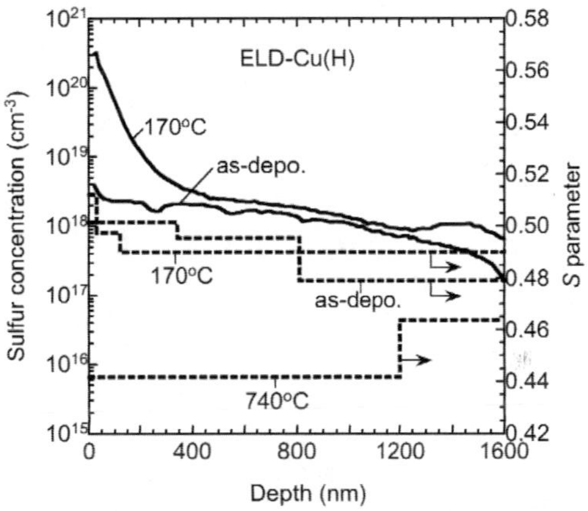


**Figure 6.** Depth distributions of sulphur and $S$ for the ELD-Cu(H). Vacancy agglomeration occurred in the region with high sulphur concentrations.

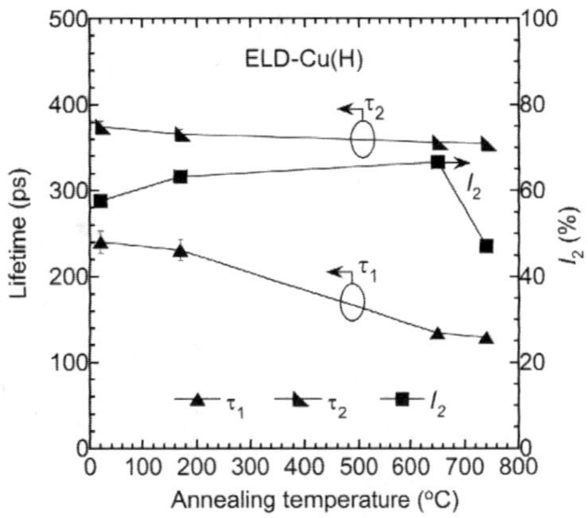


**Figure 8.** Lifetime of positrons ($\tau_1$ and $\tau_2$) and the second intensity ($I_2$) for ELD-Cu(H).

## 4. Conclusions

We have demonstrated the efficacy of positron annihilation for the characterization of vacancies in ultra-shallow region in Si and electroless deposited Cu to aid in the optimization of process parameters for those materials. For gas cluster ion implanted Si, we found that the formation of the vacancy clusters was affected by extremely high temperature and its rapid transients in the impact regions of gas cluster ions. For ELD-Cu, the close relationship between annealing behaviours of vacancy-type defects and residual impurities was observed, suggesting that the precise control of impurities in Cu is a key to suppress the atomic migration and to increase the reliability of Cu interconnects.

## References

[1] *International Technology Roadmap for Semiconductors* 2011, www.itrs.net.
[2] Uedono A, Inoue N, Hayashi Y, Eguchi K, Nakamura T, Hirose Y, Yoshimaru M, Oshima N, Ohdaira T and Suzuki R 2009 *Jpn. J. Appl. Phys.* **48** 120222
[3] Uedono A, Inoue N, Hayashi Y, Eguchi K, Nakamura T, Yoshimaru M, Oshima N, Ohdaira T and Suzuki R 2009 *Proc. 12th IEEE Int. Interconnect Tech.* 75
[4] Uedono A, Inoue N, Hayashi Y, Eguchi K, Nakamura T, Hirose Y, Yoshimaru M, Oshima N, Ohdaira T and R. Suzuki 2010 *Proc. 13th IEEE Int. Interconnect Tech. Conf.* 7.2
[5] Uedono A, Tsutsui K, Ishibashi S, Watanabe H, Kubota S, Nakagawa Y, Mizuno B, Hattori T and Iwai H 2010 *Jpn. J. Appl. Phys.* **49** 051301
[6] Uedono A, Tsutsui K, Ishibashi S, Watanabe H, Kubota S, Tenjinbayashi K, Nakagawa Y, Mizuno B, Hattori T and Iwai H 2010 *Proc. 10th Int. Workshop on Junction Technology* 149
[7] Oka Y, Uedono A, Goto K, Hirose Y, Matsuura M, Fujisawa M and Asai K 2011 *Jpn. J. Appl. Phys.* **50** 05EB06
[8] Uedono A, Moriya T, Tsutsui T, Kimura S, Oshima N, Suzuki R, Ishibashi S, Matsui H, Narushima M, Ishikawa Y, Graf M and Yamashita K 2010 *12th Int. Workshop on Junction Technology Extended Abstracts* IEEE Press 85
[9] Uedono A, Dordi Y, Li S, Mizunaga G, Tenjinbayashi K, Oshima N and Suzuki R 2012 *IEEE Int. Interconnect Tech. Conf.* Session 9.4
[10] Uedono A, Yamashita Y, Tsutsui T, Dordi Y, Li S, Oshima N and Suzuki R 2012 *J. Appl. Phys.* **111** 104506
[11] Yamada I, Matsuo J, Toyoda N and Kirkpatrick A 2001 *Mat. Sci. Eng. R* **34** 231
[12] Webb E, Witt C, Andryuschenko T and Reid J 2004 *J. Appl. Electrochem.* **34** 291
[13] Bachand J, Freytsis A, Harrington E, Gwinn M, Hofmeester N, Hautala J, Mack M E and Regan K 2002 *Proc. Ion Implantation Technology* IEEE, New York 669
[14] Vaskelis A, Norkus H J, Rozovskis G and Vinkevicius H I 1997 *Trans. IMF* **75** 1
[15] Van Veen A, Schut H, De Vries J, Hakvoort R A and Ijpma M R 1990 *AIP Conf. Proc.* **218** 171
[16] Detailed and up-to-date information of QMAS is reported in http/qmas.jp
[17] Kirkegaard P, Eldrup M, Mogensen O E and Pedersen N J 1981 *Comput. Phys. Commun.* **23** 307
[18] Maekawa M and Kawasuso A 2010 *J. Phys.: Conf. Ser.* **225** 012032
[19] Simpson P J, Knights A P, Chicoine M, Dudeck K, Moutanabbir O, Ruffell S, Schiettekatte F and Terreault B 2008 *Appl. Surf. Sci.* **255** 63
[20] Yamauchi J, Aoki N and Mizushima I. 1997 *Phys. Rev. B* **55** R10245
[21] Uedono A, Suzuki T, Nakamura T, Ohdaira T and Suzuki R 2007 *Jpn. J. Appl. Phys.* **46** 1938
[22] Mantl S and Triftshäuser W 1978 *Phys. Rev. B* ***17*** 1645
[23] Onishi T, Mizuno M, Fujikawa T, Yoshikawa T, Munemasa J, Mizuno M, Kihara T, Araki H and Shirai Y 2011 *J. Electro. Mat.* **40** 1384

Journal of Physics: Conference Series **443** (2013) 012068 doi:10.1088/1742-6596/443/1/012068

# Reflection high-energy positron diffraction: the past 15 years and the future

**Y Fukaya[1], M Maekawa[1], I Mochizuki[1], K Wada[2], T Hyodo[2] and A Kawasuso[1]**

[1]Advanced Science Research Center, Japan Atomic Energy Agency, 1233 Watanuki, Takasaki, Gunma 370-1292, Japan

[2] Institute of Materials Structure Science, High Energy Accelerator Organization (KEK), 1-1 Oho, Tsukuba, Ibaraki 305-0801, Japan

fukaya.yuki99@jaea.go.jp

**Abstract**. Reflection high-energy positron diffraction (RHEPD) is the positron counterpart of reflection high-energy electron diffraction (RHEED). Owing to the positive charge of the positron, RHEPD provides a powerful tool with which to determine the structure of the first surface layer. We have been investigating important surface systems concerning their unique electric and magnetic properties and also phase transition phenomena using positron beams (flux: $10^3$~$10^4$ $e^+$/sec) with $^{22}$Na sources. Currently, we are developing a new RHEPD apparatus with a bright and intense positron beam (flux: $10^5$ $e^+$/sec) based on the LINAC at the Slow Positron Facility, KEK. Here, we summarize the past results and the future prospects of the RHEPD study in the surface science.

## 1. Introduction

In early 1980s, low-energy positron beam was utilized to study surface structures [1]. As the results of absence of exchange interaction for positrons and the Coulomb repulsion between positrons and ion cores, in the low energy region, the positron-solid interaction becomes much less complex as compared to the electron-solid interaction. LEPD has been applied to the structure analysis of surfaces including Cu and CdSe [2,3] and has proven to be superior to low-energy electron diffraction (LEED) in a number of important cases.

Ichimiya proposed that reflection high-energy positron diffraction (RHEPD) could also have a great advantage in the surface structure analysis [4] as compared to its electron counterpart, RHEED. Positrons are repelled from crystal surface. When a high-energy (10-20 keV) positron beam is incident on the crystal surface at grazing angle, the total reflection takes place below the critical angle. Under the total reflection region, the positron beam is diffracted at the top-most surface layer. This is a distinct feature for positron diffraction, which is never observed in electron diffractions. The critical angle for total reflection is typically larger than a few degrees.

So far, we have investigated the structures, thermal vibrations, electronic excitations and phase transitions of important surfaces by using the RHEPD. In this study, we report on some representative

Journal of Physics: Conference Series **443** (2013) 012068 doi:10.1088/1742-6596/443/1/012068

results obtained to date and on ongoing research using an intense positron beam generated from a LINAC at the Slow Positron Facility in KEK.

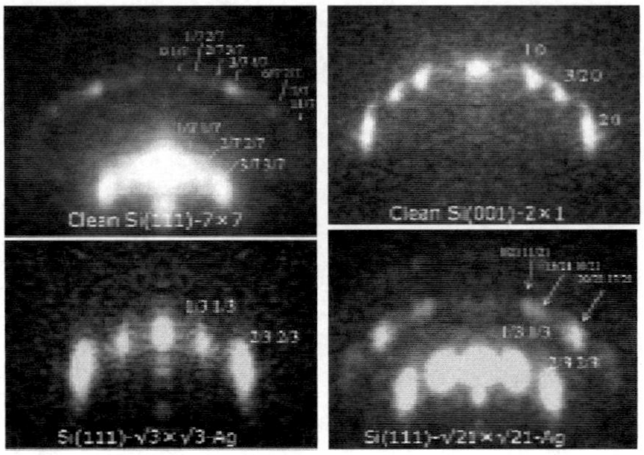


**Figure 1.** RHEPD patterns from various surface structures on crystals obtained by using a positron beam with a $^{22}$Na source. The indexes in the pattern indicate the fractional- and integer-order spots accompanied with the surface reconstruction structures.

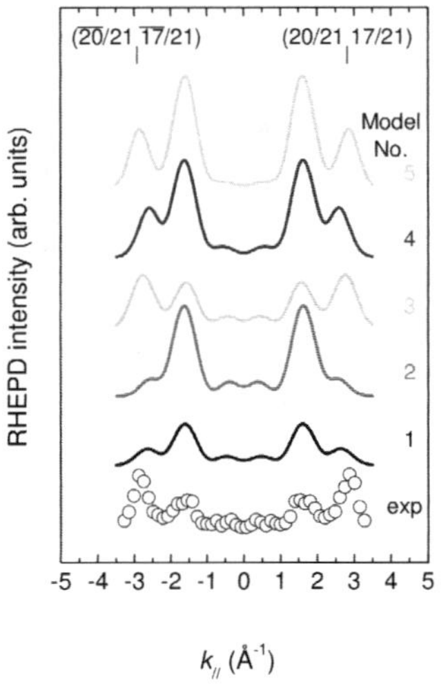


**Figure 2.** Intensity distribution in the 1/7th-Laue zone of the RHEPD pattern from the Si(111)-√21×√21-Ag surface as a function of wave number ($k_{//}$) parallel to the surface [17]. Open circles indicate the experimental line profile (Fig. 1). The line profiles denoted as solid lines represent the calculated ones for various structure models.

## 2. RHEPD study in the past 15 years

### *2.1. Early RHEPD study*

To observe clear RHEPD patterns, it is critically important that a positron beam with a sufficient long coherence length is used. In 1998, Kawasuso and Okada developed a RHEPD apparatus employing a $^{22}$Na source and electrostatic lenses, and for the first time succeeded in observing the RHEPD patterns from a Si(111)-1×1-H surface and the total reflection of positrons [5]. Afterwards, the positron beam apparatus was improved by adopting the electromagnetic lenses [6,7]. Clear RHEPD patterns from various surface structures were observed, as shown in Fig. 1.

### *2.2. Surface structure analysis*

So far, RHEPD has been applied to various surface systems such as Si(111)-√3×√3-Ag [8], Si(111)-4×1-In [9], Ge(111)-√3×√3-Sn [10], and Ge(001)-2×1-Pt [11]. Among them, the √21×√21 superstructure is reported below as one of the complicated surface structures.

The adsorption of a small amount of the metal (Ag, Au, and Cu) atoms on the Si(111)-√3×√3-Ag surface leads to the superstructure having a √21×√21 periodicity. This surface shows a high electric conductivity as compared to the Si(111)-√3×√3-Ag surface [12]. The surface structure is still controversial although it has been studied by using various surface techniques [13-16]. The Si(111)-√21×√21-Ag and the Si(111)-√21×√21-(Ag,Au) surfaces give rise to clear fractional-order spots related to the √21×√21 periodicity (see also Fig. 1). The intensity distributions from these surfaces are very similar to each other. In the structure analysis of these surfaces, we may determine the positions of the additional metal atoms on the Si(111)-√3×√3-Ag surface. However, this work was very problematic since there exist other possible ${}_{21}C_4$ structures considering the distribution of the additional metal atoms (3-4) to available adsorption sites (21). To accomplish the above work, we followed the screening method so that the true structure should reproduce both the observed rocking curves and patterns. The solid lines in Fig. 2 represent the experimental line profile of the 1/7th-Laue zone and those calculated assuming the finally remaining structural models after the screening in the rocking curve analysis. Only one structural model (#3), where the center of three adatoms corresponds to the Si trimer, reproduces the experiment [17]. The atomic positions of the Si(111)-√21×√21 surface are lately explained in terms of cohesive nature of noble metals [18,19].

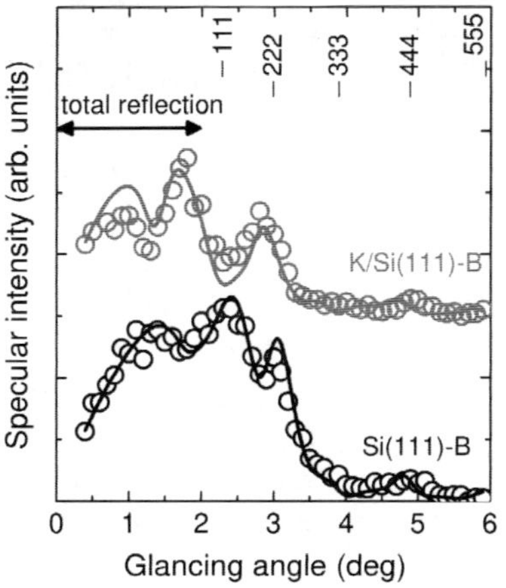


**Figure 3.** RHEPD rocking curves of specular spots from the Si(111)-√3×√3-B surface (black) and the K/Si(111)-√3×√3-B surfaces (red) under the one-beam condition. The circles indicate the measured curves. The solid lines denote the calculated curves using the optimum heights of K atoms.

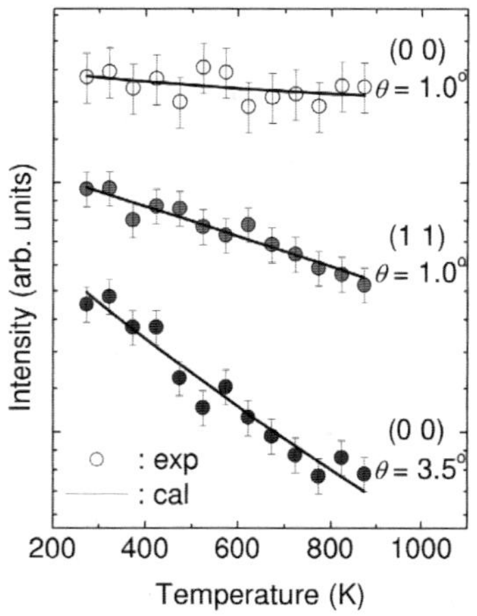


**Figure 4.** Temperature dependences of the RHPED intensities from a Si(111)-7×7 surface (circles). Solid lines represent the calculated intensities using $\Theta_S$ = 290 K and $\Theta_B$ = 600 K. The accelerating voltage of the incident positron beam is 20 kV.

### *2.3. Surface charge transfer*

Recently, the RHEPD was found to be also useful for the study on the charge transfer near the surface. When a highly B-doped Si(111) surface is saturated by the K atoms, the K/Si(111)-√3×√3-B surface is formed. This surface is known as a typical example of Mott-type insulator surfaces [20]. The electronic states of the K/Si(111)-√3×√3-B surface have been extensively investigated using angle-resolved photoemission spectroscopy [21].

The red circles in Fig. 3 represent the RHEPD rocking curves of the specular spots from the K/Si(111)-√3×√3-B surfaces at room temperature under the one-beam condition. In this condition, the rocking curve is mainly sensitive to the atomic positions perpendicular to the surface. In the total reflection region, the dip structure resulting from the adsorption of K atoms is clearly observed at 1.3°. As compared with the Si(111)-√3×√3-B surface (denoted as black circles), the position of the 111 Bragg peak is shifted to lower angle. This indicates that the mean inner potential of the Si(111)-√3×√3-B surface is dramatically changed after the adsorption of K atoms. The analyses on the basis of the dynamical diffraction theory [22] demonstrated that the above result is explained considering the charge transfer of 1.0$e^-$ per K atom to Si atoms [23].

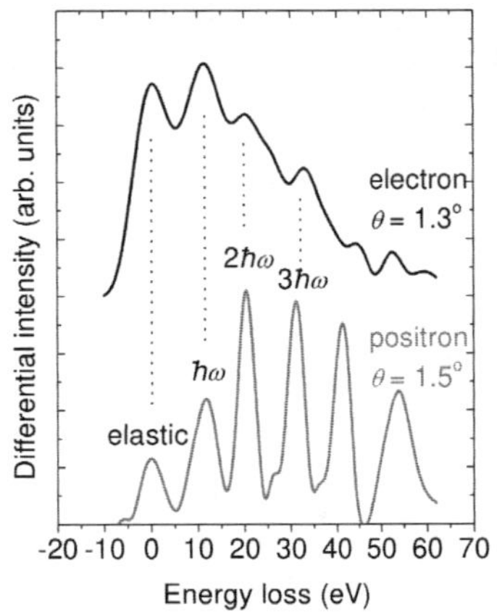


**Figure 5.** Differential curves for positrons (red line) and for electrons (black line) from the Si(111)-7×7 surface as a function of energy loss. The glancing angles are 1.5° for positrons and 1.3° for electrons. The incident azimuths are parallel to the $[11\bar{2}]$ direction for positrons and 7.5°-off oriented from the $[11\bar{2}]$ direction for electrons.

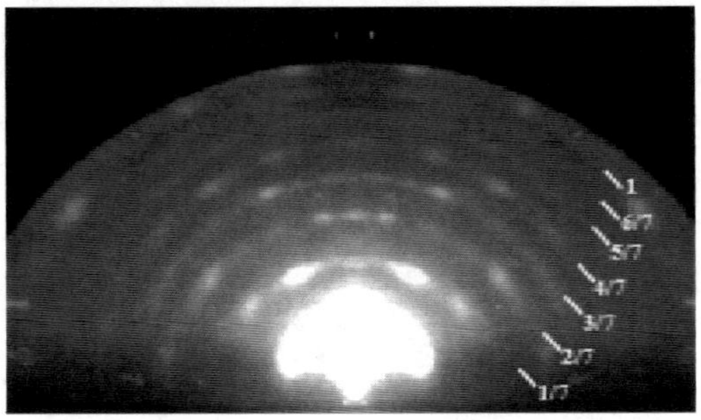

**Figure 6.** RHEPD pattern from the Si(111)-7×7 surface. The incident azimuth of the positron beam corresponds to the $[11\bar{2}]$ direction. The glancing angle is set at 2.2°. The indexes at the right hand side indicate the Laue zone. The pattern shows the superimposition of the left and right parts.

*2.4. Surface thermal vibration*

To understand the surface electrical conductivity, surface phase transition, and adsorption, the thermal vibrations of surface atoms is one of the important factors. Figure 4 shows the temperature dependence of the RHEPD intensity measured from a Si(111)-7×7 surface (see also Fig. 1). Since the incident positron energy was 20 keV, the critical angle for total reflection is estimated to be 1.4°. The RHEPD intensities of (0 0) and (1 1) spots at $\theta = 1.0°$ and (0 0) spot at $\theta = 3.5°$ satisfy the total reflection and the 444 Bragg reflection conditions, respectively. The calculations based on the dynamical diffraction theory confirmed that the RHEPD intensity in the total reflection region is not affected by the bulk Debye-temperature ($\Theta_B$). Thus, the Debye-temperature of the adatoms is solely determined under the total reflection condition. Consequently, the surface Debye-temperature ($\Theta_S$) of the adatoms on the Si(111)-7×7 surface is determined to be 290 K, which is significantly smaller as compared with that obtained by the first-principles calculations and electron diffraction. The result demonstrated that the thermal vibrational amplitude of the adatoms is much larger than those expected before [24].

*2.5. Surface plasmon*

When charged particles are incident on the crystal surface, various energy loss processes such as phonon, plasmon, and core-level electron excitations occur. Among these excitations, the surface plasmon excitation is known as a dominant process for electrons [25-28]. On account of the positive charge and the total reflection, the surface plasmon excitation due to positrons is probably different from that of electrons. To reveal this, the energy spectra of diffracted positron beam were measured.

Figure 5 shows the positron and electron energy loss spectrum of the specular spot from a Si(111)-7×7 surface. Five prominent loss peaks are observed, which are assigned to single ($\hbar\omega$) through fivefold ($5\hbar\omega$) surface plasmon loss of Si (11 eV), respectively. For positrons, the elastic peak intensity is rather weak and the peak intensities of $2\hbar\omega$ and $3\hbar\omega$ are higher than the others. On the other hand, for electrons, the peak intensity of $\hbar\omega$ is the highest and the elastic peak intensity is also relatively high. The energy loss spectra are approximately expressed by the Poisson distribution: $P(n) = n_s^n \exp(-n_s)/n!$, where $n_s$ is the mean excitation number of surface plasmon [25]. By fitting this equation to the energy loss spectra, $n_s$ = 2.6 for positrons and $n_s$ = 1.4 for electrons are obtained. Thus, the totally reflected positron beam excite more surface plasmons at surfaces than electrons [29].

## 3. RHEPD with an intense beam and future prospect

At the Slow Positron Facility in KEK, the intensity of the initial positron beam was increased to $10^7$ $e^+$/sec by the modification of converter/moderator assembly [30]. In 2011, a RHEPD apparatus was connected to this KEK positron beamline. A remoderator composed of 100 nm-thick W foil and magnetic lens were installed in front of the RHEPD chamber. The energy spread of the incident positron beam was dramatically reduced to ~8 eV and the beam radius is decreased to ~1.0 mm. As a result, we succeeded in measuring a RHEPD pattern from a Si(111)-7×7 surface (Fig. 6) which showed a significant improvement in clarity in comparison to those obtained before the use of the remoderator. In future studies, the RHEPD method utilizing the intense positron beam system will be applied to advanced surface systems that are important in future spintronics, green technology and so on. A method to directly determine the surface structure will also be developed.

## References


[1] Coleman P G (Ed.), *Positron Beams and their Applications*, World Scientific, Singapore, 2000.
[2] Mayer R, Zhang C S, Lynn K G, Frieze W E, Jona F and Marcus P M 1987 *Phys. Rev.* B **35** 3102
[3] Horsky T N, Brandes G R, Canter K F, Duke C B, Horng S F, Kahn A, Lessor D L and Mills Jr A P 1989 *Phys. Rev. Lett.* **62** 1876
[4] Ichimiya A 1992 *Solid State Phenom.* **28&29** 143
[5] Kawasuso A and Okada S 1998 *Phys. Rev. Lett.* **81** 2695
[6] Kawasuso A, Fukaya Y, Hayashi K, Maekawa M, Okada S and Ichimiya A 2003 *Phys. Rev.* B **68** 241313
[7] Kawasuso A, Ishimoto T, Maekawa M, Fukaya Y, Hayashi K and Ichimiya A 2004 *Rev. Sci. Instrum.* **75** 4585
[8] Fukaya Y, Kawasuso A and Ichimiya A 2007 *Phys. Rev.* B **75** 115424
[9] Fukaya Y, Hashimoto M, Kawasuso A and Ichimiya A 2008 *Surf. Sci.* **602** 2448
[10] Fukaya Y, Kawasuso A and Ichimiya A 2006 *Surf. Sci.* **600** 4086
[11] Mochizuki I, Fukaya Y, Kawasuso A, Yaji K, Harasawa A, Matsuda I, Wada K and Hyodo T 2012 *Phys. Rev.* B **85** 245438
[12] Hasegawa S, Tong X, Takeda S, Sato N and Nagao T 1999 *Prog. Surf. Sci.* **60** 89
[13] Ichimiya A, Nomura H, Horio Y, Sato T, Sueyoshi T and Iwatsuki M 1994 *Surf. Rev. Lett.* **1** 1
[14] Nogami J, Wan K J and Lin X F 1994 *Surf. Sci.* **306** 81
[15] Tong X, Sugiura Y, Nagao T, Takami T, Takeda S, Ino S and Hasegawa S 1998 *Surf. Sci.* **408** 146
[16] Tajiri H, Sumitani K, Yashiro W, Nakatani S, Takahashi T, Akimoto K, Sugiyama H, Zhang X and Kawata H 2001 *Surf. Sci.* **493** 214
[17] Fukaya Y, Kawasuso A and Ichimiya A 2006 *Surf. Sci.* **600** 3141
[18] Fukaya Y, Kawasuso A and Ichimiya A 2007 *Surf. Sci.* **601** 5187
[19] Fukaya Y, Matsuda I, Hashimoto M, Kubo K, Hirahara T, Yamazaki S, Choi W H, Yeom H W, Hasegawa S, Kawasuso A and Ichimiya A 2012 *Surf. Sci.* **606** 919
[20] Weitering H H, Shi X, Johnson P D, Chen J, DiNardo N J and Kempa K 1997 *Phys. Rev. Lett.*

**78** 1331
[21] Cardenas L A, Fagot-Revurat Y, Moreau L, Kierren B and Malterre D 2009 *Phys. Rev. Lett.* **103** 046804
[22] Ichimiya A 1983 *Japan. J. Appl. Phys.* **22** 176
[23] Fukaya Y, Mochizuki I and Kawasuso A 2012 *Phys. Rev.* B **86** 035423
[24] Fukaya Y, Kawasuso A, Hayashi K and Ichimiya A 2004 *Phys. Rev.* B **70** 245422
[25] Lucas A A and Šunjić M 1971 *Phys. Rev. Lett.* **26** 229
[26] Horio Y, Hashimoto Y, Shiba K and Ichimiya A 1995 *Japan. J. Appl. Phys.* **34** 5869
[27] Nakahara H, Hishida T and Ichimiya A 2003 *Appl. Suf. Sci.* **212-213** 157
[28] Tanishiro Y 2003 *Hyomen Kagaku (Journal of The Surface Science Society of Japan)* **24** 166
[29] Fukaya Y, Kawasuso A and Ichimiya A 2009 *Phys. Rev.* B **79** 193310
[30] Wada K, Hyodo T, Yagishita A, Ikeda M, Ohsawa S, Shidara T, Michishio K, Tachibana T, Nagashima Y, Fukaya Y, Maekawa M and Kawasuso A 2012 *Eur. Phys. J.* D **66** 37

Journal of Physics: Conference Series **443** (2013) 012069 doi:10.1088/1742-6596/443/1/012069

# Monte Carlo simulations of the extraction of slow positrons into gas through thin SiN windows

**B E O'Rourke, W Zhou, N Oshima, K Ito, R Suzuki**

National Institute of Advanced Industrial Science and Technology (AIST), Tsukuba, Ibaraki 305-8568, Japan

E-mail: brian-orourke@aist.go.jp, nagayasu-oshima@aist.go.jp

**Abstract.** Monte Carlo simulations have been performed of the extraction of slow positrons into air through thin silicon nitride (SiN) windows. The simulations are based on recent experiments at the positron probe microanalyzer (PPMA) in which slow positron microbeams have been used to analyze samples located outside the vacuum chamber in air. In the present study we calculated the fractional transmission of positrons for SiN windows as a function of positron energy and window thickness, and also calculated their energy and angular spectra after extraction. The effect of a thin air/gas layer between the window and sample was also simulated.

## 1. Introduction

Positrons annihilation is a unique tool for studying defects and open volumes in materials. Various experimental techniques including positron annihilation lifetime spectroscopy (PALS) have been developed and contributed to analysis and characterization over a wide range of materials [1]. Mono-energetic, low energy (slow) positrons penetrate a well defined distance into the surface layer of the sample and can be used to study thin or layered samples. Such beams of slow, monoenergetic positrons can be produced by moderation of the fast positrons produced via electron accelerators, nuclear reactors or intense radioisotope sources.

The slow positron beams are typically transported from the source to the experiment under ultra high vacuum and any samples must also be introduced into the vacuum chamber. However, for many materials it is necessary to measure outside vacuum in an environment similar to actual operational conditions. Interesting examples include wet polymer films or low-k insulating layers of electronic devices kept in humidity-controlled air or membranes in aqueous solutions. The study of powders and liquids and other samples not compatible with vacuum with low energy positrons also becomes possible.

## 2. The AIST Atmospheric Positron Probe Microanalyzer

A positron microbeam has been developed at AIST, Japan, the positron probe microanalyzer (PPMA). This pulsed, variable energy (1 keV – 30 keV) beam allows PALS measurements with a spatial resolution down to 30 μm and has been used for 2D and 3D defect distribution analysis [2]. The PPMA also allows the beam to be focussed on small, thin vacuum windows and hence be extracted into the air. Silicon nitride (SiN) windows were installed on the PPMA sample chamber and have been used to successfully extract slow positrons outside the vacuum chamber [3,4]. Three different windows have been used with the following dimensions; 30 nm × 0.6 mm × 0.6 mm, 200 nm × 1.5 mm × 1.5 mm and 500 nm × 3.0 mm × 3.0 mm. The density of the SiN windows is around 3.0 g/cm$^3$ and the compositional ratio of Si to N is approximately 1 : 1[3]. When one side of the window is evacuated to vacuum the window bends slightly towards the vacuum side. This means that a small air gap is present between the window and the sample. For the 30 nm and 200 nm windows this minimum air gap, $d_0$, was measured using a laser microscope to be 35 μm – 40 μm and 65 μm – 70 μm respectively. The deflection of a clamped film when $d_0 >> h$, can be approximated by the expression $d_0 \approx (r^4 P / Eh)^{1/3}$ [5], where $r$ is the radius, $P$ the applied pressure (1 atm), $E$ the Young's modulus and $h$ the material thickness. For CVD prepared SiN films a value of $E$ = 270 GPa has been reported [6]. Assuming a value for the radius, r, equal to half the lateral dimension, values of $d_0$ = 37, 85 and 312 μm for the 30, 200 and 500 nm films respectively are calculated, in reasonable agreement with the measured value for the thinner two films.

When the slow positron beam passes through the SiN window it will interact and some fraction of the incident positrons will be backscattered, some annihilated in the window and some fraction transmitted. The transmitted beam will emerge with an energy distribution and angular dispersion. Although stopping profiles in multilayer systems can be estimated by scaling laws [7], Monte Carlo simulations can be used to determine energy and angular distributions alongside stopping profiles. The present report aims to estimate these effects at the AIST atmospheric PPMA using the Penelope2008 code [8]. Penelope 2008 provides full support for positron interactions and can simulate processes with energies down to several hundred eV.

## 3. Monte Carlo Simulations

### *3.1. Slow positron transmission through SiN*

In the initial simulation a slow, monoenergetic positron beam at normal incidence to a thin SiN film was simulated and the positron transmission fraction and energy and angular distributions of transmitted positrons calculated.

#### *3.1.1 Positron Transmission Fraction*

A typical simulation result is shown in figure 1 where the relative fractions of transmitted, backscattered and absorbed (annihilated) positrons are plotted as a function of positron energy for the 200 nm thick SiN window. The transmitted fraction increases from 0 around 3 keV to almost 100% at 30 keV. The backscattered fraction decreases from around 10% at 5 keV to less than 1 % at 30 keV.

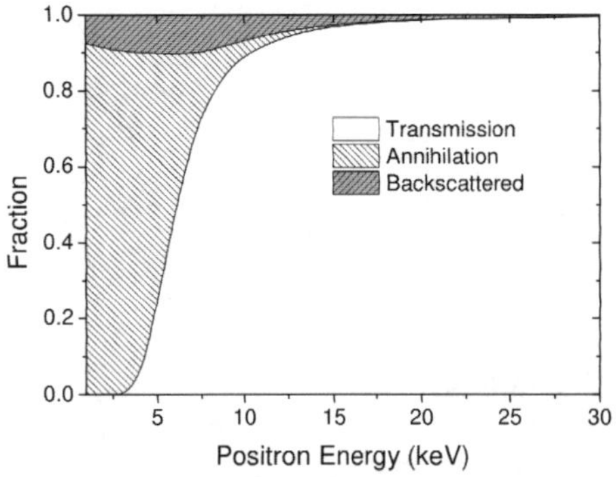


**Figure 1.** Fractional transmission, back-scattering and ab-sorption (annihilation) of positrons incident on a 200 nm SiN film as a function of positron energy.

Although other profiles have also been shown to give good agreement with measurements [9] the stopping profile, $P$, of positrons in materials is typically characterised by the Makhov profile [10];

$$P(z,E) = -\frac{\mathrm{d}}{\mathrm{d}z}\exp\left[-\left(\frac{z}{z_0}\right)^m\right] \quad \text{with} \quad z_0 = \frac{AE^n}{\rho\,\Gamma(1+1/m)} \tag{1}$$

*A*, *n* and *m* are material and energy dependant parameters [11], *E* is the energy in keV and *z* the material thickness in nm. $\rho$ is the material density (in g/cm$^3$) and $\Gamma$ is the Gamma function. Neglecting backscatter, the transmission through a film of a certain depth can be estimated by integrating this function. We can then fit our simulation results to this integral to find the parameters for SiN.
Using the transmission fractions calculated via the Monte Carlo simulation gives the following result for these parameters (with $m$ = 2); $A$ = 32.8 ± 0.6, $n$ = 1.66 ± 0.01. Figure 2 shows the simulation results for positron transmission through various thicknesses of SiN films plotted alongside curves generated with the calculated fit parameters.

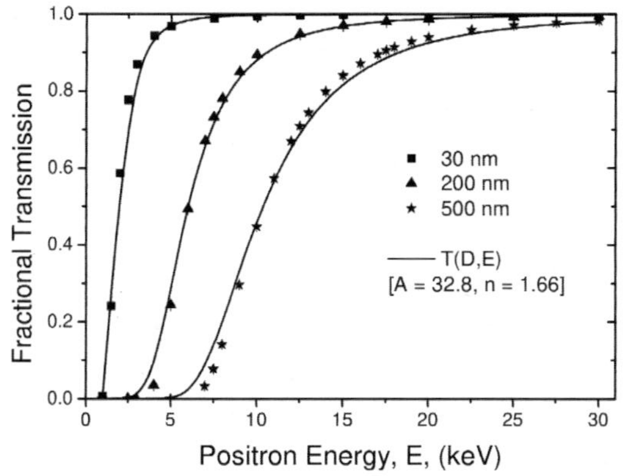


**Figure 2.** Plot of the calculated transmission of positrons through thin SiN films as a function of positron energy and film thickness and compared to curves generated by an integral of the Makhov profile with the parameters $A$ = 32.8, $n$ = 1.66, $m$ = 2.

*3.1.2 Energy and angular distribution of transmitted positrons*

The initially monoenergetic and parallel positron beam will emerge with broad energy and angular distributions. Figure 3a shows the calculated energy profile of initially 5 keV positrons transmitted through both 30 nm and 200 nm SiN films. At 5 keV the transmission through 30 nm films is 97% while that for 200 nm films is 24%. For the thicker 200 nm film it is seen that a broad energy distribution which peaks around 3.5 keV is obtained whereas for the thinner 30 nm film the distribution is more sharply peaked with a maximum at 4.9 keV. An estimate for the width of the distribution ($E_-$, $E_+$) can be obtained by finding the energy at which the probability is half of this maximum value (N[$E_\pm$] = 0.5 N[$E_{\mathrm{peak}}$]). This analysis was repeated for a range of impact energies at the three SiN film thicknesses, 30 nm, 200 nm and 500 nm and is plotted in figure 3b. The error bars in this figure represent the distribution width. As the positron energy is increased the energy distribution width narrows and approaches the initial impact energy.

The angular distribution of transmitted/backscattered positrons is shown in figure 4. The number of positrons emitted at a function of polar angle, $\theta$, is given by the probability per steridian ($\varepsilon$) multiplied by the factor $2\pi\sin(\theta)$. In figure 4a this quantity is plotted for a positron beam of 5 keV energy incident on 30 nm and 200 nm SiN films. In both cases the distribution peaks at a particular angle ($\theta_{max}$) and then decays with increasing angle to zero at 90° (perpendicular to the film). The distribution between 90° and 180° corresponds to positrons backscattered from the film. The angle where $2\pi\sin(\theta)\varepsilon$ reaches a maximum, $\theta_{max}$, is plotted as a function of positron energy for the 30 nm, 200 nm and 500 nm SiN films in figure 4b. This angle decreases with decreasing window thickness and increasing positron energy. For the 30 nm window, $\theta_{max}$ is less than 5° for E > 10 keV.

*3.1.3. Temporal and Spatial Resolution at the sample*

Using the results from the previous section it is possible to estimate the effect of the SiN window on the temporal resolution. A simple estimate is to calculate the time difference of arrival at the sample between the high energy and low energy positrons. For this analysis we define the high energy and low energy values as the energy at which the probability is half of peak value ($N[E_{\pm}] = 0.5\ N[E_{peak}]$). We also define the distance positrons travel between the SiN window and sample as equal to the minimum gap $d_0$ divided by the cosine of the peak angle $\theta_{max}$. The pulse broadening, $\Delta T$, is then given by,

$$\Delta T = \frac{d_0}{\cos\theta_{max}}\left(\frac{1}{\sqrt{2E_-/m}} - \frac{1}{\sqrt{2E_+/m}}\right), \tag{2}$$

and the results plotted in figure 5a.
It is clear that even for very low energy positron beams (≥ 1keV) the reduction in temporal resolution is only a few ps and is therefore expected to have no significant influence on the final temporal resolution.

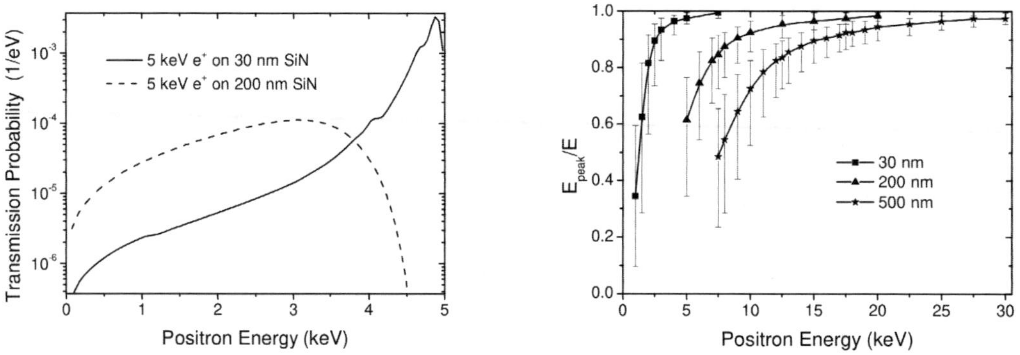


a) b)

**Figure 3. a)** Energy distribution of initially 5 keV positrons transmitted through 30 nm and 200 nm SiN films. **b)** $E_{peak}$ of transmitted positrons and distribution width ($N[E_{\pm}] = 0.5\ N[E_{peak}]$) shown by the error bars plotted as a function of incident positron energy for transmission through 30 nm, 200 nm and 500 nm SiN films.

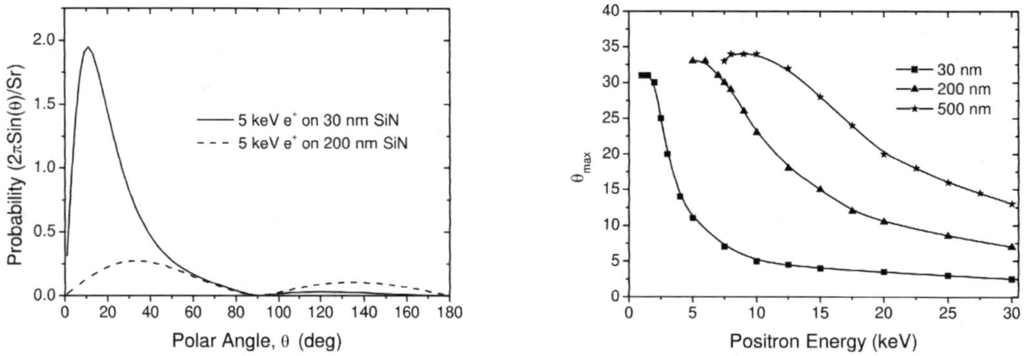


a) b)

**Figure 4. a)** Angular distribution of initially 5 keV positrons transmitted/backscattered through 30 nm and 200 nm SiN films. **b)** The angle of peak positron emission, $\theta_{max}$, plotted as a function of incident positron energy for transmission through 30 nm, 200 nm and 500 nm SiN films.

We can also estimate the positron beam size on the sample. If we assume the initial beam diameter, $w_0$, as 30 μm, the spot size on target can be estimated by; $w = w_0 + 2d_0 \tan\theta_{max}$. The result of this calculation is shown in figure 5b. The reduction in beam size resolution is estimated to be quite small, especially when the thinnest (30 nm) window is used over a wide energy range.

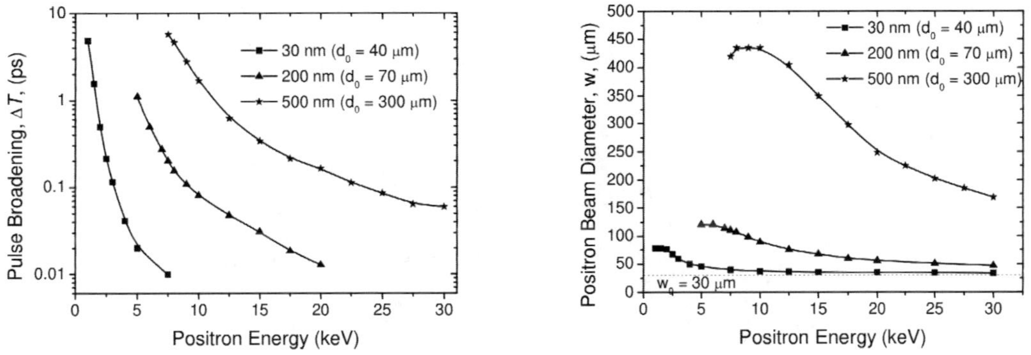


a) b)

**Figure 5.** a) Estimate of the pulse broadening, $\Delta T$, caused by positrons passing through the SiN window. $\Delta T$ is estimated using the minimum window-sample gap, $d_0$, the width of the energy distribution and the angle of maximum positron emission, $\theta_{max}$. b) Estimated size of the positron beam spot on target based on $\theta_{max}$ and the minimum window – sample gap, $d_0$. For the 30nm and 200nm windows $d_0$ is based on the value measured using a laser microscope whereas for the 500 nm window a value of 300 μm was assumed based on the calculation in section 2.

*3.2. Effect of air/gas layer*

As mentioned in section 2, due to the curvature of the SiN window when one side is under high vacuum, a thin layer of air (or gas) will be present between the SiN window and the sample. This layer will also cause a loss of transmission and broadening of the energy and angular distributions.

If we consider the SiN – Gas layer – Sample as a multilayer system, the stopping probability $\eta_i$ of positrons in the $i$-th layer is given by [1,3];

$$\eta_i = \int_{x_i}^{x_{i+1}} P_i(z,E)\mathrm{d}z \ , \quad P_i(z,E) = -\frac{\mathrm{d}}{\mathrm{d}z}\exp\left[-\left(\frac{z-\delta z_i}{z_{0i}}\right)^m\right] \ \text{with} \ z_{0i}\Gamma\left(1+\frac{1}{m}\right) = \frac{A_i E^{n_i}}{\rho_i}, \tag{3}$$

where $P_i$ is the stopping profile of positrons in the $i$-th layer and z(nm) is the depth from the surface of the top layer. The depth correction factor $\delta z_i$ is given by [3];

$$\delta z_i = \begin{cases} x_i - z_{0i}\left[-\ln\left(1-\sum_{k=1}^{i-1}\eta_k\right)\right]^{\frac{1}{m}} & \text{for } i \geq 2 \ , \\ 0 & \text{for } i = 1 \end{cases} \tag{4}$$

where $x_i$ is the depth from the surface to the interface between the $(i-1)$-th and the $i$-th layers.

Figure 6 shows the result of simulations based on a recent experiment involving a polyvinyl alcohol (PVA) film. In this experiment the 30 nm SiN window was used together with a 35 μm Kapton spacer to give a window – sample separation of 70 μm. $N_2$ gas at 1 atm was present in this region. Monte Carlo calculations based on this geometry were performed and the fraction stopped in each layer (after removal of the small fraction backscattered from the SiN window) was determined. Also shown in the figure is the calculated stopping probability (or absorbed fraction) in each layer based on the model outlined above. This model tends to slightly overestimate the absorption in the PVA sample and underestimate the absorption in the $N_2$ layer but overall agreement is good indicating that this model can be used as a guideline for future experiments.

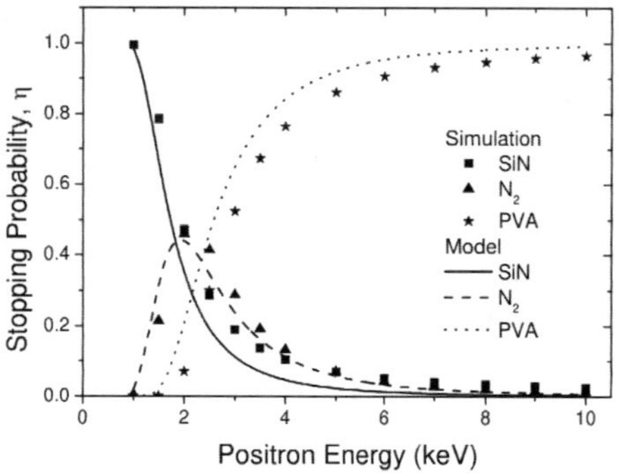


**Figure 6.** Plot of the calculated stopping probability in each layer [SiN (30 nm), $N_2$ (70 μm), PVA (2 mm)] as a function of positron energy. Also shown is the stopping probability based on the multi-layer Makhov model with typical values of the free parameters ($A$ = 40, $n$ = 1.6) for each layer [10].

## 4. Summary

Monte Carlo simulations of the transport of low energy positron beams through thin SiN windows and gas layers have been performed. The simulations were motivated by the recent development of an atmospheric positron probe microanalyzer at our institute whereby a focussed positron beam is extracted from the vacuum into air. It was found that the transmission of slow, monoenergetic positrons through SiN can be modelled using the well known Mahkov profile. The simulations also suggested that the effect of extraction through SiN windows on the temporal resolution in PALS is insignificant. For the lateral beam size on target some degradation is expected but a lateral resolution less than 100 μm should be possible using the thinner (30 μm) SiN windows over a wide energy range.

**Acknowledgement:** This work was supported by NEDO. The SiN shape measurement was supported by the AIST Nano-Processing Facility.

## 5. References

[1] Coleman P 2000 *Positron Beams and Their Applications* (Singapore: World Scientific)
[2] Oshima N, Suzuki R, Ohdaira T, Kinomura A, Narumi T, Uedono A and Fujinami M 2008 *J. Appl. Phys.* **103** 094916; 2009 *Mat. Sci. Forum* **607** 238; 2009 *Appl. Phys. Lett.* **94** 194104
[3] Oshima N, O'Rourke B E, Kuroda R, Suzuki R, Watanabe H, Kubota S, Tenjinbayashi K, Uedono A and Hayashizaki N 2011 *Appl. Phys. Express* **4** 066701
[4] Zhou W, Chen Z, Oshima N, Ito K, O'Rourke B E, Kuroda R, Suzuki R, Yanagishita H, Tsutsui T, Uedono A and Hayashizaki N 2012 *Appl. Phys. Lett.* **101** 014102
[5] Landau L D and Lifshitz E M 1986 *Theory of Elasticity 3rd Edition* (Oxford: Butterworth) pp 53
[6] Pierson H O 1999 *Handbook of Chemical Vapor Deposition (CVD): Principles, Technology, and Applications 2nd Edition* (William Andrew Publishing)
[7] Aers, G C 1994 *Appl. Phys. Lett.* **64** 661
[8] Baro J, Sempau J, Fernandez-Varea J and Salvat F 1995 *Nucl. Instrum. Methods Phys. Res. B* **100** 31; see http://www.nea.fr/tools/abstract/detail/nea-1525
[9] Baker J A, Chilton N B, Jensen K O, Walker A B, Coleman P G 1991 *J. Phys.: Condens. Matter* **3** 4109
[10] Mills A P Jr and Wilson R J 1982 *Phys. Rev. A* **26** 490; Valkealahti S and Nieminen R M 1983 *Appl. Phys. A* **32** 95-106; Vehanen A, Saarinen K, Hautojärvi P and Huomo H 1987 *Phys. Rev. B* **35** 4606-4610
[11] Algers J, Sperr P, Egger W, Kogel G and Mauer F H 2003 *Phys. Rev. B* **67** 125404

Journal of Physics: Conference Series **443** (2013) 012070 doi:10.1088/1742-6596/443/1/012070

# Positron Re-emission Studies from W (100)

**K Sudarshan[1§], V N Petrov[1, 2], S N Samarin[1], P Guagliardo[1], A H Weiss[3] and J F Williams[1]**

[1]ARC Centre of Excellence for Antimatter and Matter Studies, School of Physics, The University of Western Australia, Crawley, Perth, Western Australia 6009. Australia.
[2]St. Petersburg State Polytechnical University, Polytechnicheskaya 29, St. Petersburg 195251, Russia
[3]Physics Department, University of Texas at Arlington, Arlington, TX 76019-0059, USA

[§]E-mail: sudarshan.kathi@uwa.edu.au

**Abstract.** A tunable low energy positron beam has been established at The University of Western Australia. The positron beam can deliver approximately 700 positrons per second in the energy range of 0-1 keV. Initial studies of thermal positron re-emission from a W(100) target showed that the positron work function was 2.75 ± 0.01eV. It was also found that the positron reemission yield was sensitive to the sample temperature and surface cleanliness.

## 1. Introduction

As the positron implantation depth increases with increasing incident energy, surface physics studies require tunable positron sources in the low energy range. A detailed account of slow positron beams and the interaction of positrons with surfaces and thin films can be found in the review by Schultz and Lynn [1]. A tunable, low energy positron beam for surface studies has been established at The University of Western Australia (UWA). The salient design features of the positron beam, various experimental capabilities built around the facility, and our initial measurements of low energy positron re-emission from W(100) are described in the paper.

## 2. The variable energy positron beam at UWA

### *2.1. The positron beam set up and features*

The positron beam apparatus at UWA consists of three stages: thermal positron production, positron transport and an experimental chamber containing a sample. The positrons are produced by a 20 mCi $^{22}Na$ source (procured from iThemba labs, South Africa) and are moderated by a polycrystalline tungsten film of thickness 1 μm, in transmission geometry. The moderated positrons are electrostatically transported to the experimental chamber. The principal design of the beam transport is based on that used at the University of Texas at Arlington, USA [2]. The moderated positrons are extracted and electrostatically transported through a 90 degree bender. The potentials on various elements of the lens assembly and beam optics have been optimized using SIMION. The fast positrons from the source which were not moderated are not passed through the bent section of the beam transport. The potential between the moderator and the sample determines the beam energy.

A schematic of the experimental chamber is shown in Fig.1. The chamber is equipped with a position sensitive micro channel plate (MCP) detector at $45^0$ with respect to the incident beam. The position sensitive MCP in combination with two grids is used as retarding field analyzer (RFA) to measure the energy and intensity of positrons or electrons. Two grids are placed in front of the MCP to provide the retarding field. A W(100) sample is in the center of the chamber on a manipulator. The single crystal can be moved in 3D and the angle with respect to incident beam can be changed as required. The experimental chamber is equipped with a UHV compatible evaporator to deposit thin films of Fe, Ag and LiF on the W single crystal substrate, and a microbalance to calibrate the deposition rates. The sample can be annealed in situ. A facility to magnetize the thin ferromagnetic films and change the direction of magnetization by reversing the direction of current in a copper coil placed next to the sample was added to study magnetic films. The entire beam line is under a vacuum of $2\text{-}4\times10^{-10}$ mbar.

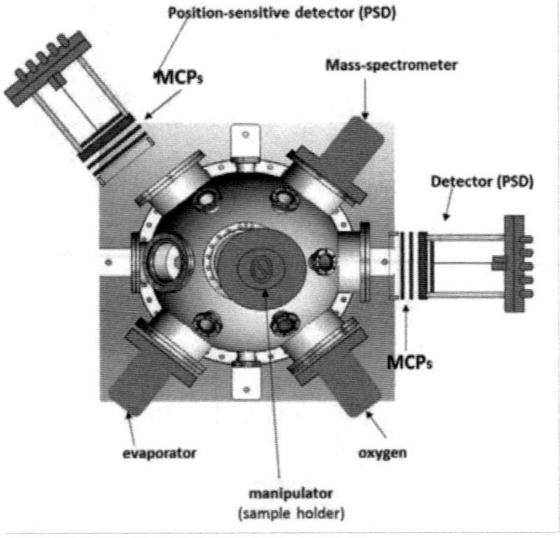


**Figure 1**. The schematic of the experimental chamber showing various capabilities of the positron beam at UWA

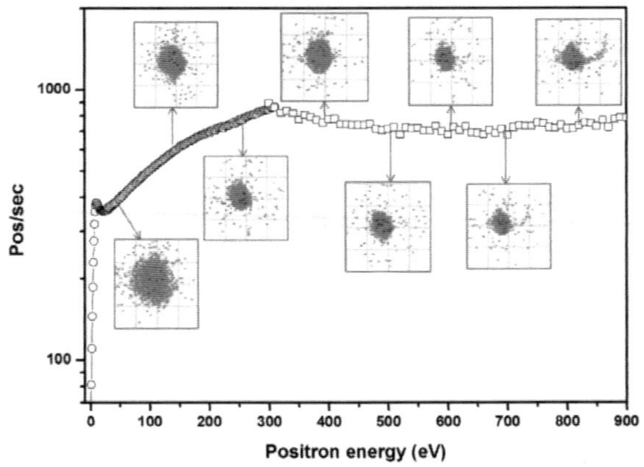


**Figure 2**. The transmitted intensity of the positrons at different energies. The insets show the maximum dimension of the positron beam.

*2.2. Characterization of the variable energy positron source*

The positron beam has been optimized and characterized in terms of the transmission intensity, focusing, and the energy distribution. To measure the beam intensity and size at different positron energies, the position sensitive MCP is moved into the direct path of the beam behind the sample. The dimensions of the MCP are such that the beam intensity and its diameter can be measured at about 10 cm downstream from the sample position. The potential of the beam exit or delivery point into the chamber is grounded and positrons are in a field free region in the chamber. The beam intensity and spot size in the energy range of 0-900 eV is shown in Fig.2. The size of the inset box containing the beam profile is 4 x 4cm. These dimensions represent the upper limit of the beam spot size at the sample position, as they are measured downstream from the sample. A measurement of the positron retarding curve of elastically scattered positrons gives a full-width-at-half-maximum (FWHM) of 0.7 eV for the elastic peak. This can be taken as a measure of the energy spread in the incident beam.

## 3. Positron re-emission from W(100)

Positrons implanted in a solid quickly thermalize. At low energies, the implantation depths are small compared to the positron mean diffusion length and hence most positrons will encounter the surface during their random walk diffusion. The surface dipole layer, which is responsible for the positive electron work function, can in some cases eject the energetic positrons. For a positron, the ground state in many metals lies higher in energy than the vacuum level thereby resulting in the spontaneous reemission of positrons from those metals. Positrons also undergo refraction at the surface during re-emission which results in the re-emission being confined to a narrow angular range about the surface normal [3].

Tungsten meshes, films and crystals are the most commonly used moderators in variable energy positron beams. Our first measurements with the newly constructed positron beam at UWA concern positron re-emission from a tungsten single crystal. Although workable procedures for the use of tungsten as a moderator have evolved over the years and are being practiced [4], the understanding of the physics of moderation and the factors influencing the moderation efficiency are far from complete.

### *3.1. Positron work function of W(100)*

Positrons with an energy given by the work function of the material are re-emitted from the surface. The positron work function is difficult to determine due to the difficulties associated with contact potential differences in determining the exact energy of reemitted positrons, and the influence of surface characteristics of the sample on the re-emission process. This is reflected in the range of work functions of W(100) reported in literature. For example, Jin et al measured the work function of W(100) foil to be -1.9±0.3 eV at a pressure of $10^{-6}$ Torr [5], measurements at $10^{-10}$ Torr by Amarendra et al [6] for W(100) foil gave -2.48±0.05 eV, and a value of -3.0±0.3 was reported by Chen et al [7] and Hugenschmidt et al [8].

The energy spectra of re-emitted positrons can be measured by applying either a positive retarding bias to the detector or a negative retarding bias to the sample. The positron retarding curve measured using a positive retarding potential on the MCP and the negative of its derivate are shown in Fig.3 as circles. The sample was cleaned by annealing in the presence of oxygen (at 1400 $^0$C) and then with electron flashes in vacuum (2400 $^0$C). The cleaning procedure adopted was based on reported procedures for obtaining clean W(100) surfaces [6]. The low energy part of the re-emitted positron energy spectrum is reported to range down to 0 eV [6-8]. However, in the present experiment no low energy tailing was observed, probably due to the cleanliness of the sample and it was difficult to estimate the contact potential difference or retarding bias where the positron counts in energy spectra approach zero. When the retarding potential is applied to the sample, and in the absence of contact potential difference between the sample and the detector, the peak position in the derivative of the retarding curve (energy spectra) is expected at a negative voltage which is equal in magnitude to the earlier case where positrons were retarded by the grid in front of the detector. The peak position of the energy spectra could be different from work function value due to contact potential difference with analyser. The retarding curve and energy spectra of re-emitted positrons measured with a negative potential on sample is also shown in Fig.3 as triangles. The overall equations governing the measurement with the incorporation of contact potential differences can be summarized as:

$$-\varphi_{e^+} = e\,V_{Ret}^{GS} + (\varphi_-^G - \varphi_-^S) \;=\; -\,e\,V_{Ret}^{SG} - (\varphi_-^G - \varphi_-^S)$$

where $\varphi_{e^+}$ is the positron work function of the sample, $V_{Ret}^{GS}$ and $V_{Ret}^{SG}$ are the retarding potentials on the grid and sample respectively, corresponding to maxima in the re-emitted energy spectra; and $(\varphi_-^G - \varphi_-^S)$ is the contact potential difference between the grid and the sample. The peak position of the energy spectra measured by both methods differs by twice the contact potential difference. The positron work function is estimated from the above measurements as:

$$-\,2\varphi_{e^+} = eV_{Ret}^{GS} - eV_{Ret}^{SG}$$

and was found to be -2.75 ± 0.01 eV.The energy spread of the re-emitted positrons was about 180 meV which is in reasonable agreement with energy spread of 250 meV reported by Amarendra et al [6].

### *3.2 Effect of temperature*

To get further insights into the thermalized positron re-emission efficiency of W(100), the positrons reemitted in the direction normal to the surface were measured under different conditions. The W single crystal surface was cleaned with electron flashes (heated to 2400$^0$C) and the re-emission yield from W was measured at regular time intervals after the electron flashing, as shown in Fig.4. The positron yield was found to increase rapidly as the sample cooled and approached a maximum after 1.5 hours. The possible lower yield at high temperature might be caused by either a broader angular distribution of remitted positrons due to phonon scattering or a reduction in the work function of the material [9], or increased trapping of positrons in defects reducing their diffusion length [10]. The reduction in the intensity of positrons was nominal for more than 20 hours. With progression in time, the temperature of the sample was reduced and the contamination of the surface increased [11]. The results show that the positron yield is lower from hot tungsten although the exact temperature of the crystal, when the maximum in yield was achieved, could not be ascertained in this study. The reasons for the marginal rise in the intensity beyond 25 hours are not clear.

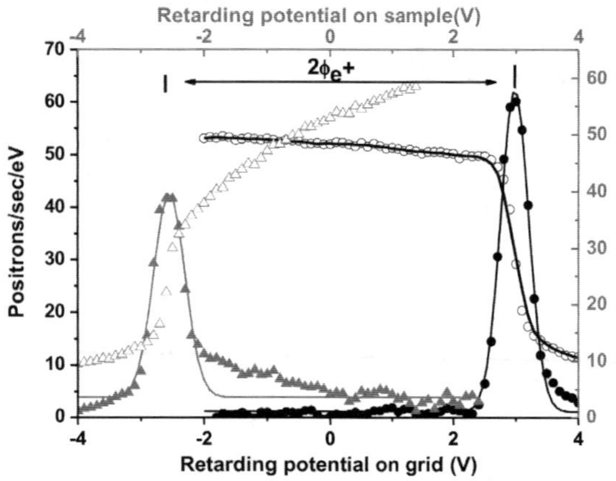


**Figure 3**. Positron counts with retarding bias applied to the MCP grid (open circles) and the negative of its derivative showing the energy spectrum (filled circles). The positron counts with retarding bias applied to the sample (open triangles) and its derivative (filled triangles).

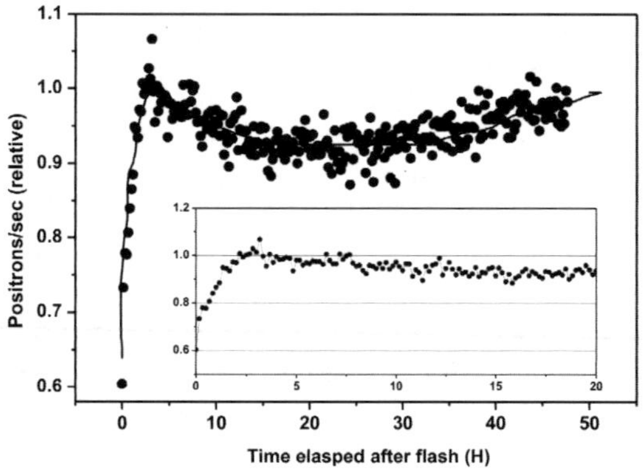


**Figure 4.** Re-emitted positron intensity from W(100) as a function of time after cleaning. The inset shows a magnified plot at initial times. Positrons in the energy range of 2-4 eV were taken as re-emitted positrons.

## 4. Conclusion

A slow positron beam which can deliver ~700 positrons/sec in the energy of 0-1keV was established and characterized at The University of Western Australia. The beam was utilized to study positron re-emission and scattering from metallic surface. The re-emitted positron measurements from W(100) have shown that the positron work function is 2.75 ± 0.01 eV. The positron re-emission process was shown to be sensitive to the temperature and cleanliness of the sample.

## 5. References

[1] Schultz P J, Lynn K G, 1988. *Rev. Mod. Phy.* **60** 701
[2] Zhou H, 1996. PhD thesis, University of Texas at Arlington,
[3] Murray C A and Mills, Jr. A P, 1980. *Sol. Stat. Commun.* **34** 789.
[4] Zecca A et al, 2010. *Nucl. Instr. Meth.* B **268** 533.
[5] Jin B et al, 1994. *Jpn. J. Appl. Phys.* **33** L1493.
[6] Amarendra G et al, 1996. *J. Appl. Phys.* **80** 4660.
[7] Chen D M et al, 1985. *Phys. Rev. B.* **31** 4123.
[8] Hugenschmidt C et al, 2002. *Appl. Surf. Sci.* **194** 283.
[9] Rosenberg I J et al, 1987. *Phys. Rev. B.* **35** 2083.
[10] Lynn K G, Schultz P J, 1985. *Appl. Phys. A.* **37** 31.
[11] Al-Qaradawi I Y, Coleman, P G, 2002. *Appl. Surf. Sci.* **194** 20.

Journal of Physics: Conference Series **443** (2013) 012071 doi:10.1088/1742-6596/443/1/012071

# In-situ (C)DBS at high temperatures at the NEPOMUC positron beam line

**Markus Reiner, Philip Pikart and Christoph Hugenschmidt**
Technische Universität München, Physik-Department E21 and FRM-II, Lichtenbergstraße 1, D-85748, Germany

E-mail: markus.reiner@frm2.tum.de

**Abstract.** We report on the current status of the sample heating stage in the CDB-spectrometer at the NEPOMUC positron beam line. The currently installed new sample heating is described in detail and various design aspects are discussed briefly. As an exemplary application, the positron diffusion at high temperatures in Ge was investigated by a depth dependent evaluation of both the Doppler broadening of the annihilation line and free Ps annihilation at the surface. It was confirmed that the temperature dependence of the positron diffusion is extraordinarily strong above 670 K.

## 1. Introduction

CDBS is outstanding among the various PAS methods due to its elemental selectivity on the vicinity of the positron annihilation site [1]. Performing (C)DBS with a monoenergetic positron beam at high temperatures is an unique experimental technique for the investigation of defect annealing and temperature driven structural changes in thin film systems [2]. However, it requires not only a highly intense positron beam in order to enable sufficiently short measurement times, but also an appropriate sample heating device, which enables the controlled heating of specimens in an UHV condition and biased to high voltages up to -30 kV. For this reason, during the recent years, several heatable sample holders have been applied at the CDB spectrometer [3] of the NEPOMUC positron beam line [4,5]. Within this contribution, we present the currently installed sample heating device as a result of various design studies and recent measurements on the temperature dependence of the positron diffusion in Ge(111).

## 2. The new sample heating device at the NEPOMUC CDB-spectrometer

The previous sample heaters at the NEPOMUC CDB-spectrometer were integrated in the standard sample holder for room temperature measurements. For this reason, design limitations constrained the maximum reachable temperature to 800 K. In order to increase the temperature further, an extra sample holder for high temperature measurements was constructed and is currently installed at the CDB-spectrometer.

In this device, the sample is heated by the light of a halogen lamp, which is focused onto the bottom of the sample by an elliptic reflector. Compared to the previous designs, the reflecting area has been more than trebled. Thus, the maximum sample temperature, which can be reached by use of the 250 W halogen lamp, will be clearly increased. Besides, in the new set-up, the sample will be movable and hence, spatially resolved measurements will also become possible at high temperatures.

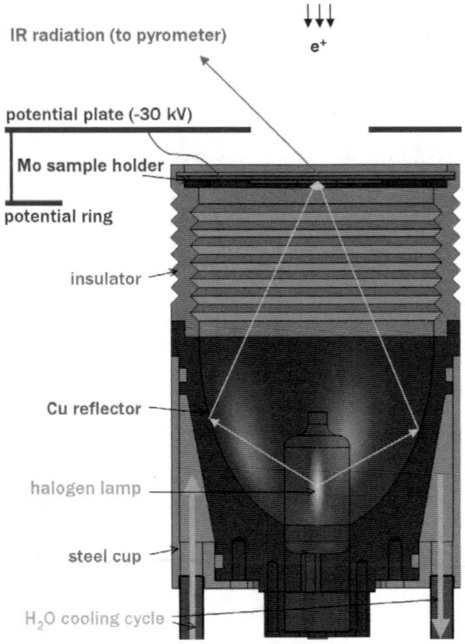


**Figure 1.** Mirror lamp assembly and heatable sample holder

The detailed sketch in figure 1 shows all relevant parts of the new sample heating device. The sample holder is located above the mirror lamp assembly, which consists of the Au-coated elliptic Cu reflector and the halogen lamp in the lower focal point. In order to run the halogen lamp at full power, a water cycle is used for cooling the mirror lamp assembly. The water will be pumped through the gasketed free volume between reflector and the steel cup, which carries the reflector. In its other focal point, the specimen is placed and hold by an assembly of Mo sheets, which enables fast sample changing. The Mo sheets are fixed on a ceramic insulation, which was designed in order to deal with the electric potential difference of up to 30 kV between reflector and specimen. The potential plate and ring, which enclose the sample holder, lead to a well defined electric potential in the vicinity of the specimen.

It was found, that the two color pyrometer METIS MQ22 from Sensortherm$^{TM}$, which can be placed outside the sample chamber, is well suited for a contact-free temperature measurement. For this, an IR transparent window is flanged at the sample chamber. Using this method, it is necessary to attach a small temperature control point consisting of colloidal graphite to the sample before heating. In this way, the sample has a well defined emissivity. The results of the pyrometer measurements are in excellent agreement with those obtained by electrical temperature sensors.

In addition, the presented design enables scanning an area of 2 cm x 2 cm on the sample with the NEPOMUC positron beam. The maximum spatial resolution reached by electrostatically focusing the beam onto the specimen will be 0.3 mm. Typical measurement times with an collinear assembly of four pairs of HPGe detectors will be less than 1 min for DBS and 2 h for CDBS. Within these short measurement times, a large variety of samples (with a typical size between 0.5 cm x 0.5 cm and 2 cm x 2 cm) can be investigated at temperatures up to around 1300 K.

## 3. High temperature behavior of positron diffusion in Ge

### *3.1. Introduction*

As an exemplary measurement, the positron diffusion in single crystalline Ge at temperatures between 673 K and 853 K was investigated. Several studies based on different experimental techniques have been published on this issue. Jorch et al. [6] observed an extraordinarily strong temperature dependence of the positron diffusion length $L_+$ between 600 and 850 K by the investigation of free Ps annihilation at the surface, as well as Uedono et al. [7] by depth dependent DBS. Soininen et al. [8] found a weaker dependence by depth dependent DBS and investigation of Ps annihilation. These results illustrate that the high temperature behavior of positron diffusion in Ge, which is related to the electron and hole mobility in the semiconductor, is a point of long-term interest and give motivation for the presented study.

### *3.2. Experimental*

A single crystalline Ge(111) sample (1.5 cm x 1.5 cm) was investigated and heated up to a maximum temperature of 853 K. Before the measurements, no surface treatment like polishing or etching has been performed in order to avoid chemical contamination. Instead, the specimen was heated up to the maximum temperature in a first heating cycle in order to remove oxide from the surface. After cooling down to room temperature, the specimen was heated for a second time and the positron diffusion was investigated.

For this, depth dependent DBS (S(E)) and investigation of the free Ps annihilation at the surface (F(E)) were performed with an incident positron energy between 1 keV and 16 keV. The S-parameter was determined by evaluating a central peak region of 511±0.825 keV and a total peak region of 511±11 keV of the broadened annihilation line. The F-parameter was determined by the evaluation of the linearized valley-to-peak ratio (valley region 450 - 500 keV, peak region 500 - 522 keV), which is sensitive to the $3\gamma$-decay of o-Ps with its continuous energy spectrum. The $3\gamma$-decay event can only take place for free Ps, which does not interact with electrons from the bulk. In case of the singlecrystalline Ge, the annihilation of free Ps can be expected to take place only at the surface.

Both the S(E)- and the F(E)-curve were used for a quantitative determination of $L_+$ by fitting both curves simultaneously. Since the total range of incident energies was accounted for these fits, also epithermal positrons were considered. For performing the fits, the VEPFIT software package [9] was applied. A more detailed description of the method of data evaluation can be found in [2].

### *3.3. Results and Discussion*

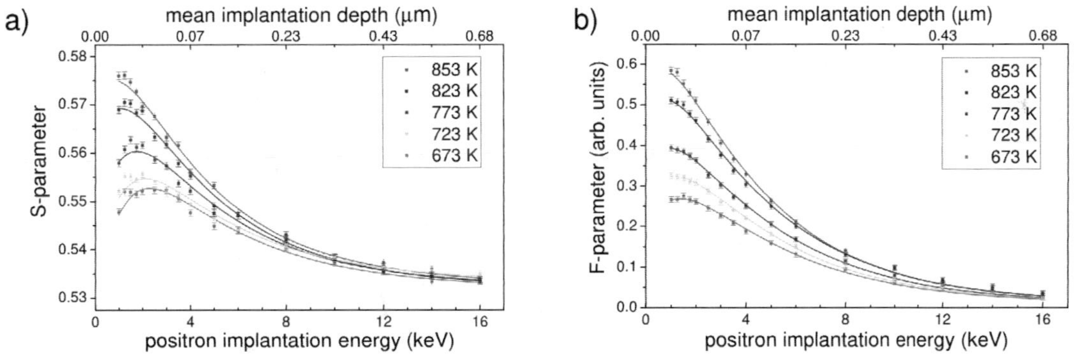


**Figure 2.** S(E) and F(E) on Ge(111) at high temperatures; solid lines are fit results.

The recorded S(E)- and F(E)-curves (measurement time 12 min per depth profile) at high temperatures are shown in figure 2. At the various temperatures, for high incident positron energies, the S-parameter saturates at a bulk value of 0.532±0.002. For the F-parameter, at all temperatures this bulk value is 0 and hence, no significant $3\gamma$-decay is found for high implantation energies, i. e. in the bulk.

For lower implantation energies, i. e. in the surface region, the F-parameter increases, since positrons, which diffuse to the surface, can desorb as Ps leading to the observed $3\gamma$-annihilation of o-Ps. The fraction of positrons that annihilates as o-Ps increases with temperature and hence, positrons may be thermally desorbed as Ps from surface states. Also the S-parameter at the surface, which takes higher values than in the bulk, increases with the temperature. This observation is directly connected to the increasing fraction of positrons annihilating as o-Ps. The formation of Ps at the surface also leads to the annihilation of p-Ps. Due to its low binding energy, at the annihilation of p-Ps only a small Doppler shift is detected resulting in a high S-parameter. For implantation energies below 3 keV, the S(E)- and F(E)-curves exhibit further features, which can be attributed to the annihilation of epithermal positrons. Here, their behavior is not treated in detail.

The solid lines in figure 2 represent VEPFIT results. The according results for $L_+$ and its temperature dependence are shown in figure 3. The decrease from 146±5 nm to 106±5 nm between 673 K and 853 K is too large in order to be explained by a $T^{-\epsilon}$ with $\epsilon = 0.25$, which corresponds to acoustical phonon scattering during positron diffusion [10]. This behavior was observed in [8], where, only implantation energies above 6 keV were evaluated in order to exclude any influence of epithermal positrons. This method of data evaluation was also tested on the current data, but led to large uncertainties in the fit results.

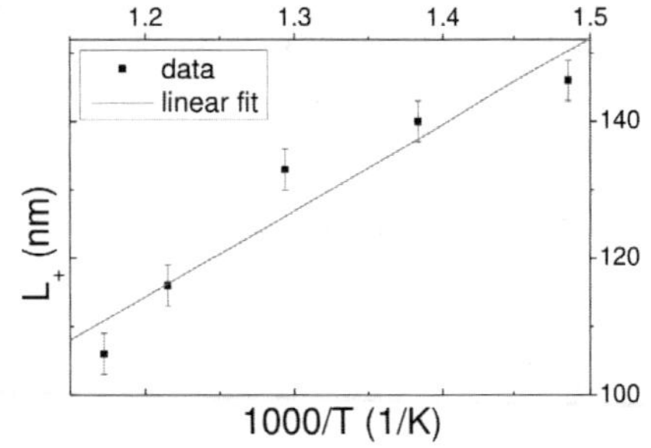


**Figure 3.** Temperature dependence of $L_+$ in Ge(111)

A stronger temperature dependence of $\epsilon$=1 can explain the results much better (see fitted line in figure 3). This is also in reasonable agreement with the results published in [6], where positron implantation energies below 5 keV were investigated and $\epsilon$=0.8 was found. In particular, the present study, based on the evaluation of S(E)- and F(E)-scans, confirms the results given in [7], where also epithermal positrons were considered and $\epsilon$=1 was found in the evaluation of S(E)-scans. Possible explanations for the observed strong temperature dependence of $L_+$ in Ge are given in [6]; there it was tentatively concluded that the positron diffusion is affected by some kind of positron-lattice coupling.

## 4. Conclusion and outlook

Within this contribution, the development and design of the new sample heater in the CDB-spectrometer at the NEPOMUC positron beam line is presented. Due to the use of the highly intense positron beam, the device will enable novel experiments at temperatures up to 1300 K. Furthermore, the high temperature behavior of the positron diffusion in Ge(111) was investigated exemplary. The depth dependent evaluation of both the Doppler broadening of the annihilation radiation and the free Ps annihilation at the surface revealed an extraordinarily strong temperature dependence of the positron diffusion above temperatures of 670 K.

## References

[1] Asoka-Kumar P, Alatalo M, Ghosh V J, Kruseman A C, Nielsen B and Lynn K G 1996 *PRL* **77** 2097
[2] Reiner M, Pikart P and Hugenschmidt C 2012 *Phys. Procedia* **35** 104
[3] Reiner M, Pikart P and Hugenschmidt C 2013 submitted
[4] Stadlbauer M, Hugenschmidt C and Schreckenbach K 2008 *Apl. Surf. Sci.* **255** 136
[5] Hugenschmidt C, Löwe B, Mayer J, Piochacz C, Pikart P, Repper R, Stadlbauer M and Schreckenbach K 2008 *Nucl. Instrum. Meth. A* **593** 616
[6] Hugenschmidt C, Piochacz C, Reiner M and Schreckenbach K 2012 *New J. Phys.* **14** 055027
[7] Jorch H H, Lynn K G and MacKenzie I K 1981 *PRL* **47** 362
[8] Uedono A, Moriya T, Komuro N, Kawano T, Tanigawa S and Ikari A 1996 *Jpn. J. Appl. Phys.* **35** 4599
[9] Soininen E, Mäkinen J, Beyer D and Hautojärvi P 1992 *PRB* **46** 13104
[10] van Veen A, Schut H, Clement M, de Nijs J M M, Kruseman A and IJpma M R 1995 *Appl. Surf. Sci.* **85** 216
[11] Puska J and Nieminen R M 1994 *Rev. Mod. Phys.* **66** 841

Journal of Physics: Conference Series **443** (2013) 012072 doi:10.1088/1742-6596/443/1/012072

# Flash lamp annealing of tungsten surfaces marks a new way to optimized slow positron yields

**W Anwand[1], J M Johnson[1,2], M Butterling[1], A Wagner[1], W Skorupa[3], G Brauer[1]**

[1] Institut für Strahlenphysik, Helmholtz-Zentrum Dresden-Rossendorf, PF 510119, 01314 Dresden, Germany

[2] Department of Physics, The University of Texas at Austin, 1 University Station C1600 Austin, Texas 78712-0264, USA

[3] Institut für Ionenstrahlphysik und Materialforschung, Helmholtz-Zentrum Dresden-Rossendorf, PF 510119, 01314 Dresden, Germany

E-mail: w.anwand@hzdr.de

**Abstract**. Tungsten in the form of a mono-crystalline foil with an optimum thickness of about 2 μm is often used as a positron moderator in mono-energetic positron beams with $^{22}Na$ positron sources. The efficiency of such a moderator strongly depends on its prior heat treatment, i.e. an annealing procedure with considerable difficulty at temperatures of about 2000 °C under vacuum conditions. Flash lamp annealing (FLA) has been tested as new method to quickly anneal W foils in order to produce easy manageable, low-cost moderators with a high efficiency. With FLA, just the surface of a W foil is heated above the melting point (3422°C) within 1 to 3 ms, i.e. without melting the whole foil volume. In this way, a surface cleaning is reached connected with a considerable increase in the positron diffusion length. Conventional polycrystalline W foils of 9 μm ± 25% thickness, heat treated by FLA, were characterized and tested as positron moderators. First promising tests result in a moderator efficiency of $\sim 3*10^{-4}$ and clearly demonstrate that FLA is also applicable to tungsten meshes.

## 1. Introduction

The worldwide scarcity of commercially available $^{22}Na$ sources and the resulting dramatic rise in its price demand new thinking about an optimal application of $^{22}Na$ as a $\beta^+$ source in mono-energetic positron beams for depth-dependent Doppler broadening measurements [1]. The positron moderator is the crucial factor for the creation of enough mono-energetic positrons for their use in such investigations. There are two main requirements for the moderator properties: a stable, high moderation efficiency and manageability at low costs. Tungsten generally meets both requirements well. Nevertheless, the commonly used 2 μm single crystalline tungsten foils are especially susceptible to radiation damage and require frequent defect annealing, which is associated with considerable difficulties. Fig. 1 shows the decrease in the count rate of such a foil in comparison to the decrease calculated with the decay law. The foil was mounted in front of a 2.4 GBq source. It becomes clear that moderator annealing is essential for an effective operation of mono-energetic positron beams. A new method was sought to replace the laborious annealing of the thin foils mostly done by means of electron heating under high vacuum conditions by a more simple method. Flash lamp annealing (FLA) seems to be viable for this purpose and experimental results regarding the first application of flash-lamp-annealed tungsten foils are presented.

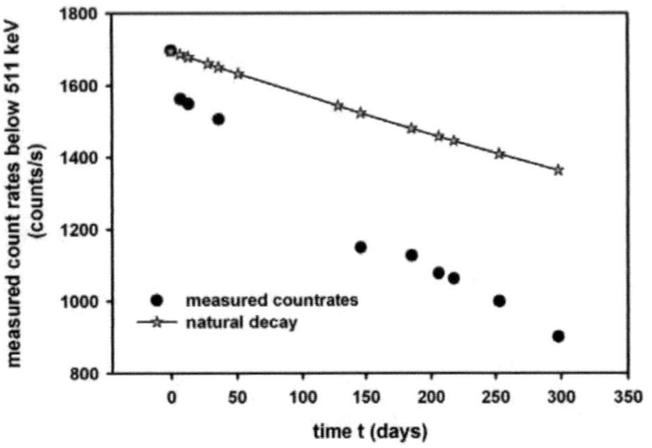


**Figure 1.** Decrease of the count rate from a 2 μm single crystalline foil in comparison with the decrease caused by the natural decay.

## 2. Efficiency of tungsten foils with different thicknesses

FLA of tungsten requires opaque foils in order to avoid local overheating, which causes the formation of holes in the foils. Opaque polycrystalline foils are commercially available with a minimum thickness of 9 μm ± 25%. However, thicker moderator foils lead to decreasing moderator efficiency. For this reason, a short calculation of the efficiency of a 2 μm foil in comparison to the thicker foil of 9 μm was made. The calculations were performed with the following assumptions: The positrons from

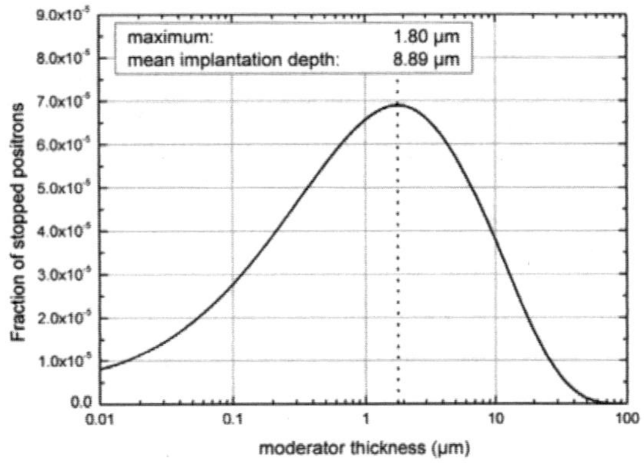


**Figure 2.** Stopped positrons in dependence on the penetration depth in tungsten.

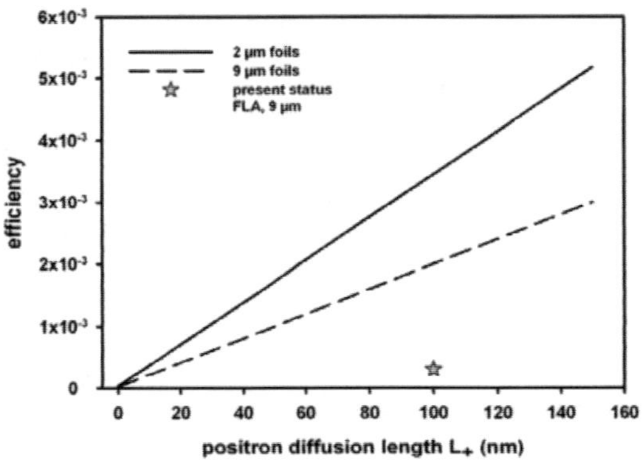


**Figure 3.** Expected moderator efficiency versus positron diffusion length.

the source capsule penetrate a 5 μm thick Ti window and enter the moderator foil, which is directly mounted behind the window. The energy distribution of the positrons only marginally changes after passing the Ti window. The fraction of the positrons F(z) which is stopped in a certain depth z within the moderator can be calculated by use of the energy distribution of the positrons N(E) and assuming a Makhovian profile P(E,z) for the depth distribution according to Eq. (1).

$$F(z) = \int_{0keV}^{545keV} N(E)P(E,z)dE \tag{1}$$

Fig. 2 shows the dependence on the penetration depth for positrons stopped in a tungsten foil. It becomes clear that the optimum thickness of a tungsten moderator foil is about 2 μm. A thickness of 9 μm correlates with the mean positron implantation depth, with only 59% of stopped positrons below the surface compared to the thinner foil. The expected moderation efficiencies for both foils are shown in Fig.3. Thereby the moderator efficiency is defined as ratio of moderated positrons divided by fast positrons entering the moderator. Assuming that the bulk diffusion length $L_{+,bulk}$ = 135 nm [2] the expected maxima of the efficiencies should be 4.5 x $10^{-3}$ for 2 μm and 2.7 x $10^{-3}$ for 9 μm thick foils. In Ref.[3], it was reported a moderator efficiency of a 6 μm thick tungsten foil of 2.6 x $10^{-4}$ after annealing at 2100 °C in vacuum. Surface properties, grain boundaries, impurities and other influences, which were not included in the rough calculations, are responsible for the considerable discrepancy between calculation and reality.

## 3. Flash lamp annealing of tungsten foils

FLA [4] was used for cleaning the surface of the foils and for defect annealing of the poly-crystalline foils. It offers a chance for the optimization of the moderator properties. This method enables the heating of the surface of the W foil above the melting point of 3422 °C in 1 to 3 ms without melting the whole volume. The very short heat input into the surface is realized by xenon-filled gas discharging lamps, which were charged by a capacitor bank with a capacitance of 4.8 mF and a maximum charging voltage of about 4 kV. The whole capacitor bank is discharged via induction coils for the adjustment of the duration of the flash. Twelve Xe flash lamps generate the short light pulse with a spectral distribution between 400 and 700 nm, peaking around 500 nm. For the FLA of the 9 μm W foils, a light pulse of 1.25 ms has been chosen with an energy density of 32 J/cm$^2$ resulting from an electrical power of 31 MW. This heat treatment was carried out stepwise in an Ar flow at normal pressure, starting with an energy density of 8 J/cm$^2$. The complete cleaning of the surface could be observed at an energy density of 12.5 J/cm$^2$, which corresponds to a surface temperature of about 1600 °C. While the surface cleaning could be obtained below 2000 °C , temperatures above the melting point had to be applied in order to obtain longer positron diffusion lengths due to grain growth and defect annealing (Fig.4a and b and Fig.5).

**Figure 4a.** Scanning Electron Microscopy (SEM) image of the as-received foil.

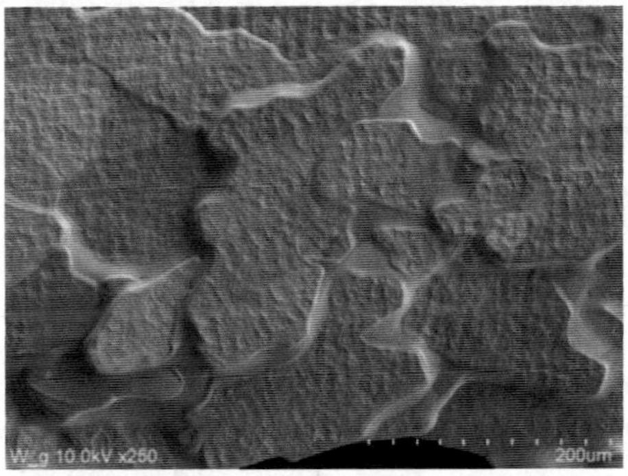


**Figure 4b.** Molten tungsten foil surface after FLA.

## 4. Characterisation and test of the moderator foils and meshes

The mono-energetic positron beam “SPONSOR” at Helmholtz-Zentrum Dresden-Rossendorf [5] has been operating with a 2.4 GBq $^{22}$Na source and a 2 μm single crystalline W moderator for about nine years. Finally, in 2011 the source had decayed to 308 MBq and the measured count rate below the 511keV line decreased from 1800 c/s at the beginning to 22 c/s. Therefore, the moderator was replaced by a flash lamp annealed 9 μm polycrystalline W foil and a nearly 20 times higher count rate of 400 c/s could be immediately reached. Taking into account the distance between Ge detector and sample, efficiency and size of the Ge detector, the number of positrons entering the moderator and other experimental influences, the efficiency of the moderator could be estimated to be about 3 x $10^{-4}$, which is marked as a star in Fig.3. Fig.5 shows the measured S parameter versus the incident positron energy E for the as-received W foil and the moderators that were used. The positron diffusion lengths $L_+$ were fitted using the VEPFIT software [6]. The fits result in very short $L_+ \sim 1$ nm for the as-received foil, in $L_+ = 48$ (8) nm for the 2 μm foil and 94 (13) and 99 (16) nm for FLA 1 and 2 foils, respectively. Because of 25% deviation in thickness of the poly-crystalline foils a complete surface melting could not be obtained and the diffusion length in the FLA foils is shorter than the expected one [2]. From the comparison of the diffusion lengths of the 2 μm foil and the poly-crystalline foil it becomes clear that the shorter diffusion length only contributes to half of the degradation of the efficiency, which is more than an order of magnitude in total. Structural changes at the surface primarily influence the moderator properties. In this context, it is worth mentioning that all W foils showed a decrease in the efficiency of about 20% under vacuum conditions of $5x10^{-7}$ Torr after one week. The primary efficiency could be restored by exposure of the foils to air for a few hours.

W meshes can be etched with a homogenous volume degradation, which leads to a uniform diameter of the W wires. For this reason, heating of the entire surface to the melting point is possible without damage by local overheating, as seen in Fig. 6. Taking into account an efficiency like that of the poly-crystalline foils and three times higher surface area of the meshes (diameter of the wire 10-15 μm), a total efficiency of $9x10^{-4}$ should be achievable using 10-12 meshes.

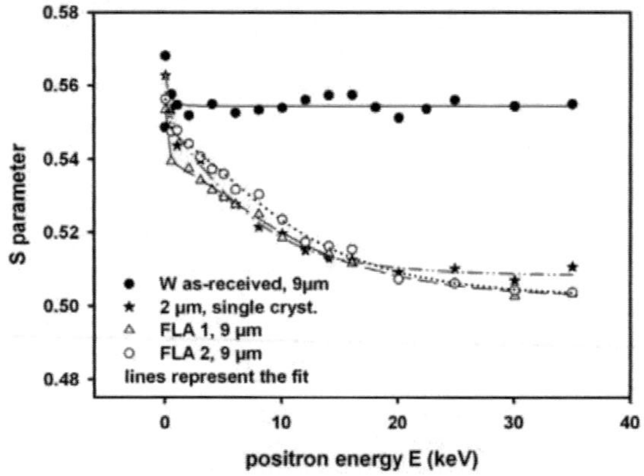


**Figure 5.** S(E) of W moderators.

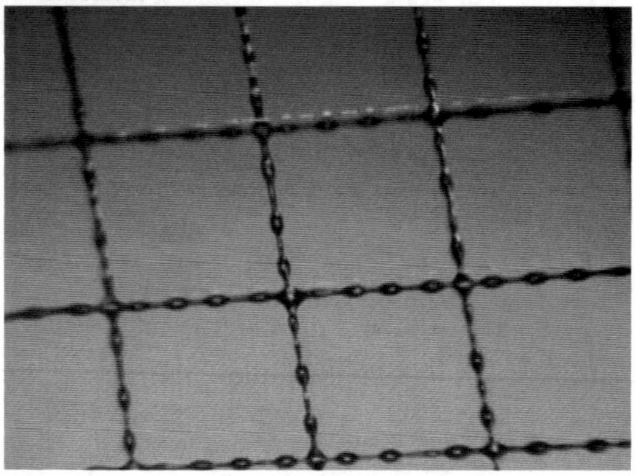

**Figure 6.** W meshes treated by FLA.

## 5. Conclusions

FLA is a suitable annealing procedure for W foils of a thickness of 9 μm. By heating the surface to near the melting point, it could not only be cleaned, but a longer positron diffusion length could be reached, which would be impossible with a thermal treatment of around 2000 °C. The obtained moderator efficiency of $3x10^{-4}$ agrees well with the published value [3], though it falls short of the optimum, and further efforts for its improvement are needed. Nevertheless, the short annealing time and stable properties of the foils even when exposed to air legitimate this method. The application of FLA to W moderator meshes is promising. The intention is to realize this as a next step.

## Acknowledgement

The support of Th. Schumann (with FLA) and of E. Christalle (with SEM) is gratefully acknowledged.

## References

[1] Coleman P G (Ed.) 2000 *Positron Beams and their applications* (Singapore: World Sci. Publ.)
[2] Vehanen A, Lynn K G, Schultz P J and Eldrup M 1983 *Appl. Phys. A* **32** 163
[3] Gramsch E, Throwe J, Lynn K 1987 *Appl. Phys. Lett* **51** 1862
[4] Skorupa W, Anwand W, Panknin D, Voelskow M, Yankov R A, Gebel T 2005 *Vacuum* **78** 673
[5] Anwand W, Brauer G, Butterling M, Kissener H-R, Wagner A, *Design and Construction of a Slow Positron Beam for Solid and Surface Investigations* in: B.N. Ganguly, G. Brauer (Eds.), *Near Surface Depth Profiling by Mono-energetic Positrons* Defect and Diffusion Forum Vol. 331 2012 pp 25-40
[6] van Veen A, Schut H, de Vries J, Haakvoort R A, Ijpma M R *Positron beams for solids and surfaces* in: P.J. Schultz, G.R. Massoumi, P.J. Simpson (Eds.), *Proceedings of the AIP Conference* vol. **218** 1990 pp 171-177

Journal of Physics: Conference Series **443** (2013) 012073 doi:10.1088/1742-6596/443/1/012073

# Optimization of growth parameters of TiO2 thin films using a slow positron beam

**M. Butterling[1], W. Anwand[1], S. Cornelius[1], K. Potzger[1], A. Smekhova[2], M. Vinnichenko[1] and A.Wagner[1]**

*[1]Helmholtz-Zentrum Dresden-Rossendorf, Bautzner Landstr. 400, D-01328 Dresden, Germany*

*[2] M.V.Lomonosov Moscow State University (MSU), Leninskie Gory 1, Moscow, 119991, Russia*

E-mail: m.butterling@hzdr.de

**Abstract**. $TiO_2$ thin films grown on fused silica were investigated using positron Doppler broadening spectroscopy at the slow-positron-beam SPONSOR [1] at the Helmholtz-Zentrum Dresden-Rossendorf. Effects of changes in different parameters like temperature or oxygen flow during film deposition on positron sensitive parameters have been investigated and first results will be presented.

## 1. Introduction

The ferromagnetic properties of transition metal doped $TiO_2$ are intensely investigated in recent years. Such dilute magnetic semiconductors (DMS) have potential applications in spintronics, i.e. due to spin-polarization of mobile charge carriers, or in magneto-optics. Ferromagnetic transition metal doped $TiO_2$ films can be created in different ways, i.e. from thin film growth or ion implantation. Until now there is an intense debate about the origin of the ferromagnetic properties and whether they are related to highly mobile spin-polarized electron currents or not. Consequently, highly sensitive structural investigations are applied which revealed that ferromagnetic properties may arise from either secondary phases [1], Co enriched anatase clusters [2] or intrinsic defects [3, 4]. Moreover, structural point defects such as vacancies can influence the ferromagnetic interaction between the transition metal dopants [5]. It was also found that Co doped $TiO_2$ can be ferromagnetic and insulating at the same time which can not be explained by the usual ferromagnetic double exchange model describing the coupling in DMS [6]. The major problem, however, is the occurrence of metallic Co secondary phases, especially in ion-implanted $TiO_2$ [7], which have the potential to mimic all of these effects and thus camouflage an intrinsic ferromagnetic coupling.

In future we will study the origin of ferromagnetic properties of V, Co or Mn doped anatase $TiO_2$ thin films. The effects of oxygen pressure and substrate temperature on the (defect) structure of the grown films are of importance concerning the film quality (for further treatments and investigations) and can be easily investigated by depth-resolving positron Doppler broadening spectroscopy.

## 2. The SPONSOR setup

The Slow Positron System of Rossendorf (SPONSOR) [8] was used for depth-resolving Doppler broadening measurements. There positrons were implanted with energies between 27 eV... 36 keV resulting in an implantation depth of several micrometers (the planned film thickness is about 300 nm). The energy resolution at the annihilation line is (1.09 ± 0.01) keV.

## 3. Choice of the substrate material

The substrate material can have a strong influence on the film structure: $TiO_2$ films are either polycrystalline, if deposited on fused silica (a), or epitaxial, if deposited on $SrTiO_3$(100) (b). From (a) it is evident that the $TiO_2$ structure can be adjusted to be either rutile or anatase depending on growth temperature. The largest electric conductivities have been achieved for $TiO_2$ anatase films either undoped or doped with Nb. Besides $SrTiO_3$, $LaAlO_3$ may also serve as a possible substrate material. Depth-profiles obtained by Doppler broadening spectroscopy at SPONSOR for as-received material (figure 1) indicate that $SrTiO_3$ is the better candidate: the diffusion length $L_+$ (obtained by the VEPFIT package [9]) is more than double the value of that for $LaAlO_3$.

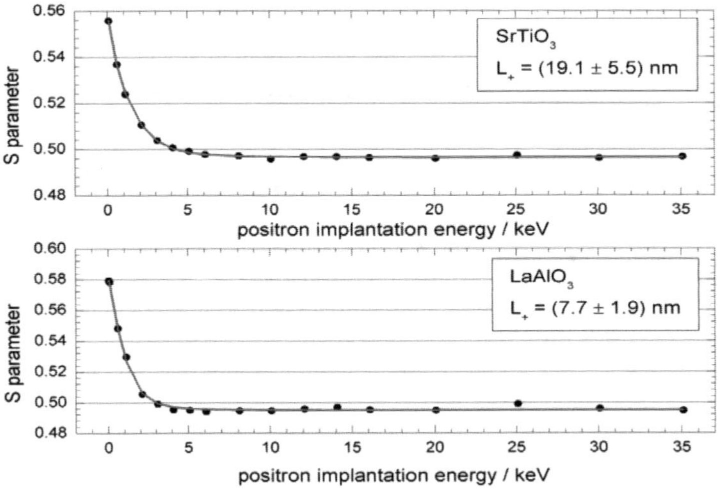


**Figure 1.** Depth profiles of substrate candidates for $TiO_2$ films. The diffusion length $L_+$ was calculated using the VEPFIT package [9].

Due to these investigations $TiO_2$ will be grown on $SrTiO_3$ for further investigations of ferromagnetic properties.

## 4. Effect of substrate temperature during film growth on the structure of $TiO_2$

Thin films of $TiO_2$ were grown by sputter deposition. The substrate temperature influences structure and defect situation of deposited films and influences defect induced room-temperature ferromagnetism. To investigate a temperature effect we used a series of thin $TiO_2$ films grown on fused silica where the films were deposited at constant partial oxygen pressure. The related profiles are shown in figure 2 also containing the profile of the substrate of fused silica (f-$SiO_2$). One can see the effect of the substrate on the S parameter but the layer is not affected by this meaning that the films are not too thin. Again the diffusion length $L_+$ and the S parameter of the$TiO_2$ film layer were calculated (figure 3). Surprisingly there is a non-monotonous behaviour of the S parameters with a maximum for a substrate temperature of around 400 °C. The small layer thickness leads to larger error bars for $L_+$ complicating the interpretation of results.

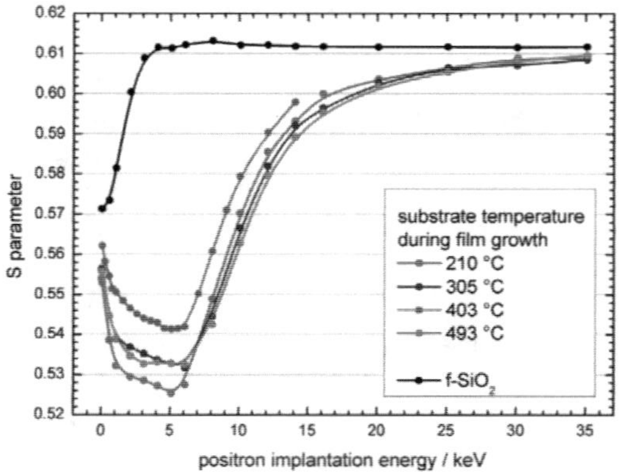


**Figure 2.** Depth profiles obtained by positron Doppler broadening spectroscopy for different substrate temperatures.

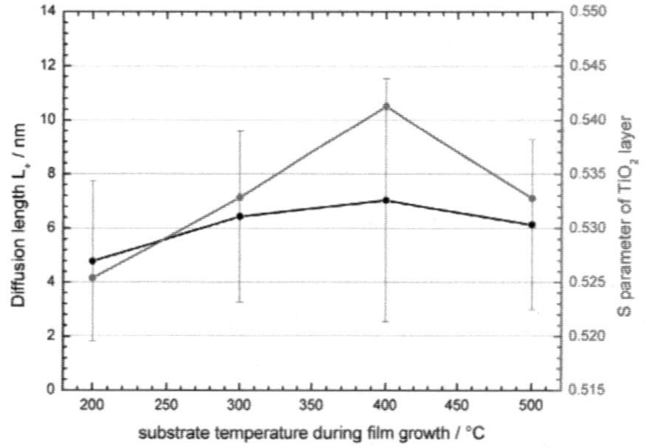


**Figure 3.** S parameter and calculated diffusion length $L_+$ (using the VEPFIT package [9]) for the $TiO_2$ films grown on fused silica (f-SiO2).

$TiO_2$ will be grown at 400 °C due to the optimum of the diffusion length for this temperature.

## 5. Influence of oxygen partial pressure during film growth on the structure of $TiO_2$

The origin of the ferromagnetic properties was explained by molecular orbitals formed from the valence electrons on the three Ti ions surrounding the oxygen vacancy and couple via direct exchange interaction. A direct relationship between the oxygen partial pressure during growth, the electronic conductivity as well as the saturation magnetic moment of anatase TiO2 films was established [3].

We used positron Doppler broadening spectroscopy to investigate the influence of oxygen partial pressure on film structure on a nano-scale. This investigated series consists also of $TiO_2$ grown on fused silica at a temperature of 400 °C at different oxygen partial pressures. The profiles (figure 4) as well as the extracted S parameters and diffusion lengths $L_+$ (figure 5) show a clear dependence on oxygen partial pressure. A simultaneous increase of S and $L_+$ indicates that with increasing pressure smaller defects begin to agglomerate, leading to larger values for S and $L_+$. The jump in $L_+$ (within the error bars) indicates a sensitive pressure range of around $4{\cdot}10^{-4}$ mbar.

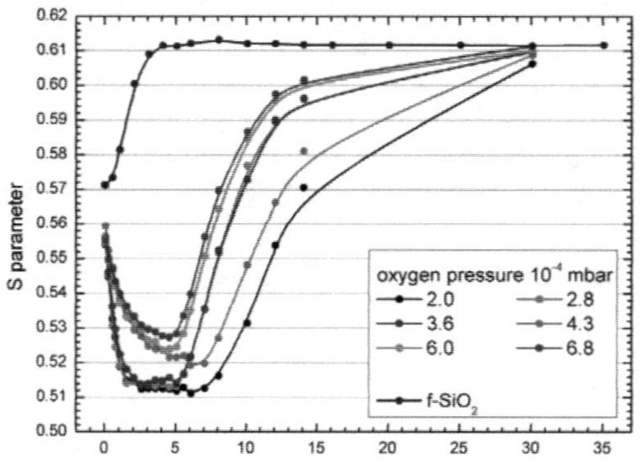


**Figure 4.** Depth profiles obtained by positron Doppler broadening spectroscopy for different oxygen partial pressures.

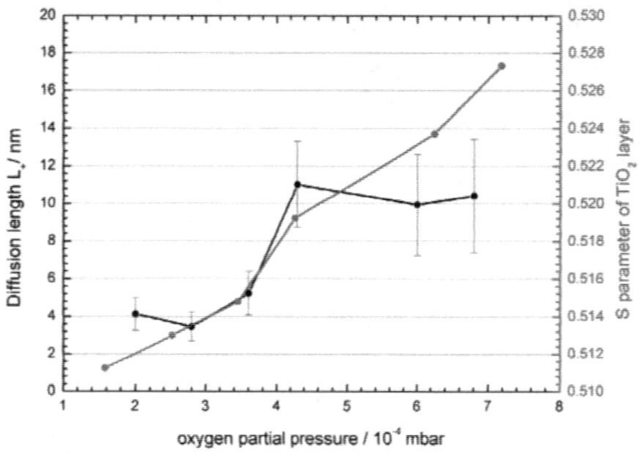


**Figure 5.** S parameter and calculated diffusion length $L_+$ (using VEPFIT [9]) for $TiO_2$ films grown on fused silica (f-SiO2) for different oxygen partial pressures.

## 6. Conclusions and Outlook

First pre-investigations for the investigation of thin $TiO_2$ films deposited on fused silica were done by using positron Doppler broadening spectroscopy. The aim was to improve the conditions during the film growth and to decide for a suitable substrate material within the framework of investigating the ferromagnetic properties of V, Co or Mn doped anatase $TiO_2$ thin films. $SrTiO_3$ was chosen as substrate material for the future film deposition. It also turned out that variations in substrate temperature and oxygen partial pressure have great influence on the defect structure of the films. Optimal parameters were found as following:

a) the substrate temperature should be in the range of 400 °C
b) at an oxygen partial pressure of around $4 \cdot 10^{-4}$ mbar present defects in the layer start to agglomerate

Next step will be the film deposition on $SrTiO_3$ at different substrate temperatures and oxygen partial pressures to investigate the effects also for this substrate material.

This work is supported by the Initiative and Networking Fund of the German Helmholtz Association, Helmholtz-Russia Joint Research Group HRJRG-314, and the Russian Foundation for Basic Research, RFBR #12-02-91321-SIG_a.

## References

[1] Zhou S, Talut G, Potzger K, Shalimov A, Grenzer J, Skorupa W, Helm M, Fassbender J, Cizmar E, Zvyagin S A and Wosnitza J 2008 *J. Appl. Phys.* **103** 083907
[2] Chambers S A, Droubay T, Wang S M, Lea A S, Farrow R F C, Folks L, Deline V and Anders S 2003 *Appl. Phys. Lett.* **82** 1257
[3] Yoon S D, Chen Y, Yang A, Goodrich T L, Zuo X, Arena D A, Ziemer K, Vittoria C and Harris V G 2006 *J. Phys. – Condens. Matter.* **18** L355
[4] Zhou S Q, Cizmar E, Potzger E, Krause M, Talut G, Helm M, Fassbender J, Zvyagin S A,Wosnitza J and Schmidt H 2009 *Phys. Rev.* B **79** 113201
[5] Kaspar T C, Heald S M, Wang C M, Bryan J D, Droubay T, Shutthanandan V, Thevuthasan S, McCready D E, Kellock A J, Gamelin D R and Chambers S A 2005 *Phys. Rev. Lett.***95** 217203
[6] Griffin K A, Pakhomov A B, Wang C M, Heald S M and Krishnan K M 2005 *Phys. Rev. Lett.* **94** 157204
[7] Akdogan N, Nefedov A, Zabel H, Westerholt K, Becker H W, Somsen C, Gok A, Bashir A, Khaibullin R, and Tagirov L 2009 *J. Phys.* D, *Appl. Phys.* **42** 115005
[8] Anwand W, Brauer G, Butterling M, Kissener H R, Wagner A Design and Construction of a Slow Positron Beam for Solid and Surface Investigations *Near Surface Depth Profiling by Mono-energetic Positrons* ed Ganguly B N and Brauer G (Switzerland: TransTech Publishers) to be published
[9] van Veen A, Schut H, de Vries J, Hakvoort R A and Ijpma M R, Analysis of positron profiling data by means of "VEPFIT" *Positron beams for solids and surfaces* ed Schultz P J, Massoumi G R, Simpson P J (NY: Amer. Inst. Phys 1990) pp 171-196

Journal of Physics: Conference Series **443** (2013) 012074 doi:10.1088/1742-6596/443/1/012074

# Optimization of a multi-ring detector for Ps time of flight measurements

**L Di Noto[1,3], M Benetti[2], S Mariazzi[1], G-F Dalla Betta[2] and R S Brusa[1]**
[1]Dipartimento di Fisica, Università di Trento and INFN, Gruppo collegato di Trento, Via Sommarive 14, I-38123 Povo, Trento, Italy
[2]Dipartimento di Ingegneria e Scienza dell'Informazione, Università di Trento and INFN, Gruppo collegato di Trento, Via Sommarive 14, I-38123 Povo, Trento, Italy

E-mail: di.noto@science.unitn.it

**Abstract.** We have designed a multi-ring detector (MRD) based on Bismuth Germanate (BGO) crystals, coupled to Silicon PhotoMultipliers (SiPM) for measuring the Ps time of flight (TOF). The set-up geometry was optimized by Monte Carlo simulations to take into account at different Ps velocities: (i) the background noise due to backscattered positrons, (ii) the cross-talk between adjacent detectors, (iii) the lifetime of Ps decay. Three parameters were defined to evaluate the different configurations and a figure of merit was obtained. This allows the choice of the best set up configuration for measuring Ps emitted with a particular energy range, optimizing the signal to noise ratio and keeping the acquisition time acceptable.

## 1. Introduction

A TOF apparatus to investigate Ps cooling by measuring the energy distribution of Ps emitted from porous materials was developed in Trento [1] and it is being set up at the NEutron induced POsitron Source MUnich (NEPOMUC) [2]. The system was realized to work with continuous $e^+$ beams. The secondary electrons ejected by the $e^+$ entering the sample are used as start signal as recorded by two channeltrons. The stop signal is given by the Ps annihilation $\gamma$ rays observed by detectors placed at a known distance $z$ from the sample surface. In order to improve the efficiency of this apparatus we designed a 4-ring detector where each ring is separated from another ring by tungsten shields. A multi-ring detector MRD allows the acquisition of simultaneously TOF spectra at different distances. Since the MRD requires small size detectors, BGO crystals, with $4 \times 4 \times 10$ mm$^3$ dimensions, coupled with SiPMs were considered [3].

## 2. Monte Carlo Simulations and results

To take into account the increasing of the noise caused by background signal from the backscattered positrons and by false counts of Ps that don't annihilate in front of the ring (*cross-talk effect* between adjacent detectors), many Monte Carlo simulations have been carried for different configurations obtained varying the thickness ($t \div 1$-9 mm) of the shield among the detector's rings together with the

[3] To whom any correspondence should be addressed.

distance ($d$÷20-90 mm) between the detectors and the chamber axis (see figure 1). Each ring covers an angle of 180° in the xy plane and the slit dimension was chosen to be $\Delta$=4 mm to fit with the BGO-SiPMs dimensions. The distance $h$ between the target and the middle of the first ring was fixed to 1 cm, while $D$=10 mm was the radius of the chamber [3]. To extract a figure of merit, three parameters $M1_j$ , $M2_j$, $M3_j$ were defined for each ring $j$, and studied at different velocities $v$ of the emitted Ps:

- $M1_j$ is the fraction of Ps detected by each ring j and $M1 = \sum_{j=1}^{4} M1_j$;
- $M2_j$ takes into account the rate of the noise events due to the background γ rays coming from $e^+$ annihilation on the chamber walls and detected in each ring *j;*
- $M3_j$ takes into account the rate of the false events due to the *cross-talk* recorded in each ring *j*.

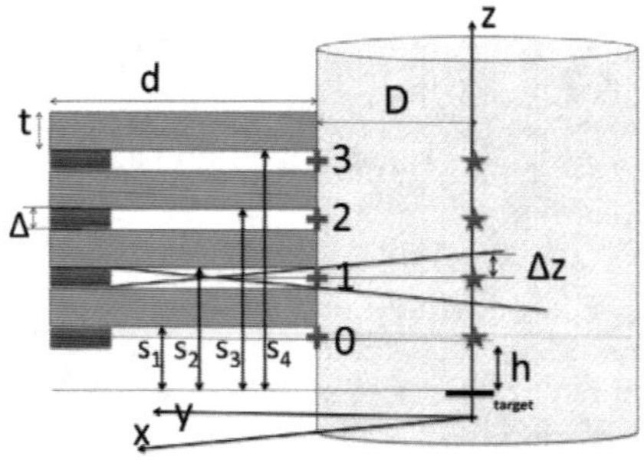


**Figure 1.** Geometry of the set up used in the Monte Carlo simulations

By considering Ps annihilating in flight along the z axis (stars in figure 1), the $M1_j$ parameters can be calculated as the product of the solid angle $\alpha_j(d) = \frac{\pi\Delta}{d+D}$ and the probability that Ps decays in front of each detector $j$. Defining the lower z-coordinate of each ring $s_i = h + \frac{\Delta}{2} + i\,(t + \Delta)$ with i=0,..,3, we obtain $M1\,(t,d,v) = \sum_{i=0}^{3} M1_j\,(t,d,v) = \sum_{i=0}^{3} \alpha_j \left(e^{-\frac{s_i}{v\tau}} - e^{-\frac{s_i+\Delta}{v\tau}}\right)$, with $\tau = 142\,ns$ the oPs lifetime. The parameter $M1$ as a function of $v$ (figure 2) increases at low $v$ until it reaches a maximum, then it decreases. Given the MRD geometry, this behavior is due to the balance of two factors: on one side $M1$ decreases if $v$ increases because the time spent by Ps in front of the slits decreases; on the other side, as $v$ increases, Ps annihilations are also recorded by the more distant rings. In fact only when the TOF required for Ps to reach the higher ring is comparable with $\tau$, the counting rate of the higher scintillator becomes significant. In figure 3 $M1$ as a function of $t$ and $d$ is reported, where $M1$ is bigger for smaller shield dimensions. The obscured zone below the *resolution* line corresponds to geometries that are not usable because the γ annihilations are directly seen by two different rings. The importance of $M1_j$ is the fact that it can be related to the measurement time. Considering that in a TOF measurement with a continuous $e^+$ beam $10^6$ $e^+$/s maximum can be used to avoid false start-stop coincidence events [3] and that at least $10^5$ events must be collected for each spectrum, we can estimate the acquisition time as $T_j = 10^{-1}/(M1_j \cdot \beta)\,[s]$ where $\beta(v)$ is the fraction of the implanted $e^+$ emitted as Ps with the desired energy [1,4].

To estimate the background noise, the distributed $e^+$ annihilations on the chamber walls were simulated with four γ beam sources [5] in front of each detector (numbered crosses in figure 1). For each shield configuration, the deposited energies $E_i$ from each source $i$ ($i$=0-4) in the ring $j$ was evaluated considering the beam sources $i$ pointing towards the ring $j$, in order to calculate:

$M2_j = \frac{\sum_{i \neq j, i=0}^{3} E_i}{\bar{E}}$ where $\bar{E}$ is the deposited energy from the source placed in front of the ring $j$.

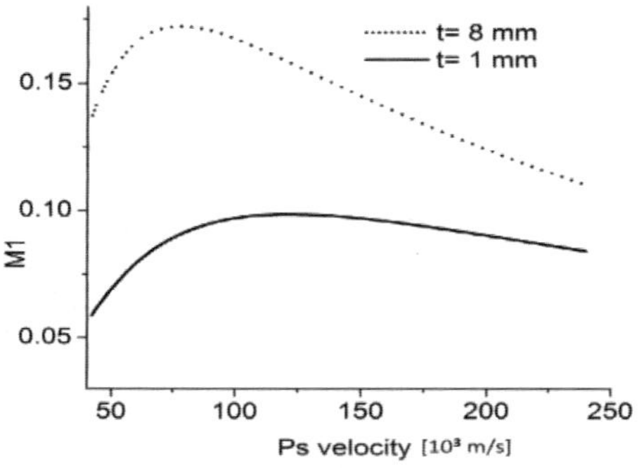


**Figure 2.** *M1* calculated with *d*=20 mm and *t*=8 mm and *t*=1 mm as a function of the Ps velocity

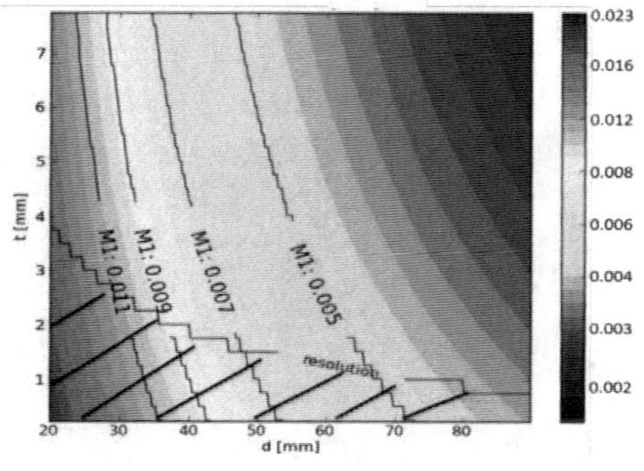


**Figure 3.** *M1* as a function of *t* and *d* calculated for Ps velocity of $v = 6 \cdot 10^4$ m/s

Similarly, the noise due to *cross-talk* was simulated with γ beam sources placed on the chamber axis z (numbered stars in figure 1) and pointing towards the considered ring $j$: $M3_j = \frac{\sum_{i \neq j, i=0}^{3} E_i^c}{\bar{E}}$, where $E_i^c$ are the deposited energies in the ring $j$ for each source $i$. $M2_j$ and $M3_j$ express the two effects that increase the noise of a MRD with respect to a single detector. In order to sum the two effects their relative weights have to be determined. Because the Ps emission yield depends on the sample properties and the Ps energy, while the intensity of the backscattered $e^+$ is connected to the $e^+$ beam, to the sample surface and to the set up characteristics, we can look at an experimental TOF spectrum (figure 3) carried out at Trento by using a single ring detector placed at *h*= 1 cm (figure 3). From this plot, for each Ps velocity we can extract a signal to background ratio $\delta\ (v)$ in order to obtain a new parameter $M_j = M3_j + M2_j / \delta\ (v)$ for each ring j. Thus $M_j$ takes into account both the background and the *cross-talk* noise.

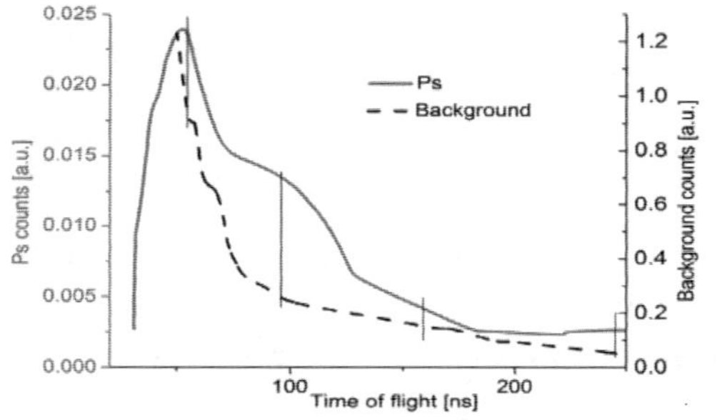


**Figure 4.** TOF measurement of escaping Ps in continuous line, background in dot. The vertical bars show the velocities considered as examples for calculations.

Because $M_j$ and $T_j \propto 1/M1_j$ depend on *v*, also the choice of the best geometry depends on the *v* value that is going to be measured. As examples, we have considered four most significant values of *v* (corresponding to values indicated by vertical lines in figure 4): 1) $v = 2 \cdot 10^5$ m/s where the background dominates over the Ps signal, 2) $v = 1 \cdot 10^5$ m/s where $M1$ reaches its maximum 3) $v = 6 \cdot 10^4$ m/s where the background and signal are comparable, 4) $v = 4 \cdot 10^4$ m/s where the background is lower than the signal. From geometric considerations it can be observed that the ring called 1 or 2 in figure 1 are more exposed to the background and cross-talk noise. For this reason, in order to size the MRD, we have considered only the second ring, where the worst conditions in term of signal to noise ratio occur. Thus the MRD geometry must be chosen looking at the signal to noise ratio

Journal of Physics: Conference Series **443** (2013) 012074 doi:10.1088/1742-6596/443/1/012074

$M1_2/M_2$, the measuring time $T_2 = \frac{10^{-1}}{M1_2 \cdot \beta}[s]$, and the spatial resolution $\Delta z = \frac{\Delta}{d}\left(\frac{d}{2} + D\right)$ (see figure 1) which is related to the resolution on the kinetic energy value $E$ extracted by a TOF measurement by the relation: $\frac{\Delta E}{E} = \frac{2\Delta z}{z}$.

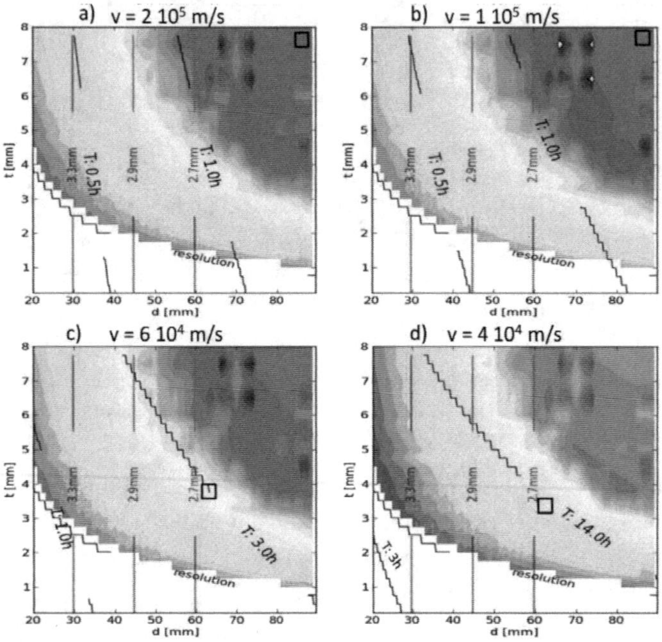


**Figure 5.** Colours express the value of the ratio $M1_2/M_2$, in log scale, as a function of t and d, for four different Ps velocities (the same indicated with bars in figure 3). The limitations imposed by the resolution (in the bottom) are shown. The resolution $\Delta z$ on the Ps annihilation position is marked by red vertical lines and the measurement time is indicated by black lines. The squares indicate the best configuration set up as explained in the text.

The behavior of $M1_2/M_2$ with the time $T_2$ and the $\Delta z$ values (expressed in the continuous black and red lines respectively) is shown in figure 5 for the selected four Ps velocities. To calculate $T_2$ the parameter $\beta$ was estimated by multiplying the yield (25%) of emitted Ps as reported in [4] to the fraction (0.08, 0.06, 0.02, 0.016) calculated considering the Ps signal in figure 4. From figure 5 we can observe that the signal to noise ratio $M1_2/M_2$ increases going towards the zone in which the shield dimensions increase, because the effect of the decreasing of noise (i.e of $M_2$) is more relevant with respect to the decrease of $M1_2$. Thus, by imposing a limitation on the energy resolution and on the measurement time (if necessary), and optimizing the signal to noise ratio, for each velocity the best configuration can be chosen. For instance, for larger velocity, for which Ps is emitted more abundant and the measurement time is always less than 3 hours, the best set up is with higher values of t and d (see figure 5.a,b). On the contrary, when the measurement time increases, the best configuration can be chosen by imposing for instance T<3 h and $\Delta z$ <2.7 mm as indicated in figure 5c. Finally, when Ps with low velocity is of interest, the measurement time increases and configurations with smaller shield are necessary (figure 5 d).

## 3. References

[1] Di Noto L, Mariazzi S, Bettonte M, Nebbia G and Brusa R S 2012 *Eur. Phys. J. D* **66** 118

[2] Hugenschmidt C 2010 *Physics with many positrons* ed R S Brusa, A Dupasquier and A P Mills Jr (Amsterdam: IOS; Bologna: SIF) p 399

[3] Mazzuca E, Benetti M, Mariazzi S, Brusa R S, Dalla Betta G-F and Piemonte C 2012 *J. Instr* **7** P05006

[4] Mariazzi S, Bettotti P and Brusa R S 2010 *Phys. Rev. Lett.* **104** 243401

[5] Gentit F-X 2002 *Nucl. Instr. Meth. Phys. Res. A* **486** 35

Journal of Physics: Conference Series **443** (2013) 012075 doi:10.1088/1742-6596/443/1/012075

# Efficient emission of positronium atoms from an Na-coated polycrystalline tungsten surface

**H Terabe[1], S Iida[1], K Wada[2], T Hyodo[2], A Yagishita[2] and Y Nagashima[1]**

[1] Department of Physics, Tokyo University of Science, 1-3 Kagurazaka, Shinjuku, Tokyo, 162-8601, Japan

[2] Institute of Materials Structure Science, High Energy Accelerator Research Organization (KEK), 1-1 Oho, Tsukuba, Ibaraki, 305-0801, Japan

E-mail: j1211706@ed.kagu.tus.ac.jp and ynaga@rs.kagu.tus.ac.jp

**Abstract.** Time-of-flight spectra for the ortho-positronium emitted from clean and Na-coated tungsten surfaces have been measured using the pulsed slow positron beam at KEK-IMSS slow positron facility. Emission efficiency of positronium from the Na-coated sample was found to be several times greater than that from uncoated tungsten surfaces.

## 1. Introduction

Recently, we reported that the emission efficiency of positronium negative ions ($Ps^-$) from tungsten surfaces bombarded with slow positrons was enhanced by alkali-metal coating [1, 2]. This enhancement is thought to be due to the surface dipole barrier reduction by alkali-metal deposition, which results in the increase in the fraction of conduction electrons available for the $Ps^-$ emission.

The $\gamma$-ray energy spectra for the investigation of $Ps^-$ emission indicated that the positronium (Ps) yields also increase by the coating. The Ps emission energy, $T_{\mathrm{Ps}}$, from the surface is expressed as

$$T_{\mathrm{Ps}} = -\phi_+ - \phi_- + E_{\mathrm{B}}, \quad (1)$$

where $\phi_+$ and $\phi_-$ are the positron and electron work functions, respectively, and $E_{\mathrm{B}}$ is the Ps binding energy (6.8 eV). The values of positron and electron work functions for the surface with alkali-metal coating can be written as

$$\phi'_+ = \phi_+ + \Delta \text{ and } \phi'_- = \phi_- - \Delta, \quad (2)$$

where $\Delta$ is the effect of surface dipole barrier reduction by the coating. Therefore, the emission energy for the Ps emission is unchanged by the coating.

In the present work, we have measured the energy distribution of ortho-positronium ($o$-Ps) emitted from Na-coated tungsten surface using a time-of-flight (TOF) technique in order to investigate the mechanism of Ps emission.

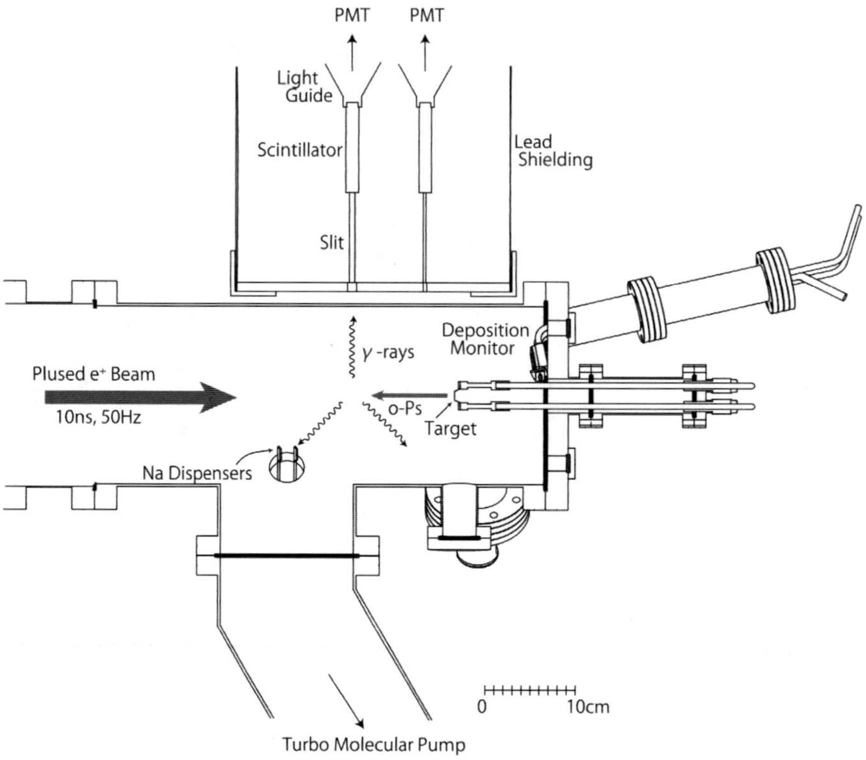


**Figure 1.** Schematic diagram of the Ps-TOF chamber.

## 2. Experimental Procedure

The measurements were performed using a pulsed slow positron beam at the KEK-IMSS slow positron facility [3]. Positrons were generated by a 55 MeV electron linac operated in a short pulse mode and moderated using a tungsten moderator. The repetition rate of the linac was 50 Hz. The pulsed slow positron beam was magnetically transported after being accelerated to an energy of 4.2 keV in order to minimize pulse widening. The pulse width near the target was 10 ns.

Figure 1 shows the schematic diagram of the measurement system. The base pressure of the target chamber was $6\times10^{-8}$ Pa.

The target was a polycrystalline tungsten foil of 25 $\mu$m thick. It was annealed *in situ* at 1800 K for 30 min by the passage of an electric current. After being cooled down to room temperature, Na atoms were deposited. The thickness of the Na coverage was 2.6 Å, where the electron work function has been reported to be the lowest [4]. The target was grounded and the positron incident energy was 4.2 keV.

Two plastic scintillation detectors can be installed to detect the $\gamma$-rays. Only the detector located 120 mm from the target was used in this measurement. The scintillator was 100 mm $\times$ 100 mm square and 10 mm in thickness and coupled on opposite side to a photomultiplier tube (HAMAMATSU H6614) by a light guide. The $\gamma$-rays were detected through a lead slit of 6 mm width. The time resolution of this measurement system including the beam pulse width was 10 ns, which was better than that used for the TOF measurement of $o$-Ps emitted from $SiO_2$ [5]. The time intervals between the linac pulses and the signals from $o$-Ps self-annihilation were measured using a digital oscilloscope with a sampling speed of 500 MS/s. The waveform data were transferred to a PC and then TOF spectra were obtained off-line.

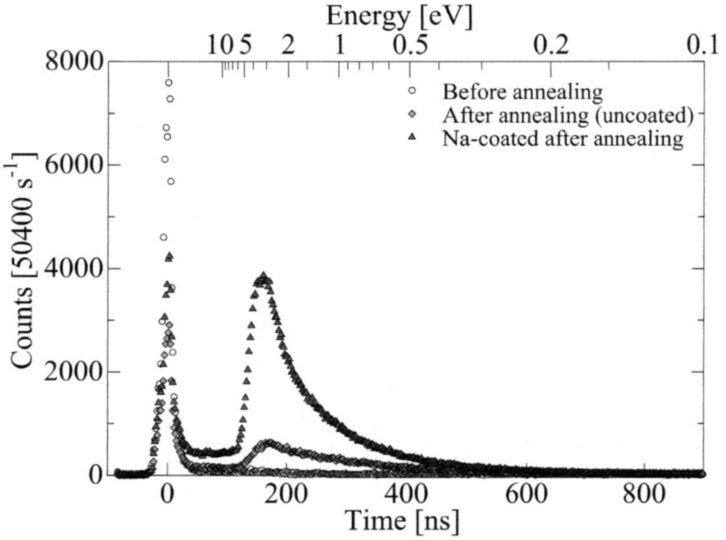


**Figure 2.** TOF Spectra for unannealed, annealed and Na-coated W surfaces.

## 3. Results and Discussion

Figure 2 shows the obtained TOF spectra for unannealed, annealed and Na-coated samples. The spectra were normalized to the measurement time. Large peaks near time zero are attributed to the prompt annihilation of the positrons in the target and self-annihilation of para-positronium. The peak becomes higher by the coating. This indicates that the fraction of re-emitted slow positrons becomes lower by changing $\phi_+$. The spectra show that the yield of *o*-Ps component increases by Na coating. For the annealed and Na-coated samples, the components appear after about 120 ns. It corresponds to the energy of 5 eV, which agrees with the Ps emission energy calculated using equation (1) and reported values of $\phi_+$ and $\phi_-$ for polycrystalline tungsten [6, 7]. This indicates that the maximum energy of Ps emission is independent of Na coating as expected.

Mills [8] and Gidley *et al.* [9] suggested that the positrons bound in their image potential well at the surface may be emitted as Ps atoms by lowering $\phi_-$ when the surface is coated with alkali-metal. According to this model, the yield of lower energy component of *o*-Ps in the TOF spectrum should increase. However, figure 2 shows that the yield of higher energy component increases further and the shape of the *o*-Ps energy distribution becomes narrower than that for the uncoated surface.

## Acknowledgments

We thank K. Michishio, R. H. Suzuki, T. Shidara, S. Osawa, M. Ikeda and other KEK staff for their generous support for the positron beam line. This work has been performed under the approval of the Photon Factory Program Advisory Committee (Proposal No. 2010S2-003). It was partly supported by a Grant-in-Aid for Scientific Research (No. 24221006) from the Ministry of Education, Science and Culture of Japan.

## References

[1] Nagashima Y, Hakodate T, Miyamoto A and Michishio K 2008 *New J. Phys.* **10** 123029
[2] Terabe H, Michishio K, Tachibana T and Nagashima Y 2012 *New J. Phys.* **14** 015003
[3] Wada K et al. 2012 *Eur. Phys. J.* D **66** 37
[4] Kiejna A and Wojciechowski K F 1981 *Prog. Surf. Sci.* **11** 293

[5] Nagashima Y, Morinaka Y, Kurihara T, Nagai Y, Hyodo T, Shidara T and Nakahara K 1998 *Phys. Rev.* B **58** 12676
[6] Weiss A H and Coleman P G 2000 *Positron Beams and Their Applications* ed P G Coleman (Singapore: World Scientific) p 129
[7] David R L (ed) 2008 *CRC Handbook of Chemistry and Physics* 89th edn (Boca Raton, FL: CRC Press)
[8] Mills A P Jr. 1983 *Positron Solid-State Physics: Proc. Int. School Phys. 'Enrico Fermi', Course 83* ed W Brandt and A Dupasquier (Amsterdam: North-Holland) p 432
[9] Gidley D W, Köymen A R and Capehart T W 1988 *Phys. Rev.* B **37** 2465

# Study of Neutron Induced Defects in Ceramics using the GiPS Facility

**S May-Tal Beck,[1] M Butterling[2], W Anwand[2], A Beck[1], A Wagner[2], G Brauer[2], A Ocherashvili[1], I Israelashvili[1], O Hen[3]**

[1] *Physics department, NRCN,P.O.Box 9001, Beer-Sheva, Israel*
[2] *Institut für Strahlenphysik, Helmholtz-Zentrum Dresden-Rossendorf, 01314 Dresden, Germany*
[3] *School of Physics and Astronomy, Tel-Aviv University, Tel-Aviv 69978, Israel*

smtbeck@gmail.com

**Abstract**. Preliminary results are presented from a study of neutron irradiation damage in Sapphire and $B_4C$, produced with a fluence of $6\times10^{18}$ n/cm$^2$ and ~$10^{15}$ n/cm$^2$, respectively. Measurements were performed at the GiPS facility and the SPONSOR beam at HZDR, and in the PAL spectrometer at NRCN. Bulk and vacancies lifetimes were identified in the Sapphire, ~150ps and ~188ps, respectively, with complete trapping in the irradiation induced vacancies. Irradiation damage in $B_4C$ found to be limited to the surface. A single lifetime of ~166ps was measured in both irradiated and non-irradiated samples, and was associated with the bulk.

## 1. Introduction

There has been an increased interest in defects within structural materials motivated by future fission and fusion reactor needs. While reactor steels are extensively studied, much more research effort is needed in order to understand radiation damage in ceramic materials and its effect on their macroscopic characteristics [1].

Sapphire and $B_4C$ are interesting to study. Sapphire is a candidate material to be used in diagnostic systems for burning plasma experiments [2], due to its transparency to a wide range of wavelengths (200-5000 nm), high melting temperature (~2300 K) and hardness close to that of a diamond. Its optical and electronic properties are expected to be affected by the harsh radiation environment. The $B_4C$ is interesting for the nuclear industry due to the high cross-section for thermal neutron capture in Boron, which produces helium inside the material. The much higher neutron flux expected in future reactors can cause swelling of the material and macroscopic crack formation.

PAS methods are suited [3-5] and applied to study neutron-induced defects in Sapphire and $B_4C$ samples. In the present paper first but preliminary results are presented.

## 2. Samples and Measuring Techniques

Two pairs of Sapphire samples were available as disks of 14 mm in diameter and a total thickness of 6 mm. One pair was neutron irradiated in the Israel Research Reactor 1 (IRR1) [6] some few years ago, to a fluence of $6\times10^{18}$ n/cm$^2$. The other was not irradiated and was annealed at 1773 K for 30 minutes before measurements. Two pairs of cubic samples of $B_4C$ were available, each having a size of $10\times10\times8$ mm$^3$. One pair had been previously irradiated to a neutron fluence of ~$10^{15}$ n/cm$^2$ in the IRR1 [6]. Despite of the high penetrability of neutrons, predictive calculations for $B_4C$ [7] show that damage can be produced only on the surface due to the high cross section for the $^{10}B(n,\alpha)^{7}Li$ process, of ~3800 barn for thermal neutrons. Based on the calculations of Ref.7, and taking into account the neutron fluence of ~$10^{15}$ n/cm$^2$ on the $B_4C$ samples, a damage of ~$10^{-4}$ -$10^{-3}$ dpa is expected to exist on their surface. A single crystalline Si sample was used as a reference.

Journal of Physics: Conference Series **443** (2013) 012076 doi:10.1088/1742-6596/443/1/012076

All Sapphire and $B_4C$ samples were measured at the GiPS facility [3]. $B_4C$ samples were measured also at the SPONSOR beam [4] since neutrons were expected to create damage only on their surface [7]. Lifetime measurements on the same Sapphire samples were performed also in the PAL spectrometer at NRCN [5]. The GiPS facility is suitable for thick samples (>= 1cm³) of solids and liquids. At the SPONSOR beam mono-energetic positrons from 0.027 keV to 37 keV are available. The PAL spectrometer at NRCN had an average time resolution ~190ps during measurements.

### 2.1. GiPS

The data from GiPS measurements are collected in list mode, namely, for each event the data from each of the detectors is recorded, making possible an offline correlative data analysis. Selection of good annihilation events was done by requiring that only one pair of ($BaF_2$, HPGe) detectors fired, and by applying energy cuts on the spectra from these detectors. The first criterion selected ~0.25 of the collected events, and energy cuts further reduced the number of good events by a factor of ~6, leaving $\sim 1.5\times 10^5$ events from each measurement for final analysis. The energy scales were not calibrated, but due to the fact that the 511 keV photons were the only ones detected, the selection criteria were defined in channel units. These criteria included events within one and five FWHM from the center of a fitted Gaussian to the 511 keV photo-peaks in the $BaF_2$ and HPGe energy spectra, respectively. As an example, figure1 shows one HPGe energy spectrum of events that passed all selection criteria from the non-irradiated $B_4C$ measurement. Lifetime data, gained from the $BaF_2$ time stamps and selected by the same set of criteria, are shown in figure 2 for the four $BaF_2$ detectors.

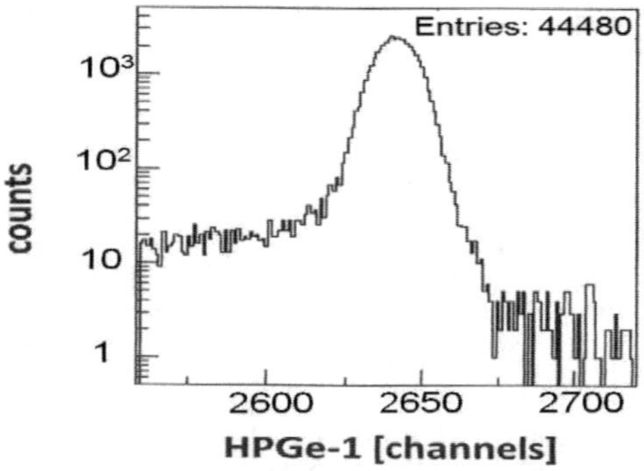


**Figure 1.** Energy spectrum from one HPGe detector, for the selected events of the non-irradiated $B_4C$ measurement.

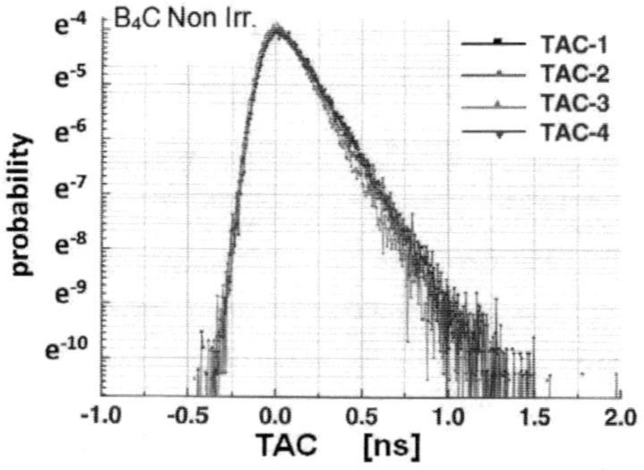


**Figure 2.** Positron lifetime spectra from the four $BaF_2$ detectors, for the selected events of the non-irradiated $B_4C$ measurement.

Mean positron lifetimes were fitted using the RESOLUTION program [8]. The time resolution for each $BaF_2$ detector was defined from the Si measurement, ~180ps. For all spectra it was possible to fit only one decaying exponential. Figure 3 presents the averaged mean values of positron lifetimes extracted from the four different detectors per each measurement. The error bars include the statistical error, as well as the spread of results from the different detectors. DB analysis is still pre-mature and is not presented here.

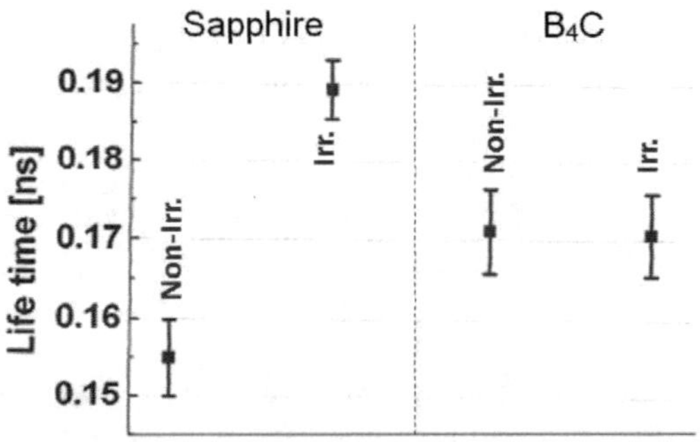


**Figure 3.** Averaged mean positron lifetimes for the four measured samples (see text for further details).

### 2.2. SPONSOR

DB spectra for $B_4C$ samples, irradiated and non-irradiated, were collected with one high-purity Germanium (HPGe) detector in varied steps of the positron energy. The energy resolution of the HPGe detector was (1.09±0.01) keV at 511 keV. The number of counts accumulated in the photo-peak was $9.5\times10^5$ per positron penetration energy. Energy regions defined for S and W calculations were $|E\gamma\text{-}511\text{keV}|<0.92\text{keV}$ and $2.3\text{keV}<|E\gamma\text{-}511\text{keV}|<3.0\text{keV}$, respectively. Figure 4a shows the S values as a function of positron penetration energies, for the two measurements. Slight differences between the irradiated and non-irradiated samples appear on the surface. Figure 4b shows the S-W plot, where this difference is more pronounced. Positron diffusion lengths, calculated using the VEPFIT program [9], were found to be identical: $L_+ = (96 \pm 4)$ nm and $L_+ = (97 \pm 6)$ nm for the irradiated and non-irradiated samples, respectively. This is expected since the irradiation damage is concentrated on the surface of the sample and therefore has little effect on the average diffusion length.

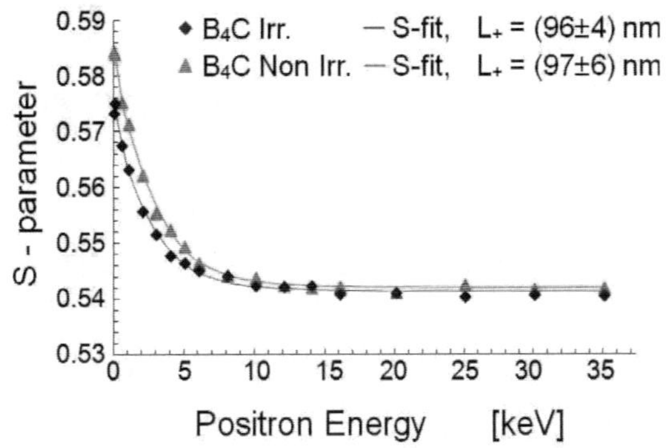


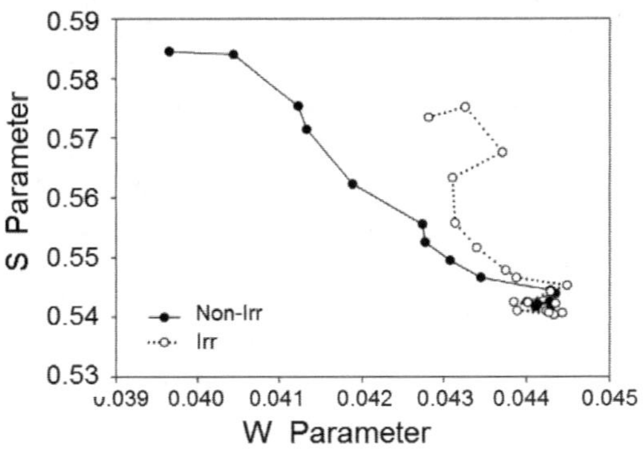


**Figure 4: (a)** S values extracted from two $B_4C$ samples, non-irradiated and irradiated to a fluence of $\sim10^{15}$ n/cm$^2$, as a function of positron beam energy. The marker size includes the uncertainties, dS=0.0006. **(b)** S-W plot for the values extracted for these two samples.

2.3. PAL spectrometer

The source and samples assembly used included two 1μm Ni foils, as beddings for the $^{22}$Na source, and an air gap between the foils and the samples for safety reasons. The Ni foils and the samples were held by Perspex housings.

Positron lifetimes in Perspex and their relative intensities were found from a Perspex measurement performed with a similar assembly (7.6 μm Kapton foils instead of Ni), and based on the simulated relative annihilation intensities in the assembly materials (done in a similar way to that described in detail in [10]). The lifetimes: 125ps (p-Ps), ~360ps and ~1770ps (o-Ps) contribute relative intensities of ~0.3 and ~1.96 of that of the long component. These values are similar to other published values for nylon materials [11]. A measurement of well annealed Fe samples was used to extract the $\tau_{e+}$(Ni)~235ps. The measurement of the Sapphire non-irradiated and annealed samples, were used to extract the relative intensity in Ni, ~8.5%, as well as the relative intensities in the Perspex components. Time resolution was defined from $^{60}$Co measurements [5], taken in between Sapphire measurements, ~190ps. Finally, for each Sapphire measurement, all source correction components were fixed, except for the intensity of the o-Ps component that might include annihilations in the air gap. The POSITRONFIT package [8] was used to analyse the lifetime spectra. The resulted mean lifetimes are 144±4ps, 159±1ps and 188±2ps for the non-irradiated and well-annealed, non-irradiated and irradiated to $6\times10^{18}$ n/cm$^2$ samples, respectively. The errors are statistical only.

## 3. Results and Discussion

The measurements presented here, of ceramic samples that were exposed to low dose of neutron irradiation, can lead to few conclusions:

The lack of background events in lifetime measurements in the GiPS facility is evident by the single decay obtained for each of the samples. The changes between the results obtained from the different detectors, and taken into account in the error estimation (figure 3), reflect the systematic uncertainty of the measurement.

For the Sapphire, two distinguished lifetime were measured for the non-irradiated and annealed, 155±5ps, and for the irradiated samples, 189±4ps, in similarity to previously reported values, and thus can be attributed to the bulk region and to trapping in vacancies [12]. The criteria applied to the data selected a clean sample of events, with relatively low statistics. More events can be selected in further analysis of the same data, containing more than one pair of detectors, but adequate background reduction criteria should be studied and applied. Positron lifetimes resulted from measurements of the same samples at the NRCN spectrometer (listed above), agree within 1$\sigma$ with the results of GiPS. While in [13] two lifetimes are reported for Sapphire irradiated to $\sim 10^{20}$-$10^{21}$ n/cm$^2$, it is evident from the present measurement that for the fluence of $6\times10^{18}$ n/cm$^2$, only one type of defect is visible, which fully traps the positrons.

Lifetime results from the $B_4C$ measurements do not show any difference between the irradiated and non-irradiated samples, and agree with published results for the non-irradiated material [14]. However, since the GiPS facility probes the entire volume of the sample, if damage has been created in few micron depth of the material, a maximum intensity of 2% is expected. This amount cannot be observed with the current low statistics. The DB measurement in the SPONSOR beam is much more sensitive to the surface of materials, and shows slight change on the surface between the two $B_4C$ samples, which is not apparent by the VEPFIT results. Since the $10^{-3}$ dpa in a depth of few microns should be visible in slow positron beam measurements, further investigation is needed to understand the measurement results in relation with the damage estimation. The two distinguished regions seen on the S-W plot, resulted mainly from different W values on the surface, might indicate on the existence of He bubbles there. Further measurements with correlated DB data should enhance the sensitivity to the W parameter and produce a better picture of the neutron irradiation induced defects.

**References**

[1] Ledermann P 2004 Session I: Overview of Relevant Irradiation Environments *Workshop on Advanced Computational Materials Science: Application to Fusion and Generation IV Fission Reactors, (Washington, D.C. 31 March – 2 April 2004) (Oak Ridge National Laboratory report ORNL/TM-2004/132)* ed. R E Stoller et al, pp 8-15.

[2] Duffy D M 2010 *Phil. Trans. R. Soc.* A **368** 3315-28

[3] Butterling M, Anwand W, Cowan T E, Hartmann A, Jungmann M, Krause-Rehberg R, Krille A and Wagner A 2011 *Nucl. Instr. Meth.* B **269** 2623-9

[4] Anwand W, Brauer G, Butterling M, Kissener H-R and Wagner A 2012 *Defect and Diffusion Forum* **331** 25-40

[5] May-Tal Beck S, Brauer G, Anwand W, Berant Z, Shahal O, Ganor M and Israelaswily I 2004 *Materials Science Forum* **445-446** 495-7

[6] Information at http://www.soreq.gov.il/default_EN.asp.

[7] Simeone D et al 1997 *J. Nucl. Mater.* **246** 206

[8] Kirkegaard P, Pedersen N J and Eldrup M 1989 *Technical Report Riso-M-2740 (Riso National Laboratory Report, Roskilde, Denmark*)

[9] Van Veen A, Schut H, De Vries J, Hakvoort R A and Ijpma M R 1991 *AIP Conf. Proc.* **218** 171-198

[10] May Tal Beck S, Anwand W, Wagner A, Brauer G, Beck A, Ocherashvili A, Hen O, Haroush S, Eisen Y and Moreno D 2012 *Defect and Diffusion Forum* **331** 95-112

[11] Surbeck H 1977 *Helv. Phys. Acta* **50** 705-21

[12] Liszkay L et al 2004 *Materials Science Forum* **445-446** 138

[13] Nagashima Y, Kawashima K, Hyodo T, Hasegawa M, Lee B T, Hiraga K, Yamaguchi S, Forster M and Schaefer H E 1995 *Materials Science Forum* **175-178** 461-4

[14] Chang-Hong Liu 2000 *Materials Transactions, JIM,* **Vol. 41, No.10** 1293

# Status of the Linac based positron source at Saclay

**J-M. Rey[a], G. Coulloux[a], P. Debu[a], H. Dzitko[b], P. Hardy[a], L. Liszkay[a], P. Lotrus[a], T. Muranaka[c], C. Noel[b], P. Pérez[a], O. Pierret[b] N. Ruiz[a], Y. Sacquin[a]**

[a] CEA/IRFU, CEA Saclay, 91191 Gif sur Yvette, France.

[b] CEA/DAM, CEA Bruyère le Chatel, 91297 Arpajon, France.

[c] CERN Beams Department, CH 1211 Geneva 23, Switzerland

E-mail: j-m.rey@cea.fr

**Abstract.** Low energy positron beams are of major interest for fundamental science and materials science. IRFU has developed and built a slow positron source based on a compact, low energy (4.3 MeV) electron linac. The linac-based source will provide positrons for a magnetic storage trap and represents the first step of the GBAR experiment (Gravitational Behavior of Antimatter in Rest) recently approved by CERN for an installation in the Antiproton Decelerator hall. The installation built in Saclay will be described with its main characteristics. The ultimate target of the GBAR experiment will be briefly presented as well as the foreseen development of an industrial positron source dedicated for materials science laboratories.

## 1. Introduction

One of the fundamental questions of today's physics concerns the action of gravity upon antimatter. A difference between the free fall of matter and antimatter in the gravitational field of Earth would signal a violation of the Weak Equivalence Principle. There are indirect limits on its validity with antimatter. However these are model dependent or controversial [1]. Projects to detect the free fall of charged antiparticles, in the gravitational field of the Earth, all failed because of the difficulty to shield the apparatus against electromagnetic fields [2]. No experimental direct measurement has ever been successfully performed with antimatter particles. The GBAR experiment, recently accepted by CERN, proposes a direct measurement with antihydrogen atoms, $\overline{H}$, by first producing the antihydrogen ion $\overline{H}^+$ [3], as suggested by J. Walz and T. Hänsch [4]. This ion can be cooled down to a few μK, by sympathetic cooling with ordinary laser-cooled ions. Velocities less than 1 m/s could be reached for a quasi-vertical free fall. An R&D program has been launched at IRFU (CEA/Saclay) to demonstrate the feasibility of the production of antihydrogen with the use of a cloud of positronium (Ps) atoms (the bound state of an electron and a positron). This target, when bombarded with antiprotons, should allow combining its positrons with the incoming antiprotons and create $\overline{H}$ atoms and $\overline{H}^+$ ions.

## 2. The electron linac

In order to produce the high positron rate needed by the GBAR experiment the use a 10 MeV / 2 mA has been foreseen. As a demonstrator for the positron production setup patented by CEA a small electron linac derived from a commercial industrial machine (Fig 1) has been purchased. Using the proper HF power source its nominal kinetic energy and average current were given to be 5.5 MeV and between 0.1 and 0.2 mA. The repetition rate is 200 Hz with bunch length of 4 µs. The acceleration length is 21 cm after which the beam diameter is 1 mm. The initial commissioning of the linac performed with a filament emitting electrons showed a beam current of 0.007 mA for energy of 5.5 MeV. An impregnated cathode as electron emitter has been installed to increase the beam current. Using it the electron beam reached the expected value of 0.14 mA. Nevertheless energy measurement was not possible in that new configuration as the beam had to exit the vacuum through a thin titanium window to dump on an aluminium wedge used to measure the energy and the beam would have burned this titanium window.

**FIGURE 1.** Side view of the electron linac.

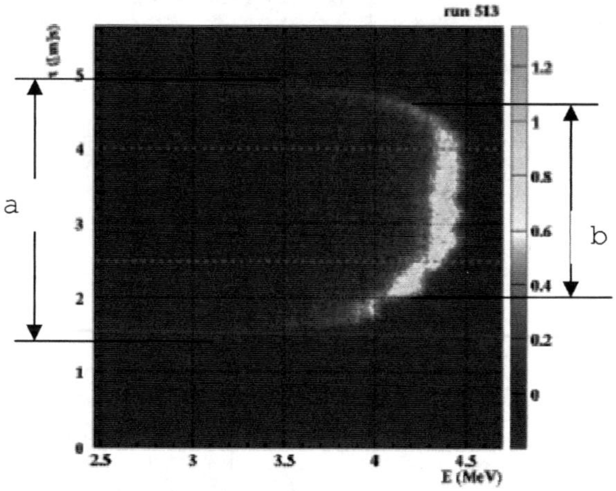


**FIGURE 2.** Time evolution of the energy spectrum of one bunch of the electron beam. a= 4 µs, bunch length as directed by HF power source. b=2.5 µs, effective bunch length of the beam structure.

As the positron production was lower than what had been simulated using a 5.5 MeV electron beam the full measurement of the electron beam characteristics had to be undertaken. This has been done using a compact beam analyzer. This device has been installed in the linac bunker and replaced temporarily the positron source.

This beam analyzer consists of a 180° bending magnet and a fast readout system based on optical fibers. It provides a bunch time analysis of the electron energy as can be seen on figure 2. One can state that for a nominal bunch length of 4 µs the effective high power emission has duration of 2.5 µs during which the beam energy reaches 4.3 MeV. Reducing the cathode emitting current leads to an increase of the peak energy. Fig. 3 shows the beam energy dependence as a function of the emitted cathode current for two repetition rates. This energy dependence as a function of the beam current reflects a beam loading effect.

In parallel calculation of the voltage gradient within the accelerating cells were done showing already critical values (locally near 20 MV/m). A further increase of the HF power source would have led to electrical breakdown within the cells. Therefore we decided to work with a lower than the expected beam current to optimize the energy and the positron production rate.

## 3. The positron source

The positrons are produced in a tungsten target by pair production. The setup shown in Fig.4. consists of a magnetic selector that extracts the positrons emitted after the water cooled target from the remaining electron flow. These positrons are then directed on the pad detector shown on Fig.5. It consists of aluminum pads connected to current measurement devices and protected by a thin metallic grid that can be kept at various voltages. Using this detector allowed to adjust the magnetic configuration to optimize the positron capture.

The average positron energy emitted from the W target has a mean value of 800 keV and ranges from 0 to 3 MeV, which is advantageous concerning moderation efficiency, compared to positrons generated by a higher energy electron beam. Based on the 0.2 mA beam of 5.5 MeV energy simulations, performed with the GEANT3 software, predict rates of $5.10^{11}$ $e^{+}.s^{-1}$ when the beam hits a 200 μm thick target at 5 degree incidence angle. This small angle provides greater heat transfer and positron emission in vacuum while keeping an effective target thickness for the incoming electrons. At such low incident energies the positrons are emitted almost isotropically from the target (with 2/5 of these positrons emitted backwards from the incident beam direction).

## 4. The moderator

Although the design is adapted to a neon cryogenic moderator the initial low energy positron measurements have been performed using a more conventional tungsten mesh moderator. The tungsten meshes have been placed after the positron production target and the magnetic field configuration of the positron source adjusted for low energy positron transport. The measurement of the low energy positron production rate can only be performed outside of the biological shield because of the high level of RF noise near the electron linac and the very low signal of moderated positrons. A 13 m long transport line consisting of a vacuum tube in which a continuous magnetic field (0.5 to 1 mT) is established by coils wound directly on the tube has been installed to guide the positrons outside of the biological protection.

Trials have been made to optimize the number of tungsten grids for moderated positron production.

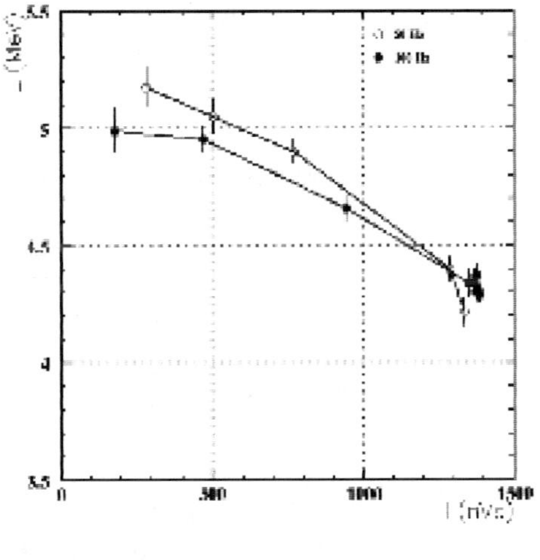

**FIGURE 3.** Beam energy (in MeV) as a function of the beam peak current (1 nVs = 200 μA) for 50 and 100 Hz repetition rate.

**FIGURE 4.** The positron source with the electron/positron separator

The optimum value is around 12 grids as shown on Fig.6. An accelerating voltage is applied across the tungsten mesh grids to extract the positrons towards the extraction line.

The positron detection is done indirectly once they are converted to 511 keV annihilation gamma rays. An aluminum target is inserted in the vacuum tube and used as a target to produce the gamma rays. The gammas are then detected using a BGO scintillator and photomultiplier.

**FIGURE 5.** The positron detector.

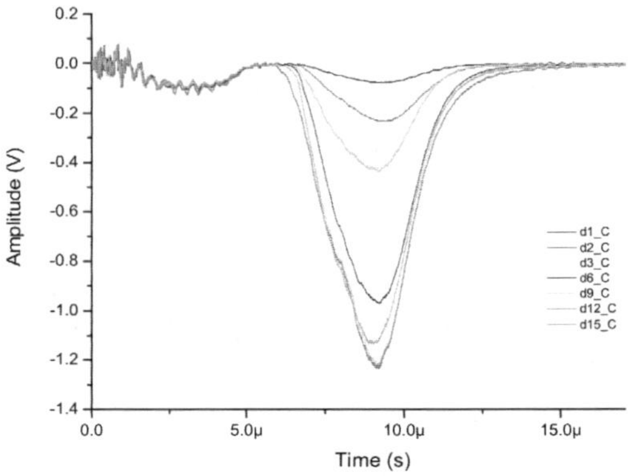


**FIGURE 6.** Positron signal for different number of W meshes in the moderator. Respectively 1, 2, 3, 6, 9, 12, and 15 meshes.

## 5. Future: the gbar project and positron sources for materials science.

The GBAR experiment[5] or "Gravitational behavior of anti-matter at rest" is the basic motivation for the development of an intense compact positron source. This experiment has been accepted by CERN in May 2012. It will produce anti hydrogen atoms and ions during interaction of antiprotons on a positronium target. A very high flux of positrons is needed to produce enough positronium for those interactions. A target value of 2.8 $10^8$ slow positrons per second is needed, to be compared to the actual 3.2 $10^6$ $s^{-1}$ achieved. Nevertheless the efforts invested in the development of this positron source will be made available for research in materials science. Compact non-radioactive slow positron sources are now available with slow positron production rates of 1.5 $10^7$ $s^{-1}$ for the low flux system and 4 $10^7$ $s^{-1}$for high flux system. A company will soon be created to commercialize these products.

## 6. Conclusion

CEA-IRFU has developed an innovative compact linac based positron source for the need of a fundamental experiment dedicated to the gravity behavior of anti-matter. This device works now efficiently after several years of fine tuning. As consequences the GBAR experiment has been approved by CERN and optimized positron sources will be made commercially available for materials science analysis.

**Acknowledgments**

This work is supported in part by Conseil Général de l'Essonne under ASTRE 2006 funding and by Research National Agency under contract number ANR-05-BLAN-0380-01.

**References**

[1] Nieto M and Goldman T 1992 *Phys. Rep.* **205** 5221-281
[2] Darling T et al. 1992 *Rev. Mod. Phys.* **64** 237
[3] Pérez P and Rosowsky A 2005 *Nucl. Inst. Meth. A* **545** 20-30
[4] Walz J and Hänsch T 2004 *General Relativity and Gravitation* **36** 561
[5] GBAR Proposal Chardin G et al CERN-SPSC-P-342 September 2011; http://cdsweb.cern.ch/record/1386684/files/SPSCP-342.pdf
[6] Pérez P and Rosowsky A 2004 *Nucl. Instr. Meth. A* **532** 523-532
[7] de Mascureau J et al 1993 Proceedings of the 1993 IEEE Particle Accelerator Conference 2115
[8] Rey J-M et al 2008 *AIP Conf. Proc.* **1099** 944-947

Journal of Physics: Conference Series **443** (2013) 012078 doi:10.1088/1742-6596/443/1/012078

# Measurement of the Positron Annihilation Induced Auger Electron Spectrum from Ag(100)

**P Joglekar[1], K Shastry[1], N G Fazleev[1,2] and A H Weiss[1]**

[1] Physics Department, University of Texas at Arlington, Arlington, TX 76019, USA
[2] Physics Department, Kazan Federal University, Kazan 420008, Russian Federation

E-mail: weiss@uta.edu

**Abstract**. Research has demonstrated that Positron Annihilation Induced Auger Spectroscopy (PAES) can be used to probe the top-most atomic layer of surfaces and to obtain Auger spectra that are completely free of beam-impact induced secondary background. The high degree of surface selectivity in PAES is a result of the fact that positrons implanted at low energies are trapped with high efficiency at an image-correlation potential well at the surface resulting in almost all of the positrons annihilating with atoms in the top-most layer. Secondary electrons associated with the impact of the incident positrons can be eliminated by a suitable choice of an incident beam energy. In this paper we present the results of measurements of the energy spectrum of electrons emitted as a result of positron annihilation induced Auger electron emission from a clean Ag(100) surface using a series of incident beam energies ranging from 20 eV down to 2 eV. A peak in the spectrum was observed at ~40 eV corresponding to the $N_{2,3}VV$ Auger transition in agreement with previous PAES studies. This peak was accompanied by an even larger low energy tail which persisted even at the lowest beam energies. Our results for Ag(100) are consistent with previous studies of Cu and Au and indicate that a significant fraction of electrons leaving the sample are emitted in the low energy tail and suggest a strong mechanism for energy sharing in the Auger process.

## 1. Introduction

In this paper, we present the energy spectrum of Auger electrons emitted from Ag(100) obtained using Positron Annihilation Auger Electron Spectroscopy (PAES). In PAES, the Auger electrons are excited by the creation of core holes through an annihilation process. It has been shown [1] that PAES is significantly more surface selective than Electron stimulated Auger Electron Spectroscopy (EAES) and X-ray induced Auger Electron Spectroscopy (XAES) due to the fact that the positrons are trapped in a surface state localized just above the surface at the time of annihilation. The localization of the positrons in the surface state ensures that almost all of the annihilation induced core hole excitations that result in Auger transitions occur in the topmost atomic layer [2]. More importantly, for the purpose of this paper, the energy of the positron beam used in PAES can be made arbitrarily low making it possible to eliminate most or all of the beam induced secondary electrons leaving the surface [3]. In contrast, the highly penetrating beam used in XAES excites Auger electrons from many inelastic mean free paths below the surface. The beam induced secondary background present in EAES and XAES (which is typically many times larger than the signal) makes the measurement of these spectral features extremely difficult. Here we present the results of measurements of energy spectra of electrons emitted as a result of positron annihilation at the surface of Ag(100). The data indicate that a significant fraction of electrons leaving the sample are emitted in the low energy tail.

Journal of Physics: Conference Series **443** (2013) 012078 doi:10.1088/1742-6596/443/1/012078

The spectral weight of the tail is several times larger than can be accounted for in terms of inelastic backscattering and inelastic processes associated with electron transport through the thin layer of electron gas covering the atomic cores in the top-most layer. This suggests that the tail is intrinsic to the Auger transition and that a significant fraction of the Auger transitions proceed through multi-electron processes in which the energy made available through filling the core hole is shared amongst 2 or more electrons.

## 2. Experiment

The measurements were performed using a high resolution PAES system at University of Texas at Arlington described more fully in Ref [4]. The incident positron beam, generated from beta decay of $Na_{22}$ source moderated through a thin (1μm) tungsten (W) foil. The beam is electromagnetically guided by 40 Gauss axial magnetic field towards the sample. The sample was cleaned daily by ion sputtering (without annealing) and maintained under UHV conditions ($6\times10^{-10}$ Torr) during the time of PAES data collection.

The time of flight of the Auger electrons is obtained by measuring the time interval between the annihilation gamma signal from a $BaF_2$-PMT detector and the signal due to the detection of the Auger electron by a channel plate detector placed after at the end of a ~ 1m flight path. The emission of the Auger electron occurs within $\sim10^{-15}$ sec of the annihilation and may be considered to be essentially simultaneous with the detection of the gamma rays emitted when the core electron is annihilated. The overall timing resolution of the system is ~ 1.9 n-sec with a corresponding energy resolution of ~ 3.2 eV at 40 eV and ~ 16.5 eV at 250 eV [4].

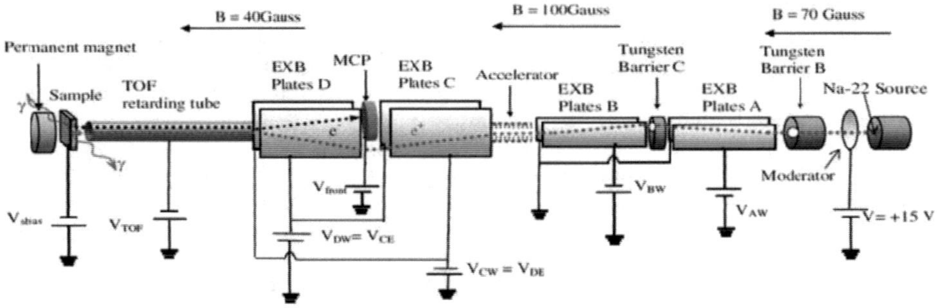


**Figure 1.** TOFPAES setup at University of Texas at Arlington. Ref [4,5].

A permanent magnet is placed behind the sample to produce a 50 mT field at the sample surface. Solenoids produce a 4mT in the transport region. The resulting field redirects the momentum of the outgoing electrons toward the beam axis [6] and reduces the time spread due to the angular divergence of the electrons leaving the sample surface. The flight path consists primarily of a ~ 0.7 m electric field free region inside the T-O-F retarding tube. The energy spectrum is determined from the time of flight spectrum using a conversion function obtained using an empirical calibration procedure [4, 5].

## 3. Results and Discussion

The energy spectrum of electrons emitted in the energy range of 20 eV to 600 eV as a result of 18 eV positrons incident on the surface of Ag(100) is shown the Figure 2. A peak may be seen at ~40 eV which is due to the annihilation induced $N_2VV$ and $N_3VV$ Auger transitions [7]. Small peaks, containing and order of magnitude less integrated intensity, may also be observed at energies of ~160 eV, ~240 eV, and ~380 eV. The origin of these peaks can be attributed to the bulk impurities Sulphur (~146 eV and ~373 eV) and Carbon (~263 eV) diffusing to the surface after sputtering. The effect on the intensity of these peaks of incorporating sample annealing into the sputtering cycle will be studied and investigated in future. Figure 3 shows the variation in the secondary electron peak as a function of the sample bias. The beam energy used in this experiment was approximately 2 eV. We can clearly see that the secondary electron peak has a broad maximum at the lagging edge which is increasing as the sample bias is increasing.

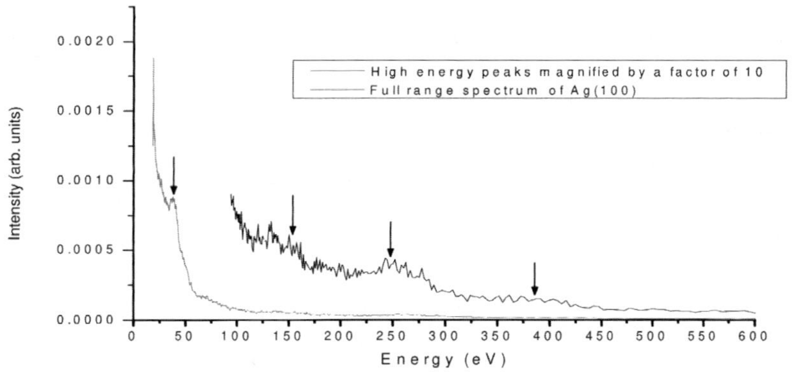


**Figure 2.** Energy spectrum of Ag(100) at 18 eV beam energy. The upper spectrum is a magnified version of the lower spectrum (×10). The arrow ~40 eV indicates the position of a peak associated with the $N_{23}VV$ Auger transition. The arrows pointing to the magnified spectrum at point to much lower intensity peaks of Sulphur at ~146 eV, ~373 eV and of Carbon at ~263 eV.

Figure 3 shows the 2 eV positron beam induced electron spectrum for positrons incident on a Ag(100) surface at different sample bias (and hence kinetic energies) of 0 V, 1 V, 5 V and 10 V. The Ag NVV Auger peak can be seen at ~40 eV in each case. The results obtained are consistent with the experimental results of Mukherjee et al. [7] for the Cu(100) surface.

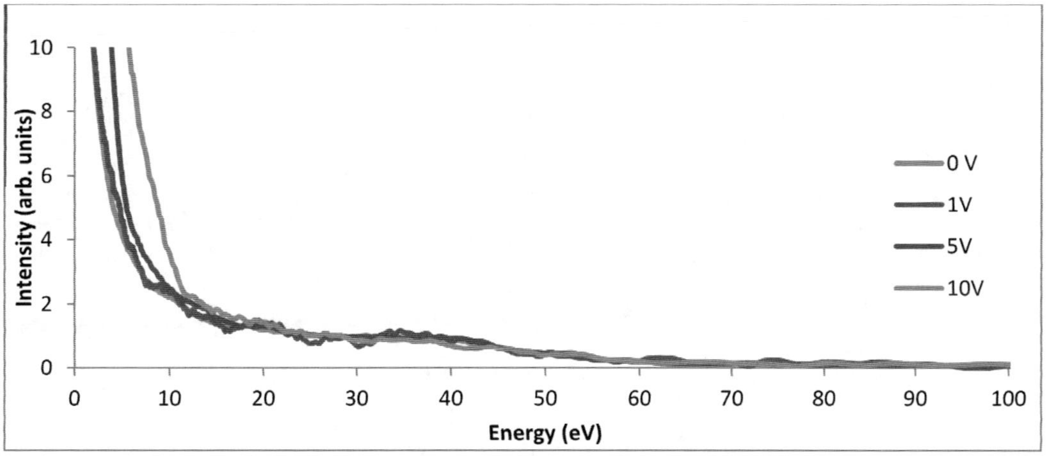


**Figure 3.** The low energy portion of the PAES spectrum obtained using a beam of positrons incident on a Ag(100) surface at energies of 2 eV, 3 eV, 7 eV and 12 eV. The spectrum has been smoothed using a 3 eV moving average [5].

The observed shift in the steep negative slope below 15 eV is in qualitative agreement with the equation for the maximum kinetic energy of the beam impact induced secondary electrons derived in ref. 8 (eq. 1) based upon consideration of the energy released when a positron makes a transition from a scattering state above the vacuum level to a bound state at the surface given by eq. 1:

$$KE_{max} = KE_{e+} + E_b - \phi^- , \quad (1)$$

where $KE_{max}$ is the maximum kinetic energy of the impact is induced secondary electrons, $KE_{e+}$ is the kinetic energy of the incident positron, $E_b$ is the positron surface state binding energy, and $\phi^-$ is the electron work function. Using eq. 1 with calculated values of $E_b = 3.12$ eV, $\phi^- = 4.5\ eV$ we obtain the upper bounds on the beam induced secondary electron energies of approximately 0 eV, 1 eV, 5 eV and 10 eV, for the incident positron beam energies of 2 eV, 3 eV, 5 eV and 12 eV respectively. It may be seen that the significant low energy tail does not disappear even as the positron beam energy is reduced to the point the incident beam can no longer directly knock out secondary electrons. Mukherjee et al. suggested that the similarly large low energy tail that was observed in Cu and Au was due to multi-electron Auger processes [9]. Another possible explanation is that valence electrons are ejected as a result of the sudden creation of the core hole via the annihilation process. In this case conservation of energy would require that the sum of the energies of the two annihilation gammas would be given by eq. 2:

$$E_{gamma1} + E_{gamma2} = 2m_0c^2 - (BE_{e^+} + BE_{core} + \textstyle\sum_i BE_{valence_i}) , \quad (2)$$

where $E_{gamma\ 1,2}$ are the energies of the two gamma rays, $m_0c^2$ is the rest mass energy of the positron and electron, $E_{e^+}$ is the total energy of the positron in the surface state, $BE_{core}$ is the binding energy of the core electron, and $\sum_i BE_{valence_i}$ is the sum of the binding energy of the low energy valence electrons emitted during the annihilation process.

**Summary**

In this paper we have presented the results of measurements of the spectrum of electrons emitted as a result of positron annihilation induced Auger transitions on an Ag(100) surface. The measurements were carried out for a series of positron beam energies ranging from 18 eV down to 2 eV. All the spectra show a prominent but broad peak at ~40 eV corresponding to the $N_{2,3}VV$ Auger transition in agreement with previous PAES studies. The presence of a large low energy tail (LET) in the spectrum below the Auger peak at the lowest beam energies is consistent with previous experiments with Cu and Au, respectively. Mukherjee et al. present arguments that extrinsic processes such as Auger electrons undergoing inelastic backscattering [9] are not sufficient to account for the majority of the LET observed. They suggest that the majority of the LET is due to intrinsic Auger processes in which the energy associated with the filling of the core hole is shared amongst 2 or more valence electrons. Another possible explanation is that a significant fraction of the tail is due to low energy electrons that are emitted during the core hole creation process. Future experiments are planned that will be aimed at determining the relative importance of these different contributions.

## 4. Acknowledgement

This work was supported in part by the National Science Foundation Grant #DMR-0907679.

**References**

[1] Weiss A, Mayer R, Jibaly M, Lei C, Mehl D and Lynn K G 1988 *Phys. Rev. Lett.* **61** 2245
[2] Mehl D et al 1990 *Phys. Rev. B* **41** 799
[3] Jensen K O and Weiss A 1990 *Phys. Rev. B* **41** 3928
[4] Xie S 2002 *PhD Dissertation* University of Texas at Arlington
[5] Nadesalingam M P 2007 *PhD Dissertation* University of Texas at Arlington
[6] Kruit P, Read F H 1983 *J Phys. E* **16** 313
[7] Jung E, Zhou H, Kim J, Starnes S, Venkataram R and Weiss A H 1993 *Appl. Surf. Sci.* **116** 318
[8] Mukherjee S, Nadesalingam M P, Guagliardo P, Sergeant A D, Barbiellini B, Williams J F, Fazleev N G and Weiss A H 2010 *Phys Rev Lett* **104** 247403
[9] Mukherjee S F, Shastry K and Weiss A 2011 *Phys. Rev. B* **84** 155109

Journal of Physics: Conference Series **443** (2013) 012079 doi:10.1088/1742-6596/443/1/012079

# The Upgrade of the Neutron Induced Positron Source NEPOMUC

**C. Hugenschmidt[1,2], H. Ceeh[1], T. Gigl[1,2], F. Lippert[1], C. Piochacz[1,2], P. Pikart[1], M. Reiner[1], J. Weber[1], S. Zimnik[1]**

[1]Physik-Department E21, Technische Universität München, James-Franck-Straße 1, 85748 Garching, Germany

[2]FRM II, Technische Universität München, Lichtenbergstraße 1, 85748 Garching, Germany

E-mail: `Christoph.Hugenschmidt@frm2.tum.de`

**Abstract.** In summer 2012, the new NEutron induced POsitron Source MUniCh (NEPOMUC) was installed and put into operation at the research reactor FRM II. At NEPOMUC *upgrade* 80% $^{113}$Cd enriched Cd is used as neutron-gamma converter in order to ensure an operation time of 25 years. A structure of Pt foils inside the beam tube generates positrons by pair production. Moderated positrons leaving the Pt front foil are electrically extracted and magnetically guided to the outside of the reactor pool. The whole design, including Pt-foils, the electric lenses and the magnetic fields, has been improved in order to enhance both the intensity and the brightness of the positron beam. After adjusting the potentials and the magnetic guide and compensation fields an intensity of about $3 \cdot 10^9$ moderated positrons per second is expected. During the first start-up, the measured temperatures of about 90°C ensure a reliable operation of the positron source. Within this contribution the features and the status of NEPOMUC *upgrade* are elucidated. In addition, an overview of recent positron beam experiments and current developments at the spectrometers is given.

## 1. Introduction

High intensity positron sources are in high demand for a large variety of positron beam experiments in solid state and surface physics as well as for fundamental studies. At electron accelerators (linacs) positrons can be created from absorption of high-energy bremsstrahlung, which is released by decelerating electrons in a target. The absorption of high energy $\gamma$-radiation from nuclear fission can be used to generate positrons by pair production at a research reactor, e.g. at the POSH beam in Delft [1]. At NEPOMUC the nuclear reaction $^{113}$Cd(n,$\gamma$)$^{114}$Cd is applied in order to provide high-energy $\gamma$-rays for pair production in Pt foils [2]. Pt is applied as positron moderator as well, since it leads to the emission of mono-energetic positrons. Similar sources have been designed at the PULSTAR reactor in the USA [3] and at the research reactor in Hamilton, Canada [4].

Up to now, NEPOMUC has been operated successfully with an intensity of about $10^9$ moderated positrons per second [5]. The design and features of the positron source as well as the positron beam facility are reported elsewhere (see e.g. [2] and references therein). After five years of reactor operation, the beam tube SR 11 had to be replaced due to the burn-up of $^{113}$Cd inside the tip of the beam tube. The main change of NEPOMUC *upgrade* is the application of highly enriched $^{113}$Cd in order to ensure a lifetime of 25 years. In order to enhance the intensity and the beam brightness, additional improvements have been made to the design of the source components [6]. In this work, the improvements of the final layout of NEPOMUC *upgrade* and the expected beam parameters are presented. Results of the calculated temperature field inside the beam tube are shown and compared with the thermal behavior during the first start-up of the reactor with the new positron source. Finally, a short overview of positron beam experiments at NEPOMUC is given.

## 2. The new positron source NEPOMUC

### *2.1. In-Pile Positron Source*

As in the previous setup, the principle of NEPOMUC *upgrade* at the research reactor FRM II is based on the emission of high-energy prompt $\gamma$-rays after thermal neutron capture in $^{113}$Cd. A structure of Pt foils converts the released high-energy $\gamma$-radiation into positron-electron pairs and leads to the emission of mono-energetic positrons (see e.g. [2]).

The lifetime of the first source, which used an converter of natural Cd with a thickness of 3 mm, was limited by the burn-up of the $^{113}$Cd after 1250 days of reactor operation at the nominal power of 20 MW. For this reason, the inclined beam tube SR11 with the in-pile positron source had to be replaced. The main task was a considerable extension of the operation time of the positron source to 25 years. Therefore, Cd enriched with 80% $^{113}$Cd, i.e. 6.5 times higher amount of $^{113}$Cd than in natural Cd, is applied at NEPOMUC *upgrade*.

As shown in figure 1, NEPOMUC *upgrade* consists of three main components: (i) the outer 'beam tube' surrounded by the $D_2O$ of the moderator tank and with Cd as converter inside the tip, (ii) the evacuated 'experimental tube' carrying the magnetic coils for positron beam transport, and (iii) the innermost 'potential tube' with the Pt foil structure and electric lenses.

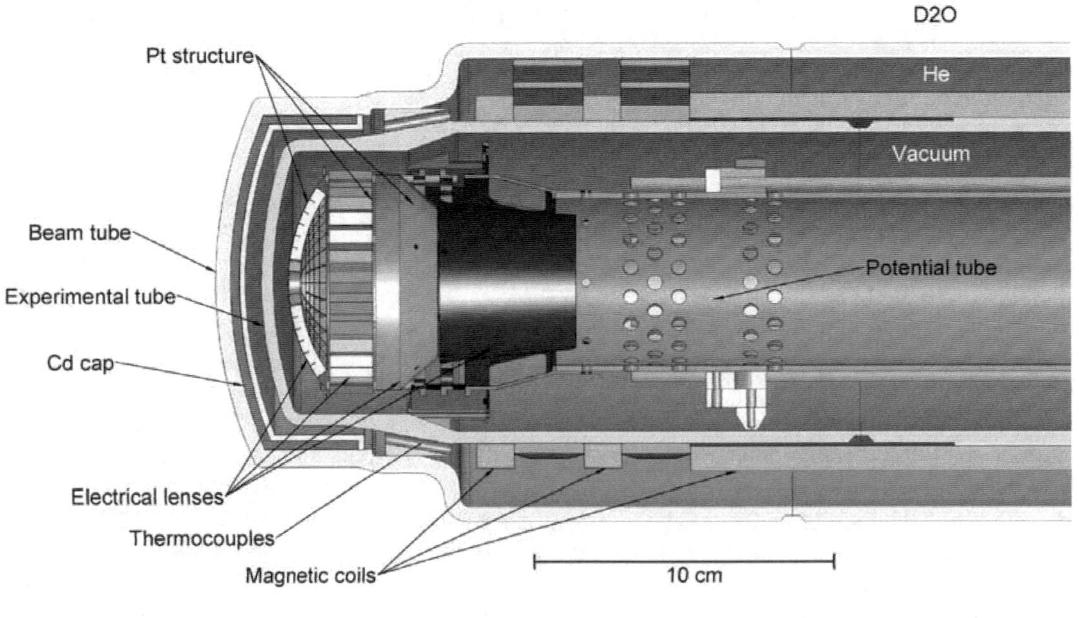


**Figure 1.** Cross sectional view of the new in-pile positron source NEPOMUC *upgrade*.

Compared with the previous position, the Cd cup is 50 mm closer to the fuel element which leads to a higher thermal neutron capture rate. In addition, the smaller Cd cup at NEPOMUC *upgrade* leads to a lower neutron flux depression. The according increase of the positron production rate in Pt is about 20%.

The design of the inner parts has been improved in order to enhance both the beam intensity and its brightness. For this reason, the electric and magnetic fields in the beam formation stage were simulated for various configurations by finite element (FE) calculations [6].

Absorption of high energy $\gamma$-quanta in the Pt front section with additional small vanes and in the electrical lenses contribute to the positron production. For the formation of a high brightness beam, only the moderated positrons from the Pt front plate with low transverse momentum are collected. A higher total moderation efficiency is expected due to the improved geometry and (inelastic) scattering of positrons from the Pt lenses onto the front moderation foil.

A combination of electric and magnetic fields is used to accelerate and to form the beam. In order to minimize transport losses the positrons should be guided adiabatically, i.e. with low kinetic energy. For this reason, the potential of the 'potential tube' and the insulated tube inside the chicane through the biological shield can be biased at any desired value to ensure a fully adiabatic beam transport. Outside the biological shield, the positron beam can be guided to a remoderator using a W(100) crystal in back reflexion geometry for brightness enhancement [7]. A new beam switching and remoderation unit allows quick toggling between the primary high-intensity and the high-brightness remoderated positron beam [6].

The $\gamma$-heating generates a thermal power in the Al/Cd/Al structure of about 1.35 kW. Dependent on the material the heat density of the inner parts in the front section is typically between 0.4 and 1 W/g leading to an additional heat power of approximately 0.6 kW and hence to a total of about 2 kW. The heat conductance between experimental tube and beam tube has to be sufficient since the heat load has to be dissipated in the surrounding heavy water. In order to ensure a high durability in the long term and for safety reasons, the temperature in the experimental tube must not exceed 150 $^{o}$C.

FE calculations have been performed in order to display the temperature field in the outer tubes at a reactor power of 20 MW (see figure 2). The heat load in the components mainly due to $\gamma$-absorption was taken from Monte Carlo calculations. The heat transmission coefficient from the beam tube to the surrounding water (45 $^{o}$C) was set to 2200 W/m$^2$K. Heat transport by He (1.2 bar) between the tubes barely contributes to the cooling. Most important is the contact area between the tubes for heat conductance. As expected, the maximum temperature is reached at the tip of the experimental tube and amounts to 110 $^{o}$C.

### *2.2. Positron beam parameters*

The positron beam is transported in a magnetic guide field of typically 7 mT with a kinetic (longitudinal) energy of $E_L$=1 keV (20 eV) of the primary (remoderated) positron beam. The relevant beam parameters such as $Y^+_{mod}$ yield of (re)moderated positrons, i.e. beam intensity, transverse energy $E_\perp$, and diameter $d^+$ of the positron beam are described and compared with the former NEPOMUC beam in [6]. The according values expected for the primary moderated and the remoderated positron beam at NEPOMUC *upgrade* are summarized in Table 2.2.

For NEPOMUC *upgrade*, a maximum intensity of about $3{\cdot}10^9$ moderated positrons per second is expected. The correct number of the positron intensity strongly depends on various effects such as amount of (in-)elastically scattered positrons, which contribute to the primary positron beam, moderation efficiency of the Pt moderator at high temperature and in the radiation field, and beam extraction and formation at the tip of the beam tube. However, the more conservative estimation of the beam parameters are given in parentheses. The total efficiency of the beam guiding and focusing at the remoderator –and hence $Y^+_{mod}$ of remoderated positrons– is expected to be higher due to the higher brightness of the primary beam at NEPOMUC *upgrade*.

### *2.3. Commissioning of the new positron source*

During the first start-up of the reactor the temperature of the experimental tube at the contact area to the outer tube was measured with four independent thermocouples in order to ensure a

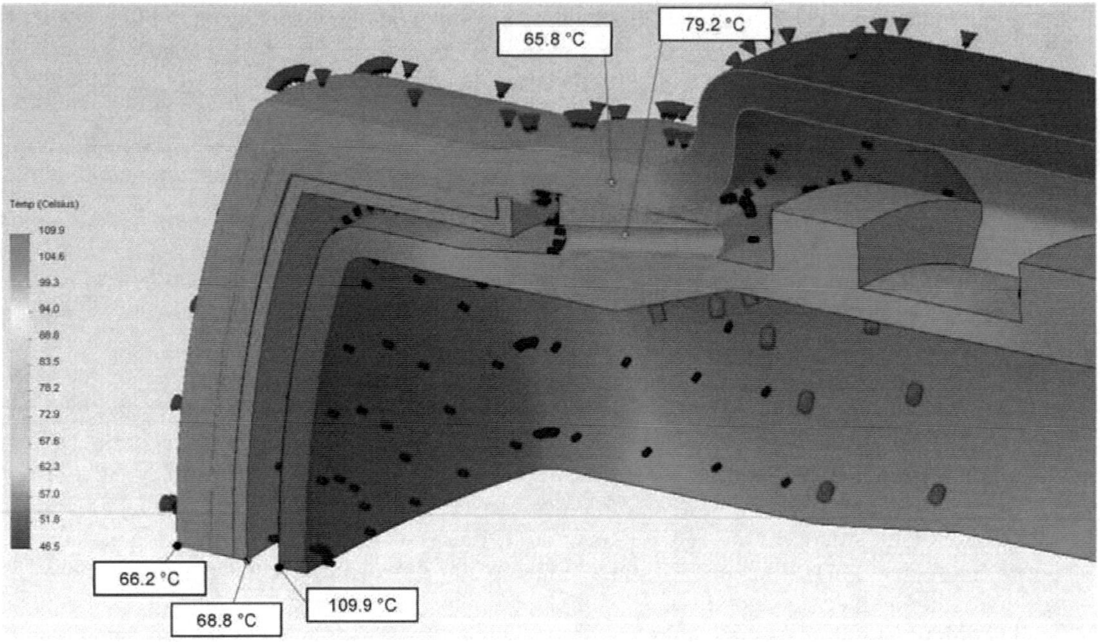


**Figure 2.** FE-calculation of the temperature field in the beam tube and experimental tube during reactor operation at 20 MW. For details see text.

| | **NEPOMUC *upgrade*** | |
|---|---|---|
| | **primary** | **remoderated** |
| $Y^+_{mod}$ $[s^{-1}]$ | $3{\cdot}10^9$ $(2{\cdot}10^9)$ | $\approx 2 \cdot 10^8$ |
| $E_{\perp,\mathrm{FWHM}}$ $[eV]$ | 0.6 (1.2) | $\approx 1$ |
| $d^+_{\mathrm{FWHM}}$ $[mm]$ | $< 7$ $(d_{full} < 22)$ | $< 1$ $(d_{full} < 2)$ |

**Table 1.** Expected beam parameters at NEPOMUC *upgrade* for the primary moderated (values from [6]) and the remoderated positron beam: $Y^+_{mod}$ yield of (re)moderated positrons, $E_\perp$ transverse energy, and $d^+$ diameter of the positron beam. Standard settings are a magnetic guide field of 7 mT, a kinetic (longitudinal) energy of $E_L$=1 keV and 20 eV for the primary and the remoderated beam respectively. (Conservative limits in parentheses.)

safe operation of SR11. The mean temperature $T_m$ and the reactor power $P$ are shown in figure 3 (left). A maximum temperature of about 90(2) °C is reached at 20 MW. During the increase to the power steps of 5, 10, 15, and 20 MW the temperature almost immediately follows the increase of the reactor power and shows a nearly linear behavior (figure 3 (right)), that indicates a sufficient good heat conduction between the two tubes. The further temperature increase during the power steps is attributed to the warming of the SR11 and the surrounding $D_2O$. Compared to the FE-simulation the measured temperature is slightly higher, which is explained by a lower heat transmission coefficient from the beam tube to the $D_2O$.

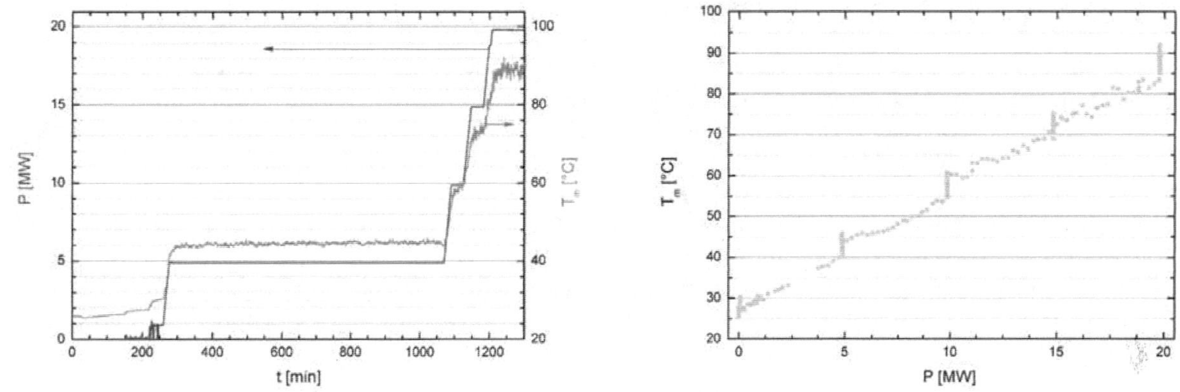


**Figure 3.** Reactor power $P$ and temperature of the experimental tube $T_m$ during start-up of the reactor (left), almost linear increase of $T_m$ as function of reactor power (right).

In August 2012, first tests confirmed the operational reliability of SR11, and first positrons were successfully guided from the source to the outside of the reactor pool. Presently, various parameters are adjusted such as potentials of the moderator, electric lenses and potential tube as well as magnetic guide and compensation fields for proper beam extraction. The shape and the intensity of the positron beam are measured at the first accessible place outside the reactor shielding.

## 3. Positron Beam Experiments

NEPOMUC *upgrade* will enable the continuation of a large variety of studies using the high-intensity positron beam as well as the development of new experiments. These experiments range from high-sensitive and elemental selective investigations in surface physics over defect spectroscopy in thin films and solids to fundamental studies in atomic and plasma physics.

### *3.1. Surfaces*

Positron annihilation induced Auger electron Spectroscopy (PAES) for element selective surface studies has several advantages over conventional AES such as suppressed secondary electron background in the range of Auger-transition energies and exceptional surface sensitivity. Only a few PAES studies have been performed so far due to the very long measurement time of up to 20 days per spectrum with conventional positron sources. After upgrading the spectrometer at NEPOMUC, we succeeded in recording PAES spectra with high energy resolution within several minutes. For example, the surface coverage and island growth were studied in the systems Cu/Pd and Cu/Fe [8]. Moreover, with first *time-dependent* PAES the theoretically predicted segregation process of Cu in Pd could be observed experimentally [9].

Currently a x-ray source is installed in order to gain element information of the surface with complementary techniques such as XAES and XPS. An additional STM is also mounted for the examination of the surface topology.

Journal of Physics: Conference Series **443** (2013) 012079 doi:10.1088/1742-6596/443/1/012079

### *3.2. Defects and thin films*

Coincident Doppler-Broadening Spectroscopy (CDBS) enables defect studies and element specific measurements in the near surface region and in the bulk of the specimen up to a few $\mu$m according to a positron energy of up to 30 keV. The sample defects can be mapped by scanning the beam in two dimensions with a lateral resolution of $<300\,\mu$m. Since depth dependent CDBS is particularly suited to study the elemental surrounding of the positron annihilation site metallic layers and (nano-)clusters embedded in Al were investigated by CDBS [10, 11]. Defect annealing in thin films [12] or in Cu and Ni after severe plastic deformation [13] can be observed by in-situ DBS as a function of time and temperature. Presently, a new sample stage is installed in order to extend the temperature range to 100...1300 K [14].

The Pulsed Low-Energy Positron System (PLEPS) was developed at the Universität der Bundeswehr (UniBW) München for depth dependent positron lifetime measurements. At NEPOMUC, lifetime spectra can be recorded within a few minutes, and the high peak-to-background ratio of $10^4$ allows the extraction of at least three lifetime components reliably [15]. Several PLEPS experiments were performed e.g. on the free volume in polymer films [16] or in bioadhesive [17], on irradiation induced defects after He implantation in InN and GaN [18] or on vacancies in a thin film perovskite oxide [19]. An interface including pulsing units and an additional remoderator for brightness enhancement was developed in order to enable the operation of the Scanning Positron Microscope (SPM) [20] at the NEPOMUC beam line [21].

### *3.3. Fundamental experiments*

Additional experimental setups can easily be connected to the positron beam line at the open beam port. Examples are, a nitrogen-filled drift chamber for the investigation of positron moderation by inelastic scattering [22], a coincidence setup for the detection of the correlated electron-electron and electron-positron emission from surfaces after slow positron impact [23] or a high precision measurements of the decay rate of the negative positronium ion $Ps^-$ [24]. Due to the limited place at the present position, it is planned to extend the positron beam facility to a new experimental hall at the east side of the reactor building. A new spectrometer for the measurement of the Angular Correlation of Annihilation Radiation (ACAR) using a $^{22}$Na source has been set up recently [25]. It is planned to move this apparatus to the new hall in order to enable depth dependent ACAR experiments for non-destructive studies of the electronic structure –in particular anisotropies of the Fermi surface– in thin films, at interfaces and surfaces. Another longterm experiment at an additional beam port will be the implementation of a stellarator setup for the creation and studies of electron–positron plasmas [26].

## 4. Summary and outlook

The main change of the new beam tube at NEPOMUC *upgrade* is the utilization of 80% enriched $^{113}$Cd in order to provide a long-life high intensity positron beam. A new design of the positron beam extraction was developed, which is expected to lead to an increase of both beam intensity and brightness. The anticipated intensity of the mono-energetic positron beam is about $3.0{\cdot}10^9$ moderated positrons per second. In addition, a new beam switching and remoderation unit was constructed in order to allow quick toggling between the primary high-intensity and the high-brightness remoderated positron beam. Presently the beam parameters are adjusted for providing a high positron intensity. In 2012, the spectrometers PLEPS, CDBS and PAES will be connected to the beam line in order to continue the positron beam experiments.

## Acknowledgements

The authors thank the group of G. Dollinger (UniBW München) for the continuous support to operate PLEPS at NEPOMUC. Funding of the projects by the DFG (project no. A3 within the Transregional Collaborative Research Center TRR 80) and by the BMBF (project no. 05K10WOB) is gratefully acknowledged.

## References

[1] Schut H, van Veen A, de Roode J and Labohm F 2004 *Mat. Sci. For.* **444-446** 507–509
[2] Hugenschmidt C, Schreckenbach K, Stadlbauer M and Straßer B 2005 *Nucl. Instr. Meth. A* **554**(1-3) 384–391
[3] Hawari A I, Gidley D W, Xu J, Moxom J, Hathaway A G, Brown B and Vallery R 2009 *AIP Conf. Proc.* **1099** 862–865
[4] 2009 Personal communication P. Mascher, McMaster University, Hamilton
[5] Hugenschmidt C, Löwe B, Mayer J, Piochacz C, Pikart P, Repper R, Stadlbauer M and Schreckenbach K 2008 *Nucl. Instr. Meth. A* **593** 616–618
[6] Hugenschmidt C, Piochacz C, Reiner M and Schreckenbach K 2012 *New Journal of Physics* **14** 055027
[7] Piochacz C, Kögel G, Egger W, Hugenschmidt C, Mayer J, Schreckenbach K, Sperr P, Stadlbauer M and Dollinger G 2008 *Appl. Surf. Sci.* **255** 98–100 ISSN 0169-4332
[8] Mayer J, Hugenschmidt C and Schreckenbach K 2010 *Surf. Sci.* **604** 1772–1777 ISSN 0039-6028
[9] Mayer J, Hugenschmidt C and Schreckenbach K 2010 *Phys. Rev. Lett.* **105** 207401
[10] Hugenschmidt C, Pikart P, Stadlbauer M and Schreckenbach K 2008 *Phys. Rev. B* **77** 092105 (pages 4)
[11] Pikart P, Hugenschmidt C, Horisberger M, Matsukawa Y, Hatakeyama M, Toyama T and Nagai Y 2011 *Phys. Rev. B* **84** 014106
[12] Reiner M, Pikart P and Hugenschmidt C 2012 *Physics Procedia* **35** 104 – 110 ISSN 1875-3892 ¡ce:title¿Positron Studies of Defects 2011¡/ce:title¿
[13] Oberdorfer B, Steyskal E M, Sprengel W, Puff W, Pikart P, Hugenschmidt C, Zehetbauer M, Pippan R and Würschum R 2010 *Phys. Rev. Lett.* **105** 146101
[14] Reiner M, Pikart P and Hugenschmidt C 2012 *J. Phys.: Conf. Ser.* **to be published**
[15] Sperr P, Egger W, Kögel G, Dollinger G, Hugenschmidt C, Repper R and Piochacz C 2008 *Appl. Surf. Sci.* **255** 35–38 ISSN 0169-4332
[16] Egger W, Sperr P, Kögel G, Wetzel M and Gudladt H J 2008 *Appl. Surf. Sci.* **255** 209–212 ISSN 0169-4332
[17] Rätzke K, Wiegemann M, Shaikh M Q, Harms S, Adelung R, Egger W and Sperr P 2010 *Acta Biomater.* **6** 2690–2694 ISSN 1742-7061
[18] Reurings F, Tuomisto F, Egger W, Löwe B, Ravelli L, Sojak S, Liliental-Weber Z, Jones R E, Yu K M, Walukiewicz W and Schaff W J 2010 *physica status solidi (a)* **207** 1087–1090 ISSN 1862-6319
[19] Keeble D J, Mackie R A, Egger W, Löwe B, Pikart P, Hugenschmidt C and Jackson T J 2010 *Phys. Rev. B* **81** 064102
[20] David A, Kögel G, Sperr P and Triftshäuser W 2001 *Phys. Rev. Lett.* **87** 067402
[21] Piochacz C, Egger W, Hugenschmidt C, Kögel G, Schreckenbach K, Sperr P and Dollinger G 2007 *phys. stat. sol. (c)* **4** 4028–4031
[22] Löwe B, Schreckenbach K and Hugenschmidt C 2008 *Appl. Surf. Sci.* **255** 96–97 ISSN 0169-4332
[23] van Riessen G A, Schumann F O, Birke M, Winkler C and JKirschner 2008 *J. Phys. Cond. Matt.* **20** 442001
[24] Ceeh H, Hugenschmidt C, Schreckenbach K, Gärtner S A, Thirolf P G, Fleischer F and Schwalm D 2011 *Phys. Rev. A* **84**(6) 062508
[25] Ceeh H, Böni P, Chioncel L, Leitner M, Weber J and Hugenschmidt C 2012 *J. Phys.: Conf. Ser.* **to be published**
[26] Pedersen T S, Danielson J R, Hugenschmidt C, Marx G, Sarasola X, Schauer F, Schweikhard L, Surko C M and Winkler E 2012 *New Journal of Physics* **14** 035010

Journal of Physics: Conference Series **443** (2013) 012080 doi:10.1088/1742-6596/443/1/012080

# Fast inorganic scintillators – status and outlook

**Rainer W. Novotny**

2nd Physics Institute, Justus-Liebig-University, Heinrich-Buff-Ring 16, 35392 Giessen, Germany

r.novotny@exp2.physik.uni-giessen.de

**Abstract**. Scintillation detectors based on inorganic materials represent one of the most widely applied instrumentation techniques in the fields of nuclear and high-energy physics as well as medical or industrial applications. Driven by experimental requirements the research has focused onto a faster response, shorter decay times and higher compactness implementing high-Z ions. The discovery of the fast core-valence luminescence in $BaF_2$, the allowed electric dipole transitions in $Ce^{3+}$, and the short radiation lengths of BGO and $PbWO_4$ have set important milestones. However, excellent time resolution is affected as well by the density of photoelectrons produced at the early stage of the signal generation and the integral light output. The paper will give a detailed overview of the present status on fast inorganic scintillators, their performance and the theoretical limitations on achievable energy and time resolutions. The results are illustrated by various applications in research. The state of the art scintillator material will be discussed based on new materials such as $LaBr_3$ or $LaCl_3$, which were doped with rare earth ions such as $Ce^{3+}$-centers to reach decay times between 20 and 40ns, respectively.

## 1. Introduction

For decades, scintillation detectors based on inorganic materials have become one of the most widely applied instrumentation techniques in physics – in particular in the fields of nuclear and high-energy physics. Their discovery and development are strongly correlated with the experimental needs in basic research and technology in physics. Visual counting of the discovered X-rays or natural radioactivity became possible with $BaPt(CN)_4$, $CaWO_4$ or ZnS at the end of the 19th century. The first construction of scintillation detectors, made possible by the development of the photomultiplier tube [1], started with the discovery of activated and pure alkali halide crystals. NaI(Tl) and CsI(Tl), introduced by Hofstadter [2,3], provide since more than 60 years efficient photon and particle detection. The continuous increase of the energy range of the probes to be detected directed crystal development towards faster response, shorter decay times and higher compactness implementing high-Z ions. In particular, the discovery of the fast core-valence luminescence in $BaF_2$ [4,5], the allowed electric dipole transitions in $Ce^{3+}$ and the short radiation lengths of BGO [6] or $PbWO_4$ [7] set important milestones in the last decades of the 20th century to find or even engineer the ideal scintillator.
In particular, the evolution in nuclear, medium and high-energy physics illustrates the impact on the detector development by asking for complex detector systems to cope with multi-particle production and high photon multiplicities. In high-energy physics detector technology had to be adapted in a similar way.

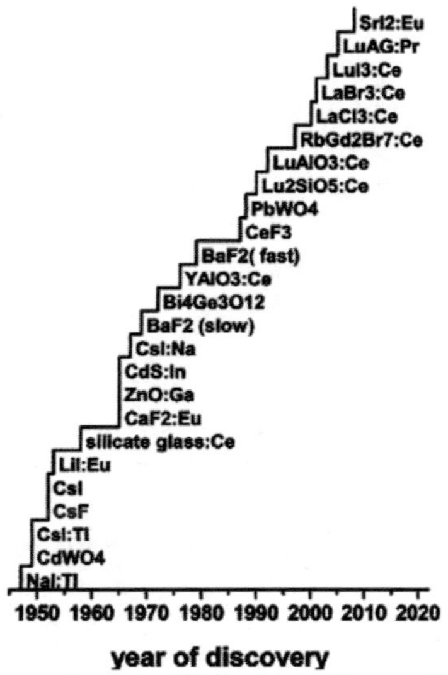


**Figure 1.** Development of inorganic scintillators starting with the availability of electronic photo sensors.

## 2. Basic properties of inorganic scintillators

The overall luminescence yield is one of the most selective criteria for the selection of a scintillator material. This holds in particular for nuclear structure studies or medical imaging when low energetic photons have to be detected with high resolution. A phenomenological approach leads to the following estimate of the primary light yield Y of scintillation photons per deposited energy given in MeV:

$$\mathbf{Y} = \frac{\mathbf{E}_\gamma}{\beta \cdot \mathbf{E}_g} \cdot \mathbf{S} \cdot \mathbf{Q} \tag{1}$$

where $\beta \cdot E_g$ is the mean energy necessary for the formation of one electron-hole pair in a medium with a forbidden zone of width $E_g$ and $E_\gamma$ is the absorbed energy. S describes the efficiency of the energy transfer to the luminescent center and Q the quantum yield of the intra-center luminescence. Assuming S and Q to be close to 100% the light yield will depend inverse proportionally on the energy $E_g$.

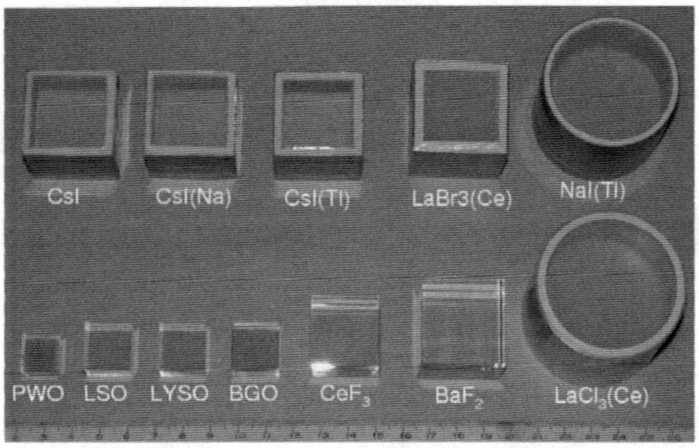


**Figure 2.** Comparison of different cubic scintillator materials shown with dimensions of $1.5X_0$. The picture is taken from [8].

However, the energy resolution finally achieved with the complete detector is a complex interplay between statistical fluctuations within the above formula, the light collection efficiency and the performance and matching of the photo sensor. The different mechanisms of interaction for neutral and charged particles within the detector medium and the completeness of energy absorption have to be included on top. Due to high count rates, coincidence requirements or particle identification via time-of-flight technique time resolution is another criteria for scintillator selection. Additional factors including the rise and decay time of the scintillation process, fluctuations of the photon transit time from the emission point within the crystal to the photo sensor and fluctuation within the photo detector affect the timing performance. Therefore, most of the present applications focus on the search for bright and fast scintillators limiting the variety of optional luminescence centers either intrinsic or created by dopants.

The detection of electromagnetic (EM) probes relies on the measurement of the energy and impact location, both with high resolution. In particular, the reconstruction of the invariant mass of neutral mesons, which decay preferentially into photon pairs, requires the simultaneous measurement of energy and the relative angle between both photon directions. Besides a full geometrical coverage with high efficiency the absorption of the total photon energy in the active material is mandatory. The asymptotic cross sections of the elementary processes determine the radiation length $X_o$ depending on the composition of the material. The subsequently created pairs of electrons and positrons, which undergo bremsstrahlung at energies above a material specific critical value, close the circle by generating again high-energy photons. In this process, leptons loose their energy on a logarithmic scale and lead to a lateral spread of the secondary products due to multiple scattering, which is quantified geometrically by the Molière-radius ($R_M$). The interplay of these processes leads to the formation of the EM shower. The present accelerator technology in medium- and high-energy physics aiming for higher beam energies and luminosities has forced the scintillator development towards extremely compact, fast and radiation hard crystals (see figure 2).

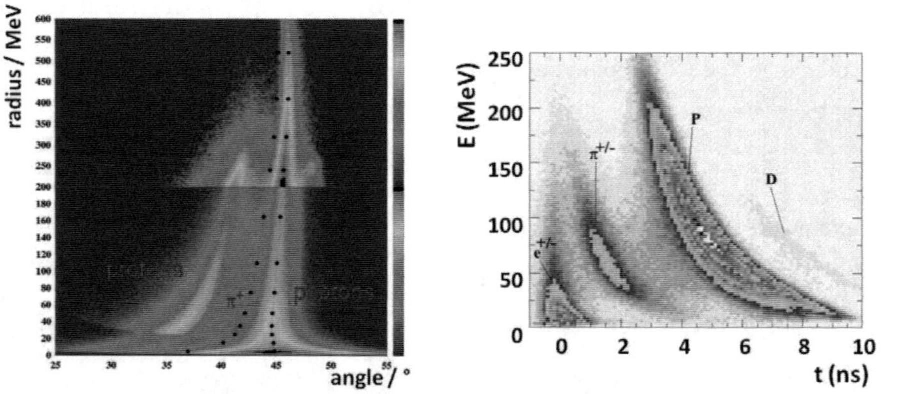


**Figure 3.** Both correlations illustrate the particle identification power of $BaF_2$ crystals exploiting the different line shape characterized by the specific ratio of the fast and slow scintillation component (left) and the correlation between measured energy and time of flight based on a time resolution $\sigma <$ 100ps. In case of the left figure, the lower part has been scaled by a factor and the black dots determine the ridge and the left valley of the photon response.

## 3. Present search for fast scintillators

In the 1980's a new opportunity has been observed, the core-valence luminescence in $BaF_2$ which showed extremely fast rise and slow times, respectively, at a level below 1ns. Since the specific band structure - showing an energy distance between the core and valence band being smaller than the usual band gap between valence and conduction band - suppressed Auger transitions leading to a temperature independent so-called cross luminescence with a moderate light yield compared to the dominant luminescence based on self-trapped excitons [9]. In addition, both mechanisms showed a strong sensitivity to the ionization density of the absorbed particle providing particle identification via pulse shape analysis in addition to the excellent time-of-flight resolution. Figure 3 illustrates both techniques. Since the ratio of the fast and slow scintillation component remains nearly constant for a given particle species, the correlation of both separately measured and calibrated components can be expressed by polar coordinates as plotted in the left part of figure 3. Unfortunately, only a few ionic crystals shows such a highly sensitive behavior.

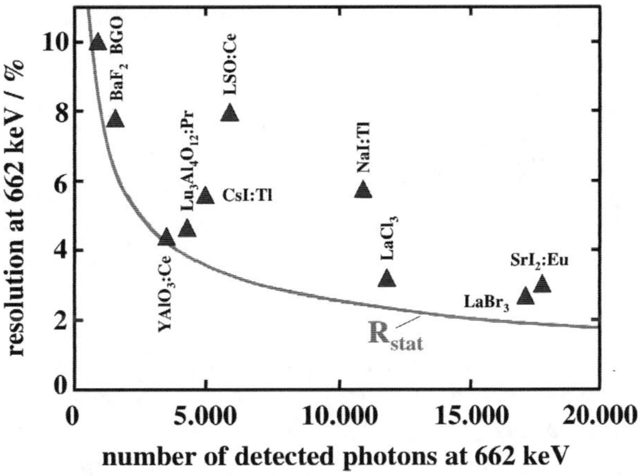


**Figure 4.** Energy Resolution (FWHM/E) measured at 662keV $\gamma$-energy with different scintillator crystals as a function of the number of detected photons in a photomultiplier tube. The red curve shows the limit expected from photon statistics [11].

About one decade ago [10] the search for bright scintillators with small band gap succeeded in growing large volume rare earth halides such as $LaBr_3$ or $LaCl_3$. In order to accomplish fast luminescence for high-rate studies or good timing resolutions these halides were doped with rare earth ions. Primarily $Ce^{3+}$-centers with an allowed 5f-4d transition provide decay times $\tau$ between 20 and 40ns depending on the host structure. Implementing $Pr^{3+}$-ions instead can further reduce $\tau$ by nearly a factor of two.

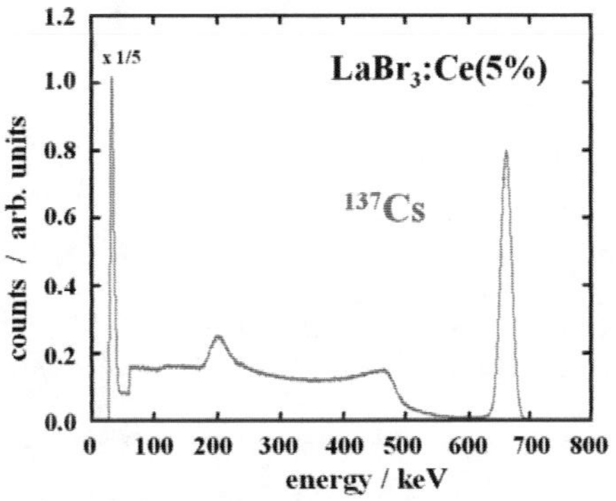


**Figure 5.** Response of a 19x19mm$^2$ $LaBr_3$:Ce crystal to a $^{137}$Cs source. The part below 60keV is scaled down by a factor of 5 to display the simultaneous detection of 32keV Ba $K_\alpha$ X-rays[11].

Exploiting the fundamental limits given by Equation (1) the achievable energy resolution should be limited only by Poisson statistics of the luminescence yield. Figure 4 illustrates the presently achieved energy resolutions (FWHM) with respect to the theoretical limit and figure 5 shows the response function to $\gamma$-rays of 662keV energy ($^{137}$Cs) measured with a small $LaBr_3$:Ce crystal. However, besides the imperfection of the crystal, of light collection or of the sensitivity of the photo sensor a not yet fully understood mechanism of a non-proportional response at energies below 100keV appears to have a strong impact. This holds in particular for scintillators like LSO, NaI(Tl) or CsI(Tl), respectively. The recently developed halides have started to replace the traditional materials like NaI(Tl) or CsI(Tl) in a wide field of application for γ-cameras, radiation protection or homeland security. Unfortunately, most of these materials are hygroscopic and therefore, have to be canned and consequently impose limits on a dense packing of multi detector systems.

In the past years Ce-doped silicate based inorganic scintillator materials were developed and heavily used for medical applications such as SPECT and PET tomography. Mass production of small samples but large quantities and the principal capability to grow also large size crystals was established for oxyorthosilicates such as $Lu_2SiO_5$ (LSO) and $Lu_{2(1-x)}Y_{2x}SiO_5$ (LYSO), both doped with cerium. The latter material combines high density ($\rho$ = 7.4g/cm$^3$), fast response (decay time $\tau$ ~40ns at420nm) and in case of high energy applications a compact EM-shower containment ($X_o$=1.14cm, $R_M$ = 2.3cm). In addition, high light yields comparable to NaI(Tl) were achieved for optimized Ce-concentrations [12].

## 4. Outlook

The engineering of future inorganic scintillation detectors and concepts has already started partly promoted by a series of new technologies in material sciences. The recently developed method of *micro pulling down* offers the consideration of optical fibers based on inorganic scintillators. Diameters between 0.3mm and 2.0mm, respectively, can be grown up to a typical length of 100cm. Production has been started successfully with doped or un-doped LuAG, YAG or LYSO samples (see figures 6 and 7). These fibers offer all advantages of inorganic materials such as high light yield, radiation hardness and fast response and provide intrinsically excellent position information. The currently developed Silicon photomultiplier, pixel arrays of avalanche diodes operated in the Geiger-mode, offer a perfectly matched readout with high quantum efficiency. First attempts for small animal PETS or position sensitive monitor detectors even for photon beams are under investigations.

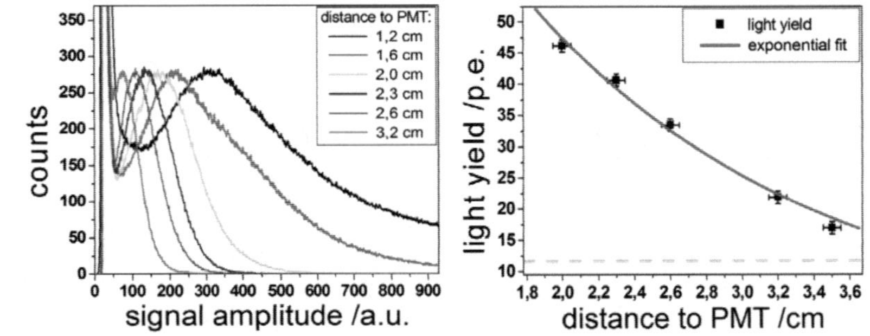


**Figure 6.** Response of a LYSO:Ce fiber (Ø=0.3mm, length L=100mm) to a $^{241}$Am α-source as a function of the source distance to the photo sensor. The fibers still show a strong light attenuation due to imperfections [14].

For the first time, crystals have been proposed to construct a homogeneous calorimeter combining the EM and hadronic part, which would eliminate disturbing dead materials and would drastically improve the energy resolution in particular for hadronic jets measurement. The concept is based on the dual readout method to measure both Cherenkov and scintillation light simultaneously. Due to the extremely large volume of the calorimeter, the crystal material must be dense and UV-transparent. Present concepts either follow a bulk scintillator approach based on materials like BGO, rare earth doped $PbF_2$ or doped PWO emitting green scintillation light with slow decay time. The fast Cherenkov component can be either separated from the scintillation light either by the difference in wavelength or decay time similar to phoswich technology [8]. Alternatively, densely packed Ce-doped and un-doped scintillation fibers based on LuAG would allow a direct readout of both light components. In addition, the fibers could be integrated into an absorber matrix operating as a sampling calorimeter.

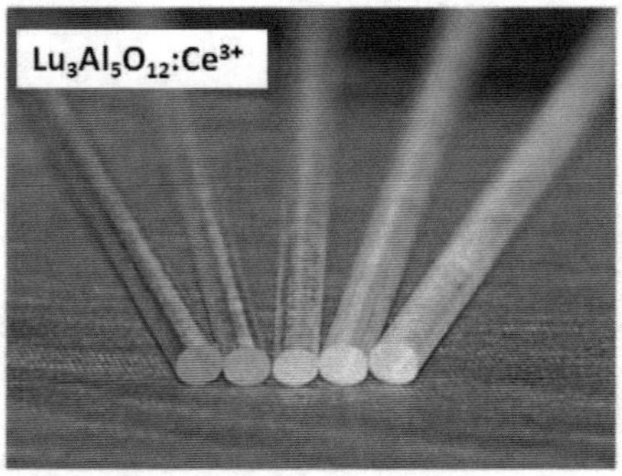


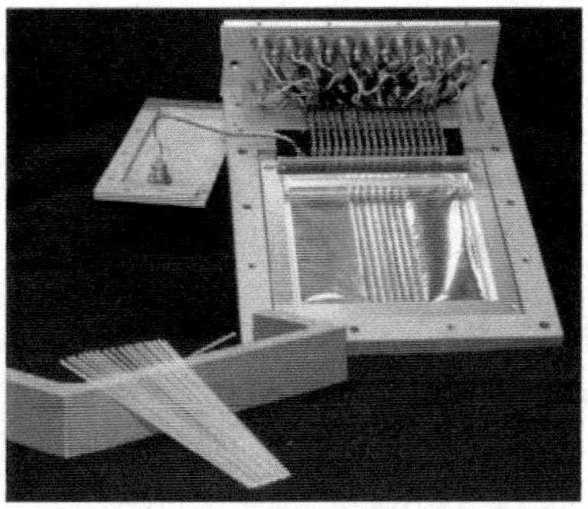

**Figure 7.** Inorganic fibres made of LuAG:Ce (left) and a first detector prototype used as 2D position sensitive monitoring detector[15].

In contrast to all previous investigations and applications of inorganic scintillators the study of the luminescence of scintillating nano-particles made of doped insulators has opened up a completely new field. Similar to the unexpected luminescence features of the quantum dots (Q-dots), made of semiconductors, the relaxation of energy, non-proportionality, surface effects and quantum and dielectric confinement can completely deviate from our understanding of a well known scintillator crystal. In spite of the tradition going back more than 60 years, inorganic scintillators are still a very vivid and expanding field of research and application.

**References**

[1] Curran S C, Baker W R 1948 *Rev. Sci. Instrum.* **19:116**
[2] Hofstadter R 1948 *Phys. Rev.* **74** 100
[3] Hofstadter R 1950 *Nucleonics* **6** 79
[4] Farukhi M R et al. 1971 *IEEE Trans. on Nucl. Sci.* **NS-18** 200
[5] Ershov N N et al. 1982 *Opt. Spectrosc.* **53** 51
[6] Weber M J et al. 1973 *J. Appl. Phys.* **44** 5495
[7] Derenzo S E et al. 1990 *IEEE Trans. on Nucl. Sci.* **37** 203
[8] Zhu Ren-Yuan 2009 *Proc. of XIII Int. Conf. On Calorimetry in High Energy Physics JOP Conf. Ser.* **160** 012017
[9] Aleksandrov Y M 1984 *Sov. Phys. Sol. State* **26** 1734
[10] Dorenbos P 2002 *Nucl. Instr. and Meth.* **A 486** 208
[11] Dorenbos et al. 2004 *IEEE Trans. on Nucl. Sci.* **51** 1289
[12] Thiel M et al. 2008 *IEEE Trans. on Nucl. Sci.* **55 142**
[13] Auffray E et al. 2009 *Conf Record IEEE Nucl. Sc. Symp., Orlando, USA, October 25-31* **N 43-2**
[14] Diehl S et al. 2011 *Conf Record IEEE Nucl. Sc. Symp., Valencia, Spain, October 23-29* **N 47-2**
[15] Novotny R W et al. 2011 *European Communit-Research Infrastructure Activity under FP7, Hadron Physics 2, Work Package* **WP 21**

16th International Conference on Positron Annihilation (ICPA-16) IOP Publishing
Journal of Physics: Conference Series **443** (2013) 012081 doi:10.1088/1742-6596/443/1/012081

# Positron beams from small accelerators and status of a novel positron storage project

**MH Weber,[1] BD Riley, CJ Baker, KG Lynn**
Center for Materials Research, Dana Hall 102
Washington State University, Pullman WA 99164-2711, USA

E-mail: m_weber@wsu.edu

**Abstract.** Positrons are generated by an intense beam of deuterons. Deuterons with energies up to 3 MeV and 300 $\mu$A intensity or 1000 W power are guided onto a carbon target. The (d,n) reaction converts $^{12}C$ into the positron emitter $^{13}N$. In one version a CVD diamond target is rotated behind a W foil moderator after irradiation. Rapid heating of the diamond and defect generation limits this method to deuteron power <100 W. At higher power $^{13}N$ molecules are released had must be condensed behind a moderator. Graphite is employed for higher power doses. External direct current or e-beam heating are used to reach temperatures in excess of 2000 K when nitrogen is released from graphite. Efforts to maximize the release of $^{13}N$ containing molecules as well as the goal to accumulate positrons in a novel trap design are discussed. Up to 75% of the activity was released at 2 MeV. This decreases with deuteron energy.

## 1. Introduction

In recent years a number of experiments have been carried out or are proposed which require very large numbers of positrons.[1] These include more ambitious positron-in-materials experiments, the formation of $Ps_2$ (Positronium) molecules,[2] Bose Einstein condensation of Ps, and the creation of anti-hydrogen atoms.[3] These are the stepping stones towards probing fundamental properties of antimatter and gravity.[4] Considerably more ambitious is the idea of using stored antimatter as a battery to deliver energy in remote locations such as outer space. At $1.8\times10^{11}$ MJ/kg (including readily available electrons) positrons are more than 10 billion times as efficient per mass than chemical fuels. Less than a nano-gram of positrons contains sufficient energy to lift 1 kg of payload beyond the gravitational pull of Earth. Three challenges require solutions: Production, storage and conversion to accelerating forces. Here, a recently proposed concept to store positrons and a positron source based on a small van de Graaff accelerator to test this idea are discussed.[5,6]

The implementation of Penning-Malmberg trap enabled on-demand delivery of $\sim10^9$ positrons in short bursts.[7] In typical cylindrical traps positrons are confined radially by a large homogeneous magnetic field of several Tesla strength. Electric repulsive potential barriers of about 1kV provide axial confinement. To retain larger numbers of positrons the required electric potential rises rapidly. Figure 1 shows an example (dashed line).

[1] Corresponding author: m_weber@wsu.edu

Almost all positron sources utilize powerful accelerators and nuclear reactors or long lived radioisotopes. To a potential user of positrons for materials research or high energy physics it is desirable to have on-demand access to positrons sufficient for several experiments without the concern of longer lived radioactive materials or the need to travel to a large facility. Potentials on the order of less than 100 V are desirable for portable traps. These could be filled at a central location and delivered to users for positron annihilation experiments.

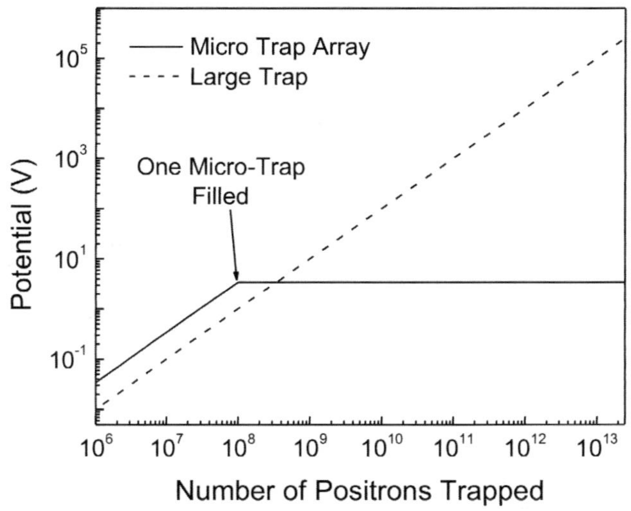


**Figure 1**: Retaining potential required to store positrons in a single trap (dashed) or an array of micro-traps (solid).

The novel concept would make portable traps practical allow for much larger numbers of stored positrons. It calls for separating the single cloud of positrons in 0.05 m diameter by 0.1 m long cylinder into N parallel cylinders. The conducting walls of each cylinder shields the repulsive fields from one cylinder to any other. The diameter of each cylinder is reduced to micrometer dimensions. The length to diameter aspect ratio is raised to about 1000 to maintain the overall length of each micro-trap. Adding more positrons requires filling more micro-traps at no added cost of increasing the confining potential (solid line in figure 1).

A "proof of principle" experimental project under way at Washington State University calls for storing $>10^{11}$ positrons in about $2\times10^4$ micro-traps of 100 $\mu$m ∅ and 0.1 m length each.[5,6] Each micro-trap will hold about $10^7$ positrons.

A number of challenges need to be met for this project.

1) The trap will be assembled from 200 50 mm ∅ wafers of silicon. Microelectromechanical systems (MEMS) technologies are used to etch 20 000 holes into each wafer. The Bosch process and cryo-etching are refined to achieve straight hole walls and homogeneous etching rates across the wafers. The wafers will be gold coated and bonded together with micrometer precision to form the array of micro-traps.[8]
2) The array must be aligned to within 1 $\mu$m deviation over its 0.1 m length inside of a 7 Tesla superconducting magnet. Vibrations must be kept to less than 10 $\mu$m.
3) Vacuum conditions must be maintained well below $10^{-10}$ torr to minimize positron molecule collisions.
4) Monoenergetic positrons will be guided onto the magnet, trapped and stored. Electric potentials are applied to segments of the trap to facilitate the stacking more and more positrons to many micro-traps in parallel. To fill the micro-trap array with $10^{11}$ positrons in an hour or less the incident positron beam must deliver more than $3\times10^7$ positrons per second, not accounting for any inefficiencies.
5) Computer simulations are ongoing to determine the highest stable densities and ideal geometry.[9]

In the following, a beam concept to deliver sufficient positrons is discussed. The nuclear reaction $^{12}C(d,n)^{13}N$ has one of the largest cross sections to produce a positron emitting isotope.[10] $^{13}N$ decays with a half-life of just under 10 min. 300 $\mu$A of 3 MeV deuterons suffice to generate a production rate of $1.2\times10^{11}$ positrons/sec. With an overall of better than $2.5\times10^{-4}$ efficiency to provide moderated positrons a filling time of 1 hour (not losses) can be reached.

## 2. Accelerator and positron production

The basics of positron production with a 3 MV deuteron accelerating van de Graaff machine have been described earlier.[11] The High Voltage Engineering (HVE) KN-3000 can deliver 0.5 to 3.0 MeV deuterons at currents from 30 to 300 $\mu$A. With a single focussing element and a pair of horizontal and vertical deflectors the beam is focussed onto a carbon target. The target is located in a 30 cm thick borated paraffin box within a 5 cm thick lead shroud to shield against neutrons and energetic gamma rays. Radiation detectors are installed to monitor the gamma rate and neutron dose at the deuteron target and at the cryogenic vacuum pump between the accelerator and the shielding igloo.

Two versions of delivering energetic positrons to a moderator have been tested. In the first case, the $^{13}$N source isotope was trapped in the diamond or graphite targets and, after a suitable irradiation time, rotated behind a W thin foil moderator. In the second version, the graphite target is heated sufficiently to release the radioisotope. $^{13}$N containing molecules can then be condensed behind the moderator. Nitrogen diffuses out of graphite at temperatures in excess of 2000 K.[12] In the case of positron emission tomography (PET) approaches to produce $^{13}$N ammonia dissolved in water or to burn graphite by adding heat and oxygen were tested.[13]

### *2.1. Rotating source*

The original concept of this positron beam called for irradiating diamond targets with the deuteron beam. Any $^{13}$N activity would be trapped in the diamond while the generated heat is transported efficiently from the diamond surface to a water-cooled copper substrate. After irradiation times of 3 half lives the target was rotated 90 degrees behind a W thin foil moderator. Figure 2 shows a beam example where the intensity is monitored as a function of angular position behind the moderator. The data were corrected for the decay of the source. This concept works well at low power as was demonstrated originally by Xie et al as well as Cassidy et al.[14, 15]

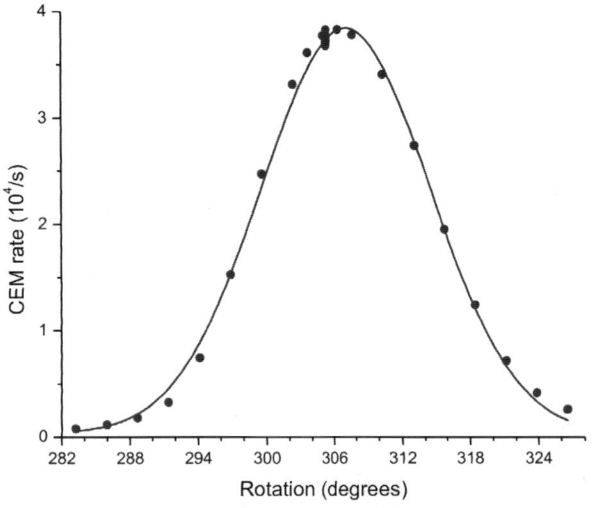


**Figure 2**: Positron beam intensity versus source rotation. At 307° the source is aligned with the moderator. The rates are corrected for source decay.

When the power of the deuteron beam was increased above approximately 100 to 200 W (50 $\mu$A at >2 MeV deuterons) and after longer operations the positron beam intensity degraded. The deuteron beam generated damage in the diamond targets to deteriorate the thermal properties of the diamond target. The target graphitized and eventually disintegrated due to thermal stresses. The vacuum in the accelerator and positron beamlines was maintained by cryopumps. The loss in positron beam intensity was accompanied by a build-up of radiation detected at the cryopumps. This was confirmed by replacing the diamond target with graphite of lower thermal conductivity. $^{13}$N containing isotopes were released at lower deuteron beam power.

### 2.2. *Thermal $^{13}N$ release and condensation*

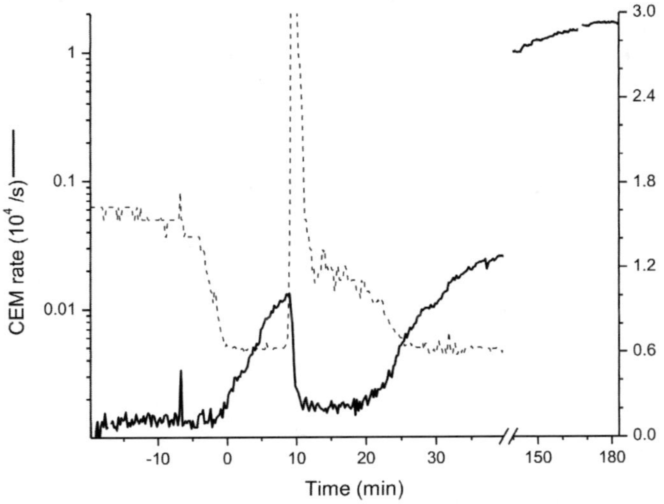


**Figure 3**: Positron intensity (solid line, left log scale) as the source cup is cooled below 20 K (near 0 min). The total pressure (dashed line, right scale) is also shown. Note the time step near 40 min.

The opposite approach was tested. The deuteron beam generates the $^{13}N$ isotope as well as provides sufficient heat to release the isotope. A copper coldfinger was installed behind the moderator to condense any gases. When the temperature drops below 20 K the positron beam intensity rises and a sudden drop in the pressure in the target chamber is observed as shown in figure 3. After about 10 minutes the temperature jumps up, the positron beam is lost before both recover. The required temperatures below 20 K suggest that a significant fraction of the $^{13}N$ molecules are $^{13}N^{14}N$.

Again, this concept shows promise at relatively low deuteron power, several issues remain to be resolved. Most importantly the thermal isolation of the deuteron target and the $^{13}N$ condenser must be improved. Second, some of the $^{13}N$ inventory never is released from the target or may condense elsewhere at higher temperature. In an environment rich in deuterium, carbon and residual water from the vacuum system $^{13}N$ tagged compounds such as $NH_3$, HCN or deuterated versions are possible.

## 3. Graphite target heating

The deuteron target was separated from the positron beam to evaluate and optimize the release of $^{13}N$ compounds. The heating of the target was separated from accelerator power heating by installing either direct current heating or electron gun heating. In the former thin graphite targets were heated with up to 250 A passing through the target. About 1000 W were provided into the target and target holder system. The electron beam system was operated with 9 keV electrons at up to 140 mA onto 0.025 m square graphite targets. In a separate test 110 mA at 7 keV (770 W) heated the graphite to 2000 K as measured with an optical pyrometer.

A residual gas analyzer (RGA) was used to observe the partial pressures to determine the released gasses. As expected, nitrogen (masses 14 and 28) was released at a power consistent with the 2000 K threshold. Other dominant gases were CO and argon as well as water, hydrogen and deuterium. Over time the partial pressure of argon and water dropped significantly. Of these, argon was not expected from graphite. However, the graphite sample was not selected for purity.

Target heating was carried out either together with deuteron bombardment or separately after the deuteron beam was shut down. The former combined the heat sources of the deuteron beam with the secondary heater and the latter permitted more careful monitoring of the $^{13}N$ isotope release from the target. Radiation monitors were installed in the paraffin and lead box and at the cryopump between the accelerator and the target. When the deuteron beam is shut down, almost all radiation is due to $^{13}N$ decays. About 1% of the initial dose is due to $^{56}Mn$ with a 2.58 hr half live. Figure 4(a) shows the sudden drop in activity on the target along with a rise of the radiation dose at the cryopump when the heater is turned on. The deuteron beam is off during this measurement and the data were corrected for the 10 min half live of $^{13}N$. In this case about 35% of the $^{13}N$ inventory was released after 3 MeV deuteron bombardment. Figure 4(b) shows the fraction of released radiation as a function of deuteron beam energy or implantation depth. Most tests were conducted at 2 MeV when the heater power was gradually increased. In both cases, by combined heating or after shutting down the deuteron beam the maximum released fraction was 75% to date. Estimates of the total positron flux generated are consistent with thick target yields for the $^{12}C(d,n)^{13}N$ reaction.

It appears that the $^{13}N$ containing molecules are released more rapidly by the direct current heating compared to e-beam heating from the opposite side of the target. The rise in radiation release precedes the observation of nitrogen (masses 14 and 28 on the RGA). In the initial experiments the accelerator deuteron beam power caused sufficient heating to release some of the $^{13}N$ inventory above a threshold power level. However, during the most recent experiments at even higher deuteron power this no longer occurs. The causes for these effects are under investigation.

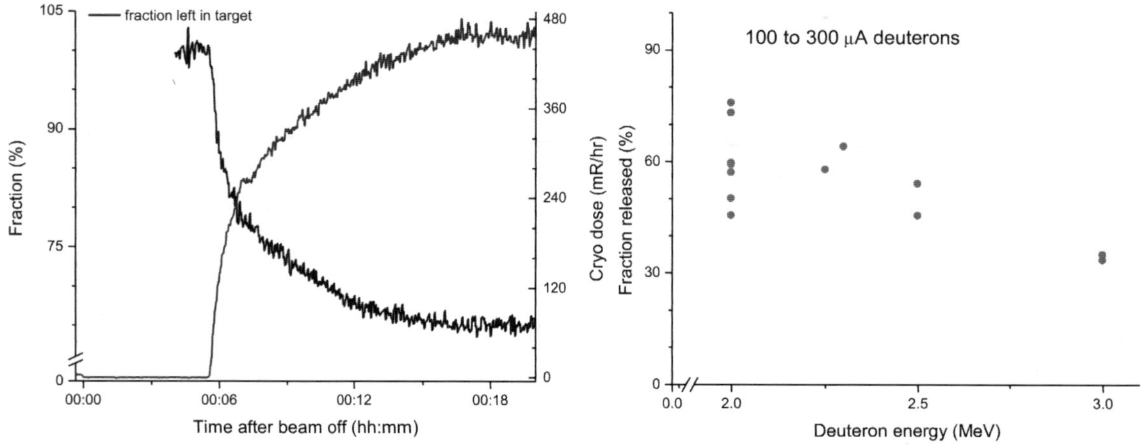


**Figure 4**: a) Source decay corrected fraction of $^{13}N$ radiation in the graphite target (left scale) and dose in the cryopump (right scale). 2 MeV deuterons were used. b) Fraction of radiation released from the target as a function of deuteron beam energy (red dots).

## Conclusions

A small 3 MV van de Graaff accelerator was commissioned to generate the $^{13}N$ positrons source from bombarding graphite with deuterons. The graphite must be heated to about 2000C completely release the isotope which is then frozen near a thin film W moderator. An initial beam was observed, however, the compatibility of the high temperature generation of $^{13}N$ and freezing the isotope containing compound close to the moderator requires further improvements. Modifications are under way. Computer simulations indicate that space charge repulsion can be reduced dramatically by replacing a single large trap with an array of long but $\mu$m diameter tubes. A test bed is under construction.

## Acknowledgments

The authors appreciate the tremendous amount of work on this project by Lloyd Pilant, Alan Hunt, M Shurtliff and A Lynn on the positron beam. A Verma, J Jennings, CJ Baker, and A Narimannezhad are contributing on the positron storage project. Financial support for the overall project is acknowledged from the WM Keck Foundation for the accelerator and the Army Research Laboratory under contract no. W9113M-09-0075. The Office of Naval Research supports the computer modelling efforts.

## References

[1] Mills A P Jr 2011 *Rivista del Nuovo Cimento* **34** 151
[2] Cassidy D B and Mills A P Jr 2007 *Nature* **449** 195
[3] Amoretti M, Amsler C, Bonomi G, *et al Nature* **419** 456
[4] Holzscheiter M H, Charlton M and Nieto M M 2004 *Physics Reports* **402** 1
[5] Baker C J, Jennings J, Verma A, Xu J, Weber M H and Lynn K G 2012 *Eur. Phys. J.* D **66** 109
[6] Xu J, Weber M H and Lynn K G 2010 *CLXXIV Enrico Fermi School on Physics with Many Positrons* ed A Dupasquier A P Mills jr R S Brusa Societa Italiana de Fisica 481
[7] Murphy T J and Surko C M 1992 *Phys. Rev.* A **46** 5696
[8] Verma A, Jenning J, Johnson R D, Weber M H and Lynn K G 2012 *J. Vac. Sci. Technol.* B **31** 32001
[9] Narimannezhad A, Xu J, Baker C J, Weber M H and Lynn K G 2012 to be published
[10] Young F C and Rose D V 1996 *Atomic Data Nucl. Data Tables* **64** 223
[11] Hunt A W, Pilant L, Cassidy D B, Tjossem R, Shurtliff M, Weber M H and Lynn K G 2002 *Appl. Surf. Sci.* **194** 296
[12] Kohl W H 1960 *Materials and Techniques for Electron Tubes* Reinhold Pub. Corp. New York
[13] Dence C S, Welch M J, Hughey B J, Shefer R E and Klinkowstein R E 1994 *Nucl. Med. Biol.* **21** 987
[14] Xie R, Petkov M, Becker D, Canter K F, Jacobsen F M, Lynn K G, Mills R and Roellig L O 1994 *Nucl. Instrum. Meth. Phys. Res.* B **93** 98
[15] Cassidy D B, Canter K F, Shefer R E, Klinkowstein R E and Hughey B J 2002 *Nucl. Insrum. Meth. Phys. Res.* B **195** 442

Journal of Physics: Conference Series **443** (2013) 012082 doi:10.1088/1742-6596/443/1/012082

# New experiment stations at KEK Slow Positron Facility

**K Wada[1], T Hyodo[1], T Kosuge[1], Y Saito[1], M Ikeda[2], S Ohsawa[2], T Shidara[2], K Michishio[3], T Tachibana[3], H Terabe[3], R H Suzuki[3], Y Nagashima[3], Y Fukaya[4], M Maekawa[4], I Mochizuki[4] and A Kawasuso[4]**

[1] Institute of Materials Structure Science, High Energy Accelerator Research Organization (KEK), 1-1 Oho, Tsukuba, Ibaraki 305-0801, Japan

[2] Accelerator Laboratory, High Energy Accelerator Research Organization (KEK), 1-1 Oho, Tsukuba, Ibaraki 305-0801, Japan

[3] Department of Physics, Tokyo University of Science, 1-3 Kagurazaka, Shinjuku, Tokyo 162-8601, Japan

[4] Advanced Science Research Center, Japan Atomic Energy Agency, 1233 Watanuki, Takasaki, Gunma 370-1292, Japan

E-mail: `ken.wada@kek.jp`

**Abstract.** Recent development of the Slow Positron Facility at the Institute of Materials Structure Science, High Energy Accelerator Research Organization (KEK) is reported. The facility, equipped with a dedicated 55 MeV linac, provides a high-intensity, pulsed slow-positron beam. The beam is produced in a production unit at a high tension of up to 35 kV and guided magnetically through a grounded beam line, and then branched using compact branching units in the experiment hall. An overview, some details of three experiments currently conducted and the outlook of the facility are described.

## 1. Introduction

This article reports recent development of the Slow Positron Facility (SPF) at the Institute of Materials Structure Science (IMSS), High Energy Accelerator Research Organization (KEK). This facility provides a slow-positron beam produced by using a dedicated linac. First, an overview and the present status of the facility is given. Next, three experiment stations in the facility, namely, those for a Ps negative ion ($Ps^-$), a reflection high-energy positron diffraction (RHEPD), and positronium time-of-flight (Ps-TOF) are reported. Then the outlook of the facility is described.

## 2. Overview

The KEK-IMSS-SPF provides a high-intensity, pulsed slow-positron beam produced by using a dedicated linac. The electron-beam energy of the linac is 55 MeV with the operation power of 600 W at the maximum. The maximum repetition rate is 50 Hz. It operates in two pulse modes: the long pulse mode of $1\,\mu s$ pulse width, and the short pulse mode of variable 1–10 ns pulse width. In October 2010, a new positron converter and moderator assembly was introduced, yielding an increase of an order of magnitude in the intensity of the beam [1]. The long pulse

Journal of Physics: Conference Series **443** (2013) 012082 doi:10.1088/1742-6596/443/1/012082

mode provides $5 \times 10^7$ slow-$e^+$/s, and the short pulse mode provides $5 \times 10^6$ slow-$e^+$/s. The initial beam diameter is about 20 mm.

The experiment hall of the facility consists of two floors. The dedicated linac and the positron converter and moderator assembly is on the basement floor. Currently, there are four beam-line branches (figure 1): two of them (SPF-A1, SPF-A3) lies on the basement floor, and one branch (SPF-A2) goes up to the deck floor and comes down vertically, and the last one (SPF-B1) goes up on the ground floor.

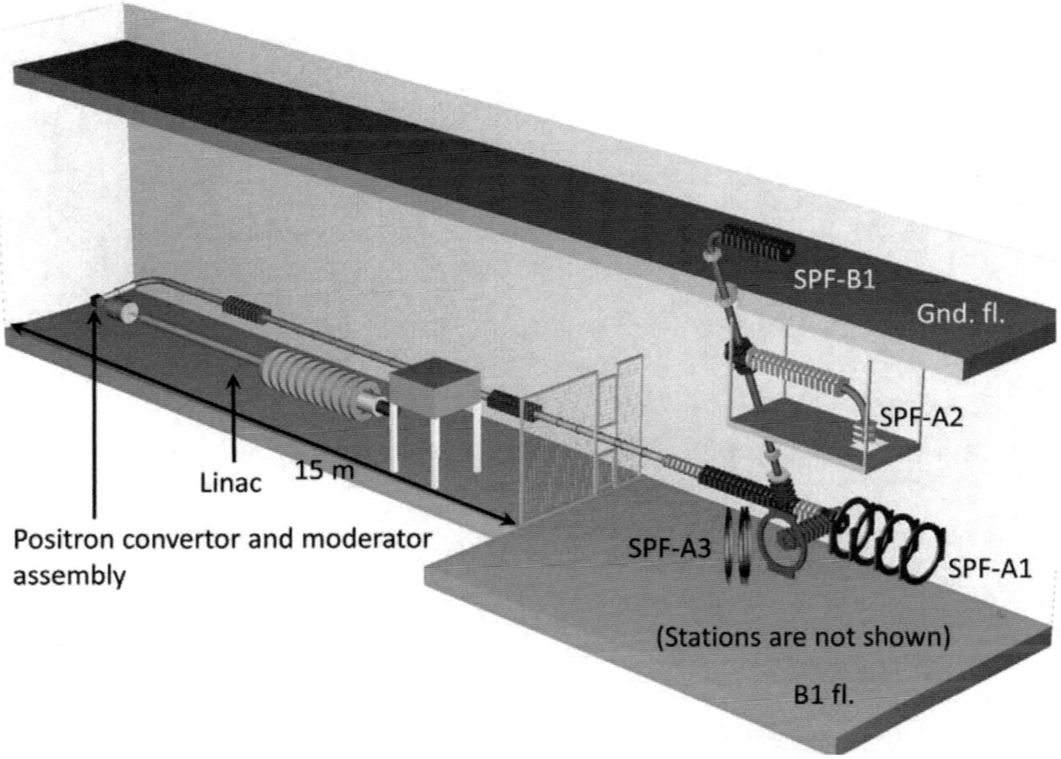


**Figure 1.** A schematic view of the KEK-IMSS-SPF beam-line.

The positron converter and moderator assembly is at high tension of up to 35 kV with reference to the grounded potential slow-positron beam line, and the tension is adjusted in accordance with the requirements of individual experiments. Standardized beam-line-branching units for the transportation energy of up to 35 keV allow flexible rearrangements of the branches.

## 3. Present experiment stations

### *3.1. Ps negative ion experiment station*

A method to efficiently produce positronium negative ion ($Ps^-$), an exotic system composed of one positron and two electrons bound through Coulomb interaction, has been developed by Nagashima et al. [2, 3]. The photodetachment of the $Ps^-$ ions produced by using the pulsed positron beam in this facility was performed [4] at the branch SPF-A1. The incident positrons hit the target of 25 $\mu$m W film coated with one ML of Na, and $Ps^-$ ions were formed on the surface. The emitted $Ps^-$ ions are then accelerated by an electrostatic field. Immediately after that, the photodetachment was accomplished by irradiation of 25 Hz pulsed (width 12 ns) Q-switched Nd:YAG laser. Since the slow positron beam is provided in 50 Hz, the data with and without the laser irradiation were obtained simultaneously. The Doppler shifted annihilation $\gamma$-rays from the accelerated $Ps^-$ detected by two Germanium detectors verified the formation of the $Ps^-$ ions and the reduction of them the photodetachment into Ps atoms.

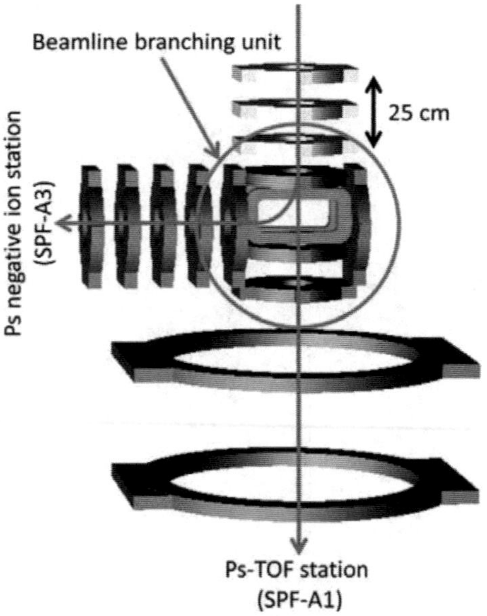


**Figure 2.** The beam-line-branching unit (within the blue circle) standardized for the transportation of the positrons with energy of up to 35-keV. Only the configuration of the magnetic coils are shown.

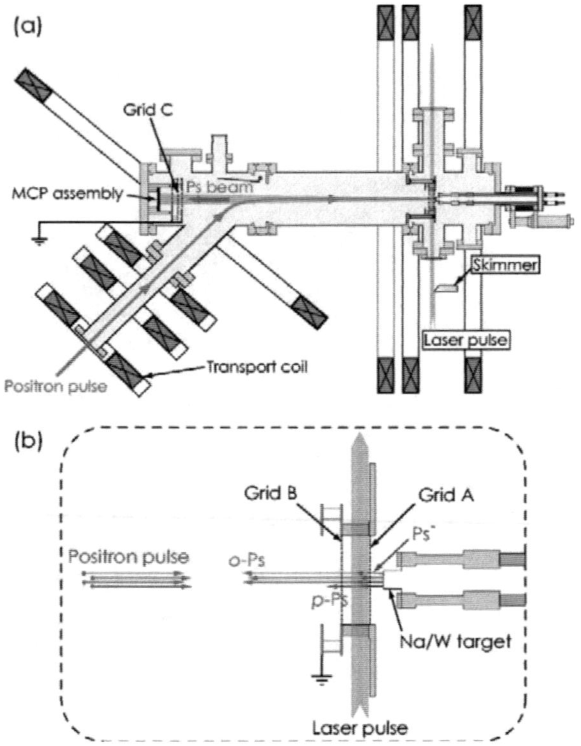


**Figure 3.** Schematic of the energy-tunable-Ps-beam production and detection system: (a) a whole view; and (b) details of the $Ps^-$ source and its photodetachment region (not to scale).

After developing the above technique, we built a new beam-line branch, SPF-A3, and an energy-tunable Ps beam was successfully produced in ultra-high-vacuum environment and detected [5]. The energy of the Ps, which is a neutral atom and hence cannot be accelerated by a static electric field directly, tuned accelerating the $Ps^-$ ions before the photodetachment.

Figure 2 shows the branching unit together with the coils around it (the beam-line duct is not shown). The branching unit is standardized and compatible with the positron energy up to 35 keV. The positrons are guided into selected direction by switching relevant Helmholtz coils. A pair of steering coils (shown in red in Figure 2) is used to suppress the perpendicular drift of the curving positron-beam trajectory.

Figure 3a shows the station at the end of the new branch. The slow-positron beam of 4.2 keV was guided by 45° along a curved magnetic field and led to the target. The $Ps^-$ ions emitted from the target were accelerated by an electrostatic potential of 0.3-2.3 kV, and then photodetached by the pulsed laser (Figure 3b). The resulting ortho-Ps atoms traveled in the direction of the acceleration of the $Ps^-$ ions, and were detected by a Microchannel Plate (MCP) with a pulse-counting mode, placed at 80cm away from the target. All the charged particles back-scattered from the target were transported back along the curved magnetic field and not detected by the MCP. Some of the $\gamma$-rays from positron annihilations in the target were detected

Journal of Physics: Conference Series **443** (2013) 012082 doi:10.1088/1742-6596/443/1/012082

by the MCP and gave information on the time of the Ps formation. The time interval of the detections of this $\gamma$-ray and the Ps clearly showed that the Ps flew with the energy expected from the electrostatic acceleration of the $Ps^-$ ions [5]. Production of an energy-tunable Ps beam from 300 eV to 1.9 keV in an ultra-high-vacuum environment was achieved.

### *3.2. Reflection high-energy positron diffraction experiment station*

Reflection high-energy positron diffraction (RHEPD) is the positron version of the reflection high-energy electron diffraction (RHEED). Since the crystal potential for the positron is positive, positrons incident at a small glancing angle are totally reflected, which is not the case for the electrons. Thus the positron diffraction is a very sensitive method for the topmost surfaces [6]. RHEPD was proposed by A. Ichimiya [7] and was brought into practical use by Kawasuso and Okada [8] with a $^{22}$Na-based slow-positron beam at Japan Atomic Energy Agency (JAEA).

In 2010, a RHEPD station was installed at the end of the beam-line branch SPF-B1 in KEK-IMSS-SPF [1]. Magnetically transported positrons with an energy of 10 keV were introduced into the magnetic-field-free region, which were isolated from the magnetic field of the transportation coils by a 3 mm thick iron plate. About a half of the transported positrons went through a magnetic lens placed in the magnetic-field-free region, a $\phi 10$ mm aperture, a set of Einzel lens, and a $\phi 3\,\mathrm{mm} \times 30\,\mathrm{mm}$ collimator, then reached the sample. The observed specular intensity was 14 times intensified [1] over the former measurements with the $^{22}$Na-based beam at JAEA.

One of the results with the installed RHEPD station is the determination of the temperature-dependent atomic configuration of well-ordered and defect-less Pt-induced nanowires on the Ge(001) surface [9]. This research was performed along with scanning tunneling microscopy (STM) and angle-resolved photoemission spectroscopy (ARPES).

The other results is the experimental evidence for the correlation of the spin splitting and the outward displacement of the Pb atoms in $Ag_2Pb$ alloy on Ag(111)/Si(111) surface [10], which shows giant Rashba effect. The shift of the position of the Pb atom for each Ag thickness was determined by using RHEPD rocking curves (specular intensity as a function of the glancing angle of the incident positrons). Then the electronic states of $Ag_2Pb$ alloy on the Ag(111)/Si(111) surface were investigated by using angle resolved photoemission spectroscopy (ARPES), while controlling the position of the Pb atom by changing the Ag(111) thickness. These results confirmed that the magnitude of the Rashba energy depends on the outward displacement of the Pb atoms as theoretically predicted [11]. They also suggest that the Rashba energy can be controlled by changing the substrate Ag thickness.

After the above experiments, a brightness-enhancement unit was installed at the magnetic-field-free region. Details of the unit will be reported elsewhere [12]. The unit consists of a magnetic lens, transmission-type remoderator of 100-nm-thick W(111) crystal, and several extraction electrodes. The W crystal was annealed *in situ* by the passage of an electric current. The incident positrons with a beam energy of 15 keV were focused by a magnetic lens on the W crystal at an applied voltage of 10 kV. The diameter of the beam spot focused on the crystal is 3 mm at FWHM. The remoderated beam was transported through a set of Einzel lens to the sample, where the beam diameter is less than 1 mm. The energy spread of the 10 keV beam was improved from 110 eV at FWHM without the remoderation to 8 eV with it. The observed diffraction patterns from the Si(111)-7×7 surface with the remoderated beam shows the fractional index spots due to the 7×7 super structure, which were not clearly identified with the former [12].

The obtained RHEPD rocking curve with remoderated beam yielded 45 times brighter than that with $^{22}$Na-based slow-positron beam (Figure 4).

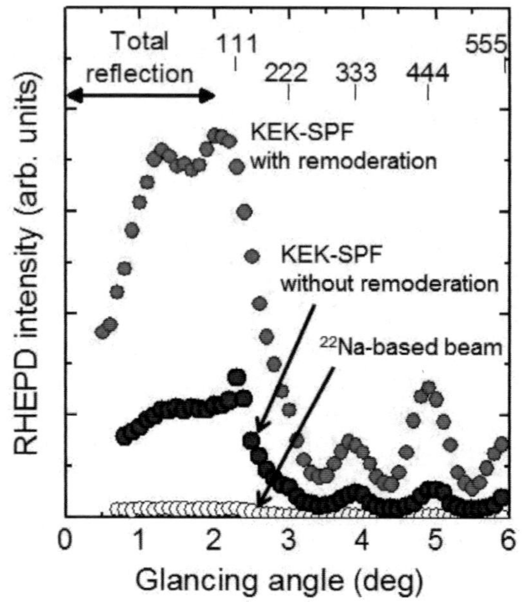


**Figure 4.** RHEPD rocking curves from a Si(111)-7×7 surface observed with 10 keV positron beams. The data obtained by the linac-based beams with and without remoderation at KEK-IMSS-SPF, and that by a $^{22}$Na-based beam are presented.

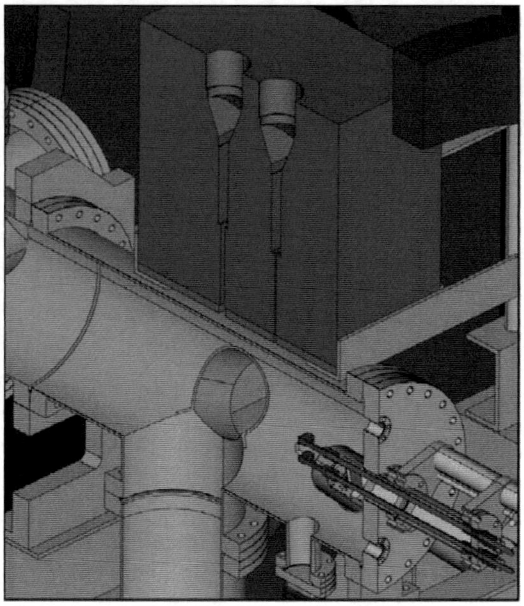

**Figure 5.** A cross section of the revised positronium time-of-flight detection system. The annhilation $\gamma$-rays are viewed by detectors through two sets of lead slits.

*3.3. Positronium time-of-flight experiment station*

The Ps time-of-flight experiment station was revised. More compact detection system gives a better time resolution: the revised detection system gives a time resolution of 14 ns, while that of the old one was 23 ns. The detailed configuration of the apparatus is reported elsewhere [14].

The incident positrons hit the target sample and the ortho-Ps atoms emitted from the surface annihilate in flight. The annihilation $\gamma$-rays were viewed through two sets of lead slits by detectors and the time of the detection is recorded. One pair of the lead slits, of 2 mm width, is located at 40 mm away from the target and the other, of 6 mm width, at 120 mm. The detector consists of set of plastic scintillators of $100 \times 100 \times 10\,\mathrm{mm}^3$, light guide, and mesh-type photomultiplier tubes. The prompt peak in the TOF spectrum caused by the $\gamma$-rays from the annihilations in the sample gives information of the time zero. With this revised system, an enhancement of the emission efficiency of the Ps was observed for Na-coated W-film [14].

## 4. Outlook

The RHEPD experiments with remoderated beam are starting from Oct. 2012. The low-energy positron diffraction (LEPD), which is the positron version of low-energy electron diffraction (LEED), will be installed at the end of the beam-line branch SPF-A2 in the year 2013.

A DC-beam section, which convert the pulsed slow-positron beam into the DC-beam using a penning trap with a pulsed voltage supply, will be installed next year. Then a coincidence-Doppler-experiment station will be arranged.

The user beam time in this facility is about 180-days in the years 2010 and 2011. Currently, the facility provide user beam time for two experiments, Ps$^-$ and RHEPD. From October 2012, it is going to provide user beam times for five experiments, two Ps-TOF experiments and one positron-impact-induced ion-desorption experiment, as well as the continuing two experiments.

About 100 current sources for about 200 sets of magnetic coils for the slow-positron transportation must be optimized for each transportation energy and for each beam-line branches. For accommodating increasing needs for the readjustments of the currents, a client-server control system is being installed. With this system, current sources are controlled remotely from a client PC through the server and the programmable logic controller (PLC). About 30 current sources are already under control of the system. The system will be completed in October 2012. Then sets of optimized values of all the magnetic-coil current for one experiment will be switched smoothly to the next.

Any researcher can use the KEK-IMSS-SPF beam line through approval by the proposal approval committee in Photon Factory, KEK. The deadline of the application for experiments for a period from October to March is the first Friday in May of the same year, and for those for a period from May to July is the first Friday in November of the year before. Further information on the application for using the beam line will be provided by the corresponding author in response to inquiries.

## 5. Summary

The KEK-IMSS-SPF provides the slow-positron beam produced by using a dedicated linac. It is operated in a long-pulse mode providing $5 \times 10^7$ slow-$e^+$/s and a short-pulse mode providing $5 \times 10^6$ slow-$e^+$/s in 50 Hz at maximum. The positron converter and moderator assembly is at high tension of up to 35 kV and the entire slow-positron beam line is at ground potential, which allows one to perform an experiment just as is done in common laboratories. Beam-line branches are flexibly rearranged with a standardized branching unit compatible with a transportation energy up to 35 keV.

Production of an energy-tunable Ps beam with an energy variable from 300 eV to 1.9 keV in an ultra-high-vacuum environment was achieved. Recent results with the RHEPD station include temperature transition of Pt-induced nanowires on the Ge(001) surface and the correlation of the spin splitting and the outward displacement of the Pb atoms in $Ag_2Pb$ alloy on Ag(111)/Si(111) surface.

With a brightness enhancement unit, the RHEPD specular intensity was significantly intensified and the resolution of the rocking curve was improved. The Ps-TOF station was revised, yielding a better time-resolution.

## References

[1] K Wada, T Hyodo, A Yagishita, M Ikeda, S Ohsawa, T Shidara, K Michishio, T Tachibana, Y Nagashima, Y Fukaya, M Maekawa and A Kawasuso, Eur. Phys. J. D **66**, 37 (2012).
[2] Y Nagashima, T Hakodate, A Miyamoto and K Michishio, New J. Phys. **10**, 123029 (2008).
[3] H Terabe, K Michishio, T Tachibana and Y Nagashima, New J. Phys. **14**, 015003 (2012).
[4] K Michishio, T Tachibana, H Terabe, A Igarashi, K Wada, T Kuga, A Yagishita, T Hyodo and Y Nagashima, Phys. Rev. Lett., **106**, 153401 (2011).
[5] K Michishio, T Tachibana, R H Suzuki, K Wada, A Yagishita, T Hyodo and Y Nagashima, Appl. Phys. Lett., **100**, 254102 (2012).
[6] S Y Tong, Surf. Sci. **457**, L432 (2000).
[7] A Ichimiya, Solid State Phenom., **28/29**, 143 (1992/93).
[8] A Kawasuso and S Okada, Phys. Rev. Lett., **81**, 2695 (1998).
[9] I Mochizuki, Y Fukaya, A Kawasuso, K Yaji, A Harasawa, I Matsuda, K Wada and T Hyodo, Phys. Rev. B, **85**, 245438 (2012).
[10] Y Fukaya, M Maekawa, I Mochizuki, A Kawasuso, I Matsuda, M Ogawa, R Yukawa, K Wada and T Hyodo, presented at ICPA-16, also submitted to Phys. Rev. Lett.
[11] G Bihlmayer, S Blugel and E V Chulkov, Phys. Rev. B, **75**, 195414 (2007).
[12] M Maekawa, et al, in preparation.
[13] K Ito, R Yu, K Sato, K Hirata and Y Kobayashi, J. Appl. Phys., **98**, 094307 (2005).
[14] H Terabe, S Iida, K Wada, T Hyodo, A Yagishita and Y Nagashima, submitted to J. phys.: Conf. Ser.

Journal of Physics: Conference Series **443** (2013) 012083 doi:10.1088/1742-6596/443/1/012083

# The Bristol HIDAC 2D-ACAR Spectrometer

**S.B. Dugdale, J. Laverock, C. Utfeld, M.A. Alam, T.D. Haynes, D. Billington and D. Ernsting**

H.H. Wills Physics Laboratory, University of Bristol, Tyndall Avenue, Bristol BS8 1TL, UK

E-mail: `s.b.dugdale@bristol.ac.uk`

**Abstract.** We describe the Bristol spectrometer for positron annihilation Fermi surface studies utilising high-density avalanche chambers (HIDACs) as position sensitive detectors. Measurements on $\alpha$-$SiO_2$ show, through the momentum distribution of para-positronium, the substantial improvement in resolution compared to Anger cameras. Measurements of the Fermi surface of V are used to determine the resolution. The new spectrometer is found to have an efficiency of $12.5 \pm 0.6$ % and a (coincidence) contribution to the position resolution of $0.96 \pm 0.1$ mm.

## Introduction

Together with the closely related technique of Compton scattering [1, 2, 3], positron annihilation has established itself as a powerful probe of the Fermi surface (FS) of a variety of materials [4]. With its ability to deliver a direct image of the occupied momentum states (through the Lock-Crisp-West (LCW) folding of the measured momentum distribution back into the first Brillouin zone [5]), and hence the FS, two-dimensional angular correlation of positron annihilation radiation (2D-ACAR) experiments have been able to reveal Fermi surfaces whose particular shapes are thought to be responsible for a wide range of phenomena from the magnetic ordering of $f$-electron magnetic moments in Gd-Y alloys [6] and ferromagnetic shape-memory alloys such as $Ni_2MnGa$ [7], through weak ferromagnets close to quantum critical points [8, 9] to borocarbide superconductors with nested FS sheets [10, 11]. The unique properties of the positron have even been exploited to uncover the FS of nanoscale precipitates in Al-Li alloys [12].

## Position Sensitive Detectors

The 2D-ACAR technique utilises the positron as a probe and relies on the coincidence measurement of the two annihilation photons' positions [4]. Thus the angular deviation, which is directly related to the electron momentum, is resolved and the momentum density can be mapped out. Experimentally, the spectrometer assembly comprises two position sensitive detectors (PSDs) (almost) symmetrically positioned with respect to the source/sample chamber. The four main objectives a detector system for 2D-ACAR needs to meet are good spatial resolution (if instrument resolution function deconvolution is to be avoided [13, 14]), high detection efficiency for $\sim 511\,\text{keV}$ photons, sufficient resolving time for the coincidence measurement and long term stability, as normal acquisition times are of the order of weeks. These technological requirements are, however, hard to meet simultaneously and a trade-off between them usually forces compromise.

Journal of Physics: Conference Series **443** (2013) 012083 doi:10.1088/1742-6596/443/1/012083

Historically, different detector technologies have been used, but three types have emerged. The first 2D-ACAR machine built by Berko [15] utilised multi-element scintillation counters and such an approach has been followed recently by others [16]. The second type of PSDs are so-called Anger Cameras, which have been adapted from medical imaging and use a Tl-ativated NaI crystal optically coupled to a hexagonal closed-packed array of photomutiplier tubes [17]. Pioneered by West, the first spectrometer was built in the late 1970s [18], and a similar setup was in operation in Bristol for nearly two decades. Anger cameras offer modest resolution ($< 4$ mm, although some recent approaches should improve on this [19]), exceptional electronic stability, good resolving times of the order of nanoseconds ($\sim 50$ ns) and coincidence efficiencies of $\sim 10\,\%$. Amongst all PSDs the last type, namely the High Density Proportional Chambers (HDPC), stands out due to its excellent resolution. The concept of HDPCs being a Multi-Wire Proportional Chamber (MWPC) embedded in Pb photon-electron converters is due to Jeavons [20, 21, 22]. However, although the resolution of 1.0 mm was exceptional, the weak point resided in relatively long coincidence resolving times ($> 100$ ns) causing appreciable background contributions due to accidental coincidences. Moreover, the comparatively low coincidence efficiencies ($\sim 8\,\%$) demanded more technological developments in order to compete with the Anger Camera [18, 22, 23].

In recent years the HDPC has mostly been applied to the medical imaging technique of Positron Emission Tomography (PET), which stimulated further developments and culminated in various technological improvements tackling its efficiency [24]. In particular the newest generation of the so-called HIDAC (where the acronym stands for HIgh Density Avalanche Chamber) provides sub-millimetre position resolution combined with nanosecond coincidence timing and a high efficiency of $\sim 11\,\%$ [24, 25]. Consequently, such a system is ideally suited for application in a 2D-ACAR spectrometer, since it does not only result in shorter acquisition times than the older versions, but also permits FS investigations with unprecedented precision, and will allow systems with small Brillouin zones (e.g. complex systems with large unit cells such as $Lu_5Ir_4Si_{10}$ [26]) to be studied.

**Figure 1.** The current HIDAC design as utilised in the Bristol spectrometer includes a stack of seven modules with an active area of $290 \times 280\,\mathrm{mm}^2$.

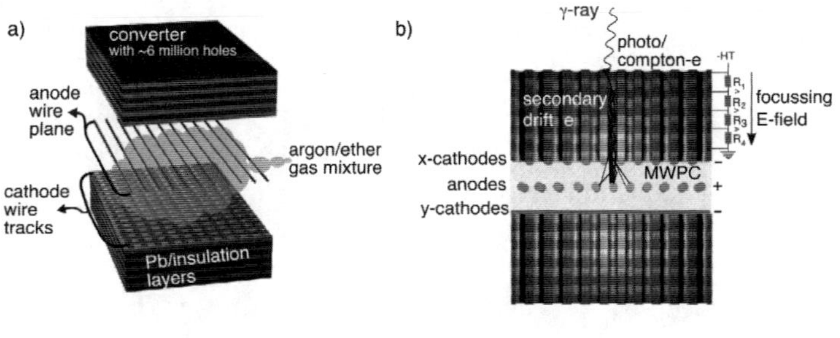


**Figure 2.** (a) One HIDAC module consists of a gas-filled MWPC with electron-photon converters either side. The converters constitute alternating layers of Pb-Sb alloy and insulation, drilled with a matrix of holes. (b) The HIDACs operate by incident photons releasing electrons which enter an adjacent hole and avalanche towards the MWPC in which they are detected. The position information is then extracted via the printed cathode tracks.

MWPCs possess a dense 2D wire grid, where each wire acts as a proportional chamber resolving the position of the incident particle [27]. The use of gas-filled MWPC for $\gamma$-rays has been proven useful up to energies of 8 keV, above which the photoelectric capture cross-section in the gas becomes strongly suppressed and the detection efficiency drops dramatically. The logical consequence of such behaviour led to the introduction of a denser medium in order to increase the capture efficiency. These so-called converters serve to convert the non-(gas-)ionising photons to ionising particles which can subsequently be detected by the MWPC. The (2.5 mm) converters comprise 12 stacked sheets of a Pb-Sb alloy (50 $\mu$m) separated by fibre-glass reinforced epoxy-resin interlayers (140 $\mu$m). A hole-matrix with 0.4 mm diameter and 0.5 mm pitch is drilled into the converter, which would ideally deliver sub-millimetre intrinsic spatial resolution while providing a greater surface area and thus higher efficiency. Fig. 2a depicts a HIDAC module (comprising MWPC and converters). Photons interact with the Pb-Sb alloy and liberate fast electrons via the photoelectric (or Compton) effect. The free electrons (Fig. 2b) then escape via an adjacent hole, where in order to guarantee efficient extraction an additional electric field focuses the photo-electrons onto the centre of the hole and the gradient accelerates them resulting in an avalanche effect from gas ionisation as they travel to the MWPC. More details can be found in Ref. [28].

**The Bristol HIDAC spectrometer**

The HIDAC system utilised in the Bristol 2D-ACAR spectrometer has been developed, constructed and installed by Oxford Positron Systems and the design follows broadly their most recent 3D HIDAC-PET [24, 25]. In the current 2D-ACAR version, one detector consists of seven stacked individual HIDAC modules. Each module constitutes an individual unit including its own MWPC and set of two solid converters. The sensitive area is rectangular in shape and the dimensions of (290 x 280) $\text{mm}^2$ give an acceptance angle of $\pm 14$ mrad at a sample-detector distance of 10 m. In order to minimise any resolution degradation due to small misalignments of individual modules relative to each other within one HIDAC detector and relative to the other detector, an iterative post-processing procedure is applied to the list-mode data.

**Spectrometer Resolution**

The characteristic spectrum of $\alpha$-$SiO_2$, featuring thermalised para-positronium (p-Ps) peaks at $\mathbf{p} = 0$ and 'Umklapp' (higher-momentum components) at the reciprocal lattice points, provides the most obvious way to compare resolution functions, since these peaks are assumed to be $\delta$-functions convoluted with the experimental resolution. Fig. 3a shows such spectra measured with the new HIDAC spectrometer at low temperature $(20 - 35\,\text{K})$ and a sample-detector distance of 10 m. For comparison, Fig. 3b shows the equivalent spectrum measured in the old Anger Camera system at similar temperature and a sample-detector distance of 12 m. The difference in the resolution underlying these two spectra is clearly evident in the respective widths of the $p$-Ps peaks in the $p_x$ direction (where the only additional contribution to the intrinsic detector resolution is due to the temperature broadening). The substantial improvement in resolution is also clearly evident in the ratio of the central peak height to the broad background coming from electron-positron (rather than p-Ps) annihilation.

When the intrinsic detector resolution only provides a small contribution to the total angular resolution, this kind of measurement is governed by various difficulties. For example, at the level of resolution expected for the new HIDAC system, many-body effects may start to broaden the p-Ps peaks (e.g. [29]). In order to gain a more accurate picture of the coincidence resolution an alternative method based on the momentum distribution of a metallic system was employed. Here a well studied and simple metal with strong FS features that are well described by first-principles calculations serves as a reference measurement for the new, high-resolution 2D-ACAR apparatus. Such an approach then allows the resolution to be modelled by electronic structure calculations. The ideal candidate for such a reference measurement is the [100]-projection 2D-ACAR spectrum of V. The projected FS along this direction is dominated by the strong features due to the projected $N$-hole pockets [30, 31], and the electronic structure is well described by theoretical predictions from first-principles calculations [32, 33].

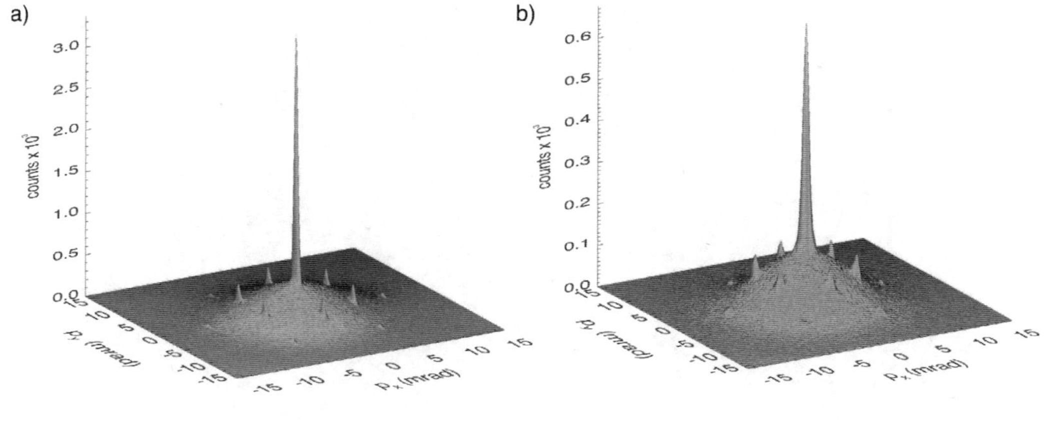


**Figure 3.** $SiO_2$ spectra measured with the (a) HIDAC and (b) Anger Camera spectrometers.

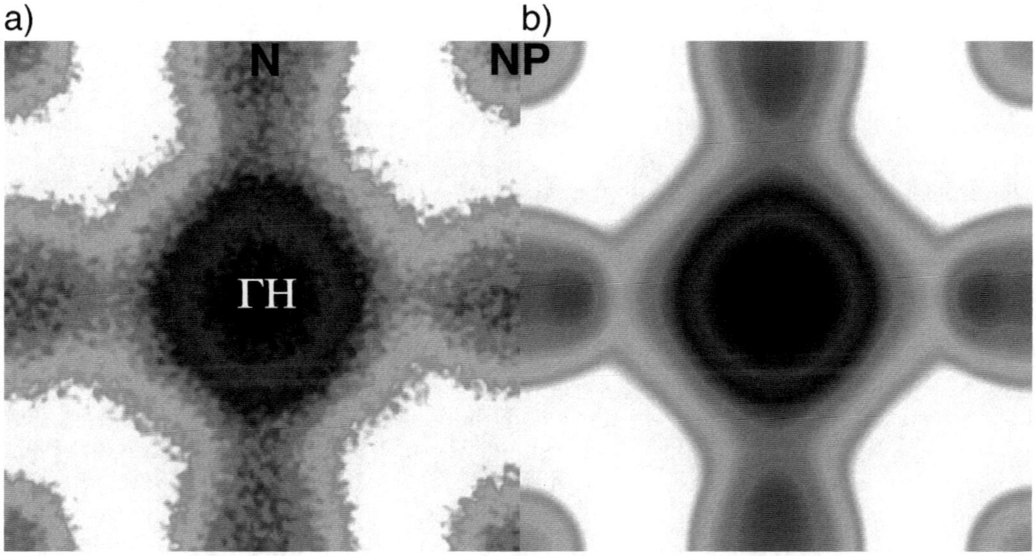


**Figure 4.** (a) The LCW of V ([100] projection) as measured on the HIDAC spectrometer (b) The convoluted results of the optimised LMTO calculation (Ref. [33]).

The high-statistics data (300 million counts) were processed as described in Ref.[28] and histogrammed on a (1024 × 1024) mesh with a calibration of 0.05 mrad/channel. The detectors were located at an average distance of 8 m, with one detector offset by $\Delta z = 0.3$ m with respect to the other which ensures a smooth, flat-topped momentum sampling function [28]. The theoretical result convoluted with the fitted resolution is compared to the experimental data in Fig. 4. The FWHM of the fitted resolution function corresponds to a resolution of $\Gamma_x = 1.25 \pm 0.05$ mm and $\Gamma_y = 8.10 \pm 0.08$ mm, where the $y$-resolution is completely dominated by the finite source spot on the sample. Finally, $x$-resolution can be decomposed into a temperature-dependent part and the intrinsic detector coincidence resolution. The temperature contribution at $T = 4$ K, assuming an effective mass of $m^* \sim 1m_e$, is $\Gamma_{e^+} = 0.8$ mm. Thus the intrinsic *coincidence* contribution to the detector resolution is $\Gamma_c = 0.96 \pm 0.1$ mm (including estimates

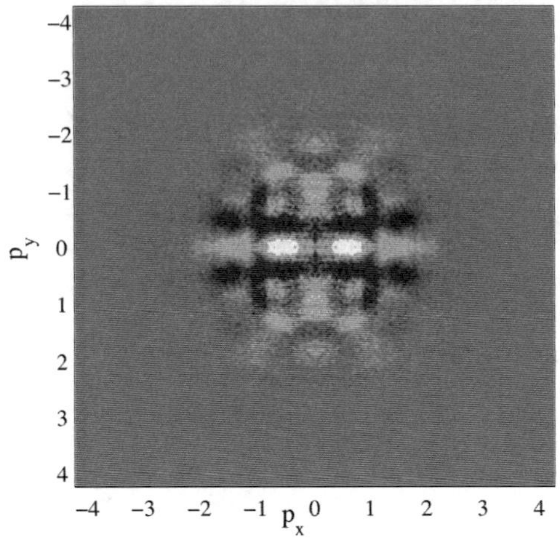


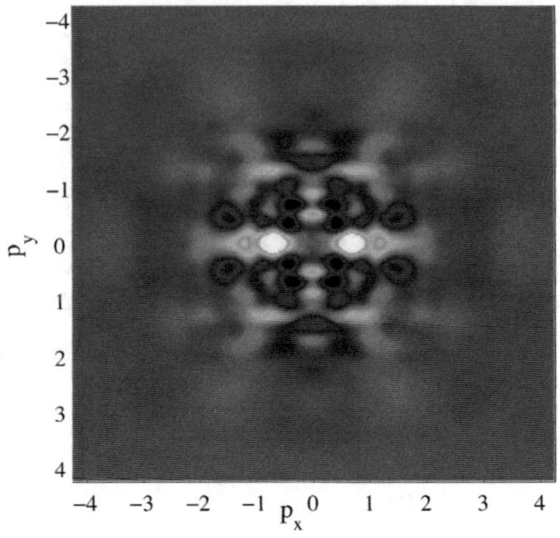


**Figure 5.** Radial anisotropies (at 20K) for the $[01\bar{1}0]$ direction in $NbFe_2$ from experiment (**left**) and theory (**right**).

of the temperature contribution and the error of the fit), which is very similar to the results obtained in PET applications [24]. Finally, the single detector position resolution is estimated to be $\Gamma_i = 0.68\,\mathrm{mm}$, which is (perhaps unsurprisingly) in close agreement with the converter hole pitch of 0.5 mm, particularly taking into account that the processing electronics and the data treatment itself (e.g. LCW-folding) will contribute some broadening as well. One should also note, that here the coincidence detector resolution is determined by the intrinsic detector resolution in quadrature (i.e. $\Gamma_c = \sqrt{2} \cdot \Gamma_i$). However, as observed for Anger Camera systems, where the intrinsic position resolution is well defined via for instance single camera images of a phantom, in the case of HIDAC such measurements are precluded due to the required accuracy of the phantom at the current level of resolution, and determinations via the coincidence detector resolution in practice result in a factor between $\sqrt{2}$ and 2 [18]. Hence, the estimate given here can be viewed as an upper limit on the intrinsic spatial resolution of the HIDAC. As a result, this number quantifies the last performance characteristic of the new spectrometer and proves the high level of resolution reached in the new 2D-ACAR system utilising HIDAC detectors.

### Detector Efficiency

In the case of HIDAC detectors the major factors influencing their efficiency are the size and number of modules as well as the converter's hole matrix. The small hole sizes and the dense hole grid in the converters affect not only the resolution but also provide a larger effective internal surface area for interactions, increasing the probability of electron escape from the Pb-Sb alloy. From measurements of count rates in coincidence and anti-coincidence (the latter corrected for non-sample-related background), the efficiency of a single module was calculated to be $\epsilon_s \sim 2.5 \pm 0.3\%$ [28]. The total detector efficiency, of all seven modules (including the attenuation factor of the converters) was calculated to be $\sim 12.5 \pm 0.6\%$.

### Paramagnetic phase of $NbFe_2$

Finally, in order to demonstrate the new spectrometer on a 'new' material, in Fig. 5 the radial anisotropy of a $[01\bar{1}0]$ projection of $NbFe_2$ is shown alongside theoretical calculations [32]. There is clear agreement in the shape of the anisotropy between experiment and calculation. This measurement is part of a wider study to understand the role of electronic structure in governing the magnetic properties of the quantum-critical $Nb_{1-y}Fe_{2+y}$ system [34, 35, 36, 37, 38].

## Conclusions
The Bristol HIDAC 2D-ACAR spectrometer is a powerful tool for Fermiology. Future work will focus on determining the FS in electronically complex materials, and, for example elucidating its role in driving charge-density waves in materials like $R$NiC$_2$[39, 40].

## Acknowledgments
This work was supported by EPSRC(UK). We are indebted to Alan Jeavons, the late Roy Chandler and Chris Dettmar of Oxford Positron Systems.

## References
[1] Utfeld C *et al.* 2010 *Phys. Rev. B* **81**(6) 064509
[2] Laverock J *et al.* 2007 *Phys. Rev. B* **76**(5) 052509
[3] Dugdale S B *et al.* 2006 *Phys. Rev. Lett.* **96** 046406
[4] Dugdale S B 2007 *Physica Status Solidi C* **4** 3851–3856
[5] Lock D G, Crisp V H C and West R N 1973 *Journal of Physics F: Metal Physics* **3** 561
[6] Crowe S J *et al.* 2004 *Europhysics Lett.* **65** 235–241
[7] Haynes T D *et al.* 2012 *New Journal of Physics* **14** 035020
[8] Major Z *et al.* 2004 *Phys. Rev. Lett.* **92** 107003
[9] Hamid A S *et al.* 2011 *Phys. Rev. B* **84**(23) 235107
[10] Dugdale S B *et al.* 1999 *Phys. Rev. Lett.* **83**(23) 4824–4827
[11] Dugdale S B *et al.* 2009 *Supercond. Sci. Technol.* **22** 014002
[12] Laverock J *et al.* 2010 *Phys. Rev. Lett.* **105** 236401
[13] Dugdale S B *et al.* 1994 *Journal of Physics: Condensed Matter* **6** L435–L443
[14] Fretwell H M *et al.* 1995 *Europhysics Lett.* **32** 771–776
[15] Berko S, Haghooie M and Mader J 1977 *Phys. Lett.* **63A** 335
[16] Inoue K *et al.* 2002 *Nucl. Instr. and Meth. in Phys. Res. A* **487** 471
[17] Anger H O 1963 *Nucleonics* **21** 56
[18] West R N, Mayers J and Walters P A 1981 *J. Phys. E: Sci. Instrum.* **14** 478
[19] Leitner M, Ceeh H and Weber J A 2012 *New Journal of Physics* **14** 123014
[20] Jeavons A, Charpak G and Stubbs R 1975 *Nucl. Instr. and Meth.* **124** 491
[21] Jeavons A P 1978 *Nucl. Instr. and Meth.* **156** 41
[22] Manuel A, Samoilov S, Fischer O, Peter M and Jeavons A 1979 *Helv. Phys. Acta* **52** 255
[23] Bisson P *et al.* 1982 *Helv. Phys. Acta* **55** 100
[24] Jeavons A P, Chandler R A and Dettmar C A R 1999 *IEEE Trans. Nucl. Sci.* **43** 468
[25] Missimer J, Madi Z, Honer M, Keller C, Schubiger A and Ametamey S M 2004 *Phys. Med. Biol.* **49** 2069
[26] Mansart B *et al.* 2012 *Proceedings of the National Academy of Sciences* **109** 5603–5608
[27] Charpak G 1970 *Ann. Rev. Nucl. Sci.* **20** 195
[28] Utfeld C 2010 *Probing the electron momentum density: The Fermi surface and spin polarisation of electronically complex materials* Ph.D. thesis University of Bristol
[29] Barbiellini B and Platzman P 2009 *Phys. Status Solidi C* **6** 2523
[30] Parker R and Halloran M 1974 *Phys. Rev. B* **9** 4130
[31] Hughes R J *et al.* 2004 *Phys. Rev. B* **69**(17) 174406
[32] Barbiellini B, Dugdale S B and Jarlborg T 2003 *Comp. Mater. Sci.* **28** 287–301
[33] Laverock J, Haynes T D, Alam M A and Dugdale S B 2010 *Phys. Rev. B* **82** 125127
[34] Brando M *et al.* 2008 *Phys. Rev. Lett.* **101**(2) 026401
[35] Subedi A and Singh D J 2010 *Phys. Rev. B* **81**(2) 024422
[36] Tompsett D A, Needs R J, Grosche F M and Lonzarich G G 2010 *Phys. Rev. B* **82**(15) 155137
[37] Alam A and Johnson D D 2011 *Phys. Rev. Lett.* **107**(20) 206401
[38] Haynes T D *et al.* 2012 *Phys. Rev. B* **85** 115137
[39] Shimomura S *et al.* 2009 *Phys. Rev. Lett.* **102**(7) 076404
[40] Laverock J, Haynes T D, Utfeld C and Dugdale S B 2009 *Phys. Rev. B* **80**(12) 125111

Journal of Physics: Conference Series **443** (2013) 012084 doi:10.1088/1742-6596/443/1/012084

# Spin-polarized positron annihilation spectroscopy for spintronics applications

**A. Kawasuso, Y. Fukaya, M. Maekawa, I. Mochizuki, and H. Zhang**
Advanced Science Research Center, Japan Atomic energy Agency, 1233, Watanuki, Takasaki, Gunma, 370-1292 JAPAN

E-mail: kawasuso.atsuo@jaea.go.jp

**Abstract.** Spin-polarized positron annihilation spectroscopy will be useful in studying spintronics materials. Here, we summarize some fundamental aspects of this method for the future spintronics studies.

## 1. Introduction

Spin-polarized positron annihilation spectroscopy (SP-PAS) has so far been used for studying the polarized band structures of bulk ferromagnets [1–16]. Considering the facts that important spin-related phenomena such as magnetoresistance, spin current, spin accumulation and so on occur near surfaces/interfaces and spin device metarials such as spin injection electrodes are utilized in thin fims, by developing low-energy spin-polarized positron beam, SP-PAS will be further applied to the spintronics field. As for the sensitivity of positrons to vacancy defects, SP-PAS will also be an important tool to study vacancy-induced magnetism.

From the above viewpoints, we have been developing spin-polarized positron beams. We performed SP-PAS experiments with the Doppler broadening of annihilation radiation (DBAR) measurements on some ferromagnets. For studying surface spin phenomena, we revive the spin-polarized positronium (Ps) annihilation method based on the DBAR measurements, of which the original principle was established by the Michigan group using annihilation lifetime measurement [17]. In this paper, we briefly report the above two experiments.

## 2. SP-PAS for ferromagnetic systems

The positron-electron momentum densities for the *ith* majority (spin-down (↓)) and minority (spin-up (↑)) bands, $\rho_i^{\downarrow(\uparrow)}(\boldsymbol{p})$, of a ferromagnet are given by

$$\rho_i^{\downarrow(\uparrow)}(\boldsymbol{p}) = \left| \int e^{-i\boldsymbol{pr}} \Psi_+(\boldsymbol{r}) \Psi_i^{\downarrow(\uparrow)}(\boldsymbol{r}) \sqrt{\gamma[n_-(\boldsymbol{r})]} d\boldsymbol{r} \right|^2, \tag{1}$$

where $\Psi_+(\boldsymbol{r})$ is the positron wave function, $\Psi_i^{\downarrow(\uparrow)}(\boldsymbol{r})$ is the electron wave function and $\gamma[n_-(\boldsymbol{r})]$ is the enhancement factor. According to Berko [5,6], DBAR spectra (convolutions of apparatus function and 1D-ACAR spectra) under the fields parallel and anti-parallel to the motion of the positrons are given by

$$N_\pm(p_z) = \frac{\lambda_S}{4} \sum_{i=1}^{occ} \left[ \frac{(1\pm P) N_i^{\downarrow}(p_z)}{\lambda^{\Uparrow}} + \frac{(1\mp P) N_i^{\uparrow}(p_z)}{\lambda^{\Downarrow}} \right], \tag{2}$$

where summation is over all occupied states, $\lambda_S = 4\pi r_e^2 c$ ($r_e$: the classical electron radius), $P$ is the longitudinal polarization of positrons, $N_i^{\downarrow(\uparrow)}(p_z)$ is the double integral of $\rho_i^{\downarrow(\uparrow)}(\boldsymbol{p})$ and $\lambda^{\Uparrow(\Downarrow)}$ is the total annihilation rate of spin-up (spin-down) positrons:

$$\lambda^{\Uparrow(\Downarrow)} = \frac{1}{2}\sum_{i=1}^{occ}[\lambda_S w_i^{\downarrow(\uparrow)} + \lambda_T(w_i^{\downarrow(\uparrow)} + 2w_i^{\uparrow(\downarrow)})]. \tag{3}$$

Here, $\lambda_T = \lambda_S/1115$ and $w_i^{\downarrow(\uparrow)}$ is the overlap integral between the positron and *ith* majority (minority) spin band wave functions ($= \int\int\int \rho_i^{\downarrow(\uparrow)}(\boldsymbol{p})d\boldsymbol{p}$). Equation 2 suggests that $N_+(p_z)$ and $N_-(p_z)$ exhibit the characters of more majority and more minority spin bands, respectively. Their difference is given by

$$N_+(p_z) - N_-(p_z) = \frac{\lambda_S P}{2}\sum_{i=1}^{occ}\left[\frac{N_i^{\downarrow}(p_z)}{\lambda^{\Uparrow}} - \frac{N_i^{\uparrow}(p_z)}{\lambda^{\Downarrow}}\right]. \tag{4}$$

The summation of partial differential spectra $(N_i^{\downarrow}(p_z) - N_i^{\uparrow}(p_z))$ is given by

$$\begin{aligned}\sum_{i=1}^{occ}[N_i^{\downarrow}(p_z) - N_i^{\uparrow}(p_z)] &\propto \Delta N(p_z) + P\frac{\lambda^{\Uparrow} - \lambda^{\Downarrow}}{\lambda^{\Uparrow} + \lambda^{\Downarrow}}\Sigma N(p_z) \\ &\approx \Delta N(p_z) + P^{3\gamma}\Sigma N(p_z),\end{aligned} \tag{5}$$

where $\Delta N(p_z) = N_+(p_z) - N_-(p_z)$, $\Sigma N(p_z) = N_+(p_z) + N_-(p_z)$ and $P^{3\gamma} = (N_+^{3\gamma} - N_-^{3\gamma})/(N_+^{3\gamma} + N_-^{3\gamma})$, where $N_\pm^{3\gamma}$ denotes the three-photon annihilation intensities in positive and negative fields. Therefore, either through eq.(4) or eq.(5), one can extract the components related to magnetic electrons.

From eq. (3), the positron annihilation lifetime spectrum of a ferromagnet is composed of two different lifetime components even if $P$=0. The lifetime spectra in positive and negative field are given by

$$L_\pm(t) = \frac{\lambda_S}{4}\sum_{i=1}^{occ} w_i^{\downarrow}(1\pm P)exp(-\lambda^{\Uparrow}t) + \frac{\lambda_S}{4}\sum_{i=1}^{occ} w_i^{\uparrow}(1\mp P)exp(-\lambda^{\Downarrow}t). \tag{6}$$

Figure 1(a) shows the experimental set-up of DBAR measurement in magnetic field. Figure 1(b) shows the differential DBAR spectra obtained for polycrystalline Fe, Co and Ni samples under magnetic field using a $^{68}$Ge-$^{68}$Ga source [18]. These spectra exhibit that 3*d* electrons having broader momentum distribution are responsible for polarization effects. The magnitude of field-reversal asymmetry reflects the difference of magnetization of these ferromagnets.

### 3. SP-PAS for surface spin systems

A fraction of positrons implanted near a metal surface are emitted as Ps by picking up surface electrons near the Fermi level. Ps annihiltion characteristics depends on the relative direction of positron and electron spin polarizations. The energy spectrum and annihilation lifetime of three-photon annihilation of spin-triplet Ps are separated from those of two-photon annihilation due to spin-singlet Ps and free positron annihilation. To detect spin-polarized electrons through Ps annihilation, it is convenient to observe the three-photon annihilation event. The Michigan group first demonstrated this experiments based on lifetime measurement [17]. We show below that the similar experiment is possible with DBAR measurement.

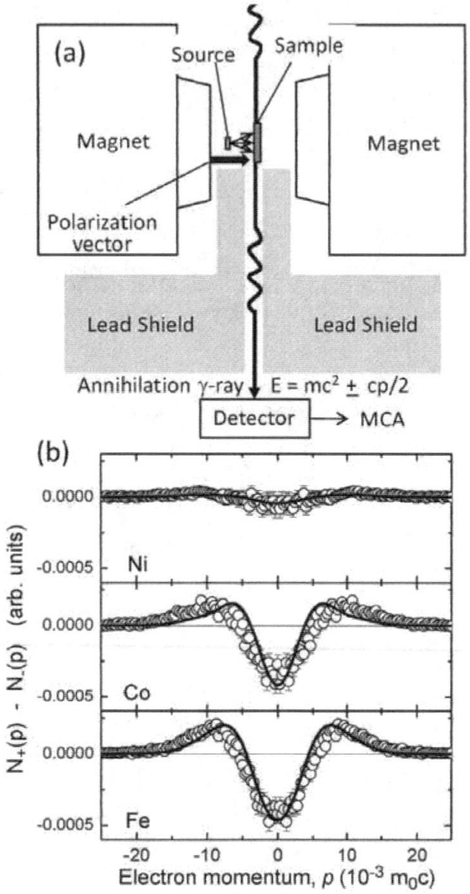


**Figure 1.** (a)Experimental set-up of DBAR measuerments under magnetic field. (b)Differential DBAR spectra obtained for polycrystalline Fe, Co and Ni samples under magnetic fields of ± 1 T.

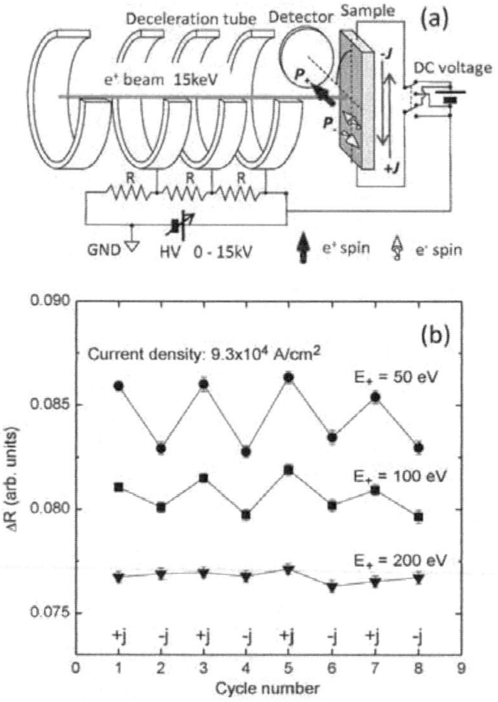


**Figure 2.** (a)Experimental set-up for detecting surface spin polarization under direct current. (b)Current reversal (+j ↔ -j) oscillation of $\Delta R$ due to spin-triplet positronium obtained for a Pt thin film surfaces at different incident positron energies ($E_+$).

Fractions of each spin state of Ps are given by

$$F_{|00\rangle} = (1 - P_+P_-cos\varphi)/4, \quad (7)$$

$$F_{|10\rangle} = (1 - P_+P_-cos\varphi)/4, \quad (8)$$

$$F_{|11\rangle} = (1 + P_+ + P_-cos\varphi + P_+P_-cos\varphi)/4, \quad (9)$$

$$F_{|1-1\rangle} = (1 - P_+ - P_-cos\varphi + P_+P_-cos\varphi)/4, \quad (10)$$

where the subscript $|\, Sm\rangle$ of $F$ denote total spin $S$ and magnetic quntum number $m$, $P_\pm$ is the positron and electron polarizations, respectively, and $\varphi$ is the angle between the positron and electron polarization directions. The total three-photon annihilation fraction is given by

$$F_{Ps}^{3\gamma} = \epsilon(1)(F_{|11\rangle} + F_{|1-1\rangle}) + \epsilon(0)F_{|10\rangle}, \quad (11)$$

where $\epsilon(1)$ and $\epsilon(0)$ are the detection efficiencies of annihilation photons from $m$=±1 and $m$=0 states, respectively [19]. When the positron and electron polarization directions are the same, the $|11\rangle$ state is preferentially formed and the fractions of the other states are relatively lower, resulting in the increase of $F_{Ps}^{3\gamma}$. Conversely, when the positron and electron polarization directions are opposite, the fractions of the $|11\rangle$ and $|1-1\rangle$ states are relatively lower as compared with the $|10\rangle$ and $|00\rangle$ states. Hence, $F_{Ps}^{3\gamma}$ decreases. Ps three-photon annihilation is traditionally characterized by a quantity $R$ [20],

$$R = (T - U)/U = \frac{(1 - F_{Ps}^{3\gamma})R_0 + F_{Ps}^{3\gamma}R_1U_1/U_0}{1 - F_{Ps}^{3\gamma} + F_{Ps}^{3\gamma}U_1/U_0}, \tag{12}$$

where $T$ and $U$ denote the area under the entire intensity curve and under the two-photon annihilation spectrum and the subscripts 0 and 1 denote the cases of 0% and 100% positronium emission, respectively. Although $R$ is a complicated function of $F_{Ps}^{3\gamma}$, for a small $F_{Ps}^{3\gamma}$, $\Delta R$=$R$-$R_0$ is approximately proportional to $F_{Ps}^{3\gamma}$. Thus, by using Eqs.(8) to (12), the asymmetry of $\Delta R$ upon spin flip ($+P_- \leftrightarrow -P_-$) is

$$A = \frac{\Delta R(+P_-) - \Delta R(-P_-)}{\Delta R(+P_-) + \Delta R(-P_-)} = \frac{2\epsilon(1) - \epsilon(0)}{2\epsilon(1) + \epsilon(0)}P_+P_-cos\varphi. \tag{13}$$

Therefore, from the observation of $\Delta R$ by changing the relative direction between positron and electron spin polarizations, polarized electrons are detected.

Figure 2(a) shows the experimental set-up for the current-induced spin polarization on metal surfaces. Figure 2(b) shows the $\Delta R$ obtained for a Pt thin film (30 nm thick) under direct current as a function of successive current reversal at different incident positron energies. From this result, excess electron spins are accumulated on the Pt surface under direct current.

**Acknowledgement**
This work was supported by JSPS KAKENHI Gant Number 24310072.

## References

[1] Hanna S S, Preston R S 1957 *Phys. Rev.* **106** 1363, **109** 716
[2] Mijnarends P E and Hambro L 1964 *Phys. Lett.* **10** 272
[3] Mijnarends P E 1973 *Physica* **63** 248
[4] Berko S, Zuckerman J 1964 *Phys. Rev. Lett.* **13** 339
[5] Berko S 1967 *Positron Annihilation*, ed A T Stewart and L O Roellig (New York: Academic Press) p 61
[6] Berko S and Mills A P 1971 *J. Phys., Colloq.* **32** C1
[7] Hohenemser C, Weingart J M and Berko S 1968 *Phys. Lett. A* **28** 41
[8] Mihalisin T W and Parks R D 1966 *Phys. Lett.* **21** 610, 1967 *Phys. Rev. Lett.* **18** 210, 1969 *Solid State Commun.* **7** 33
[9] Shiotani N, Okada T, Sekizawa H, Mizoguchi T and Karasawa T 1973 *J. Phys. Soc. Jpn.* **35** 456
[10] Šob M, Szuszkiewicz S and Szuszkiewicz M 1984 *Phys. Stat. Sol. (b)* **123** 649
[11] Jarborg T, Manuel A A, Mathys Y, Peter M, Singh A K and Walker E 1986 *J. Magn. Magn. Mater.* **54-57** 1023
[12] Szuszkiewicz S, Šob M and Szuszkiewicz M 1986 *J. Magn. Magn. Mater.* **62** 202
[13] Genoud P, Singh A K, Manuel A A, Jarlborg T, Walker E, Peter M and Welle M 1988 *J. Phys. F* **18** 1933
[14] Genoud P, Manuel A A, Walker E and Peter M 1991 *J. Phys.: Condens. Matter* **3** 4201
[15] Kondo H, Kubota T, Nakashima H, Kawano T and Tanigawa S 1992 *J. Phys.: Condens. Matter* **4** 4595
[16] Hanssen K E H M and Mijnarends P E 1986 *Phys. Rev. B* **34** 5009
[17] Gidley D W, Köymen A R and Capehart T W 1982 *Phys. Rev. Lett.* **49** 1779
[18] Kawasuso A, Maekawa M, Fukaya Y, Yabuuchi A and Mochizuki I 2011 *Phys. Rev. B* **83** 100406(R)
[19] Drisko R M 1956 *Phys. Rev.* **102** 1542
[20] Lynn K G and Welch D O 1980 *Phys. Rev. B* **22** 99

Journal of Physics: Conference Series **443** (2013) 012085 doi:10.1088/1742-6596/443/1/012085

# A TOF-PET Detector based on Quadrant-Sharing PMTs and Optimized Leading-edge Timing Method

**Yingjie Wang[1, 2], Junhui Liu[1, 2], Daowu Li[1, 2], Tianbao Zhang[1, 2], Zhiming Zhang[1, 2], Xingzhong Cao[1, 2], Baoyi Wang[1, 2], Long Wei[1, 2]**

[1]Key Laboratory of Nuclear Analysis Techniques, Institute of High Energy Physics, Chinese Academy of Sciences, Beijing 100049, China
[2]Beijing Engineering Research Center of Radiographic Techniques and Equipment, Beijing 100049, China
19B Yuquan Lu, Shijingshan District, Beijing, China

E-mail: weil@ihep.ac.cn

**Abstract**. A time-of-flight positron emission tomography (TOF-PET) detector was developed based on a 6×9 LYSO array and four single channel photomultipliers (Hamamatsu R9800). Leading-edge timing circuit with optimized parameter was used instead of the constant fraction discriminator. The results showed that all 54 elements in the flood histogram could be identified clearly. The average coincidence resolving time was 402 ps FWHM.

## 1. Introduction

Positron emission tomography (PET) is a powerful functional imaging tool widely used in the early diagnosis of tumors. Time-of-flight (TOF) technique improves the signal-to-noise ratio of a PET image and consequently reduces dose and shortens scanning time.

PS-PMT and single channel PMT are both studied for TOF-PET by various researchers. Kim acquired 477 ps timing resolution by coupling LSO to Hamamatsu PS-PMT H8500 [1]. Szczesniak compared a series of fast PMTs and suggested that PMTs for TOF-PET should have low time jitter, high quantum efficiency (QE), and the best anode pulse shape quality [2]. The coincidence resolving time (CRT) of 422 ps was reported using PMT-quadrant-sharing LYSO detector and constant fraction discriminator (CFD) [3].

In this study, we developed a detector with LYSO array and quadrant-sharing PMTs and then incorporated the detector to a leading edge discriminator (LED). Good CRT was acquired.

## 2. Detector and electronics

The detector module consisted of crystals, a light guide, and PMTs. They were packaged in a detector box together with front-end electronics.

### 2.1. Crystals

Scintillators for TOF-PET application should exhibit high Z, high density, high light yield, and fast decay time. $BaF_2$, $LaBr_3$:Ce, and LYSO are all good candidates for TOF-PET. However, $LaBr_3$ deliquesce easily. The fast light output of $BaF_2$ peak at 220 nm, which requires PMTs with quartz window. Thus, LYSO was chosen as the scintillator of our detector based on the light output, stopping power, packaging technology, and cost.

According to a previous study on LSO scintillators for TOF-PET, the longer the LYSO crystal, the higher the detecting efficiency will be, although with worse timing resolution, and vice versa [4]. Therefore, an LYSO crystal with length of 20 mm was used to balance timing resolution and detecting efficiency.

### 2.2. PMTs

Hamamatsu R9800 was a single channel PMT with good timing ability. The anode pulse rising time was 1.0 ns, and the transmit time spread was 270 ps. The QE peak was at 420 nm, which matched well with the LYSO emission spectrum.

The QE of the photocathode border area of R9800 was relatively low because of the manufacturing techniques. Therefore, avoiding the use of the border area is important to obtain good timing resolution.

### 2.3. Detector Module Design

Four R9800 (2 × 2) were arranged in a light tight aluminum box. LYSO crystals each measuring 4.2 mm × 5 mm × 20 mm were wrapped together by 6 × 9 to form a scintillator block. The size of the block was 25.2 mm × 45 mm × 20 mm. The block was placed in the area outlined by the bold rectangular line in Figure 1. In fabricating large detectors based on this design, in the $x$-direction, the block covers half of the PMT sensitive area. In the $y$-direction, the block does not cover the entire two PMTs to avoid the border area of PMTs.

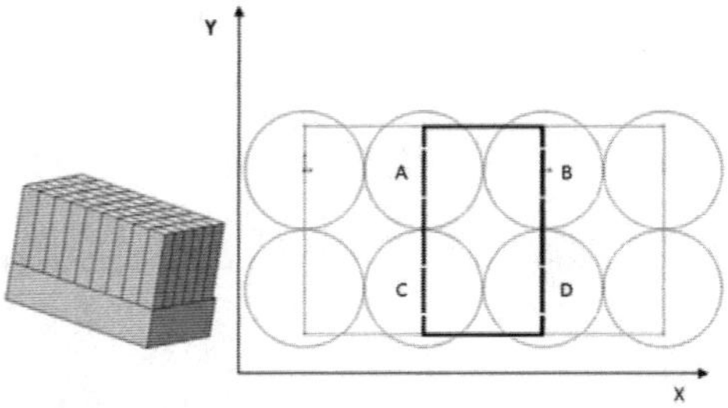


**Figure 1.** Schematic of the LYSO-R9800 detector module.

**Figure 2.** Photograph of the detector module.

As demonstrated in Figures 1 and 2, a light guide, made of K9 glass, with dimensions 45 mm × 45 mm × 9 mm, and all sides polished, was placed between the scintillator array and PMTs.

### 2.4. Electronics

The anode signals of four R9800s were processed individually. Each signal was amplified and shaped to quasi-Gaussian shape, 300 ns width. Leading-edge timing method was adopted in the timing electronics. The threshold of the comparer in the circuit largely influenced the timing performance. The optimum threshold of the circuit was acquired and described in section 3.2.

## 3. Measurement setup

### 3.1. Flood histogram and CRT measurement setup

We only fabricated one R9800-LYSO detector in this work. Therefore, another detector based on XP2020Q and $BaF_2$ with known timing resolution of 159 ps was used to build a coincidence time measurement system (Figure 3).

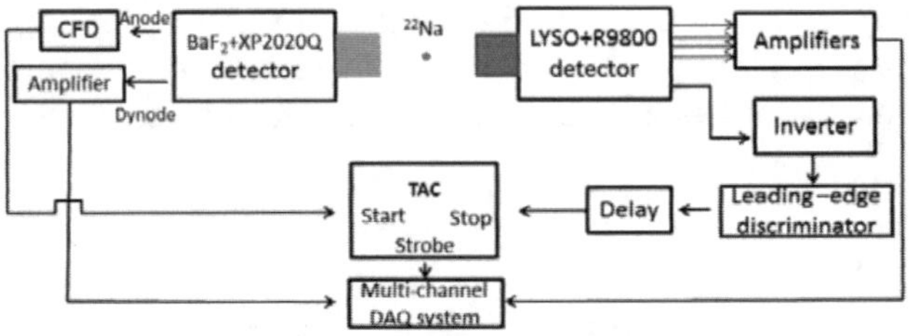


**Figure 3.** $BaF_2$-LYSO timing resolution measurement setup.

The dynodes of four R9800 were combined to do the timing. The inverted pulse was amplified, and then, fed to the self-developed LED. The output of the LED was delayed and fed to a time-analog converter (TAC) to act as the stopping channel. The anode signal of XP2020Q-$BaF_2$ detector was conducted to a CFD (ORTEC 935) to do the timing. The timing signal was fed to the TAC to act as the starting channel. The coincidence signals of the two detectors were used to strobe the multi-channel data acquisition (DAQ) system. The anode signals of R9800 and XP2020Q were used to select 511 keV gamma photon events.

The amplified anode signals (S1, S2, S3, and S4) of four R9800s (A, B, C, and D) were fed to the DAQ system. The two-dimensional position information of the crystal array can be obtained by X (S1+S3), Y (S1+S2), and E (S1+S2+S3+S4) [5]. A flood histogram of the scintillator array was displayed on a LabWindows user interface. The flood histogram was utilized to create the position and energy tables of every single LYSO crystal in the scintillator array. Then, the CRT between every crystal and the $BaF_2$ detector ($TR_s$) was measured. After deducting the timing resolution value of the XP2020Q-$BaF_2$ detector, we obtained the timing resolution of R9800+LYSO detector ($TR_L$) and the CRT of two R9800-LYSO detectors ($TR_C$).

### 3.2. Best threshold of leading-edge timing circuit

In our previous studies, we found that the timing resolution is largely influenced by the threshold of the comparer in the timing circuit. The special value of the threshold for various detectors should be investigated.

A similar system with CRT measurement in section 3.1 was used to study the optimum threshold of the timing circuit. The scintillator array was replaced by an LYSO crystal fixed in the middle of the light guide. The crystal measured 4 mm × 4 mm × 22 mm, with five sides wrapped with Teflon. The timing resolution of the single LYSO detector at FWHM to 511 keV gamma photon was recorded along with the changes in the LED threshold.

## 4. Results

### 4.1. Special threshold of LED

The relationship between timing resolution and threshold of LED is plotted in Figure 4. The best timing resolution of 218 ps was obtained when the threshold was 15.6 mV. Accordingly, we can conclude that the CRT of two such detectors was $218 \times \sqrt{2} = 309$ ps. This result was better than using ORTEC 935 CFD with 351 ps CRT. These results indicate that LED may be more appropriate for an LYSO detector than CFD to obtain better timing resolution.

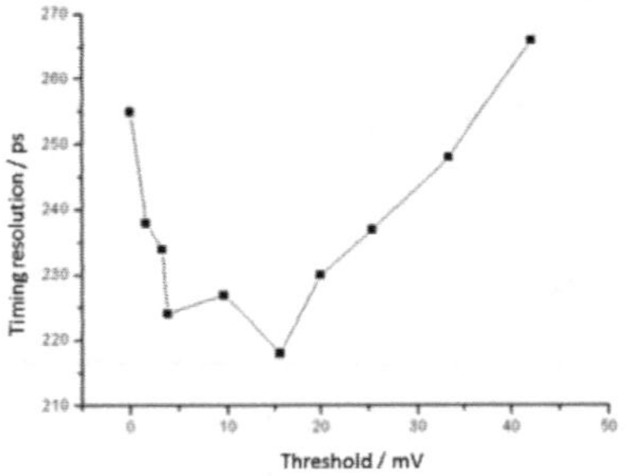


**Figure 4.** Relationship between timing resolution and the LED threshold.

4.2. Spatial resolution and timing resolution results

All 6 × 9 elements of the LYSO array can be resolved clearly in the flood histogram (Figure 5). The peak-to-valley ratio was better than 4:1. The results show that the detector had good spatial resolution ability for human-scale TOF-PET.

Using the system mentioned in section 3.1 and the optimized parameter of timing circuit, the average CRT with $BaF_2$ detector of the 54 LYSO crystals was measured as 326 ps. The coincidence time spectrum is shown in Figure 6. Therefore, the CRT of two identical R9800-LYSO detectors developed in this work could be calculated according to equation 1:

$$CRT_{LYSO\text{-}LYSO}=\sqrt{2}\times\sqrt{CRT^2_{LYSO-BaF_2}-TR^2_{BaF_2}}=\sqrt{2\times(326^2-159^2)}=402\text{ ps} \quad (1)$$

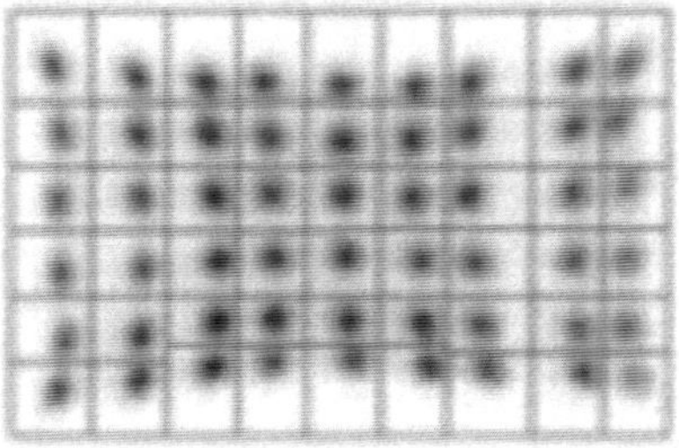

**Figure 5.** Flood histogram and one-dimensional projection.

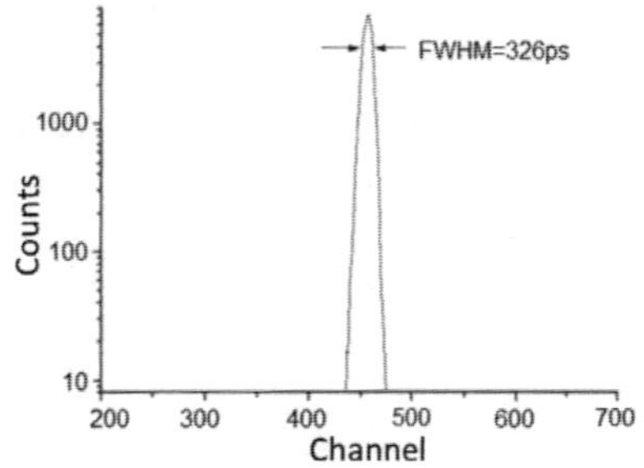


**Figure 6.** CRT spectrum between LYSO-R9800 and $BaF_2$-XP2020Q.

## 5. Summary

We developed a TOF-PET detector based on an LYSO array and four R9800 PMTs. All the LYSO crystals measuring 4.2 mm × 5 mm × 20 mm in a matrix was identified. CRT of 402 ns was obtained using an LED. The detector developed in this work can be easily extended in one direction, with high cost effectiveness and large detecting area. We are currently developing whole body TOF-PET detector modules based on this study.

The LYSO crystals are obtained from a local factory. The average light output is less than 80% of the output of Saint Gobain LSO crystals. An improved timing resolution will be acquired if all the LYSO crystals are replaced with better ones.

## Acknowledgement

This work is supported by the National Natural Science Foundation of China (Grants Nos. 11175200 and 10805049), Main Direction Program of Knowledge Innovation of Chinese Academy of Sciences, Grant (Grant No. KJCX2-EW-N06), and Special-funded Program on National Key Scientific Instruments and Equipment Development (Grant No. 2011YQ120096).

## Reference

[1] Kim C L 2006 *Ieee Nuclear Science Symposium Conference Record* **6** 1919
[2] Szczesniak T, Moszynski M, Swiderski L, Nassalski A, Lavoute P and Kapusta M 2009 *Ieee Trans. on Nucl. Sci.* **56** 173
[3] An S, Li H, Liu S, Ramirez R, Zhang Y, Wang C, Baghaei H and Wong W 2011 *Ieee Trans. on Nucl. Sci.* **58** 2155
[4] Moses W W, and Derenzo S E 1999 *Ieee Trans. on Nucl. Sci.* **46** 474
[5] Anger H O 1958 *Rev. Sci. Instrum.* **29** 27

# QtCDB2 software for coincidence Doppler broadening measurement system

**M. Petriska, . Slugeň, . Sabelová, S. Soják and Jana eterníková**

Institute of Nuclear and Physical Engineering, Slovak Technical University, Bratislava, 812 19 Slovakia

E-mail: `martin.petriska@stuba.sk`

**Abstract.** A digital system for the coincidence measurement of Doppler Broadening of positron annihilation (CDB) has been changed to one which is fully software based. Spectrometer is based on two HpGe detectors, HV sources and PC with PCI-9820D digitizer. Detector pulses are digitized directly at the HpGe detector pre-amplifier outputs. The previous external trigger chain was replaced by data processing system with a software trigger. All pulses from detectors are processed by a trapezoid filter. Spectrometer performance was tested and compared in various conditions.

## 1. Introduction

In 2011, a new spectrometer for coincidence Doppler broadening (CDB) of the positron annihilation peak was built and tested at PALS laboratory in Bratislava [1]. The spectrometer has new advantages like electrical Cryo - Cooling and digital pulse processing system. As a digitizer, 14bits Adlink PCI9820 was used. The external triggering chain was based on NIM modules: constant fraction discriminators and fast coincidences. The system was tested by measurement of Fe samples with He and H implantation. These measurements showed that the system has a long-term instability caused by temperature changes in the laboratory. To make the system even simpler, the software was rewritten to process all input data without the need of external coincidence trigger signal. Real-time transfer using PCI-9820 on-board 3k-sample FIFO was used to achieve this possibility . This feature is possible only if the data throughput from the digitizer is less than the available PCI bandwidth, therefore a lower sampling speed 20MSps was applied instead of the maximum 60MSps applicable to this device.

## 2. Hardware setup

Our new CDB system consists of detectors, their power supply and computer with acquisition card, see Fig. 1. As HpGe detectors we used Canberra Gc2019 detectors with relative efficiency 20 %. Both detectors are the same type, but their HV supply is different. The HV source used is Dual NIM based Canberra 3125.

## 3. Trapezoid filter

Analogous to the old QtCDB software, QtCDB2 uses the same trapezoidal filter [2] for energy extraction from digitized pre-amplifier outputs signals. To see results from the filter, QtCDB2 (Fig. 4) software based on Qt and Qwt libraries can show pulse histograms from output of the

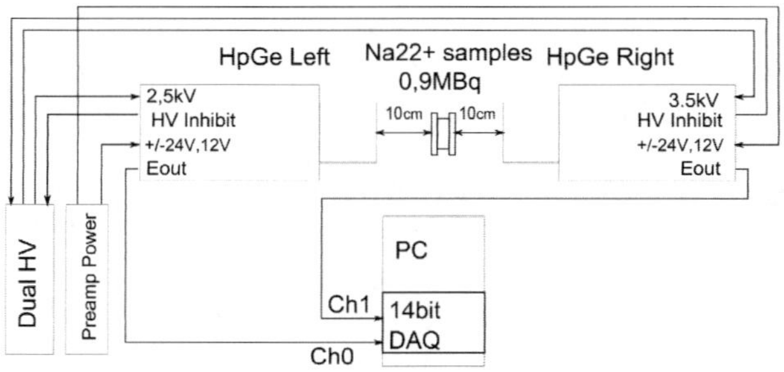


**Figure 1.** CDB schematic diagram, HPGe Detectors Canberra GC2019, HV Source Canberra 3125, DAQ Adlink PCI-9820, PC Intel Core Quad 2.6Ghz, Win7 64bit, Preamp Power from NIM Bin

trapezoid filter, input of the filter, input of the filter with shifted baseline to same start value, and timing signal which is used to select rise time of the trapezoid pulse (see Figs. 2 and 3). To suppress pile ups shorter than trapezoid length, a shape filter based on controlling flat top deviation from linear shape and pulse height in end of pulse level was used. Pile ups longer than trapezoidal filter (K+L) are compensated by the trapezoid filter. Pulses whose heights are below the LLD level are discriminated from the processing.

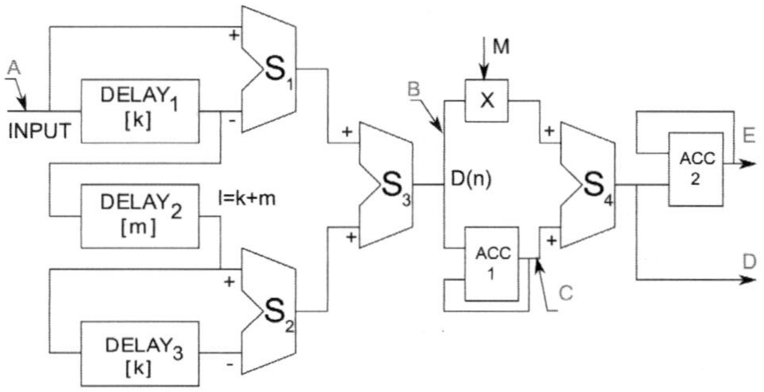


**Figure 2.** Trapezoidal filter block scheme [2]. Signals in points A..E are shown in Fig. 3,
A: input v(n)
B: D(n)=v(n)-v(n-k)-v(n-l)+v(n-k-l)
C: integrated B
D: timing output
E: energy output

## 4. Data acquisition mode

The QtCDB2 software supports two data acquisition modes of the PCI-9820 card.

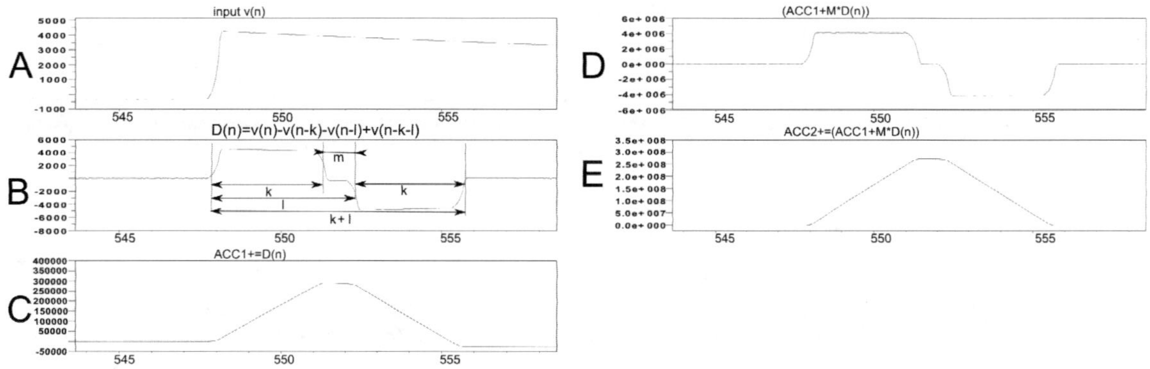


**Figure 3.** Signals from trapezoidal filter. A,B,C,D,E are points in parts of the filter. (See Fig. 2) The first vertical axis has scale in output values from 14bits ADC, the other vertical axis are in values as computed by filter, horizontal axis are in μs.

Journal of Physics: Conference Series **443** (2013) 012086 doi:10.1088/1742-6596/443/1/012086

### *4.1. Two steps of data acquisition mode*

The old QtCDB code used two steps of data acquisition - DAQSTEPPED mode with external triggering. One pulse processing used in the QtCDB needs a small buffer (about 3kB). Its size was appropriate to trapezoidal pulse size. In the new code we process a long buffer with lot of pulses. The maximum buffer size is determined by internal 128MB SDRAM and card driver. For one measurement we need 2*14bits, i.e. 4 bytes, therefore in the buffer should be stored 4.2 million samples. DAQSTEPPED acquisition is made in two steps and needs restarting ADC after each data reading from the buffer. This takes death time to acquisition. However in this mode the card should work properly in all available sampling rates supported by the DAQ card.

### *4.2. Double Buffer mode*

This mode applies two acquisition data buffers. The first buffer is filled with data from the ADC, while the second one is used to send data from the card to trapezoidal filter. Role of the buffers is periodically switched. Double buffer mode allows to process all data continuously without the need of ADC restart. The mode is suitable for the ADC sampling rates below 20MSps, due to PCI transfer throughput.The lower sampling speed means few data points in the trapezoid filter with the same shaping time. This means that the resolution is worse than in the previous mode with maximum sampling rate of 60MSps. The obtained resolution with 20MSps, 0.890keV is good compromise between FWHM and count-rate. The results from different sampling rate measurements with the same trapezoidal filter parameters are shown in Table 1. The FWHM was calculated from $E_{left}$-$E_{right}$=0 diagonal cut Gaussian extrapolation. Shaping time was chosen according to supplied specification of the GC2019 detectors and was same in all measurements - shaping time of 4µs and flat top length of 1.1µs.

**Table 1.** Obtained FWHM, count rates with different PCI-9820 ADC sampling rates and acquisition modes. Shaping time 4µs and flat top length 1.1µs

| Sampling Rate [MSps] | FWHM[keV] | Count rate [cps] | (K+L) [samples] | ADC Mode |
|---|---|---|---|---|
| 60 | 0.883 | 109 | 546 | DAQSTEPPED |
| 30 | 0.886 | 216 | 273 | DAQSTEPPED |
| 20 | 0.890 | 623 | 182 | Double Buffer |
| 15 | 0.915 | 855 | 136 | Double Buffer |
| 12 | 0.937 | 962 | 109 | Double Buffer |
| 10 | 0.957 | 1045 | 91 | Double Buffer |
| 5 | 1.159 | 1771 | 45 | Double Buffer |

## 5. Long term measurement stability

The previous CDB system [1] had a problem with the HV source stability. This instability was seen in the CDB spectra as a worsening in the FWHM and it depends on temperature changes in the room. The used HV source Canberra 3125 is a dual HV in one NIM module. Temperature dependent errors were seen in both HpGe outputs signal in the same level. Even it was not observed in energy difference, which is the main part of CDB spectra, we decided to eliminate this error. Spectrum shift was eliminated by continuous spectra recalibration. The values computed of trapezoidal filters, which are proportional to gamma energies are stored in buffers.The buffer size is adjustable, but usually it can save energies from about 1 million coincident events. In the first step, from these 1 million values, two one-dimensional 8000 channel

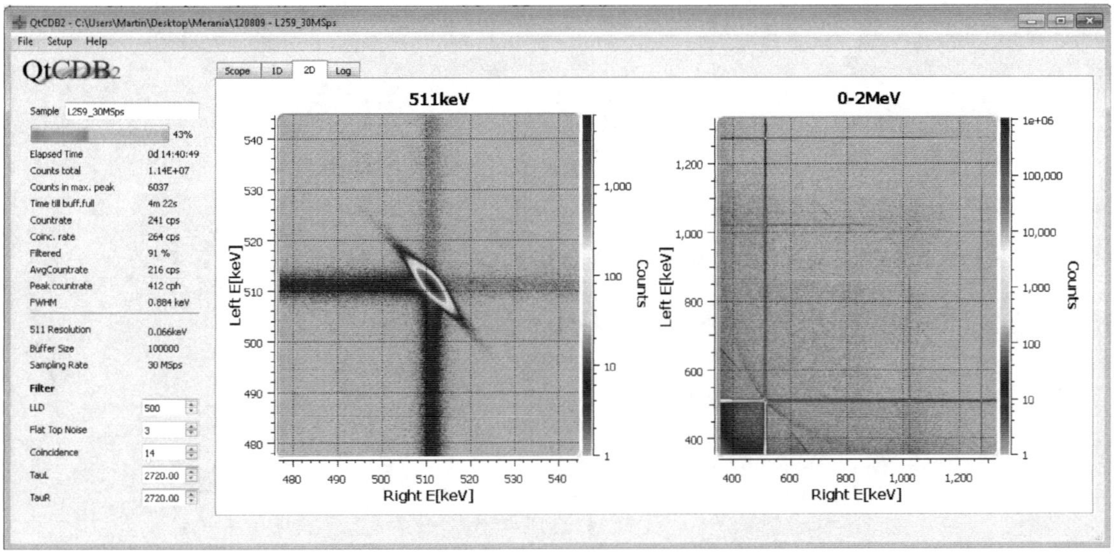

**Figure 4.** QtCDB2 user interface with 511keV peak for CDB analysis and 0-2MeV spectrogram for positron annihilation in flight analysis [3]

energy histograms for both detectors are made. These collected histograms are calibrated from 511keV and 1274keV peaks by the Gaussian peak extrapolation. In the second step, computed calibration coefficients from these histograms are used to calculate calibrated energy values from the stored events in the buffer. To track the 511keV peak movement even precisely, calibration coefficients are changed linearly from the previous calibration coefficients across the buffer. The calibrated energy values are used for 2D spectrograms (Fig. 4) with desired energy scales. The first spectrogram with 511keV peak in middle is used for the CDB analysis. The second spectrogram with energy coverage from 0 to 2MeV should be used for positron annihilation in flight analysis [3].

## 6. Conclusion

At the time when we started to build our CDB system (2010), NIM DSP modules used in [4] were discontinued and the multiparameter analogue MCA was expensive. The proposed solution with software trapezoidal filtering realized by PC and automatic spectra stabilisation based on continual recalibration from sodium-22 energy peaks gives good long-term stability and resolution (0.883keV) with the used HpGe detectors even if laboratory temperature is not well stabilized. The QtCDB2 software contains GPL libraries and remain open source. The software sources are available for the positron annihilation community at [5].

## References

[1] Petriska M, Sabelova V, Slugeň V, Sojak S, Stacho M and Veterníková J 2012 *Physics Procedia* **35** 117–121 ISSN 1875-3892

[2] Jordanov V T and Knoll G F 1994 *Nuclear Instruments and Methods in Physics Research Section A: Accelerators, Spectrometers, Detectors and Associated Equipment* **345** 337–345 ISSN 0168-9002

[3] Čížek J, Vlček M and Procházka I 2012 *New Journal of Physics* **14** 035005 ISSN 1367-2630

[4] Van Petegem S, Van Waeyenberge B, Segers D and Dauwe C 2003 *Nuclear Instruments and Methods in Physics Research Section A: Accelerators, Spectrometers, Detectors and Associated Equipment* **513** 622–630 ISSN 0168-9002

[5] 2012 URL `http://sourceforge.net/projects/qtcdb2/`

Journal of Physics: Conference Series **443** (2013) 012087 doi:10.1088/1742-6596/443/1/012087

# Software for analysis of waveforms acquired by digital Doppler broadening spectrometer

**M. Vlček[1], J. Čížek[1], I. Procházka[1]**

[1] Faculty of Mathematics and Physics, Charles University in Prague, V Holešovičkách 2, Prague 8, CZ-18000, Czech Republic

E-mail: marian.vlcek@gmail.com

**Abstract.** High-resolution digital spectrometer for coincidence measurement of Doppler broadening of positron annihilation radiation was recently developed and tested. In this spectrometer pulses from high purity Ge (HPGe) detectors are sampled in the real time by fast digitizers and subsequently analyzed off-line by software. We present description of the software routines used for pulse shape analysis in two spectrometer configurations: (i) semi-digital setup in which detector pulses shaped in spectroscopic amplifiers (SA's) are digitized; (ii) pure digital setup in which pulses from detector pre-amplifiers are digitized directly. Software developed in this work will be freely available in the form of source code and pre-compiled binaries.

## 1. Introduction

Measurement of Doppler broadening of positron annihilation radiation in coincidence is a well established experimental technique capable of characterizing chemical composition of materials in vicinity of open volume defects. Recent advances in this technique include development of a high-resolution digital spectrometer [1]. Excellent performance of this spectrometer was demonstrated by precise measurement of positron annihilation-in-flight [2]. Digital spectrometer exhibit not only superior energy resolution and signal-to-noise ratio over traditional analogue spectrometers but represent also more economical alternative. It employs real time sampling of pulses from HPGe detectors by fast 12-bit digitizers and their subsequent off-line analysis by software routines. This approach provides several advantages compared to traditional analogue setup:

(i) all detector signals are recorded and accessible for latter analysis,

(ii) data analysis can be repeated several times to optimize its parameters and to filter out pulses damaged by pile-up or by ballistic deficit,

(iii) tedious adjustment of the analogue NIM devices is not necessary anymore.

However, a proper design and implementation of software routines for waveform analysis is crucial for performance of the digital spectrometer.

In this work, algorithms for pulse shape analysis in two spectrometer configurations are described:

(i) semi-digital setup in which detector signals are firstly processed by pseudo-Gaussian filter in SA's in order to improve the signal-to-noise-ratio and the output SA signals are digitized,

(ii) pure digital setup where pulses from detector pre-amplifiers are digitized directly.

## 2. Data analysis

### *2.1. Semi-digital setup*

In semi-digital configuration, waveforms has pseudo-Gaussian shape shown in figure 1(a). We use shaping with time constant of 4 $\mu$s and the waveforms consist of 1000 points taken with sampling period of 20 ns. The analysis of sampled waveforms is performed by a code CMFIT [3] in sequence of two runs. In the first run, maximum of the waveform is determined by parabolic fitting. Base-line level prior to the pulse is fitted by a constant line. The onset of the pulse is defined as a point, where derivative of the waveform exceeds a preselected threshold value. If RMS of the base-line or $\chi^2$ value of the parabolic fit exceed preselected levels, waveform is discarded. Pulse height is calculated as a difference between maximum obtained by parabolic fitting and estimated base-line level prior to the pulse. Calculated value is added to the histogram of pulse heights which is then calibrated using known energies of the photo-peaks in the spectrum.

Waveform is further normalized by shifting position of maximum to a common reference point on the horizontal scale and by vertical scaling to a common height. Normalized waveform is finally added to two-dimensional histogram of pulse shapes. Ideal shape of the waveform shown in figure 1(a) is determined using the most frequent values in the two-dimensional histogram of pulse shapes for each channel. Upper and lower bound around the ideal pulse shape (see inset in figure 1(a)) are determined at points where the counts in the two-dimensional histogram fall to $\frac{1}{10}$ of the maximum for each channel.

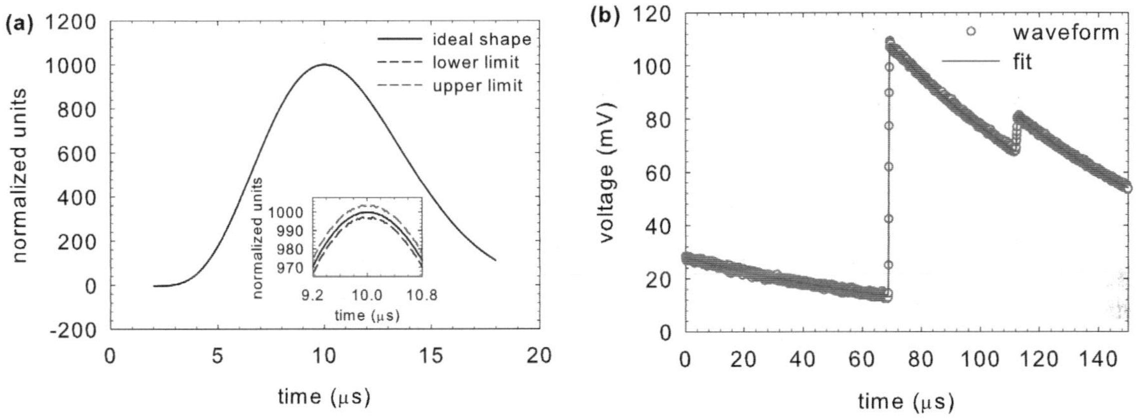


**Figure 1.** (a) "Ideal waveform shape" (blue) in the semi-digital configuration. The inset presents detail of lower (green) and upper (red) bounds applied by shape filter. (b) Typical waveform measured in pure-digital configuration; solid line shows fit of the signal by model function given by equation 1.

In the second run the shape of each waveform is examined by the shape filter constructed in the first run, i.e. shape of the normalized waveform is compared with the ideal shape. Only the waveforms which fall everywhere into the corridor between the lower and the upper bound are accepted. Time interval between maxima of the pulses from first and second detector is calculated. The waveforms are accepted only if their time difference is smaller than 100 ns. Finally, at the end of second run, two-dimensional histogram of $E_1 + E_2 - 2m_0c^2$ versus $E_1 - E_2$ and its one-dimensional vertical and horizontal cuts are constructed.

### *2.2. Pure digital setup*

Typical waveform acquired in pure digital setup consisting of 2500 points with sampling period of 60 ns is shown in figure 1(b). Pulse shape analysis in pure digital setup is performed in two runs.

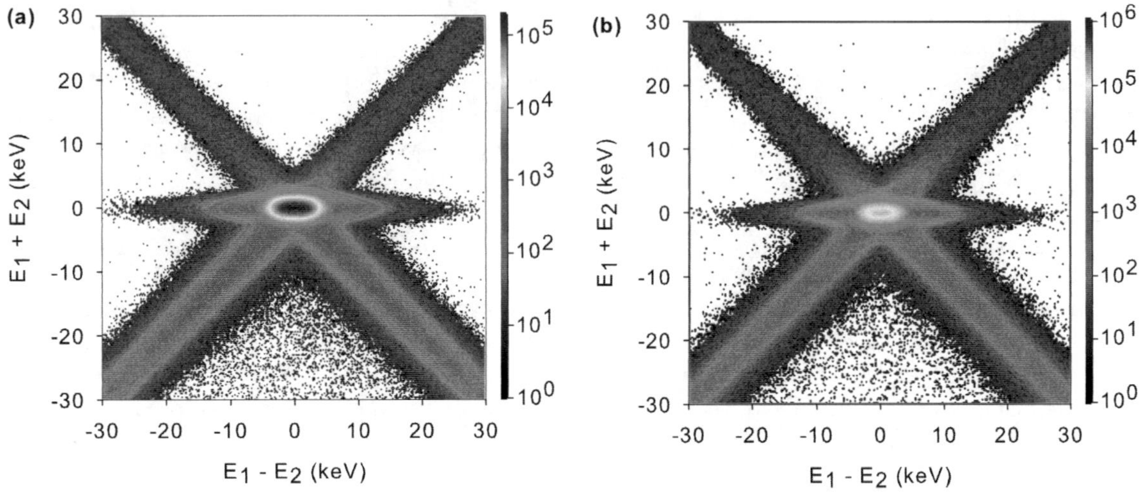


**Figure 2.** Two-dimensional CDB spectra measured on a well annealed pure Al (99.9999%) reference sample in (a) semi-digital configuration, (b) pure-digital configuration.

In the first run, waveforms are analyzed by DCDB MPI program [3]. At first, waveform is convoluted with differentiating smoothing window. Convoluted waveform is then used to determine number of pulses in the waveform and raw initial values of their amplitudes and positions.. Waveforms containing zero or more than two pulses are discarded. Original waveform is then fitted by model function

$$f(t) = f_{main}(t) + f_{pile-up}(t) + f_{prec}(t) + bcg. \tag{1}$$

Main pulse $f_{main}$ is modeled as product of Heaviside step function $H_s$ and exponential decay function with amplitude $\beta_0$, decay constant $\beta_1$ and time origin $\beta_2$. The product is than convoluted with a Gaussian with standard deviation $\beta_3$ to approximate finite rise-time of the sampled signal.

$$f_{main}(t) = \left[\frac{1}{\beta_3\sqrt{2\pi}} \exp\left(-\frac{t^2}{2\beta_3^2}\right)\right] \star [\beta_0 H_s(t-\beta_2)\exp(-\beta_1(t-\beta_2))]. \tag{2}$$

Possible additional pile-up pulse occurring on the decaying tail of the main pulse is described by formally identical equation

$$f_{pile-up}(t) = \left[\frac{1}{\beta_3\sqrt{2\pi}} \exp\left(-\frac{t^2}{2\beta_3^2}\right)\right] \star [\beta_4 H_s(t-\beta_5)\exp(-\beta_1(t-\beta_5))], \tag{3}$$

where $\beta_4$ and $\beta_5$ are amplitude and time origin of pile-up pulse, respectively. Exponential tail produced by another possible pulse with amplitude $\beta_6$ preceding the main pulse is approximated by an exponential decay

$$f_{prec}(t) = \beta_6 \exp(-\beta_1 t). \tag{4}$$

Baseline level $bcg$ is assumed to be constant. To check the goodness of fit the $\chi^2$ value is calculated and stored together with fitted parameters $\beta_i$ in an output file.

The second processing run is performed by DCDB Hist code [3] which performs energy calibration and constructs one and two-dimensional spectra. The $\chi^2$ value and the fitted parameters $\beta_i$ are read from outpu file created in the first run. Histograms of the amplitudes

$\beta_0$ of the main pulses from the first and the second detector are constructed and time interval between the main pulses from first and second detector is calculated. Waveform is discarded if time interval exceeds preselected value of 200 ns. Histogram of pulse amplitude is calibrated using known energies of two gamma peaks, e.g. energy of annihilation peak (511 keV) and $^{22}$Na start photon (1274 kev). The waveforms where fitting resulted in too high $\chi^2$ values are discarded. Finally, two-dimensional histogram of $E_1 + E_2 - 2m_0c^2$ versus $E_1 - E_2$ and its one-dimensional cuts in horizontal and vertical direction corresponding to the Doppler broadened profile of annihilation peak and resolution function of the spectrometer, are constructed.

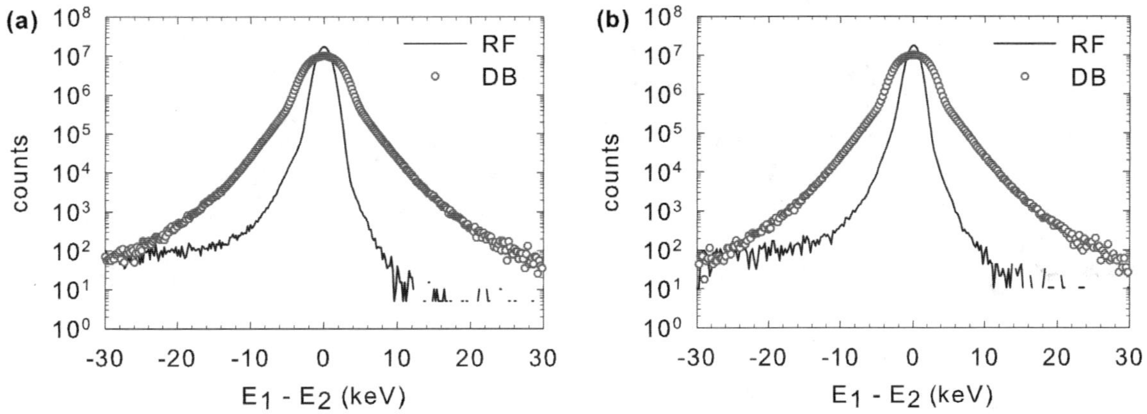


**Figure 3.** One-dimensional cuts from CDB spectra in figure 2 corresponding to the Doppler broadened profile (points) and the resolution function (lines): (a) semi-digital configuration, (b) pure-digital configuration.

Results of benchmark measurement of reference 99.9999% Al sample are shown in figure 2 and 3. CDB spectra with quality improved in comparison to the traditional analogue configuration were obtained both in the semi-digital and the pure-digital setup. Semi-digital configuration exhibits slightly lower background, while pure-digital setup has better energy resolution.

## 3. Conclusion

Software for off-line analysis of pulse signals from HPGe detectors has been developed. Two configurations were considered: (i) semi-digital setup where pulses shaped in SA are sampled; (ii) pure digital setup where pulses from detector pre-amplifiers are digitized directly. The algorithms employed in software analysis are described. Software developed in this work is freely available in the form of source code and pre-compiled binaries at `http://physics.mff.cuni.cz/kfnt/us/groups/pas/software.html`.

## Acknowledgement

This work was supported by Grant Agency of Charles University (Project no. 566012), by the grant SVV-2012-265303 and by the Academy of Sciences of the Czech Republic (under project KAN 300100801).

## References

[1] Čížek J, Vlček M and Procházka I 2010 *Nucl. Instrum. Meth. A* **623** 982 – 994
[2] Čížek J, Vlček M and Procházka I 2012 *New Journal of Physics* **14** 035005
[3] Čížek J and Vlček M URL `http://stacks.iop.org/1367-2630/14/i=3/a=035005`

Journal of Physics: Conference Series **443** (2013) 012088 doi:10.1088/1742-6596/443/1/012088

# First Experiments with MePS

**M Jungmann[1], J Haeberle[1], R Krause-Rehberg[1]**
**W Anwand[2], M Butterling[2], A Wagner[2], J M Johnson[2], and T E Cowan[2]**

[1]Univ. Halle, Inst. für Physik, Von-Danckelmann-Platz 3, 06108 Halle, Germany
[2]Institut für Strahlenphysik, Helmholtz-Zentrum Dresden-Rossendorf, PF 510119, 01314 Dresden, Germany

E-mail: Reinhard.Krause-Rehberg@physik.uni-halle.de

**Abstract:** MePS (Mono-energetic Positron System) is part of the EPOS system (ELBE Positron Source) in the HZDR (Helmholtz-Zentrum Dresden-Rossendorf). It is one of the installations at ELBE (Electron LINAC for beams with high Brilliance and low Emittance), which supplies a 40-MeV electron beam. MePS makes use of the excellent time structure of the primary electron beam of ELBE (repetition frequency up to 26 MHz; bunch length < 5ps) to produce a pulsed, intense slow positron beam to allow positron lifetime spectroscopy. In order to avoid spurious signals, which, in other systems, are often obtained by positrons reflecting from the sample surface, a bent tube (45°) was added between accelerator and sample chamber. The MePS system has been used to study the pore system of a series of low-k dielectric layers.

## 1. Introduction

MePS is a user-dedicated, highly-intense, bunched beam system using mono-energetic positrons for materials research with positrons and positronium. The MePS system is now almost finished. We started the setup at ELBE in December 2007 and we preliminary finished it in September 2011. After changing the moderator from a Pt foil to a stack of tungsten foils with 10 tungsten meshes on top, we gained a factor of 20 in intensity. The count rate at the output of the photomultiplier (lifetime signal rate at energy window of 50%), which is located close to the sample (separated only by the glass of a view flange), was 25000 cps (25 μA at 30 MeV electron beam power = 750 W; 40 × 10 mm $BaF_2$ scintillator on Hamamatsu H3378-50 PMT). At ELBE's maximum power of 40 kW (1mA at 40 MeV), a count rate of several $10^5$ cps is to be expected. MePS is designed to use this power level (sufficiently strong water-cooling of target and beam dump). The chopper pulse generator is not yet available, so the time resolution is only about 500 ps. However, this time resolution is sufficient for porosimetry investigations. First experiments were done to study low-k dielectric layers.

## 2. The setup of the MePS system

The ELBE electron beam has a unique time structure: very short electron bunches (< 5 ps) hit the electron-positron converter in cw-mode with a frequency of $26/2^n$ MHz (n = 0, 1, 2,...) in Cave 111b. The repetition frequency can be adjusted according to the requirements: 77 ns is used for positron lifetimes, while very long lifetimes can be measured by 616 ns repetition time (porosimetry with positronium). In the latter case, a higher bunch charge can be applied, so that the intensity will not decrease. A scheme of the system is shown in figure 1. The converter consists of a stack of 50 tungsten foils (0.1 mm each), which is water-cooled directly. A beam dump made from 5N-Al stops the electrons after they have passed the converter. The moderator is located in front of it and parallel to the entrance window of an Einzel lens. The moderated positrons are extracted by the transport energy of 2 keV. The lens focuses the positron bunch into a longitudinal magnetic guidance field. More than 30 steering coil pairs allow the adjustment of the beam inside the beamline at any position. The beam is fed from the cave 111b through a cable tunnel into the positron lab (cave 111d). The tunnel lies beneath a 3.2 m shielding of heavy concrete. Although the electron and thus the positron bunch is very small after pair production, when the bunch arrives in the lab 6 m away it is broader (about 10 ns). This is due to the moderation process, inhomogeneity in the acceleration field, and different paths in the beamline (three bends). Thus, a combination of chopper and double-slit buncher are needed to improve the timing. The chopper cuts a 4 ns window into the bunch, and the buncher will produce the time focus at the sample position.

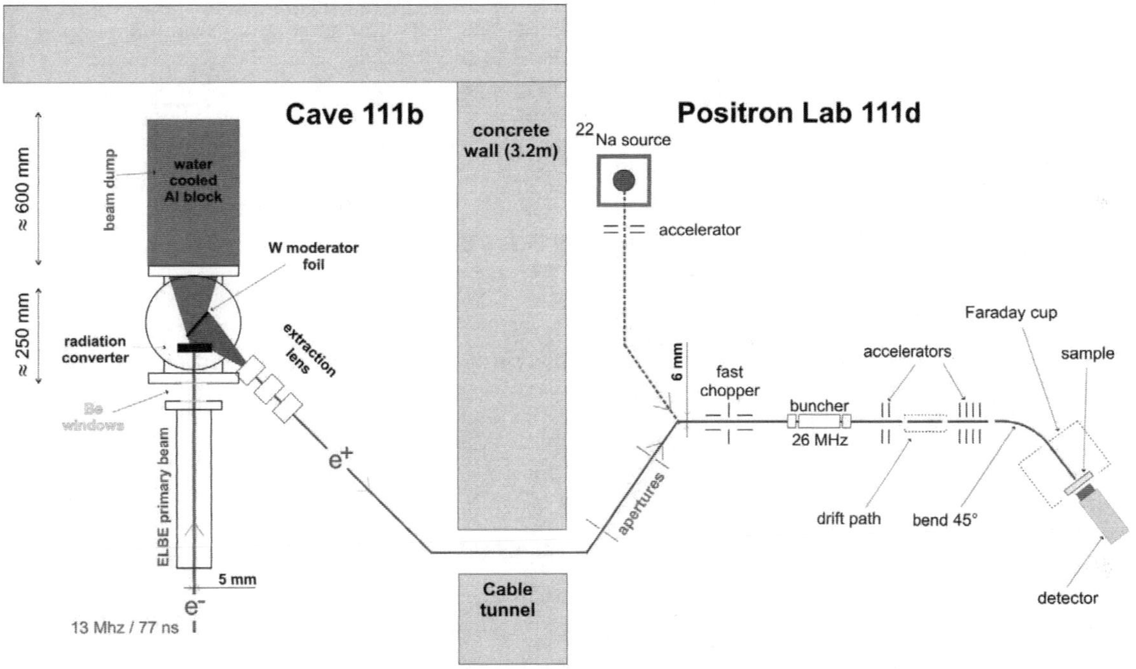


**Figure 1.** The scheme of MePS. Positron production in Cave 111b and the positron lab (Cave 111d) are separated by a 3.2 m concrete wall. The beam length is about 12 m [1,2]. The 22-Na source is a future extension of the system to allow experimental work when the electron beam is not available.

The shift of the time focus due to different acceleration voltages is compensated by a different velocity in a drift path and also by different RF-amplitudes in the buncher. After an additional beam bend of 45° (see discussion below) the sample is mounted in a Faraday cup. The PMT (Hamamatsu H3370-50) is located close to the sample but outside the vacuum, separated only by the glass of a viewing window.

### 3. Solving the problem of backscattered positrons

For high-z sample material, a large fraction of positrons is scattered back into the beamline [3]. They are guided by the longitudinal field until they reach the accelerator. There, they will be re-accelerated to the sample and, thus, will produce side peaks in the lifetime spectrum. Usually, an additional E×B filter in the beamline shall suppress these artifacts. However, this suppression is never complete. We used another approach: a 45° bend after the accelerator is equipped with steering coils. They guide the bunch through the bend to the sample, but they will not allow the backscattered positrons to reach the accelerator. They will hit the wall away from the detector. Figure 2 shows the effect of this filter: in

Journal of Physics: Conference Series **443** (2013) 012088 doi:10.1088/1742-6596/443/1/012088

the left panel the lifetime spectrum of a Cu sample is shown in a straight setup (no bend). The accelerator is working and one can clearly see a strong side peak about 70 ns after the main spectrum. In the right panel, the bend is mounted, and the sample was measured under otherwise identical conditions. The effect is clearly visible: there are no longer any spurious signals, and, in addition, the background is distinctly reduced.

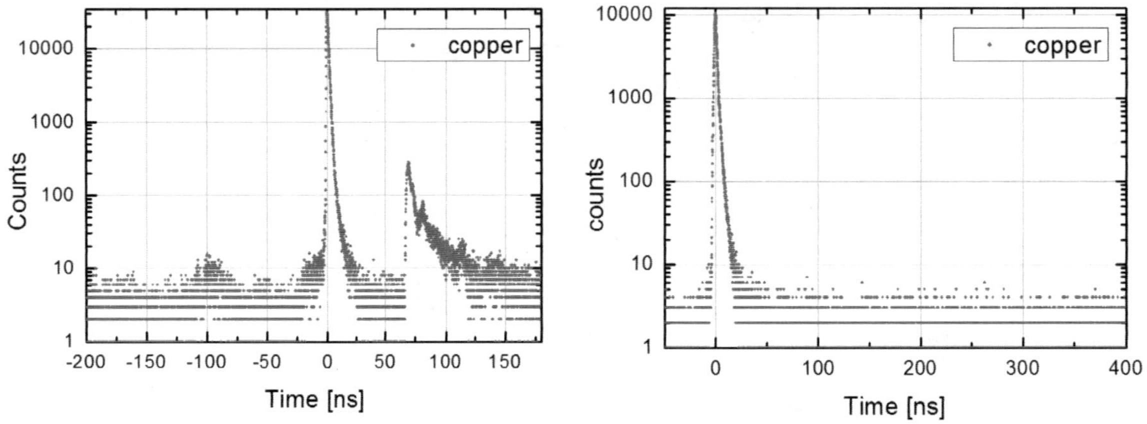


**Figure 2.** Lifetime spectra of Cu in a straight setup (left panel) and after assembly of a 45° bend into the beamline between accelerator and sample (right panel).

**4. Simulation of the required chopper pulse**

In order to specify the necessary pulse width for the chopper pulse, we performed an over-all time-dependent simulation with SIMION 8 from the moderator to the PMT.

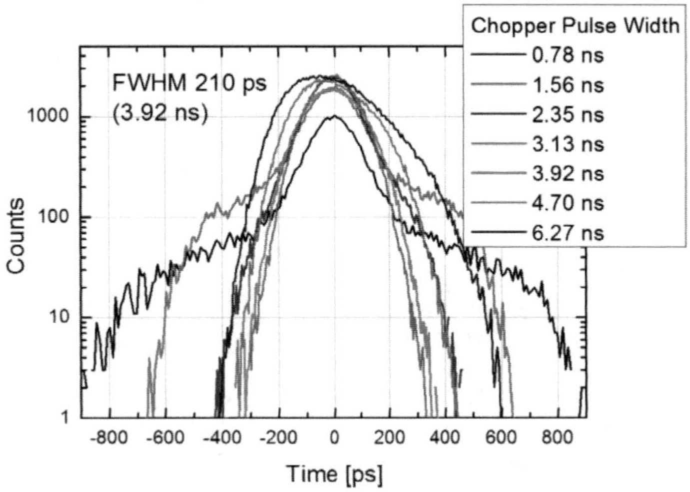


**Figure 3.** The simulated time structure at sample position as function of the width of a Gaussian chopper pulse. A pulse width of 4 ns is still acceptable.

Figure 3 shows the simulated time structure of the positron beam at the sample position for different chopper pulse widths. For a width of 3.9 ns, a time resolution of 210 ps is to be expected (contribution of PMT is included). Thus, the pulse generator must provide pulses with 13 MHz, 100V amplitude and a width of 4 ns and a nearly Gaussian shape. This demand should be realizable. In case of problems due to the high repetition frequency, one could use 616 ns (1.625 MHz) as standard mode for the lifetime measurements. For high count rates (> $10^5$), however, the intensity of long lifetimes can be decreased due to multiple hits of the detector during one bunch. Certainly, this disturbance can be corrected.

## 5. First use of MePS: Study of low-k dielectric layers

Low-k insulating layers are used to reduce the product R×C in ULSI integrated circuits in order to allow for higher clock frequencies. As a first experiment at MePS, we studied a set of such low-k dielectric layers, which were prepared variously.

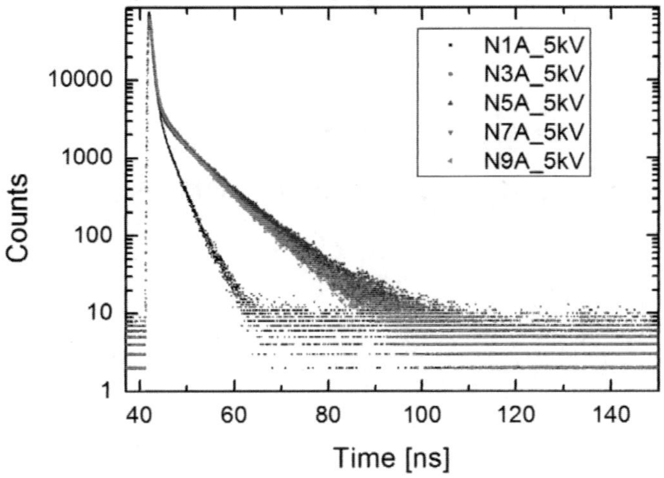


**Figure 4.** Positron lifetime spectra of 1 μm low-k layers measured at 5 keV (about 450 nm positron penetration depth). The samples were variously treated: as-received (N1), variously plasma treated (N2, N3) and with an additional TiN cap layer (N7, N9).

The samples were CVD grown with a low-k layer thickness of about 1 μm. Figure 4 shows the comparison of the spectra accumulated at 5 keV corresponding to a mean positron penetration depth of 450 nm. The spectra show hardly any spurious signals in the background. The peak-to-background ratio is about $10^4$. The spectra were recorded without the chopper (chopper plates grounded). Thus, the original pulse has only been treated by the buncher. The time resolution is 500 ps (FWHM), which is sufficient for porosimetry. The observed pore size at this particular depth was 0.68 nm for the untreated sample and 1.12 to 1.18 nm for the treated samples [4].

## 6. Acknowledgement

The authors would like to acknowledge the support of the BMBF under project number 05K10NHA. The help of R. Ecke and N. Ahner, Fraunhofer ENAS Chemnitz, is also acknowledged.

## 7. References

[1] Krause-Rehberg R, Brauer G, Jungmann M, Krille A, Rogov A and Noack K 2008 Applied Surface Science 255 22
[2] Krause-Rehberg R, Jungmann M, Krille A, Butterling M, Anwand W, Brauer G, Rogov A, Noack K, Cowan T E 2010 *Proceedings of the International School of Physics "Enrico Fermi"*, Course CLXXIV "Physics with many positrons", IOS Press Amsterdam 463-479
[3] Mäkinen J, Palko S, Martikainen J and Hautojärvi P 1992 J. Phys. Cond. Mat. 4 L503
[4] S. J. Tao 1972 J. Chem. Phys. 56 5499

16th International Conference on Positron Annihilation (ICPA-16) IOP Publishing
Journal of Physics: Conference Series **443** (2013) 012089 doi:10.1088/1742-6596/443/1/012089

# Determination of electron binding energies in two-dimensional Coincident Doppler Broadening spectra

**Philip Pikart, Christoph Hugenschmidt**
Technische Universität München, Physikdepartment E21, Lichtenbergstr. 1, 85747 Garching
E-mail: philip@pikart.de

**Abstract.**
In Coincident Doppler Broadening Spectroscopy (CDBS) the 1022 keV sum energy of the annihilation photons is utilized to validate an undisturbed detection of the two-gamma electron-positron decay. Due to conservation of energy the sum energy is lowered if the annihilating particles, in particular the electron, are in a bound state. Usually this effect is neglected because the binding energies of annihilated electrons are small compared to the energy of the annihilation photons. A novel data evaluation algorithm allows to clearly identify the influence of binding energies in the CDBS spectrum. Exemplary measurements are presented and compared to calculated spectra.

## 1. Introduction

As described in widely known publications [1–3] the low background of coincident Doppler broadening measurements is based on the evaluation of the sum energy $\hat{E}$ of an annihilating electron positron pair. It is regarded to be $\hat{E} = 2 \cdot m_0c^2 = 1022\,\text{keV}$ in the case of the detection of a two-$\gamma$ electron-positron-decay in two collinear detectors. However, the sum energy is not exactly 1022 keV, it is reduced by the binding energies of the annihilating particles $E_\text{B}$. While the binding energy of the diffusing positron is only a few eV [4], the binding energy of core electrons ranges up to many keV. Within the scope of this work, $\hat{E}$ is precisely evaluated in the CDBS spectra and the influences of the binding energy are analyzed.

## 2. Data evaluation

$E_1$ and $E_2$ are the energies of a collinearly emitted photon pair, which are both recorded in a two-dimensional multi-channel-analyzer (MCA) matrix. $\Delta E$ designates the Doppler shift of both annihilation, which is representing the longitudinal movement of the electron-positron pair:

$$\hat{E} = E_1 + E_2 = 2 \cdot m_0c - E_\text{B} = 1022\,\text{keV} - E_\text{B} \quad \Delta E = \frac{E_1 - E_2}{2} \tag{1}$$

Figure 1 illustrates the contributions of different electron orbitals of iron to different areas in the acquisition matrix for ideal data recorded with infinitely high resolution. The Doppler broadening leads to diagonal lines in the spectrum which are limited by the maximal momentum of the electron in the direction of the detector axis. All events on one diagonal line have the

Journal of Physics: Conference Series **443** (2013) 012089 doi:10.1088/1742-6596/443/1/012089

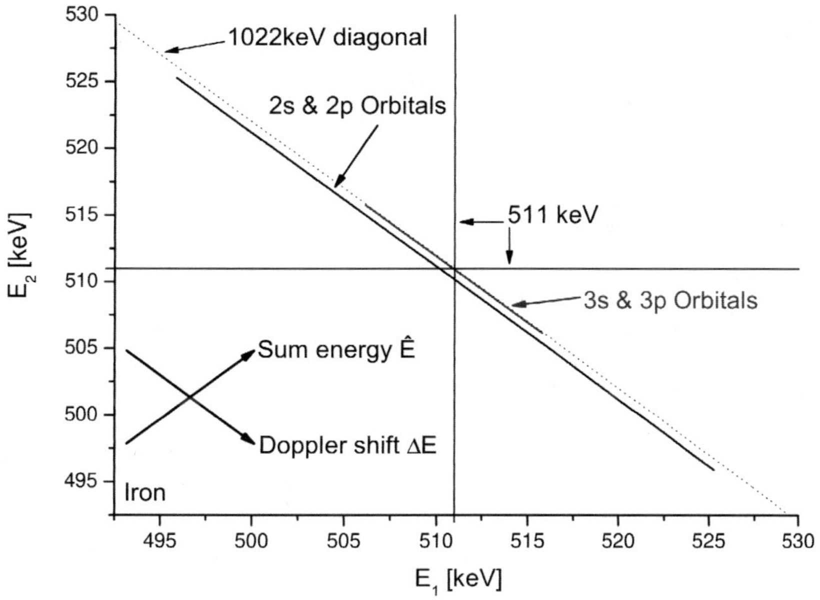


**Figure 1.** The schematic distribution of annihilation events from different orbitals in the CDB spectrum, exemplarily for iron. The diagonal arrows designate the axis of sum energy and Doppler shift. The distance of an event to the 1022 keV diagonal results from the binding energy $E_{\mathrm{B}}$.

same sum energy according to equation 1. The sum energy varies due to the electron binding energies. With increasing $E_B$ the core annihilation probability $P$ decreases and is negligible for $1s$-electrons (not shown).

In real measurements, these event lines widely overlap because of the finite energy resolution of $\approx$ 1.3 keV of HPGe-Detectors [5]. Consequently, the lines of different sum energy are not detected separately, but the variation of the mean sum energy $\hat{E}$ over the Doppler shift $\Delta E$ is expected to be observable.

For this task, a novel data evaluation algorithm is used. It determines the mean sum energy $\hat{E}(\Delta\, E)$ by fitting the Gaussian-shaped detector function to areas in the matrix with constant Doppler shift $\Delta\, E$. For details about the used algorithms, see [6].

## 3. Experiment

For the present experiment, annealed iron was chosen as sample material due to its 2s and 2p orbitals, which annihilate with positrons with a probability $P$ of 0.034 % and 0.076 % [7]. The binding energy of these orbitals is 845 eV and 720 eV respectively. In the acquisition matrix, the according events are dominating in the outer wings of the Doppler broadened photo peak - i.e. for $|\Delta E| > 5$ keV - as it is visible in figure 1, whereas the 3p and 3s orbitals practically overlap with the 1022 keV center peak due to their low binding energy of $<$ 100 eV. Although the annihilation rate with the 2s and 2p orbitals is below 0.1%, the background in the relevant quadrants of the spectrum is in the range of $10^{-5}$ compared to the photopeak, so that a measuerment of 2p and 2s electron annihilations is feasible. The mean sum energy $\hat{E}$ in the range of high Doppler shifts $|\Delta E| > 5$ keV is expected to be significantly reduced by the electron binding energy.

## 4. Calculations

For the calculation of the sum energy as a function of the Doppler shift $\hat{E}(\Delta E)$, the contributions from the relevant orbitals are calculated separately. The intensity $I_n(\Delta E)$ of the single orbitals $n$, which represent one diagonal line in figure 1, is approximated by a parabolic function due to the projection of an isotropic electron momentum distribution to one dimension. The intensity of each line is set to the respective core annihilation probability $P_n$. $E_{\max,n}$ is the maximal Doppler shift regarding the orbital $n$.

$$I_n(\Delta E) = \frac{\sqrt{(\Delta E_{\max,n})^2 - (\Delta E)^2}}{\frac{4}{3} \cdot (\Delta E_{\max,n})^3} \cdot P_n \tag{2}$$

The spectra are folded with a Gaussian function with a $E_{\mathrm{FWHM}}$ of 1.05 keV which represents the coincident detector resolution $Res(E)$.

$$Res(E) = \frac{1}{\sqrt{\frac{\pi \cdot E_{\mathrm{FWHM}}}{2 \cdot ln2}}} \cdot e^{-\frac{1}{2}\left(\frac{E \cdot 2\sqrt{2ln(2)}}{E_{\mathrm{FWHM}}}\right)^2} \tag{3}$$

$\bar{I}_n(\Delta E)$ is the intensity distribution function of orbital $n$ after convolution with the Gaussian detector function.

$$\bar{I}_n(\Delta E) = (I_n(\Delta E) * Res(E(\Delta E))) \tag{4}$$

The resulting $\hat{E}(\Delta E)$ is the average sum energy of all events as a function of the specific Doppler shift $\Delta E$. $n_{max}$ is the number of orbitals included in the calculation. $E_{\mathrm{B}\,n}$ denotes the binding energy of the respective orbital.

$$\hat{E}(\Delta E) = \frac{\sum_{n=1}^{n_{max}} \bar{I}_n(\Delta E) \cdot (1022keV - E_{\mathrm{B}\,n})}{\sum_{n=1}^{n_{max}} \bar{I}_n(\Delta E)} \tag{5}$$

## 5. Results

In figure 2 the measured and the calculated sum energy are shown. The statistical error of the measured values is calculated by variance analysis of neighbored but independent evaluation points.

In the momentum area of $15 \cdot 10^{-3} m_0 c < |p| < 30 \cdot 10^{-3} m_0 c$ a significant decrease of $\hat{E}$ is visible. This sum energy deficit is in good agreement with the calculated data. The deviation from the calculated data can be explained by the strongly simplified momentum distribution of the single electron orbitals. For $|p| > 30 \cdot 10^{-3}\, m_0 c$ the measured values scatter largely since the number of events in the evaluated area was to small to detect a stable center of the peak.

## 6. Conclusion and Outlook

The measurement of the binding energies provides an additional element-specific parameter compared to conventional evaluation of Doppler broadening only. It was shown that the two-$\gamma$ sum energy as a function of the Doppler shift can be extracted from a two-dimensional CDBS acquisition matrix with a standard event number of $\approx 1.5 \cdot 10^7$ counts. For future applications, this may help interpreting ambiguous CDB results which occur when alloys or elements with a similar CDB-spectrum are measured.

To get more exact $\hat{E}(\Delta E)$-spectra, more experiments with higher statistics and better energy resolution have to be performed. Furthermore material combinations, like iron and cobalt are to be investigated. They carry an almost identical CDBS shape, but may be distinguished by different binding energies.

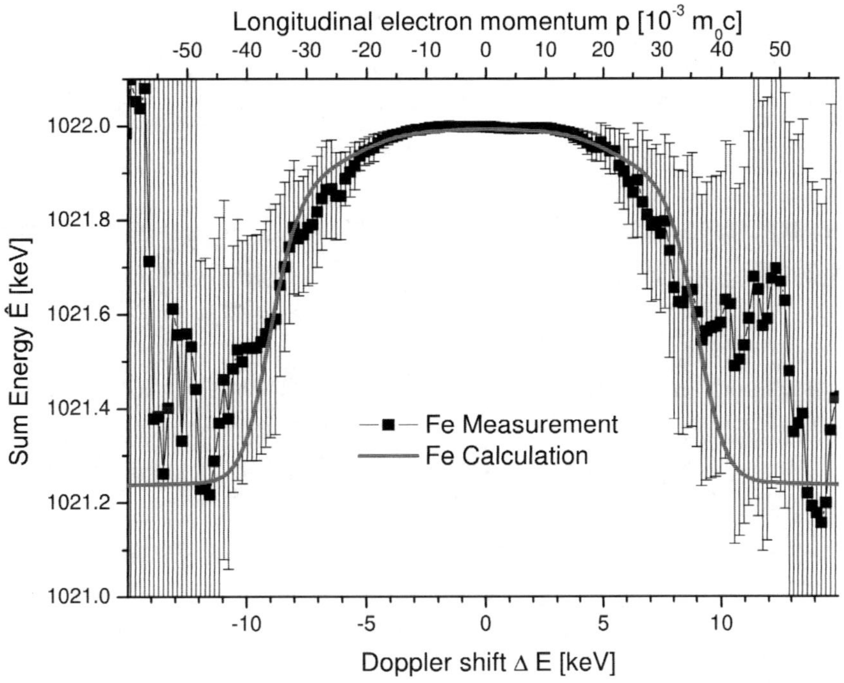


**Figure 2.** For each Doppler shift $\Delta E$ the center of weight $\hat{E}$ of the contributing events is calculated. The sum energy deficit is caused by the binding energy of the annihilated electron.

## 7. Acknowledgments

The authors sincerely want to thank the BMBF (Bundesministerium für Bildung und Forschung) for funding of the HPGe-detector equipment and especially Prof. Peter Böni and the E21 chair of the Physics Department at the Technische Universität München for the possibility to complete the first authors PhD-thesis about CDBS.

[1] K. G. Lynn, J. E. Dickman, W. L. Brown, M. F. Robbins, and E. Bonderup, 1979, *Phys. Rev. B*, 20:3566–3572.
[2] P. Asoka-Kumar, M. Alatalo, V. J. Ghosh, A. C. Kruseman, B. Nielsen, and K. G. Lynn, 1996, *Phys. Rev. Lett.*, 77(10):2097–2100.
[3] P. E. Mijnarends, A. C. Kruseman, A. van Veen, H. Schut, and A. Bansil, 1998, *J. Phys.: Condens. Matter*, 10:10383–10390.
[4] B. L. Brown, W. S. Crane, and Jr. A. P. Mills, 1986. *Appl. Phys. Lett.*, 48(11):739–741.
[5] G. F. Knoll, 2000, *Radiation Detection and Measurement.* John Wiley & Sons, Inc..
[6] Philip Pikart, 2012, Dissertation, Technische Universität München, Lichtenberstr. 1 85747 Garching Germany.
[7] K. O. Jensen and A. Weiss, 1990, *Phys. Rev. B*, 41(7):3928–3936.

Journal of Physics: Conference Series **443** (2013) 012090 doi:10.1088/1742-6596/443/1/012090

# Development of a sample chamber with humidity control for an atmospheric positron probe microanalyzer

**W Zhou[1,*], N Oshima[1], Z Chen[2], K Ito[2], B E O'Rourke[1], R Kuroda[1], R Suzuki[1], H Yanagishita[3], T Tsutsui[4], A Uedono[4] and N Hayashizaki[5]**

[1]Research Institute of Instrumentation Frontier (RIIF), National Institute of Advanced Industrial Science and Technology (AIST), Tsukuba, Ibaraki 305-8568, Japan
[2]Metrology Institute of Japan (MIJ), National Institute of Advanced Industrial Science and Technology (AIST), Tsukuba, Ibaraki 305-8565, Japan
[3]Research Institute for Innovation in Sustainable Chemistry (ISC), National Institute of Advanced Industrial Science and Technology (AIST), Tsukuba, Ibaraki 305-8565, Japan
[4]Division of Applied Physics, Faculty of Pure and Applied Science, University of Tsukuba, Tsukuba, Ibaraki 305-8573, Japan
[5]Research Laboratory for Nuclear Reactors, Tokyo Institute of Technology, Meguro, Tokyo 152-8550, Japan

zhou-w@aist.go.jp

**Abstract.** In order to perform positron lifetime measurements on thin films under atmospheric conditions, a slow positron microbeam was extracted into air using silicon nitride thin films (30 nm and 200 nm) as a vacuum window. Even the thinner window (30 nm) was found to reliably withstand a differential pressure of 1 atm under various stress tests. By placing the sample in an enclosed chamber through which gas with a fixed, controllable relative humidity (RH) was continuously passed, the RH dependence of the *ortho*-positronium lifetime for bulk fused silica was examined.

## 1. Introduction

The positron annihilation technique, based on keV-energy positron beams, is a powerful tool for investigating atomic defect and sub-nanometer scale pores of thin films. In practical measurements slow positrons are generated in a high vacuum, therefore, it is necessary to place samples in a vacuum chamber. Recently, extraction of slow positrons into air has been achieved with the help of a pulsed positron microbeam, produced for the positron probe microanalyzer (PPMA) in AIST[1-3]. This method is promising to be apposite for exploring nanoscopic holes of thin films under practical conditions. In this paper, we report details of reliability tests of the vacuum window and the humidity controller, specially designed for PPMA, and preliminary results of variable-energy positron lifetime measurements for bulk fused silica, performed at various humidity conditions using the developed system.

---

[*] To whom any correspondence should be addressed.

## 2. Atmospheric PPMA

A schematic view of the atmospheric PPMA[4,5] in AIST is shown in Fig. 1. An intense positron beam produced by an electron linear accelerator (LINAC) is focused by a magnetic coil on a remoderator. The brightness enhanced positrons are transferred from the remoderator and are extracted to air through a vacuum window. The window is a silicon nitride (SiN) thin film on a silicon substrate, providing a tiny opening typically less than 1 $mm^2$ for extraction of positrons to air through the SiN film. Then the positron microbeam is focused by an objective lens down to a final beam size of 200 $\mu$m[1] at the vacuum window. The vacuum window is placed on the central hole of the terminal metal flange of PPMA, and is fixed by epoxy glue[5]. The SiN film is a fragile part in the vacuum system, thus accidental breakage could cause serious damage to the vacuum system due to the rapid leak of air. In consideration of various working conditions, for instance, high humidity and/or temperature, the SiN film should not break for at least 10 h to achieve the total number of counts during positron lifetime measurements under high relative humidity (RH) at 30 $^{\circ}$C. It is therefore necessary to ensure the reliability of the SiN windows.

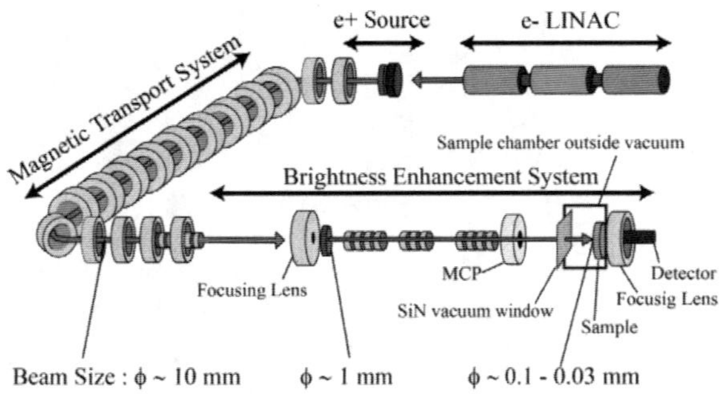


Fig. 1. Schematic illustration of the atmospheric PPMA[1].

## 3. Strength tests for SiN vacuum window

Materials such as Be, Ti and SiN are often used as vacuum windows. In our apparatus, the areal density of the window should be as small as possible to optimize the efficiency of extracting low energy positrons into air. Based on this condition SiN was chosen for the window. SiN films with thicknesses of 30 nm and 200 nm (density ~ 3 g/$cm^3$) were purchased from Silson Company. For the strength tests we built a simple vacuum system, consisting of a scroll pump, a leak valve and a vacuum gauge that are connected by a flexible metal vacuum tube with each other (Fig. 2). The SiN vacuum window was fixed at the end of this system. Three different strength tests were performed; (a) water adsorption, (b) heat treatment, and (c) sample contact, and each test was performed at least 10 times for each SiN film. During the tests the degree of vacuum was kept constant around 5.3 Pa. The details and a summary of the strength tests of SiN windows are shown in Fig. 3 and Table 1, respectively.

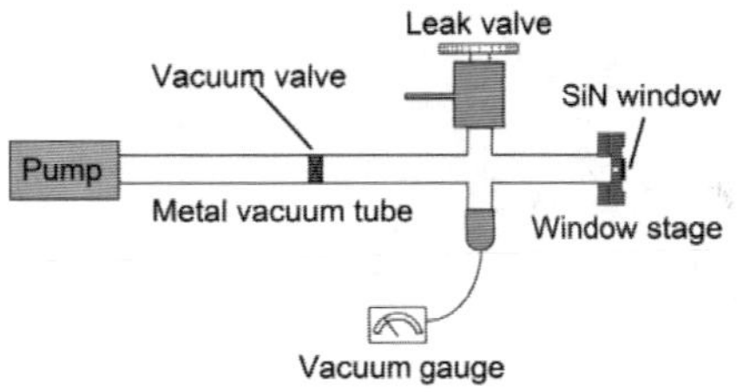


Fig. 2. The vacuum system for strength tests.

Table 1. Summary of the strength tests for SiN windows

| Properties | 30-nm thick | 200-nm thick |
|---|---|---|
| Lateral size | 0.6 mm×0.6 mm | 1.5 mm×1.5 mm |
| Vacuum degree (Pa) | 5.3 | 5.3 |
| Displacement of the window center | 35 $\mu$m ~40 $\mu$m | 65 $\mu$m ~70 $\mu$m |
| Water resistance (immerse in water for 10 h) | Succeed | Succeed |
| Temperature resistance (90 $^{\circ}$C~110 $^{\circ}$C for 20 min) | Succeed | Succeed |
| Sample directly contact | Failed | Succeed |
| Sample mounted with Kapton spacers | Succeed | Succeed |

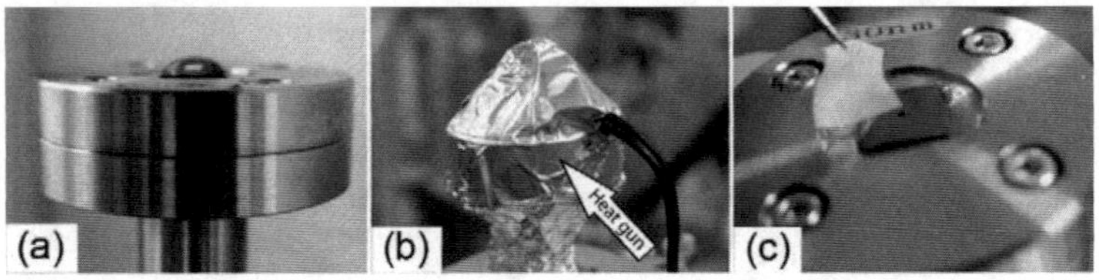


Fig. 3. Strength tests for the SiN vacuum window. (a) The SiN window was immersed in distilled water completely and then left to dry by evaporation at room temperature. The SiN window worked well under the effect of the surface tension of a water-drop; (b) The vacuum window and the stage were covered by 4 layers of aluminium foil. A thermal sensor was placed near the SiN window to monitor the temperature. The window was heated by a heat gun to an enough high temperature from the direction shown by arrow. The SiN window also worked normally; (c) A piece of a soft rubber film was put onto the thin vacuum window directly. In this test, the 30-nm SiN window was broken.

The 30-nm-thick SiN window failed the direct sample contact test. For this reason, 35-μm Kapton foils were placed between the sample and the vacuum window as a spacer when the 30-nm-thick SiN was used for the positron measurements. The kapton spacers keep the SiN window away from sample and protect the window preventing direct sample contact. After these three strength tests, both the 30-nm and 200-nm-thick SiN films were confirmed applicable to the vacuum window of the atmospheric PPMA.

## 4. Sample chamber with humidity controller

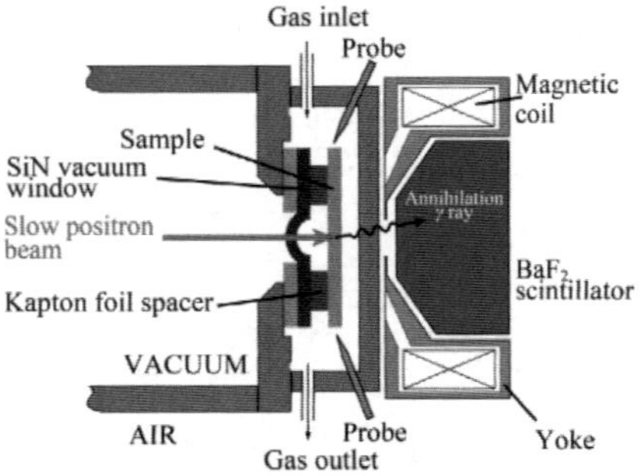


Fig. 4. Schematic illustration of the sample chamber with humidity controller (not to scale).

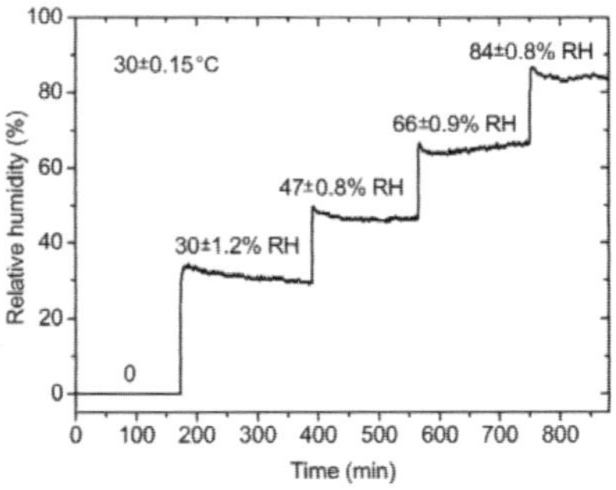


Fig. 5. Response and stability of the RH in sample chamber.

A sample chamber equipped with a humidity controller (Makuhari Rikagaku Garasu, Inc.), specially designed for the atmospheric PPMA, was installed on the air side of the SiN window (Fig. 4). A $BaF_2$ scintillator was placed outside the sample chamber to detect the annihilation $\gamma$-rays of the positrons. A magnetic coil was used as an objective lens to regulate the positron beam diameter[2]. The beam diameter at the sample was adjusted to be smaller than the lateral size of the SiN window. The slow positrons will reach the sample surface after crossing the SiN film and air. The transmission efficiency of positrons in SiN, the positron beam properties emerging from the SiN thin window and depth profile of positrons in a sample are discussed elsewhere[6]. The humidity controller and sample chamber were electrically floated from the ground level to supply the accelerating bias for adjusting the positron injection energy into the sample. A metal lid was used to seal the sample chamber, and a nitrogen gas with different levels of humidity were passed continuously through the chamber, allowing a fixed and controllable relative humidity and keeping the pressure constant at 1 atm. Two probes were used to monitor the humidity and temperature inside the chamber. By controlling the flow rates of a dry nitrogen gas and a nitrogen gas saturated with water vapor, RH inside the chamber can be varied from 0 to over 90%. The response and stability of RH in sample chamber is shown in Fig. 5. RH in the chamber stabilizes within 2 min with a standard deviation not more than 1.2%, when changing from one value to another value. Note that, based on Fick's law of diffusion, the time of water molecules at 30 °C to pass through the channel between the SiN window and the sample is estimated as ~10 s, which is much shorter than the time to get the stability of the humidity condition in the chamber (~2 min).

## 5. Performance of the atmospheric PPMA

In order to check if the *ortho*-positronium (*o*-Ps) lifetime is correctly measured even if the relative humidity varies, bulk fused silica with an *o*-Ps lifetime of ~1.6 ns[5] was selected as a standard sample. The RH dependence of the *o*-Ps lifetime for bulk fused silica is shown in Fig. 6. The 30-nm-thick SiN film was selected as the vacuum window for the measurements. The positron incident energy $E$ was 4 keV and the beam diameter was 200 $\mu$m. The total count in each spectrum was $1\times10^6$. The thickness of the fused silica sample was 0.6 mm with the lateral size of 1.5 cm × 1.5 cm. At high RH, the thickness of adsorbed water layer on the surface of fused silica is reported to be less than 1 nm[8], so that the effect of the water layer on the lifetime measurements at $E$ = 4 keV is negligible. As in Fig. 6, the measured *o*-Ps lifetimes for the fused silica (~1.6 ns) show no dependence on RH (0, 43% RH and 85% RH). This result indicates that the apparatus worked well during the lifetime measurements.

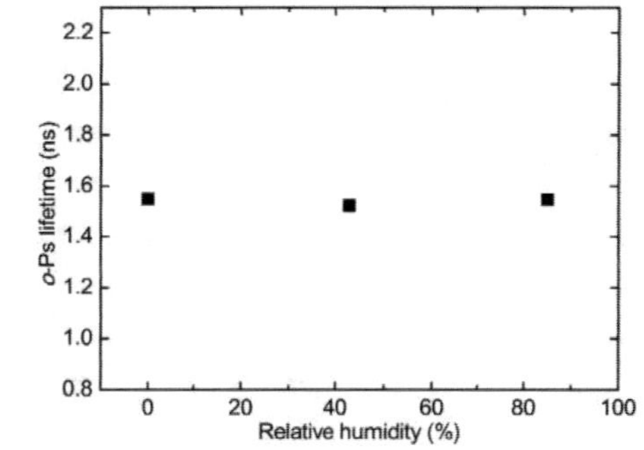


Fig. 6. Relative humidity dependence of the *o*-Ps lifetime for the fused silica. The relative measurement uncertainty was evaluated not more than 5%.

## 6. Conclusion

A sample chamber equipped with a humidity controller for the atmospheric PPMA has been developed. The SiN thin films (thickness 30 nm and 200 nm, lateral size 0.6 mm × 0.6 mm and 1.5 mm × 1.5 mm, respectively) were confirmed suitable for the vacuum window of the atmospheric PPMA. The humidity controller was constructed to provide variable and stable relative humidity conditions in the sample chamber. The RH dependence of the *o*-Ps lifetimes for the fused silica was measured to confirm the normal operation of the atmospheric PPMA.

This work was supported by NEDO. The SiN shape measurement was supported by the AIST Nano-Processing Facility.

## References

[1] Oshima N, Suzuki R, Ohdaira T, Kinomura A, Narumi T, Uedono A, and Fujinami M 2009 *Radiat. Phys. Chem.* **78**, 1096

[2] Oshima N, Suzuki R, Ohdaira T, Kinomura A, Narumi T, Uedono A, and Fujinami M 2009 *Appl. Phys. Lett.* **94**, 194104

[3] Oshima N, Suzuki R, Ohdaira T, Kinomura A, Narumi T, Uedono A, and Fujinami M 2008 *J. Appl. Phys.* **103**, 094916

[4] Zhou W, Chen Z, Oshima N, Ito K, O'Rourke B E, Kuroda R, Suzuki R, Yanagishita H, Tsutsui T, Uedono A, and Hayashizaki N 2012 *Appl. Phys. Lett.* **101**, 014102

[5] Oshima N, O'Rourke B E, Kuroda R, Suzuki R, Watanabe H, Kubota S, Tenjinbayashi K, Uedono A, and Hayashizaki N 2011 *Appl. Phys. Express* **4**, 066701

[6] O'Rourke B E, et al., this proceedings

[7] Doty P M, Aiken C, and Marh H 1944 Analytical Edition **16**, 686

[8] Arima K, Jiang P, Deng X Y, Bluhm H, and Salmeron M 2010 J. Phys. Chem. C **114**, 14900

Journal of Physics: Conference Series **443** (2013) 012091 doi:10.1088/1742-6596/443/1/012091

# Position-resolved Positron Annihilation Lifetime Spectroscopy

**A Wagner[1], M Butterling[1], F Fiedler[1], F Fritz[1,2], M Kempe[1,2] and T E Cowan[1,2]**

[1]Institute for Radiation Physics,
Helmholtz-Zentrum Dresden-Rossendorf, POB 510119, D-01314 Dresden, Germany
[2]Technische Universität Dresden, D-01062 Dresden, Germany

E-mail: a.wagner @hzdr.de

**Abstract.** A new method which allows for position-resolved positron lifetime spectroscopy studies in extended volume samples is presented. In addition to the existing technique of in-situ production of positrons inside large ($cm^3$) bulk samples using high-energy photons up to 16 MeV from bremsstrahlung production, granular position-sensitive photon detectors have been employed. A beam of intense bremsstrahlung is provided by the superconducting electron linear accelerator ELBE (Electron Linear Accelerator with high Brilliance and low Emittance) which delivers electron bunches of less than 10 ps temporal width and an adjustable bunch separation of multiples of 38 ns, average beam currents of 1 mA, and energies up to 40 MeV. Since the generation of bremsstrahlung and the transport to the sample preserves the sharp timing of the electron beam, positrons generated inside the entire sample volume by pair production feature a sharp start time stamp for positron annihilation lifetime studies with high timing resolutions and high signal to background ratios due to the coincident detection of two annihilation photons. Two commercially available detectors from a high-resolution medial positron-emission tomography system are being employed with 169 individual $Lu_2SiO_5$:Ce scintillation crystals, each. In first experiments, a positron-lifetime gated image of a planar Si/$SiO_2$ (pieces of 12.5 mm × 25 mm size) sample and a 3-D structured metal in Teflon target could be obtained proving the feasibility of a three dimensional lifetime-gated tomographic system.

## Introduction

Positron Annihilation Lifetime Spectroscopy (PALS) is a powerful tool which allows for studies of crystal lattice defects on the nanometer scale and at low defect concentrations, open volumes in polymers, porosity and others. Especially, lattice defects like dislocations, mono vacancies and vacancy clusters, as well as open-volume cavities are perfectly suited for investigations using positrons. Several techniques have been developed over the last decades which make use of either kinematical observables of the annihilation radiation of positrons with electrons from the sample materials and of the annihilation lifetime of positrons after injection into the sample material. Especially, information about the electron density at the annihilation site is generated through Positron Annihilation Lifetime Spectroscopy (PALS) which correlates the distribution of annihilation lifetimes with the concentration and the type of defects.

Widespread standard techniques make use of radioactive positron sources like $^{22}$Na where the positron emission is accompanied by electromagnetic transitions from excited states in the daughter nuclei. Because photon emission is prompt with respect to the positron emission the time resolution of the additional photon detector adds to the overall achieved accuracy in lifetime measurements. This additional contribution can in turn be avoided using an accelerator-based source of positrons which furthermore offers the advantage of possible adjustments of the source strength and the time structure. We will describe in the following chapters, how a versatile source of positrons using a superconducting electron accelerator has been realized and used in position-resolved positron annihilation lifetime spectroscopy.

**The accelerator facility ELBE**

Starting shortly after the dawn of the new millennium a superconducting electron accelerator at the Helmholtz-Zentrum Dresden-Rossendorf came online [1]. Being the result of a development based on superconducting accelerating structures in the scope of the Tesla collaboration [2] a powerful new electron accelerator has been constructed which currently offers beam currents of up to 1.6 mA at maximum beam energies of 40 MeV. The facility is named Electron Linear Accelerator with high Brilliance and low Emittance (ELBE) and its high-intensity electron beam is being used as a driver for various types of secondary radiation e.g. coherent infrared light, THz radiation, photo-neutrons, bremsstrahlung, and positrons. A layout of the facility and the various end stations are shown in Fig. 1. Employing superconducting technology allows easy adjustment of the electron beam time structure which is crucial for time-of-flight experiments or lifetime experiments as described here. The micro-pulse repetition rate can be selected as $2^n$ divisors of 26 MHz with n=0…8. Typically, the micro-pulse repetition rate for positron annihilation lifetime experiments is chosen to be 26 MHz or 13 MHz with micro-pulse intervals of 38 ns or 77 ns, respectively. The micro-pulse width has been measured to be less than 5 ps using electro-optical sampling.

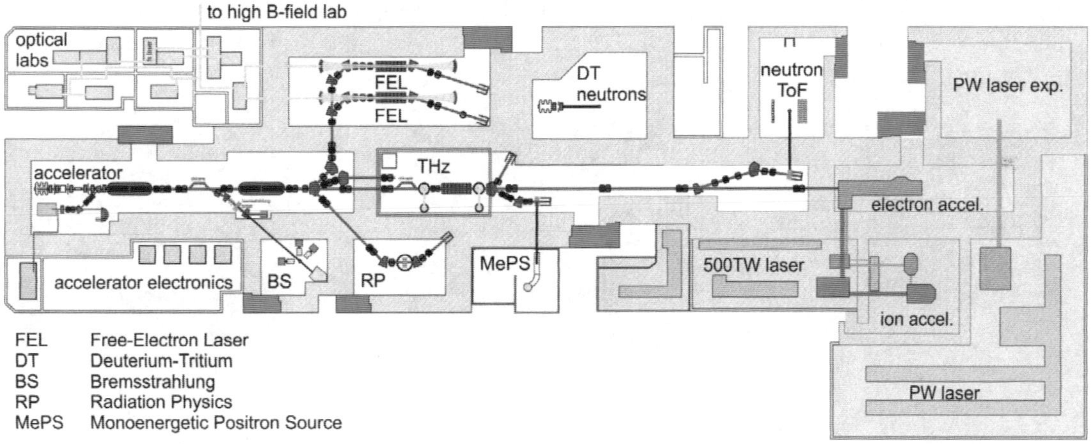


**Figure 1.** Layout of the ELBE centre for high power radiation sources. The overall width is about 100 m. Experiments discussed here took place at the bremsstrahlung facility labelled BS.

**Position-resolved Positron Annihilation Lifetime Spectroscopy**

The gamma-induced positron annihilation lifetime facility [3,4] has been extended by a set of position-sensitive photon detectors which will allow reconstructing a three-dimensional image of the distribution of positron lifetimes inside bulk samples. Two pixelated photon detectors each made from 13 × 13 crystals of $Lu_2SiO_5$ of 4 × 4 × 20 $mm^3$ volume [5] have been set up at the facility. Figure 2 shows a sketch of the setup with both detectors fixed perpendicular to the incoming photon beam. Each detector is equipped with 4 photomultiplier tubes. Dedicated preamplifier and discriminator electronics have been developed in-house in order to achieve a high time resolution of 450 ps per detector. The photon energy deposition is calculated using the sum of all four charge-integrated signals and individually calibrated crystal responses. The obtained energy resolution is 12.1% (FWHM) and 13.3% (IFWHM) at 511 keV photon energy for both detectors, respectively. Signal partitioning between the four photomultiplier tubes of one detector allows identification of the crystal in which the photon has interacted.

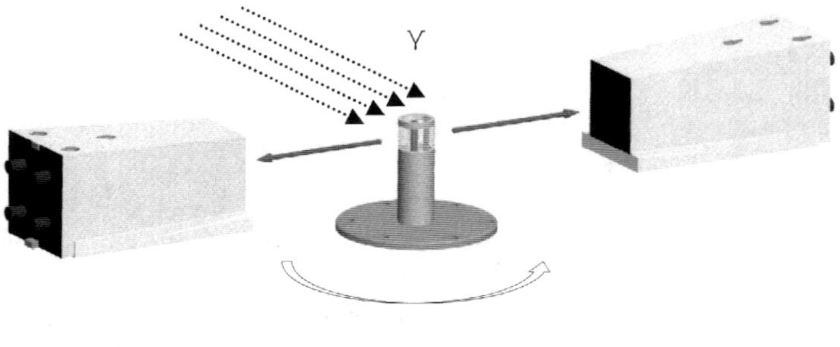


**Figure 2.** Sketch of the positron annihilation lifetime tomography setup. The bremsstrahlung beam hits the sample which is mounted on a rotational stage allowing for 3D image reconstruction. Two pixelated $Lu_2SiO_5$ detectors detect both annihilation quanta in coincidence.

The new system has been successfully tested with a two-dimensional structure made from Si/$SiO_2$ which had not been rotated during the experiment. Equal-sized pieces of monocrystalline Silicon and microscope slides of dimensions 12.5 × 25 × 0.8 $mm^3$, see Fig. 3, have been fixed in between two thin-walled Kapton sheets and mounted parallel to the beam direction. Fig. 4 shows the projected images derived from correlated events between both detectors. The annihilation lifetime has been derived from the time difference between the mean time of all 8 photomultiplier signals and the radio-frequency of the accelerator.

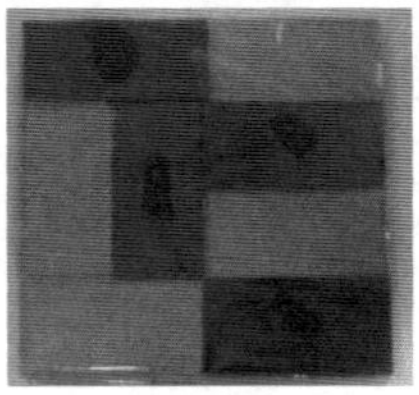

**Figure 3.** 2D target consisting of Si wafer material (dark) and microscopy slides (light) wrapped into Kapton foils.

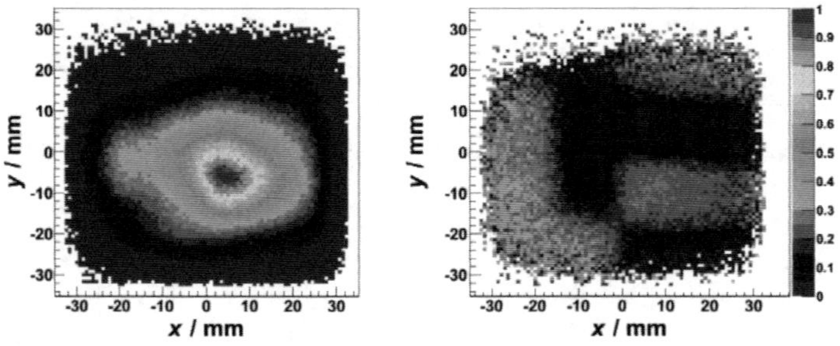


**Figure 4.** Two-dimensional distribution of two-photon annihilation integrated over all positron annihilation lifetimes (left) and the ratio of intensities gated for annihilation lifetimes in excess of 220 ps (1 σ) by all lifetimes (right).

The lifetime-integrated distribution shows no distinct features except an enhancement in the centre of the image due to the increased solid angle for correlated detection. Gating on positron lifetimes in excess of 220 ps (1 σ of the timing resolution ) clearly discriminates for regions inside the sample with enhanced formation of o-Ps, namely $SiO_2$. Both materials have been selected in order not to emphasize areas with different pair production yields. The timing resolution has not been corrected for the inaccuracy of time-zero in the direction of the beam (x-axis in Fig. 4) which amounts to about 48 ps (RMS). Perpendicular to the beam, taking the mean timing of both detectors the annihilation lifetime inaccuracy due to different positron production sites cancels. From Fig. 4 a lateral position resolution

of 2.8 mm has been obtained by fitting one of the boundaries with the Gauss error function. Further improvement and the reduction of image artefacts will be subject of a future publication.
As a fully three-dimensional example, a sample has been constructed consisting of a 25 mm diameter Teflon cylinder with embedded slabs made from Copper, Iron, and Aluminium having the same volume (12 mm$^2$ × 25 mm) but different geometrical shapes, see Fig. 5.

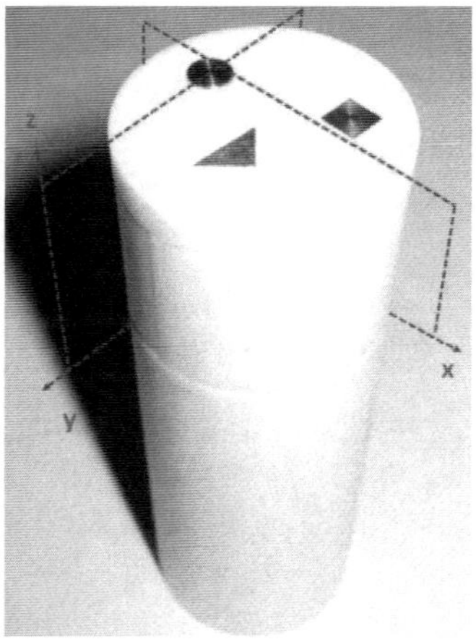


**Figure 5.** Three-dimensional sample made from Teflon with embedded slabs made from Cu, Fe, and Al. Dashed lines indicate the projections shown in Fig. 6.

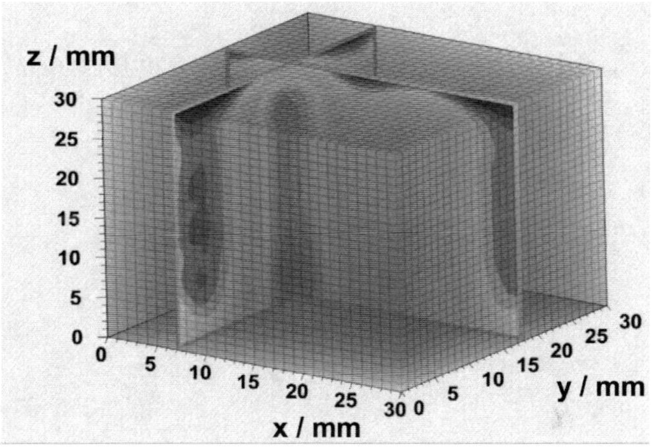


**Figure 6.** 3-D reconstruction of positron annihilation events gated on positron lifetime inside the sample shown in Fig. 5. Red colour indicates an enhancement of shorter annihilation lifetimes.

Figure 6 shows two perpendicular cuts through the 3-D reconstructed positron lifetime distribution. Clearly, the regions with shorter annihilation lifetimes can be identified. The newly developed system complements earlier developments of a positron annihilation microprobe which enabled high-resolution defect analysis at surfaces [6] or a high-energy positron beam for PALS studies in bulk material [7]. Further developments of the presented system aim at improving the position resolution by using smaller scintillator crystals and improved timing resolution.

Acknowledgements
The authors like to thank M. Sobiella, A. Hartmann, and K.Heidel for technical support and the ELBE accelerator crew for stable operations.

---

References:

[1] Gabriel F, Gippner P, Grosse E, Janssen D, Michel P, Prade H, Schamlott A, Seidel W, Wolf A and Wünsch R 2000 *Nucl. Inst. Meth. Phys. Res.* **B 161** 1143
[2] Koepke K, 1995 *Nucl. Inst. Meth. Phys. Res.* **B 99** 706
[3] Butterling M, et al., 2010 *Phys. Status Solidi* **A 207** 334
[4] Butterling M, Anwand A, Cowan T E, Hartmann A, Jungmann M , Krause-Rehberg R, Krille A and Wagner A, 2011 *Nucl. Inst. Meth. Phys. Res.* **B 269** 2623
[5] Siemens: Biograph® PET scanner: www.medical.siemens.com
[6] Haaks M, Bennewitz K, Bihr H, Männig U, Zamponi C, Maier K 1999 *Appl. Surf. Sci.* **149** 207
[7] Oshima N, Kuroda R, He C, Kinomura A, Ohdaira T, Toyokawa H, Suzuki R, Hayashizaki N, Hattori T 2007 *Phys. Status solidi* **C4** 4023

16th International Conference on Positron Annihilation (ICPA-16) IOP Publishing
Journal of Physics: Conference Series **443** (2013) 012092 doi:10.1088/1742-6596/443/1/012092

# First 2D-ACAR Measurements on Cu with the new Spectrometer at TUM

**J A Weber[1], P Böni[1], H Ceeh[1], M Leitner[2,3] and Ch Hugenschmidt[1,2]**

[1] Physik-Department E21, Technische Universität München, James-Franck-Straße 1, 85748 Garching, Germany

[2] FRM II, Technische Universität München, Lichtenbergstraße 1, 85748 Garching, Germany

[3] Physik-Department E13, Technische Universität München, James-Franck-Straße 1, 85748 Garching, Germany

E-mail: josef-andreas.weber@frm2.tum.de

**Abstract.** The two-dimensional measurement of the angular correlation of the positron annihilation radiation (2D-ACAR) is a powerful tool to investigate the electronic structure of materials. Here we report on the first results obtained with the new 2D-ACAR spectrometer at the Technische Universität München (TUM). To get experience in processing and interpreting 2D-ACAR data, first measurements were made on copper. The obtained data are treated with standard procedures and compared to theoretical calculations. It is shown that the measurements are in good agreement with the calculations and that the Fermi surface can be entirely reconstructed using three projections only.

## 1. Introduction

When a positron annihilates in a solid with an electron the resulting annihilation radiation carries the momentum information of the electron-positron pair. This information is contained in a Doppler shift from the mean value of 511 keV, and the angular deviation from exact antiparallel directions of the two $\gamma$ quanta. With 2D-ACAR the latter is measured, which yields a specific projection of the two $\gamma$ momentum density $\rho^{2\gamma}$. Since a positron thermalizes within a few ps [1] the angular deviation is caused mainly by the electron momentum in the solid.

The low momenta that are measured with 2D-ACAR are due to occupied states near the Fermi surface in the case of a metal. In these states, any electron is in a delocalised Bloch state. If we represent the according wave function in the basis of momentum eigenfunctions, it has to be expressed by a sum over different momentum components. Mathematically $\rho^{2\gamma}$ is given by

$$\rho^{2\gamma}(\mathbf{p}) \propto \sum_{j,\mathbf{k}:E(j,\mathbf{k})<E_\mathrm{F}} \left| \int e^{-i\mathbf{p}\cdot\mathbf{r}} \Psi^{\mathrm{ep}}_{j,\mathbf{k}}(\mathbf{r},\mathbf{r})\mathrm{d}\mathbf{r} \right|^2 \quad (1)$$

where the sum goes over all the states $\mathbf{k}$ below the Fermi level in the $j$th band. The wave function of the positron and an electron in the band $j$ and state $\mathbf{k}$ is given by

$$\Psi^{\mathrm{ep}}_{j,\mathbf{k}}(\mathbf{r},\mathbf{r}') = \Psi^{\mathrm{e}}_{j,\mathbf{k}}(\mathbf{r})\Psi^{\mathrm{p}}(\mathbf{r}')\sqrt{\gamma_{j,\mathbf{k}}(\mathbf{r},\mathbf{r}')} \quad (2)$$

Journal of Physics: Conference Series **443** (2013) 012092 doi:10.1088/1742-6596/443/1/012092

with a general enhancement factor $\gamma_{j,\mathbf{k}}(\mathbf{r},\mathbf{r}')$, which accounts for the interaction of the positron with the electron. If we assume an independent particle model, i.e. $\gamma_{j,\mathbf{k}}(\mathbf{r},\mathbf{r}') = 1$, and put in the Fourier expansion $\Psi^{\mathrm{e}}_{j,\mathbf{k}}(\mathbf{r})\Psi^{\mathrm{p}}(\mathbf{r}) = \sum_{\mathrm{G}} C_{j,\mathbf{G}} \exp\left(i(\mathbf{k}+\mathbf{G})\mathbf{r}\right)$ with reciprocal lattice vectors $\mathbf{G}$ we get:

$$\rho^{2\gamma}(\mathbf{p}) \propto \sum_{j,\mathbf{k}} \Theta(E_{\mathrm{F}} - E_{j,\mathbf{k}}) \sum_{\mathbf{G}} \left| C^{j}_{\mathbf{G}}(\mathbf{k}) \right|^2 \delta_{\mathbf{p}-\mathbf{k},\mathbf{G}} \tag{3}$$

From this equation we see that filled bands give a continuous distribution and bands crossing the Fermi level yield breaks which are distributed through the reciprocal space. An even more descriptive result is achieved if we define the following function $n(\mathbf{P})$ of a vector in reciprocal space $\mathbf{P}$, assuming a constant positron wave function:

$$n(\mathbf{P}) = \sum_{\mathbf{G}} \rho^{2\gamma}(\mathbf{P}+\mathbf{G}) \propto \sum_{j,\mathbf{k}} \Theta(E_{\mathrm{F}} - E_{j,\mathbf{k}}) \,. \tag{4}$$

The operation described by equation (4) is called the Lock-Crisp-West (LCW) procedure [2]. With latter equation the $\rho^{2\gamma}$ of equation (3) can be represented in the reduced zone scheme. The LCW procedure is applicable for $\rho^{2\gamma}$ and its projections in the main symmetry directions, i.e. the measured 2D-ACAR spectra.

The method used here to calculate the full three dimensional $\rho^{2\gamma}$ is based on the maximum entropy algorithm (MEA) [3]. Although a MEA to de-convolve ACAR spectra is known to have positive effects on reconstructing $\rho^{2\gamma}$ [4], it was just recently, that a MEA has successfully been applied for reconstructing the 3D momentum density from 2D-ACAR measurements [5]. The idea behind the MEA can be found, e.g, in [6]. The main purpose of the reconstruction is to get a three dimensional density $D$ which agrees within the 2D-ACAR spectra $M_i^{\alpha}$ (where $i = 1...N^2$ runs over all pixel) for all measured angles $\alpha$ with the statistical error $\sigma_i$. It could be realized by minimizing the following expression, where $R_i^{\alpha}(D)$ is the Radon transformation, i.e. the projection, of the density $D$ at an angle of $\alpha$:

$$\chi^2 = \sum_{\alpha}\sum_{i} \frac{\left(R_i^{\alpha}(D) - M_i^{\alpha}\right)^2}{\sigma_i^2} \tag{5}$$

If $M$ is of size of $N^2$, then $D$ has a size of $N^3$. Consequently, the minimization problem is highly under-determined. This is the reason why the MEA is employed. In order to solve the minimization an additional criterion is used: The resulting density $D$ should be flat. For this purpose, a Lagrange function with $\chi^2$ and an entropy function $\sum_j D_j \ln(D_j)$ is constructed by the use of the Lagrange multiplier $\lambda$:

$$L(D) = \sum_{j} D_j \ln(D_j) - \lambda\chi^2 \tag{6}$$

$L$ has to be maximized in order to obtain $D$. The solution to this problem can be written in an iterative form [7]

$$D_i^{(n+1)} = A \frac{D_i^{(n)} \exp\left(-\lambda \frac{\partial\chi^2}{\partial D_i}\right)}{\sum_j D_j^{(n)} \exp\left(-\lambda \frac{\partial\chi^2}{\partial D_j}\right)} \tag{7}$$

with the normalisation constant A.

In case of $\rho^{2\gamma}$, the quality of the reconstruction can be improved and the calculation time is reduced if the full symmetry of the reciprocal lattice is taken into account. Doing this, the MEA becomes a fast and reliable method to recreate the three dimensional $\rho^{2\gamma}$.

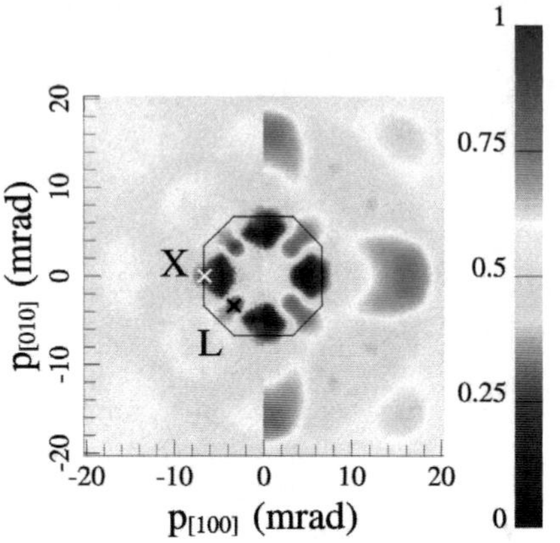


(a) Measured (left) and calculated (right) anisotropy of the 2D-ACAR spectrum.

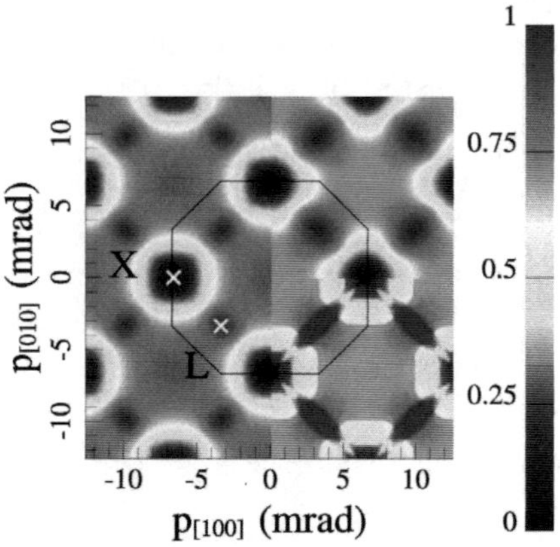


(b) LCW folded density in the [001] direction: measurement (left), calculation (right bottom) and calculation convolved with the experimental resolution (right top).

**Figure 1.** [001] projection of copper (colour online).

## 2. Experiments

Measurements were performed at the new 2D-ACAR spectrometer at the TUM in Garching. The details of the set-up are reported in detail by H. Ceeh et al. [8]. A single crystalline copper sample was prepared from commercially available copper single crystal by orienting and cutting. To remove lattice defects produced by cutting, the sample was electrolytically polished.

During the measurement the sample was cooled below 10 K. A spectrum with more than $10^8$ counts was recorded for projections along the main symmetry axes [100], [110] and [111]. All data were symmetrised according to their point group in order to enhance statistics. For the simulation of 2D-ACAR spectra, the Munich SPR-KKR package, version 5.4 by H. Ebert et al. was used [9].

## 3. Results and Discussion

In figure 1(a) the anisotropy of the measured (left) and the calculated data (right) is displayed. Obviously the innermost first order features are in very good agreement: There is an increased density in [110] direction at the L-point due to the necks of the Fermi surface and a decreased density in [100]. The minimum along the latter direction corresponds to the X-point of the Brillouin zone.

This attribute is even more explicit in the LCW folded density on the left side of figure 1(b): The highest density is located at the projected L-points, where the Fermi surface extends to the boundary of the Brillouin zone, whereas the lowest density is again at the X-points. In the case of copper, there are no electronic states at the X-point in the band which crosses the Fermi level and hence a low $\rho^{2\gamma}$ is measured. The right side of figure 1(b) displays the theoretical LCW density. For a better comparison, the upper right corner shows the calculation convolved with the experimental resolution. Again, the congruency is very good. The only small difference concerns the size of the necks at the L point.

To calculate the three dimensional $\rho^{2\gamma}$ a MEA was applied using the three measured projections and all the equivalent directions. The data were then folded into the first Brillouin zone by the LCW procedure for three dimensions. In figure 2 the isosurface dividing the Brillouin

**Figure 2.** Isosurface plot of the reconstructed 3D $\rho^{2\gamma}$ of copper from only three projections. The density for the isosurface is chosen in such a way, that the resulting isosurface divides the Brillouin zone in two equal parts.

zone in equal parts is plotted, which has the well known shape of the Fermi surface of copper (see e.g. [10, 11]). Although the qualitative agreement with previous measurements is good, three projections are not sufficient to extract reliable quantitative data.

## 4. Conclusion and Outlook

We have performed first measurements with the new 2D-ACAR spectrometer at TUM in Garching on the model system copper. The validity of the obtained data could be proven using ab-inito calculations together with LCW folding and MEA. Even though only three projections were recorded, the 3D $\rho^{2\gamma}$ could be obtained.

In the near future, it is planed to measure more projections of copper to compare our results quantitatively. Further investigations will then aim at more complex materials, especially correlated electron systems. Spectra will be collected as a function of temperature in order to investigate phase transitions. In addition we plan to compare the measurements with results obtained by complementary techniques such as, e.g., quantum oscillatory methods.

## Acknowledgements

The authors would like to thank the technical staff of the MLL accelerator laboratory and the directorial board, especially L. Beck for his kind support. This project is funded by the Deutsche Forschungsgemeinschaft through the Transregio TRR80.

## References

[1] Schultz P J and Lynn K G 1988 *Rev. Mod. Phys.* **60**(3) 701
[2] Lock D G, Crisp V H C and West R N 1973 *J Phys.* F Met. Phys. **3** 561
[3] Gull S and Daniell G 1978 *Nature* **272** 686
[4] Fretwell H M *et al.* 1995 *Europhys. Lett.* **32** 771
[5] Pylak M, Kontrym-Sznajd G and Dobrzyski L 2011 *Appl. Phys.* A **104**(2) 587
[6] Burch S F, Gull S F and Skilling J 1983 *Comput. Vision Graph.* **23** 113
[7] Skilling J and Bryan R K 1984 *Mon. Not. R. Astron. Soc.* **211** 111
[8] Ceeh H, Weber J A, Leitner M, Böni P and Hugenschmidt C 2013 *Rev. Sci. Instrum.* **84** 043905
[9] Ebert H, Ködderitzsch D and Minár J 2011 *Rep. Prog. Phys.* **74** 096501
[10] Burdick G A 1961 *Phys. Rev. Lett.* **7**(5) 156
[11] Kondo H *et al.* 1993 *Phys. Status Solidi* B **177** 345

Journal of Physics: Conference Series **443** (2013) 012093 doi:10.1088/1742-6596/443/1/012093

# A novel setup for the pulsing and energy enhancement of a positron beam

**C Piochacz**[1,2], **C Hugenschmidt**[1,2]

[1] Physik-Dep. E21, Technische Universität München, James-Franck-Str. 1, D-85748 Garching
[2] FRM II, Technische Universität München, Lichtenbergstr. 1, D-85748 Garching

E-mail: christian.piochacz@frm2.tum.de

**Abstract.** We propose a detailed concept to raise the total energy of a positron beam by an amount of at least 10 keV. The setup consists of a combination of a pre- and a main-buncher which is able to convert the continuous remoderated positron beam of NEPOMUC into pulses of about 2.5 ns width with a repetition rate of 5 MHz. This pulsed beam is suited for a new setup which elevates the beam energy by rf-fields. In contrast to existing rf-accelerators, the new setup has only negligible influence on the transversal beam phase space.

## 1. Introduction

By tuning the implantation energy of a positron beam it is possible to perform depth resolved measurements from the surface into the bulk of the investigated sample. This is one of the prime advantages of monoenergetic positron beams over the bulk studies performed with $\beta^+$ emitters. The beam energy is usually adjusted by biasing the source, the sample or both according to the desired implantation energy. However, in some cases this method is not possible for technical reasons.

The remoderated positron beam at NEPOMUC has an kinetic energy of about 20 eV [1] with respect to ground potential. Hence the implantation energy at the instruments has to be tuned by the sample potential. This is e.g. not possible at the PAES system [2] at NEPOMUC, as there the (re-) emitted low energy charged particles are in the scope of interest and the detector can not be biased to the same potential as the sample.

A completely different approach is to raise the total beam energy by using time dependent fields. In common rf-accelerators the kinetic energy is enhanced due to the potential difference between two succeeding electrodes which are charged by an appropriate rf. These accelerators are designed to provide charged particle beams of several MeV or more and substantial precautions are made to ensure the transversal stability of the beam. As the implantation energy in depth resolved positron measurements is usually limited to energies below about 30 keV, a more compact design, which has a much lower effect on the transversal phase space of the beam, is proposed here.

## 2. Experimental setup

The presented setup can be divided into three parts with distinct assignments (see Fig. 1). Firstly, the continuous beam is converted into a pulsed beam by a combination of a pre-buncher and a double gap main-buncher. In the second step the potential energy is raised by using a rf-electrode and in the last section the potential energy is transformed into kinetic energy by an electrostatic accelerator. As the second component enhances the potential energy it is called positron energy *elevator* to emphasize the difference to a classic rf-accelerator.

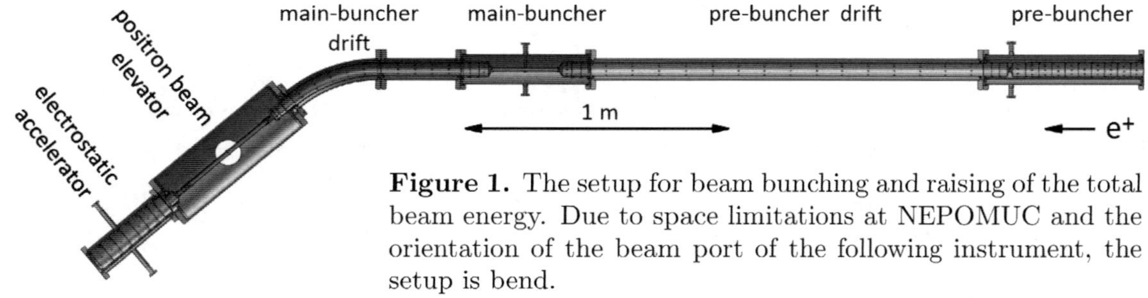


**Figure 1.** The setup for beam bunching and raising of the total beam energy. Due to space limitations at NEPOMUC and the orientation of the beam port of the following instrument, the setup is bend.

*2.1. Positron beam bunching units*

The bunching is based on the time-dependent modulation of the beam energy. The energy modulation occurs at the gap between two electrodes where e.g. on the second one a time dependent potential is applied. In the common double gap approach a third electrode follows and hence a time dependent field appears again[1]. If the modulation function has a rising and a falling edge and the length of the center electrode is matched to the mean energy of the positrons, the energy modulation at both gaps leads to a compression of the positron beam. It can be shown, that the time dependency of the energy modulation $eU_{\mathrm{m}}(t)$ at one gap has to have the following form:

$$eU_{\mathrm{m}}(t) = eU_0 \left( \left( \frac{1}{1 - t/t_0} \right)^2 - 1 \right) \tag{1}$$

where $e$ is the elementary charge, $eU_0$ the kinetic energy of the reference particle and $t_0$ the time it needs for the drift $d$ after which the time focus appears. If the time $t_0 = d/\sqrt{2\, {}^{e}\!/_{m}\, U_0}$ is sufficiently long this expression can be approximated linearly to:

$$eU_{\mathrm{m}}(t) \approx 2\, eU_0 \frac{t}{t_0} = \sqrt{8/m}\, (eU_0)^{3/2}\, \frac{t}{d} \tag{2}$$

A sine wave has a rising and a falling edge which are approximately linear and is easy to generate as well as to amplify. Thus it is often used as a modulation function despite the fact that only a fraction of the function leads to a correct energy modulation of the positrons and therefore a high background is generated.

*2.1.1. Pre-Buncher* To avoid the background, a pre-buncher was developed, which modulates only at the second gap. The potential difference at the first gap is spread over several gaps formed by a number of electrodes called equalization drift (ED). If the transit time $\tau_{\mathrm{pb}}$ equals the reciprocal of the repetition frequency $f_{\mathrm{pb}}$, the energy modulation of the positrons during the pass of the ED is averaged to zero. Using such a buncher, the modulation function does not have to have a rising and a falling edge and an arbitrary function can be used which approximates the function given by equation (1) as exactly as possible or even accounts for higher order effects not regarded by this equation. Although such a function is obviously better suited for bunching purposes, it is more demanding to generate and to amplify. However, a high amplitude is necessary to minimize the aberrations and to attain a sharp time focus. Hence the pre-buncher

[1] Single gap bunchers which are based on $\lambda/4$-resonators are several meters long for frequenzies below 50 MHz.

is only used to compress as many positrons as possible into a bunch which is just short enough to fit the linear parts of the following sine wave buncher. One design object was to keep the repetition rate $f_{\mathrm{pb}}$ low. $f_{\mathrm{pb}}$ depends on the ED length $l_{\mathrm{ed}}$ and the mean energy of the positrons $e\bar{U}_{\mathrm{pb}}$ is as follows:

$$1/f_{\mathrm{pb}} = \tau_{\mathrm{pb}} = \frac{l_{\mathrm{ed}}}{\sqrt{2e\bar{U}_{\mathrm{pb}}/m}} \tag{3}$$

As all electrodes are attached to the rf, the total capacity should be small and hence $l_{\mathrm{ed}}$ should not be larger as about 500 mm. By regarding about 15 eV as a lower limit for $e\bar{U}_{\mathrm{pb}}$, a minimum repetition rate of 5 MHz is feasible. With equation (2) and assuming a modulation energy of about $eU_{\mathrm{pb}} = 10\,\mathrm{eV}$ and an unchanged mean energy the drift length is about $d_{\mathrm{pb}} = 1.4\,\mathrm{m}$. Rearranging equation (2) and substituting the energy spread of the primary beam $\Delta E_{\mathrm{p}}$ for the modulation energy $eU_{\mathrm{m}}$ leads to the chromatic aberration $\tau_{\mathrm{c}}$:

$$\tau_{\mathrm{c}} = \frac{\Delta E_{\mathrm{p}}\, d_{\mathrm{pb}}}{\sqrt{8/m}\,(eU_0)^{3/2}} \tag{4}$$

The chromatic aberration is in this case the main reason for the finite extension of the time focus and amounts to about 20 ns if a reasonable energy spread of 1 eV is used.

*2.1.2. Main-Buncher* The main-buncher is built up in the same way as those used in the PLEPS [3] or in the Munich SPM [4]. It consists of three electrodes, where the first and the last are biased to a fixed potential $U_{\mathrm{b}}$. The center electrode forms together with a coil a resonating circuit and oscillates with the frequency $f_{\mathrm{b}}$ and an amplitude $\hat{U}_{\mathrm{b}}$ around the potential $U_{\mathrm{b}}$. In order to keep the chromatic aberrations low, the mean energy of the positrons after the time focus of the pre-buncher and therefore also during the pass through the main-buncher has to be considerably higher as the maximum energy modulation at the pre-buncher. To ensure both a high mean energy $eU_{\mathrm{b}}$ and a short center electrode, the main-buncher is operated at a multiple of the frequency of the pre-buncher. By choosing 20 MHz each fourth main-buncher cycle is used for further beam compression. Using a 250 mm long center electrode leads to a pass energy of about 284 eV. By the resonant amplification, a peak to peak voltage of 100 V can be easily achieved. However, since the beam is pre-bunched, the maximum potential difference which acts onto the positron beam at the first gap, shrinks to about 95 eV. Inserting this value into equation (4) and regarding $eU_{\mathrm{pb}} = 10\,\mathrm{eV}$ as an upper limit for the energy spread in front of the main-buncher, the chromatic aberration $\tau_{\mathrm{cb}}$ of the main-buncher amounts about 1.5 ns. The deviations of the sine wave from equation (1) lead to an additional, the so called spherical aberration $\tau_{\mathrm{sb}} \approx 0.63\ f_{\mathrm{b}}^2 \Delta t_{\mathrm{b}}^3$ . It is dependent from the time frame $\Delta t_{\mathrm{b}}$ during which the modulation field acts onto the pre-bunched beam. Inserting the estimated width of the pre-bunched pulses results in 2 ns for $\tau_{\mathrm{sb}}$ and therefore to a total aberration $\tau_{\mathrm{b}} = \sqrt{\tau_{\mathrm{cb}}^2 + \tau_{\mathrm{sb}}^2}$ of about 2.5 ns.

*2.2. Positron beam energy elevator*

The energy elevator is built up in the same way as the main-buncher above. In contrast to the main buncher, the parameters of the elevator and the beam are matched to each other, so that the gaps are field free in first order approximation when the positrons pass them. Hence, the kinetic energy of the positrons is not altered there and the transversal phase space keeps almost unaffected. The passing through the center electrode lasts half of one period and therefore the positron gains a potential energy equal to twice of the amplitude of the applied sine wave.

To attain the aspired amplitude of $\hat{U}_{\mathrm{e}} = 5\,\mathrm{kV}$ the power losses in the resonating circuit have to be very low. Too high losses would overload the primary sine wave generator or amplifier and would further result in an exceeding heating of the resonating components, especially of those within the vacuum. By using the identity $1/2\,L\hat{I}^2 = 1/2\,C\hat{U}_{\mathrm{e}}^2$ of the maximal stored energy in the coil with the inductance $L$ and in the capacitor with the capacity $C$, the effective power $P_{\mathrm{eff}}$, which the amplifier has to provide, can be calculated:

$$P_{\mathrm{eff}} = \frac{1}{2}R\hat{I}^2 = \frac{1}{2}R\frac{C}{L}\hat{U}_{\mathrm{e}}^2 = \frac{1}{2}\omega_0^2 RC^2\hat{U}_{\mathrm{e}}^2 \qquad (5)$$

$\hat{I}$ and $R$ are amplitude of the current and the total resistance, respectively. Hence, to attain a high amplitude with low power $R$ and $C$ have to be small. By finite element simulations and preliminary measurements it could be shown, that a total resistivity of about $0.3\,\Omega$ and a capacity of about 20 pF are reasonable assumptions. With these values and an repetition frequency of again $f_{\mathrm{e}} = 20\,\mathrm{MHz}$ an amplitude of 5 kV can be achieved with a 24 W amplifier.

To minimize the pulse length at both gaps and therefore the perturbation of the transversal and longitudinal phase space, the beam is accelerated in front of the elevator and passes it with about 730 eV. Due to the higher energy, the center electrode has to have a length of about 400 mm. Nevertheless, the pulse width is at the gaps elongated from the formerly calculated 2.5 ns to about 2.7 ns. This finite width results in an unintended energy modulation of about 18 eV at each gap. If these energy modulations and those of the bunchers are treated as independent from each other, a total energy spread can be estimated by the sum of their squares to about 98 eV. It should be emphasized, that the low potential difference of maximal 18 eV at the gaps has a completely negligible influence onto the transversal phase space of the beam.

### *2.3. Electrostatic acceleration*
To preserve the pulse width after the energy elevation, the positrons have to be accelerated immediately after the energy elevation. Therefore, a first acceleration of a few hundred eV is made right after a short exit electrode of the elevator. However, in order to keep the perturbation of the beam low, the main acceleration to about 10 keV is done over seven electrodes spaced over a total distance of about 320 mm.

## 3. Conclusion
It could be shown, that the combination of the novel pre-buncher and the double gap main-buncher is in principle suited to generate highly efficient short pulses of about 2.5 ns with a repetition rate of 5 MHz. Due to the shortness of the pulses, the proposed positron beam elevator is able to raise the total energy of the beam by at least 10 keV with negligible influence onto the longitudinal and the transversal phase space. The calculations here are based on formulas widely used in the field of pulsed particle beams. A deeper insight of the system can only be gained by trajectory calculations leading to a detailed phase space analysis which have already been started.

## Acknowledgments
The authors would like to thank Gottfried Kögel for the manifold input and fruitful discussions.

## References
[1] Piochacz C, Kögel G, Egger W, Hugenschmidt C, Mayer J, Schreckenbach K, Sperr P, Stadlbauer M and Dollinger G 2008 *Applied Surface Science* **255** 98–100
[2] Mayer J, Hugenschmidt C and Schreckenbach K 2010 *Surface Science* **604** 1772–7
[3] Bauer-Kugelmann W, Sperr P, Kögel G and Triftshäuser W 2001 *Materials Science Forum* **363-365** 529–31
[4] Triftshäuser W, Kögel G, Sperr P, Britton D, Uhlmann K and Willutzki P 1997 *Nucl. Instr. and Meth. B* **130** 264–9

16th International Conference on Positron Annihilation (ICPA-16) IOP Publishing
Journal of Physics: Conference Series **443** (2013) 012094 doi:10.1088/1742-6596/443/1/012094

# First measurements with the Munich 2D-ACAR spectrometer on Cr

**Hubert Ceeh[1], Josef Weber[1], Christoph Hugenschmidt[1,2], Michael Leitner[2] and Peter Böni[1]**

E-mail: hubert.ceeh@frm2.tum.de
[1]Technische Universität München, Lehrstuhl E21, James-Franck Straße, 85747 Garching, Germany
[2]FRM II, Technische Universität München, Lichtenbergstraße 1, 85747 Garching, Germany

**Abstract.** The Munich 2D-ACAR spectrometer at the Maier-Leibnitz accelerator laboratory in Garching has recently become operational. In the present implementation a 2D-ACAR spectrometer is set up, with a baseline of 16.5 m, a conventional $^{22}$Na positron source and two Anger-type gamma-cameras. The positrons are guided onto the sample by a magnetic field generated by a normal conducting electromagnet. The sample can be either cooled by a standard closed-cycle-cryostat to low temperatures or heated by a resistive filament to temperatures up to 500 K. We present the key features of this new 2D-ACAR spectrometer and, in addition, discuss first measurements on the pure metal system Cr. The 2D-ACAR measurements have been performed on Cr at different temperatures: at 5 K and at room temperature in the anti-ferromagnetic phase and at 318 K slightly above the paramagnetic phase transition.

## 1. Introduction

Almost all properties of a material are defined by its electronic structure. For a metal, the most important characteristic is the Fermi surface, namely the boundary between occupied and unoccupied states in reciprocal space. The traditional experimental method for studying the Fermi surface is the de Haas-van Alphen effect, which is limited to low temperatures, and therefore cannot be utilized for the study of certain paramagnetic materials, for example Cr, since the Néel temperature of Cr is 311 K [1]. For this reason, numerous ACAR investigations, both experimental [2, 3, 4] and theoretical [5, 6, 7] have been performed on paramagnetic Cr. Studying the fermiology of the paramagnetic phase of Cr would allow to gain insight to the driving mechanism of the anti-ferromagnetic ordering of Cr below 311 K, since a nesting feature of the Fermi surface gives a possible explanation. However, several 2D-ACAR investigations performed in the paramagnetic [8, 9] phase of Cr revealed significant discrepancies between the measured and the calculated two-photon-momentum distribution, which cannot be explained in terms of an unexpected topology of the Fermi surface. These discrepancies are most probably due to $e^+$-$e^-$-correlations and/or many-body effects not accounted for by theory. Recently, an experimental approach to this problem, the utilisation of a state-dependent enhancement has been presented in[10]. With this method the enhancement is directly obtained by fitting the contribution of the individual bands to the data. This way, the Fermi surface nesting vector was deduced via the computation of the static susceptibility from the fitted LMTO calculations. In our study we focussed on the differences in the Fermi surface topology between paramagnetic and anti-ferromagnetic Cr. Therefore, 2D-ACAR measurements at different temperatures have been performed.

## 2. Experimental setup

The 2D-ACAR spectrometer consists of two main components, the source-sample chamber including the sample environment, and the detector assembly consisting of two Anger-type gamma cameras, which were obtained from the positron group of Bristol university. The Anger cameras are positioned symmetrically at a distance of 8.25 m to the source-sample chamber. The Munich 2D-ACAR spectrometer features two interchangeable sample holders, with an available temperature range between 5 K and 500 K, both of which were used for this study. The details of the Munich 2D-ACAR spectrometer and the performance characteristics can be found in [11]. For this study, four 2D-ACAR measurements have been conducted, each with the integration direction chosen along [100]. Two measurements were performed with the coolable sample holder in the anti-ferromagnetic phase of Cr, one at 5 K with a total of $1.10 \cdot 10^8$ counts and the other one at room temperature with a total of $1.84 \cdot 10^8$ counts. Two additional measurements were performed with the heatable sample holder, one at room temperature, in order to ensure reproducibility, with a total of $2.09 \cdot 10^8$ counts and the other one at 318 K with a total of $2.63 \cdot 10^8$ counts. Before each of the two measurement campaigns the orientation of the Cr crystal was checked by means of Laue diffractometry and aligned to $\pm 1°$. Before the sample was mounted the surface was etched with concentrated hydrochloric acid in order to remove the oxide layer from the surface. The crystal was cut by means of spark erosion, therefore it is assumed that the crystal is defect free. Positron lifetime measurements are underway to investigate the defect concentration.

The recorded ACAR spectra were corrected for the momentum sampling function and then folded back to **k**-space following the LCW-procedure[12]. This way features in the projection of the momentum distribution attributed to the Fermi surface are enhanced and easier to identify. The LCW-folded data were corrected for background contributions and scaled from 0–1 so that the dynamic range is identical for all three measurements as well as for theoretical calculations: 0 represents unoccupied states while 1.0 represents the occupied states.

In addition, first principle band structure calculations have been performed for paramagnetic Cr using the Munich SPR-KKR-package [13] and the expected two-photon momentum distribution was deduced. For the calculations the independent particle model was applied, therefore $e^+$–$e^-$–correlation effects are not included in the theoretical data.

## 3. Results

Two of the four measurements were performed at room temperature in order to check if the ACAR spectra are consistent before and after the change of the sample holder. Since both spectra are compatible within the statistical limits only the spectrum with the higher statistics is presented here (see figure 1). The comparison of the low temperature measurement and the room temperature measurement shows the effect of decreasing resolution with increasing temperature. However, besides the lower resolution in the room temperature measurement both spectra are comparable, therefore only one is shown.

The features at the N-points appear to be occupied in the anti-ferromagnetic phase (High symmetry points of the Brillouin zones of both phases, *simple cubic* for the anti-ferromagnetic phase and *bcc* for the paramagnetic phase are indicated in figures 1–5). This picture changes when the sample temperature is increased above the Néel temperature (see figure 2). A small difference in temperature drives the sample into the paramagnetic state and yields a drastic change in the LCW-folded data. The N-hole pockets become more pronounced, which can be seen comparing cuts along Γ-N-Γ for both phases (see figure 4). Furthermore, the electron pocket structure at the X-points connecting the N-points is smaller than in the anti-ferromagnetic

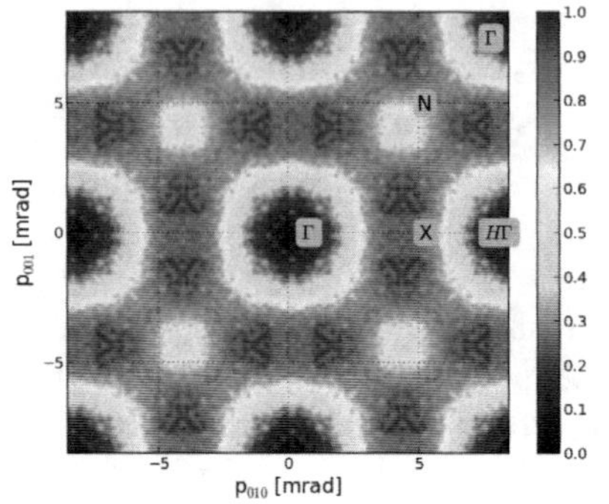


**Figure 1.** LCW distribution for anti-ferromagnetic Cr [100] at room temperature.

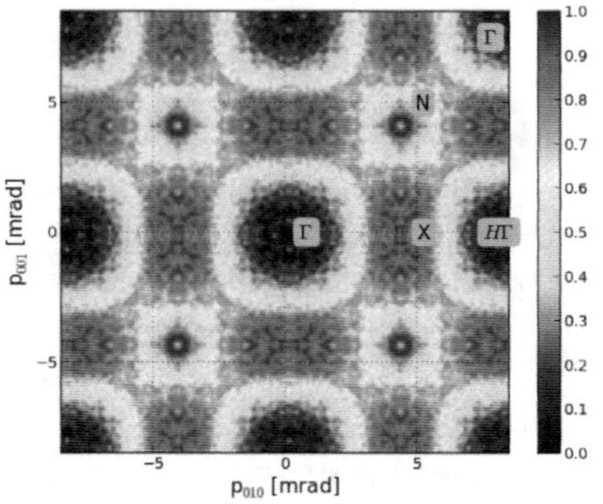


**Figure 2.** LCW distribution for Cr [100] at 318 K. Slightly above the Néel temperature for Cr of 311 K.

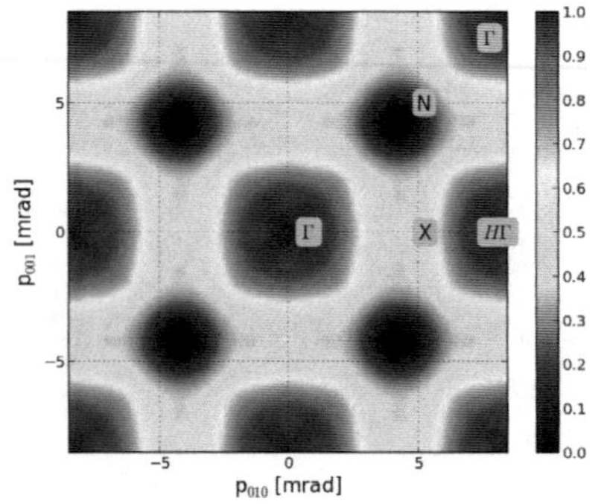


**Figure 3.** LCW distribution for paramagnetic Cr [100] obtained from first principle calculations.

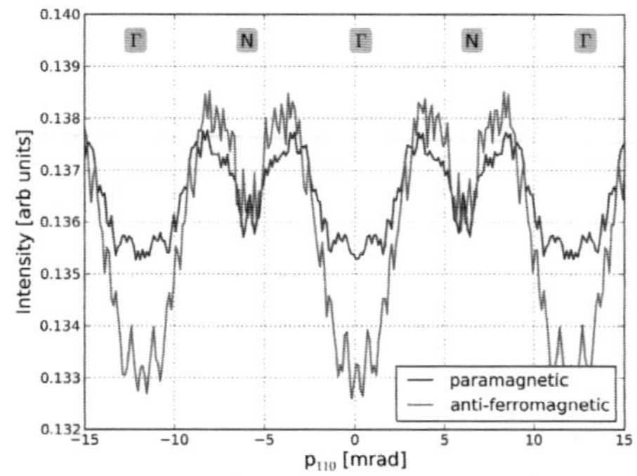


**Figure 4.** Cut through the LCW-folded projections along Γ-N-Γ ($p_{001}$ $=p_{010}$) for the paramagnetic and the anti-ferromagnetic phase. The initial LCW-projections were normalized to the same amount of counts.

phase. Generally, the agreement of our measurement with previous 2D-ACAR measurements (see ref.[8]) on paramagnetic Cr is considerable. However, the discrepancies between our data and the SPRKKR calculations are quite substantial. This has also been reported from different laboratories [8, 9]. The relative intensities in the calculation for the occupied and the unoccupied states can not be reproduced in the LCW-folded data (see figure 3).

## 4. Discussion

The experimental resolution of our spectrometer lies between 1.54 mrad and 1.64 mrad [11], which is only slightly less than the extent of the N-point features in our data. However, the changes in the LCW-folded data when the sample is driven over the magnetic phase transition cannot be attributed to resolution effects, since a temperature increase of 20 K results in a degradation of the overall resolution of less than 0.01 mrad.

During data analysis the LCW-folding was applied to the isotropic data, which was obtained by azimuthal averaging of the original data. The isotropic data does not contain the full information

on the Fermi surface topology. With the isotropic data alone features at the X-point, similar to those described above, can be produced by the LCW-folding, as can be seen in figure 5.
A possible explanation would be that the positron annihilation with a high lying flat band close to the Fermi energy is strongly enhanced. Also it has to be considered the application of the LCW-theorem is only valid if the positron wave function can be characterized as a plane wave, which is not true if strong correlation effects are present, like it is the case for Cr [14].
To summarize, we were able to reproduce the results from previous measurements[8] we also could show, that the 2D-ACAR spectrum is drastically different in the anti-ferromagnetic and in the paramagnetic phase. However, the interpretation of our data with respect to the calculation remains an unsolved challenge.

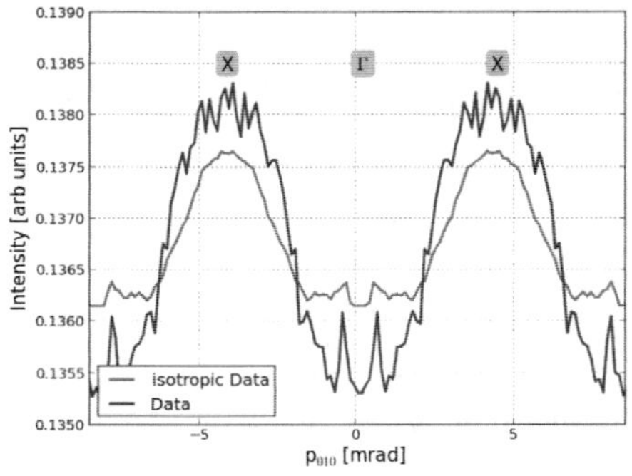


**Figure 5.** Cut through the LCW-folded data along $X\Gamma X$ ($p_{001} = 0$) in comparison with the same cut for a LCW-folding of the isotropic data. The position of high symmetry points is indicated.

## 5. Outlook

For the interpretation of the 2D-ACAR data of Cr a solid understanding of the enhancement is needed. We plan on recording multiple 2D-ACAR projections of Cr in the paramagnetic phase in order to perform a full 3D reconstruction of the momentum density, respective measurements are under way. This way, we hope to get a better grasp of the enhancement in Cr and with that gain insight to the electronic structure.

## Acknowledgements

This project is funded by the Deutsche Forschungsgesellschaft (DFG) within the Transregional Collaborative Research Center TRR 80 “From electronic correlations to functionality”.

## References

[1] Steinitz M O 1986 *J. Magn. Magn. Mater.* **60** 137
[2] Shiotani N et al. 1977 *J. Phys. Soc. Jpn.* **43** 4
[3] Singh A K and Singru R M 1983 *J. Phys. F: Met. Phys.* **13** 2189
[4] Hughes R J et al. 2004 *Phys. Rev. B* **69** 174406
[5] Singh A K et al. 1988 *Europhys. Lett.* **6** 67
[6] Sundamrajant V, Kanheret D G and Singru R M 1992 *J. Phys.: Condens. Matter* **4** 89754988
[7] Rubaszek A et al. 2001 *Phys.Rev. B* **65** 125104
[8] Dugdale S B et al. 1998 *J. Phys.: Condens. Matter* **10** 10367
[9] Biasini M 2000 *Physica B* **275** 285
[10] Laverock J et al. 2010 *Phys.Rev. B* **82** 125127
[11] Ceeh H et al. 2013 *Rev. Sci. Instrum.* **84** 043905
[12] Lock D G, Crisp V H C, West R N 1973 *J. Phys. F: Met. Phys.* **3** 561
[13] Ebert H, Kodderitzsch D, Minár J 2011 *Rep. Prog. Phys.* **74** 096501
[14] Rabou LP L M and Mijnarends P E 1984 *Solid State Commun.* **52** 933

Journal of Physics: Conference Series **443** (2013) 012095 doi:10.1088/1742-6596/443/1/012095

# Development of a new time and position resolving detector for the pulsed low energy positron system PLEPS

**U Ackermann[1], W Egger[1], P Sperr[1], B Löwe[1], L Ravelli[1], G Kögel[1], G Dollinger[1] and O Jagutzki[2]**

[1] Universität der Bundeswehr München, Institut für Angewandte Physik und Messtechnik, Werner-Heisenberg-Weg 39, Neubiberg, 85577, Germany
[2] Universität Frankfurt, RoentDek GmbH c/o Institut für Kernphysik, Frankfurt am Main, 60438, Germany

ulrich.ackermann@unibw.de

**Abstract.** The pulsed low energy positron system PLEPS at the Munich research reactor FRM-II is a user facility for depth resolved positron lifetime measurements. Besides positron lifetime measurements 2D-AMOC (Two Dimensional Age Momentum Correlation) experiments are also possible. 2D-AMOC provides in coincidence the lifetime of the positron and the longitudinal momentum distribution of the annihilated electron. It would be of great scientific concern to measure simultaneously the entire 3D-momentum distribution of the electron annihilating with the positron and the corresponding lifetime of the positron (4D-AMOC). To perform 4D-AMOC measurements a time and position resolving detector is required in coincidence with a pixelated Germanium detector. Therefore a time and spatially resolving detector is currently developed at our institute with envisaged time resolution of 100 ps (FWHM) and a spatial resolution of about 2.6 mm (FWHM) over an area of 12 cm². First test measurements have been carried out with a 25 mm diameter MCP (Micro Channel Plate) image intensifier and with special delay-line anode readout for the spatial information. Up to now 178 ps (FWHM) time resolution and on average 3.4 mm (FWHM) position resolution have been achieved with $BaF_2$ as scintillator material and a $^{60}Co$ source.

## 1. Introduction

The pulsed low energy positron system [1] (PLEPS) at the high intensity neutron induced positron source [2] (NEPOMUC) at the Munich research reactor FRM-II is a very powerful user facility to perform depth-dependent high-quality positron annihilation lifetime spectroscopy (PALS) measurements. To enhance its defect identification potential a setup to perform two dimensional age momentum correlation (2D-AMOC) measurements has been recently commissioned [1]. In addition to the lifetime of the positron 2D-AMOC measurements provide information about the correlated longitudinal momentum of the electron annihilating with the positron and thus about the chemical environment at the location of annihilation. It would be of great scientific concern to measure the full three dimensional momentum distribution of the electron annihilating with the positron together with the positron's lifetime (4D-AMOC).

In the 2D-AMOC setup the two almost collinearly emitted gamma quanta from the annihilation process are registered with a photomultiplier tube (PMT) to measure the lifetime of the positron in coincidence with a Germanium detector to measure the longitudinal momentum of the electron.

4D-AMOC measurements require a time and position resolving detector operating in coincidence with an energy and position resolving detector. In our 4D-AMOC setup we use a commercially available pixelated Germanium detector from Canberra as position and energy resolving detector. The time and spatially resolving detector is currently under development. The pixelated Germanium detector has an energy resolution of 1.35 keV (FWHM) at 662 keV. For the time and position resolving detector a time resolution of about 100 ps (FWHM) is envisaged. The required angular resolution for both detectors is 2.6 mrad (FWHM). This corresponds to 2.6 mm position resolution (FWHM) at a distance of 1 m from the specimen.

To meet these constraints, a new approach for the time and position resolving detector has been chosen using a specially designed MCP image intensifier with a $BaF_2$ scintillator and a capacitively coupled delay-line anode readout providing the spatial information. First time and position resolution tests have been made with a prototype. The results are presented in this paper.

## 2. Experimental setup and results

To figure out the optimum design parameters for the new time and position resolving detector, we used an MCP image intensifier from Proxitronic (active diameter 25 mm) with a fused silica entrance window and a bialkali photocathode evaporated on the inner side of the window. For the time resolution measurements a cylindrical $BaF_2$ scintillator of 15 mm diameter and 10 mm height was coupled to the entrance window. For the position resolution measurements two identical $BaF_2$ scintillators (4 mm diameter, 8 mm height) were mounted at a distance of 7.5 mm. The time signal was taken directly from the MCPs. The electron cloud produced in the channels of the MCP image intensifier is collected on a so-called resistive screen anode which is capacitance-coupled to a commercial position-encoding delay-line anode (RoentDek GmbH) placed outside the MCP image intensifier. Figure 1 shows a schematic drawing of the MCP detector setup.

### 2.1. Time resolution measurement

A conventional fast-slow configuration was used for the time resolution measurement (see figure 2). Start and stop signals for the time to amplitude converter were derived from a PMT (Photonis, XP2020 URQ) and the MCP image intensifier using constant fraction triggers. In the measurements a $^{60}Co$ (1.17 MeV and 1.33 MeV) point source was placed between the MCP image intensifier and the PMT.

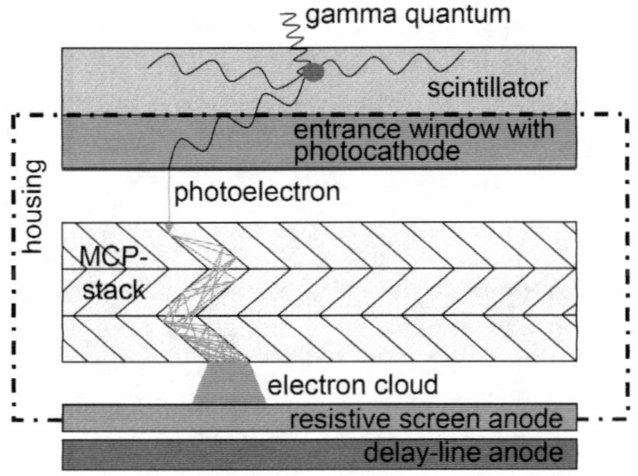


**Figure 1.** Principle drawing of the MCP detector setup.

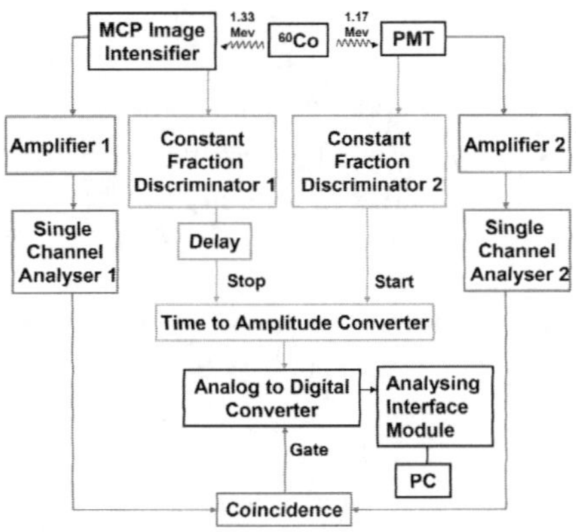


**Figure 2.** Scheme of the time resolution measurement setup.

The energy window was set on the 1.17 MeV energy peak for the PMT and on the 1.33 MeV energy peak for the MCP image intensifier. The measured coincidence resolving time was 206 ps ± 1 ps (FWHM). From the known time resolution of the PMT the MCP image intensifier time

Journal of Physics: Conference Series **443** (2013) 012095 doi:10.1088/1742-6596/443/1/012095

resolution of 178 ps ± 1 ps (FWHM) could be deduced. In figure 3 the recorded time spectrum is shown.

### 2.2. Position resolution measurement

The 2D-position information of the interaction of the gamma quantum in the scintillator is deduced from a RoentDek delay-line anode readout. The delay-line anode in this setup consists of a three layer structure with LC-delay configuration. A detailed description of the delay-line anode could be found in [3]. An electron cloud hitting the resistive screen anode will induce a signal on the three layers of the delay-line anode. This signal goes to the endings of the respective LC-delays leading to six signals - two signals per layer - with different arrival times. From the differences of the arrival times relative to the signal taken directly at the MCPs the centroid information of the electron cloud and hence the 2D-position of the interaction of the gamma quantum in the scintillator is derived.

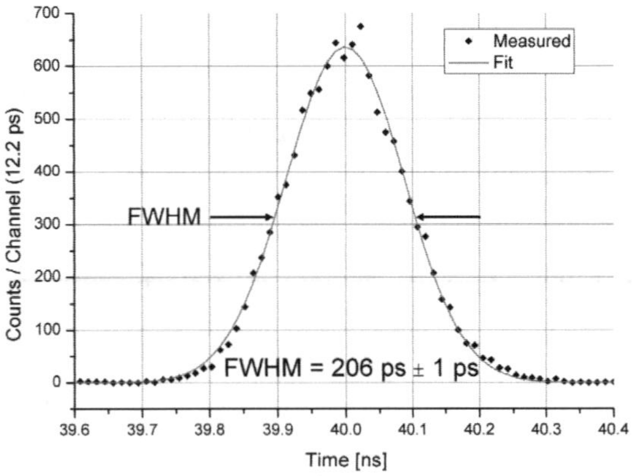


**Figure 3.** Observed coincidence resolving time spectrum with the MCP image intensifier and PMT.

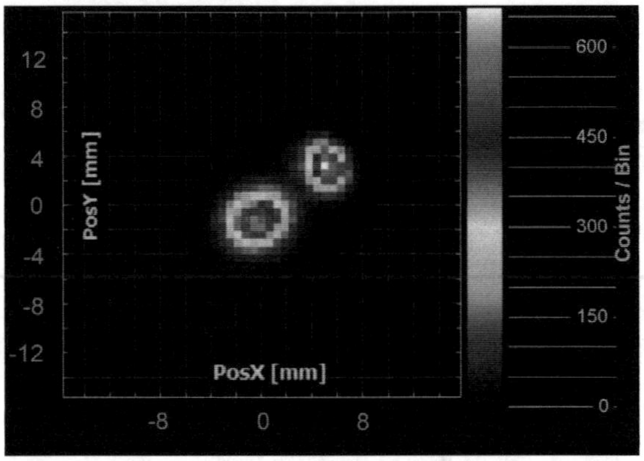


**Figure 4.** Recorded detector signals as function of position. The two $BaF_2$ crystals can be clearly separated.

Two identical $BaF_2$ scintillator cylinders (4 mm diameter, 8 mm height) were mounted with a plastic holder on the entrance window of the MCP image intensifier, one in the center of the entrance window, the other in a distance of 7.5 mm (center to center). The $^{60}Co$ point source was placed on top of the plastic holder between the scintillators. Figure 4 shows the recorded detector signals as a function of position. The position of the two $BaF_2$ cylinders is clearly resolved. The projection of the counts to the x-axis results in an average position resolution (FWHM) of approximately 3.3 mm and for the projection to the y-axis of 3.4 mm, respectively.

## 3. Conclusion and outlook

A new detector approach with a MCP image intensifier and additional delay-line anode readout for reaching the goal of 100 ps time resolution and 2.6 mm position resolution, FWHM respectively, was presented in this paper. The aim is to upgrade PLEPS in the near future with a position and time resolving detector fulfilling the above mentioned constraints and thus enabling in coincidence with a pixelated Germanium detector 4D-AMOC measurements. We have reached with a 25 mm active diameter MCP image intensifier a time resolution of 178 ps (FWHM) and on average a 3.4 mm position resolution (FWHM) in our first experiment with $BaF_2$ as scintillator material. This result seems very promising to achieve the above stated required position and time resolution. Tests will be made with a MCP image intensifier with 40 mm active diameter and delay-line anode readout and different shaped scintillators.

The so far measured time resolution with the 25 mm active diameter MCP image intensifier is in the order of the until now achieved time resolution at PLEPS. Already at this moment it would be possible to implement the MCP image intensifier at PLEPS and to perform time resolution measurements. Due to the fact that we used for the time resolution measurements already at our institute existing instrumentation the goal of 100 ps (FWHM) seems feasible with especially for the MCP image intensifier adapted electronics. Also the position resolution of 2.6 mm (FWHM) should be obtainable by performing further experimental position resolution tests with different shaped scintillator materials.

**Acknowledgement**
Funding from the BMBF (Project 05K10WNA-Posimethod) is gratefully acknowledged.

**References**

[1] Egger W 2010 Pulsed low-energy positron beams in materials sciences *Proc. Int. School of Physics Enrico Fermi Course CLXXIV* ed R S Brusa, A Dupasquier and A P Mills Jr (Amsterdam: IOS Press) pp 419–449

[2] Hugenschmidt C 2010 Positron sources and positron beams *Proc. Int. School of Physics Enrico Fermi Course CLXXIV* ed R S Brusa, A Dupasquier and A P Mills Jr (Amsterdam: IOS Press) pp 399–418

[3] Schössler S, Bromberger B, Brandis M, Schmidt L Ph H, Tittelmeier K, Czasch A, Dangendorf V and Jagutzki O 2012 Time and position sensitive single photon detector for scintillator read-out *J. Instrum.* 2012 JINST 7 C02048

16th International Conference on Positron Annihilation (ICPA-16) IOP Publishing
Journal of Physics: Conference Series **443** (2013) 012096 doi:10.1088/1742-6596/443/1/012096

# Geant4 simulation of the effect of backscattered positrons on the lifetime spectra of PLEPS

**L Ravelli, B Löwe, W Egger, G Kögel, P. Sperr and G Dollinger**
Universität der Bundeswehr München, Werner-Heisenberg-Weg 39, 85577 Neubiberg, Germany
E-mail: luca.ravelli@unibw.de

**Abstract.** The Pulsed Low Energy Positron System (PLEPS) allows the measurement of positron lifetime spectra of very high quality with peak-to-background ratios up to $3 \cdot 10^4$. At those peak-to-background ratios small structures appear in the lifetime spectra due to backscattered positrons. Despite their small overall contribution - less than 2 % of the total events in the spectrum even with backscattering coefficient as high as 40 % - those satellite structures can render the data analysis difficult. To understand the origin of those satellite structures and to further improve the performance of the system, comprehensive simulations of the target chamber of PLEPS have been undertaken. The results reproduce fairly well the background of the lifetime spectrum. It is now possible to identify the origin of the background structures and also plan some additional countermeasures.

## 1. Introduction

Positron Annihilation Lifetime Spectroscopy (PALS), when combined with a pulsed beam of variable energy, yields a very powerful technique capable of providing depth profiles of open volume defects e.g. in layered structures [1] and implanted samples [2] which are inhomogeneously damaged.

The Pulsed Low Energy Positron System [3] (PLEPS) at the high intensity positron source NEPOMUC [4] (NEutron-induced POsitron source MUniCh) at the research reactor FRM-II is a user facility to perform high quality PALS measurements in very short times. Count rates up to $10^4$ counts per second have been achieved, which results in short measuring time (about 10 minutes per lifetime spectrum with $> 3 \cdot 10^6$ counts) and high peak to background ratios (up to $3 \cdot 10^4$). This allows routinely to separate up to four lifetimes.

New and ever more complicated layered materials require further improvements of the ability of PLEPS to separate more lifetimes. To this purpose two key factors have to be considered: *a*) time resolution and *b*) peak to background ratio.

(a) High time resolution is of particular importance for the separation of short–lived lifetime components. The time resolution of a pulsed positron beam is determined by the pulse width and the detector response. With small ($< 1\,\mathrm{cm}^3$) scintillators, detector responses in the 100 ps range can be achieved. The trade off is a dramatic loss in event rate and consequently a high price to pay in terms of measurement time.

(b) High peak-to-background ratio with PLEPS may be achieved either by using two detectors face-to-face in coincidence or a single detector mounted immediately below the target.

Unfortunately the excellent peak-to-background ratio obtainable in the former case is outweighted by the dramatic loss in event rate. For the single detector, at high peak-to-background ratios, structures due to backscattered positrons appear in the lifetime spectra.

Already with the present setup the amount of events coming from backscattered positrons is less than 2 % of the total counts in the lifetime spectrum even for high Z targets and high implantation energies (backscattering coefficient as high as 40 %). These backscattered positrons partly increase the homogeneous background, whereas some cause satellite structures in the spectra (see figure 2). Note that even the most pronounced satellite structure contain less than 0.2 % of the total counts in the lifetime peak. In spite of their small overall contribution, those satellite structures can render the analysis of the spectra difficult. Eliminating these satellite structures is presently the most promising way to improve the quality of the lifetime spectra and to augment the number of lifetimes, which can be separated with PLEPS. To this aim comprehensive simulations of the backscattered positrons in PLEPS have been performed.

## 2. Simulations

### *2.1. Electric and magnetic field modeling with COMSOL Multiphysics*

PLEPS can be conceptually divided into two main parts: the pulsing section - consisting of a prebuncher, a chopper and a buncher operating at a frequency of 50 MHz, which corresponds to a time window of 20 ns - followed by the acceleration section - consisting of a drift tube, an accelerator and the target chamber. The target chamber consists of a Faraday cage and the sample holder, which is kept at the negative high voltage (0.5 kV - 20 kV). The potential of the Faraday cage is set to -2 kV with respect to the sample. This ensures that, in combination with the variable drift tube potential, the pulse shape is independent of the implantation energy and consequently the time resolution varies little over the full energy range. The positrons are magnetically guided through the system in a constant solenoidal field. For a detailed description of the entire system see [3]. In this work, we focus on positrons backscattered from the target. Therefore only the target chamber of PLEPS was considered. The electric and magnetic fields were determined by COMSOL Multiphysics [5].

The simulation of the magnetic field has been done using rotational symmetry and periodic boundary conditions. In figure 1 (*a*) the section of the target chamber parallel to the beam direction and the coils that generate the magnetic field are shown. The magnetic flux density is plotted as surface- and streamline-plots as a function of the position. The magnetic field is homogeneous along the beam path, except for the last 30 mm where the field increases due to the presence of the mu-metal shielding of the detector, and it is consistent with the measured values. In the Faraday cage the rotational symmetry is broken and, therefore, a three dimensional simulation of the electric field was necessary. The complicated fan-like structure of tungsten blades covered with Kapton in the top of the Faraday cage was replaced by a simple plate. Figure 1 (*b*) shows the section of the target chamber parallel to the beam direction, in which the electric potential as contour-plot and the electric field as arrow field are shown.

### *2.2. Simulation with Geant4 and results*

With the electric and magnetic fields obtained by COMSOL, the trajectories of the backscattered positrons were simulated using Geant4 (release 9.5 patch 01) [6] and the PENELOPE (PENetration and Energy LOss of Positrons and Electrons and photons) low-energy electromagnetic model, including ionization, bremsstrahlung and multiple scattering. Geant4 takes into account the slowing down of the positrons inside the specimen. A range cut value of 1 nm and an energy threshold limit of 50 eV were set. To register only events that are actually seen by the detector, also the tungsten shielding of the detector was considered and the 511 keV annihilation quanta that deposited less than 300 keV were rejected like in the real experiment.

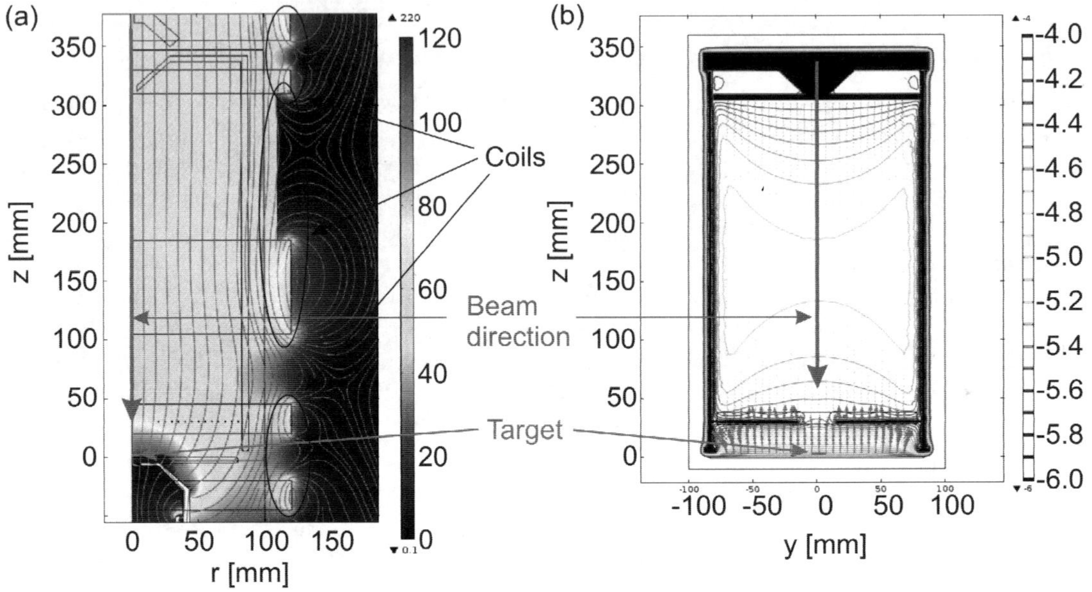


**Figure 1.** Section of the target chamber parallel to the beam direction showing the magnetic flux density as surface- and streamline-plots on the left side (*a*) and the electric potential as contour-plot and the electric field as arrow field on the right (*b*).

For each simulation $10^8$ positrons were implanted into the sample to obtain approximately the same number of events in the measured and in the simulated spectra. The annihilation position and the time between the implantation of the positron and the arrival of the gamma quantum in the detector were saved. We choose a gold target because due to its high backscattering coefficient the satellite structures are more pronounced than for other samples. The simulations have been performed for different positron implantation energies and the influence of the following possible countermeasures was studied: a) extending time window, b) using different materials in the top of the Faraday cage and c) varying the solenoidal magnetic field.

As an example, in figure 2 the comparisons between the measured spectra and the results obtained from the simulations at 4 keV positron implantation energy with time windows of 20 ns (left) and 40 ns (right) are shown for two different material in the top of the Faraday cage. From both the measured and the simulated spectra the constant background value was subtracted.

At 4 keV implantation energy (figure 2 (left)), the experimental measurement shows a very pronounced structures centered at 10 ns and a second one at the end of the time window and partly re-entering the time window from the opposite end. The simulation with tungsten in the top of the Faraday cage agrees very well with the experimental spectrum. Using a low Z material like Kapton instead of tungsten in the simulation, suppresses completely the folded structure under the lifetime peak.

Extending the time window (figure 2 (right)), independently of the material in the top of the Faraday cage, the fold-over of the second structure is completely avoided and this disturbance removed from the region under the peak of the lifetime spectrum and shifted to a position where the lifetime spectrum is already decayed. A first experimental test with a 40 ns time window has confirmed this result. The reduction of the second peak (centered at 22 ns) by using different materials could be of interest for samples with long lifetimes. At 4 keV implantation energy the

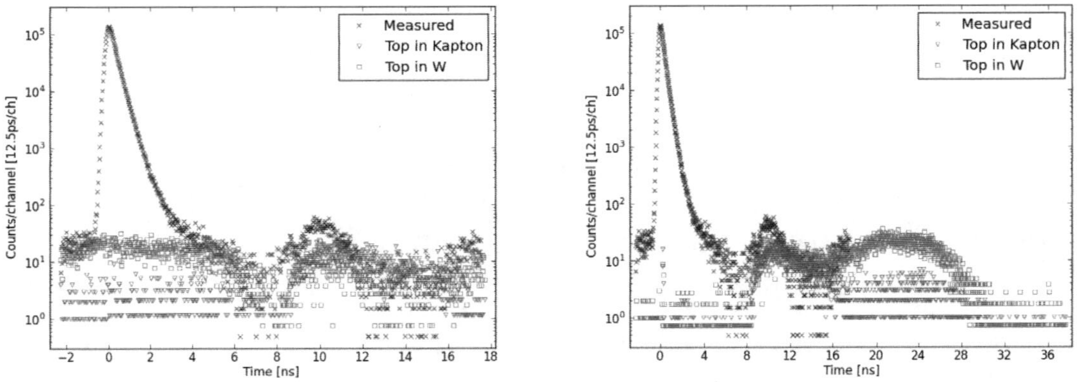


**Figure 2.** Measured lifetime spectrum of a gold target at 4 keV positron implantation energy and simulated background structure with two different materials in the top of the Faraday cage and time window of 20 ns (left) and of 40 ns (right).

influence of the magnetic field on the background structures was negligible. However, at higher implantation energies a small part of the backscattered positrons annihilates on the side walls of the target chamber and this results in a sharp structure centered at about 4 ns. Increasing the magnetic field in the target region prevents these positrons to annihilate on the side walls and thus allows to suppress this structure.

## 3. Conclusions

Comprehensive simulations of positrons backscattered from the sample and annihilating in the target chamber of PLEPS have been performed and compared to the measured lifetime spectra showing very good agreement. The simulations have shown three possible pathways to reduce the structures in the background: shifting the background structures outside the region of interest by a) extending the time window and b) increasing the magnetic field and c) reducing the backscattering coefficient of the top of the Faraday cage by using low Z materials.

Moreover, the possibility to almost quantitatively reproduce the behavior of backscattered positrons in the target chamber opens up a new way to check and determine the implantation profile on layered systems with depth-dependent Z-values by comparing the measured backscattered background as a function of the implantation energy with the simulated one.

## Acknowledgment

Funding from the BMBF (Project 05K10WNA-Posimethod) is gratefully acknowledged.

## References

[1] Keeble D J, Wicklein S, Dittmann R, Ravelli L, Mackie R A and Egger W 2010 *Phys. Rev. Lett.* **105** 226102
[2] Brusa R S, Mariazzi S, Ravelli L, Mazzoldi P, Mattei G, Egger W, Hugenschmidt C, Löwe B, Pikart P, Macchi C and Somoza A 2010 *Nucl. Instr. and Meth.* B **268** 3186
[3] Egger W 2010 *Proc. Int. School of Physics Enrico Fermi Course CLXXIV* ed Brusa R S, Dupasquier A and Mills A P (Amsterdam: IOS Press) p 419
[4] Hugenschmidt C 2010 *Proc. Int. School of Physics Enrico Fermi Course CLXXIV* ed Brusa R S, Dupasquier A and Mills A P (Amsterdam: IOS Press) p 399
[5] Comsol multiphysics website: URL `http://www.comsol.com`
[6] Agostinelli S *et al.* 2003 *Nucl. Instr. and Meth.* A **506** 250 URL `http://geant4.cern.ch/`

Journal of Physics: Conference Series **443** (2013) 012097 doi:10.1088/1742-6596/443/1/012097

# Positron annihilation lifetime spectroscopy based on secondary electron emission from carbon foil

**C K Ng, B Yang and S Fung**

Department of Physics, The University of Hong Kong, Pokfulam Road, Hong Kong, People's Republic of China

E-mail: data@hku.hk

**Abstract.** Using the emission of secondary electron (SE) from a thin carbon foil in transmission geometry guided by a strong radial electric field in the region between the foil and the sample under investigation, a new variable energy positron annihilation lifetime spectroscopy (PALS) system is being implemented. To this end, the SE emission from a carbon foil of 30nm thick has been investigated in transmission geometry. Considerable emission of SE from the foil with peak energy of about 5eV is observed. We have found that both the energy loss and dispersion of the positrons after transmission through the carbon foil are small enough for the proposed positron lifetime-depth profiling PALS system to be realised. These experimental results and the timing simulations of Cai et. al. [1] indicate that the proposed variable PALS based on secondary electrons generated by carbon foil have a high time resolution.

## 1. Introduction

The major difficulty of developing variable energy Positron Annihilation Lifetime Spectroscopy (PALS) is the lack of a natural start signal of the spectroscopy to indicate that the positron entered the sample [2]. In the previous work [3], our group suggested an innovative design of variable energy positron annihilation lifetime spectroscopy based on the emission of secondary electron (SE) from a carbon foil. The design is of cylindrical symmetry along the y-axis and involves 4 parts. Figure 1 shows the schematic layout of our proposed design of the variable energy PALS. In the first part, a layer of carbon foil is mounted on a grounded tungsten mesh, which is perpendicular to the direction of the positron beam. The second part is a needle mounted behind the centre of the carbon foil and the tungsten mesh. The needle just touches the surface of the sample to ensure that the latter is electrically grounded. This needle-sample arrangement provides a radial electric field which deflects outwards the secondary electrons from the carbon foil. To enhance the effect of the radial electric field, an annular electrode at 2kV is provided around the sample as the third part of the system. The last part is a microchannel plate (MCP), placed behind the needle-sample arrangement to detect the secondary electrons that pass through the space between the sample and the annular electrode. The secondary electrons so detected give rise to the start signal for the variable energy PALS.

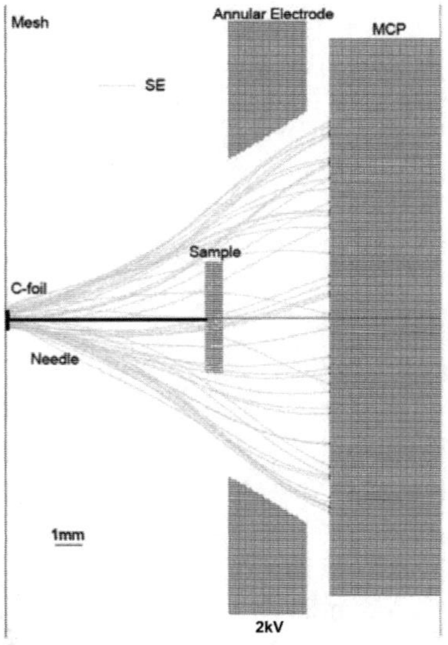


**Figure 1.** Schematic layout of our proposed design of the variable energy PALS

## 2. Experimental

To investigate the feasibility of using the secondary electrons generated from carbon foil as a start signal for the spectroscopy, an experiment is set up as shown in figure 2. A variable energy incident positron beam is produced by the slow positron beam facility at the University of Hong Kong. A 30nm thick carbon foil from Arizona Carbon Foil Co., Inc. was mounted on a grounded tungsten mesh. Two retarding grid analyzers for biasing off the positrons and generated secondary electrons were located behind the carbon foil. The phosphor screen visualized the profile of positrons and generated secondary electrons. A digital camera was remote-controlled to take photos of the positron beam and generated secondary electrons.

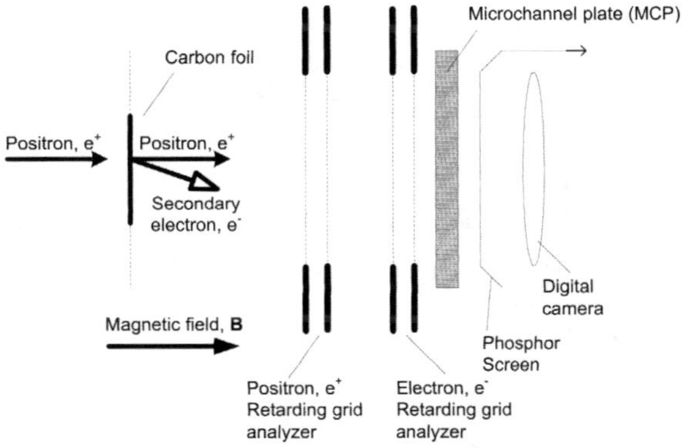


**Figure 2.** Schematic layout showing the experimental setup

## 3. Results and discussion

The energy spectrum of the generated secondary electrons with different positron beam energies is obtained as shown in figure 3. A peak energy around 5eV is observed and it undergoes a rapid drop in intensity so that nearly no contribution above 40eV is seen. The energy loss and dispersion for 2keV, 5keV and 10keV of positron are shown in figure 4. For the beam energy less than 10keV, there is a significant energy loss for positrons. There are 1keV energy loss for 2keV positron beam and 0.5keV for 5keV positron beam respectively. When the positron beam energy increases, the energy loss decreases significantly. The reason is due to the high transmission rate of energetic positrons. The energy dispersion also exhibits a similar pattern. Although there is little energy loss and energy dispersion for the 10keV positron beam, the energy of transmitted positron beam is still close to its incident energy so that we could still make use of the beam profile for depth profiling.

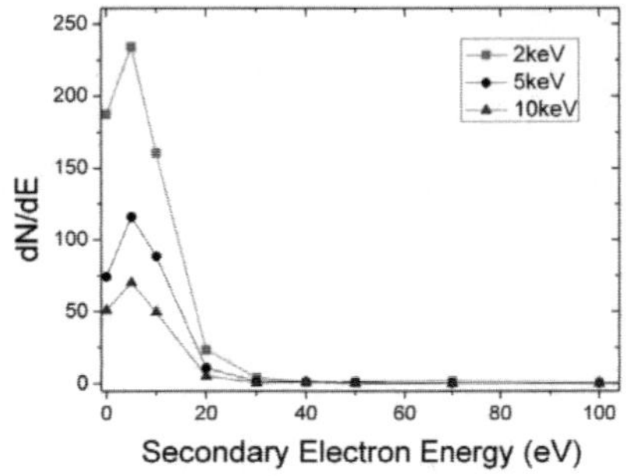


**Figure 3.** Energy spectrum of secondary electrons for incident positron beam energies of 2keV, 5keV and 10keV.

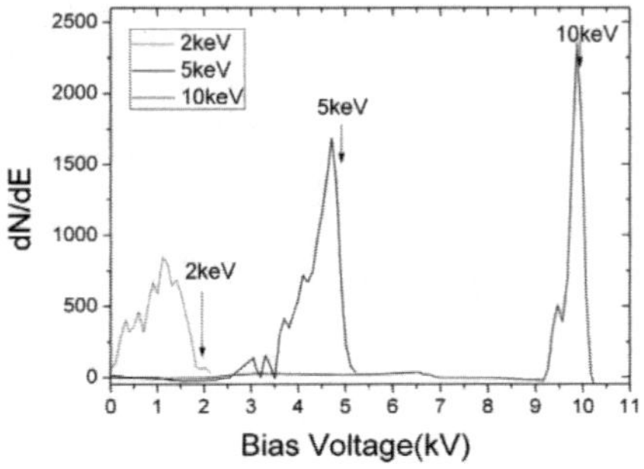


**Figure 4.** Energy spectrum of the transmitted positron beam for incident energies of 2keV, 5keV and 10keV.

There is a difference of the positron beam profiles with and without carbon foil respectively for positrons with an incident energy of 2keV. When the positron beam energy increases, there is less scattering of positron from the beam axis. For calculating the scattering angle of the positrons, the normal distribution $I(r) = A\exp[-\ln 2\frac{r^2}{r_i^2}]$ is used to fit the curves of beam profiles where $r_1$ and $r_2$ are the half widths at half maximum intensity for the non-scattered and scatted beams respectively.

The resultant beam profile is assumed to be a convolution of the unscattered beam profile and the scattering disc of radius $d\tan\theta$, $\theta$ is the average scattering angle and $d$ is the distance from the carbon foil to the microchannel plate. As a result,

$$\theta = \tan^{-1}\left(\frac{\sqrt{r_2^2 - r_1^2}}{d}\right) \qquad (1)$$

The calculated results for 2keV, 5keV and 10keV positron beams are shown in table 1.

**Table 1.** Average positron scattering angles and maximum secondary electrons emission

| Positron beam energy | Energy loss | Fraction of energy loss | Transmitted fraction | Calculated positron scattering angle | Maximum Secondary electrons emission |
|---|---|---|---|---|---|
| 2 keV | 1 keV | 50% | 40% | 4° | 61° |
| 5 keV | 0.5 keV | 10% | 64% | 7° | 60° |
| 10 keV | 0.13keV | 1.3% | 76% | 5° | 64° |

To calculate the emission angle of the secondary electrons, the spiral motion of the secondary electrons in the axial magnetic field of the beam has to be taken into account because these are low energy electrons. The cyclotron radius $\frac{mv_\perp}{Be}$ is convoluted in giving the maximum angle $\theta_{max}$ of emission as:

$$\theta_{max} = \tan^{-1}\frac{v_\perp}{v_{//}} = \tan^{-1}\left(\frac{Be\sqrt{r_3^2 - r_1^2}}{\sqrt{2mE}}\right) \quad (2)$$

Where $v_\perp$ is maximum transverse velocity of secondary electrons and $v_{//}$ is the beam axis velocity which can be obtained from the average forward positron beam energy of 5eV. The calculated results are listed in table 1. According to the work from Sigmund, the secondary electrons follow a cosine $\theta$ distribution [4]. As a result, we assume that the average scattering angle can be calculated as 45°.

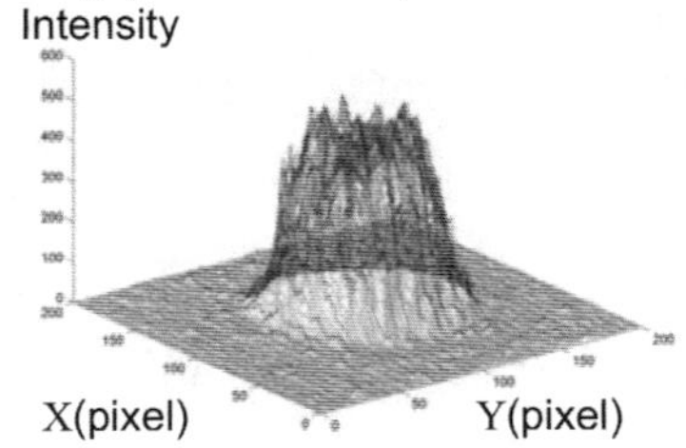


**Figure 6.** Beam profile without carbon foil

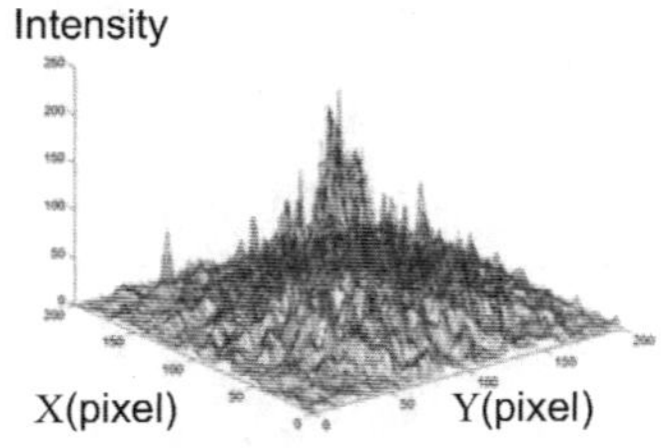


**Figure 7.** Beam profile with carbon foil

A simulations using SIMION [5] for testing the resolution of the variable energy PALS was performed by Cai et. al [1]. Timing simulations were carried out using typical set of 5eV SEs with uniformly distributed scattering angles up to 65°. The results of the simulation are shown in figure 8. A full width at half maximum (FWHM) dispersion in the time of flight (TOF) spectrum of 40ps is observed. The transmitted positrons were taken to be uniformly scattered within an angle of 10°. The time of flight spectrum for a 5keV beam is shown in figure 9 where the full width at half maximum dispersion is only 15ps.

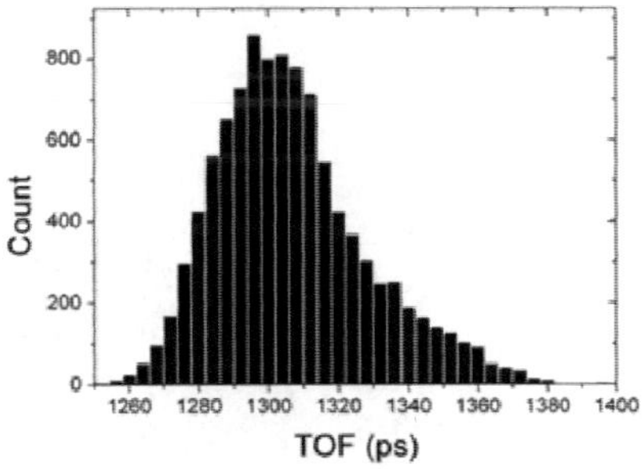


**Figure 8.** Simulation result of time of flight (TOF) for secondary electrons with kinetic energy 5eV

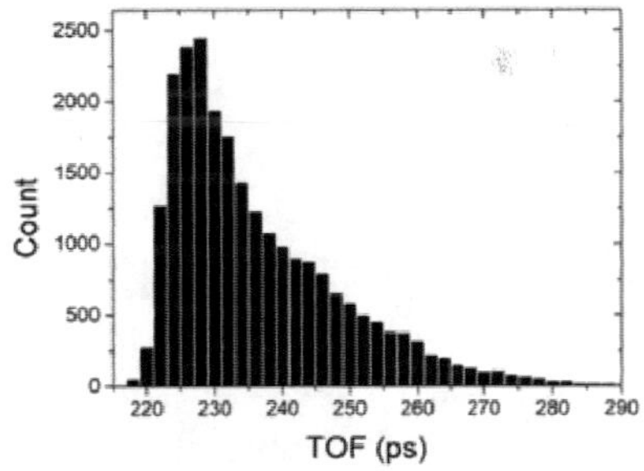


**Figure 9.** Simulation result of time of flight (TOF) for positrons with an incident energy 5keV

In conclusion, the results from calculations and simulations strongly suggest that the proposed variable PALS based on secondary electrons generated by carbon foil is able to achieve a time resolutions of less than 200ps.

## References

[1] L H Cai, B Yang, C C Ling, C D Beling and S Fung 2011 *J. Phys. Conf. Ser.* **262,** 012009
[2] P G Coleman 2000 *Positron Beams and their applications* (World Scientific)
[3] D Chen, J D Zhang, C C Cheng, C D Beling and S Fung 2008 *Appl. Surface Sci.* **255**, 122-124
[4] P Sigmund 1969 *Phys. Rev.* **184**, 383-416
[5] D A Dahl 2000 *SIMION 3D Version 7.0* (Idaho National Engineering Environmental lab)

Journal of Physics: Conference Series **443** (2013) 012098 doi:10.1088/1742-6596/443/1/012098

# A position sensitive germanium detector for the measurement of angular deviation of annihilation radiation

**Benjamin Löwe[1], Markus Reiner[2], Werner Egger[1], Christoph Hugenschmidt[2] and Günther Dollinger[1]**

[1] Universität der Bundeswehr München, Institut für Angewandte Physik und Messtechnik (LRT 2), Werner-Heisenberg-Weg 39, 85577 Neubiberg, Germany
[2] Technische Universität München, FRM II, Lichtenbergstr. 1, 85747 Garching, Germany

E-mail: benjamin.loewe@unibw.de

**Abstract.** To improve electron momentum sensitivity in Coincidence Doppler Broadening Spectroscopy (CDBS) measurements it is envisaged to measure the angular correlation of annihilation radiation along with the energy of both annihilation photons. For this purpose two position sensitive 36-fold pixelated, planar germanium detectors will be utilized. The position sensitivity of one of those detectors has been tested with a collimated gamma source. A data acquisition system consisting of 37 sampling analogue-to-digital converters with PC based on-line/off-line processing has been installed. A position sensitivity of 1.6 mm has been achieved.

## 1. Introduction

Currently two established positron annihilation spectroscopy (PAS) methods exist by which the momentum of the electron the positron annihilates with can be measured. Both measure different projections of the electron momentum. For Doppler Broadening Spectroscopy this is the longitudinal component derived from the energy of the annihilation quanta. Angular Correlation of Annihilation Radiation (ACAR) measures the angular deviation of both photons from 180 ° and therefore the full transverse component of the momentum.

Recent measurements with two germanium double-sided strip detectors (DSSD) by Williams et al. [1, 2] yielded a position resolution of approx. 0.5 mm (FWHM) and showed that the coincident acquisition of the angular deviation along with the photon energy is possible with high resolution (0.5 mrad for the angular resolution and 1.63 keV for the energy resolution).

Milechina et al. [3] showed that it is also possible to achieve a position resolution of 0.5 mm with a germanium detector whose front side has been pixelated into contacts of a size of 5 mm × 5 mm each.

To get a position resolution smaller than the size of a pixel it is necessary to analyse the induced charges that form especially in the next neighbour pixels during the time the charge carriers are travelling through the germanium crystal.

In this contribution we report on a test of the position resolution of a 36-fold pixelated, planar HPGe detector.

Journal of Physics: Conference Series **443** (2013) 012098 doi:10.1088/1742-6596/443/1/012098

## 2. Experimental set-up

The germanium detector (Canberra EGPS 48*48*20-36 PIX) consists of a crystal with an active size of 48 mm × 48 mm and a thickness of 20 mm, which is located 5 mm behind a 1 mm thick aluminium window. On the front side the crystal is segmented into 6 × 6 pixel contacts (8 mm × 8 mm each). An additional contact on the backside of the crystal covers the full volume. Every single pixel as well as the full volume contact is connected to one of the 37 built-in, resistive, charge sensitive pre-amplifiers (Canberra PSC823C), with a decay constant of 50 µs. The pixels are specified with an (FWHM) energy resolution of 1.30 to 1.37 keV (mean value: 1.33 keV), and 2.80 keV for the full volume contact, at a photon energy of 662 keV. The pixel detector is operated at a bias voltage of + 2500 V.

To determine the spatial resolution that can be achieved with the detector a collimated $^{137}$Cs gamma source (photon energy: 661.7 keV) mounted onto a x-y positioning stage with a travel range of 70 mm in both dimensions has been constructed. The collimator consists of two tungsten discs of a respective thickness of 20 mm and 10 mm stacked behind each other. A hole with a diameter of 0.8 mm has been eroded into the centre of both of them. The collimated source is mounted on a double axis stepper motor system (OWIS LTM 80) and can be positioned in front of the detector with a relative accuracy of 20 µm.

## 3. Electronic set-up

A fast data acquisition system has been set up for reading all 37 detector outputs and processing their signals at high count rates. The heart of the system is the 8-channel flash-ADC Struck SIS3302 whereof five modules are in operation for this detector. Each channel of the SIS3302 is sampling at a rate of 100 MHz, with a resolution of 16 bit per sample. These five modules are installed in a VME bus crate and can be accessed by a PC over a PCI Express to VME interface (SIS1100e/3104).

The signal height, and therefore the energy, of an event is calculated on a built-in FPGA on the ADCs using a moving average window algorithm with correction for the pre-amplifier's decay time [4]. The trigger generated by the full volume channel is distributed to all other modules to trigger the pixel channels. Each channel then writes its event data to one of its two buffers. The data per event contain a time-stamp, a configurable length of raw data (0 to 1024 samples) and the calculated signal height. While the ADC is acquiring data into one buffer the other buffer can be read-out by the computer in the meantime. The maximum event rate feasible with this set-up depends on the number of raw samples saved per event. At 100 samples per event and channel approx. 9 kcps can be reached. By setting the trigger level of the full volume channel high enough to reject photons that didn't deposit their full energy into the crystal, the use of the available bandwidth can be further optimized.

The data acquisition software on the computer is constantly polling all modules to check if a data buffer is full and then switching all modules through a VME broadcast at the same time to the other buffer. This ensures that the events in the channel buffers are synchronous to each other.

## 4. Data analysis

The acquired data is calibrated and processed on the computer. An in-house developed software analyses events that deposited energy in only one pixel and whose energies at the full volume contact and the pixel are in a range of 640 keV to 680 keV. These events are then analysed on-line but can also be stored for later off-line analysis.

When determining the position of the photon interaction in the detector one has to exclude pixels at the border of the active area of the crystal. Therefore only the inner 4 × 4 pixels are considered.

The acquired raw samples of all eight neighbours of the pixel of interest are then smoothed with a moving average algorithm. From the maximum value of such a smoothed signal its first value is then subtracted, to correct its offset. This difference corresponds to the height of the influence signal in the eight next neighbour pixels.

To get the position within the pixel in x direction the height of the influenced signals of the left ($I_l$) and the right ($I_r$) neighbour is used, whereas for the y direction the corresponding signals from the top ($I_t$) and bottom ($I_b$) neighbour pixel is taken.

The x and y positions within the pixel are calculated as the logarithm of the ratio of the relevant neighbour pixel intensities according to:

$$x = log\left(\frac{I_r}{I_l}\right) \text{ and } y = log\left(\frac{I_t}{I_b}\right). \quad (1)$$

The positions are then added to a two-dimensional histogram when the sum of all eight next neighbour pixels exceeds an adjustable threshold.

## 5. Measurements

To test the position resolution, data has been acquired with the collimated source at several positions in front of the detector. Fig. 1 shows the two-dimensional histogram of events analysed with the algorithm described above. The sub-pixel coordinates x and y have been scaled with a constant factor before being added to the global coordinates X and Y of the respective pixel. This has been done to prevent overlapping of two adjacent pixels. In fig. 1 the collimated source has been positioned near the centre of pixel 22. When comparing the counts in pixel 22 and its next neighbours a background of 20 % can be attributed to photons not being stopped in the tungsten.

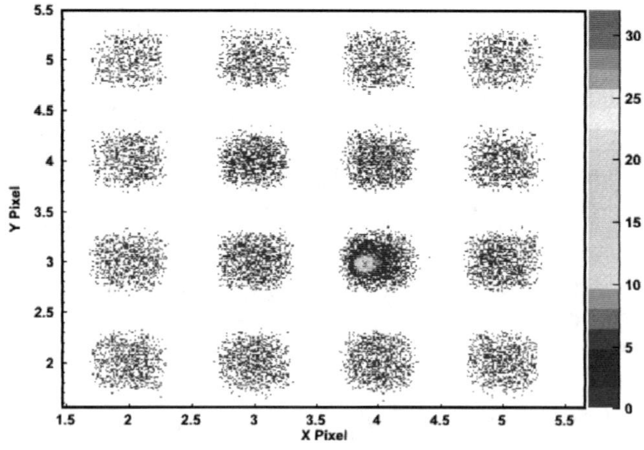


**Figure 1.** Two-dimensional histogram of analysed positions in the detector with the source in front of pixel 22 (X=4, Y=3)

Fig. 2 shows in detail two two-dimensional histograms of events within pixel 22. For both histograms the source was at the same y position but the x positions differ by 2.0 mm. Fig. 3 shows the projection of the events to the x axis.

For estimating the lateral resolution a Gaussian has been fitted to both projected histograms. With the known distance of 2.0 mm between both positions and the width of the Gaussian fit, a position uncertainty of $\Delta x = 1.6$ mm (FWHM) can be calculated.

Simulations with Geant 4 suggest a spread of the photon interactions in the crystal of approximately 1.07 mm. Assuming a gauss-like distribution would result into an x resolution of the detector of $\Delta x = 1.2$ mm.

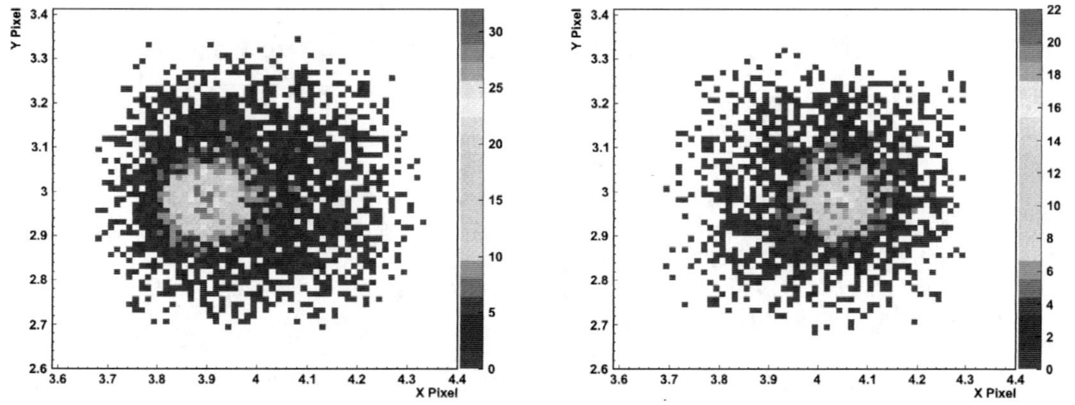


**Figure 2.** Pixel 22, x positions differ by 2 mm

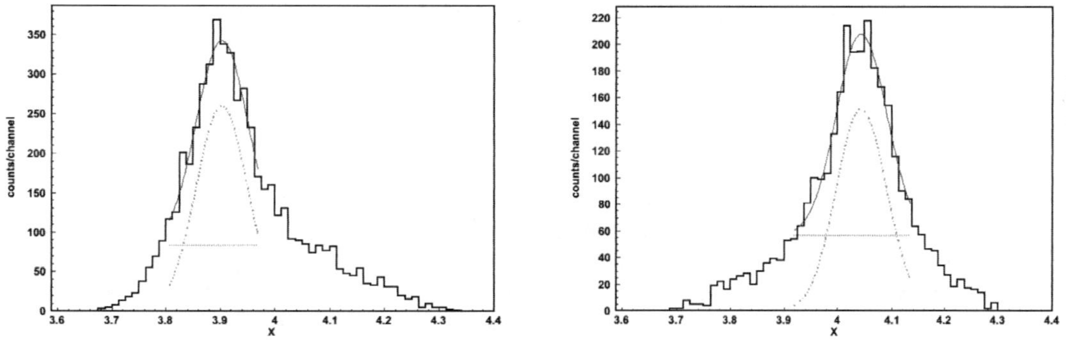


**Figure 3.** Projections of fig. 2 to the x axis

## 6. Conclusion

A data acquisition system for a 36-fold pixelated HPGe detector has been developed and successfully tested. It enables sub-pixel resolution by analysing the influenced charges in the next neighbour pixels. The test measurements with a collimator with a relatively large aperture have shown that a lateral resolution of 1.6 mm can be reached. The actual value of the detector's position resolution should be even better.

Used in coincidence with the second pixelated detector, this resolution will already be sufficient to improve the momentum information contained in CDBS experiments.

## Acknowledgments

The authors want to thank Wolfgang Engl and Ricardo Riedel for the installation of the translation stage and for the help at constructing the collimator holder.

Funding from the BMBF (Project 05K10WNA-Posimethod) is gratefully acknowledged.

## References

[1] Williams C, Baker W, Burggraf L, Adamson P and Petrosky J 2010 *Nuclear Science, IEEE Transactions on* **57** 860–869 ISSN 0018-9499

[2] Williams C, Burggraf L, Adamson P and Petrosky J 2011 *Nucl. Instr. and Meth.* **A 629** 175–184

[3] Milechina L and Cederwall B 2005 *Nucl. Instr. and Meth.* **A 550** 278–291 ISSN 0168-9002

[4] Lauer M 2004 *Digital Signal Processing for segmented HPGe Detectors, Preprocessing Algorithms and Pulse Shape Analysis* Ph.D. thesis

16th International Conference on Positron Annihilation (ICPA-16) IOP Publishing
Journal of Physics: Conference Series **443** (2013) 012099 doi:10.1088/1742-6596/443/1/012099

# Optimization of the Scintillator Size for Positron Lifetime Measurements

M. Yamawaki[1], Y. Kobayashi[1], K. Hattori[2], Y. Watanabe[2]
[1] National Institute of Advanced Industrial Science and Technology (AIST), Tsukuba, Ibaraki 305–8565 & 8568, Japan,
[2] TOYO SEIKO Co., Ltd, 3-195-1 Umaganji, Yatomi-cho, Aichi 490–1412, Japan,
E-mail: yamawaki.masato@aist.go.jp

**Abstract.** Optimization of the measurement condition for positron annihilation lifetime spectroscopy (PALS) is discussed from the viewpoint of non-destructive "on-site" material inspection. Numerical analysis based on a least-squares estimation and experiments with various sizes of $BaF_2$ scintillators by conventional PALS suggested that the use of relatively large $BaF_2$ scintillators enables on-site material inspection by PALS with reasonable accumulation time and time resolution.

## 1. Introduction

On-site inspection of building structures, in-line flow products, etc. by positron annihilation lifetime spectroscopy (PALS) requires reliable measurements in a reasonably short time [1], preferably, several minutes or less. In conventional positron lifetime measurements it is necessary to sandwich a sodium-22 positron source with two specimens cut out from the material to be inspected and several hours are generally required to accumulate annihilation events for deducing positron lifetimes of the materials. Recently we have developed a positron lifetime measurement system based on an anti-coincidence method [2]. This system enables measurements to be performed without taking out sample pieces from the target materials; it is, in principle, applicable to non-destructive "on-site" inspection of various structures. The use of smaller scintillators in the anti-coincidence system provides a better time resolution but reduces the count rate, as in the conventional positron lifetime measurements.

We discuss analytical relations among the count rate, time resolution, and lifetime evaluation accuracy based on the least-squares estimation method. Using the results of the numerical analysis and the PALS experiments performed with different sizes of $BaF_2$ scintillators, it is shown that the use of larger scintillators can shorten the time required for the positron lifetime measurements, aimed for non-destructive on-site material inspection.

## 2. Mathematical modeling and analytical consideration

The effect of the deviations of different parameters on the deviation of the fitting function can be written as the following equation.

$$\delta y(\tau_i) = K_{N1}\delta N_1 + K_{\lambda 1}\delta\lambda_1 + K_{N2}\delta N_2 + K_{\lambda 2}\delta\lambda_2 + K_\sigma\delta\sigma + K_{t_0}\delta t_0 + \delta n_b \quad i = 1,\cdots,m \tag{1}$$

Here $y(\tau_i)$ is the mathematical model of the lifetime histogram which consists of two components having different lifetimes $1/\lambda_1$ and $1/\lambda_2$

$$y(\tau_i) = \frac{N_1\lambda_1\Delta\tau}{\sqrt{2\pi}\sigma}e^{-\lambda_1\tau}\int_{-\infty}^{\tau_i} e^{-\frac{(t-t_0)^2}{2\sigma^2}}e^{\lambda_1 t}dt + \frac{N_2\lambda_2\Delta\tau}{\sqrt{2\pi}\sigma}e^{-\lambda_2\tau}\int_{-\infty}^{\tau_i} e^{-\frac{(t-t_0)^2}{2\sigma^2}}e^{\lambda_2 t}dt + n_b \quad i = 1,\cdots,m \tag{2}$$

where $\tau_i$ is the time corresponding to the $i$-th channel and $n_b$ is the background noise. The coefficients in the right hand side of differential equation (1) are given by the following equations:

$$K_{Nk} = \frac{\lambda_k\Delta\tau}{\sqrt{2\pi}\sigma}J_{0k}(\tau_i) \qquad k = 1,2 \tag{3}$$

$$K_{\lambda k} = \frac{N_k \Delta\tau}{\sqrt{2\pi\sigma}}[(1-\lambda_k\tau_i)J_{0k}(\tau_i) + \lambda_k J_{1k}(\tau_i)] \qquad k=1,2 \tag{4}$$

$$K_\sigma = \sum_{k=1}^{2} \frac{N_k\lambda_k \Delta\tau}{\sqrt{2\pi\sigma^2}}[-(1-\frac{t_0^2}{\sigma^2})J_{0k}(\tau_i) - \frac{2t_0}{\sigma^2}J_{1k}(\tau_i) + \frac{1}{\sigma^2}J_{2k}(\tau_i)] \tag{5}$$

$$K_{t_0} = \sum_{k=1}^{2} \frac{N_k\lambda_k \Delta\tau}{\sqrt{2\pi\sigma^3}}[-t_0 J_{0k}(\tau_i) + J_{1k}(\tau_i)] \tag{6}$$

where $N_k$ is the count of the $k$-th component. $J_{0k}(\tau_i)$, $J_{1k}(\tau_i)$ and $J_{2k}(\tau_i)$ are defined as follows.

$$J_{0k}(\tau_i) = \int_{-\infty}^{\tau_i} e^{-\frac{(t-t_0)^2}{2\sigma^2}} e^{-\lambda_k(\tau_i - t)} dt \qquad k=1,2 \tag{7}$$

$$J_{1k}(\tau_i) = \int_{-\infty}^{\tau_i} t e^{-\frac{(t-t_0)^2}{2\sigma^2}} e^{-\lambda_k(\tau_i - t)} dt \qquad k=1,2 \tag{8}$$

$$J_{2k}(\tau_i) = \int_{-\infty}^{\tau_i} t^2 e^{-\frac{(t-t_0)^2}{2\sigma^2}} e^{-\lambda_k(\tau_i - t)} dt \qquad k=1,2 \tag{9}$$

$\sigma$ is the standard deviation of the system time resolution, $t_0$ is the time when the start signal is generated, $n_b$ is the background noise and $\Delta\tau$ is the time width of the frequency channel.

According to the least-squares method, standard deviations $s_{N1}$, $s_{N2}$, $s_{\lambda 1}$, $s_{\lambda 2}$, $s_\sigma$, $s_{t0}$ and $s_{nb}$ of parameters $N_1$, $N_2$, $\lambda_1$, $\lambda_2$, $\sigma$, $t_0$ and $n_b$ are obtained from the element $[(\boldsymbol{M}^T\boldsymbol{M})^{-1}]_{ii}$ of the diagonal matrix $(\boldsymbol{M}^T\boldsymbol{M})^{-1}$. Here $\boldsymbol{M}$ is a matrix defined by the following equations:

$$\boldsymbol{M} = \boldsymbol{W}\boldsymbol{H} \tag{10}$$

$$\boldsymbol{W} = \begin{bmatrix} 1/\delta n_1 & 0 & \cdots & 0 \\ 0 & 1/\delta n_2 & \cdots & 0 \\ \vdots & \vdots & \ddots & \vdots \\ 0 & 0 & \vdots & 1/\delta n_m \end{bmatrix} \tag{11}$$

$$\boldsymbol{H} = [\boldsymbol{h}_1 \quad \boldsymbol{h}_2 \quad \cdots \quad \boldsymbol{h}_m]^T \tag{12}$$

$$\boldsymbol{h}_i = [K_{N1}(\tau_i) \quad K_{N2}(\tau_i) \quad K_{\lambda 1}(\tau_i) \quad K_{\lambda 2}(\tau_i) \quad K_\sigma(\tau_i) \quad K_{t0}(\tau_i) \quad 1] \qquad i=1,2,\cdots,m \tag{13}$$

$\delta n_i$ is the weight used in the least-squares fitting and is described as shown below, considering that positron annihilation is random events which follow the Poisson process.

$$\delta n_i = [y(\tau_i) + \delta n_w^2]^{1/2} \qquad i=1,2,\cdots,m \tag{14}$$

where $y(\tau_i)$ is the term equivalent to the square of the standard deviation of count errors which occurred in Poisson process and $\delta n_w$ is the standard deviation of white noise assumed to be contained uniformly in every channel.

The standard deviation of $\lambda_1$ is given as follows:

$$s_{\lambda 1} = ([(\boldsymbol{M}^T\boldsymbol{M})^{-1}]_{22})^{1/2} \tag{15}$$

The standard deviation $s_{\tau 1}$ of the averaged lifetime $\tau_1$ is denoted by the following formula:

$$s_{\tau 1} = s_{\lambda 1} / \lambda_1^2 \tag{16}$$

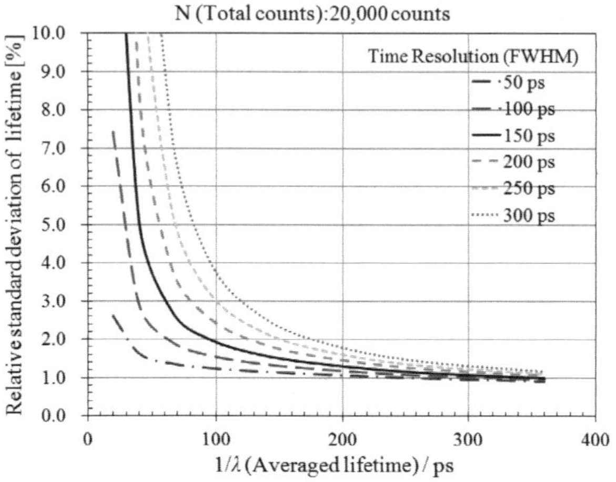


**Figure 1 Variations of relative standard deviations of positron lifetime for various time resolutions as a function of the average lifetime**

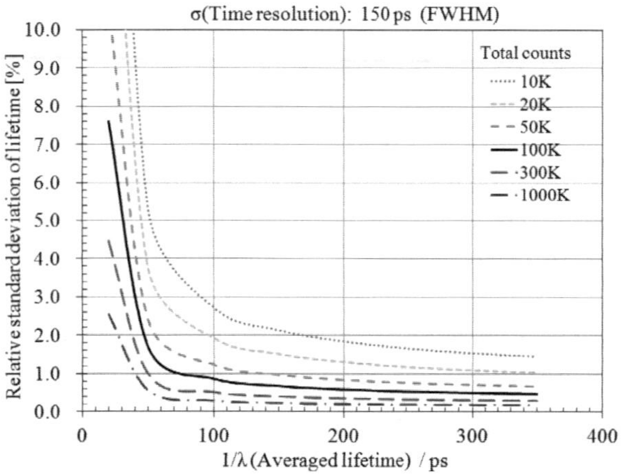


**Figure 2 Variations of relative standard deviations of positron lifetime for various total counts as a function of the average lifetime**

The above-mentioned mathematical model is useful to obtain the numerical relation among the standard deviation of the lifetime, the system time resolution and the total counts of lifetime data. It is also useful to seek the possibility of the existence of additional lifetime components. Considering the typical condition of the application to metallic material inspection, it was assumed in the present numerical analysis that the lifetime spectrum consists of two components: the first component with the average lifetime ranging from 100 ps to 250 ps and the second component with an average lifetime of 380 ps (the positron lifetime of Kapton to sandwich sodium-22) and 17% frequency (relative intensity). The total number of counts and the time resolution were fixed to $2\times10^4$ and 150 ps, respectively.

Figure 1 shows an example of the results obtained by numerical analysis under constraint of $\tau_2$=380 ps. The number of total counts was fixed to $2\times10^4$. It is seen that the measurement error of the 150 ps lifetime is 1-2% when the time resolution is 150 ps (FWHM) and the improvement of the time resolution reduces the lifetime estimation errors whatever the lifetime is, but it is not so effective unless the lifetime of the first component is less than 100 ps. Also, it is clear that the lifetime measurement near 50 ps or less will bring about the rapid increase of lifetime estimation errors with time resolution.

Figure 2 shows the variations of the estimation errors of lifetime $1/\lambda_1$ = 150 ps as a function of the average lifetime for total counts raging from 10K to 1000K. Despite the value of the lifetime ($1/\lambda_1$) the lifetime estimation error decreases in proportion to the reciprocal of the square root of the total counts of the lifetime data. For example, the error of $1/\lambda_1 \cong$ 150 ps is about 2% for the total counts of 10K but it is as low as 0.2% for 1000K counts.

Moreover, another analysis indicates that the use of the second lifetime as a known parameter is effective for the improvement in accuracy. Under the fixed second lifetime of 380 ps and intensity of 17%, the first lifetime of around 150 ps can be determined with several picoseconds errors if the total counts are larger than $2\times10^4$.

## 3. Experiment and consideration

We performed positron lifetime measurements of Si single crystal to study the influence of the scintillator size on time resolution and count rate. A sodium-22 (850 kBq) source was sandwiched between two pieces of a Si wafer. The PAL measurement system used in this experiment was mainly composed of $BaF_2$ scintillators made by Ohyo Koken Kogyo Co.,LTD, photomultiplier tubes H3378-

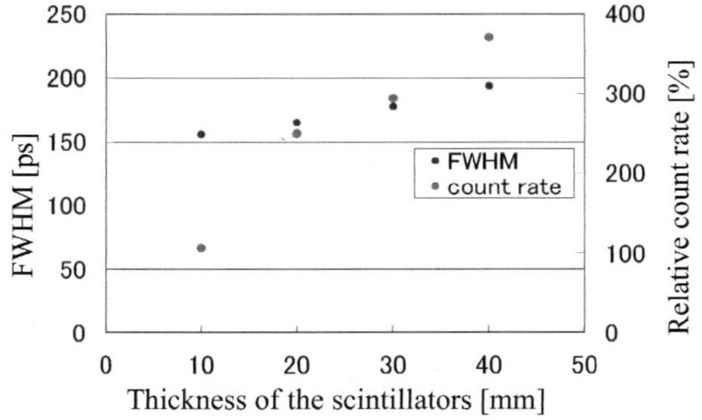


**Figure 3 Deviation of FWHM and Relative count rate versus thickness of scintillator's**

51 produced by Hamamatsu Photonics K.K. and a digital oscilloscope, Wave surfer 44MXs-B, produced by LeCroy corporation. Positron lifetimes were measured by determining the time intervals between 1.27 MeV start $\gamma$-rays and 0.511 MeV stop $\gamma$-rays. $BaF_2$ scintillators used for the measurements were 30 mm in diameter on top, 40 mm in diameter on bottom and 10, 20, 30 and 40 mm in thickness. Figure 3 shows variations of time resolution in FWHM and relative count rate with thickness of the scintillators. The relative count rate is expressed as percentage to the count rate ($2.0\times10^4$/5min) obtained with two 10 mm thick scintillators. We see that with increase of the size of scintillators the count rate rapidly increases but the time resolution deteriorates rather slowly. With the help of the mathematical model in section 2 and eq.(16) we evaluated standard deviations of lifetime $1/\lambda_1$=150 ps (a typical positron lifetime for a metal) based on the experimentally determined time resolutions and count rates in Fig. 3 (the positron lifetime in Si is ~220 ps and much longer than 150 ps). The evaluated standard deviations of the 150 ps lifetime for 5 min accumulation are listed in table 1. The standard deviation decreases continuously from 2.31% for the 10 mm thick scintillators to 1.39% for the 40 mm thick scintillators. This result indicates that the increased count rate obtainable with a pair of larger scintillators effectively improves the accuracy of the positron lifetime measurements in spite of some worsening of the time resolution. The use of large scintillators can shorten the accumulation time as required by on-site material inspection by PALS.

**Table 1 Estimated standard deviations of the 150 ps lifetime for 10, 20, 30 and 40 mm thick (5 min accumulation)**

| Thickness of scintillators [mm] | 10 | 20 | 30 | 40 |
|---|---|---|---|---|
| Standard deviation of 150 ps lifetime [ps] | 2.31 | 1.55 | 1.48 | 1.39 |

## 4. Conclusion

With the help of the least-squares estimation theory, which uses the fitting function derived from the linearization of the mathematical model of the positron lifetime histogram, standard deviations of the 150 ps lifetime were evaluated from experimental data of Si obtained with various sizes of $BaF_2$ scintillators. Our conclusion is that the use of larger scintillators can improve the accuracy of the positron lifetime determination and shorten the time required for the positron lifetime measurements, in particular, those aimed for non-destructive on-site material inspection.

## Acknowledgments

This research was supported in part by a grant from Chubu Bureau of Economy, Trade and Industry METI. We gratefully acknowledge helpful discussions with Dr. Kenji Ito (AIST), Dr. Nagayasu Oshima (AIST) and Prof. Hamdy F.M. Mohamed (Minia University, Egypt).

## References

[1] Oshima N, Suzuki R, Ohdaira T, Kinomura A, Kubota S, Watanabe H, Tenjinbayashi K, Uedono A, Fujinami M, 2011 *Journal of Physics: Conference Series* **262** 012044
[2] Yamawaki M, Kobayashi Y, Hattori K, Watanabe J, 2011 JJAP. **50** 086301